REVIEWS in MINERALOGY

Chemical Weathering Rates of Silicate Minerals

Editors:

Arthur F. White

U.S. Geological Survey, Menlo Park, CA

Susan L. Brantley

Pennsylvania State University

Front cover: Denticulated margin on naturally weathered hornblende formed by side-by-side coalescence of lenticular etch pits. From a weathered corestone developed on the Carroll Knob Complex in the Appalachian Blue Ridge near Otto, North Carolina. Field of view = 60 μm wide. Courtesy of M.A. Velbel, Michigan State University. See Velbel (1993) *American Mineralogist* 78:405-414.

Back Cover: SEM image of a naturally weathered (001) cleavage surface of an alkali feldspar from gravels formed from the granite at Shap, northern England. Scale bar: 2 μm. The paired triangular and trapezoidal etch pits have developed at the outcrop of edge dislocations. These form extended loops around perthitic albite lamellae in orthoclase. The lamellae extend parallel to *b*, the long axis of the micrograph. Courtesy of M.R. Lee and I. Parsons (see Chapter 8).

Series Editor: **Paul H. Ribbe**

Department of Geological Sciences
Virginia Polytechnic Institute & State University
Blacksburg, Virginia 24061 U.S.A.

Mineralogical Society of America
Washington, D.C.

MINERALOGICAL SOCIETY OF AMERICA
Printed by BookCrafters, Inc., Chelsea, Michigan.

REVIEWS IN MINERALOGY

(Formerly: SHORT COURSE NOTES)

ISSN 0275-0279

Volume 31

Chemical Weathering Rates in Silicate Minerals

ISBN 0-939950-38-3

ADDITIONAL COPIES of this volume as well as others in the series
may be obtained at moderate cost by writing or phoning:

MINERALOGICAL SOCIETY OF AMERICA
1015 Eighteenth Street, NW, Suite 601
Washington, D.C. 20036 U.S.A.

Telephone: (202) 775-4344

FOREWORD

This is the twenty-second year of a nearly continuous series of short courses sponsored by the Mineralogical Society of America in the subjects of mineralogy, petrology, crystallography, and geochemistry. Most have been given in conjunction with the Society's annual meeting together with the Geological Society of America, but recently there has been a move to presenting short courses at the semi-annual meetings of the American Geophysical Union as well. The printed volumes that accompany these courses, mixed together with several monographs, now total thirty-one, with two more due this year (on *silicate melts* and the element *boron* in mineralogy and geochemistry).

This volume, *Chemical Weathering Rates of Silicate Minerals*, was edited by Art White (U.S. Geological Survey, Menlo Park) and Sue Brantley (Penn State University), who, in addition to their own feedback to the authors, sent all chapters out for peer review and did their best to keep things on schedule. In the latter effort they failed, but through no fault of their own, and the result was some hurried and not up-to-standards copy editing—for which I apologize. Nonetheless, the subject matter is as impeccable as the authors who deal with it, and the reader is left to judge the results. This book is a follow-on to Volume 23 (1990), *Mineral-Water Interface Geochemistry*, co-edited by Mike Hochella and Art White, and still available at a reasonable price from MSA.

I am particularly greatful to Dr. Jodi Rosso for unscrambling computer software problems with good humor and limitless patience. At the moment she is my nominee to succeed me at this post. Margie Strickler, now having contributed to her nineteenth RiM volume, has been a significant help, as was my colleague, Arthur Snoke. I lost my paste-up man, Brett Macey, who started well but thought the M.S. thesis he was doing under my supervision took precedence over work on this volume. I agreed and thus got stuck with the job myself; so if any figures or pages are misplaced, there is no one to blame but ...

Paul H. Ribbe
Series Editor
Blacksburg, VA

ACKNOWLEDGMENTS

We gratefully acknowledge the Mineralogical Society of America for agreeing to sponsor a short course on Chemical Weathering Rates of Silicate Minerals and to publish this book. This format provides an ideal vehicle for both integrating existing scientific information and for proposing future research topics and directions. These objectives are meet in a timely and economic manner by the *Reviews in Mineralogy* Series. We greatly appreciate the outstanding effort of Paul Ribbe, Series Editor, and his staff, who overcame the maze of computer platforms and software idiosyncrasies to intergrate the contributions into the final book. We also acknowledge the support of Alex Speer and his staff at the MSA office for assisting in organizing the short course and Laura Serpa of the University of New Orleans for handling the associated GSA symposium and theme sessions. We also thank Steve Newton who worked tirelessly on proofing manuscripts and the contributors themselves who gave a significant amount of their time and effort in the last several months to bring this effort to fruition. Finally we thank Gerald Boulmay of the Bourbon Orleans Hotel for the inspirational suggestion that "Darlin', in New Orleans, anything is possible."

Art F. White
Menlo Park, California

Susan L. Brantley
University Park, Pennsylvania
September 11, 1995

Chemical Weathering Rates of Silicate Minerals

TABLE OF CONTENTS, VOLUME 31

Chapter 1 A. F. White & S. L. Brantley

CHEMICAL WEATHERING RATES OF SILICATE MINERALS: AN OVERVIEW

Chapter 2 A. C. Lasaga

FUNDAMENTAL APPROACHES IN DESCRIBING MINERAL DISSOLUTION AND PRECIPITATION RATES

Chapter 3   **W. H. Casey & C. Ludwig**

SILICATE MINERAL DISSOLUTION AS
A LIGAND-EXCHANGE REACTION

Chapter 4   **S. L. Brantley & Y. Chen**

CHEMICAL WEATHERING RATES OF
PYROXENES AND AMPHIBOLES

DISSOLUTION AND PRECIPITATION KINETICS
OF SHEET SILICATES

Chapter 6   P. M. Dove

KINETIC AND THERMODYNAMIC CONTROLS ON SILICA REACTIVITY IN WEATHERING ENVIRONMENTS

Chapter 7   **A. E. Blum & L. L. Stillings**

FELDSPAR DISSOLUTION KINETICS

Chapter 8   M. F. Hochella, Jr. & J. F. Banfield

CHEMICAL WEATHERING OF SILICATES IN NATURE:
A MICROSCOPIC PERSPECTIVE
WITH THEORETICAL CONSIDERATIONS

CHEMICAL WEATHERING RATES OF SILICATE MINERALS IN SOILS

WEATHERING RATES IN CATCHMENTS

Chapter 11   **H. Sverdrup and P. Warfvinge**

ESTIMATING FIELD WEATHERING RATES
USING LABORATORY KINETICS

Chapter 12   **R. F. Stallard**

RELATING CHEMICAL AND PHYSICAL EROSION

Chapter 13   **Robert A. Berner**

CHEMICAL WEATHERING AND ITS EFFECT
ON ATMOSPHERIC CO_2 AND CLIMATE

Chapter 1

CHEMICAL WEATHERING RATES OF SILICATE MINERALS: AN OVERVIEW

Art F. White

U. S. Geological Survey
Menlo Park, CA 94025 U.S.A.

Susan L. Brantley

Department of Geosciences
Pennsylvania State University
University Park, PA 16802 U.S.A.

"The ruins of an older world are visible in the present structure of our planet... The same forces are still destroying, by chemical decomposition or mechanical violence, even the hardest rocks and transporting the materials to the sea."   James Hutton

INTRODUCTION

Chemical weathering of silicate minerals results from the differences in the thermodynamic conditions that existed at the time of mineral formation and that of ambient conditions at the earth's surface. The solid state characteristics of silicate minerals, generally established at high temperature and/or pressures, continue to be reflected during weathering through mineral compositions, crystallographic structures and the petrographic fabric and textures of rocks. The ambient processes which most influence silicate weathering are associated with the flow and chemistry of water at the earth's surface. The term "weathering" implies strong dependency on processes associated with the hydrosphere, atmosphere, and biosphere.

Chemical weathering also encompasses an immense range of physical scale. Ultimately weathering reactions consist of the detachment and attachment of atoms at a mineral surface. The rates of these reactions are dependent on solid substrates and aqueous boundary layers which must be measured on the scale of Ångströms and which are generally not chemically comparable to bulk mineral nor fluid compositions. During weathering the chemistry of these interfaces is affected by the surrounding micro-environments controlled by diverse processes ranging from diffusion of reactants into micropores and along grain boundaries to biogeochemical interactions on microbial substrates and adjacent to plant roots. These micro-environments, in turn, comprise larger systems such as soils and watersheds which are commonly studied in terms of natural weathering rates. Final "up-scaling" leads to the consideration of global weathering rates and the linkage to global climate history and the development of life on the earth. The scope and scale of these chemical and physical processes provide both significant barriers to the quantification of weathering mechanisms and rates as well as fascinating opportunities for multi-disciplinary research.

Historical beginnings

The importance of chemical weathering has long been considered in scientific inquires. The early geologic literature from the time of Hutton made reference to the disintegration of rocks to form soils. During that time, numerous proposals were advanced to explain chemical weathering. Extensive lateritization observed in Brazil was attributed by Darwin (1876) to submarine processes while feldspathic decomposition in granite and the origin of kaolin was attributed by Brongniart (1839) to electric currents resulting from the contact of heterogeneous rock types. Fournet (1833) and Hartt (1853), writing even earlier, more correctly attributed chemical weathering to "the efficacy of water containing carbonic acid" in promoting the decomposition of igneous rocks.

In accord with current global climate studies, Hunt (1873) reported that chemical weathering profiles in the Blue Ridge Mountains of Virginia were "effected at a time when a highly carbonated atmosphere, and a climate very different from our own, prevailed." Antecedent to the recent interest in the role of vegetation on chemical weathering, Belt (1874) observed that the most intense weathering of rocks in tropical Nicaragua was confined to forested regions and attributed this effect to " the percolation through rocks of rain water charged with a little acid from decomposing vegetation." Chamberlin (1899) proposed that enhanced rates of chemical weathering associated with major mountain building episodes would result in a draw down of atmospheric CO_2 levels which would lead to global cooling. Early attempts in estimating the age of the earth based on salt accumulation in the oceans (Jolly 1898) also recognized the role of weathering in the chemistry of runoff and river discharge.

In addition to field observations, experiments were also initiated at an early date to determine the relationship between observed mineral weathering and solution chemistry. In 1848, Rogers and Rogers published a paper in the American Journal of Science entitled On the decomposition and partial solution of minerals and rocks by pure water and water charged with carbonic acid. In parallel sets of experiments, "quantities (40 grains) of the finely powered minerals (including feldspars, mica and analcime) were placed with a certain volume (10 cubic inches) of liquid in green glass bottles and agitated from time to time over a prescribed period. The liquid, separated by filtration, was evaporated to dryness ... and is submitted to quantitative analysis." In commenting on these results, Bischof (1863) postulated that, " If now 40 grains of hornblende, unpowered, the surface is only one millionth of the powered, were treated in the same way and the water was renewed every two days, the time required for perfect (total) solution would be somewhat more than six million years." This was perhaps the first realistic estimate of the time scales involved in weathering processes at the earth's surface based on an experiential study. It is reasonable to conclude that by 1897, when Merrill published his milestone volume Rocks, Rock Weathering and Soils, the major processes controlling chemical weathering rates at the earth's surface had been identified and that many of the fundamental approaches used to study weathering today had been defined.

Recent advances

The major advances since that time have been in the understanding of the fundamental chemical processes which control mineral weathering and in the efforts to quantify weathering rates on both the laboratory and field scales. These advances parallel the development of other branches of chemistry and physics, including the development of fundamental theories based on statistical and quantum mechanics which have permitted insights into what has often appeared to be insurmountably complex chemical reactions involving minerals and water. One of the first major reviews of this topic was presented in 1981 in Reviews in Mineralogy, Vol. 8, *Kinetics of Geochemical Processes*. Likewise, the application of ligand exchange theory, initially developed to explain ion exchange processes on metal oxide surfaces, has lead to the ability to isolate the effects of specific aqueous species on dissolution and precipitation reaction on the mineral surfaces. This subject was systematically addressed in 1990 in Reviews in Mineralogy, Vol. 23, *Mineral-*

Water Interface Geochemistry. Finally, the relatively recent development of new analytical instrumentation has permitted, for the first time, direct observation of the solution-mineral interface. Some of the useful techniques applied to mineral weathering include Auger electron and X-ray photoelectron spectroscopy, X-ray absorption spectroscopy and tunneling and atomic force microscopy. The application of such techniques was presented in Reviews in Mineralogy, Vol. 18, *Spectroscopic Methods in Mineralogy and Geology.* Recognition must also be given to the exponential increase in computation power and speed without which analysis, data reduction and interpretation would not be possible.

Along with the development of fundamental techniques to characterize weathering reactions has come the recognition of the fundamental importance of weathering in terms of the earth's ecosystems and interactions with biology, hydrology, and climate. Chemical weathering has gained recognition in environmental studies related to water quality, watershed acidification, nutrient cycling related to issues such as deforestation, and the feedback between chemical weathering and global CO_2 budgets and greenhouse warming. These issues provide both the scientific incentive and funding to conduct long term hydrochemical monitoring programs which have provided greater details on processes and rates of natural chemical weathering. Methodologies include characterization of weathering environments based on soil water chemistry and hydrology as well as long term records of solute fluxes in watersheds. Also, weathering studies are being conducted over a much wider change of geochemical environments and geographic localities. Efforts have also been advanced to more accurately characterize global weathering flux rates and to develop global weathering models.

PRACTICAL IMPORTANCE OF CHEMICAL WEATHERING

One of the oldest direct applications of weathering was to economic geology, in particular to the formations of laterites, bauxite deposits and supergene enrichment of metal deposits (Brimhall and Dietrich 1987). The recognition of importance of weathering in soil development predates even mining interest and studies on the interactions between soil fertility and mineral weathering are still of great importance in agriculture. Additional emphasis on chemical weathering in the last quarter of this century parallels the recognition by society of the importance of the total environment and the increasing strain industrial development places on ecosystems. Several examples of the application of silicate weathering rates to important societal concerns are now discussed.

Sources of inorganic nutrients in soils

Chemical weathering contributes macro nutrients such as Mg, Ca, K and PO_4 in addition to micronutrients such as Fe, Mn and B to soils underlain by silicate rocks. Ultimately these elements are used and recycled by plant communities (see Marschner, 1995, for a detailed review of mineral nutrition of plants). Recent studies of soil chronosequences, comprising soils of different ages developed from similar silicate rock types, provide interesting insights on relationships between soil fertility and the degree of chemical weathering. Figure 1a shows a compilation for three chronosequences (Jennifer Hardin, 1995, pers. comm.) which indicate that the mass for organic carbon   (g cm^{-3}) initially increases and thereafter decreases as a function of soil age. In such a plot, organic carbon can be used as an indicator of soil productivity and soil age as an indicator of the degree of soil weathering. Initial increases in fertility relate both to release of inorganic nutrients from silicate weathering and to the increase in clay content which increases water retention. However, the end products of extensive weathering found in the older soils ultimately result in the depletion of the sources for these inorganic nutrients, in addition to progressively poorer permeability and drainage in soils, resulting in a decrease in soil fertility.

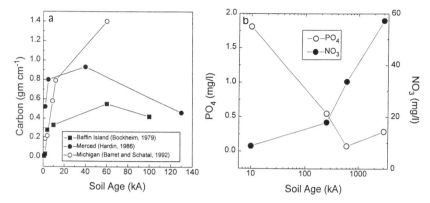

Figure 1. Trends in soil fertility with age (a) change in solid organic content for indicated soil chronosequences and (b) average changes in dissolved PO_4 and NO_3 content in soil waters for the Merced soil chronosequence, California (after White et al., 1995).

Variations in inorganic nutrients in soil waters also exhibit significant trends with soil age such as shown for PO_4 and NO_3 for the Merced chronosequence (Fig. 1b; also see White, this volume). Clearly the younger soils produce a surplus of available phosphate while older soils apparently become phosphate limited. This trend is the opposite of that for nitrate, which increases with soil age as a result of a decline in plant and microbial productivity. The principal source of phosphate in crystalline rocks is apatite, which occurs as accessory phases and inclusions in silicate minerals. As shown by the SEM photograph of an apatite inclusion in a feldspar grain from a 250 kA Merced soil (Fig. 2), only minimal etching of the crystal surface has occurred. This relatively pristine condition suggests that the apatite has only been recently exposed to the grain surface as the enclosing feldspar

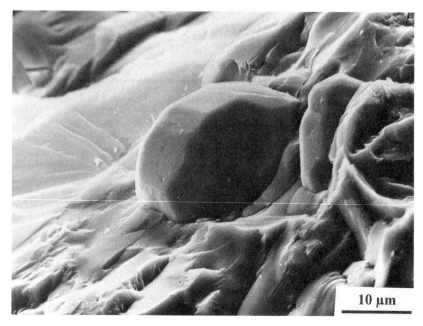

Figure 2. SEM photo of apatite inclusion in K feldspar grain from the 250 kA Merced soil profile.

grain was gradually dissolved. In the case of other major inorganic nutrients such as Mg, Ca and K, the rate of supply to the biologic community is directly dependent on the dissolution rates of the residual silicate phases.

Role of chemical weathering in buffering watershed acidification

Acidification of soils, rivers and lakes is a major environmental concern in many parts of North America and Northwestern Europe. Areas at particular risk are uplands where silicate bedrock, resistant to chemical weathering such as granites and schists, is overlain by thin organic layers. Although atmospheric deposition of hydrogen and sulfur is the most recognized factor in watershed acidification, land use practices also create acidification. An example is the effects of conifer reforestation in areas of Wales and Scotland (Farley and Werritty 1989). Chemical weathering of silicates is the only process by which acidity can be neutralized over long time periods in such environments. Cation exchange in the soil zone can cause neutralization only if base cations are replenished by weathering.

As pointed out by Drever (1988; also see this volume), the relationship between acid input and weathering rate has major practical importance. A reasonable environmental objective is to decrease atmospheric emissions to a level where the input of acidity is equal to or less than the rate of weathering in sensitive watersheds. The principal unresolved question is the feedback between the two processes. For example, if the weathering rate decreases as acid input decreases, a greater reduction in emissions would be required to bring acid loading into balance with weathering. One of the most direct methods to assess such affects has been based on experimental investigations of the effect of pH on silicate dissolution as discussed in the present volume. The dissolution rates of different silicate phases respond differently to variations in pH. As indicated in Figure 3a, the dissolution rate for plagioclase feldspar (albite) is generally insensitive to pH at near neutral values but increases with the -0.5 power of pH in acidic solutions (Blum, 1994) (also see this volume). Extrapolation of this relationship suggests that the weathering rates may increase by a factor of approximately 3 from pH 4 to 5, a range critical in the onset of watershed acidification.

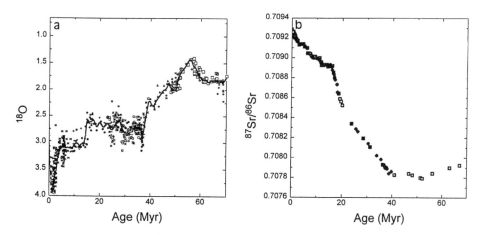

Figure 3. Isotope trends in marine carbonates over the last 70 M yr. (A) ^{18}O (after Raymond and Ruddiman, 1992) and (B) $^{87}Sr/^{86}Sr$ (after Edmond, 1992).

Data on the effects of acidification on natural weathering rates are not conclusive. A continuous 25 year record of the chemistry of bulk precipitation and surface water at the Hubbard Brook Experimental Forest is probably the best long term record for acidification

(Driscoll et al. 1989). The increase in pH (Fig. 3b) from approximately 4.1 to 4.4 correlates with a decrease in atmospheric sulfate loading with time. Silica concentrations during the same interval, however, remain essentially constant, suggesting that no discernable decrease in weathering rates has occurred with the decline in watershed acidification. In contrast, in comparing mineral and elemental depletion in soil profiles with input/output budgets in acidified Adirondack watersheds, April et al. (1986) suggested that present weathering rates may be accelerated by a factor of 3 relative to past long term rates. As suggested by Sverdrup (1990; also see this volume) the effects of pH alone can not be considered in applying measured dissolution rates to natural acidification problems. For example, decreasing pH may also increase aqueous Al concentrations in soil and surface water which may retard dissolution. Clearly, the problems associated with acidification are a subset of the overall challenge of characterizing weathering rates in watersheds.

The impact of chemical weathering on long term climate change

The intensive interest in past and present global climate changes has renewed efforts to quantitatively understand feedback mechanisms between climate and chemical weathering. On time scales longer than a million years, atmospheric CO_2 levels are primarily controlled by the balance between the rate of volcanic input from the Earths interior and the rate of output through chemical weathering of silicates at the earth's surface. Such chemical weathering through silicate hydrolysis reactions can buffer atmospheric CO_2, thus moderating large increases and decreases in global temperature and precipitation through the greenhouse effect. This balance has been significantly disturbed by the burning of large quantities of fossil fuel. A major uncertainty in climate models is the sensitivity of weathering rates to these climatic variables. If small changes in climate cause large changes in weathering, the feedback is strong, coupling in the models is tight, and predicted climate variability is limited. If weathering is relatively insensitive to changes in temperature and precipitation, then the feedback is weak and wider excursions in global climate are predicted.

The most detailed records of climate change (temperature) and chemical weathering during the Cenozoic are based on the respective $^{18}O/^{16}O$ and $^{87}Sr/^{86}Sr$ ratios of marine sediments (Fig. 4a,b). The overall increase in ^{18}O, attributed both to concentration of ^{18}O in continental ice sheets and temperature-dependent enrichment of ^{18}O in calcite, implies an overall decline in the earth's temperature over the last 70 million years. The seawater $^{87}Sr/^{86}Sr$ ratio reflects a balance between input of radiogenic Sr from chemical weathering of the continental silicate rocks and nonradiogenic Sr from hydrothermal activity associated

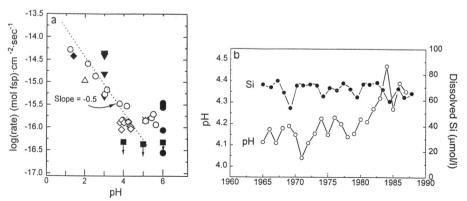

Figure 4. Effect of pH on chemical weathering rates. (a) Experimental rates for albite dissolution at 25°C (after Blum, 1994). (b) Variation in mean annual pH and Si discharge from Hubbard Brook, New Hampshire.

with seafloor spreading. To a first approximation, changes in the rates of sea floor spreading have probably been minor in the last 30 to 40 myr, implying that the increase in $^{87}Sr/^{86}Sr$ is related to increased weatherability of crystalline silicate rocks on the land.

If the increase in marine $^{87}Sr/^{86}Sr$ ratios reflect increases in weathering fluxes of continental rocks, the corresponding decrease in temperature represented by the marine ^{18}O record indicates a lack of any influence of temperature on weathering during Cenozoic time. Raymo and Ruddiman (1992) and Edmond (1992), among others, have proposed that chemical weathering rates are controlled by tectonics and that rapid weathering can occur in cool mountainous environments due to glaciation and orographic effects producing high precipitation and runoff (see Stallard, this volume). This theory argues that the potential Cenozoic "icehouse" effect created by CO_2 uptake by increased silicate hydrolysis is avoided by negative feedbacks from a decrease in the burial rate of organic carbon, and from weathering and precipitation of secondary minerals associated with submarine basalts.

In contrast, Berner et al. (1983, 1994; also see this volume) evokes a strong global feedback between increasing temperature associated with increased atmospheric CO_2 and enhanced chemical weathering. This temperature connection appears to be supported by increases in measured chemical fluxes in smaller scale watersheds in warmer climates (Velbel, 1993; White and Blum, 1995). In the climate driven scenario, the marine $^{87}Sr/^{86}Sr$ record is explained, at least in part, not by changes in absolute weathering rates but changes in the sources of weatherable rocks. For example, enhanced radiogenic Sr signature can be imparted by uplift and weathering of silicates with unusually high $^{87}Sr/^{86}Sr$ ratios, such as presently found in the Himalayan headwaters (Edmond, 1992).

A number of important issues related to weathering rates need to be addressed to resolve this controversy. The mechanisms controlling Sr release from specific mineral phases must be determined. The chemical weathering rates of bedrock relative to physically eroded sediments resulting from glaciation is also important, as is the role of topography on weathering processes.

PURPOSE AND CONTENT OF PRESENT VOLUME

This book reviews current thinking on the fundamental processes that control chemical weathering of silicates, including the physical chemistry of reactions at mineral surfaces, the role of experimental design in isolating and quantifying these reactions, and the complex roles that water chemistry, hydrology, biology, and climate play in weathering of natural systems. The chapters in this volume are arranged to parallel this order of development from theoretical considerations to experimental studies to characterization of natural systems. Secondly, the book is meant to serve as a reference from which researchers can readily retrieve quantitative weathering rate data for specific minerals under detailed experimental controls or for natural weathering conditions. Toward this objective, the authors were encouraged to tabulate available weathering rate data for their specific topics. Finally this volume serves as a forum in which suggestions and speculations concerning the direction of future weathering research are discussed.

The comprehensive nature of the volume provides opportunities to address important temporal and spacial issues that often separate the work and thinking of investigators working on specific aspects of chemical weathering. As has become apparent in assembling this volume, a number of important issues related to chemical weathering are unresolved. No effort was made to reach a consensus on these issues. Divergences in opinion were accepted between various authors and are apparent in the chapters of this volume. The following sections briefly describe the contents of each chapter and attempt to integrate these contributions into the overall theme of the volume.

Fundamental approaches in describing mineral dissolution and precipitation rates

Fundamental knowledge of the chemical and physical laws governing chemical reactions is necessary to understanding the kinetics of geochemical processes. Chapter 1, *Fundamental approaches in describing mineral dissolution and precipitation rates*, provides an overview of the approach by which scientific knowledge is integrated from many different spatial and temporal scales to understand chemical weathering. A general form for the rate law for dissolution or precipitation of heterogeneous mineral surface reactions is proposed which comprises three terms, wherein the effects of pH, temperature, and affinity can be separated. In many cases, the affinity term is the least understood.

The ultimate goal of a kinetic study is to understand the atomic mechanisms controlling the reaction. To relate general rate equations to actual mechanisms, it is often necessary to simplify the mineral structure and assume a simple surface speciation model or a surface density of specific sites such as kink sites. While most of the present geochemical database for mineral dissolution derives from flow-through reaction cells using mineral powders, new in situ optical techniques allow mineral surfaces to be imaged during dissolution. In either case, in order to extrapolate rate measurements made in the laboratory to the field, it is necessary to estimate a value for the mineral-water interfacial area. Where careful experiments can be made, and all the relevant parameters in the rate equation can be measured, surface area can then be calculated as the ratio between the observed flux and the predicted flux per unit area.

One of the newest and most exciting areas of geochemical kinetics research is the use of ab initio methods. With the exponential increase in computer power, the size of the geochemical system that can be studied from first principles has increased to the point of becoming relevant to complex issues of mineral weathering. Much of the work on bonding in mineral surfaces focuses upon zeolites and clays, where molecular clusters are chosen so as to mimic the chemical features of the phase. With larger computers, larger clusters are now modeled, allowing investigation of longer range forces. Cluster calculations have been completed for gibbsite, kaolinite, quartz, and feldspar. For example, ab initio modeling shows that adsorption of a proton onto a bridging oxygen in a tectosilicate leads to a lengthening of bond distance, which in turn lowers the activation energy for dissolution. Calculated activation energies for low pH dissolution of feldspar agree with experimental measurements. Under basic conditions, ab initio calculations suggest that the formation of an adsorption 5-fold coordinated Si species controls the activation energy and the rate of dissolution of quartz. The speed of new computers has also aided in the development of more sophisticated Monte Carlo models for dissolution, which are useful to bridge the gap between ab initio models, surface complexation models, and the overall observed rates.

Silicate mineral dissolution as a ligand-exchange reaction

Chapter 2, *Fundamental approaches in describing mineral dissolution and precipitation rates,* notes that the reactivity trends and mechanisms by which metals are released from a dissolving mineral surface are similar to the trends and mechanisms of ligand exchange around dissolved metal complexes. The correlation between mineral dissolution rate and the ligand exchange constant for the metal ion is especially compelling for metal oxides and for the orthosilicate minerals. However, for more polymerized silicates, the reactivity trends are not as easily discerned. The most important factor determining reactivity is the metal-oxygen bond strength. While adsorption of protons or ligands affects reactivity, the effect is small compared to the relative rates associated with different metal bonds.

The similarity between mineral dissolution and ligand exchange can be explained by noting several factors. The bond distances for metal-oxygen bonds in minerals are similar to the bond distances in solution. Coordination numbers are also similar between solid and solution. It is therefore useful to correlate the dissolution rate of a mineral with the acid-base properties of the surface and the acid-base properties of the metal solute ion. In addition, it can be shown that the same ligands that increase the reactivity of bonds between metals and hydration waters in solution should also increase metal oxide dissolution. Of course, solution pH is a complicating factor in that surfaces complexes may protonate differently than solution complexes. The general model of silicate mineral dissolution as a ligand-exchange reaction may be very useful in predicting reactivity trends and mechanisms, especially for silicates without extensive polymerized structures (i.e. the orthosilicates). In some cases, it may be possible to derive linear free energy relations between rate coefficients and equilibrium constants.

Chemical weathering rates of pyroxenes and amphiboles

Natural and laboratory rates of dissolution of the inosilicates are summarized in Chapter 4, *Chemical weathering rates of pyroxenes and amphiboles*. Interpretation of rates measured in experimental reactors are equivocal due to such problems as changing solution and surface chemistry, secondary phase precipitation, transport limitations, surface area changes, nonstoichiometric dissolution, and changing kinetics as a function of run duration. Comparison of natural etching rates measured for soil hornblendes yields mineral dissolution rates which are several orders of magnitude lower than laboratory rates, as observed for mass balance studies (see White, this volume): some researchers have suggested that the mechanisms of weathering in the field and in the lab reactors are distinctly different (see Hochella and Banfield, this volume). Understanding the rates of weathering of the Ca- and Mg-containing inosilicates is especially important considering the critical role that the weathering of these minerals play in the feedback between temperature and carbon dioxide over long time scales .The activation energy of these weathering reactions is probably between 80 and 120 kJ mol^{-1}.

Early experiments for inosilicates, in the absence of oxygen, documented parabolic dissolution rates; however, subsequent research has shown after an initial transient period, rates became independent in time (linear kinetics). Researchers have not reported linear dissolution kinetics of Fe-containing inosilicates in the presence of oxygen above pH 3. Even after thousands of years, dissolution of hornblende in natural systems may still not attain linear steady state with respect to surface reactions. Coupled oxidation reduction reactions have also been documented to proceed at the surface of several Fe-containing inosilicates.

Significant discrepancies exist in the literature among dissolution rates, and no rate model is generally accepted. Rate measurements at pH 5 range from 10^{-8} to 10^{-14} mol m^{-2} s^{-1}. Less research has been completed on the dissolution kinetics of inosilicates, and the relative effects of solution components are poorly understood compared to other silicate phases (see Nagy; Dove, and Blum and Stillings, this volume). Where dissolution data is unavailable, dissolution rates of endmember compositions may be predicted using the observed proportionality between log rate and connectedness (i.e. the number of bridging oxygens/tetrahedral unit) and the trend observed for dissolution rate of any silicate of a given connectedness to increase with increasing metal-solvent exchange.

Plots of log rate versus pH for amphiboles and pyroxenes differ from the other silicates (see Nagy; Dove, and Blum and Stillings, this volume). Dissolution rates tend to decrease with increasing pH above neutral. Some research suggests that an alkaline acceleration of dissolution would be observed in carbon dioxide-free systems. As predicted by the ab initio calculations (see Lasaga, this volume), dissolution rates of silicates are expected to increase at higher pH. The lack of experimental corroboration is therefore a contradiction to the theoretical calculations. In general, the slope of the log rate-pH curve for most Fe-free inosilicates is low, while the slope may be steeper for Fe-and/or Al-containing phases.

The weathering of sheet silicates

Unlike most of the other silicates, the sheet silicates characteristically both dissolve and precipitate during weathering. In natural systems, nucleation of phyllosilicates occurs naturally both homogeneously and heterogeneously, and growth occurs both epitaxially and topotaxially. In Chapter 4, *Dissolution and precipitation kinetics of sheet silicates*, the state of knowledge concerning laboratory investigation of reaction rates of phyllosilicates is summarized. In general, much more is known about dissolution rates than precipitation rates, largely because of the difficulties inherent in precipitation experiments. While the time scale for attaining equilibrium between pure water and aluminosilicate clays in closed systems in the laboratory is on the order of 2 to 4 years, almost any clay can be synthesized in the laboratory at low temperature from oxy-hydroxide gels. Precipitation from dilute solution, however, is more difficult.

Dissolution and growth of phyllosilicates occurs mainly on the basal and the edge surface. The most active sites on phyllosilicates are protonated or deprotonated metal groups at edges, and it is the net proton surface charge on edges which is thought to predominantly control pH-dependent dissolution of clay minerals under far from equilibrium conditions. Although this surface charge may also control precipitation, little data have been collected to support this hypothesis. Considerable controversy continues concerning the exact identity of the surface species which control reaction rates. Because of the reactivity of edge sites, sheet silicates preferentially dissolve inward from edge to core.

For phyllosilicates and gibbsite, dissolution rates on the order of 10^{-13} mol m^{-2} s^{-1} have been observed at pH 5 and 25°C in the laboratory. Aluminum-free phyllosilicates such as talc and chrysotile dissolve slightly faster, while brucite dissolves several orders of magnitude faster. In general, like the orthosilicates and feldspars, rates are near constant as a function of pH near neutral, and increase at both high and low pH. The absolute value of the slope of the log rate-pH curve (often referred to as the reaction order with respect to H$^+$) under acid conditions for most phyllosilicates and for the aluminum and magnesium oxides varies from small values to as high as 1.6. Under basic conditions, the absolute value of the reaction order may be closer to unity, although less data is available. Some experimental evidence suggests that the absolute value of the reaction order increases with temperature; however, the true temperature dependence is not well understood (see Lasaga, this volume). Activation energies reported for sheet silicates range from 7.1 to 88 KJ mol^{-1} under a variety of conditions.

Although the effect of pH on dissolution of phyllosilicates has been the most extensively studied, several inhibitors (such as aluminum) and catalysts (such as organic ligands) of dissolution have also been identified. Ligand exchange with surface hydroxyl groups (see Casey, this volume) is thought to cause ligand-promoted dissolution to occur in parallel to proton-promoted dissolution. Increased dissolved salt concentrations also enhance the rate of dissolution of some sheet silicates, in agreement with observations for minerals such as quartz (see Dove, this volume). Several investigators have parameterized the effect of affinity on reaction, using both simple linear models and complex models relating both Al inhibition and affinity.

Controls on silica reactivity in weathering environments

Comprising 20% of the volume of the exposed crust, quartz is found in almost every subaerial environment. As the endmember oxide to the varied classes of polymerized silicate minerals, it has been well-studied under a variety of conditions of temperature and pressure. In Chapter 5, *Kinetic and thermodynamic controls on silica reactivity in weathering environments*, the complexity of both thermodynamic and kinetic controls on silica reactivity is explored, especially under ambient conditions.

The complexity of silica chemistry is exemplified by the large number of polymorphs formed. In weathering environments, quartz is present mainly as a clastic constituent; however, opaline silica is commonly found in soils, and is typically attributed to biogenic processes, such as the formation of phytoliths in some grasses. Interpretation of the grain surface textures of quartz in soil has yielded information concerning the processes of disaggregation, fragmentation, dissolution, and precipitation. However, many grains have silica coatings, which are typically amorphous and which obscure underlying textures.

Solubility of silica polymorphs generally increases with decreasing order and density of the structure. Solubility is roughly constant as a function of pH up to pH 8.5, where the silicic acid molecule dissociates in solution and the solubility increases. In water, the quartz surface is dominated by surface hydroxyl (silanol) groups which can protonate and deprotonate. The charge of the quartz surface is controlled by these surface acid and base reactions, where no specifically adsorbing ions are present in solution. Modeling the quartz-water interface using the triple layer surface model suggests that hydrogen ions coordinate at the innermost layer (as inner-sphere complexes) while sodium and other similar ions adsorb in one of the two outer layers.

The rate of quartz dissolution has been shown to follow a rate law which is zero order in silicic acid, while the rate of precipitation follows a first order rate dependence upon silicic acid in solution. For such conditions, then, the net rate of dissolution of quartz follows a simple dependence upon the difference $(1-)$ where it is the saturation state of the solution with respect to quartz. Although this dependence has been observed under hydrothermal conditions with only water, when Pb or Na are added to solution, deviations from this simple behavior occur. The interaction between alkali cations and the silica interface also causes an increase in dissolution rate of quartz and amorphous silica under ambient conditions. The magnitude of the dissolution rate effect due to salt decreases with higher pH and increases with increasing temperature. The effect of temperature on dissolution of quartz is characterized by an activation energy in the range of 60 to 76 kJ mol^{-1}. Rate of dissolution of amorphous silica and quartz show a minimum at pH 2, the point of zero net proton charge.

The presence of organic acids in weathering solutions may accelerate the dissolution of quartz, while the presence of dissolved aluminum or ferric iron may inhibit dissolution. The presence of iron coatings may also contribute to bacterial adhesion upon quartz surfaces, although the net effect of such adhesion is unknown. Quartz dissolution has been successfully modeled using surface complexation models over a wide range of temperature, pH, and salt concentrations. In the future, this model will be extended to include Al, Fe, and organic components. This model has also been extended to predict the subcritical fracture kinetics of quartz, and thus is useful in understanding the water-promoted disaggregation of quartz-cemented rock.

Dissolution rates of feldspars

As discussed earlier in this overview, great interest exists concerning the rates of weathering reactions which can neutralize acidity generated by S and CO_2 from burning of fossil fuels; this interest especially drives research into the rates and mechanisms of dissolution of feldspars, the most common minerals in the crust. The most studied of the silicates, feldspars and their reaction rates are summarized in Chapter 7, *Feldspar dissolution kinetics*. However, despite the sustained experimentation with feldspars, extrapolation of the laboratory weathering rate to the field often overpredicts natural weathering. This discrepancy may be related to variability in factors such as reaction duration, surface area and coatings, defect density, unsaturated versus saturated flow, saturation state, and biology.

Researchers agree with the observation of increasing feldspar dissolution rate above and below neutral pH, and this is usually attributed to the protonation and deprotonation of the mixed oxide surface. The trough exhibited in the log rate-pH curve has also been observed for orthosilicate minerals (see Brantley, this volume), phyllosilicate minerals (see Nagy, this volume), and quartz (see Dove, this volume), but it is poorly documented for inosilicates (see Brantley, this volume). Interpretation of surface protonation of feldspars is made difficult due to fast proton-cation exchange and leaching of the feldspar surface. Although most workers assume that dissolution rate is a function of protonation of aluminol or silanol groups, recent investigations and ab initio calculations have suggested that protonation of bridging oxygens may control dissolution rate, especially under acidic conditions. The many models for feldspar dissolution range from atomistic, to surface speciation to leached layer/diffusion, and to macroscopic surface topographic models.

At extreme pH values of 1 or 2, the leached feldspar surface consists of a variably polymerized silicon-rich amorphous layer. Natural feldspar surfaces, weathered for up to several hundreds of thousands of years under higher pH conditions show shallow leaching of Na, Ca, and K. In general, in those regions of pH space where Si-rich layers due form, the dissolution rate of feldspars exceeds that of amorphous silica, and diffusion through the leached layer is not thought to be rate-limiting

Most dissolution rate measurements for albite and potassium feldspar suggest rates on the order of 10^{-12} to 10^{-13} mol m^{-2} s^{-1} over the range pH 5 to 8, with an apparent activation energy for dissolution of about 50kJ mol^{-1}. Anorthite dissolution rates are faster, and the apparent activation energy is close to 80 kJ mol^{-1}. A few workers have documented that, like the sheet silicates and quartz, dissolved Al inhibits the dissolution of alkali feldspars under far-from-equilibrium conditions. However, no dependence on aluminum has been documented for anorthite dissolution. Organic chelates have also been observed to strongly increase the dissolution rate of feldspars, in agreement with observations for other silicates. Most workers believe that a correlation exists between the ability of a ligand to complex aluminum and its capacity to accelerate dissolution. Workers have also documented that feldspar dissolution, again like that of the phyllosilicates, also decreases as the absolute free energy of reaction decreases. Unlike the sheet silicates and quartz, however, salt-solutions decrease feldspar dissolution kinetics under acid conditions.

Microscopic approaches to weathering

The last half of this volume describes silicate weathering rates based on observations of natural weathering processes. As mentioned in the introduction, the range of spacial scales is a paramount issue in quantify weathering rates in natural systems and the arrangement of chapters reflect the order of increasing scale. Chapter 8, on microscopic approaches to studying weathering, addresses important but often unrecognized issues related to microscopic scale weathering in the natural environment. Specific surface-related issues include the role of surface morphology on reaction rates, surface free energies, the correlation between physical and reactive surface areas and the role of specific reaction sites such as step and dislocation structures in determining reaction rates.

The chapter places significant emphasis on distinguishing between the roles that internal and external surface play in weathering reactions. Unlike experimental studies which consider short term reactions far from equilibrium, many natural weathering reactions involving primary minerals, as documented by transmission electron microscopy (TEM), occur along internal reaction fronts with secondary phases forming in close proximity to these features. In such cases, it is the microtextural characteristics of the dissolving mineral that are the most important underlying factor in determining what chemical reactions will come into play and on the rates of chemical transport to and away from such sites. In additionto solid state differences, the properties of water and associated dissolved species under such conditions are different depending on the physical dimensions of the confined space and properties of the substrate.

A model is presented which is used to explore the role of microenviroments on chemical weathering. Internal weathering in this approach is defined as the space in which the flux of material to and from the primary/secondary interface is controlled by diffusional processes. Diffusion in turn is influenced by viscosity and chemical differences of water in very small pore spaces. Due to slow movement of product species out of such spaces, primary mineral dissolution occurs at near-saturated conditions while rates of secondary mineral formation is limited by diffusion of required reactants (if any) into the pore spaces. Secondary phase nucleation may be augmented by the semi-coherent boundaries in which secondary minerals directly inherit structural components from the primary phases. Such processes are important in overcoming nucleation barriers in weathering reactions.

Weathering rates in soils

The observation that soils cover most of the earth's land surface, coupled with the early recognition that rocks decompose to form soils, has produced a vast literature describing weathering features in soils. However, the modern discipline of soil science has been chiefly concerned with the interaction of weathering products, such as clays, with the biosphere and has not extensively investigated weathering rates of primary minerals that ultimately lead to soil development. As discussed Chapter 9, *Weathering rates of silicate minerals in soils*, the soil environment presents several unique features in terms of chemical weathering. Soil profiles are physically accessible in term of weathering processes and can be often to spatially characterized in a single vertical dimension.

The concept of soil sequences, defined in terms of one dependent or variable soil property and a number of independent or constant properties, also has great potential for simplifying our understanding of weathering processes. Variable soil properties include parent material, climate, topography, age and biology. Examples of soil sequences include chronosequences developed on river and marine terraces which weather under similar conditions but for variable lengths of time and climosequences, which define soils influenced principally by differences in precipitation and temperature. A unique aspect of soils is that weathering rates can be addressed from two perspectives, long term rates based on observed chemical and mineralogical changes in the soil, and short term rates reflecting solute transport in soils. If other soil properties remain constant, the integration of the short term rates over time should result in solid state mass balances which define the long term rates. A lack of agreement would suggest changes in other soil properties with time.

Chapter 9 presents a number of methods and examples by which quantitative chemical weathering rates are determined. Approaches are demonstrated for calculating weathering velocities in saprolite profiles and in documenting the effects of climate on rates of soil development. Use of soil solute fluxes in calculating mineral weathering rates involves solving mass balances which relate the array of solute species to dissolution and precipitation of specific mineral phases. Solute mass fluxes require the determination of the fluid flux through the soil, based on either estimates of hydraulic conductivity or on input/output balances for conservative tracers such as chloride.

Calculation of weathering rates requires normalization of mass balances to specific mineral surface areas. Approaches to surface area characterization include macroscopic geometric estimates and microscopic BET measurements. Surface areas determined by these approaches progressively diverge as soil mineral surfaces roughen and porosity increases with weathering intensity. Other reasons for variations in rate constants include the effects of soil age on mineral surface reactivity, the role of hydrologic heterogeneity on fluid residence time and reactive mineral surfaces, the effects of soil solution pH, speciation and reaction affinities, and finally the impact of vegetation and climate.

Weathering rates in catchments

A catchment or watershed is generally defined in terms of a hydrologic unit which contributes to measurable surface water discharge. Generally the greater the homogeneity of a catchment in terms of geology, hydrology, topography and climate, the greater the usefulness in terms of characterizing weathering processes. Major advances in catchments studies in recent years have followed examples such as the Coweeta and Hubbard Brook studies, which utilized integrated ecosystem approaches including chemical weathering, nutrient cycling, hydrologic processes such as storm runoff events and climate. Most such studies have involved areas of moderate to high relief in vegetated temperate climates. Many of these studies have also been initiated to answer specific environmental concerns such as acid deposition, changing agricultural practices and deforestation. Affected catchments cannot be considered pristine in terms processes which may impact chemical weathering.

As discussed in Chapter 10, *Weathering rates in catchments*, one of the principal geochemical objectives in catchment studies is to define elemental input and output fluxes. The chemical weathering rate, normalized to mass flux per unit area of catchment per year, is the net difference between the input fluxes from precipitation, dry deposition, adsorption, biomass degradation and output fluxes from stream discharge, desorption and biomass degradation. The extent to which each of these terms can be accurately measured or estimated in individual catchments is also discussed as are parameters which produce differences in chemical weathering rates in different watersheds. Such differences are attributed to climate, which is dependent principally on temperature and precipitation, lithology, which influences the relative dissolution rate of different mineral phases and topographic relief, which controls the relative importance of chemical versus physical weathering. Other important parameters, generally more difficult to quantify, are the effect of vegetation type and density and the distribution and depth of catchment soils.

One of the central issues discussed is the methods for converting watershed fluxes to quantitative dissolution rates for specific mineral phases. As in the case for soils, this problem is directly related to the surface areas of specific minerals and the extent and duration of contact between mineral surfaces and infiltrating meteoric water. The chapter points out that these issues are, at least, partially responsible for the apparent discrepancies observed between experimental dissolution rates and weathering rates calculated for minerals based on watershed fluxes. Other issues are related to the aging of mineral surfaces, the formation of secondary protective layers, the presence of solutes that inhibit dissolution and the approach of field solutions to thermodynamic saturation.

Methods for reconciling experimental and soil and catchment weathering rates

As previously indicated, chemical weathering rates are important when considering a number of environmental issues. As discussed in Chapter 11, *Estimating field weathering rates using laboratory kinetics*, these concerns foster efforts to develop geochemical models, principally on the soil and catchment scale, which can be used both to explain presently observed (static) hydrochemical conditions or to predict possible changes (dynamic) based on future environmental conditions. The extent and detail to which such watershed models incorporate weathering kinetics and rates vary significantly depending on the goal, formulation and complexity of the models. Three levels of development include (1) assignment of a constant weathering rate based on the net difference between input and output fluxes for a specific study site, (2) assignment of standard rates, either arbitrary or determined as a function of soil properties and modified by solution pH, and (3) the development of a geochemical weathering model which incorporates mineral rates constants determined from experimental data.

Chapter 11contains a detailed discussion of criteria needed for an experimental kinetic data base applicable for modeling of weathering reactions. Also discussed are methods by which transition state theory can be used to separate out the effects of individual chemical parameters such as base cations, Al, CO_2, and pH on reaction rates and how mass balance equations are constructed which tie chemical weathering to fluxes related to cation exchange, nutrient uptake, soil chemistry and hydrology. These processes are incorporated into an integrated computer model which serves as a tool for comparing and evaluating rates predicted from the experimental data base but which are corrected for natural soil and catchment conditions. These predicted weathering rates are compared to rates determined from catchment budgets, Sr isotopes and soil mineral losses. An important conclusion derived from the model is that the consideration of isolated chemical parameters often produces inaccurate estimates of their relative impact on weathering. For example, watershed acidification, which decreases pH, would may be expected to increase weathering. However such acidification also mobilizes significant Al, which at the same time, tends to retard the dissolution rates and therefore buffer the reaction rates. Consideration of an integrated model results in much closer agreement between experimental dissolution rates and natural weathering rates.

Chemical versus physical weathering

One of the significant problems in quantifying the rates of denudation and erosion in natural systems is the intricacy of the coupling between physical and chemical weathering. Chapter 12, *Relating chemical and physical erosion,* explicitly emphasizes that two classes of material move through watersheds: solid products of weathering and dissolved solute loads. Although most geochemical researchers do not address the interplay between physical and chemical weathering, the importance of solid transport in altering landscapes has been studied more fully in the geological literature.

Chemical weathering enhances physical removal of material by disaggregating rock, and physical weathering enhances chemical dissolution by increasing access of water to rock material and increasing mineral surface area. Two regimes of erosion are defined: in weathering-limited erosion, the rate of transport processes exceed the rate of chemical weathering (and concurrent generation of loose material); in transport-limited erosion, the rate of chemical weathering exceeds the rate of transport, and soils continue to develop until access of water to fresh bedrock is restricted. Some lithologies or environments cannot experience both regimes: for example, unconsolidated deposits do not need chemical weathering to disaggregate the rock material. Such deposits, especially where unprotected by vegetation, often produce the greatest yields of sediment per watershed area. Similarly, glacial erosion produces large sediment yields in which the chemical disaggregation of rock appears to play an insignificant role.

Although a landscape surface may not attain steady state, the concept of steady state erosion allows the development of simple models which relate chemical and physical weathering. In these simple models, the sum of solid and dissolved fluxes leaving a water shed must equal the mass of bedrock degraded within the watershed. Quantification of these fluxes is highly dependent upon the ability to measure fluxes, which can only be accomplished by long term collection of representative solute and solid load fluxes. The dissolved load is simpler to measure, although achieving accurate separation of dissolved and suspended load is difficult and time-consuming, and often must simply rely on operational definitions. Improvements in collection of suspended sediments now enable researchers to collect velocity-weighted average samples in order to estimate fluxes. However, estimatesof the bed-material transport are still a significant source of error in river discharge studies.

Models relating the solid and solute flux for rivers can be used to allow prediction of the comparative importance of silicate and carbonate weathering, the effect of runoff, and the implications for landscape evolution. At low runoff, silicate weathering is more effective, while at high runoff carbonates weather more rapidly. The model also predicts that silicate fluxes (solid and solute) are relatively insensitive to runoff, while carbonate fluxes are a strong function of this variable. Such model results are consistent with observations of erosion-resistant carbonate landforms in arid regions, and the general absence of such landforms in the tropics. Future use of these steady state physical and chemical erosion models will have direct application to studies of land usage and landscape evolution.

Chemical weathering and its effect on atmospheric CO_2 and climate

As previously discussed, significant research is being directed at the feedback processes between silicate weathering, global CO_2 changes and long term effects on the earth's climate. Chapter 13 reviews the fundamental concepts underlying these linkages. If atmospheric CO_2 increases, there is global warming due to the greenhouse effect. But if the rate of CO_2 uptake by silicate weathering also increases with increasing temperature, a negative feedback is provided. Conversely, if CO_2 decreases, there is global cooling and the rate of CO_2 uptake by weathering decreases. Other factors affecting weathering rates over geologic time are also considered. The rise and fall of global sea levels by several hundred meters can cause significant variations in the continental land masses, which affects the weathering ratios of silicate versus non-silicate (carbonate) rocks. Active tectonics enhances physical erosion and allows greater exposure of primary silicates. Mountain uplift also increases granulation due to glaciation and produces orographic effects which can increase precipitation.

A final factor influencing weathering is the role of long term variations in the extent and type of vegetation. Vascular plants, which have evolved over the last 400 million years, may affect long term global weathering in a number of ways. Higher plants accelerate weathering in order to obtain nutrients from rocks via the secretion of organic acids from their roots. In contrast, initial results show little evidence of accelerated weathering related to primitive plants such as lichens. Organic litter from plants decomposes to CO_2 and organic acids which provide additional components for weathering. Plants also anchor soils against erosion allowing retention of water. And on a regional scale, plants recirculate water through transpiration permitting increased rain fall.

The preceding weathering processes, along with estimates of other parameters such as solar radiation, are quantified and incorporated into a model used to predict CO_2 levels in the earth's atmosphere over the last 600 my. Sensitivity analyses are presented which demonstrate the importance of the rise of vascular land plants. Also the sensitivity of CO_2 in the global climate models (GCM), which predict temperature feedback, is discussed with results indicating the correlation is fairly robust regardless of the exact model used.

WEATHERING AT DIFFERENT SCALES

Philosophy of kinetic models

As pointed out by Lasaga (this volume), the assignment of a rate equation to express mineral weathering—whether at the atomic, reactor, soil, bedrock, watershed, or global scale—must be compatible both with experimental observation and with fundamental concepts of chemical kinetics. It is the hope that a rate equation, properly parameterized for a well-understood and carefully controlled system, will be useful in extrapolating to other systems. For rate equations based upon elementary reactions, where reactions occur between molecules exactly as written, predicted rates should extrapolate to new systems where the mechanism of reaction is the same Laidler (1987). Thus, great effort is expended in measuring chemical kinetics in well-controlled systems and developing theoretical models to express the rate and mechanism of mineral reaction. For a rate equation such as those discussed in this volume, several terms must be parameterized, including the rate constant, an activation energy and pre-exponential factor, an affinity term, a surface complex term, and a surface area term.

Several philosophical approaches are emerging with respect to this general endeavor of measuring and extrapolating rates from one system to another. On one side, the "optimistic kineticist" believes that careful measurement of laboratory dissolution kinetics, surface complexation, and mineral surface areas will yield predictive models for mineral weathering at a variety of scales from the atomic surface to the global. Hovering on the other side is the "pessimistic kineticist" who wonders if accurate parameterization of these factors can ever be attained in a way that will be usefully predictive. These latter geochemists believe that natural weathering kinetics will never be fully quantified or easily predicted, but they know that the attempt at quantification is a useful process by which understanding emerges.

In fact, every geochemical modeler implicitly chooses a level of complexity for their proposed model, based upon the scale at which they address the weathering problem (Sverdrup, this volume). The level of complexity dictates the parameters which must be measured for their system. As models attempt to simulate systems at larger scales, models typically incorporate parameters measurable at the scale of interest, and ignore or lump other parameters which are measured at lower scales. For example, a mass balance model of a watershed is typically normalized per unit land area (i.e. per hectare), whereas a mass balance model of dissolution in a chemical reactor is typically normalized per unit area of mineral surface.

By the time models reach the scale of global or continental weathering (Berner, this volume), weathering is analyzed as a system of equations describing fluxes, normalized by subaerial land area, between major reservoirs. This systems approach, while unable to reproduce exact rates or elucidate fundamental mechanisms, yields insight into the sensitivity of the global system to changes in the significant variables. The modeling of Berner (this volume) thus successfully teaches the researcher about the flow of carbon dioxide through the earth's litho-, hydro-, and atmospheric system over long time scales, without parameterizing all of the complexity of the weathering reactions at the atomic scale. However, such a model is subject to the constraint that it utilize mathematical expressions which characterize the fluxes in agreement with the underlying chemical and physical principles. An example would be the assumption that weathering rates on a global scale obey the Arrhenius equation which describes the effects of temperature in reaction rates on an experimental scale. Furthermore, research must always continue to investigate whether other variables or fluxes or reservoirs may be more important than those which have been explicitly attributed to weathering processes.

Relating models at different scales

Much of the general areas of disagreement expressed within this volume concern the philosophical problem of relating models for weathering developed at different scales. While ab initio calculations can teach us about fundamental reactions which occur at the atomic scale, such calculations cannot be made with enough atoms so as to simulate dissolution at kinks at the mineral surface where dissolution is presumed to actually occur. Similarly, although we can start to measure the density of kinks using new microscopic techniques, we do not as of yet have models which can relate the measurement of BET surface area to the kink density or reactive surface area. Thus surface complexation models for dissolution (see Casey and Ludwig, this volume) typically are normalized to BET surface area, because it is measurable, rather than reactive surface area, which remains at this time unmeasurable. To scale from chemical reactors to watersheds, models usually assume weathering occurs predominantly at the soil grain surface, which is typically estimated from geometric surface area based upon acreage of land, soil porosity, and grain geometry (White, this volume).

However, Hochella and Banfield (this volume) point out that there is much evidence for weathering which occurs in natural samples along internal porosity of grains and microcracks in bedrock, which may or may not be accessed during measurement of nitrogen adsorption for BET surface area, and which is certainly ignored in the assessment of geometric surface area. Thus the question of scaling from the laboratory to the watershed becomes a problem in deciding which flow path contributes the predominant flux of the component of interest to the overall mass balance, deciding how to measure in the laboratory the rate of dissolution by the mechanism which dominates that flowpath, and determining how the mineral surface area along that flow path can be modeled.

Mineral surface area

For a system where flowpath is defined, the parameter of mineral surface area therefore represents the scaling parameter that is used to extrapolate from the atomic surface to the chemical reactor to the field. Unfortunately, this parameter is perhaps the least well-understood parameter in the commonly used rate equations. Although most experimentalists use the BET surface area to normalize dissolution kinetics, Brantley and Chen (this volume) point out that for many phases, no linear relationship has been demonstrated between rate and surface area. Thus, although almost every rate equation assumes a direct dependence of rate upon BET surface area, little evidence has been collected to support this assumption, and even less work has investigated the difference between reactive and total surface area. Similarly, watershed mass balance studies (Drever, this volume) and global models (Berner, this volume) are typically normalized by land surface area, although the proportionality between flux and land area may not be simple and direct. Again, for a watershed mass balance study, the flowpath of the water through the system, and the mineral surface area / water volume ratio along that flow path will determine the total flux of dissolved components.

Although mineral surface area has a thermodynamic definition (Lewis and Randall, 1961), as a parameter it must be defined operationally. If we follow White and Peterson (1990) and use the dissolution flux measured in any field system to calculate a "reactive" surface area based upon assumed laboratory reaction kinetics, our calculated reactive area for any field system is almost invariably less than the geometric or BET surface area (Fig. 5a). Thus the scaling parameter, surface area, has a lower value when measured in a field system of spatial dimension of kms as compared to the laboratory system of spatial dimension of cms.

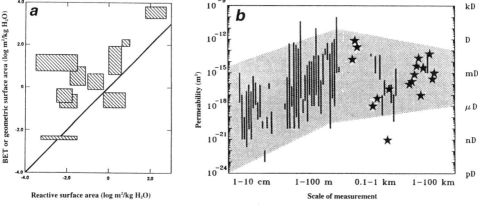

Figure 5. (a) Comparison of estimated geometric or BET surface area of soil or aquifer materials undergoing dissolution or precipitation in ten field experiments plotted versus the reactive surface area calculated by dividing the observed flux in the field by the assumed laboratory reaction rate constant (after White and Peterson, 1990). Each box represents the estimated areas, with associated error. All data sources are summarized in White and Peterson (1990). The one study which plots with reactive area > BET/geometric area was completed for a basaltic glass characterized by low reaction rates and low estimated surface area; the one study which plots directly on the line indicating reactive area = BET/geometric area summarizes the only study for which precipitation (rather than dissolution) was modelled, and the only study for which carbonates rather than silicates were investigated. (b) Ranges in values of permeability (bars) along with single analyses of permeability (stars) plotted against the characterisitc scale of measurement (from Clauser, 1992). Measurements completed at the 1 to 10 cm scale are generally laboratory measurements, while measurements made *in situ* in boreholes or determined regionally plot at the 20m to 1 km and 1 to 100 km scales respectively. Geographic locations and references for all measurements are summarized in Clauser (1992). Shaded area emphasizes the general trend of the data.

A similar phenomenon has been observed with the Darcy permeability, which also varies as a function of the scale of measurement (Fig. 5b): typically the Darcy permeability is greater when measured at the large field system scale than when measured for a small hand sample in the laboratory (Clauser, 1992). A general explanation for the discrepancy in permeability is that the larger the scale of measurement, the larger the heterogeneity of flow path which can be accessed by the flow system measurement, and in particular, the more macropores or fractures can act as "short-circuits" for flow. In this respect, the general increase in permeability measured at larger scales can partially explain the decrease in reactive surface area observed at larger scales in kinetic estimates: in moving from the laboratory to the field, the heterogeneity of flow paths increases and the effective surface area for dissolution will decrease as more water flows along "short-circuit" paths.

However, an important additional complication for dissolution reactions is the nonlinearity of chemical reactions inherent in the affinity term: reactions are inhibited as saturation state of solution increases. Thus, along precisely those flow paths where mineral surface area / water volume is highest, reaction rate will perhaps become the slowest due to the affinity effect. Investigation of this affinity effect is thus of great importance (see Blum and Stillings, this volume).

Coupling of transport and reaction

Because of the coupling between transport and reaction, accurate Quantification of mineral weathering rates in field systems will have to rely upon numerical models developed in the future which will couple fluid advection, diffusion, and chemical reaction. Such codes will evaluate the relative flux contribution from dissolution along internal porosity, along micro- and macro-pores, and along fractures. Only with such codes will we be able to simultaneously calculate flow path, mineral surface area/water volume, fluid residence time, and reaction progress—as of yet, even the simplest of these parameters cannot be successfully measured simultaneously in the field in any adequate manner (e.g. Konikow, 1986). Codes which attempt to couple fluid flow and reaction are becoming available, albeit for use over limited time and spatial scales (see, for example, Steefel and Van Cappellen, 1990).

Although full parameterization of such codes may remain impossible on any spatially extensive scale, we will need to continue to collect both field and laboratory measurements to provide the "best" inputs for the basic physical and chemical variables in the systems of interest. Even with the best input parameters such codes cannot be validated (Bredehoeft and Konikow, 1993), however, and our understanding will only continue to advance by continued design and implementation of field experiments which carefully extract one variable at a time to isolate and compare with the coupled numerical models. In this category are such field experiments as Drever and Zobrist (1992), which analyzed the effect of altitude, and the experiment of Brantley et al. (1993) which isolated the effect of drainage on weathering, and detailed calulations of weathering rates in soil chrono-sequences (White et al., 1995).

Transient versus steady state

Inherent in the foregoing discussion is the assumption that steady state dissolution rates can be measured in the laboratory or in the field over manageable time frames. Many experimenters have asserted that they have measured dissolution kinetics at steady state for slow-dissolving silicate phases. Analysis of these experiments, however, often reveals that dissolution rates, which initially started high, slowed throughout the experiment, and never reached steady state. As summarized by Brantley and Chen (this volume) and White (this volume), evidence has also been summarized suggesting that the rates of weathering of some silicate minerals in the field may also not reach a steady state. For example, Mogk and Locke (1988) have asserted that even after 100,000 y the surfaces of hornblende grains weathering in soils have not reached a steady state composition, suggesting that weathering of the grains continues to slow with time.

Similarly, researchers investigating dissolution fluxes from watersheds have noticed that long duration transient changes in solute flux occur after an environment is perturbed; indeed, it is unknown how long is necessary for a weathering landscape to reach steady state (see Stallard, this volume). As Stallard points out, the concept of steady state must entail understanding of both physical and chemical weathering for every land surface, and while some regimes may reach steady state, others may always represent transient responses to outside forcing functions such as climate. In this regard, the coupling between biological processes and weathering also introduces a timescale which is not well understood. Thus, another reason for the discrepancy between laboratory and field may result from scaling over large time scales: if steady state is generally not achieved for silicate weathering, then the longer the dissolution experiment (or the soil age), the slower the rate of weathering for any given mineral surface. Such a phenomenon may partly explain the slow natural dissolution rates observed for many silicate minerals.

CONCLUSIONS

By providing in this volume a summary and guide to the literature of chemical weathering kinetics, we have attempted to bridge the scales of theory, experimental measurement, and natural observation. The inconsistencies among chapters or among approaches is thus indicative of the problems inherent in understanding chemical weathering at all scales. We feel that many new approaches toward investigating mineral weathering currently underway will serve to push our understanding forward in the next decade.

REFERENCES

April RR, Newton R, Coles LT (1986) Chemical weathering in two Adirondack watersheds: Paststst and present-day rates. Geol Soc Am Bull 97:1232-1238

Barrett, LR, Schatal RJ (1992) An examination of podzolization near Lake Michigan using chronofunctions. Canad. J. Soil Sci. 72:527-541

Belt, T (1874) The Naturalist in Nicaragua. Chicago, University of Chicago Press, 326 p

Berner, RA (1994) 3GEOCARBII:A revised model of atmospheric CO_2 over phanerozoic time. Am J Sci 294:56-91

Berner RA, Lasaga AC, Garrels RM (1983) The carbonate-silicate geochemical cycle and its effect on atmospheric carbon dioxide over the past 100 million years. Am J Sci 283:641-683

Bischof G (1863) Lehrbuch der Chemischen und Physikalischen Geologi. Bonn, Adoph Marus 232 p

Blum AE (1994) Feldspars in weathering. In: Parsons I (ed) Feldspars and Their Reactions. Netherlands, Kluwer, Dordrecht, Netherlands, p 595-629

Bockheim JG (1979) Properties and relative ages of soils of SW Cumberland peninsula, Baffin Island. Arctic and Alpine Res 11:289-306

Brantley SL, Blai AC, Cremeens, DL, MacInnis I, Darmody, RG (1993) Natural etching rates of feldspar and hornblende. 55:262-272

Brimhall GH, Dietrich WE (1987) Constitutive mass balance relations between chemical composition, volume, density, porosity, and strain in metasomatic hydrochemical systems: Results on weathering and pedogenesis. Geochim Cosmochim Acta 51:567-587

Bredehoeft JD Konikow LF (1992) Ground-water models cannot be validated. Adv Water Res 15:75-83

Brongniart A (1839) Observations sur la structure intérieure du Sigillaria elegans compareé à cells des Lepidodendron et des Stigmaaaria et à celle des végétaux vivants. Archives du Muséum d'Histoire Naturelle, Paris, Tome I, p 405-461

Chamberlin TC (1899) An attempt to frame a working hypothesis of the cause of glacial periods on an atmospheric basis. J Geology 7:545-584

Clauser, C (1992) Permeability of crystalline rocks. EOS Trans Amer Geophys Union 73:233, 237-238

Darwin C (1876) Geological Observations on the Volcanic Islands and Parts of South America Visited during the Voyage of the HMS 'Beagle.' London, Smith and Elders, 239 p

Drever JI (1988) The Geochemistry of Natural Waters. New Jersey, Prentice Hall, 359 p

Drever JI, Zobrist J (1992) Chemical weathering of silicate rocks as a function of elevation in the southern Swiss Alps. Geochim Cosmochim Acta 56:3209-3216

Driscoll CT, Likens GE, Hedin LO, Eaton JS, Bormann FH (1989) Changes in the chemistry of surface waters. Envir Sci Technol 23:137-142

Edmond JM (1992) Himalayan tectonics, weathering processes and the strontium isotope record in marine carbonates. Science 258:594-1597

Farley DA, Werritty A (1989) Hydrochemical budgets for the Loch Dee experimental catchments, southwest Scotland (1981-1985) J Hydrology 109:351-368

Fournet MJ (1833) Memoire sur la Decomposition des Minerais d'origine ignee et leur conversion en Kaolin. Ann. de Chimie et Physique LV:240-256

Harden JW (1987) Soils Developed in Granitic Alluvium near Merced, California. U.S. Geol. Surv. Bull 1590-A 68 p

Hartt CF (1853) Geologia e geografia fidsca do Brasil. Sao Paolo, Companhia Editoria Nacional, 329 p

Hunt TS (1874) Proceedings Boston Society Natural History 16:115-117.

Konikow LF (1986) Predictive accuracy of a ground-water model: Lessons from a post audit. Ground Water 24:173-184

Jolly J (1898) An estimate of the geological age of the earth. Sci Reans Royal Dublin Soc 7 (ser 2):23-66

Laidler KJ (1987) Chemical Kinetics. New York, Harper and Row, 531 p

Lewis GN, Randall M (1961) Thermodynamics (Revised by Pitzer KS, Brewer L). New York, McGraw-Hill, p.482

Marschner H (1995) Mineral Nutrition of Higher Plants. London, Academic Press, 529 p.

Merrill GP (1906) A Treatise on Rocks, Rock Weathering and Soils. New York, MacMillan

Meybeck M (1987) Global chemical weathering of surficial rocks esimated from dissolved river loads. Am J Sci 287:401-428

Mogk DW, Locke WW III (1988) Application of auger electron spectroscopy (AES) to naturally weathered hornblende. Geochim Cosmochim Acta 52:2537-2542

Raymo ME, Ruddiman WF (1992) Tectonic forcing of late Cenozoic climate. Nature 359:117-122

Rodgers WB, Rodgers RE (1848) On the decomposition and partial solution of minerals and rocks by pure water and water charged with carbonic acid. Am J Sci 5:401-405

Steefel CI, Van Cappellen P (1990) A new kinetic approach to modeling water-rock interaction: The role of nucleation, precursors, and Ostwald ripening. Geochim Cosmochim Acta 54:2657-2677

Sverdrup HU (1990) The Kinetics of Base Cation Release Due to Chemical Weathering. Lund, Sweden, Lund University Press

Velbel MA (1993) Temperature dependence of silicate weathering in nature: how strong a feedback on long-term accumulation of atmospheric CO_2 and global green house warming? Geology 21:1059-1062

White AF, Peterson ML (1990) Role of reactive-surface area characterization in geochemical kinetic models. In: DC Melchior, RL Bassett (eds), ACS Symp Ser 416:461-475, Am Chem Soc, Washington, DC

White AF, Blum AE (1995) Effects of climate on chemical weathering rates in watersheds. Geochim Cosmochim Acta 59:1729-1747

White AF, Blum AE, Schulz MS, Bullen TD, Harden JW, Peterson ML (1995) Chemical weathering of a soil chronosequence on granitic alluvium 1:Reaction rates based on changes in soil mineralogy. Geochim Cosmochim Acta (in press)

Chapter 2

FUNDAMENTAL APPROACHES IN DESCRIBING MINERAL DISSOLUTION AND PRECIPITATION RATES

Antonio C. Lasaga

Department of Geology and Geophysics
Yale University
New Haven, CT 06511 U.S.A.

INTRODUCTION

This chapter emphasizes the fundamental approaches to the study of water-rock interactions and their ramifications in weathering, global change, environmental concerns and other crustal processes. The basic point is that major advances in our understanding of the geochemistry of Earth's surface processes can only ultimately come about from a sound understanding of the chemical and physical laws that govern these processes. It is useful to carry out the typical geochemist's scatter diagram to extract some sort of correlation between two of many variables from an assortment of field data. In fact, that is usually the important first step. But it is only a first step. True quantification of the rates of surface processes requires that a much deeper understanding follow these scatter diagrams.

Perhaps the most important task to discuss in this first chapter is the need and adequacy of integrating the scientific knowledge from many different spatial and temporal scales as well as the fundamental physics and chemistry of the mineral reactions in a full treatment of chemical weathering, which includes the global effects of such weathering. In this short course, the weathering process will be viewed from the atomistic point of view, the submicron point of view, the micron surface chemistry point of view, the laboratory point of view, the watershed point of view and the continental point of view.

A good place to start is with the general rate law for mineral precipitation and dissolution. This rate law is most directly applicable to the laboratory scale, i.e. the rate law is derived from the many and varied experimental data on mineral-water interactions, Hence, the rate law is obtained from data at the middle of the spatial range. On the one hand, then, there is a need to relate the laboratory rate law to the atomic scale phenomena. This must be done if any fundamental understanding of the kinetics is desired. This linkage provides important meaning to the various terms which may appear in the rate law. On the other hand, the laboratory rate law is then also linked to observations in the field, whether at the outcrop, watershed or global level. This short course embarks in both of these directions.

The importance of understanding the fundamental chemical processes at mineral surfaces has been well appreciated by numerous workers in recent years. In particular, the role that these processes play in determining the overall kinetic rate of dissolution/precipitation or adsorption/desorption reactions has been analyzed. Two factors that have received a lot of attention have been temperature and solution pH. Certainly, these two factors exert a pivotal role in the overall rate law for dissolution or precipitation of minerals. However, any hope of modeling the behavior of the solid crust in the presence of water requires that the possibility of achieving equilibrium be included in the model. This inclusion cannot be carried out if the rate law for the various mineral-water reactions does not incorporate the critical change of the rate with deviation from equilibrium. Understanding this variation is, in fact, essential to properly understand also the pH, temperature or ionic strength effects previously studied. As will be discussed further below, one of the most interesting discoveries of our recent work has been the non-trivial variation of the rates with deviation from equilibrium. Of course, this result has a number of important implications for our modeling of chemical weathering and geochemical cycles, implications that will be explored also in later chapters.

There are many factors that govern the rate of growth and dissolution of minerals in contact with fluids in the crust. Therefore, a very large effort has been made in recent years to formulate a general framework that can be used to extract the overall rate laws for the geologically important minerals. In particular, this effort has found that the structure and dynamics of the mineral surface must be fully understood to make substantive progress towards the kind of rate law needed in the modern-day hydrogeologic, hydrothermal, and weathering models. The kinetic study of mineral surfaces requires a combination of kinetic theory, macroscopic experiments, field observations and atomic scale observations. Indeed, the atomic scale observations of mineral surface dynamics needed to make fundamental changes in our kinetic paradigms have been quite limited.

A general form of the rate law for heterogeneous mineral surface reactions that generalizes much simpler rate laws that have been used by numerous workers and that we have been using is given by:

$$Rate \ = \ k_0 \, A_{min} \, e^{-E_a/RT} \, a_{H^+}^{n_{H^+}} \, g(I) \, \prod_i a_i^{n_i} \, f(\Delta G_r) \qquad (1)$$

k_o is an intensive rate constant with units of moles/cm^2/sec and incorporates all the pre-exponential factors involving the mineral surface dynamics. A_{min} is the reactive surface area of the mineral and E_a is the apparent activation energy of the overall reaction. At this point, it should be pointed out that the value of the "reactive surface area" to use in equation (1) and how to measure it is a major topic of debate today. This surface area enigma is a problem for both the field-based studies of watersheds as well as the laboratory measurements of mineral-fluid kinetics. We will return to it in a later section. Another much studied variable, the pH dependence of the dissolution/precipitation reactions, has been explicitly taken out by the term $a_{H^+}^n$ in Equation (1), where a_{H^+} is the activity of the hydronium ion in solution. The focus here has been on the physical chemistry governing the adsorption of H^+ and OH^- ions on mineral surfaces. A possible dependence on the ionic strength of the solution, I, in addition to that obviously entering through the activities of ions (e.g. through the interactions with mineral surface charges), is also indicated by the g(I) term. The terms in (1) involving the activities of other species in solution, a_i, incorporates other possible catalytic or inhibitory effects on the overall rate (Note that the symbol $\prod$ stands for multiplication of all such terms). Such catalysis or inhibition is kinetically quite distinct from the final term, $f(\Delta G_r)$, which accounts for the important variation of the rate with the deviation from equilibrium ($\Delta G_r = 0$).

It is important to emphasize the *form* of the equation. By writing all the various terms as products, the kinetic rate law is asserting what amounts to a "separation of variables" statement. Thus, the pH effect would be considered separately from the

Arrhenius term. An extreme case opposite to our rate law would have a term such as:

$$Rate \ = \ k_o \, A \, e^{-E_a(a_{H+})/RT} \, f(\Delta G_r) \tag{2}$$

where the *main* effect of pH is to change the exponential in the Arrhenius equation (and hence to vary the rate even at constant temperature). A-priori, there is no reason to reject Equation (2). However, kinetic theory supports (1). This is a basic statement: rate laws must be based on a good understanding of the mechanisms and atomic dynamics. As an example, there were a dozen very different functions for the temperature dependence of the rate widely in use by chemists and which fit the limited data quite well until atomic theory nailed the Arrhenius form that is used today (Lasaga, 1996). It should be noted that some of the water-rock kinetic work has suggested that the activation energy may vary with pH. We shall return to this effect; however, in this case, the assertion is not (2) but rather:

$$Rate \ = \ k_0 \, A_{min} \, e^{-E_a(a_{H+})/RT} \, a_{H+}^{n_{H+}} \, g(I) \, \prod_i a_i^{n_i} \, f(\Delta G_r) \tag{3}$$

One might make an important comment on (3). Often the a_{H+}^n term has been obtained from the pH dependence of the rate. However, if the activation energy varies with pH, the variation of $\exp(-E_a(a_{H+})/RT)$ with pH at *constant* temperature will dominate and the a^n term will exhibit quite different behavior (in fact, n may go from positive to negative in the acid range).

Some of the recent work (Kline and Fogler, 1981; Furrer and Stumm, 1986; Blum and Lasaga, 1988; 1991; Casey and Sposito, 1992; Brady and Walther, 1992) has pointed out the importance of writing the kinetic rate laws using the concentration (or activity) of adsorbed reactants on the mineral surface. Following their lead, it would be better to write our general rate law as (Lasaga et al., 1994):

$$Rate \ = \ k_0 \, A_{min} \, e^{-E_a/RT} \, X_{H+,ads}^{n_{H+,ads}} \, g(I) \, \prod_i X_{i,ads}^{n_i} \, f(\Delta G_r) \tag{4}$$

where for most reactions, equilibrium can be assumed between the mole fraction of adsorbed species on the mineral surface, $X_{i,ads}$, and the solution composition. For any given mineral, then, $X_{i,ads}$ will be a function of pressure, temperature and the solution composition. Alternatively, recent workers have rewritten Equation (4) using an *adsorption isotherm* to describe the relation between $X_{H+,ads}$ or $X_{i,ads}$ and a_{H+}, a_i in solution. For example, a Langmuir adsorption isotherm has been used by Blum and Lasaga (1991), Stillings and Brantley (1994), Gautier et al., 1994). Note that E_a is *not* a function of pH now, i.e. the effect of temperature on the H^+ adsorption isotherm is taken out in the $X_{H+,ads}(T)$ term.

The functional dependence of the rate on the overall ΔG_r is essential to any attempts to apply the experimental kinetic data to natural processes. Our ignorance of the role of this function in the interpretation of mineral dissolution kinetic data is a major flaw in current kinetic work. For example, differentiating catalytic or inhibitory effects (e.g. Al effects on feldspar dissolution) from the effects of $f(\Delta G_r)$ are very important in further interpretation of the kinetics. The most important point to make, however, is that this function, $f(\Delta G_r)$, (a) *must* be present in any overall rate law and (b) is most likely to be non-trivial (based on all the previous chemical and physical studies of simpler inorganic salts). $f(\Delta G_r)$ needs to be in the rate law because thermodynamics requires that the overall rate be modulated in such a way that $f(0) = 0$. Therefore, the rate law must depend ultimately on the ΔG_r,

Figure 1. Comparison of calculated output Al concentrations with observations for column experiments involving bauxitic gibbsite. The x-axis is the ratio L/v, where L is the length of the column and v is the flow rate. For fast flow rates, (small L/v), the concentration of Al would be linearly proportional to L/v (see text). The excellent agreement was not obtained by fitting to the data but follows from a direct application of Equation (1).

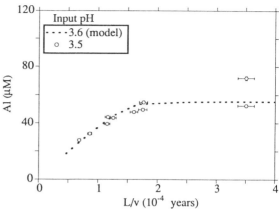

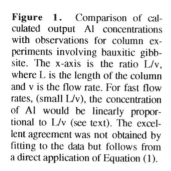

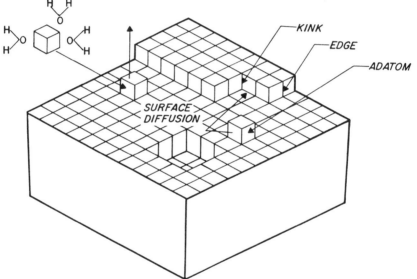

Figure 2. Basic surface processes during crystal growth or dissolution. The dynamics of the surface are controlled by the movement of steps, which, in turn, depend on the motion of adsorbed species (adatoms), kink sites, and defects on the surface. The attachment and detachment processes will also depend on hydration and dehydration of the water molecules surrounding the particular species being added or removed from the mineral surface.

which is not the same as any dependence on individual species in solution—whether or not these species are involved in the calculation of ΔG_r. As a result, a logical fundamental way to approach rate laws in water-rock kinetics is to first study the $f(\Delta G_r)$ and then work out the various terms in front of (1) or (4). Historically, the reverse procedure has taken place.

If, furthermore, the principle of detailed balancing is applicable to the overall reaction, then the derivative of f at equilibrium, $f'(0)$, is well-defined (i.e. has the same value as equilibrium is approached from undersaturation ($\Delta G_r \leq 0$) or supersaturation ($\Delta G_r \geq 0$)). The existence of $f'(0)$ renders valid the application of the theory of irreversible thermodynamics, which requires linear relations between "fluxes" and "forces" (e.g. deGroot and Mazur, 1962; Fisher and Lasaga, 1981). However, in the field of crystal growth, the theory of irreversible thermodynamics may not always be applicable even very close to equilibrium between a mineral and the composition of a solution, especially when non-equilibrium defects in the mineral play a major role in the kinetics (e.g. see Lasaga, 1981a). The shape of the function f will be discussed further below.

The overall form of Equation (1) implies an important degree of transferability from experimental work at certain conditions to other conditions. An interesting test of such transferability was carried out using gibbsite. Nagy and Lasaga (1992) carried out an experimental study of the $f(\Delta G_r)$ function for gibbsite at 80^oC and pH 3. If Equation (1) is correct, then, an adjustment of the rate for pH (with the a_{H+}^n term) and for temperature (with the $\exp(-E_a/RT)$ term) would lead to a correct overall rate under different conditions including conditions near equilibrium. In other words, the $f(\Delta G_r)$ should be transferable. Such a test was conducted using column experiments at 25^oC and at pHs of 3 to 4.2 (Mogollon et al., 1995). The rate law was adjusted for the different pH and temperature by carrying out experiments very far from equilibrium. The input solution into the columns had essentially no aluminum concentration. As the solution flowed through the column, the concentration of aluminum would vary depending on the reaction rate between the solution and the gibbsite. For slow enough flow rates, the concentration of aluminum would reach values that would approach equilibrium with the gibbsite. The output aluminum concentration (and the output pH), therefore, should be calculable with a flow and reaction model, if the rate law is accurately known. For conditions far from equilibrium the dissolution rate of the gibbsite in the column is nearly constant and, therefore, the aluminum concentration will be inversely proportional to the flow rate. However, as equilibrium is approached and the variation in the function $f(\Delta G_r)$ becomes important, the output aluminum concentration curve will begin to deviate significantly from such an inverse relation. The important assumption, based on (1), is that the *same* $f(\Delta G_r)$ function from the Nagy and Lasaga (1992) work could be transferred to the model of the column experiments at 25^oC and different pH. The comparison between the measured output of the column experiments and the calculations using such a scheme is shown in Figure 1. Note that the model predictions were not in any way calibrated to the actual experimental data exhibited in Figure 1. After calibrating the rate law for pH and T from independent experiments, the full rate law was fixed and used to obtain the dashed line in Figure 1. The very good agreement in the region of curvature (i.e. where the $f(\Delta G_r)$ is exerting an important influence), attests to the validity of the form of Equation (1), at least in this case.

A simple rate model

It is illuminating to show how the form of Equation (1) (or (4)) can arise from a well-defined underlying kinetic mechanism. The simplest of these mechanisms involves attachment and detachment at kink sites. We should first recall the properties of kink sites.

Kink Sites. The central focus of step growth occurs at sites, such as number 2 in Figure 2, which have *half* the number of bonds present as in the bulk phase. Such sites are special because they self-generate (i.e., an atom arriving at such a site creates a new similar site); they are given the name *kinks*, because of their obvious appearance. Kink sites are important sites for modeling growth and, in particular, we can show that *at* equilibrium the impingement rate of atoms to kink sites, k_+, is equal to the kink site dissolution rate, k_-,

$$k_+^{kink} = k_-^{kink} . \tag{5}$$

Note that Equation (5) is *not* true of other sites (e.g. adatoms) at equilibrium.

For the simple case of a solid containing N atoms, with each atom having s nearest neighbors in the bulk solid, it can be shown that (Lasaga, 1990) the net change in energy upon dissolution of the entire crystal is

$$\Delta E = \frac{s}{2} N \left(\phi_{ss} + \phi_{ff} - 2 \phi_{sf} \right) , \tag{6}$$

where only short-range interactions have been taken into account and where
- ϕ_{ss} = solid-solid interaction energy
- ϕ_{sf} = solid-fluid interaction energy
- ϕ_{ff} = fluid-fluid interaction energy

This result can also be written as

$$\Delta E = \frac{s}{2} N \Phi ,$$

where

$$\Phi \equiv \phi_{ss} + \phi_{ff} - 2 \phi_{sf}$$

The energy per solid particle is then given by

$$\Delta E = \frac{s}{2} \Phi . \tag{7}$$

Let us now contrast this calculation with the process whereby a particle on the *surface* with n solid bonds (i.e., n nearest solid neighbors and (s − n) fluid neighbors) dissolves. The dissolution process can be thought of as an exchange of a fluid molecule with the molecule on the surface. In this case, the energy before the dissolution (focusing only on the particles to be exchanged) is given by

$$E_{before} = -s \phi_{ff} - n \phi_{ss} - (s - n) \phi_{sf} .$$

After the exchange, the energy is given by:

$$E_{after} = -s \phi_{sf} - n \phi_{sf} - (s - n) \phi_{ff} .$$

Therefore, the net change in energy on dissolution is given by

$$\Delta E = n \left(\phi_{ff} + \phi_{ss} - 2\phi_{sf} \right) , \tag{8}$$

$$\Delta E = n \Phi , \tag{9}$$

Comparing Equations (7) and (9), it is clear that the energy change upon bulk dissolution of the solid equals the change upon detachment of a surface species, *only* when the number of bonds, n, equals s/2. This surface species is precisely what is termed a *kink* site. For example, for a cubic material (s = 6), the kink site would have 3 neighbor bonds (see Fig. 1 above). This is the fundamental link of kink sites to the thermodynamics of surfaces. Basically, the 1/2 comes into play because in the bulk we count bonds twice. This is not true of dissolution of surface species and so the dissolution of a kink site will lead to the same energy change as the energy change for bulk dissolution.

Kinetic Model. Because of the importance of establishing the *form* of the overall rate laws, it is very useful to link the form to the underlying surface and atomic scale processes by considering atomic mechanisms, which is the ultimate goal of a kinetic study. For this purpose, let us simplify the mineral structure considerably and assume a monoatomic structure such as in Figure 3. Imagine that the A* site is a kink site. Assume, furthermore, that the crystal dissolves and grows dominantly by attachments and detachments at these special kink sites. Then, if N_{kink} is the number of kink sites per unit area of mineral, the rate of addition of atoms, A, (in moles/cm²/sec) to the kink sites (Fig. 3) is:

$$R_+ = k_+ C_A N_{kink} \qquad (10)$$

where the assumption is that the rate is proportional to the collision of dissolved A atoms with the kink sites on the surface, which, in turn, is proportional to the number of A units dissolved in solution near the surface kink sites (C_A). Similarly, the rate of detachment of A units from kink sites is:

$$R_- = k_-^{kink} N_{kink} \qquad (11)$$

Both of these rates, R_+ and R_-, are in units of moles/cm² mineral/sec. At this point, fundamental use of microscopic reversibility is used. Thus, the physical concept is that at equilibrium there is no *net* rate of growth or dissolution:

$$R_{net} = R_+ - R_- = 0$$

Therefore, because the concentration of A in solution at equilibrium is C_A^{eq} (which, in fact, will be the same in the bulk as well as adjacent to the surface—even if such a difference existed in other non-equilibrium cases), we have that

$$k_+ C_A^{eq} = k_-^{kink} \qquad (12)$$

Furthermore, from the definition of free energy of a reaction:

$$\Delta G_r = RT \, ln \, \frac{C_A}{C_A^{eq}} \qquad (13)$$

or

$$C_A = C_A^{eq} \, e^{\Delta G_r / RT} \qquad (14)$$

Inserting this last expression for C_A in the equation for R_+ and using (12), we have that the total rate is given by:

$$R_{net} = R_+ - R_- = k_-^{kink} N_{kink} \left(e^{\frac{\Delta G_r}{RT}} - 1 \right) \qquad (15)$$

To extend this equation to the case that a *catalyst* is involved, let us assume that B, in solution, can adsorb onto the kink sites of mineral A (Fig. 4). In this case, the attachment and detachment rates of A units at the kink sites in the presence of B would obey the rate laws:

$$R_+ = k'_+ C_A N'_{kink} \qquad (16)$$

$$R_- = k'^{\,kink}_- N'_{kink} \qquad (17)$$

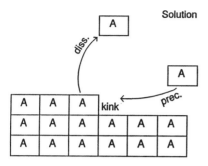

Figure 3 (top). Simple model of the growth and dissolution of a mineral by attachment and detachment at kink sites.

Figure 4 (bottom). Simple model of the growth and dissolution of a mineral by attachment and detachment of A units at kink sites in the presence of a catalyst B. The kinetics now depends also on the interaction between surface B species and dissolved B molecules.

where N'_{kink} is now the number of kink sites per unit area that have B molecules chemisorbed to them. k'^{kink}_{-} would now be the detachment rate of A units in the presence of the chemisorbed B unit (which, if B catalyzes the process by lowering the activation energies, would be much greater than k^{kink}_{-}). By the same principle of microscopic reversibility, at equilibrium:

$$k'_{+} \, C^{eq}_{A} \; = \; k'^{kink}_{-} \tag{18}$$

Furthermore, considering the reaction:

$$Kink \; + \; B^{near surface} \; = \; Kink - B$$

one obtains,

$$N'_{kink} \; = \; K_{B,eq} \, N_{kink} \, C_{B} \tag{19}$$

where C_B is now the concentration of B near the mineral surface (which in some cases will be the same as that in the bulk) and $K_{B,eq}$ represents an equilibrium adsorption constant for B on the mineral surface. Proceeding just as before, therefore, the net rate of the catalyzed reaction will be:

$$R_{net} \; = \; k'_{+} \, C^{eq}_{A} \, e^{\frac{\Delta G}{RT}} \, N'_{kink} \; - \; k'^{kink}_{-} \, N'_{kink}$$

or

$$R_{net} \; = \; K_{B,eq} \, k'^{kink}_{-} \, C_{B} \, N_{kink} \left(e^{\frac{\Delta G}{RT}} \; - \; 1 \right) \tag{20}$$

If the total rate is required then we simply multiply by the area,

$$R_{tot} \; = \; K_{B,eq} \, k'^{kink}_{-} \, A_{min} \, C_{B} \, N_{kink} \left(e^{\frac{\Delta G}{RT}} \; - \; 1 \right) \tag{21}$$

These results have the same functional dependence as Equations (1) or (4). Note that the $f(\Delta G_r)$ function takes on the classical TST form, in this case,

$$f(\Delta G_r) = e^{\Delta G_r/RT} - 1$$

which is linear with ΔG_r close to equilibrium and approaches a constant value (dissolution plateau) of -1 for very undersaturated solutions ($\Delta G_r \to -\infty$), which can be rewritten as

$$f(\Delta G_r) = e^{\Delta G_r/RT} \left(1 - e^{-\Delta G_r/RT}\right)$$

CAVEAT - TST FORMS OF THE FUNCTION $f(\Delta G_r)$

The type of function, $f(\Delta G_r)$, obtained from models involving elementary reactions at particular surface sites (such as in the previous section) needs some further discussion. This type of TST function has been discussed by (Lasaga 1981; Aagaard and Helgeson, 1982) and used by Nagy et al. (1991), for example, although with a sign convention change as

$$f(\Delta G_r) = 1 - e^{\Delta G_r/RT}, \tag{22}$$

It is important to stress that the often quoted generalization:

$$f(\Delta G_r) = 1 - e^{\Delta G_r/\sigma RT}, \tag{23}$$

is **not** on the same footing. Nagy et al. (1991) have already shown the problems with justifying Equation (23), even for a very simple case of two elementary reactions in sequence. In fact, Hollingworth (1957), who did not advocate Equation (23), also made similar points. Equation (23) is almost certainly not fundamentally valid for $\sigma \neq 1$, but has been used in recent work (e.g. Oelkers et al., 1994). It is important, therefore, to point out the fatal error in the only derivation of this equation, namely a two-page proof by Boudart (1976) (quoted in Aagaard and Helgeson, 1982). Boudart "proved" a version of Equation (23) by considering the overall reaction to consist of a sequence of s steps (e.g. elementary reactions) each with a forward, r_i, and a reverse, r_{-i}, rate:

$$1 \Longleftrightarrow 2 \Longleftrightarrow 3 \Longleftrightarrow 4 \cdots \Longleftrightarrow s$$

where 1, 2, 3,...stand for the reactants (and products) of the sequence of elementary reactions. Boudart introduces a tautology:

$$(r_1 - r_{-1})r_2 r_3 ... r_s + r_{-1}(r_2 - r_{-2})r_3 ... r_s + \tag{24}$$

$$+ r_{-1}r_{-2}(r_3 - r_{-3})...r_s + ... + r_{-1}r_{-2}r_{-3}....(r_s - r_{-s})$$

$$= r_1 r_2 r_3 ... r_s - r_{-1}r_{-2}r_{-3}...r_{-s}$$

Because an *overall rate* can only be well-defined if the entire sequence is in full steady state (e.g. Lasaga, 1981, 1996), Boudart assumes such a steady state and an overall rate, r, such that the net rate of each elementary reaction, $r_i - r_{-i}$, is the same throughout the system except for a stoichiometric number:

$$r_i - r_{-i} = \chi_i r \tag{25}$$

where χ_i is the stoichiometric number of step i. Given these assumptions, it is easy to show that

$$r = \frac{r_1 r_2 r_3 ... r_s - r_{-1}r_{-2}r_{-3}...r_{-s}}{\chi_1 r_2 r_3 ... r_s + r_{-1}\chi_2 r_3 ... r_s + ... + r_{-1}r_{-2}...\chi_s} \tag{26}$$

Boudart next introduces the overall forward rate, $\vec{r}$, and the overall reverse rate, $\bar{r}$, by letting any one individual reverse step, e.g. $r_{-i} = 0$, and any one individual forward step, e.g. $r_j = 0$. The fatal flaw sneaks in at the next step. Boudart uses the relation $r_{-i} = 0$ to obtain:

$$\vec{r} = \frac{r_1 r_2 r_3 \ldots r_s}{P} \tag{27}$$

where P is the denominator of Equation (26). Similarly, he uses the relation $r_j = 0$ to write

$$\bar{r} = \frac{r_{-1} r_{-2} r_{-3} \ldots r_{-s}}{P} \tag{28}$$

from which he concludes that

$$\frac{\vec{r}}{\bar{r}} = \prod_{i=1}^{s} \frac{r_i}{r_{-i}} \tag{29}$$

This is *wrong*. The mistake is that while the numerator in Equations (27) and (28) was properly adjusted to the different conditions (forward and reverse), the denominator was simply labeled P and then cancelled. However, as the reader can easily show, the numerator, P, is quite different in Equation (27) and in Equation (28)! This mistake completely invalidates the rest of the "proof." Of course, if one analyzes the problems that even two elementary reactions can give in obtaining (at steady state) something even remotely resembling Equation (23) (e.g. see Hollingworth, 1957 or Nagy et al., 1991), then it is not surprising that a general "proof" of (23) should have been very suspect. The main point to make here is that, unless one can make a good case for Equation (22), the function $f(\Delta G_r)$, as discussed below should be left as a very general function to be determined experimentally and related to reaction mechanisms based on careful observation of the mineral surfaces.

OTHER FORMS OF $f(\Delta G_r)$

The function $f(\Delta G_r)$ for overall reactions is difficult to predict a-priori (other than $f(0) = 0$ which guarantees that the kinetics will be fully compatible with thermodynamics). It is also true that far away from equilibrium (very negative ΔG_r), the rate ultimately becomes **independent** of ΔG_r (the dissolution plateau). This behavior is very much akin to the Henry's Law limiting behavior for dilute solutions. However, only careful experimental work can ascertain just how "far" is needed to reach the dissolution plateau in the case of any one mineral. For example, it cannot be ascertained that $\Delta G/RT < -1$ is a sufficient condition for the dissolution plateau. On the other hand, $f(\Delta G_r)$ reflects *directly* the growth/dissolution mechanism dominating the kinetics. In fact, the enormous work on the theory of crystal growth is replete with many such functions, $f(\Delta G_r)$, arising from a multitude of surface (or transport) kinetic reaction mechanisms (e.g. Ohara and Reid, 1973; Lasaga, 1996). Furthermore, these $f(\Delta G_r)$ are often quite different from the type of $f(\Delta G_r)$ in Equation (23). These mechanisms all postulate a particular behavior of the mineral surface as rate controlling (e.g. diffusion of a bulk ion to the surface, the movements of steps arising from dislocation defects, the nucleation of surface clusters, the density and motion of steps on ordered surfaces...see Lasaga, 1996). For example, some of the type of functions that can be derived for precipitation (so ΔG_r is positive) are (Lasaga, 1996):

$$f(\Delta G_r) = \Delta G_r^{1/2} \, e^{-\frac{B}{\Delta G_r}} \tag{30}$$

or

$$f(\Delta G_r) = \Delta G_r^{5/6} \, e^{-\frac{B}{\Delta G_r}} \tag{31}$$

or

$$f(\Delta G_r) = \Delta G_r^2 \, tanh(\frac{C}{\Delta G_r}) \tag{32}$$

where B and C are constants that include a variety of terms such as surface free energies and temperature. The first two equations refer to nucleation and growth mechanisms (for steps) and the last one is the classical BCF mechanism based on steps arising from screw dislocation defects on the mineral surface. The verification and validation of these functions, $f(\Delta G_r)$, necessitate a thorough understanding of the atomic structure, bonding (see *ab initio* section) and dynamics of the mineral surfaces.

SURFACE AREA

A fundamental variable in any rate law that is central whether at the atomic or the field scale, is the reactive surface area. This topic will recur throughout this book. A fundamental approach to the study of reactive surface area requires *in situ* measurements of rates at an atomic scale. This is a new and exciting area of experimental research in the field of water-rock interactions and weathering, in particular.

Most of our present kinetics experiments use conventional flow-through reaction cells and mineral powders. The use of mineral powders gives the advantage of reaction at a large number of surfaces of the mineral, thereby providing an average reaction rate. The factors in the general rate law (4) can be tested without concern for the heterogeneity of mineral surface sites. However, these factors can be studied in more detail through direct observation of the mineral surface in the process of reacting. *In situ* optical techniques for studying the mechanisms of crystal growth and dissolution have been pioneered by K. Tsukamoto, I. Sunagawa and co-workers in the mineralogy group at Tohoku University, Japan (Maiwa, et al., 1990; Ohmoto et al., 1991; Onuma et al., 1991; Onuma et al., 1993; MacInnis et al., 1993; Tsukamoto et al., 1993; Onuma et al., 1994).

The *in situ* method is useful in kinetic studies because it bypasses much of the problem of the heterogeneity of reactive surface area, very much complementing the BET-oriented powder experiments. In powder experiments, the BET surface area measurement combines the area of less reactive sites, such as on flat terraces, with more reactive sites such as ledges. For comparison, measurements of the surface-normal rate at features such as the walls of etch pits could be made with the *in situ* method. Referring to the steps of Figure 5, at a given location on the profile over the period between t_0 and t_1, the surface height change can be measured using interferometric methods to be Δh. Therefore, the surface-normal rate of dissolution, v_n, is simply

$$v_n = \frac{\Delta h}{t_1 - t_0} \tag{33}$$

Note right away that this measured surface rate (in cm/sec) can be directly related to the bulk surface rate (i.e. the rates measured in flow-through systems) if the part of the surface studied *in situ* reflects the dominant bulk dissolution rate and *if* the BET surface area is indeed correctly related to the true reactive surface area. This conclusion follows from the relation:

$$v_n = k_+ \bar{V} \tag{34}$$

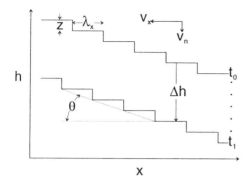

Figure 5. Schematic plot showing steps in the wall of an etch pit in the process of dissolution at times t_0 and subsequently at t_1. The pit wall is sloped at an angle θ related to the step height, z, and the step spacing, λ_x. The steps are moving to the left and consequently the surface has moved down by a distance Δh in the time between t_0 and t_1. The lateral rate of step movement is v_x and the surface-normal rate of dissolution is v_n.

where k_+ is the dissolution (growth) rate (moles/cm^2/sec) normally extracted from the bulk dissolution (or growth) experiments and where $\bar{V}$ is the molar volume of the mineral (Lasaga, 1984). Therefore, such *in situ* equipment will be able to extract independent estimates of k_+. However, k_+ is normally extracted from the flow-through experimental data (e.g. Nagy and Lasaga, 1992) by the relation:

$$k_+ = \frac{BulkRate}{A_{BET}} \qquad (35)$$

where Bulk Rate refers to the moles of mineral dissolved (or grown) in the sample cell for a given unit of time and where A_{BET} is the BET measured surface area of the mineral. Therefore, comparison of these different k_+ will provide one of the few crucial tests of the "reactive surace area-BET surface area" relationship.

On the other side of the weathering spatial scales, one may take a pessimistic outlook on the laboratory rate data and suggest that it cannot be directly applied to field problems because of lack of data on the field surface areas. From this point of view, the surface area term is very much akin to the permeability in the field of hydrogeology. In fact, hydrogeologists use Darcy's Law:

$$flow \; v = \frac{k}{\mu}(-\nabla P + \rho \vec{g}) \qquad (36)$$

where v is the fluid velocity in a porous medium, μ is the viscosity of the fluid, $\bar{g}$ is the gravity force, and ∇P is the pressure gradient acting on a fluid, to obtain the permeability, k, from flow rate measurements. It is important to note that in this case, the rest of the physics in (36) is known and thus measurement of flow rate can indeed extract k. If the the rest of the flow equation was not known, not much could be done. Similarly, in the rate equation:

$$Rate = k_0 \, A_{min} \, e^{-E_a/RT} \, a_{H^+}^{n_{H^+}} \prod_i a_i^{n_i} \, f(\Delta G_r) \qquad (37)$$

if all the chemical terms are obtained from good experimental/theoretical studies, then the term A_{min} can be extracted and studied in field measurements of the rate. For example, carefully planned column experiments (e.g. Mogollon et al., 1995) combined with laboratory flow-through kinetic studies can be very useful in extracting this kind of information.

ACTIVATION ENERGIES

The temperature dependence of the dissolution and precipitation rates is significant and is typically expressed using the Arrhenius law:

$$k = A\, e^{-E_{app}/RT} \tag{38}$$

where A is the pre-exponential factor and E_{app} is the *apparent activation energy* of the mineral-fluid reaction. Because several elementary reactions may be involved in determining the size of E_{app}, we have used the label "apparent" to distinguish it from the classical activation energy of an elementary reaction, such as treated by chemical collision theory (scattering theory) or by transition state theory. The exponential nature of Equation (38) makes temperature very important in affecting the rates. In addition, the effects of temperature can provide unique insights into the reaction mechanisms (Lasaga and Gibbs, 1990).

Recent studies have obtained apparent activation energies of mineral-solution reactions (see Table 1). The general range of apparent activation energies from recent work still leads to an average around 15 kcal/mole as suggested by Lasaga (1984). Note that an $E_{app} = 15$ kcal/mole will produce an increase of the rate by one order of magnitude between 0°C and 25 °C and another increase of an order of magnitude between 25°C and 55°C.

As the rate law in E dependence of the overall dissolution and precipitation mineral reactions may be more complex than the temperature dependence of an elementary reaction. Hence, we have used the term **apparent activation energy**, E_{app}, in Equation (38). In fact, some of the recent work (Carroll and Walther, 1990; Casey and Sposito, 1992; Brady and Walther, 1992) has focused on the variation of the apparent activation energy with pH. Because adsorption phenomena play such an important role in the dissolution/precipitation reactions, it follows that the apparent activation energy would pick up a contribution from the enthalpy of the adsorption process as discussed in Lasaga (1981b). Using Equation (4), one can write:

$$-\frac{\partial ln Rate}{\partial(1/T)} = \frac{E_{app}}{R} = \frac{E_a}{R} - n_{H+,ads}\frac{\partial ln X_{H+,ads}}{\partial(1/T)} - \sum_i n_{i,ads}\frac{\partial ln X_{i,ads}}{\partial(1/T)} - \frac{\partial ln f(\Delta G_r)}{\partial(1/T)} \tag{39}$$

Casey and Sposito (1992) have justified the pH dependence of the apparent activation energy from the second term in Equation (39). They obtain a pH dependence from the influence of the surface charge on the equilibrium constant for adsorption of hydrogen ions. Because the overall rate depends on a variety of factors (e.g. Eqn. (1) or (4)), it is not surprising that the apparent activation energy could vary with pH. However, further experiments must be carried out to ascertain whether the variation is only due to pH or to the other factors in Equation (4), particularly, the last term.

The importance of the last term in (39) has not been fully appreciated. For example, if a series of experiments at different temperatures are carried out and the ionic strength and pH are fixed, the $f(\Delta G_r)$ term can still undergo non-trivial variation because the solubility of the mineral, the dissolution rate and the solution speciation all vary with temperature. For example, the recent experiments by Casey et al. (1993) on tephroite show the approach to equilibrium of the solutions as the pH increased in the dissolution experiments. The effect of such a variation of $f(\Delta G_r)$ on the calculated activation energy can be rather large. For example, for conditions typical of those used by many workers, one can calculate a change in "activation

Table 1. Activation energies of minerals.

Mineral	E_a (Kcal/mole)	pH	References
Albite	13	Neutr	Knauss & Wolery, 1986
Albite	7.7	Basic	Knauss & Wolery, 1986
Albite	28	< 3	Knauss & Wolery, 1986
Albite	17.1	1.4	Rose, 1991
Andalusite	11.5	1	Carroll, 1989
Andalusite	5.8	2	Carroll, 1989
Andalusite	1.8	3	Carroll, 1989
Epidote	19.8	1.4	Rose, 1991
Kaolinite	16.0	1	Carroll & Walther, 1990
Kaolinite	13.3	2	Carroll & Walther, 1990
Kaolinite	10.3	3	Carroll & Walther, 1990
Kaolinite	7.7	4	Carroll & Walther, 1990
Kaolinite	2.3	6	Carroll & Walther, 1990
Microcline	12.5	3	Schweda, 1989
Prehnite	20.7	6.5	Rose, 1991
Prehnite	18.1	1.4	Rose, 1991
Quartz	17.0	7	Dove & Crerar, 1990
Sanidine	12.9	3	Schweda, 1989
Tephroite	12.9	2.5	Casey et al., 1993
Tephroite	6.3	3.5	Casey et al., 1993
Tephroite	5.7	4.2	Casey et al., 1993
Tephroite	1.1	5.1	Casey et al., 1993
Wollastonite	18.9	3 to 8	Rimstidt & Dove, 1986

energy" from 7 kcal/mole at low pH to 1 kcal/mole near neutral pH in experiments on kaolinite in the 25°C to 80°C range (Cama et al., 1995). Not only are the changes large but the predicted pattern is such as to always decrease the activation energy in the intermediate pH conditions. This type of change and pattern is precisely what has been observed. This problem would be minimal only in the "far-from-equilibrium" region. Unfortunately the definition of "far-from-equilibrium" can only be known after experiments on the particular minerals have been carried out. For example, the albite data (Burch, Nagy and Lasaga, 1993) have shown that the ΔG_r needed to be "far-from-equilibrium" exceeded -10 kcal/mole for the reaction written as in Equation (1). *Only* if the shape of the $f(\Delta G_r)$ function is known, can other effects such as catalysis, inhibition, or the effect of temperature be properly handled.

pH DEPENDENCE

A large number of recent studies on the dissolution of minerals have analyzed the pH dependence of the rates (Chou and Wollast, 1984; Helgeson et al., 1984; Furrer and Stumm, 1986; Knauss and Wolery, 1986; Tole et al., 1986; Bloom and Erich, 1987; Blum and Lasaga, 1988; Carroll-Webb and Walther, 1988; Knauss and Wolery, 1988; Wieland et al., 1988; Brady and Walther, 1989; Knauss and Wolery, 1989; Schweda,

1989; Carroll and Walther, 1990; Stumm and Wieland, 1990; Blum and Lasaga, 1991; Rose, 1991; Wogelius and Walther, 1991; Brady, 1992; Wieland and Stumm, 1992; Xie and Walther, 1992, 1994; Casey et al., 1993; Hellmann, 1994; Mogollon et al., 1994; Stillings and Brantley, 1995; Stillings et al., 1995). Under acidic conditions, the dissolution rate of many minerals is found to be proportional to a fractional power of the hydrogen ion activity, a_{H+}^n, where n, the order of the reaction, is in the range of $0 < n < 1$.

The reason for the fractional non-linear dependence of the rate on a_{H+} has been partly resolved by considering the dependence of the rate on the activity of H^+ adsorbed to the surface rather than on the bulk a_{H+} in the solution (Furrer and Stumm, 1986; Blum and Lasaga, 1988) . The amount adsorbed has been measured by acid-base titration of the mineral surface. Such titration has been applied to oxides (Furrer and Stumm, 1986), to kaolinite (Carroll-Webb and Walther, 1988), to quartz (Brady and Walther, 1992), and to albite and olivine in our lab (Blum and Lasaga, 1988; 1991). These titrations yield information on the amount of hydrogen or hydroxyl ion adsorbed at the surface of minerals as a function of solution pH. The adsorption isotherms can then be correlated with the pH-dependence of the dissolution rates. Modeling of the adsorption isotherms and dissolution data as exchange reactions between surface and solution species has resulted in identification of surface charged species which are involved in the rate-limiting dissolution step at least far from equilibrium (e.g. Blum and Lasaga, 1991). In many cases, the rate of the proton-promoted dissolution turns out to be linearly proportional to the activity of the protonated surface species (Blum and Lasaga, 1988; Brady and Walther, 1992). Certainly the experimental data have validated the form of (1) and especially of (4).

In addition, the dependence of the rate on pH at *different* temperatures will have to be further studied to verify the form of Equation (4) and to study the controversial problem of pH-dependent activation energies. As already mentioned, an important task is to differentiate the $f(\Delta G_r)$ effects in Equation (1) from *catalytic effects* due to H^+. More on this subject will be discussed in the section on ionic strength effects.

Ganor, Mogollon and Lasaga, 1995, have reexamined the pH effect on the dissolution rate of kaolinite at 25°C, 50°C and 80°C and on the activation energy of the kaolinite dissolution reaction. These new data demonstrate that the pH dependence of the dissolution rate at 25°C is stronger than previously determined. Based on the new pH dependence of the rates, the calculated activation energies for the kaolinite dissolution reaction do not exhibit a strong pH-dependence.

IONIC STRENGTH AND SOLUTION COMPOSITION

Other variables, such as the solution ionic strength, Al^{3+} concentration, or Na^+ content could affect the rates of dissolution and precipitation independently of solution saturation state. Solution ionic strength has been shown to accelerate dissolution and precipitation rates of quartz (Dove and Crerar, 1990; Dove, 1994, and references within; Hosaka and Taki, 1981a, b; Corwin and Swinnerton, 1951; and Laudise, 1958). Dove and Crerar (1990) and Dove (1994) have also shown that the nature of the cation in solution also affects the dissolution rate according to the reaction path for Si detachment in the presence of the particular cation.

The importance of factoring out the $f(\Delta G_r)$ term can be further illustrated with some recent data on the ionic strength effect ($g(I)$ in (1)) on the gibbsite dissolution rate (Mogollon, pers. comm.). Chemists have found that based on simple transition

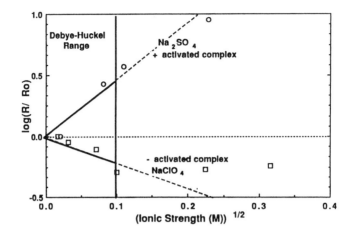

Figure 6. Ionic strength effect on the dissolution rate of natural gibbsite at 25°C and pH 3.5. The solution saturation state in the experiments was ensured to be in the dissolution plateau (Mogollon, pers. comm.).

state theory, if ions are involved in the formation of the activated complex, the rate should vary as $\sqrt{I}$, where I is the ionic strength (Lasaga, 1981, 1996). Several workers have tried to sort out the ionic strength effect on mineral dissolution rates but have obtained mixed results. Mogollon, by ensuring that the data were collected in the "dissolution plateau", has indeed obtained very interesting results that show this $\sqrt{I}$ dependence as illustrated in Figure 6. More work is needed to sort out the meaning of these data.

Similarly, the concentration of total aluminum in solution has been discussed as an important catalyst/inhibitor to mineral rates. For example, Al inhibition of dissolution rates has been proposed for albite (Chou and Wollast, 1985; Oelkers et al., 1994). For Al, it is difficult to distinguish catalysis/inhibition effects from a contribution to $f(\Delta G_r)$. Catalysis/inhibition effects can be effectively isolated by conducting experiments on the "dissolution plateau", where the rate is independent of ΔG_r.

A very useful method to separate out the catalytic/inhibitory effects of species, such as Al^{3+}, in solution from the overall role of $f(\Delta G_r)$ in the rate law involves the use of **isotach plots**, i.e. plots of constant rates contoured on an activity-activity diagram (Nielsen and Toft, 1984, Burch *et al.*, 1993). An important result of the form of Equation (1) is that if the rate law is *only* a function, $f(\Delta G_r)$, of the free energy of reaction then the *same* growth (or dissolution) rate would be obtained under various solution compositions as long as the activity product remained the same. If the rates are constant for constant ΔG_r, then the isotach plots should be straight lines with the slope determined solely by the stoichiometry of the dissolution reaction. This type of plot works very well for simple inorganic salts.

Preliminary isotach plots for albite, analcime and clinoptilolite are shown in Figure 7. Only pAl and pSi were used in these isotach plots because a_{Na^+} was maintained constant in the experiments. The dashed lines, with stoichiometric slopes of

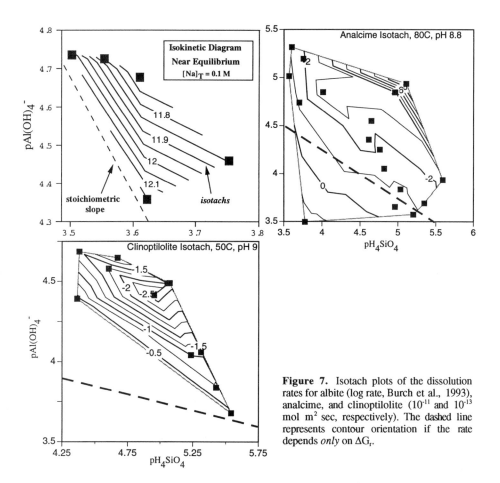

Figure 7. Isotach plots of the dissolution rates for albite (log rate, Burch et al., 1993), analcime, and clinoptilolite (10^{-11} and 10^{-13} mol m^2 sec, respectively). The dashed line represents contour orientation if the rate depends *only* on ΔG_r.

-1/3, -1/2 and -1/5 for albite, analcime and clinoptilolite, respectively, correspond to exact control by $f(\Delta G_r)$. If Al catalysis or Si catalysis were the important factors, the isotach contours would be either horizontal or vertical lines. The preliminary plots for albite and analcime suggest that the $f(\Delta G_r)$ function indeed dominates the kinetics. In contrast, preliminary results for clinoptilolite suggest Al inhibition at lower temperatures. More experiments over a broader range of Al and Si concentrations are required to further test the isotach method.

Ab INITIO APPROACHES

The use of *ab initio* (quantum mechanical) methods is steadily increasing in the earth sciences. Certainly, the nature of the bonding in minerals and mineral surfaces and the changes in atomic interactions during adsorption and reaction are one of the most fundamental approaches available to understand not only mineral weathering reactions but many other geochemical processes. One of the main driving forces in the importance of *ab initio* studies to geochemistry as a whole, is the continuing

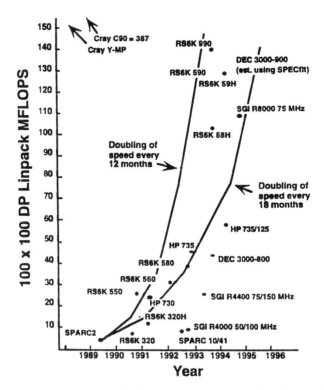

Figure 8. Representative improvements in floating point performance for various work-stations (from EMSL Benchmark Report, 1995).

exponential increase in computer power.   Figure 8 shows that the microprocessor performance is doubling every 12 to 18 months and the price of the latest hardware is often less than the price of the equipment being replaced (EMSL article, 1995). Therefore, the systems of geochemical interest that can be studied from first principles are getting bigger and more directly applicable to fundamental questions, such as those being asked here under the physical chemistry of weathering.

The previous article by Lasaga (1992) has reviewed some of the fundamental *ab initio* approaches and the first applications to mineral weathering reactions. Therefore, we focus here on more recent developments in the application of *ab initio* or first-principles methods to adsorption and chemical reaction on mineral surfaces.

A new and powerful *ab initio* method that is enabling calculations on large systems is the *density functional method* or *DFT*. In the usual *ab initio* methods, the stationary Schrodinger equation is solved in the Born-Oppenheimer approximation:

$$H \Psi = E \Psi \tag{40}$$

where H is the Hamiltonian of the system and is a differential operator constructed from the kinetic and potential energy operators (for fixed nuclei) and E is the molecular electronic energy. Ψ is the n-electron wave function. The usual solutions to (40) involve a separation of variables treatment, which write Ψ as a product of one-electron functions (molecular orbitals). The resulting one-electron equations are the so-called Hartree-Fock equations for the molecular orbitals, ϕ_i and the molecular orbital energies, ϵ_i:

$$F \phi_i(r) = \epsilon_i \phi_i(r) \tag{41}$$

where the one-electron Hamiltonian, F, involved averaging over the other electrons in the system. The problem with the HF method is that in averaging, the method neglects the instantaneous repulsion between electrons, a problem that is at the origin of the so-called *electron correlation energy*. To correct for electron correlation, the product of one-electron molecular orbitals must be augmented by additional products in what is termed a configuration interaction or CI calculation. These calculations (or similar ones involving perturbation theory, e.g. the MP2 method) are rather time-consuming, though necessary for calculating reaction energies and accurate adsorption energies.

For a system with N electrons, the Schrodinger equation (or its simplification in the Hartree-Fock equations) solves for the full wavefunction, $\Psi(r_1, r_2, ..., r_N)$. This is a function with 3N variables. From the full wavefunction, the electron density of the system can be readily computed. Because $|\Psi|^2$ is the probability of finding the N electrons at positions, r_1, r_2 ...., the total electron density is:

$$\rho(r) = N \int |\Psi|^2 \, dr_2 \, dr_3 \dots dr_N \tag{42}$$

ρ is a function only of the position, i.e. 3 variables, x,y,z. The total electron density has been used by numerous workers to understand the nature of bonding in molecules and solids (Gibbs et al., 1994). It also forms the basis for the various calculations of ionic charge on atoms. In the DFT approach, the tables are turned and the wavefunction as well as all other properties of the system are given as functions of the electron density, ρ, of the system. Hohenberg and Kohn (1964) showed that the ground state energy (and wavefunction) of a system is uniquely determined by its electron density. Furthermore, if we write $E(\rho)$ to indicate the relation between the energy and the electron density, Hohenberg and Kohn were able to prove that the energy is a minimum when the electron density is that exactly given by the ground state electron density. Therefore, a powerful variational principle was established for $E(\rho)$, which could be used to extract the density and then the ground state energy of the system.

The next step needed a specification of the actual density functional, $E(\rho)$. That is the biggest problem with DFT; while the HK theorem proves that the functional is unique (and a minimum at the exact ground state density) it does **not** give what that functional is! Therefore, approximations need to be made to come up with an appropriate $E(\rho)$, which can then be minimized to obtain the ground state electron density. One of the most successful approaches has been that of Kohn and Sham (1965), who approximated the kinetic energy of the N-electron system as if the electrons were non-interacting particles (the reader may have noticed this maneuver, which is very similar to the one used in the Hartree-Fock or in the separation of variables solution to geochemical diffusion problems, as a universal simplification by scientists of complex mathematical problems). The total energy functional, then, becomes:

$$E = E_{kin} + \int V(r)\rho(r)\, dr + \frac{1}{2} \int \int \frac{\rho(r)\,\rho(r')}{|r - r'|} \, dr\, dr' + E_{xch} \tag{43}$$

The first term is just the kinetic energy of the individual electrons. The second and third terms are merely the interaction of the electron density with an external potential (e.g. the nuclei so that V(r) is the coulomb potential between each electron and each nuclear charge) and the coulomb interaction between the electrons. The last term is the quantum mechanical exchange energy and also includes the electron correlation for the electron gas. This last inclusion is very important; the Hartree-Fock method does not include this electron correlation (which is a result of the instantaneous Coulomb interactions between the various electrons). Therefore, the normal HF methods need to be later corrected by a new calculation that adds in the electron correlation (usually done by a perturbation method, e.g. MP2 methods). The DFT approaches already have this electron correlation (at least approximately) included in the original calculation.

By minimizing the energy functional (Eqn. (43)), KS were able to obtain equations very similar to the Hartree-Fock equations:

$$[-\frac{1}{2}\nabla^2 + \phi(r) + \mu_{xch}(\rho)]\psi_i = \epsilon_i\,\psi_i \tag{44}$$

where

$$\phi(r) = V(r) + \int \frac{\rho(r')}{|\,r - r'\,|}\,dr'$$

and μ_{xch} is the electron exchange and electron correlation energy density. Both ϕ and μ_{xch} are functions of the electron density, ρ, and in later advances for non-uniform electron gases (i.e. molecules and solids), functions also of the electron density gradient, $\nabla\rho$. When these eigenvalue equations are solved, the density is obtained from

$$\rho = \sum_{i=1}^{N} |\,\psi_i\,|^2 \tag{45}$$

Because $\phi(r)$ and μ_{xch} depend on the electron density, which is calculated from Equation (45), the KS eigenvalue equations have to be solved iteratively until a self-consistent elecron density is found, again very similar to the approach to solving the Hartree-Fock equations (see Lasaga, 1992). Note, however, that these ψ_i are not molecular orbitals! The hamiltonian, H_{KS}, is pretty much the same as the Hartree-Fock hamiltonian as far as the kinetic energy and the electron-nuclear or electron-electron Coulomb interaction; it differs, however, in its treatment of the exchange term and the electron correlation. The problem is that nobody knows the correct relation between the density and the KS hamiltonian, H_{KS}. However, there are fairly good approximations for the KS hamiltonian that deal not only with the exchange term but also include a term for electron correlation right in the one-electron equations. In particular, DFT replaces the exchange-correlation energy by a one-electron integral involving local electron spin densities as well as the gradients in these densities. The exchange part is either that due to Slater (based on the behavior of a uniform density electron gas) or to Becke, the latter including a gradient correction. The electron correlation is included using local spin densities (LSD) from the work of Vosko, Wilk, and Nusair (VWN), which again is based on exact uniform electron gas results, or is treated using the gradient-corrected functional of Lee, Yang and Parr (LYP). These various DFT approaches have been now incorporated in the new releases of Gaussian 92 and Gaussian 94; therefore, making DFT calculations quite readily available to the chemical and geochemical community.

Recent comparisons of the DFT methods with the other *ab initio* methods are now available (e.g. Johnson et al., 1992). One of the DFT methods, the BYLP method, includes corrections for the gradient of the electron density (in the energy functionals) as well as electron correlation. Johnson et al. (1992) compared the BYLP

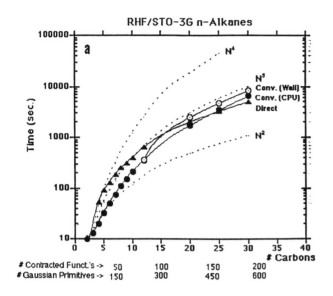

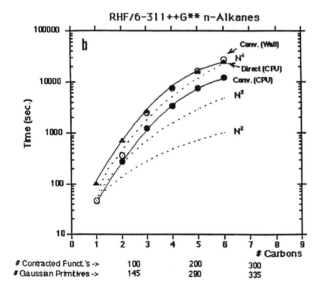

Figure 9. (a) Growth in CPU times as a function of the number of carbon atoms in a hydrocarbon chain. The calculations are Hartree-Fock calculations using a small STO-3G basis set. Results were obtained with Gaussian 92 on a DEC AXP 300. The number of basis functions and the total number of gaussians used to represent them is also given on the x-axis. (b) Same as (a) but now the basis set is a much more extensive 6-311++G** basis set (see Hehre et al., 1992 for notation; from EMSL Benchmark Report, 1995).

results with Hartree-Fock and with the electron correlation MP2 method, **all** using the same 6-31G* basis set on 32 molecules. For example, the bond lengths were found to be $0.011\mathring{A}$ too small for HF, and $0.010\mathring{A}$, $0.020\mathring{A}$ too long for the MP2 and BYLP methods, respectively. On the other hand, the vibrational frequencies are calculated more accurately by the BYLP method than the MP2 method (with BYLP having a mean absolute deviation of 73 cm^{-1} for all frequencies (compared to the harmonic frequencies) and MP2 having a mean absolute deviation of 99 cm^{-1}.

An important property of the various *ab initio* methods, is the manner in which the computational time (or cost) scales with the size of the system. Typically, many books claim that the Hartree-Fock method scales as N^4, where N is the number of atomic basis functions required to properly describe the system. The N^4 arises because the electron repulsion integrals depend on the product of 4 atomic orbitals. However, as mentioned by Johnson et al., (1992), the computational cost of HF methods for large systems is not N^4 (N = number of atomic basis functions) but rather N^2, for well-written codes. This difference arises because many of the electron repulsion integrals have values near zero and a smart program would not compute these integrals. Similarly, the computational cost of DFT methods is not N^3 but is somewhere between N and N^2. Actually, the scaling can vary depending on the system and the basis set. For example, Figure 9 shows the scaling properties of Hartree-Fock calculations on n-alkanes as n (and therefore N) increases. For this small basis set (STO-3G), the time needed indeed does scale as N^3 rather than as N^4 (using the efficient Gaussian 92 program). However, the same calculation using a very diffuse basis set, 6-311++G** leads to much less savings in the electron integrals and hence scales closer to N^4. Nonetheless, one of the advantages of DFT *ab initio* methods is that they can deal with large systems at a level that includes electron correlation (e.g. equivalent to MP2-HF methods) more efficiently that the *ab initio* methods based on Hartree-Fock. A comparison of the results for H_2O and H_4SiO_4 is given in Table 2. The table also gives the CPU time on a DEC Alpha-3000 computer.

The ability to handle larger systems is useful for the general studies of surface complexes and surface reactions, as will be evident in the discussion below.

Ab initio applications

Much of the recent work on bonding in mineral surfaces has focused on zeolites and clays. Zeolites are framework aluminosilicates which contain networks of channels and cavities that can host a wide array of molecules. Substitution of aluminum for silicon in the tetrahedral sites leads to a negative charge on the framework, often compensated by sodium ions, but sometimes compensated by a proton. In the latter case, some of the protons lead to the formation of a bridging hydroxyl group such as Si-OH-Al, which becomes a Bronsted acidic site (Fig. 10). The hydroxyl groups in zeolites play a major role in the catalytic properties used in numerous industrial processes. The catalysis of cracking, isomerization and alkylation of hydrocarbons occurs by the proton transfer from the zeolite to the carbon molecules forming carbenium or carbonium ions. The proton transfer itself, depends on the ability of the lattice to accomodate the extra negative charge. Thus, the acidity of the proton on zeolites depends on both the state before reaction as well as the stability of the so-called "Zwitter-ion" state, generated upon proton transfer to the reacting molecule (e.g. van Santen, 1994). Therefore, much focus has been done on the adsorption characteristics at these sites. For example, the adsorption of methanol on Bronsted acid sites in zeolites is the initial step in the methanol-to-gasoline process and as such has been the subject of *ab initio* work (Haase and Sauer, 1995; Shchegolev, 1994). The structure of the Si-OH-Al group has also been studied with vibrational spectroscopy (Fig. 11). Typical values of the OH stretch frequencies vary between 3650 cm^{-1} and 3550 cm^{-1} (compared to the silanol, SiOH, stretch frequency at 3745 cm^{-1}, van Santen, 1994). Experimental and theoretical values of the stretch, in-plane bending and out-of-plane bending vibrational frequencies of these acidic OH sites are in close agreement (e.g. Sauer, 1989).

Many studies have focused on molecular clusters that mimic the various features of zeolites, including the acid sites. Recent calculations involving *ab initio* quantum

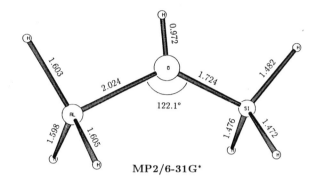

MP2/6-31G*

Figure 10. Key acid catalytic site in zeolites. The hydrogen ion is bonded to the bridging oxygen between one Al and one Si tetrahedron forming the Bronsted acidic group. This hydrogen ion site is very acidic. The structure of the cluster was carried out using an MP2/6-31G* *ab initio* calculation (Xiao and Lasaga, 1994).

Table 2. Comparison of *ab initio* methods.

H₂O

	6-31G*	MP2	Density Functional SVWN	BLYP	expt
r_{OH}	0.947Å	0.969	0.975	0.980	0.958
θ_{HOH}	105.5°	104.0°	103.7°	102.7°	104.5°
CPU time[a]	45 sec	50 sec	64 sec	70 sec	

H₄SiO₄

	6-31G*	MP2	Density Functional SVWN	BLYP
Si-O	1.6289	1.6533	1.667	1.641
O-Si-O-1	106.4°	105.8°	105.9°	105.6°
O-Si-O-2	115.7°	117.1°	116.8°	117.5°
O-Si-O-3	106.4°	105.8°	105.9°	105.6°
OH	0.947	0.970	0.978	0.975
CPU time[a]	34 min	134 min	73 min	79 min

a - CPU time on a DEC AXP 3000/800 workstation

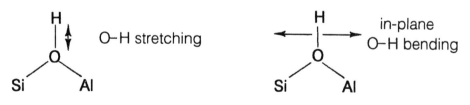

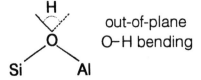

Figure 11. The three vibrational modes of the Bronsted acidic group.

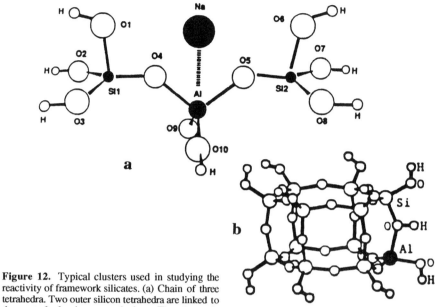

a

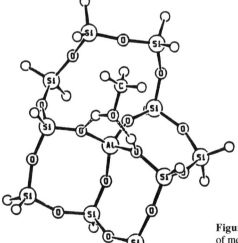

b

Figure 12. Typical clusters used in studying the reactivity of framework silicates. (a) Chain of three tetrahedra. Two outer silicon tetrahedra are linked to the central aluminum tetrahedron. The charge is balanced by a Na⁺ ion (from Uchino et al., 1993). (b) Large cluster used to simulate a zeolite. All the tetrahedra in the framework cluster contain silicon atoms except for one, which contains Al. Note the Bronsted acid hydrogen ion between one of the silicon tetrahedra and the aluminum tetrahedron. The large open spheres are silicon atoms and the smaller open spheres are oxygens. All the outer oxygens are terminated by hydrogens. The solid sphere is Al. *Ab initio* calculations even with this large a cluster were carried out using an extensive basis set, TZ2P equivalent to 6-31G** (from Limtrakul et al., 1995).

Figure 13. Equilibrium structure of the adsorption of methanol onto faujasite. The *ab initio* calculations used an extensive basis set, double-ζ^+ polarization basis DZP (from Haase and Sauer, 1995).

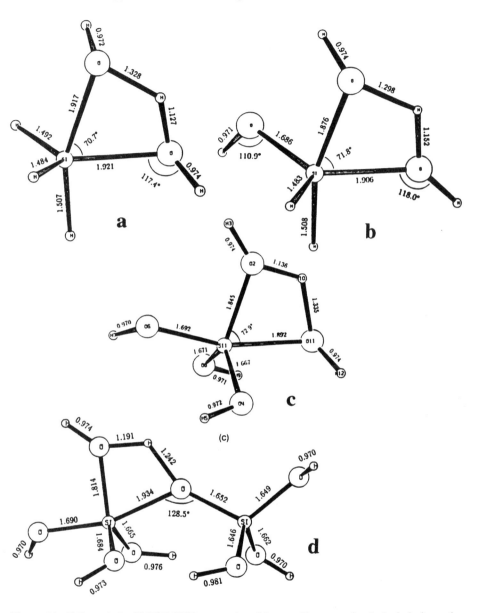

Figure 14. Fully optimized MP2/6-31G* geometries of the transition states for the hydrolysis reaction of water on (a) H_3SiOH; (b) $H_2Si(OH)_2$; (c) H_4SiO_4; and (c) $H_6Si_2O_7$ (from Xiao and Lasaga, 1994).

mechanical solutions to the Schrodinger equation have begun to use **large clusters**. Figure 12 gives a range of clusters used recently to investigate the various adsorption reactions on the surfaces of zeolites and clays. Limtrakul et al. (1995) and Haase and Sauer (1995) have employed large enough clusters to mimic the cage-like units in zeolites. For example, Figure 13 gives a cluster used in *ab initio* calculations which is typical of the faujasite zeolite structure (Haase and Sauer, 1995). Thus, a comparison

Table 3. Hydrolysis transition states activation energies (kcal/mol)

	HF/3-21G*	HF/6-31G*	MP2/6-31G*	ΔE^{expt}
$\Delta E_a^{\ddagger 1}$	22.74	44.12	30.43	
$\Delta E_a^{\ddagger 2}$	20.10	40.44	25.62	
$\Delta E_a^{\ddagger 3}$	23.83	42.42	28.17	
$\Delta E_a^{\ddagger 4}$	-	45.90	31.98	
$\Delta E_a^{\ddagger 5}$	19.71	42.32	28.51	16-21[a]

[1]Based on $H_3SiOH + H_2O^* = H_3SiO^*H + H_2O$
[2]Based on $H_2Si(OH)_2 + H_2O^* = H_2SiOHO^*H + H_2O$
[3]Based on $H_4SiO_4 + H_2O^* = H_4SiO_3O^* + H_2O$
[4]Based on $H_6Si_2O + H_2O^* = H_3SiO^*H + H_3SiOH$
[5]Based on $H_6Si_2O_7 + H_2O^* = H_4SiO_3O^* + H_4SiO_4$
a - Rimstidt and Barnes, 1980; Dove and Crerar, 1990; Dove, 1994

Table 4. SCF optimized structural parameters computed for different models.

System	Method	r(Å)			Angle(deg)	
		Si-O(H)	Al-O(H)	OH	SiO(H)Al	SiOH
$H_3SiOHAlH_3$	3-21G	1.732	1.930	0.967	128.9	120.1
	DZ	1.753	1.970	0.963	128.6	120.9
	DZP	1.706	2.027	0.951	132.3	117.8
	TZ2P	1.684	2.021	0.943	132.0	118.0
$(OH)_3SiOHAl(OH)_3$	3-21G	1.721	1.844	0.964	118.0	120.6
	DZ	1.741	1.889	0.963	119.3	122.5
	DZP	1.709	1.908	0.950	120.7	120.2
	TZ2P	1.684	1.928	0.943	121.5	120.3
$Si_6Al_6O_{30}H_{18}$	3-21G	1.716	1.839	0.971	134.1	116.4
	DZ	1.745	1.879	0.968	133.3	116.6
	DZP	1.725	1.887	0.954	136.4	114.1

DZ - double zeta basis set
DZP - double zeta basis set with polarization (e.g. d) basis functions
TZ2P - triple zeta basis set with polarization functions

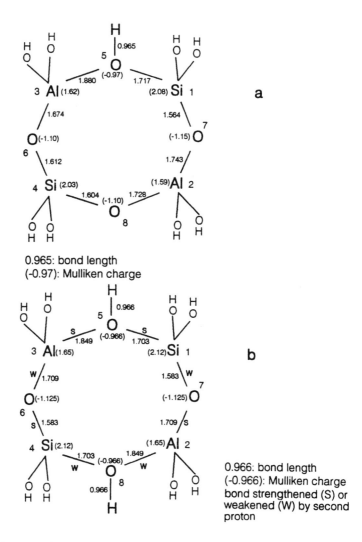

Figure 15. Four-ring cluster containing two silicon and two aluminum atoms and one (a) or two (b) acidic protons. Structure (a) is the deprotonated version of structure (b) (from van Santen, 1994). Note the changes in next-to-nearest neighbor bonds in going from (a) to (b), which just adds a hydrogen ion at position 8.

of the structural differences arising from the limitations of small clusters can be systematically studied. For example, Figure 14 compares the various transition states obtained for the hydrolysis by water of the Si-O-Si bond using a variety of clusters (including H_3SiOH, where one H is simulating the other silicon in the bridge). Table 3 compares the activation energies obtained (Xiao and Lasaga, 1994). Both the structures and energetics are rather similar. Another comparison is given in Table 4 from Limtrakul et al. (1995). Note the strong similarity arising from the covalency and local nature of the bonding in aluminosilicates, as has been repeatedly pointed out (e.g. Lasaga and Gibbs, 1987, 1991).

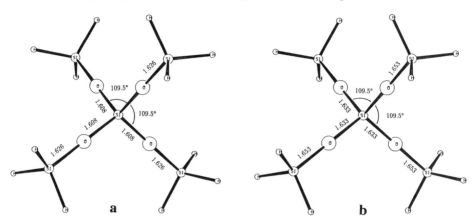

Figure 16. (a) Structure of a cluster of 5 silicon tetrahedra optimized using a 3-21G* basis set. The outer silicons are terminated by hydrogens. Note the linear Si-O-Si angles. Previous work (Lasaga and Gibbs, 1991) have found that additional polarization functions are needed on the bridging oxygens to yield a correct bent Si-O-Si angle. (b) Same as (a) but now adding electron correlation using density functional theory (BLYP). The Si-O-Si angle is still linear. (Xiao and Lasaga, 1996a).

It is interesting to point out, however, that the ability to carry out large clusters enables the next-order effects to be ascertained. There are some interesting differences that arise from longer range interactions. For example, Figure 15 shows the optimized geometry of two clusters containing 4 tetrahedral linkages in a ring (similar to that in feldspars). Both have two Al and two Si tetrahedra. However, in (a) there is an Si-OH-Al linkage, while (b) has two such linkages (i.e. cluster (a) is a deprotonated (b) cluster). Note that the Al-O and Si-O bond distances adjacent to the OH group (Si-4 and Al-2 in the figures) are much longer (1.849 and 1.703 versus 1.728 and 1.604 Å). But also the Al-O bond lengths of Al-3 (see figure for notation), which is not involved in the protonation-deprotonation, are different in the two cases. Thus, we have some indication of longer range chemical effects, albeit not as large as those in the region where the bonding is changing. Nicholas and Hess (1994) used periodic Hartree-Fock theory to compare cluster calculations with *ab initio* calculations on an infinite lattice. They also stressed that while the clusters and the infinite lattice gave very similar results, there were some significant differences, indicating the need to explore these longer range effects in the future.

The use of pentameric tetahedral units has become quite feasible at a high level of theory. For example, some of our recent optimizations of two pentameric structures are given in Figures 16, 17 and 18. The pentameric silica cluster in Figure 16 terminates the tetrahedra with H atoms. This smaller cluster yields equal Si-O bond lengths for the inner silicon atom and Si-O-Si angles of 180°, a typical problem, if extra d orbitals are not put on the bridging oxygens (e.g. Lasaga and Gibbs, 1991). This result holds true even with the calculation using density functional theory including electron correlation. Figure 17 shows the results for the bigger pentameric cluster, now using SiO_4 tetrahedra on all four tetrahedra. It is interesting to note that the problematic Si-O-Si angle, which comes out linear with the same level calculation with smaller clusters, now has an angle of 151° for the HF calculation and 160° or 131° for the density functional calculation (expt is 144°). Moreover, the bond lengths around the central Si atom now have two values, as in quartz, namely 1.614 and 1.633 Å for the DFT calculation (1.60 Å and 1.61 Å in quartz). The pentamer (Fig. 18) with one Al tetrahedra at the center exhibits Al-O-Si angles of 133° (130°

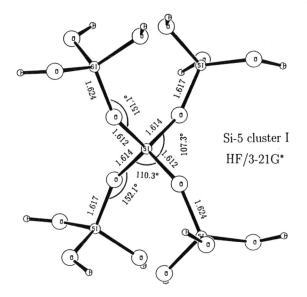

Si-5 cluster I

HF/3-21G*

Figure 17. More extensive 5-tetrahedral cluster with oxygens added to all the outer silicons. Calculations employ a 3-21G* basis set. Now the Si-O-Si angle is 151°C, much closer to experiment (144°C). The inner SiO bond lengths are also split into two different values, similar to what are observed in quartz (Xiao and Lasaga, 1996a).

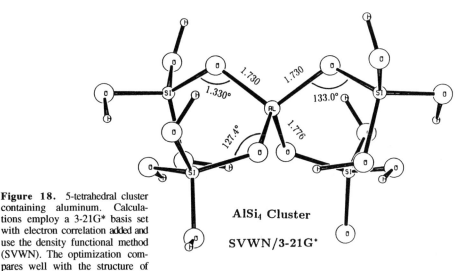

Figure 18. 5-tetrahedral cluster containing aluminum. Calculations employ a 3-21G* basis set with electron correlation added and use the density functional method (SVWN). The optimization compares well with the structure of feldspar.(Xiao and Lasaga, 1996a).

AlSi₄ Cluster

SVWN/3-21G*

in feldspar) and two Al-O bond lengths of 1.73 and 1.77 Å.

Other examples that are relevant to current research on the weathering of oxides and clays are the cluster calculations shown in Figures 19 and 20 for the minerals **gibbsite** and **kaolinite** respectively (Xiao and Lasaga, 1996a). As can be seen, the large clusters can reproduce within the experimental uncertainties rather accurately the structure of the distorted edge-sharing aluminum octahedra in gibbsite and of the linkage between the Si tetrahedral sheet and the Al octahedral sheet in kaolinite.

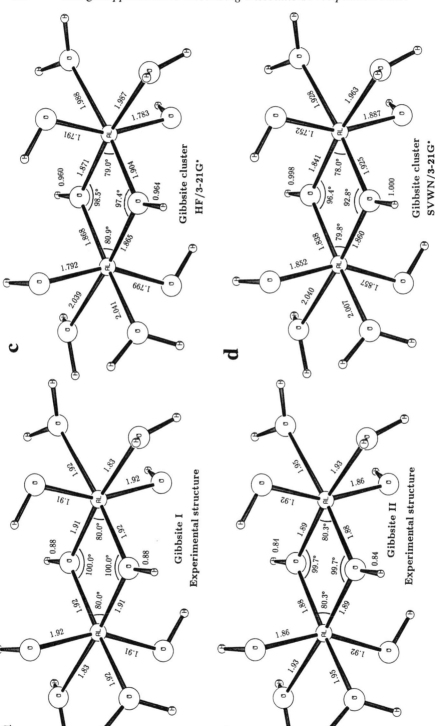

Figure 19. (a,b) Experimental structure of the two edge-sharing distorted aluminum octahedra in the gibbsite structure (data from Saalfeld and Wedde (1974)). (c) *Ab initio* results for the optimization of the edge-sharing aluminum octahedra using 3-21G* basis set. Note the favorable comparison with (a) or (b). (d) *Ab initio* results now adding electron correlation using the density functional method (SVWN in Gaussian 92) (from Xiao and Lasaga, 1996a).

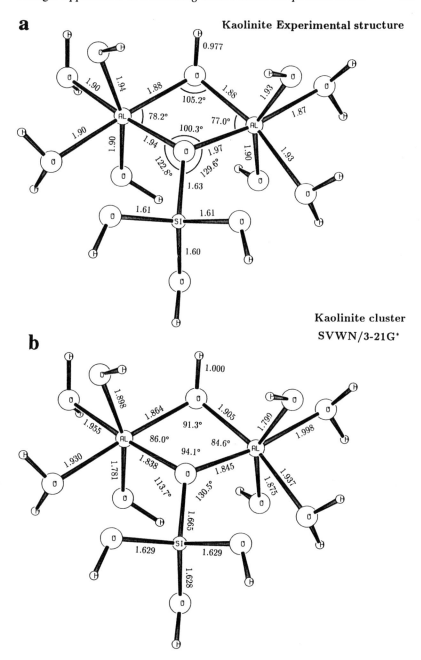

Figure 20. (a) Experimental structure of the edge-sharing distorted aluminum octahedra and the linking of the silica tetrahedra in the kaolinite structure (from Bish 1993). (b) *Ab initio* results for the optimization of a kaolinite cluster using a 3-21G* basis set and electron correlation from a density functional method (SVWN in Gaussian 92). Note the favorable comparison between (a) and (b) (from Xiao and Lasaga, 1996a).

For example, the range of Al-O bond lengths in the gibbsite distorted octahedra has a wide variation from 1.83 $\overset{\circ}{A}$ to 1.95 $\overset{\circ}{A}$ experimentally and the *ab initio* calculations yield 1.78 $\overset{\circ}{A}$ to 2.04 $\overset{\circ}{A}$ for 3-21G* and 1.75 $\overset{\circ}{A}$ to 2.04 $\overset{\circ}{A}$ for the density functional calculation including electron correlation. Note that the outer Al-O bonds will not be as accurately calculated because of the truncation of the cluster. Both the Al-O-Al and O-Al-O angles in the shared edge of the octahedra are very accurately predicted by the *ab initio* calculations. Similarly, the kaolinite cluster predicts a slightly longer Si-O bond length for the Si-O-Al$_2$ portion in accordance with observation. The Al-O bond lengths in the kaolinite cluster calculation are also very similar, although the Al-O-Al angles are not as closely reproduced as with gibbsite. We will return to the fundamental studies of reactions involving gibbsite and kaolinite below.

Much of the earlier work has used the smaller cluster, disiloxane, H_6Si_2O, to model the bridging oxygen on framework silicates (Shchegolev, 1994) The structure of disiloxane has been investigated recently (Teppen et al., 1994; Nicholas et al., 1992; Luke, 1993) as well as the acidity of the hydroxyl groups, Si-OH-Al and Si-OH-Si. Comparisons to the higher cluster disilicic acid, $H_6Si_2O_7$, have also been done recently (Teppen et al., 1994).

The adsorption energetics and structure of a variety of molecules on aluminosilicates has been studied. These include water (Pelmenschikov and Santen, 1992; 1993; Xiao and Lasaga, 1994) CO (Bates and Dwyer, 1993), CH_3OH (Haase and Sauer, 1995 NH_3 (Kyrlidis et al., 1995); proton transfer (H_3O^+) (Marchese et al., 1993; Kassab et al., 1993; Sauer et al., 1990; Xiao and Lasaga, 1994). Polymerization reactions have also been studied (Chiang et al, 1993). One of the important results of the theoretical work has been to emphasize the need to carry out electron correlation calculations to properly take into account dispersion effects in adsorption energetics.

The adsorption energy for methanol onto the faujasite cluster was calculated to be -83 kJ/mole compared to a measured value of -110 to -120 kJ/mole (Haase and Sauer, 1995).

Recent work has also focused on the local structure of and surface reaction of steam with sodium aluminosilicate glasses (Uchino et al., 1992, 1993) using clusters involving three silica or alumina tetrahedra linked in a chain.

Ab initio studies and dissolution/precipitation rates

The main interest from the point of view of weathering is not only the adsorption of possible catalysts (e.g. H^+, OH^-, organics) but also the actual reaction pathways that lead to hydrolysis or polymerization of bonds. Lasaga (1992) has already reviewed some of the earlier work on the fundamental approaches to the main atomic processes in weathering reactions. It will be useful to include here some of the recent results in this field. Xiao and Lasaga (1994, 1995), have analyzed the role of H^+ and OH^- in the catalysis of aluminosilicate weathering reactions. The adsorption of a proton onto a bridging oxygen, whether in an Si-O-Si or Al-O-Si linkage, leads to dramatic lengthening of the bond lengths, which produce much lower activation energies for reaction. Figure 21 shows the structural changes for the Al-O-Si linkage (e.g. feldspars and zeolites). Table 5 compares the energetics of the hydrolysis of Al-O-Si with water and when catalyzed by protons. Note the very good agreement with the measured feldspar activation energies for the proton-catalyzed reaction. The situation under basic conditions is a bit more complicated and discussed further below.

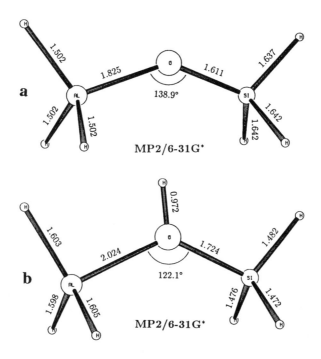

Figure 21. Effect on the SiO and AlO bond lengths upon the adsorption of a H^+ ion on the bridging oxygen. Calculations were done at the MP2/6-31G* level. This dramatic lengthening of the bonds leads to the important catalysis of the dissolution reaction of feldspars and zeolites by acids (from Xiao and Lasaga, 1994).

Table 5. Hydrolysis of feldspar.

$H_6SiOAl + H_2O$ (Kcal/mol)

	HF/3-21G*	HF/6-31G*	MP2/6-31G*	ΔE^{expt}
ΔH_o	22.76	34.83	26.27	17-29[a]

$H_6SiOAl + H_3O^+$ (Kcal/mol)

	HF/3-21G*	HF/6-31G*	MP2/6-31G*	ΔE^{expt}
ΔH_o^b	19.99(20.16)	16.27(16.40)	15.87(15.95)	17-29[a]

a - Helgeson et. al, 1984; Knauss and Wolery, 1986; Rose, 1991
b - numbers in () are based on $H_6AlOSi + H_3O^+$

The main conclusions from the *ab initio* studies of acid-catalyzed dissolution reactions are:

Dissolution Under Acid Conditions.
(1) The kinetics of quartz and feldspar dissolution at low pH involves two mechanisms, i.e., H_2O hydrolysis and $H^+(H_3O^+)$ catalysis.

(2) There is only **one** unique adsorption site for both water and the hydronium ion onto the Si–O–Al surface unit. Thus the adsorption occurs at distances roughly equal to both the Si and Al atoms of the bridging unit.

(3) The adsorption of $H^+(H_3O^+)$ onto the bridging oxygen sites of Si–O–Si and Si–O–Al linkages plays the *key role* in catalyzing the dissolution process. Calculated activation energies agree well with experimental data.

(4) The $H^+(H_3O^+)$ catalysis reaction pathway leads to major changes in the potential surface and the reaction coordinate. These changes lead to not only lower activation energies but also to smaller kinetic isotope effects, much closer to experimental data at low temperatures.

(5) The activation energies for the H^+-catalyzed reactions have been found to be distinctly lower than those for water hydrolysis, $E_a(H_3O^+) < E_a(H_2O)$. This result is in the opposite direction to the pH-dependent "apparent" experimental activation energies. The *ab initio* results show that the effects of other factors (e.g. adsorption or ΔG factors) are quite significant in the overall activation energies obtained for feldspar dissolution.

(6) The acid-catalyzed hydrolysis of the Si–O–Al bond is calculated to be preferred over the acid-catalyzed hydrolysis of the Si–O–Si bond, which explains, using first principles, the observed formation of leached layers at low pH in feldspar dissolution studies.

(7) There still remain big inconsistencies between the experimental and the theoretical kinetic isotope effects, both for quartz and feldspar dissolution. Because the isotope effects are sensitive to the reaction path, this discrepancy is important and needs to be studied further in the future.

Dissolution under basic conditions.   The mechanism under basic conditions has turned out to be somewhat surprising, emphasizing the power of *ab initio* methods to uncover unforeseen kinetic approaches. The generally accepted mechanisms of quartz dissolution in basic pH solutions can be summarized as (a) Direct attack by H_2O on a negatively charged surface site; (b) Catalysis by hydroxide ion (OH^-) on a neutral surface site. In order to test these proposals and to understand the full dynamics of the dissolution processes from first principles, Xiao and Lasaga (1995) have carried out high level ab initio molecular orbital calculations to study the following reactions:

$$(HO)_3Si–O–Si(OH)_2O^- + H_2O \longrightarrow (HO)_3Si–OH + {}^-O–Si(OH)_3$$
$$(HO)_3Si–O–Si(OH)_3 \quad + OH^- \longrightarrow (HO)_3Si–OH + {}^-O–Si(OH)_3$$

In this recent study, disilicic acid, $(HO)_3Si-O-Si(OH)_3$, and its -1 charged conformation, $(HO)_3Si-O-Si(OH)_2O^-$, have been chosen (see Fig. 22) to simulate the neutral

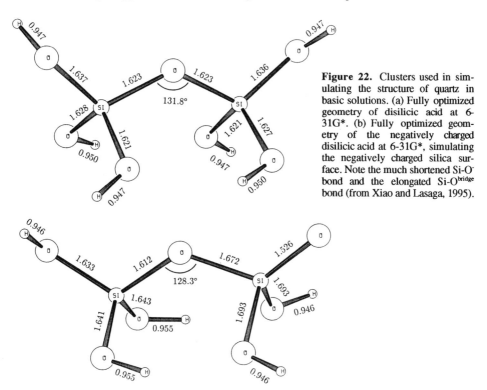

Figure 22. Clusters used in simulating the structure of quartz in basic solutions. (a) Fully optimized geometry of disilicic acid at 6-31G*. (b) Fully optimized geometry of the negatively charged disilicic acid at 6-31G*, simulating the negatively charged silica surface. Note the much shortened Si-O⁻ bond and the elongated Si-Oᵇʳⁱᵈᵍᵉ bond (from Xiao and Lasaga, 1995).

and negatively charged quartz surface site, respectively. Their reactions with OH⁻ and H_2O, which lead to the hydrolysis of the Si-O-Si bond, have then been thoroughly investigated. Figure 23 summarizes the energy changes involved in the overall reaction that leads to the rupture of a Si-O-Si bond. The work of Xiao and Lasaga (1995) has discovered some interesting consequences: (1) OH⁻ attack on $(HO)_3Si$-O-$Si(OH)_3$ will lead to the deprotonation of $(HO)_3Si$-O-$Si(OH)_3$ by transferring one H to the OH⁻, resulting in a H-bonded H_2O adsorption onto a negatively charged Si-O⁻ site, $(HO)_3Si$-O-$Si(OH)_2O^-\cdots H_2O$. This process is equivalent to H_2O attack on the negatively charged $(HO)_3Si$-O-$Si(OH)_2O^-$. (2) The next step involves the formation of a 5-fold coordinated Si species, $(HO)_3Si$-O-$Si(OH)_4^-$; this step has to overcome a large energy barrier (18.91 kcal/mol). The 5-fold coordinated Si species has almost the same potential energy as the reaction precursor - the hydrogen bonded H_2O adsorption minimum, but it significantly weakens the Si-O-Si bond; (3) The final step is the rupture of the Si-O-Si bond to form $(OH)_3Si$-OH $\cdots$ ⁻O-$Si(OH)_3$, with a much smaller energy barrier (4.49 kcal/mol). Therefore, their *ab initio* results uncover *two* adsorption minima in series before the cleavage of the Si-O-Si bond and suggest that, it is *the formation of the adsorption 5-fold coordinated Si species, not the actual hydrolysis step, that controls the activation energy and the rate of quartz dissolution in basic pH solutions*. The calculated activation energy, 18.91 kcal/mol, is similar to the one (24 kcal/mol) in H_3O^+ catalyzed hydrolysis of Si-O-Si bond, while much smaller then that (29 kcal/mol) in pure H_2O hydrolysis (see Xiao and Lasaga, 1994), indicating the role that OH⁻ as well as H_3O^+ play in catalyzing quartz dissolution. The normal mode analysis of the relevant transition states indicate that, even though

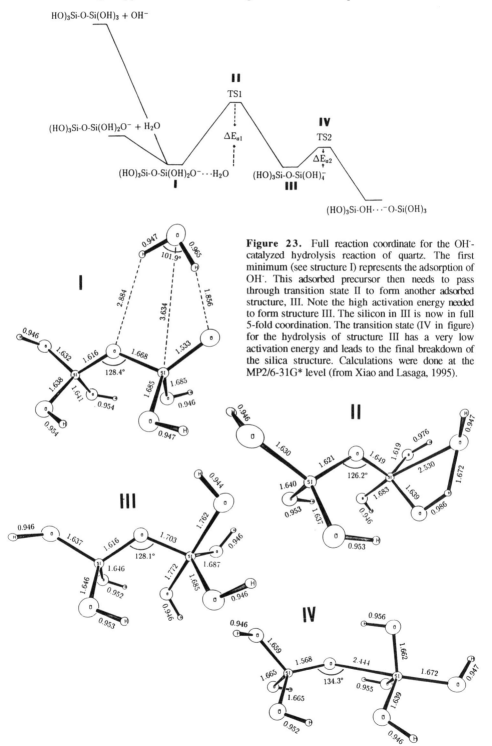

Figure 23. Full reaction coordinate for the OH⁻-catalyzed hydrolysis reaction of quartz. The first minimum (see structure I) represents the adsorption of OH⁻. This adsorbed precursor then needs to pass through transition state II to form another adsorbed structure, III. Note the high activation energy needed to form structure III. The silicon in III is now in full 5-fold coordination. The transition state (IV in figure) for the hydrolysis of structure III has a very low activation energy and leads to the final breakdown of the silica structure. Calculations were done at the MP2/6-31G* level (from Xiao and Lasaga, 1995).

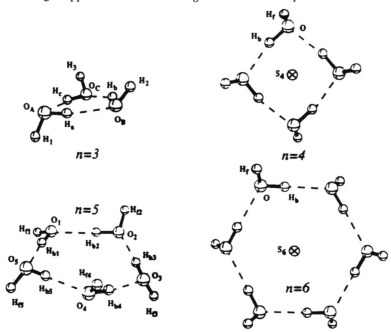

Figure 24. (a) Optimal structures of water clusters, $(H_2O)_n$ for n = 3 to 6. Calculations (Xantheas, 1995) employ electron correlation (MP2) and an extensive triple-zeta basis set.

H transfer has been involved in the overall reaction, the dominant mode in the rate determining step is the forming and breaking of the Si-O bond. This will lead to a small hydrogen kinetic isotope effect, which agrees well with recent experimental results at basic pH.

The overall conclusions from the *ab initio* results in basic solutions are:

(1) The kinetics of quartz dissolution at basic pH might involve two mechanisms, i.e., H_2O hydrolysis and OH^- catalysis.

(2) Both OH^- attack on a neutral quartz surface and H_2O attack on a negatively charged quartz surface will lead to a H-bonded H_2O adsorption on a $Si-O^-$ site. Therefore, these two mechanisms are essentially the same.

(3) The hydrolysis process includes two elementary steps: a) The formation of a 5-fold coordinated reaction intermediate; and b) The breakdown of the Si-O-Si bond. The calculated barriers for these two steps are 18.91 kcal/mol and 4.49 kcal/mol, respectively. This result suggests that it is the formation of the 5-fold coordinated species, not the actual hydrolysis step, that determines the activation energy and rate of quartz dissolution in basic pH solutions.

(4) The OH^- catalyzed reaction pathway leads to major changes in the potential surface and the reaction coordinate. These changes lead to not only lower activation energies but also to smaller kinetic isotope effects, much closer to experimental data at basic pH.

(5) The activation energies for the OH^- catalyzed reactions have been found to

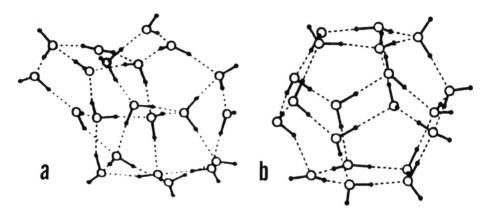

Figure 25. (a) Dodecahedron structure of $(H_2O)_{20}$. Hydrogen bonds are indicated by dashes (from Laasonen and Klein, 1994). (b) Lowest energy structure of the hydronium ion containing cluster: $(H_2O)_{20}H^+$ (from Laasonen and Klein, 1994).

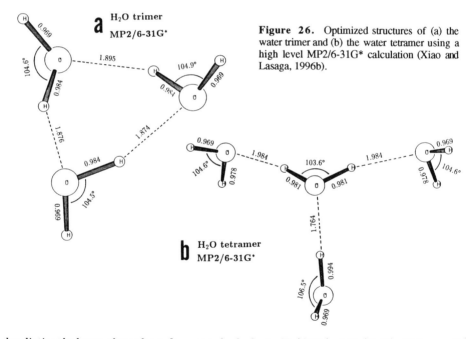

Figure 26. Optimized structures of (a) the water trimer and (b) the water tetramer using a high level MP2/6-31G* calculation (Xiao and Lasaga, 1996b).

be distinctly lower than those for water hydrolysis, $E_a(OH^-) < E_a(H_2O)$. This result is in the opposite direction to the pH-dependent "apparent" experimental activation energies. The ab initio results indicate that the effects of other factors (e.g. adsorption or ΔG factors), as discussed earlier in the chapter, may be significant in the overall activation energies obtained for quartz, feldspar or other aluminosilicate dissolution reactions.

(6) There still remain some inconsistencies between the experimental and the theoretical kinetic isotope effects. Because the isotope effects are sensitive to the reaction path, this discrepancy is important and needs to be studied further in the future.

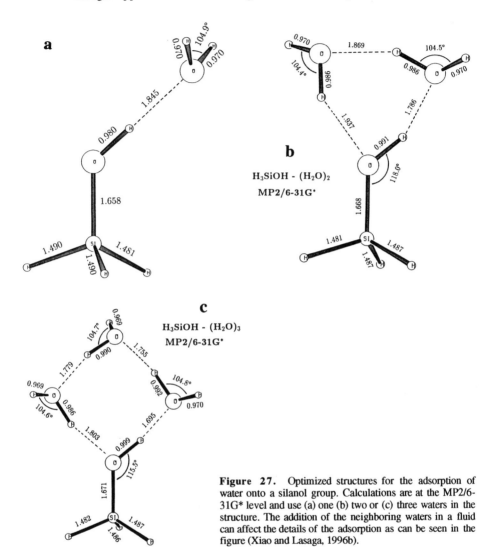

Figure 27. Optimized structures for the adsorption of water onto a silanol group. Calculations are at the MP2/6-31G* level and use (a) one (b) two or (c) three waters in the structure. The addition of the neighboring waters in a fluid can affect the details of the adsorption as can be seen in the figure (Xiao and Lasaga, 1996b).

Effect of hydration layers

A new important direction of study with *ab initio* techniques is the role of several water molecules in the adsorption and reaction schemes discussed earlier. In other words, to what extent do the nearby water molecules in solution modify the adsorption and hydrolysis reactions of both water and ions such as H^+, OH^- etc. Recent studies have been carried out on the structure of $(H_2O)_n$ clusters, $n = 1$ to 6 (Xantheas and Dunning, 1993; Xantheas, 1994, 1995) and even $(H_2O)_{20}$ and $(H_2O)_{21}H^+$ (Laasonen and Klein, 1994) (see Figs. 24 and 25). These studies have been extended to include the hydration of ions, e.g. Na^+ (Hashimoto and Morokuma, 1994); Li^+, Na^+, F^-, Cl^- (Combariza and Kestner, 1995) and Li^+, Na^+, K^+, Rb^+ (Glendening and Feller, 1995).

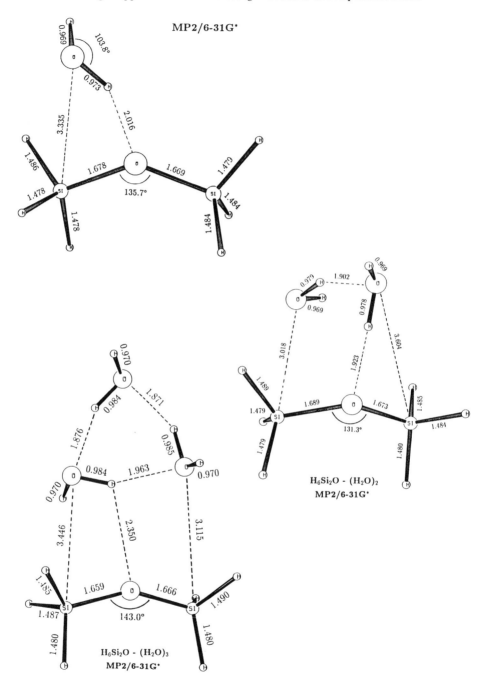

Figure 28. Optimized structures for the adsorption of water onto a bridging Si-O-Si site. Calculations are at the MP2/6-31G* level and use (a) one (b) two or (c) three waters in the structure. The addition of the neighboring waters in a fluid can affect the details of the adsorption as can be seen in the figure (Xiao and Lasaga, 1996b).

The ability to investigate the hydration of ions to a very high level using *ab initio* is important. Several recent studies (Sverjensky, 1992; Dove, 1994; Ludwig et al., 1995) have related the dynamics of ion hydration to the reactions occurring on the surfaces of minerals. To the extent that one can simplify mineral reactions with these relations, *ab initio* studies of the details of the ion-water interactions or organics-water interactions can provide useful information.

Figure 26 shows the structure of the clusters $(H_2O)_3$ and $(H_2O)_4$ at a high level calculation (MP2/6-31G*). Figure 27 compares the adsorption of water onto a silanol group including nearest-neighbor water molecules from solution (Xiao and Lasaga, 1996b). Figure 28 does the same comparison for the disiloxane cluster. The effect of including the extra waters on the adsorption structure and the energetics is non-trivial. For example, the ΔE_{ads} for water adsorbing onto the silanol group in H_3SiOH:

$$(H_2O)_n + H_3SiOH \implies H_3SiOH - (H_2O)_n$$

is -5.77 kcal/mole for the case of 2 waters (n = 2) and -8.19 kcal/mole for three waters (n = 3). Similarly, for disiloxane, the adsorption energy (for the precursor complex to hydrolysis - e.g. Lasaga and Gibbs, 1990) increases from 8.72 kcal/mole (n = 2) to 10.04 kcal/mole (n = 3).

Inclusion of the nearest water molecules in solution in studies of adsorption and surface reactions should be an important direction of study in the near future. Preliminary calculations suggest that the isotope effects and the activation energies may come much more in line with experiment for the quartz reaction, once these outside waters of hydration are taken into account. Future research will definitely involve sorting out these important hydration effects, not only for the water hydrolysis and the pH-catalyzed reactions but also for ionic and organic effects.

REACTION MECHANISMS AND RATE LAWS

The next level from the atomic bond-forming and bond-breaking studies of *ab initio* calculations, is the formulation of kinetic reaction mechanisms and their relation to observable rate laws in the lab or in the field.

The mechanism of quartz dissolution involves the adsorption ideas discussed in the last section as well as the principles utilized in the discussion of homogeneous kinetics. The quartz structure consists of a three dimensional framework of interlocking silica (SiO_4) tetrahedra. Each tetrahedron is linked to each of four other tetrahedra by means of an oxygen atom that they share. Thus the key link between tetrahedra is the Si-O-Si unit that is shown in Figure 29. Water molecules may adsorb on the surface of silica. These adsorbed water molecules can react with the Si-O-Si bonds on or near the surface. Based on the suggestion in Lasaga and Gibbs (1990a,b) we will postulate that the rate of dissolution depends on the rate of hydrolysis of the Si-O-Si unit by adsorbed water molecules. Figure 29 gives the proposed mechanism. As each of the four links of any given tetrahedron are hydrolyzed by water, the tetrahedron is "loosened" from the surface. After all four anchoring Si-O-Si have been broken, the tetrahedron becomes an $Si(OH)_4$ (or H_4SiO_4), which can easily desorb from the surface and enter the solution as $H_4SiO_4(aq)$. If we break the mechanism into each step, the reaction mechanism would be (see Lasaga, 1990):

$$H_2O + \equiv Si- \overset{K_1}{\rightleftharpoons} H_2O_{ads} \bullet Si \overset{k_1}{\to} \equiv Si - OH$$

$$H_2O + \equiv Si - OH \overset{K_2}{\rightleftharpoons} H_2O_{ads} \bullet Si - OH \overset{k_2}{\to} = Si(OH)_2$$

$$H_2O + \; = Si(OH)_2 \stackrel{K_3}{\rightleftharpoons} H_2O_{ads} \bullet Si(OH)_2 \stackrel{k_3}{\rightarrow} -Si(OH)_3 \qquad (46)$$

$$H_2O + -Si(OH)_3 \stackrel{K_4}{\rightleftharpoons} H_2O_{ads} \bullet Si(OH)_3 \stackrel{k_4}{\rightarrow} H_4SiO_4^{ads}$$

$$H_4SiO_4^{ads} \stackrel{k_5}{\rightarrow} H_4SiO_4(aq)$$

The adsorption of water molecules is assumed to occur fast enough that the concentration of adsorbed molecules reaches equilibrium. The K's refer to the equilibrium constants for H_2O adsorption onto each particular site. The slow hydrolysis steps have individual rate constants given by k_i. If we apply steady state to $\equiv$Si-OH we have that

$$k_1[H_2O_{ads} \bullet Si] + K_{-2}[H_2O_{ads} \bullet SiOH] = K_2[\equiv Si - OH]$$

Applying steady state to $H_2O_{ads}\bullet$SiOH,

$$k_2[H_2O_{ads} \bullet Si - OH] + K_{-2}[H_2O_{ads} \bullet Si - OH] = K_2[\equiv Si - OH]$$

Thus, from these two equations, it is easy to obtain that $k_1 [H_2O_{ads}\bullet Si] = k_2 [H_2O_{ads}\bullet$ Si-OH]. Similarly, applying the principle of steady state to each of the various intermediates in the reaction mechanism (46), the reader can verify that

$$k_1[H_2O_{ads} \bullet Si] = k_2[H_2O_{ads} \bullet Si(OH)] = k_3[H_2O_{ads} \bullet Si(OH)_2] \qquad (47)$$

$$= k_4[H_2O_{ads} \bullet Si(OH)_3] = k_5[H_4SiO_4^{ads}] \; .$$

Because this reaction is heterogeneous, the concentrations of these species are given as moles/cm^2 surface area of quartz. For a given amount of quartz, there is a finite total reactive surface area. Of course, by introducing the term "reactive" we are dealing with only those surface sites which are actively involved in the reactions displayed above. If the total number (moles/cm^2) of reactive sites is labeled, S_{tot}, and if the brackets [ ] now stand for the surface concentration of any species (again in moles/cm^2) then conservation of surface sites requires that:

$$S_{tot} = [\equiv Si] + [\equiv Si(OH)] + [= Si(OH)_2] + [-Si(OH)_3] \qquad (48)$$

$$+ [H_4SiO_4^{ads}] + [H_2O_{ads} \bullet Si] + [H_2O_{ads} \bullet Si(OH)]$$

$$+ [H_2O_{ads} \bullet Si(OH)_2] + [H_2O_{ads} \bullet Si(OH)_3] \; .$$

The next step is to rewrite the concentrations of the various surface species in terms of the concentration of "free" surface sites, [$\equiv$Si]. This step can be accomplished from Equation (47) and the equilibrium assumption for the H_2O adsorbed species. For example, from the first two terms of (47) :

$$k_1[H_2O_{ads} \bullet Si] = k_2[H_2O_{ads} \bullet Si(OH)]$$

Substituting the equilibrium relations for the two adsorbed species:

$$k_1 K_1 a_{H_2O} [\equiv Si] = k_2 K_2 a_{H_2O} [\equiv Si(OH)]$$

Therefore, we obtain,

$$[\equiv Si(OH)] = \frac{k_1 K_1}{k_2 K_2} [\equiv Si]$$

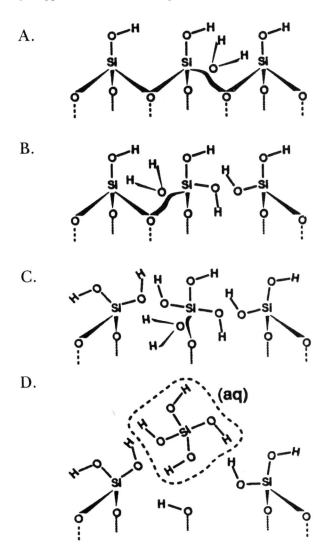

Figure 29. Reaction mechanism (A-D) for the hydrolysis of quartz by water molecules.

Similarly,

$$[= Si(OH)_2] = \frac{k_1 K_1}{k_3 K_3} [\equiv Si]$$

and

$$[-Si(OH)_3] = \frac{k_1 K_1}{k_4 K_4} [\equiv Si]$$

And the equilibrium reactions themselves lead to:

$$[H_2 O_{ads} \bullet Si] = K_1 \, a_{H_2 O} [\equiv Si]$$

Again returning to the relation,

$$k_1[H_2O_{ads} \bullet Si] = k_2[H_2O_{ads} \bullet Si(OH)]$$

it follows that

$$[H_2O_{ads} \bullet Si(OH)] = \frac{k_1}{k_2}[H_2O_{ads} \bullet Si]$$

or equivalently that

$$[H_2O_{ads} \bullet Si(OH)] = \frac{k_1}{k_2} K_1 a_{H_2O} [\equiv Si]$$

Proceeding similarly with the rest of the concentrations and factoring out the common factor, $[\equiv Si]$, Equation (48) may be rewritten as:

$$S_{tot} = [\equiv Si] D \tag{49}$$

where the parameter D is given by:

$$D = 1 + \frac{k_1 K_1}{k_2 K_2} + \frac{k_1 K_1}{k_3 K_3} + \frac{k_1 K_1}{k_4 K_4} + K_1 a_{H_2O} + \frac{k_1}{k_2} K_1 a_{H_2O} + \tag{50}$$

$$\frac{k_1}{k_3} K_1 a_{H_2O} + \frac{k_1}{k_4} K_1 a_{H_2O} + \frac{k_1}{k_5} K_1 a_{H_2O} \quad .$$

The overall rate law (i.e., the release rate of H_4SiO_4 to solution) is given by

$$Rate = k_5[H_4SiO_4^{ads}] = k_1[H_2O_{ads} \bullet Si]$$

$$= k_1 K_1 a_{H_2O} [\equiv Si] \quad ,$$

where, once more, Equation (47) and the equilibrium assumption were used. Substituting from Equation (49) yields:

$$Rate = k_1 K_1 \frac{a_{H_2O}}{D} S_{tot}$$

Finally, it is usual to relate the total number of reactive sites to the total reactive surface area:

$$S_{tot} = \alpha A_{reactive}$$

where A_{tot} is the total surface area of "active" sites (i.e., where the water can adsorb) and is related to S_{tot} by a simple geometric factor, α. If the total "reactive" surface area is some fraction β of the total surface area of quartz:

$$A_{reactive} = \beta A_{tot}$$

then the final expression for the rate law becomes:

$$Rate = k_1 K_1 a_{H_2O} \frac{\alpha \beta A_{tot}}{D}$$

which can be rewritten (dividing top and bottom by k_1K_1) as,

$$Rate = \frac{k_{net} a_{H_2O} \alpha \beta A_{tot}}{K' + a_{H_2O}} \tag{51}$$

where

$$\frac{1}{k_{net}} \equiv \frac{1}{k_1} + \frac{1}{k_2} + \frac{1}{k_3} + \frac{1}{k_4} + \frac{1}{k_5} \tag{52}$$

and where

$$K' \equiv \frac{\frac{1}{k_1 K_1} + \frac{1}{k_2 K_2} + \frac{1}{k_3 K_3} + \frac{1}{k_4 K_4}}{\frac{1}{k_1} + \frac{1}{k_2} + \frac{1}{k_3} + \frac{1}{k_4}} \tag{53}$$

Note the averaging of the rate constants and the adsorption constants that is done in obtaining the final rate law. If only one of the k_i is slow (rate-determining), then $k_{net} = k_{slow}$ and $K' = 1/K_{slow}$. If, on the other hand, all the k_i are similar ($= k$) (except for k_5, which is assumed to be fast) and all the K_i are similar ($= K$) then

$$Rate = \frac{\frac{1}{4}k\, a_{H_2O}\, \alpha\, \beta\, A_{tot}}{1/K + a_{H_2O}} \tag{54}$$

Equation (51), then, simplifies to the product of the hydrolysis rate constant (assumed similar in all steps) and the equilibrium constant for adsorption.

Note that the dependence of the rate on the activity of water predicted by Equation (54) is very similar to the Michaelis-Menten rate law i.e.

$$Rate = \frac{k'\, a_{H_2O}}{K' + a_{H_2O}} \tag{55}$$

If K' in Equation (55) were less than 1 (which it may not be), then the rate should go from independence on a_{H_2O} to linear dependence on a_{H_2O} as the activity of water drops (e.g. by adding some other solvent such as ethanol or methanol).

The activation energy, E_{act}, in the case that $1/K > a_{H_2O}$ in (54), will be given by:

$$E_{act} = E^{\ddagger}_{hydrolysis} + \Delta H_{ads} \quad , \tag{56}$$

where $E^{\ddagger}$ is the activation energy for the hydrolysis reaction activated complex and ΔH_{ads} is the standard state enthalpy of the adsorption equilibrium constant for water onto a silica surface (Lasaga and Gibbs, 1990).

Another reaction mechanism recently postulated involved the dissolution of kaolinite (Ganor et al., 1995). It is important to establish a link between the observed reaction order for pH and a molecular reaction mechanism. To do this, one must postulate the actual bond breaking and bond forming steps that occur within a kaolinite structure and that lead eventually to the release of orthosilicic acid and dissolved hydrated Al^{3+}. The overall dissolution reaction may be rewritten:

$$AlSiO_{5/2}(OH)_2 + 3\,H^+ \rightarrow Al^{3+} + H_4SiO_4 + \frac{1}{2}\,H_2O \tag{57}$$

Partly based on the pH variation of the rate, it seems that the surface breakdown reactions are controlled by the processes involving the aluminum atoms. Ganor et al. (1995) then focused on the mechanism to release aluminum into solution. In order to release an Al from an octahedral site in kaolinite into solution, two Al-O-Si bonds and four Al-OH-Al bonds must be broken. Furrer and Stumm (1986) argued that if the formation of the protonated surface species involved several protonation

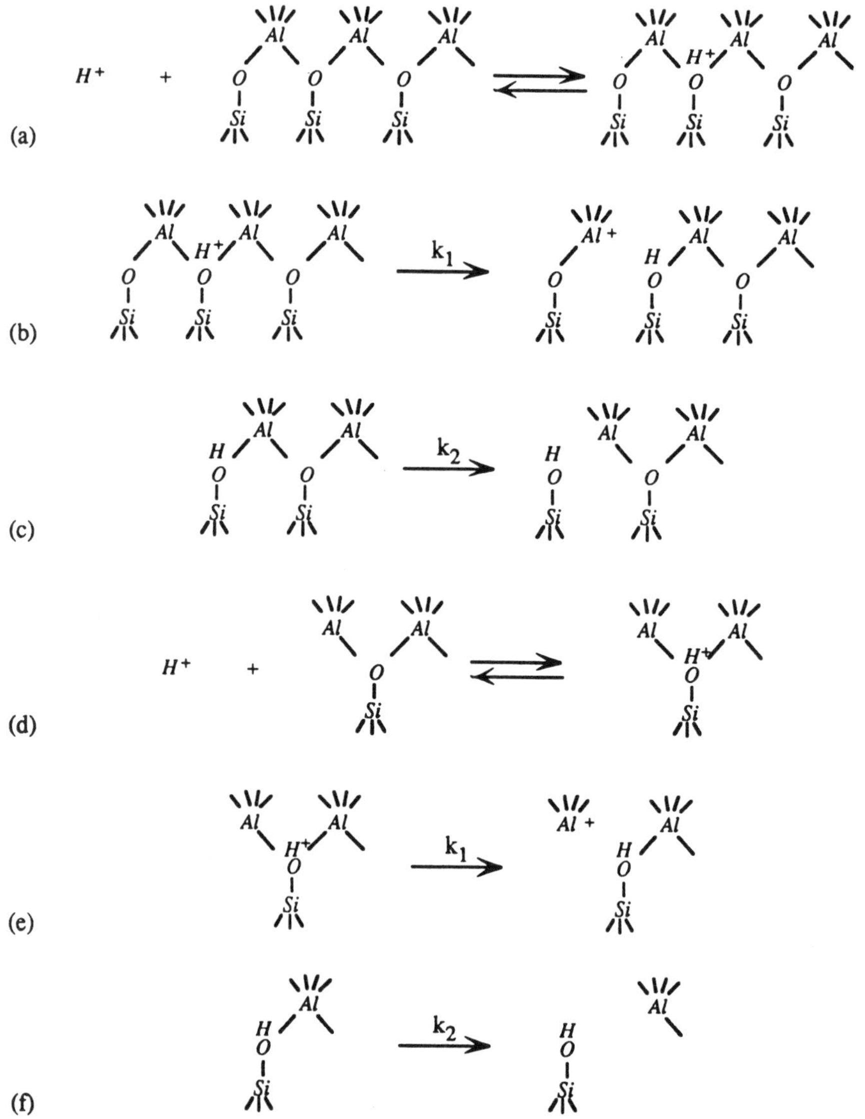

Figure 30. Suggested reaction mechanism for the acid-catalyzed dissolution of kaolinite (from Ganor et al., 1995).

steps (more than one proton is adsorbed to each surface site) then the order of the reaction with respect to the surface proton concentration will be equal to the number of protonation steps. Following the adsorption of all protons to the surface, such a mechanism requires the simultaneous breaking of all the bonds in the protonated surface complex as the key elementary reaction in the mechanism. Kinetically, such a concerted rupture would be less likely than the step-wise breaking of bonds.

Ganor et al. (1995) suggest a reaction mechanism that consists of a sequence

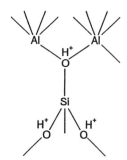

Figure 31. Triply protonated adsorption complex required to produce a high reaction order of the kinetics with respect to the amount of H^+ adsorbed on the surface of kaolinite.

of slow hydrogen-ion-mediated hydrolysis steps. As with the earlier model of quartz, the release of Al and Si atoms takes place after the sequential clipping of the covalent bonds anchoring them to the surface. In their model (Fig. 30), the rate-determining steps are the breakdown of the two Al-O-Si bonds around each Al atom. After the Al-O-Si bonds are broken the other Al-OH-Al and Si-O-Si bonds are much more quickly hydrolyzed resulting in dissolved Al^{3+} and H_4SiO_4. In the next section, a much more complex (Monte Carlo) model is introduced.

Figure 30 illustrates the sequence of proton adsorption (e.g. stages a and d) and hydrolysis of Al-O-Si bonds (stages b,c,e,f), that lead to an unzipping of the two sheets making up the kaolinite structure. In their model, the rupture of the Al-O-Si bonds is followed by fast release of Al and Si into solution:

$$4\,H^+ + 2\,Al^+ \overset{fast}{\Longrightarrow} 2\,Al^{3+} + 4\,H_2O \tag{58}$$

$$3\,H_2O + 2\,\overset{H}{O} \overset{fast}{\Longrightarrow} 2\,H_4SiO_4 \tag{59}$$

The mechanism in Figure 30 and these fast reactions account for the stoichiometry of the reaction. Because each Al in kaolinite shares four OH with four other Al, each Al has on average two OH; similarly each Si-OH shares 3 oxygens with 3 other silicons (the hydrolyzed Si-O-Al bond being now an Si-OH) leading to 1.5 oxygens per silicon. Thus, two Al^+ can be represented by $(=Al^+)_2(OH)_4$ and two Si-OH as $\equiv Si_2O_3(OH)_2$. Using these representations, the fast reactions above can be combined into:

$$4\,H^+ + (=Al^+)_2(OH)_4 + \equiv Si_2O_3(OH)_2 \overset{fast}{\Longrightarrow} 2\,Al^{3+} + 2\,H_4SiO_4 + H_2O \tag{60}$$

The combined reactions in Figure 30 and Equation (60) require six protons and release two Al and two Si into solution, in agreement with the overall reaction (57).

In their model, equilibrium is assumed for H^+ adsorption onto each particular site. The adsorption would be onto the Si-O-Al bridging oxygens, just as discussed in the earlier section on *ab initio* studies. The slow hydrolysis steps have individual rate constants given by k1 and k2. Applying the principle of steady state for the various reaction intermediates, in the same manner as done earlier for silica, the overall rate law (i.e. the release of Al to solution) is given by:

$$Rate = k_1 \left[\equiv Si - OH^+ - Al_2\right] = k_2 \left[\equiv Si - -OH - -Al\right] \tag{61}$$

where the square brackets denote the surface concentration of the species. Note that the rate law obtained from Equation (61) predicts a **linear** relation between the rate and the surface concentration of the protons adsorbed onto the Al-O-Si sites. Note that, in the end, 3 protons were adsorbed onto the surface and were involved in the reaction; however, because the steps occur in sequence we do not get a cubic rate law. If, for example, the key hydrolysis step required the surface complex denoted in Figure 31, then the rate would be proportional to the concentration of these species. In this case, the equilibrium

$$3\,S - H^+ \leftrightarrow Surface\,Complex(3H^+)$$

where S stands for any of the kaolinite surface sites, would lead to a rate law that depends on $[S\text{-}H^+]^3$ (as discussed by Stumm and Wollast, 1990). *Ab initio* results suggest that such a highly charged complex would not offer a preferable path to the sequential hydrolysis in Figure 30.

The experimental results of Ganor et al. (1995) indeed show a linear dependence of the rate on the **total** proton concentration on the kaolinite surface, i.e., the sum of the concentrations of all protonated sites. As discussed by them, the equilibrium between the various protonated sites:

$$H_{ads}^+ - S_j + S_i \leftrightarrow S_j + H_{ads}^+ - S_i \tag{62}$$

leads to a proportionality between the total surface proton concentration and the concentration in Equation (61) , as long as activity coefficients and the concentration of surface sites remain reasonably constant. Therefore, the reaction mechanism can explain the observed pH dependence of the rate law and also makes a direct relation between an elementary reaction (the hydrolysis of the Si-O-Al bond) and the overall rate, the former amenable to *ab initio* studies and the latter available from experiments. To proceed even further, the detailed structure of the minerals must be better taken into account and this is the topic of the next section.

MONTE CARLO METHODS AND RATE LAWS

With the advent of fast computers, it is important to relate the *ab initio*-type of results with the overall rate law and mechanism of adsorption-crystal growth-dissolution and eventually with the mechanism of weathering in nature. We have been working on a new Monte Carlo scheme that directly involves the crystal structure as opposed to the Ising-like box Monte Carlo models.

The Monte Carlo scheme enables us to study the effects of complex dissolution mechanisms on the overall dissolution rate. In this approach to kinetics, the rate constants are converted into probabilities (see Dixon and Schaefer, 1973). For example, in the case of the chain reaction

$$A \xrightarrow{k_1} B \xrightarrow{k_2} P,$$

there are two rate constants, k_1 and k_2. Let us assume that k_2 is greater than k_1. Suppose that a time unit is picked such that:

$$\Delta t = p\,\frac{1}{k_2} \tag{63}$$

where p is some number less than 1. Given some number of B atoms, N_B, then the number of reactions to form P in this time interval is p N_B (i.e. $k_2\,N_B\,\Delta t$). The

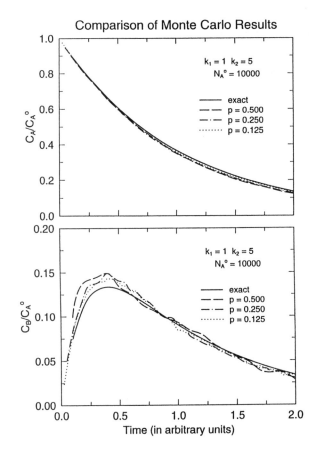

Figure 32. Comparison of the exact solution to the chain reaction (solid line) with various Monte Carlo simulations. The rate constants are $k_1 = 1$ and $k_2 = 5$ in arbitrary time units. The probability of the fast reaction (the B → P) is varied between 0.125 and 0.5; see text. (top) Plot of C_A/C_A° with time (in same arbitrary units) and (bottom) plot of C_B/C_A° with time, where C_A° is the initial amount of A.

same result would be obtained if, for sufficiently large numbers of B, we stated that the **probability** of any B atom reacting to produce P, P_B, is p. Similarly, for the same time unit, the probability that an A atom would react would be given by

$$P_A = p\,\frac{k_1}{k_2} \tag{64}$$

which is even less than p by our initial assumption. Thus, the original problem of the decay reaction can be completely replaced by the stochastic process where the reaction is followed in a sequence of time steps (of size Δt) such that at any step the probability of an A atom reacting is $P_A = p\,k_1/k_2$ and the probability of a B atom reacting is $P_B = p$. A simple computer program would then pick each A and B atom and select a random number between 0 and 1. If the atom is an A atom, then the program would check if the random number is less than P_A; if it is, then the A atom is removed and a B atom is added. If it is greater, then the A atom is left alone. The same calculation is done for all the B atoms. This sequence is repeated for as many steps as desired. Figure 32 compares the Monte Carlo method with the exact solution for the chain reaction (e.g. Lasaga, 1996).

A good illustration of the ideas behind these Monte Carlo methods is the disso-

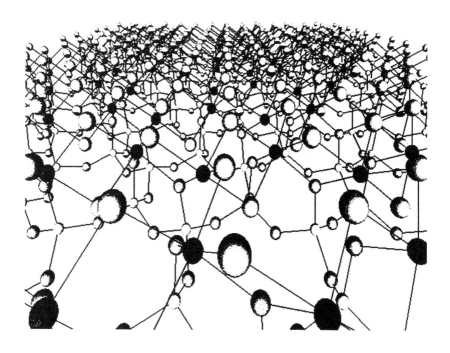

Figure 33. Single sheet in the kaolinite structure. The tetrahedral silicon layer is below and the octahedral aluminum layer is on top. Light small atoms are silicon, dark small atoms are aluminum, grey atoms are oxygens and light large atoms are OH.

lution of the clay mineral, kaolinite (Fig. 33). An atomic picture of the dissolution mechanism for kaolinite must be based on the crystal structure (of course, the surface of kaolinite will be a modified version of the bulk crystal structure). The silicon atoms are tetrahedrally coordinated by oxygens and linked in a two dimensional sheet and the aluminum atoms are octahedrally coordinated and form the other sheet in the structure.

In a more realistic Monte Carlo model of the dissolution of kaolinite, the kinetic model would track the hydrolysis reaction of all the possible surface complexes. Figure 34 illustrates the three main types of hydrolysis reactions in kaolinite leading to the dissolution of the clay sheet. The reactions may also be written as:

$$Si--O--Si + H_2O \rightarrow Si-OH + Si-OH \qquad (65)$$

$$Si--O--Al_2 + H_2O \rightarrow Al-OH-Al + Si-OH \qquad (66)$$

$$Al-OH-Al + H_2O \rightarrow Al-OH + Al-OH_2 \qquad (67)$$

The previous section dealt with the kinetic model for kaolinite dissolution by treating these reactions as independent i.e. occurring in isolation of their environment. To some extent the *ab initio* work (Xiao and Lasaga, 1994, 1995) has supported the local nature of kinetic reactions involving aluminosilicates. However, the actual rates of these hydrolysis reactions may depend on neighbor interactions (e.g. recall the discussion on the *ab initio* study of the effects of protonation on the structure of the

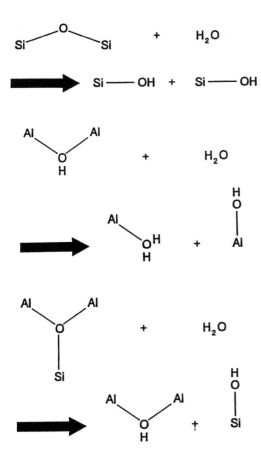

Figure 34. Three basic hydrolysis reactions involved in the dissolution of kaolinite.

4-tetrahedral ring). Thus, the hydrolysis rate constant for Si–O–Si may depend on whether the Si-O-Al$_2$ bonds on each of the two silicon atoms have been hydrolized and also on whether the other Si–O–Si units (two other ones on each of the Si) have already been hydrolized to Si-OH. So the rate constant for (65) is written as k_{ij}^{SiOSi}, where i = 1 to 3 depending on whether none, one or two of the Si-O-Al$_2$ units have been hydrolyzed and where j = 1 to 5 depending on whether there are none, 1, 2, 3, or 4 Si-OH groups attached to the two silicon atoms involved in the Si–O–Si bond being hydrolyzed. In all, there are, therefore, 15 possible different rate constants, k_{ij}^{SiOSi}, for the Si–O–Si hydrolysis.

Similarly, the rate constant for the hydrolysis of the Si-O-Al$_2$ unit, k_{ij}^{SiOAl2}, depends on whether the other Al-OH-Al bridges (see Fig. 35) have been previously hydrolyzed or not (11 possibilities). The rate can also depend on whether the adjacent Si–O–Si bridges have been hydrolyzed (4 possibilities). Hence there are 44 rate constants, k_{ij}^{SiOAl2}, i = 1-4 (0, 1, 2, 3 Si-OH units) and j = 1-11 (0,1,2,3...10 Al-OH or Al-OH$_2$ units).

The rate constant for the hydrolysis of the Al-OH-Al, k_{ij}^{AlOHAl}, will depend on

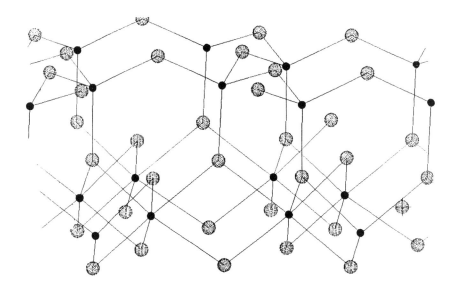

Figure 35. Close-up of the kaolinite structure showing the interactions between the various Si-O-Si, Si-O-Al, Al-O-Al and Al-OH-Al units. The rates of hydrolysis of these bonds can vary with changes in the nearby bonds as discussed in the text.

whether the adjacent bridging Al-OH-Al bond (i.e. the other part of the shared edge between octahedra) has been hydrolyzed or not (2 possibilities) and on whether the adjacent Al-OH-Al bonds have been hydrolyzed (9 possibilities), which yields 18 possible different k_{ij}^{AlOHAl}. Finally, the desorption rate constants, k_d and k'_d, for the fully hydrolyzed H_4SiO_4 and $Al(OH)_3(H_2O)_3$ surface species are needed to complete the picture. This new Monte Carlo model for kaolinite, then, would have a total of 79 rate constants! Of course, the model can be simplified (e.g. by making many of the rates similar); but, more importantly, these models can now be linked to the new *ab initio* studies of surface reactions using large clusters, as discussed earlier.

It is important to underline the difference between this new Monte Carlo approach to mineral dissolution and growth and earlier Monte Carlo models (e.g. Gilmer, 1976, 1977, 1980; Liu et al., 1982, 1989; Blum and Lasaga, 1987; Wehrli and Stumm Albano et al., 1989, Cheng et al., 1989; Uchida, 1990; Cheng, 1993; Uebing and Gomer, 1994). In the earlier models the focus was on the molecular units or "blocks" that attach or detach to the surface as well as on the surface diffusion of these "blocks". Thus, the detachment of a "block" would depend on the number of nearest neighbors bonded to it (Blum and Lasaga, 1987; Lasaga, 1990). In the current Monte Carlo scheme, the focus is on the bonds themselves. Thus the breaking of a particular bond depends on the nature of the neighboring bonds, as discussed above. By shifting to this "bond-centered" scheme, the Monte Carlo simulations can deal with actual mineral structures, such as done here for kaolinite (see Fig. 36). Previous models have almost exclusively dealt with simple cubic structures or anisotropic perturbations of a simple cubic structure. This bond-centered Monte Carlo scheme is also very compatible with the type of fundamental results coming out of the *ab initio* work, as discussed earlier on in the chapter.

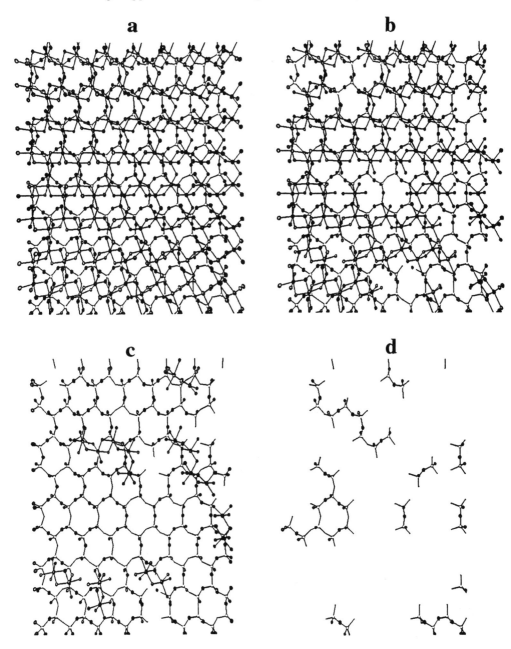

Figure 36. Monte Carlo simulation of the dissolution of kaolinite. The simulation uses the parameters of Model I in Table 6. The Al-OH hydrolyzed bonds are shown by light spheres. (a) Initial dissolution involves dominantly the hydrolysis of the Al-OH-Al linkages to produce many Al-OH and Al-H₂O bonds. (b) The aluminum octahedral layer begins to decompose. (c) The tetrahedral silicon layer is now beginning to break apart. (d) Nearly full dissolution of the sheet. Only a few polymerized silicon tetrahedra are left.

Given these rate constants, the number of bonds being hydrolyzed will depend on the population of the different species (i.e. the k are first order rate constants). Thus, in a time interval Δt the number of bonds of type i being broken is

$$Number\ hydrolyzed\ =\ k_i\,N_i\,\Delta t$$

where k_i is the actual rate constant (one of the 79) appropiate for the surface unit and N_i is the number of such surface species present (per unit area of mineral). Because the calculation will be carried out in a Monte Carlo scheme, the time interval is adjusted so that:

$$\sum_{i=1,j=1}^{3,5} k_{ij}^{SiOSi}\,N_{ij}^{SiOSi}\,\Delta t\ +\ \sum_{i=1,j=1}^{4,11} k_{ij}^{SiOAl2}\,N_{ij}^{SiOAl2}\,\Delta t\ +\ \sum_{i=1,j=1}^{2,9} k_{ij}^{AlOHAl}\,N_{ij}^{AlOHAl}\,\Delta t\ +$$

$$+\ k_d\,N_{H_4SiO_4}\,\Delta t\ +\ k_d'\,N_{Al(OH)_3(H_2O)_3}\,\Delta t\ =\ 1 \tag{68}$$

In this fashion, for each time step (of size Δt) one event is ensured to take place with a probability given by the size of the individual $k_iN_i\Delta t$ (e.g. see Blum and Lasaga (1987) for further discussion). The time evolution proceeds in a manner very similar to that discussed for the chain reaction above.

One note about the time units. These k_i are all first-order rate constants for the reaction of a particular surface unit and so have units of time^{-1}. To relate these units to the usual rate constants in units of moles/cm^2/sec, one need only multiply by a surface concentration of Si or Al in units of moles/cm^2. This connection will be clarified further below.

There are some interesting results from these kinds of models. Figure 36 shows the dissolution of kaolinite for the case that the $k_{SiOSi} = 1$, $k_{SiOAl2} = 5$, $k_{AlOHAl} = 20$ and $k_{desorp} = 200$, with **no** subtleties added for neighbor interaction. Note that, in this case, the alumina octahedral sheet is decomposing faster than the silica tetrahedral sheet. As a result the aluminum from the kaolinite sheet comes off first, followed by an unraveling of the silica framework. In this simulation, the rate is definitely non-stoichiometric.

Figure 37 quantifies the evolution of the kinetics. The units of the rate constants are arbitrary (only their relative magnitude affects the overall dissolution) and so the plots shown in Figure 37 are also given in these same dimensionless time units. Note that in this model the hydrolysis of the Si-O-Si bonds is taken as the slowest rate (e.g. rate determining). Figure 37a shows that indeed the Al atoms are released faster than the Si atoms. The speciation of the surface species is also seen in Figure 37b. Only the silicon speciation is shown for clarity. The figure shows the sequential dominance of Si-OH, Si-(OH)$_2$ and Si-(OH)$_3$ groups during the dissolution of one kaolinite layer. All three species actually reach a steady state region but not at the same time, which would cause complications in earlier kinetic models. Note the initial period of "non-steady state" similar to that observed in flow-through experiments. This time period is the type of process invoked by Chou and Wollast (1985) in discussing the effect of changing pH on the mineral surface structure. The data allow also a calculation of the actual dissolution rate that an experimentalist would measure. This is shown in Figure 37c. Note that this model would predict a substantial non-stoichiometry for the dissolution, although a rough steady state rate (for a flow-through experiment) would be reached.

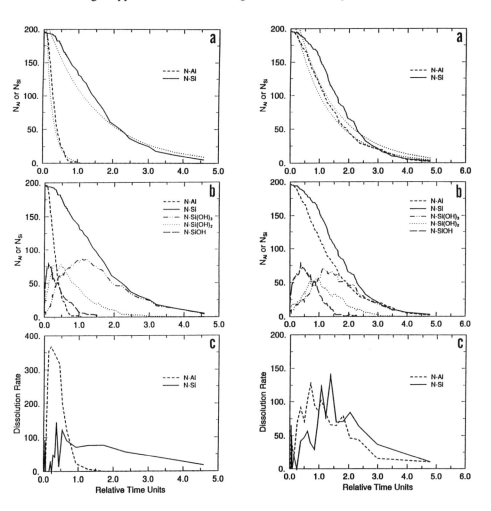

Figure 37 (left). Time evolution of the kinetics of dissolution of the kaolinite sheet. Monte Carlo simulation uses parameters in Model I of Table 6. (a) The number of Si or Al atoms remaining in the sheet are plotted versus time in dimensionless units (the time unit is tied to the value of the individual rate constants as discussed in the text). The dashed line represents the best fit of a first-order decay to the kinetics (see text). (b) Silicon surface speciation for the simulation. (c) The "dissolution rate" of Si and Al (i.e. the observable in a flow-through experiment) calculated by averaging over several time steps in the Monte Carlo simulation. Ultimately all rates go to zero because the entire sheet is dissolved.

Figure 38 (right). Same as Figure 37 but for the parameters in Model II (Table 6).

An alternative way of analyzing the overall result of the dissolution is to fit the decay of Al and Si atoms on the surface to an effective first-order rate constant. This is shown by the dotted lines in Figure 37a. It is interesting to compare the overall first-order rate constants:

$$k_{Si} = 0.69 \quad k_{Al} = 4.86$$

(again in whatever time units the original rate constants were couched). As expected,

Table 6. Monte Carlo model parameters.

MODEL I

i		k_{ij}^{SiOSi}								
j	1.0	1.0	1.0	1.0	1.0					
	1.0	1.0	1.0	1.0	1.0					
	1.0	1.0	1.0	1.0	1.0					

i		k_{ij}^{SiOAl2}								
j	5.0	5.0	5.0	5.0	5.0	5.0	5.0	5.0	5.0	5.0	5.0
	5.0	5.0	5.0	5.0	5.0	5.0	5.0	5.0	5.0	5.0	5.0
	5.0	5.0	5.0	5.0	5.0	5.0	5.0	5.0	5.0	5.0	5.0
	5.0	5.0	5.0	5.0	5.0	5.0	5.0	5.0	5.0	5.0	5.0

MODEL II

i		k_{ij}^{SiOSi}				
j	0.5	0.5	1.0	1.0	1.5
	0.5	1.0	1.0	1.5	1.5
	1.0	1.0	1.5	1.5	2.0

i		k_{ij}^{SiOAl2}								
j	1.0	1.0	1.0	1.0	1.0	1.0	1.0	1.0	1.0	1.0	1.0
	1.0	1.0	1.0	1.0	1.0	1.0	1.0	1.0	1.0	1.0	1.0
	1.0	1.0	1.0	1.0	1.0	1.0	1.0	1.0	1.0	1.0	1.0
	1.0	1.0	1.0	1.0	1.0	1.0	1.0	1.0	1.0	1.0	1.0

MODEL III

i		k_{ij}^{SiOSi}				
j	0.5	0.5	1.0	1.0	1.5
	2.0	2.0	3.0	3.0	3.0
	4.0	4.0	5.0	5.0	5.0

i		k_{ij}^{SiOAl2}								
j	0.1	0.1	0.15	0.15	0.2	0.2	0.2	0.2	0.2	0.2	0.2
	0.15	0.15	0.2	0.2	0.2	0.2	0.2	0.2	0.25	0.25	0.25
	0.2	0.2	0.2	0.25	0.25	0.25	0.25	0.25	0.25	0.25	0.25
	0.2	0.2	0.2	0.25	0.25	0.25	0.25	0.25	0.25	0.25	0.25

All models have $k_{ij}^{AlOHAl} = 20.0$ (for all i,j)

and $k_{desorp}^{H4SiO4} = k_{desorp}^{Al(hyd)} = 200.0$

the Al overall dissolution rate is much faster than the Si. However, note also that the net first-order rate constant is not the same as any of the individual rate constants. In fact, k_{Si}, which involves several hydrolysis steps, is slightly lower than k_{SiOSi}, the

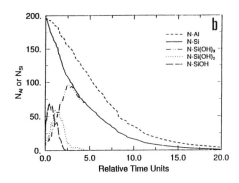

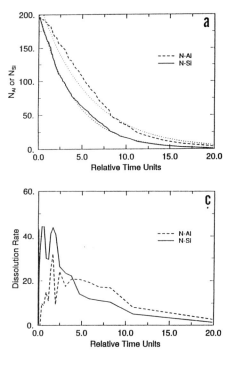

Figure 39. Same as Figure 37 but for the parameters in Model III (Table 6).

slowest elementary reaction rate constant (the Si-O-Si hydrolysis). k_{Al} is similar to the slowest rate for Al reactions ($k_{SiOAl2} = 5$).

At this point, it should be noted that these k_{Si} and k_{Al} would correspond to an experimental rate constant (in moles/cm^2/sec), if the k were multiplied by the number (moles/cm^2) of surface Si and Al atoms in a kaolinite sheet (a well-defined number). Of course, it is these **averaged** rate constants that are measured in an experiment.

A critical parameter is the linkage between the Si-O-Al$_2$ hydrolysis and the subsequent hydrolization of the Si-O-Si linkages in the tetrahedral layer. An alternative set of rate constants would allow for variations in the Si-O-Si rate constant and decrease the SiOAl2 rate constant to 1 (see Model II in table 6). The kinetic behavior is shown in Figure 38. In this case, the resulting overall dissolution is much more stoichiometric (i.e. similar to that observed in experiments - Ganor, Mogollon and Lasaga, 1995). This stoichiometric dissolution is also observed in the calculated rate (Fig. 38c) and in the overall rate constants ($k_{Si} = 0.74$, $k_{Al} = 0.83$ - see Fig. 38a). Once again, note that the overall rate constant is not simply related to the slowest elementary reaction.

A different model (such as invoked by Ganor et al., 1995) would have the hydrolysis of the Si-O-Al2 bonds as the slowest. The kinetic behavior of such a model (Model III in table 6) is exhibited in Figure 39. Note that, again, the dissolution is fairly stoichiometric after an initial non-steady-state period. The aluminum overall dissolution rate is now slightly *slower* than that of the silica, reflecting the two Si-O-Al bonds that each Al atom has (versus the sole Si-O-Al per silicon). The stoichiometric behavior is consistent with the data of Ganor et al. (1995).

These new Monte Carlo calculations can be carried out adding surface steps and other defects. For example, multiple sheets of different sizes could be considered (leading to a step between them). Future work can incorporate these variations. In addition, an important next step is to enable the adsorption of Al or Si species and the possibility of condensation reactions (e.g. olation or oxolation - see Livage, 1994) so that growth can be studied. With the addition of defects and steps and both forward (e.g. hydrolysis) and reverse (e.g. olation or oxolation) reactions present, a detailed study of the very important $f(\Delta G_r)$ function may be undertaken.

SUMMARY

Fundamental approaches needed to understand and apply the rates of mineral reactions with fluids to the many processes in nature, where these reactions play a central role, have been discussed in this chapter. These approaches link studies and observations from the atomic scale to the laboratory scale and ultimately the global scale. The chapter has emphasized the need to view the rates of mineral-fluid reactions from the point of view of a fundamental overall rate law. Therefore, the chapter has discussed the possible form of the overall rate, the meaning of the various terms in such a rate law, and the methods to obtain the values of these terms from the laboratory, nature or theory. In particular, atomic scale studies, both theoretical and experimental, are providing important new insights into the most basic dynamics that govern the various weathering and alteration processes. These studies include *ab initio* methods which can solve the quantum mechanical equations very efficiently on large clusters of atoms and can explore the adsorption and the reaction pathways and energetics of complex overall reactions. It has been shown that these methods can lead to surprising results such as the recent work on the behavior of silica surfaces in basic solutions. Other important atomic scale methods include the molecular dynamics and Monte Carlo methods. These methods link the atomic dynamics to the macroscopic observables e.g. overall rates and mineral surface structures. A new variation on the Monte Carlo method has been introduced to focus on the integration of atomic scale reactions into a reaction mechanism and to study the relation between individual elementary reactions and the overall rate. A particularly important application of these methods is the elucidation of the form of the variation of the overall rate with the saturation state of the solution, the $f(\Delta G_r)$ function. The chapter has emphasized the need to propose and study reaction mechanisms, not just individual elementary reactions, if fundamental advances are to take place in our quantification of water-rock kinetics. Experimentally, the ability to observe the dynamics of surfaces at atomic scale resolution provides the other main input to the understanding of rates at this level. These technologies will provide new information into important aspects of the overall rate law including the clarification of the "reactive surface area". These experimental methods will also supply important input into the models of surface dynamics that aim to justify the laboratory and nature scale rate laws. Later chapters will discuss many of these latter topics.

ACKNOWLEDGMENTS

The author gratefully acknowledges financial support through the Branch of Energy Science of the Department of Energy (grant DE-FG02-90-ER14153) and funding from the National Science Foundation (NSF EAR 9017976 and 9219770). The author would also like to gratefully acknowledge the technical help of Yitian Xiao in the preparation of the manuscript and the thorough reviews by Ian MacInnis, Susan Brantley, Yitian Xiao and Josep Soler.

REFERENCES

Aagard P, Helgeson HC (1982) Thermodynamic and kinetic constraints on reaction rates among minerals and aqueous solutions, I. Theoretical Considerations. Am J Sci 282:237-285.

Albano EV, Binder K, Heermann DW, Paul W (1989) Absorption on stepped surfaces: A monte carlo simulation. Surface Science 223:151-178

Armhein C, Suarez DL (1988) The use of a surface complexation model to describe the kinetics of ligand-promoted dissolution of anorthite. Geochim Cosmochim Acta 52:2785-2793

Bales RC, Morgan JJ (1985) Dissolution kinetics of chrysotile at pH 7 to 10. Geochim Cosmochim Acta 49:2281-2288

Barron EJ, Sloan JL, Harrison CGA (1980) Potential significance of land-sea distribution and surface albedo variations as a climatic forcing factor: 180 my to the present. Paleogeog Paleoclim Paleoecol 30:17-40

Bates S, Dwyer J (1993) *Ab initio* study of CO adsorption on zeolites. J Phys Chem 97:5897-5900

Berner RA, Morse JW (1974) Dissolution kinetics of calcium carbonate in sea water; IV, Theory of calcite dissolution. Am J Sci 274:103-134

Bish DL (1993) Rietveld refinement of the kaolinite structure at 1.5K. Clays Clay Minerals. 41:738-744

Bish DL, von Dreele RB (1989) Rietveld refinement of non-hydrogen atomic positions in kaolinite. Clays Clay Minerals 37:289-296

Bloom PR, Erich MS (1987) Effect of solution composition on the rate and mechanisms of gibbsite dissolution in acid solutions. Soil Sci Soc Am J 51:1131-1136

Blum AE, Lasaga AC (1987) Monte Carlo simulations of surface reaction rate laws, in W. Stumm, ed., Aquatic Surface Chemistry: Chemical Processes at the Particle-Water Interface, John Wiley & Sons, New York, p 255-292

Blum AE, Lasaga AC (1988) Role of surface speciation in the low-temperature dissolution of minerals: Nature 331:431-433

Blum AE, Yund RA, Lasaga AC (1990) The effect of dislocation density on the dissolution rate of quartz. Geochim Cosmochim Acta 54:283-297

Blum AE, Lasaga AC (1991) The role of surface speciation in the dissolution of albite. Geochim Cosmochim Acta 55:2193-2201

Bohn HL, McNeal BL, O'Connor GA (1979) Soil Chemistry, Wiley-Interscience, New York, 230 p

Boudart M (1976) Consistency between kinetics and thermodynamics. J Phys Chem 80:2869-2870

Brady PV (1991) The effect of silicate weathering on global temperature and atmospheric CO_2. J Geophys Res 96:18101-18106

Brady PV (1992) Silicate surface chemistry at elevated temperatures. Geochim Cosmochim Acta 56:2941-2946

Brady PV, Walther JV (1989) Controls on silicate dissolution rates in neutral and basic pH solutions at 25°C. Geochim Cosmochim Acta 53:2823-2830

Brady PV, Walther JV (1990) Kinetics of quartz dissolution at low temperatures. Chem Geol 82:253-264

Brady PV, Walther JV (1992) Surface chemistry and silicate dissolution at elevated temperatures. Am J Sci 292:639-658

Brantley SL, Crane SR, Crerar DA, Hellmann R, Stallard R (1986) Dissolution at dislocation etch pits in quartz. Geochim Cosmochim Acta 50:2349-2362

Burch TE, Nagy KL, Lasaga AC (1993) Free energy dependence of albite dissolution kinetics at 80°C, pH 8.8. Chemical Geology 105:137-162

Burton WK, Cabrera N, Frank FC (1951) The growth of crystals and the equilibrium structure of their surfaces. Phil Trans Royal Soc London A243:299-358

Cama J, Ayora C, Lasaga AC (1995) The Effect of deviation from Equilibrium on the dissolution rate and on apparent variations in the activation energy. (submitted to Geochim Cosmochim Acta)

Carroll SA, ms (1989) The dissolution behavior of corundum, kaolinite, and andalusite: a surface complexation reaction model for the dissolution behavior of aluminosilicate minerals in diagenetic systems. PhD dissertation, Northwestern University, Evanston, IL

Carroll-Webb SA, Walther JV (1988) A surface complex reaction model for the pH-dependence of corundum and kaolinite dissolution rates. Geochim Cosmochim Acta 52:2609-2623

Carroll-Webb SA, Walther JV (1990) Kaolinite dissolution at 25, 60 and 80°C. Am J Sci 290:797-810

Casey WH, Carr MJ, Graham RA (1988) Crystal defects and the dissolution kinetics of rutile. Geochim Cosmochim Acta 52:1545-1556

Casey WH, Sposito G (1992) On the temperature dependence of mineral dissolution rates. Geochim Cosmochim Acta 56:3825-3830

Casey WH, Hochella MF, Westrich HR (1993) The surface chemistry of manganiferous silicate minerals as inferred from experiments on tephroite (Mn_2SiO_4). Geochim Cosmochim Acta 57:785-793

Cheng VKW, Tang CM, Tang TB (1989) The Kinetic asymmetry between nucleation growth and dissolution: a Monte Carlo study. J Crystal Growth 96:293-303

Cheng VKW (1993) A monte carlo study of moving steps during crystal growth and dissolution. J Crystal Growth 134:369-376

Chiang CM, Zegarski BR, Dubois LH (1993) First observation of strained siloxane bonds on silicon oxide thin flims. J Phys Chem 97:6948-6950

Combariza JE, Kestner NR (1995) Density functional study of short-range interaction forces between ions and water molecules. J Phys Chem 99:2717-1723

Chou L, Wollast R (1984) Study of the weathering of albite at room temperature and pressure with a fluidized bed reactor. Geochim Cosmochim Acta 48:2205-2217

Chou L, Wollast R (1985) Steady-state kinetics and dissolution mechanisms of albite. Am J Sci 285:963-993

Corwin JF, Swinnerton AC (1951) The growth of quartz in alkali halide solutions. J Am Chem Soc 73:3598-3602

Cygan RT, Casey WH, Boslough MB, Westrich HR, Carr MJ, Holdren GR (1989) Dissolution kinetics of experimentally shocked silicate minerals. Chem Geol 78:229-244

de Groot SR, Mazur P (1962) Non-Equilibrium Thermodynamics. North-Holland Publishing Co., Amsterdam, 510 p

Dixon DA, Schaefer RH (1973) Computer simulation of kinetics by the Monte Carlo technique. J Chem Ed 50:648-650

Dove PM, Crerar DA (1990) Kinetics of quartz dissolution in electrolytesolutions using a hydrothermal mixed flow reactor. Geochim Cosmochim Acta 54:955-969

Dove PM, Elston SF (1992) Dissolution kinetics of quartz in sodium chloride solutions: Analysis of existing data and a rate model for 25°C. Geochim Cosmochim Acta 56:4147-4156

Drever JI (1988) The Geochemistry of Natural Waters, 2nd ed. Prentice Hall, Englewood Cliffs, NJ, 435 p

Feller D, Kendell RA, Brightman MJ (1995) A Measure of hardware and software preformance in the area of electronic structure methods. The EMSL *ab initio* methods benchmark report

Fisher GW, Lasaga AC (1981) Irreversible thermodynamics in petrology. In Lasaga AC, Kirkpatrick RJ (eds), Kinetics of Geochemical Processes, Rev Mineral 8:171-209

Furrer G, Stumm W (1986) The coordination chemistry of weathering: I Dissolution kinetics of σ-Al_2O_3 and BeO. Geochim Cosmochim Acta 50:1847-1860

Gaffin S (1987) Ridge volume dependence on seafloor generation rate and inversion using long term sealevel change. Am J Sci 287:596-611

Ganor J, Lasaga AC (1994) The effects of oxalic acid on kaolinite dissolution rate. Mineral Mag 58A:315-316

Ganor JW, Mogollon JL, Lasaga AC (1995) The effect of pH on kaolinite dissolution rates and on activation energy. Geochim Cosmochim Acta 59:1037-1052

Gautier JM, Oelkers EH, Schott J (1994) Experimental study of K-feldspar dissolution rates as a function of chemical affinity at 150°C and pH 9. Geochim Cosmochim Acta 58:4546-4560

Gibbs GV, Downs JW, Boisen MB (1994) The elusive SiO bond. In Silica, Heaney PJ, Prewitt CT, Gibbs GV (eds) Rev Mineral 29:331-363

Gilmer GH (1976) Growth on imperfect crystal faces. I. Monte Carlo growth rates. J Crystal Growth 35:15-28

Gilmer GH (1977) Computer simulation of crystal growth. J Crystal Growth 42:3-10

Gilmer GH (1980) Computer models of crystal growth. Science 208:355-363

Glendening ED, Feller D (1995) Cation-water interaction: The $M^+(H_2O)_n$ cluster for alkali metals, M = Li, Na, K, Rb and Cs. J Phys Chem 99:3060-3067

Grandstaff DE (1986) The dissolution rate of forsteritic olivine from Hawaiian beach sand. In: Colman SM, Dethier DP, (eds) Rates of Chemical Weathering of Rocks and Minerals. New York, Academic Press, 41-59.

Haase F, Sauer J (1995) Interaction of menthanol with Bronsted Acid sites of zeolite catalysts: An *ab initio* study. J Am Chem Soc 117:3780-3789

Hashimoto K, Morokuma K (1994) *Ab initio* molecular orbital study of $Na(H_2O)_n$ (n=1-6) clusters and their ions. Comparison of electronic structure of the "surface" and "interior" complexes. J Am Chem Soc 116:11436-11443

Helgeson HC, Murphy WM, Aagaard P (1984) Thermodynamic and kinetic constraints on reaction rates among minerals and aqueous solutions. II. Rate constants, effective surface area, and the hydrolysis of feldspar. Geochim Cosmochim Acta 48:2405-2432

Hohenberg P, Kohn W (1964) Inhomogeneous electron gas. Physical Review 136:B 864-B 871

Holland HD (1978) The Chemistry of the Atmosphere and Oceans. John Wiley & Sons, New York, 240 p

Hollingworth CA (1957) Kinetics and equilibria of complex reactions. J Chem Phys 27:1346-1348

Hosaka M, Taki S (1981a) Hydrothermal growth of quartz crystals in NaCl solution. J Crystal Growth 52:837-842

Hosaka M, Taki S (1981b) Hydrothermal growth of quartz crystals in KCl solution. J Crystal Growth 53:542-546

Jin JM, Ming NB (1989) A comparison between the growth mechanism of stacking fault and of screw dislocation. J Crystal Growth 96:442-444

Johnson BG, Gill PMW, Pople JA (1992) The performance of a family of density functional methods. J Chem Phys 98:5612-5623

Kassab E, Fouquet J, Allavena M, Evleth EM (1993) *Ab initio* study of protron-transfer surfaces in zeolite models. J Phys Chem 97:9034-9039

Kline WE, Fogler HS (1981a) Dissolution kinetics: catalysis by strong acids. J Colloid Int Sci 82:93-102

Kline WE, Fogler HS (1981b) Dissolution kinetics: catalysis by salts. J Colloid Int Sci 82:103-112

Knauss KG, Wolery TJ (1986) Dependence of albite dissolution kinetics on pH and time at $25°C$ and $70°C$. Geochim Cosmochim Acta 50:2481-2497

Knauss KG, Wolery TJ (1988) The dissolution kinetics of quartz as a function of pH and time at $70°C$. Geochim Cosmochim Acta 52:43-53

Knauss KG, Wolery TJ (1989) Muscovite dissolution kinetics as a function of pH and time at 70 C. Geochim Cosmochim Acta 52:1493-1502

Kohn W, Sham LJ (1965) Self-consistent equations including exchange and correlation effects. Phys Rev 140:A 1133-A 1140

Kramer GJ, van Santen RA, Emels CA, Nowak AK (1993) Understanding the acid behaviour of zeolites from theory and experiment. Nature 363:529-531

Kyrlidis A, Cook SJ, Chakraborty AK, Bell AT, Theodorou DN (1995) Electronic structure calculation of ammonia adsorption in H-ZSM-5 zeolites. J Phys Chem 99:1505-1515

Laasonen K, Klein ML (1994) Structural study of $(H_2O)_{20}$ and $(H_2O)_{21}H^+$ Using density functional methods. J Phys Chem 98:10079-10083

Lasaga AC (1981a) Transition state theory. In: Lasaga AC, Kirkpatrick RJ, (eds), Kinetics of Geochemical Processes. Rev Mineral 8:135-169

Lasaga AC (1981b) Rate laws in chemical reactions, In: Lasaga AC, Kirkpatrick RJ (eds) Kinetics of Geochemical Processes. Rev Mineral 8:1-68

Lasaga AC (1984) Chemical kinetics of water-rock interactions. J Geophys Res 89:4009-4025

Lasaga AC (1990) Atomic treatment of mineral-water surface reactions. In: Hochella MF, White AF (eds) Mineral-Water Interface Geochemistry. Rev Mineral 23:17-85

Lasaga AC (1992) *Ab initio* methods in mineral surface reactions. Rev Geophys 30:269-303

Lasaga AC (1996) Kinetic Theory in Earth Sciences. Princeton University Press, Princeton, NJ, 400 p

Lasaga AC, Blum AE (1986) Surface chemistry, etch pits and mineral-water reactions. Geochim Cosmochim Acta 50:2363-2379

Lasaga AC, Gibbs GV (1987) Applications of quantum mechanical potential surfaces to mineral physics calculations. Phys Chem Minerals 14:107-117

Lasaga AC, Gibbs GV (1990a) *Ab initio* quantum mechanical calculation of surface reactions:A new era? In: Aquatic Chemical Kinetics - Reaction Rate Processes in Natural Waters. Stumm W (ed) Wiley InterScience, New York p 259-289

Lasaga AC, Gibbs GV (1990b) *Ab-Initio* quantum mechanical calculations of water-rock interactions: adsorption and hydrolysis reactions, Am J Sci 290:263-295

Lasaga AC, Gibbs GV (1991) Quantum mechanical potential surfaces and calculations on minerals and mineral surfaces II. $6 - 31G^*$ results. Phys Chem Min 17:485-491

Lasaga AC, Soler JM, Ganor J, Burch TE, Nagy KL (1994) Chemical weathering rate laws and global geochemical cycles. Geochim Cosmochim Acta 58:2361-2386

Laudise RA (1958) Kinetics of hydrothermal quartz crystallization. J Am Chem Soc 81:562-566

Limtrakul J, Yoinuan J, Tantanak D (1995) The structural and chemical influence on catalytic properties of zeolite clusters. J Mol Struc 332:151-159

Lin FC, Clemency CV (1981) The kinetics of dissolution of muscovites at $25°C$ and 1 atm CO_2 partial pressure. Geochim Cosmochim Acta 52:143-165

Liu GZ, Van Der Erden JP, Bennema P (1982) The opening and closing of a hollow dislocation core: A monte carlo simulation. J Crystal Growth 58:152-162

Liu J, Jin JM, Ming NB, Ong CK (1989) Monte Carlo simulation of F.C.C. crystal growth with an anisotropic variable bond model. Solid State Communications 70:763-765

Livage J (1994) Sol-get chemistry and molecular sieve synthesis. In: Advanced Zeolite Science and Applications. Jansen JC (ed) Elsevier, New York, p 1-42

Luke BT (1993) An *ab initio* investigation of the lowest potential energy surface of disiloxane. J Phys Chem 97:7505-7510

MacInnis IN, Brantley SL (1992) The role of dislocations and surface morphology in calcite dissolution. Geochim Cosmochim Acta 56:1113-1126

MacInnis IN, Brantley SL (1993) Development of etch pit size distributions on dissolving minerals. Chem Geol 105:31-49

MacInnis IN, Onuma K, Tsukamoto K (1993) *In situ* study of the dissolution kinetics of calcite using phase shift interferometry. Proc. 6th Topical Meeting on Crystal Growth Mechanism, Awara, Japan 233-238

Maiwa K, Tsukamoto K, Sunagawa I (1990) Activities of spiral growth hillocks on the (111) faces of barium nitrate crystals growing in an aqueous solution. J Crystal Growth 102:43-53

Marchese L, Chen J, Wright PA, Thomas JM (1993) Formation of H_3O^+ at the Bronsted site in SAPO-34 catalysts. J Phys Chem 97:8109-8112

Mogollon JL, Ganor J, Soler JM, Lasaga AC (1995) Column experiments and the complex dissolution rate law of gibbsite. Am J Sci, in press.

Mukesh D (1991) Oscillations during carbon monoxide oxidation: a monte carlo simulation. J Catalysis 133:153-158

Nagy KL, Steefel CI, Blum AE, Lasaga AC (1990) Dissolution and precipitation kinetics of kaolinite: Initial results at 80 C with application to porosity evolution in a sandstone. In: Meshri ID, Ortoleva, PJ (eds) Prediction of Reservoir Quality through Chemical Modeling. Am Assoc Petroleum Geologists, Memoir 49:85-101

Nagy KL, Blum AE, Lasaga AC (1991) Dissolution and precipitation kinetics of kaolinite at 80 C and pH 3: The dependence on solution saturation state. Am J Sci 291:649-686

Nagy KL, Lasaga AC (1992) Dissolution and precipitation kinetics of gibbsite at 80 C and pH 3: The dependence on solution saturation state. Geochim Cosmochim Acta 56:3093-3111

Nagy KL, Lasaga AC (1993) Simultaneous precipitation kinetics of kaolinite and gibbsite at 80°C and pH 3. Geochim Cosmochim Acta 57:4329-4335

Nahon DB (1991) Introduction to the Petrology of Soils and Chemical Weathering, Wiley-Interscience, New York, 320 p

Nicholas JB, Winans RE, Harrison RJ, Iton LE, Curtiss LA, Hopfinger (1992) An *ab initio* investigation of disiloxane using extended basis sets and electron correlation. J Phys Chem 96:7958-7965

Nicholas JB, Hess AC (1994) *Ab initio* periodic Hartree-Fock investigation of a zeolite acid site. J Am Chem Soc 116:5428-5436

Nielsen AE (1984) Electrolyte crystal growth mechanisms. J Crystal Growth 67:289-310

Nielsen AE, Toft JM (1984) Electrolyte crystal growth kinetics. J Crystal Growth 67:278-288

Oelkers EH, Schott J, Devidal JL (1994) The effect of aluminum, pH, and chemical affinity on the rates of aluminosilicate dissolution reactions. Geochim Cosmochim Acta 58:2011-2024

Ohara M, Reid RC (1973) Modeling Crystal Growth Rates from Solution, Prentice-Hall, Englewood, New Jersey, 272 p

Ohmoto H, Hayashi K, Onuma K, Tsukamoto K, Kitakaze A, Nakano Y, Yamamoto Y (1991) Solubility and reaction kinetics of solution-solid reactions determined by in situ observations. Nature 351:634-636

Onuma K, Tsukamoto K, Sunagawa I (1991) Dissolution kinetics of K-alum crystals as judged from the measurements of surface undersaturations. J Crystal Growth 110:724-732

Onuma K, Tsukamoto K, Nakadate S (1993) Application of real time phase shift interferometer to the measurement of concentration field. J Crystal Growth 129:706-718

Onuma K, Kameyama T, Tsukamoto K (1994) In situ study of surface phenomena by real time phase shift interferometry. J Crystal Growth 137:610-622

Pelmenschikov AG, Morosi G, Gamba A (1992) Quantum chemical molecular models of oxides. 3. The Mechanism of water interaction with the terminal OH groups of silica. J Phys Chem 96:7422-7424

Pelmenschikov AG, van Santen RA (1993) Water adsorption on zeolites: *ab initio* interpretation of IR Data. J Phys Chem 97:10678-10680

Rees LVC (1982) When is a zeolite not a zeolite? Nature 296:491-492

Rimstidt JD, Barnes HL (1980) The kinetics of silica-water reactions. Geochim Cosmochim Acta 44:1683-1700

Rimstidt JD, Dove PM (1986) Mineral/solution reaction rates in a mixed flow reactor: wollastonite hydrolysis. Geochim Cosmochim Acta 50:2509-2516.

Rose NM (1991) Dissolution rates of prehnite, epidote, and albite. Geochim Cosmochim Acta 55:3273-3286 van Santen RA (1994) Theory of Bronsted acidity in zeolites. Advanced Zeolite Science and Applications 85:273-294

Sauer J (1989) Molecular models in *ab Initio* studies of solids and surfaces: from ionic crystals and semiconductors to catalysts. Chem Rev 89:199-255

Sauer J, Horn H, Haser M, Ahlrichs R (1990) Formation of hydronium ions on Bronsted sites in zeolitic catalysts: a quatum-chemical *ab initio* study. Chem Phys Lett 173:26-32

Schott J, Brantley S, Crerar D, Guy C, Borcsik M, Willaime C (1989) Dissolution kinetics of strained calcite. Geochim Cosmochim Acta 53:373-382

Schweda PS (1989) Kinetics of alkali feldspar dissolution at low temperature. in Water-Rock Interaction. Miles E (ed) Balkema, Rotterdam 609-612.

Shchegolev BF (1994) *Ab initio* investigation of the equilibrium geometry and electronic structure of the $(SiH_3)_2O$ and $(SiH_3)_2NH$ molecules. J Molecular Struc 315:209-212

Stallard RF (1988) Weathering and erosion in the humid tropics. In: Lerman A, Meybeck M (eds) Physical and Chemical Weathering in Geochemical Cycles. Kluwer Academic Publishers, Dordrecht, Netherlands, p 225-246

Stallard RF, Edmond JM (1983) Geochemistry of the Amazon 2. The influence of geology and weathering environment on the dissolved load. J Geophys Res 88:9671-9688

Stave MS, Nicholas JB (1993) Density functional study of cluster models of zeolites. 1. Structure and acidity of hydroxyl groups in disiloxane analogs. J Phys Chem 97:9630-9641

Stillings LL, Brantley SL (1994) Feldspar dissolution at 25°C and pH 3: Reaction stoichiometry and the effect of cations. Geochim Cosmochim Acta 59:1483-1496

Stumm W, Furrer G, Kunzl B (1983) The role of surface coordination in precipitation and dissolution of mineral phases. Croat Chem Acta 56:593-611

Stumm W, Furrer G, Wieland E, Zinder B (1985) The effects of complex-forming ligands on the dissolution of oxides and aluminosilicates. In: Drever JI (ed) The Chemistry of Weathering. Reidel Publishing Co., Dordrecht, Netherlands, p 55-74

Stumm W, Wieland E (1990) Dissolution of oxide and silicate minerals: rates depend on surface speciation. In: Stumm W (ed) Aquatic Chemical Kinetics. Wiley, New York, p 367-400

Stumm W, Wollast R (1990) Coordination chemistry of weathering: kinetics of surface-controlled dissolution of oxide minerals. Rev Geophys 28:53-69

Sverjensky DA (1993) Physical surface complexation models for sorption at the mineral-water interface. Nature 364:776-780

Teppen BJ, Miller DM, Newton SQ, Schafer L (1994) Choice of computational techniques and molecular models for *ab initio* calculations pertaining to solid silicates. J Phys Chem 98:12545-12557

Titloye JO, Parker SC, Mann S (1993) Atomistic simulation of calcite surfaces and the influence of growth additives on their morphology. J Crystal Growth 131:533-545

Tole MP, Lasaga AC, Pantano C, White WB (1986) The kinetics of dissolution of nepheline (NaAlSiO4). Geochim Cosmochim Acta 50:379-392

Tsukamoto K, Onuma K, MacInnis IN (1993) Application of real time phase shift interferometry to the slightly dissolved crystals in water. Proc. 6th Topical Meeting on Crystal Growth Mechanism, Awara, Japan 227-231

Uchida T, Sato F, Wada, K (1990) Kinetics of crystal growth on the solid-on-solid model. J Crystal Growth 99:116-119

Uchino T, Sakka T, Ogata Y, Iwasaki M (1992) Mechanism of Hydration of sodium silicate glass in a steam environment: ^{29}Si NMR and *ab initio* molecular orbital studies. J Phys Chem 96:7308-7315

Uchino T, Sakka T, Ogata Y, Iwasaki M (1993) Local structure of sodium aluminosilicate glass: an *ab initio* molecular orbital study. J Phys Chem 97:9642-9649

Uebing C, Gomer R (1994) Diffusion on stepped surfaces III. Enhanced diffusion along step edges. Surface Science 317:165-169

Velbel MA (1985) Geochemical mass balances and weathering rates in forested watersheds of the southern Blue Ridge. Am J Sci 285:904-930

Wieland EB, Wehrli B, Stumm W (1988) The coordination chemistry of weathering:III. A generalization on the dissolution rates of minerals. Geochim Cosmochim Acta 52:1969-1981

Wogelius RA, Walther JV (1991) Olivine dissolution at 25°C: Effects of pH, CO_2 and organic acids. Geochim Cosmochim Acta 55:943-954

Xantheas SS, Dunning TH jr. (1993) *Ab initio* studies of cyclic water clusters $(H_2O)_n$, n = 1-6 I. Optimal structures and vibrational spectra. J Phys Chem 99:8774-8792

Xantheas SS (1994) *Ab initio* studies of cyclic water clusters $(H_2O)_n$, n = 1-6 II. Analysis of many-body interactions. J Phys Chem 100:7523-7534

Xantheas SS (1995) *Ab initio* studies of cyclic water clusters $(H_2O)_n$, n = 1-6 III. Comparison of density functional with MP2 results. J Phys Chem 102:4505-4517

Xiao Y, Lasaga AC (1994) *Ab initio* quantum mechanical studies of the kinetics and mechanisms of silicate dissolution: $H^+(H_3O^+)$ catalysis. Geochim Cosmochim Acta 58:5379-5400

Xiao Y, Lasaga AC (1995) *Ab initio* quantum mechanical studies of the kinetics and mechanisms of quartz dissolution: OH^- catalysis. Geochim Cosmochim Acta, in press

Xiao Y, Lasaga AC (1996a) *Ab initio* studies of the structure, adsorption and hydrolysis of gibbsite and kaolinite. To be submitted to Phys Chem Minerals

Xiao Y, Lasaga AC (1996b) *Ab initio* studies of the effect of several water molecules on molecular models of mineral surface reactions. To be submitted to Geochim Cosmochim Acta

Xie Z, Walther JV (1992) Incongruent dissolution and surface area of kaolinite. Geochim Cosmochim Acta 56:3357-3363

Chapter 3

SILICATE MINERAL DISSOLUTION
AS A LIGAND-EXCHANGE REACTION

William H. Casey and Christian Ludwig

Department of Land, Air and Water Resources
and *Department of Geology*
University of California
Davis, CA 95616

INTRODUCTION

In this chapter we advance the thesis that the mechanisms by which metals are released from a dissolving mineral surface are profoundly similar to the mechanisms of ligand exchange around dissolved metal complexes. Mineral weathering resembles depolymerization of dissolved multimers (e.g. $[(FeOH)_2(H_2O)_8^{4+}(aq)]$). While reactions commonly proceed much slower at surfaces than in solution, the reactivity trends and ranges are commonly similar. The rates depend upon the acid-base properties of surface oxygens, the characters of the incoming and outgoing ligands, and the metal-oxygen and metal-ligand bond strengths. The similarity exists because general features of the metal coordination at surfaces resembles the corresponding ion in solution. The Mg-O bond length in the $Mg(H_2O)_6^{2+}(aq)$ ion, for example, is 2.10 Å and the metal is hexacoordinated to oxygens (Table 1). In the minerals periclase [MgO(s)] and forsterite [$Mg_2SiO_4(s)$], the metals are similarly hexacoordinated to oxygens and the Mg-O bond length is virtually constant at 2.11-2.13 Å (Table 1).

Table 1. Measured metal-oxygen distances (Å) for hydrated cations, oxide minerals and orthosilicate minerals. Distances for most dissolved species are given in Burgess (1988). Distances in the oxide and orthosilicate minerals are compiled in Casey (1991) and Bish and Burnham (1984), respectively. Distances for $Ga(H_2O)_6^{3+}$ and $Be(H_2O)_4^{2+}$ are inferred from Raman spectral data (Kanno, 1988).

Ion		Oxide		Silicate	
$Ca(H_2O)_6^{2+}$	2.39-2.46	CaO	2.405	Ca_2SiO_4	2.346-2.392
$Mg(H_2O)_6^{2+}$	2.10	MgO	2.11	Mg_2SiO_4	2.101-2.127
$Be(H_2O)_4^{2+}$	1.67	BeO	1.649	Be_2SiO_4	1.645
$Zn(H_2O)_6^{2+}$	2.08-2.17	ZnO	1.95	Zn_2SiO_4	1.92
$Mn(H_2O)_6^{2+}$	2.18-2.20	MnO	2.22	Mn_2SiO_4	2.185-2.227
$Co(H_2O)_6^{2+}$	2.05-2.08	CoO	2.13	Co_2SiO_4	2.123-2.134
$Ni(H_2O)_6^{2+}$	2.04-2.10	NiO	2.095	Ni_2SiO_4	2.076-2.102
$Al(H_2O)_6^{3+}$	1.87-1.97	Al_2O_3	1.86-1.97		
$Ga(H_2O)_6^{3+}$	2.02	Ga_2O_3	[IV]Ga: 1.80-1.85		
			[VI]Ga: 1.95-2.08		
$In(H_2O)_6^{3+}$	2.15	In_2O_3	1.93-2.18		

We illustrate these similarities with case studies that compare rates of reactions at the mineral surface and rates of similar reactions in solution. As the reader is certainly aware, mineral surface reactions can be devilishly complicated, as surfaces can be modified by leaching and topotactic conversion. These processes have little in common with reactions in the aqueous phase. Nevertheless, the simple hypothesis stated above will help organize observations about disequilibrium surface reactions even in cases where no complete analogue exists.

LIGAND-EXCHANGE KINETICS

What is a ligand-exchange reaction?

By definition, number of ligand atoms in the inner-coordination sphere of a metal is unchanged by a ligand-exchange reaction. The inner-coordination sphere consists of those molecules directly contacting the metal, as opposed to outer-sphere molecules that are separated from the metal by at least one molecule. The number of molecules packed into the sphere is determined by the relative sizes of the metal and coordinated molecules. For simplicity, we lump ligand substitution with true ligand exchange.

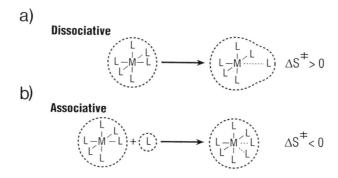

Figure 1. An illustration of dissociative (a) and associative (b) pathways for ligand exchange. A similar, albeit more complicated, figure could be created for destruction of a bridging oxygen at a mineral surface. In this case, a ligand is added to coordinate to one of the two previously bridged metals. Adapted from Burgess (1988).

For our purpose, there are only a few classes of ligand-exchange reactions to distinguish (Fig. 1). In *associative* (A) exchange, the coordination number increases by one ligand in the transition state. In a pure A mechanism (which is quite rare) one expects the incoming ligand to influence the reaction rate by distorting the stable coordination sphere of the metal. The transition-state complex has an extra ligand in the inner-coordination sphere:

$$Ni(OH_2)_6^{2+}(aq) + NH_3(aq) \;\overset{K_{eq}^{\ddagger}}{=}\; [Ni(OH_2)_6NH_3^{2+}(aq)]^{\ddagger} \tag{1}$$

The unstable seven-coordinated Ni(II) in the transition-state decomposes unimolecularly to return to octahedral coordination of ligands to the Ni(II) in the new complex. The transition-state complex decomposes unimolecularly:

$$[Ni(OH_2)_6NH_3^{2+}(aq)]^{\ddagger} \xrightarrow{\;slow\;} (Ni(OH_2)_5(NH_3)^{2+}(aq) + H_2O(l) \tag{2}$$

The fundamental assumption of transition-state theory is that there exists equilibrium between the reactants and a transition-state complex, so Equations (1) and (2) can be summarized:

$$Ni(OH_2)_6^{2+}(aq) + NH_3(aq) \underset{k_{w,A}}{\overset{k_{-w,A}}{\rightleftharpoons}} Ni(OH_2)_5(NH_3)^{2+}(aq) + H_2O(l) \qquad (3)$$

In *dissociative* (D) exchange, the coordination number decreases by one ligand in the transition state. The rate-controlling step is dissociation of the metal-ligand bond and the electronic structure of the incoming ligand exerts no influence on this dissociation. Consider the following simple reaction path, where a ligand-metal complex dissociates and the unassociated adduct reacts:

$$Ni(OH_2)_6^{2+}(aq) \underset{k_{w,D}}{\overset{k_{-w,D}}{\rightleftharpoons}} Ni(OH_2)_5^{2+}(aq) + H_2O(l) \qquad (4)$$

$$Ni(OH_2)_5^{2+}(aq) + NH_3(aq) \underset{k_{-NH_3,D}}{\overset{k_{NH_3,D}}{\rightleftharpoons}} Ni(OH_2)_5(NH_3)^{2+}(aq) \qquad (5)$$

The metal-oxygen (M-O) bond strength controls a wide range of ligand-exchange reactions that proceed via a D mechanism. Once rates are known for this elementary dissociation step, a wide range of ligand-exchange reactions are quantifiable, as we will see. Remember also the lesson of Table 1 that bond lengths and coordination numbers of metals in many silicate minerals resemble closely those in solution.

Most reactions are neither pure A or D, but have some character of each. A class of these intermediate reactions are referred to as *interchange* (I), which can be further subdivided into subclasses with a dissociative (I_D) or associative (I_A) character. Many dissociative reactions actually proceed via a two-step mechanism [the Eigen-Wilkens mechanism, Burgess (1990)]. Prior to bond dissociation, which is the rate-controlling elementary step reaction, an outer-sphere complex forms between the stable complex and the incoming ligand. This ion pair forms so rapidly that it is present at an equilibrium concentration, which is predictable from conventional electrostatic theory (e.g. the Fuoss equation). The elementary reaction is first-order in the concentration of this ion pair:

$$Ni(OH_2)_6^{2+}(aq) + NH_3(aq) \overset{K_{os}}{=} Ni(OH_2)_6 \bullet (NH_3)^{2+}(aq) \qquad \text{(ion pair)} \quad (6)$$

Rapid ion-pair formation is followed by slow dissociation of an M-O bond to water:

$$Ni(OH_2)_6^{2+} \bullet NH_3(aq) \underset{k_{w'}}{\overset{k_{-w'}}{\rightleftharpoons}} Ni(OH_2)_5(NH_3)^{2+}(aq) + H_2O(l) \qquad (7)$$

Such a mechanism explains the observation (Table 3) that second-order rate coefficients for divalent metals vary considerably with the incoming ligand, even in cases where D or I_D mechanisms are strongly suspected. When the second-order rate coefficient is divided by the predicted concentration of the ion pair at equilibrium (Table 3), the rate constant becomes virtually independent of the incoming ligand, as expected for a D or I_D mechanism. Furthermore, the rate coefficients became nearly equal to the pseudo-first order rate constant for water exchange, again suggesting that dissociation of these M-O bonds to water is the important step reaction. This theme, that rates of water movement are important, will reappear later in the chapter.

Table 2. First-order rate coefficients (k, in units of s^{-1}) and activation parameters for water exchange on some hexaaqua and monohydroxypentaaqua metal ions (Merbach and Akitt, 1990). The ratio k_{OH}/k corresponds to the ratio of rate coefficients for water exchange around the first hydrolysis product (e.g. Fe(OH)$^{2+}$(aq)) to that for the fully hydrated ion (e.g. Fe(H$_2$O)$_6$$^{3+}$(aq)). The activation parameters $\Delta H^{\ddagger}$, $\Delta S^{\ddagger}$, and $\Delta V^{\ddagger}$ were derived from the temperature and pressure variations of the reaction rates.

M^{z+}(aq)	k s^{-1}	$\Delta H^{\ddagger}$ kJ/mol	$\Delta S^{\ddagger}$ kJ/mol	$\Delta V^{\ddagger}$ cm^3/mol	pK	k_{OH}/k
Ga(H$_2$O)$_6$$^{3+}$	4.0×10^2	67.1	+30.1	+5.0	3.9	275
Ga(OH)$^{2+}$	1.1×10^5	58.9	---	+6.2		
Fe(H$_2$O)$_6$$^{3+}$	1.6×10^2	64.0	+12.1	-5.4	2.9	750
Fe(OH)$^{2+}$	1.2×10^5	42.4	+5.3	+7.0		
Cr(H$_2$O)$_6$$^{3+}$	2.4×10^{-6}	108.6	+11.6	-9.6	4.1	75
Cr(OH)$^{2+}$	1.8×10^{-4}	110.0	+55.6	+2.7		
Ru(H$_2$O)$_6$$^{3+}$	3.5×10^{-6}	89.8	-48.2	-8.3	2.7	170
Ru(OH)$^{2+}$	5.9×10^{-4}	95.8	+14.9	+0.9		
Rh(H$_2$O)$_6$$^{3+}$	2.2×10^{-9}	131.2	+29.3	-4.2	3.5	19100
Rh(OH)$^{2+}$	4.2×10^{-5}	103.0	----	+1.3		

The two-step mechanism has an analogue at mineral surfaces in that the important rate-controlling step in dissolution commonly follows an adsorption step that is so rapid that the surface concentration of adsorbate can be described with an equilibrium isotherm. The rate laws for dissolution include terms to account for the equilibrium concentration of sorbate. Oxalate, for example, is a geochemically important exudate of plants and microbes and can reach inventory concentrations of 90 g/m^2 in soil (Cromack et al., 1979; Graustein et al., 1977; Malujczuk and Cromack, 1982; Lapeyrie, 1988, Cromack et al., 1979). Carboxylic acids in general (see Jauregui and Reisenauer, 1982; Pohlmann and McColl, 1986; Lundström and Öhman, 1990) and oxalate in particular (see Amrhein and Suarez, 1988; Mast and Drever, Bennett et al., 1991; many others) considerably enhance the weathering rates of soil minerals. The adsorption of small carboxylic acids (i.e. oxalate) to a mineral surface is virtually complete in seconds (Ikeda et al., 1982), so that the slow step in dissolution is cleavage of an M-O bond.

Practically, the A, I and D character of a reaction is assessed through the dependency of the rate law on the concentration of the incoming ligand. If one supposes that the overall reaction proceeds via both pathways simultaneously, one pathway involving an outer-sphere ion-pair precursor and the other pathway involving dissociation, the overall rate law can be expressed in a simple form:

$$\frac{d[Ni(OH_2)_5(NH_3)^{2+}]}{dt} = k_f[Ni(OH_2)_6^{2+}][NH_3] - k_b[Ni(OH_2)_5(NH_3)^{2+}][H_2O] \quad (8)$$

with

$$k_f = K_{os}k_{-w'} + \frac{k_{-w}k_{NH_3}}{k_w + k_{NH_3}[NH_3]} \quad (9)$$

Table 3. The second-order rate coefficients (k_2) for exchange of each ligand around the aquated metal can be reduced to a psuedo-first-order rate coefficient (k_1) through division by the estimated equilibrium constants for formation of the outer sphere complexes (K_{os}) (from Burgess, 1988).

Ligand	k_2 $(M^{-1}s^{-1})$	K_{os} (M)	k_1 (s^{-1})
	$Ni^{2+}(aq)$		
N-Methylimidazole⁺	2.3×10^2	0.02	1.2×10^4
Imidazole H⁺	3×10^2	0.02	1.5×10^4
Ammonia	5×10^3	0.15	3.3×10^4
Hydrogen fluoride	3×10^3	0.15	2.0×10^4
Imidazole	$2.8\text{-}6.4\times10^3$	0.15	$1.9\text{-}4.3\times10^4$
1,10-Phenanthroline	4.1×10^3	0.15	2.6×10^4
Diglycine	2.1×10^4	0.17	1.2×10^4
Fluoride⁻	8×10^3	1	0.8×10^4
Acetate⁻	1×10^5	3	3.0×10^4
Glycinate⁻	2.0×10^4	2	1.0×10^4
Oxalate H⁻	5×10^3	2	0.3×10^4
Oxalate²⁻	7.5×10^4	13	0.6×10^4
Malonate²⁻	4.5×10^5	95	0.5×10^4
Methylphosphate²⁻	2.9×10^5	40	0.7×10^4
Pyrophosphate³⁻	2.1×10^6	88	2.4×10^4
Tripolyphosphate⁴⁻	6.8×10^6	570	1.2×10^4
Water exchange			3.2×10^4
	$Mg^{2+}(aq)$		
Oxine	1.3×10^4	0.2	0.7×10^5
Oxinate⁻	6.0×10^5	2.1	2.9×10^5
Fluoride⁻	5.5×10^4	1.6	0.4×10^5
5-Nitrosalicylate⁻	7.1×10^5	2	3.6×10^5
Bicarbonate⁻	5.0×10^5	0.9	5.6×10^5
Carbonate²⁻	1.5×10^4	3.5	0.4×10^5
pyrophosphateH₂²⁻	5.4×10^5	13	0.4×10^5
ADPH²⁻	1.0×10^6	9	1.1×10^5
ATPH³⁻	3.0×10^6	30	1.0×10^5
ATP⁴⁻	1.3×10^7	1200	1.1×10^5
Water exchange			1.0×10^5

and
$$k_b = k_{w'} + \frac{k_w k_{-NH_3}}{k_w + k_{NH_3}[NH_3]} \tag{10}$$

Quantities in square brackets [ ] correspond to concentrations. The rate coefficient approximately equals the first term, $k_f = K_{os}k_{-w}$, if reaction is dominantly via an outersphere complex. The second term is important for reaction along the dissociative pathway.

These methods are equivocal, however, and the most diagnostic evidence comes from $\Delta H^{\ddagger}$, $\Delta S^{\ddagger}$, and $\Delta V^{\ddagger}$ values determined spectroscopically (Merbach and Akitt, 1990). Ligand exchange via an A mechanism has a small or negative activation volume and negative activation entropies (Table 2). Reactions via a D mechanism have a positive activation entropy, a small positive activation volume, and a decreased coordination number in the activated state (Fig. 1).

Solvent exchange rates vary with M-O bond strengths

A typical rate-controlling step in ligand exchange is dissociation of a water molecule from a solvated metal. The rates of these elementary reactions have been tabulated and are observed to vary by a factor well over 10^{18} (Fig. 2). Many of alkaline metals exchange hydration waters so fast that every molecular collision transfers the ligand; that is, the rates are limited by solute diffusion. Conversely, rates of oxygen exchange around some oxyanions are slow even on a geological time scale.

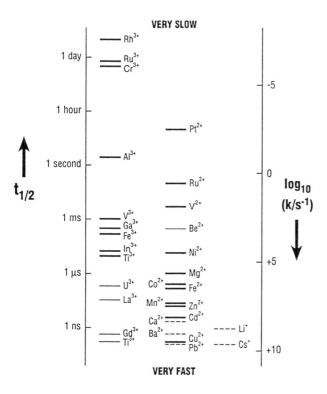

Figure 2. The first-order rate coefficient $(\log_{10}(k/s^{-1}))$ for water exchange around the corresponding hydrated metal ion (from Burgess, 1988) and mean residence times $(t_{1/2})$ for water molecules in the inner-coordination spheres of the metal. Data are for 298 K. Octahedral species are indicated by thick lines, non-octahedral species by thin lines and dashed lines indicate estimates derived from complexation studies. Note that the rates of water exchange around $Fe^{3+}(aq)$ are 1000 times slower than for $Fe^{2+}(aq)$. Electron-exchange resulting in metal oxidation greatly reduces the reactivity of some metal-oxygen bonds.

The rate coefficients tend to correlate inversely with the ionic potential (z/r) of the metal, where z is the metal valence state and r is the ionic radius (see Margerum et al., 1978). Rates decrease with increased charge on the metal and decreased ionic radius, particularly for alkaline-earth metals and for metals of the sp-block of the Periodic Table. Ion size are particularly important in exchange mechanisms because large ions accommodate the extra ligand in the activated state more easily than small ions. The mechanisms of water exchange into the hydration sphere of $Al(H_2O)_6^{3+}(aq)$,

$Ga(H_2O)_6^{3+}$(aq), and $In(H_2O)_6^{3+}$(aq) increase from dissociative to associative as the ionic radii increase from 0.51-0.81 Å (Burgess, 1988). Water exchange around $Al(H_2O)_6^{3+}$(aq) is 30,000 times slower than around $In(H_2O)_6^{3+}$(aq) (Fig. 2).

Rates for first-row divalent transition metals (Ni(II), Co(II), etc.) vary with the crystal-field stabilization energies relative to a transition state complex. These energies are given by the number of d-electrons and the ligand-field splitting parameters for the ligand in appropriate coordination to the metal. Rates of exchange for these metals are largely independent of ionic radius. Likewise, the mechanism of ligand exchange becomes increasingly dissociative with increasing number of d-electrons in the first row of transition metals. Water exchange around octahedrally coordinated Mn(II) is associative and fast. Water exchange around Ni(II) is of I_D character and slow because the octahedrally coordinated Ni(II) has strong directional bonding. The stable coordination sphere resists the distortion that accompanies ligand exchange more than metals with low ligand-field stabilization energy and spherically symmetric bonding, such as the octahedral (high-spin) Mn(II). Stated simply, there exist linear-free-energy relations (LFER) to predict these reactivities.

Metals with ionic bonds to oxygen are more reactive in water than metals that covalently bind to oxygen, but reactivity changes dramatically with the metal valence state. Rates of solvent exchange around the $Fe(H_2O)_6^{2+}$(aq) complex, for example, are a factor of 1000 times more rapid than around $Fe(H_2O)_6^{3+}$(aq) (Fig. 2). The reactivity of the Fe-O bond is lowered considerably by removing an electron, thereby oxidizing ferrous iron to the ferric state. Drawing an analogy to mineral surfaces, one then anticipates that similar changes in bond strength can be induced by electron exchange, leading to *oxidative* and *reductive* paths to metal removal from mineral surfaces. Geologically, these pathways are important because the rates of exchange of oxygen isotopes around some oxyanions, such as sulfate, are immeasurably slow. Exchange proceeds via a *reductive path* where the sulfur valence state is reduced, commonly to sulfide, and made more reactive. Subsequent re-oxidation of the sulfide to sulfate yields a new set of oxygens in the oxyanion.

The silicate polymer in most rock-forming minerals is quite resistant to hydrolysis and there is little known about the rates and mechanisms of oxygen exchange in dissolved silicate tetrahedra. There is only a single published study of ^{17}O exchange in dissolved silicate species (Knight et al., 1989), although other works are underway (Kinraide, in prep.). Knight et al. (1989) report that some sites in dissolved silica (bridging oxygens between two silicons?) took over a year to reach equilibrium, which is consistent with the dissolution rates of simple chain-silicate minerals. Depolymerization of the silicate polymer is apparently quite slow, as is covered in the chapter by S. Brantley (this volume). It is clear, however, that silicate polyhedra can link and dissociate rapidly in an aqueous solution (e.g. Kinraide and Swaddle, 1988a, 1988b; Knight et al., 1988; 1989), consistent with rapid repolymerization of silanol groups on the surfaces of some leached silicate minerals (e.g. Casey et al., 1993a). Other workers (S. Kinraide, pers. comm., 1993) find that the rates of oxygen exchange around silicate tetrahedra are uniformly rapid. By analogy with other oxyanion clusters (e.g. Richens et al, 1989), the reactivity of a particular oxygen will change dramatically with the number of coordinated metals and the accessibility of the site to solvent molecules.

Brønsted reactions commonly proceed to equilibrium and affect rates

Protons ($r \approx 10^{-13}$ cm) are so much smaller than metal ions ($r \approx 10^{-8}$ cm) that protonation of metal complexes dramatically modifies the distribution of electronic charge in the complex. Such charge redistribution can directly increase the rates of reaction by weakening the important bonds, or indirectly by causing structural changes in the metal-ligand complex that allow access to the metal site. For this reason there is intense interest in the location of Brønsted sites in multimers (e.g. Day et al., 1987; Furrer et al., 1992).

Protonation can modify the reactivity of existing ligands. Protonation of hydroxide ion, for example, converts it to a water molecule, which is generally much weaker Lewis base. Comparison of the rates of solvent exchange around metals and their hydrolysis products illustrate this point. Rates of ligand exchange, and the A or D character of the reaction, change as a coordinated water molecule is converted to a hydroxide ligand (Table 2). Removal of a proton from $Ru(H_2O)_6^{3+}$(aq), for example, causes the rate of exchange to increase by a factor of ≈ 170. Typically, the rates increase by a factor of $\approx 10^2$ to 10^3 as protons are removed from a hydration water (Table 2).

These protonation reactions are generally much more rapid than the rates of subsequent bond dissociation and they introduce a distinct pH-dependence to an otherwise slow, pH-independent reaction. Consider the reactions that depolymerize the simplest analogue for mineral dissolution: dissociation of a dissolved metal dimer. The measured rate coefficient for depolymerization of the $Fe_2(OH)_2(H_2O)_8^{4+}$(aq) dimer (e.g. Lutz and Wendt, 1970) depends linearly on hydrogen-ion concentration in solution:

$$k_{obs} = k_1 + k_2[H^+] \tag{11}$$

The psuedo-second-order term (k_2, M^{-1} s^{-1}) corresponds to a pathway where a proton first associates with, and weakens, bonds to one of the bridging hydroxide ions before the complex dissociates:

$$Fe_2(OH)_2(H_2O)_8^{4+} + H^+ \overset{k_{eq}}{=} Fe(OH)Fe(H_2O)_9^{5+} \ (equilibrium\ protonation) \tag{12}$$

$$Fe_2(OH)(H_2O)_9^{5+} + 2H_2O \overset{k_2}{\rightarrow} Fe(H_2O)_6^{3+} + Fe(OH)(H_2O)_5^{2+} \tag{13}$$

$$Fe_2(OH)_2(H_2O)_8^{4+} + 2H_2O \overset{k_1}{\rightarrow} 2Fe(OH)(H_2O)_5^{2+} \tag{14}$$

The psuedo-first-order term (k_1, s^{-1}) corresponds to the rate of dissociation into mono-hydroxide species.

These reaction pathways proceed simultaneously and one could refer to a '*proton-promoted*' pathway consisting of Equations (12) and (13), and a pathway where ligand exchange is via nucleophilic attack by the water molecule, inducing dissociation (Eqn. 14). Note that, although the rate of proton transfer is always fast, the *extent* of the acid-base reaction varies considerably with solution pH. An equilibrium protonation reaction introduces a distinct pH dependence to an otherwise slow reaction.

The parallel reaction pathways introduce considerable complexity into the interpretation of kinetic parameters. Application of the standard Arrhenius rate law

$$[k = Ae^{\frac{-E_a}{RT}}]$$

to observed rates of dissociation yields an apparent activation energy [E_a] that describes the weighted average of the reaction along two pathways. One then must expect that the temperature dependence of the observed k varies dramatically with solution pH since pH-dependent and pH-independent pathways are compounded. Such a complicated temperature dependence to the reaction rate is commonly observed in mineral dissolution studies (Carroll, 1989; Carroll-Webb and Walther, 1988; Brady and Walther, 1992; Casey et al., 1992) for exactly the same reason; the protonation reactions contribute enthalpy to the activation energy and this contribution varies with pH (e.g. Casey and Sposito, 1992). This point is illustrated for orthosilicate minerals in a later section.

Ligands can promote reaction

Other ligands can affect the dissociation rate of the above dimers. Fluoride ion, for example, could displace a bridging hydroxyl ion and introduce yet a third pathway for dissociation. Note the fundamental difference in the way that protons and the fluoride ligand affect the reaction. Protons modify the charge distribution in a pre-existing ligand associated to the metal (e.g. changing OH⁻ to H_2O). The fluoride ion directly associates with the metal and is a wholly new ligand. Thus, there exist *proton-promoted* and *ligand-promoted* pathways for dissociation of the dimer. The ligand affects rates by reducing the metal charge and hence the strengths of bonds to distal oxygens (i.e. the H_2O) molecules.

Chemical bonding at all sites in a ligand-metal complex is commonly affected by replacement or modification of one ligand, as we saw for the conversion of a coordinated H_2O to a hydroxyl ion. We saw already that the reactivities of distal M-O bonds to water increase by a factor of $\approx 10^2$ as the coordinated water deprotonates and converts to a hydroxide ion (Table 2). Rates of water exchange around Ni(II)(aq) also change appreciably as carboxylic acids, amines, ammonia and many other ligands enter the coordination sphere (Margerum et al., 1978; Wilkens, 1991).

Such a relation is not universal, however, but depends on the Lewis basicity of each ligand. Note, for example, that coordination of nitrogen from ammonia and the aliphatic amines (en, dien, trien) ligands to Ni(II) increases the rates of water exchange in the inner-coordination sphere (Fig. 3). This trend in reactivity is typically explained as resulting from reduced charge on the metal, which weakens the distal bonds to oxygen. This explanation is facile but not correct. Coordination of nitrogen in the pyridines (bypy, terpy) reduces the charge on transition metals (see p. 297-299 in Schindler, 1990), but the bonds to distal oxygens are strengthened (Griesser and Sigel, 1970). For this reason, bypy forms especially strong ternary complexes (e.g. L-Cu-bypy). The coordination of pyridines to Ni(II) has little effect (Table 4) on the rates of exchange of distal waters; the donation of electronic charge by the pyridine ligands is apparently compensated by other affects.

The increase in reactivity is also progressive (Fig. 3). For a wide range of ligands, the reactivity of distal M-O bonds to waters increase with the *number* of other electron-donating ligands in the inner-coordination sphere of the metal. The rates of water exchange around $Ni(H_2O)_{6-x}(NH_3)_x^{2+}(aq)$, for example, increase by a factor of $\approx 10^2$ as x varies from 0 to 3 (Table 4). Likewise, the dissociation rate constants for water in polydentate Ni(II)-ethylenediamine (en) complexes increases as more amines groups coordinate [e.g. $Ni(H_2O)_4(en)^{2+}(aq) < Ni(H_2O)_2(en)_2^{2+}(aq)$]. If multidentate ligands are involved, the effect increases with the ligand denticity or number of coordinated functional groups. Therefore the reactivity of the bonds to water (our proxy for bonds to oxygens at a mineral surface) reflect both the type and number of other ligands in the coordination sphere.

Ligands have acid-base properties that affect reactivities

Protons are intimately involved in the mechanism of ligand-promoted ligand exchange. Above we showed that rates of water exchange can be accelerated by removing a proton from a coordinated water to form a hydroxyl ion from a water molecule. The effect is not limited to water molecules and one expects that any ligand that can attract protons

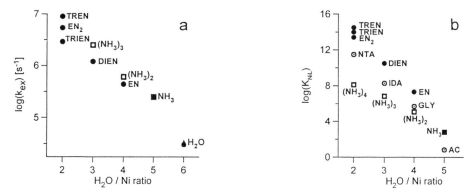

Figure 3. (a) The first-order rate coefficients at 298.2 K for exchange of water between the bulk aqueous solution and waters in the inner-coordination sphere of Ni(II)-ligand complexes as a function of the $H_2O/Ni(II)$ ratio (from Wilkens, 1991; Margerum et al., 1978; Rowland, 1975). The total coordination number of the metal (water molecules + coordinated atoms from the ligands) is six for all of the complexes shown. Ammonia complexes are identified according to the number of ammonia molecules coordinated to the Ni(II). The title $(NH_3)_2$, for example, corresponds to the $Ni(H_2O)_4(NH_3)_2^{2+}$(aq) complex.
Ligand abbreviations are: EN = ethylenediamine $[NH_2(CH_2)_2NH_2]$;
                           DIEN = diethylenetriamine $[NH_2(CH_2)_2NH(CH_2)_2NH_2]$;
                           TRIEN = $[NH_2(CH_2)_2NH(CH_2)_2NH(CH_2)_2NH_2]$;
                           TREN = $[N((CH_2)_2NH_2)_3]$.
The point marked 'H_2O' is the rate of solvent exchange around $Ni(H_2O)_6^{2+}$(aq). The complex denoted NH_3 corresponds to the $Ni(H_2O)_5NH_3^{2+}$(aq) complex, which can be assigned to both sets of homologous ligands.
(b) Metal-ligand stability constants for equilibrium in the reaction $Ni^{2+}(aq) + pL^q(aq) = NiL_p^{2-pq}(aq)$ decrease as a function of the $H_2O/Ni(II)$ ratio in the complex for ammonia and mixed amino-carboxylate complexes (acetic acid = AC; glycine = GLY, iminodiacetic acid = IDA; nitrilo-triacetic acid = NTA).

Table 4. Rate coefficients for exchange of a single water molecule from around Ni(II)-ligand complexes (from Rowland, 1975; Burgess, 1988; Wilkens, 1990; Margerum et al., 1978).

Ligand	k (s⁻¹)	Ligand	k (s⁻¹)
[a]$Ni(H_2O)_6^{2+}$	0.32×10^5	[a]$Ni(NH_3)(H_2O)_5^{2+}$	2.5×10^5
[a]$Ni(NH_3)_2(H_2O)_4^{2+}$	6.1×10^5	[a]$Ni(NH_3)_3(H_2O)_3^{2+}$	25×10^5
[b]$Ni(en)(H_2O)_5^{2+}$	4.4×10^5	[b]$Ni(dien)(H_2O)_3^{2+}$	18×10^5
[b]$Ni(trien)(H_2O)_2^{2+}$	29×10^5	[b]$Ni(en)_2(H_2O)_2^{2+}$	54×10^5
[c]$Ni(bpy)(H_2O)_4^{2+}$	0.49×10^5	[c]$Ni(bpy)_2(H_2O)_2^{2+}$	0.66×10^5
[c]$Ni(tpy)(H_2O)_3^{2+}$	0.52×10^5		
[d]$Ni(ida)(H_2O)_3^{2+}$	2.4×10^5	[d]$Ni(H_2O)(edta)^{2-}$	7×10^5
[e]$Ni(H_2O)_5Cl^+$	1.5×10^5	[e]$Ni(H_2O)_3(NCS)^{3-}$	11×10^5
[f]$Ni(2,3,2-tet)(H_2O)_2^{2+}$	40×10^5	[f]$Ni((12[ane]N_4)(H_2O)_2^{2+}$	200×10^5

[a]ammonia, [b]tertiary amines, [c]pyridines, [d]carboxylates, [e]inorganic ligands, [f]macrocycles

from other coordinated ligands can potentially change the reactivity of the remaining M-O bonds to water. Removal of a proton from an ammonia (NH_3) coordinated to Ni(II), for example, creates a highly reactive amide (NH_2^-) ligand that dissociates from the complex rapidly (Wilkens, 1991). This *conjugate-base* mechanism illustrates the point that Brønsted properties of a ligand-metal complex may differ from the ligand and metal considered separately. Proton-promoted and ligand-promoted ligand exchange are not always independent.

There also exists an internal conjugate base mechanism for polydentate ligands (Fig. 4) where a proton associates with an incoming ligand and is not fully removed from the complex. The first step is formation of a weak, monodentate, outer-sphere complex that proceeds to equilibrium (equilibrium constant = K_{os} in Fig. 4). The coordinated functional group (think of $-NH_2$) withdraws a proton from a coordinated water, thereby weakening bonds between the metal and a *distal* water molecule. This weakening leads to a slow replacement of the water by the uncoordinating functional group (rate coefficient = k' in Fig. 4). Once coordinated directly to the ligand, bonds to distal waters are further weakened and the second functional group can flop into place (rate coefficient = k" in Fig. 4).

M-O bond dissociation can control rates of chelation

Even the formation of multidentate chelate complexes can be controlled by rates of dissociation of M-O bonds to water. If the first functional group in a multidentate ligand coordinates via a D or I_D mechanism, one expects similar rate coefficients for homologous ligands. The rate coefficients for forming the monodentate aluminum-citrate (a tricarboxylic acid) complex, for example, is k = 80±10 M^{-1} s^{-1} (Lopez-Quintela et al., 1984). The rate coefficient for forming the monodentate citrate complex is experimentally identical to the rate coefficient for forming the formate complex (90±20 M^{-1} s^{-1}), which is the simplest monocarboxylic acid. Such similarity is expected if dissociation of the M-O bond to water controls the rate of the coordinating the first carboxyl functional group to the metal:

$$H_2Cit^-(aq) + Al(H_2O)_6^{3+}(aq) \xrightarrow{k_1} AlH_2Cit(H_2O)_5^{2+}(aq) + H_2O(l) \qquad (15)$$

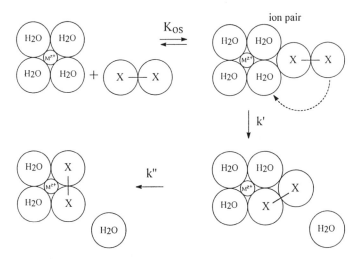

Figure 4. A reaction pathway illustrating the Internal Conjugate Base mechanism where a bidentate ligand, such as enthylenediamine, exchanges for two water molecules in the inner-coordiantion sphere of a metal.

$$HCit^{2-}(aq) + Al(H_2O)_6^{3+}(aq) \xrightarrow{k_2} AlHCit(H_2O)_4^+(aq) + 2H_2O(l) \qquad (16)$$

If the rates of solvent exchange are much slower than the rates of association of the additional functional groups in a multidentate complex, the rates of the overall chelation reaction will scale with the rates of solvent exchange.

At the other extreme, the rates of solvent exchange may be much faster than the rates of association of the ligand functional group. In this case, the first ligand functional group may not easily coordinate to the metal because of competition with the back-reacting water molecules. The rates of formation of the monodentate complex will vary considerably for different ligands and rates of ring closure can be much slower than the rates of metal desolvation. Rate constants for $Cu^{2+}(aq)$ reacting with monodentate ligands, for example, range from 2×10^8 to 20×10^8 M^{-1} s^{-1} (Burgess, 1988). Rate coefficients for forming the bidentate ring complexes are considerably smaller and range from 10^5 M^{-1} s^{-1} to 10^8 M^{-1} s^{-1}. These reactions are sterically controlled and depend on the protonation state of the functional groups.

MINERAL SURFACE REACTIONS

General comments

The features of ligand exchange that are important to the subsequent discussion of mineral dissolution are: (1) The M-O bond distances and coordination number of metals at a mineral surface are similar to those of the solvated metal in solution; (2) The rate-controlling step in ligand exchange is commonly dissociation of the M-O bonds to solvent waters. The reactivities of these elementary reactions vary by many orders of magnitude (Fig. 2); (3) Ligand exchange proceeds via several pathways that can be distinguished depending upon how protonation affects bond dissociation, whether incoming ligands displace coordinated waters, and whether the valence states of metal and ligand change during the reaction; (4) Protonation reactions and ion-pair formation are usually much more rapid than bond dissociation in ligand exchange; (5) Unexchanged ligands in a complex can affect the reactivity of distal ligands by reducing the charge on the metal; and 6) Bases can move protons from ligands in the inner-coordination sphere of the metal, thereby weakening distal bonds and enhancing the rates of exchange in a so-called 'Conjugate Base' mechanism.

Rate laws for surface reactions

Many of the scientists studying mineral corrosion have noted that the rates are properly expressed as proportional to the adsorbed concentration of rate-enhancing ligands (e.g. Koch, 1965; Wirth and Gieskes, 1979; Furrer and Stumm, 1983; Stumm and Wollast, 1990). Furrer and Stumm (1983), Pulfer et al. (1984) and Furrer and Stumm (1986) codified this dependence into a rate law for surfaces that included adsorbed concentrations of protons, hydroxyl ions and other rate-enhancing ligands. Far from equilibrium, the rates of dissolution are given by the sum of elementary reactions, yielding a rate law of the general form:

$$Rate = \sum_{p,l,o,i} k_{p,l,o,i} x_{s_{p,l,o,i}}^{p+l+o} + k_{H_2O} \qquad (17)$$

and $k_{p,l,o,i}$ and k_{H_2O} are rate coefficients for dissolution via the proton- (p), hydroxide (o), and ligand-promoted (l) pathways and nucleophilic attack by water. The variables are the mole fractions (any concentration variable could substitute) of surface complexes 'i'. This variable also includes a description of the concentrations of adsorbed protons and hydroxide ions; for example, if $l = 0$, $o = 0$ and $p \neq 0$, the rate of dissolution via the proton-promoted path alone is described. The rate coefficients must have units suitable to give the rate in terms of mol cm^{-2} s^{-1}. Rate laws such as Equation (17) assume that reactions along the various pathways are independent, and therefore additive. This assumption may fail upon close examination, but is useful.

The summation deserves elucidation. First, because these are elementary reactions, the conformation and protonation states of different surface complexes must be distinguished, which is experimentally difficult. Secondly, protons may be involved in ligand-promoted dissolution causing different surface structures and denticities. This point was clearly established by Zinder et al. (1986) who showed that oxalate-induced dissolution of goethite varies with both adsorbed oxalate and adsorbed proton concentrations. In addition, the protonation states of a ligand (and the denticity) will vary with the solution pH. The subscript p accounts for the number of protons involved in the rate-controlling step. The subscript l accounts for the number of adsorbed ligands; the subscript i accounts for ligands of similar stoichiometry but different structures, including protonation states. In this formalism, a distinction is made between protons that are adsorbed onto the ligand (think of singly deprotonated, adsorbed, oxalate; HOOCCOO⁻) and those protons that adsorb onto the mineral surface in addition to the ligand. In an extreme example, ligands may only weaken metal-oxygen bonds at the surface, making them more susceptible to proton-promoted dissolution. Distinction is also made for surface ligands with protons configured to different ligand atoms; these are different surface complexes for the purpose of the rate law.

More caveats are in order. Rate laws such as Equation (17) do not explicitly include the concentration of water, which is probably involved in the reaction, because these concentrations are usually high and constant in an aqueous solution. In fact, there is good evidence that movement of water molecules in hydration, followed by detachment of surface complexes, controls the rate of metal removal from a surface (e.g. Casey and Westrich, 1992; Ludwig et al., 1995). Rates laws such as Equation (17) require that the rate-controlling step reaction is the disruption of bonds at or near the mineral surface, and not movement of the reactive solutes through the bulk aqueous phase to the sites of reaction.

Consider the example shown in Figure 5, where we portray removal of a surface Ni(II) atom through combined actions of an oxalate ion (ox) and protons that adsorb onto the surface of NiO(s). We imagine that the elementary step reactions 1-6 are all shown in Figure 5 (we recognize that some may be missing or wrong) and that only one step controls the rate of the overall reaction. The first steps are so rapid that they proceed to equilibrium; these steps include the displacement of water molecules from the surface site. The second step, which is fast, is closing of the oxalate ring by displacement of another water molecule. The third and fourth steps are fast protonations of the underlying oxygens and the fifth and sixth steps are hydration. We speculate that the fifth or sixth steps, hydration of the metal, are slow (we do not know which is slowest or if both proceed simultaneously). This hydration(s) leads to detachment of the surface complex and is the slowest step-two waters must configure to the detaching Ni(II) if the stable Ni(OH$_2$)$_4$(ox)(aq) species is the one that detaches.

Finally and most importantly, note that we have recaptured the initial surface. A second term in the summation to Equation (17) would appear if, for example, a monodentate oxalate accelerates dissolution, or if protons adsorb onto the oxalate rather than to underlying Ni-O bonds. These surface complexes have different structures (if not stoichiometries) and must be treated separately in Equation (17).

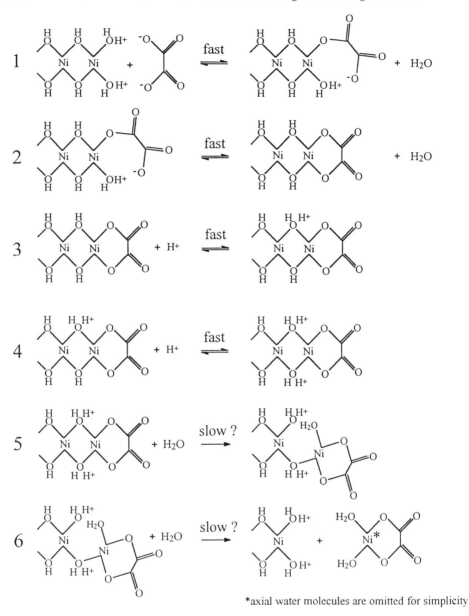

*axial water molecules are omitted for simplicity

Figure 5. A hypothetical scheme describing one pathway for the ligand-promoted dissolution of NiO(s) with oxalate and protons.

Note in this scheme, and in the previous discussion, that metal-oxygen bonds between the metal and water are commonly formed or destroyed in the rate-controlling step reaction. In our above example, the slow step in the reaction is hydration of the detaching Ni(II)-oxalate species at the surface. If this is true, one then expects that there should be similarity between the rates of solvent exchange around the metals in solution and the rates of metal removal from the corresponding surface.

Solute adsorption/desorption is usually faster than dissolution

We saw in the earlier discussion that a common first step in ligand-exchange is the formation of a electrostatically bonded ion pair. These ion pairs form much more rapidly than the subsequent exchange of ligands. Such pre-equilibration is also commonly observed at mineral surfaces although the rates of reaction are much slower than in solution. The adsorption of protons and hydroxyl ions to form surface charge commonly proceeds at time scales of 0.001 to 10 s (e.g. Astumian et al., 1981), rather than at time scales of $<10^{-6}$ s for protonation in solution.

The rates of protonation-deprotonation reactions have been characterized for simple oxide solids in electrolyte solutions via relaxation spectroscopy. This method entails disturbing an equilibrium state by instantaneous changes in pressure, temperature or energy field (see Sparks, 1989). The resulting spectral signal (usually conductance) is caused by the relaxation of the chemical system back to equilibrium. Rate coefficients for protonation-deprotonation of some oxide functional groups are compiled in Table 5.

Table 5. Rate coefficients for formation and dissociation of surface charge on metal-oxide mineral surfaces. The second-order rate coefficients (k_a) for formation of surface charge by adsorption of protons. The first-order rate coefficient (k_d) corresponds to dissociation of hyperprotonated surface groups (from Astumian et al., 1981; Sasaki et al., 1987).

Metal oxide	k_a $(M^{-1}s^{-1})$	pK(int) (M)	k_d (s^{-1})	τ_d (s)
-SO₃H groups on latex	8×10^5	0.96	9×10^4	1×10^{-5}
silica-alumina	0.3×10^5		46.	0.02
TiO₂	6.2×10^5	4.7	13.	0.08
Fe₃O₄	1.4×10^5	7.1	0.34	2.9
Fe₂O₃	2.4×10^5	8.4	0.16	6.2

First note that the rate coefficients (and derived τ_d values) scale like the acidity of the surface. The strongest acids dissociate most rapidly and are therefore proportional to the pK(int) values. This behavior is expected from our knowledge of the dissociation rates of a homologous set of dissolved acids. The activation energy for dissociation can be no smaller than the standard-state change in enthalpy at equilibrium and is commonly similar in size. Similarity between the activation energies (here expressed as τ_d values) and equilibrium constants is an example of a linear-free-energy correlation (LFER), which can be used to predict unknown rate parameters. The characteristic times for dissociation can be simply estimated as: $t \approx 1/k_d$ and are generally ten seconds or less. A characteristic lifetime for metals at the surface of a dissolving mineral can be similarly estimated and compared to Table 5. For minerals with a rocksalt or wurtzite structure, the unit cell dimensions are $\sim4\times10^{-8}$ cm and each unit cell surface exposes a single metal site. This site density corresponds to $\sim6\times10^{14}$ sites cm^{-2} or $\sim1\times10^{-9}$ moles cm^{-2} of metal. The characteristic lifetimes of metals at the surface is estimated by dividing the site density by the specific dissolution rate. For a dissolution rates of 10^{-7}, 10^{-10} and 10^{-15} mol cm^{-2} s^{-1}, for example, the characteristic lifetimes of metals at the surface are $\tau \approx 0.01$, ≈ 10, and $\approx 10^6$ s, respectively (Table 6).

A second important point in Table 5 is that characteristic metal lifetimes at the exposed mineral surfaces are generally longer than the time required to adsorb or desorb reactive ligands or protons. The rule is, however, not inviolate. The adsorption kinetics of potential rate-enhancing ligands have not been widely studied, although some data exist for oxyanion adsorption (Zhang and Sparks, 1989; 1990a,b), acetate (Ikeda et al., 1982), and metals (Hachiya et al., 1979; 1984). The relaxation times typically range in the 0.001 s to 1 s interval, which overlap somewhat with the lifetimes of metals at the more reactive mineral surfaces (Table 6). These highly reactive surfaces will dissolve via a diffusion-controlled mechanism unless experimental care is taken to shorten the diffusion length scales.

Even for minerals where $\tau \approx 1$ s, however, the rates of ligand and proton adsorption are probably fast because dissolution and adsorption rates correlate if the rate-controlling step in adsorption is replacement of a surface hydroxyl or water molecule at the surface. These reactions tend to scale like rates of solvent exchange around the corresponding metal (Hachiya et al., 1984). In other words, adsorption onto NiO(s) may be slow, but so is the rate of dissociation of the surface Ni(II)-O bonds required by dissolution. Likewise, adsorption/desorption is probably fast on rapidly dissolving minerals. The rapid adsorption reaction can be considered analogous to the rapid ion-pair formation that defines the Eigen-Wilkens mechanism.

Brønsted reactions on mineral surfaces

A connection between the acid-base properties of a surface and the dissolution rate was made by Wirth and Gieskes (1979) (see also Budd, 1961; Koch, 1965; Furrer and Stumm, 1986) who showed that rate of SiO$_2$(s) dissolution depended on the equilibrium concentration of negative charge. For surfaces, however, the protonation/deprotonation

Table 6. Estimated lifetimes for metals at the surface of a series of simple oxide and orthosilicate minerals undergoing surface-controlled dissolution. The rate data are from Casey (1991), Casey and Westrich (1992), Westrich et al., 1993; Plettnick et al., 1994, Carroll-Webb and Walther, 1988; and Brady and Walther, 1989). We assume an average site density of 1×10^{-9} moles cm^{-2} and, for most minerals, that rates decrease by a factor of 3 for every unit increase in pH.

Oxide	$\tau(s)$ pH=2	$\tau(s)$ pH=5	Silicate	$\tau(s)$ pH=2	$\tau(s)$ pH=5
CaO	0.016	0.14	Ca$_2$SiO$_4$	.33	1.8
MgO	0.06-1.0	0.6-9.0	Mg$_2$SiO$_4$	1.3×10^3	1.2×10^4
BeO	1×10^6	9×10^6	Be$_2$SiO$_4$	3.3×10^6	3×10^7
CuO	0.01	0.09	---	---	---
ZnO	0.08	0.72	Zn$_2$SiO$_4$	3.13	28
MnO	0.25	2.3	Mn$_2$SiO$_4$	2.5	22
CoO	16-80	144-720	Co$_2$SiO$_4$	360	1000
NiO	1.6×10^5	14×10^6	Ni$_2$SiO$_4$	5×10^5	4.5×10^6
SiO$_2$(am.)	3.2×10^7	1×10^5			
α-Al$_2$O$_3$	5×10^7	1×10^8			

reactions are of uncertain stoichiometry. The reactions are generally written to involve a neutral 'average' hydroxyl functional group (see Sposito, 1984; Schindler and Stumm, 1987):

$$\triangleright SOH_2^+ = \triangleright SOH + H^+ \tag{18}$$

$$\triangleright SOH = \triangleright SO^- + H^+ \tag{19}$$

where $\triangleright SOH$ represents a model hydroxyl functional group.

No information about microscopic acid-base reactions is provided by these equations and the choice of a diprotic model for the acid-base reactions is arbitrary. By microscopic reactions we mean the actual protonation of coordinatively distinct oxygens at the mineral surface. Oxygens on the surface of bromellite, for example, are associated with one, two or three beryllium ions depending upon how the BeO_4^{6-} structural unit is truncated. These oxygens can be distinguished spectrally (Boehm and Knozinger, 1983) and, by analogy with oxygens in dissolved metal-oxide clusters (e.g. Furrer et al., 1992; Richens et al. 1989; Day et al., 1987), have distinctly different acid-base properties and reactivities. Equations (18) and (19) ignore these subtleties.

Even at this broad level, the acid-base chemistry of a particular mineral surface is characteristic of its composition and structure, just as is the hydrolysis chemistry of a dissolved metal. Silicic acid [$H_4SiO_4(aq)$] does not deprotonate at the same pH as the hydrated beryllium ion [$Be(H_2O)_4^{2+}(aq)$]; likewise, the hydroxylated surface of silica [$SiO_2(s)$] is not neutral at the same pH as the surface of bromellite [$BeO(s)$]. The rapid acid-base reactions impart a net surface charge to the mineral surface. The condition of net neutrality is termed the Point of Zero Net Proton Charge (PZNPC, Sposito, 1984).

An equilibrium relation can be defined for the formation of positive or negative surface charge on these generic functional groups. For Reaction (18), for example:

$$K = \frac{f_{SOH_2^+}\, x_{SOH_2^+}}{a_{H^+}\, f_{SOH}\, x_{SOH}} \tag{20}$$

where f_i and x_i are the rational activity coefficient and mole fraction of the ith surface species, respectively, and a_{H^+} is the activity of protons in the aqueous phase. The parameter that is most commonly measured, however, is a conditional equilibrium constant:

$$K' = \frac{K\, f_{SOH}}{f_{SOH_2^+}} = \frac{x_{SOH_2^+}}{a_{H^+}\, x_{SOH}} \tag{21}$$

which is a function of temperature and solution composition, including ionic strength.

These conditional equilibrium constants vary with the extents of reaction and the range of variation distinguishes them from most aqueous reactions. One does not expect factors of 100 to 1000 variation in conditional equilibrium constants describing say, removal of a second proton from phosphoric acid, yet this range of variation is not rare for surface protonation reactions (Fig. 6). It arises because of the large and varying electrostatic field at a surface. Sophisticated models (e.g. Constant Capacitance Model, Schindler and Stumm, 1987) have been derived to account for these large corrections for surface reactions. The situation is a little more complicated for minerals which have a structural charge due to uncompensated cation substitutions (e.g. Al(III) for Si(IV) in clays) but is similar.

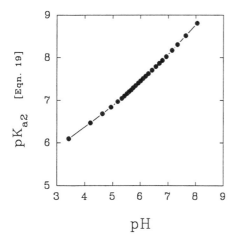

Figure 6. Variation in the conditional equilibrium constant with pH for deprotonation of the surface of $SiO_2(s)$. Data from Fürst (1976).

Proton-promoted dissolution

From the similarity in oxygen coordination chemistry (Table 1), one expects a strong correlation between mineral dissolution rates at well-defined conditions and rates of water exchange about the corresponding dissolved cation *if* the dissociation step controls the reaction rate. Well defined conditions mean: (i) that minerals are compared where the M-O bonds that are chosen for study are important to the reaction rate; (ii) that experiments are conducted in similar solutions; and (iii) that the valence state of metals at the mineral surface is unaltered.

In order to evaluate this prediction we must first assemble a set of minerals that have no covalent silicate framework so that dissociation of the M-O bonds can be observed and that are comparably protonated at a given solution pH. The simple oxide and orthosilicate minerals satisfy these criteria. The Brønsted acid-base properties of these minerals are also comparable. Although the K' values (Eqn. 21) vary by factors of 10 to 100 with pH, the variations are similar for many oxide minerals once differences in the PZNPC values are taken into account (Fig. 7). The PZNPC values of minerals within the orthosilicate class are quite similar. Therefore, at a given pH, there are comparable concentrations of adsorbed protons.

For dissolution at conditions where positive surface charge predominates (pH < PZNPC) and far from equilibrium, the empirical rate law (Eqn. 17) can be simplified to:

$$Rate = k_{H^+} x^n_{SOH_2^+} \tag{22}$$

The hypothesis is that these rates of dissolution will differ considerably with isostructural changes in mineral composition, but in a way that is predictable from the solution chemistry (e.g. Fig. 2). Such a correlation is shown in Figure 8 for the simple orthosilicate minerals at pH = 2 and 25°C. Note that the general reactivity trends for solvent-exchange kinetics are also observed for mineral dissolution.

The dissolution rates of silicates containing alkaline-earths vary with ionic size of the cation. Calcium can be more quickly removed from a surface site than magnesium; beryllium is very resistant. For minerals containing first-row transition metals, rates vary

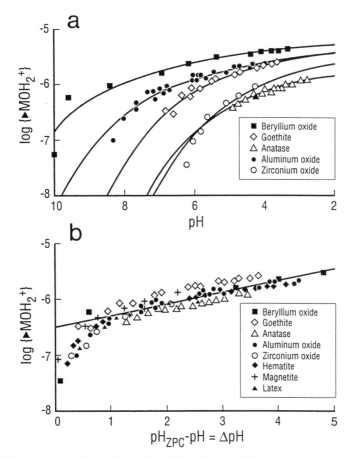

Figure 7. (a) Surface protonation isotherms from Wieland et al. (1988). Symbols correspond to titration curves at ionic strength (I) of 0.1 M, except for hematite where I = 0.2 M. The concentrations of protonated sites ($\triangleright MOH_2^+$) are given in moles/m^2. (b) Surface concentrations from Figure 7a as a function of $DpH = PZNPC - pH$.

with the number of valence d-electrons. The dissolution rate of liebenbergite [$Ni_2SiO_4(s)$] is much, much slower than tephroite [$Mn_2SiO_4(s)$] although the M-O bond lengths differ only slightly (Table 1). Clearly, dissociation of a divalent M-O bond controls the rate of dissolution of these compounds and the variation in rate spans a range of 10^5.

Ligand-promoted dissolution

By hypothesis, the same ligands that increase the reactivity of bonds between dissolved metals and hydration waters should also increase the rates of dissolution of the metal-oxide mineral. The ligands can affect dissolution rates via either the conjugate-base mechanism or by acting as a strong Lewis base. This hypothesis could be tested by examining ligand-promoted dissolution of a mineral such that a conspicuous reactivity trend in ligand-exchange is reproducible. Earlier (Table 4), for example, we showed that increased numbers of certain ligands (some amines and carboxylates) enhance the rates of water exchange around Ni^{2+}(aq). Our hypothesis, above, could be restated such that the

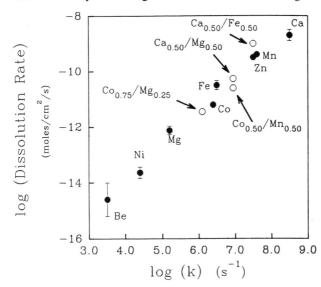

Figure 8. The dissolution rates of end-member orthosilicates at pH = 2 and 25°C plotted against the first-order rate coefficient, k (s^{-1}), for water exchange from the solvent into the hydration sphere of the corresponding dissolved cation (from Casey and Westrich, 1992; Westrich et al., 1993). The mineral and ions are identified by the divalent cation (for example, Mn^{2+} refers to the dissolution rate of tephroite (Mn_2SiO_4) and rate of solvent exchange around $Mn(H_2O)_6{}^{2+}$(aq).

rate coefficients for bunsenite [NiO(s)] dissolution will increase predictably with the number of these functional groups coordinated to a surface Ni(II).

Before conducting these experiments, however, it is necessary to evaluate the competing reactions and to simplify the rate law. Empirical information allows us to pare down the complexity of Equation (17). First, most ligands are so large that the molecularity of the reaction is likely to be unity. Very small ligands such as fluoride may be exceptional (Pulfer et al., 1984). Secondly, ligand-promoted dissolution will only be interesting if the rates are much more rapid than via the proton- and hydroxyl-promoted pathways. We can simplify the rate law by conducting experiments near the PZNPC value of NiO(s), so that the rates of proton- and hydroxyl-promoted dissolution are relatively small.

We could also simplify the rate law if only a single surface complex is stable at the experimental conditions. This means that we must choose to study ligands where the protonation state and denticity do not vary within the experimental pH range. Experiments with aminocarboxylate ligands allow us to test the hypothesis most cleanly because coordination of a surface Ni(II) to one amino and one carboxyl group for the ligands (GLY, NTA, IDA) forms a five-member ring regardless of the molecule size. We can conduct experiments with these ligands at pH = 6 and pH = 8.5. These pH conditions bracket many natural environments and also allow one to compare the relative effects of carboxyl and amino groups on the rates. Carboxylates should be more important than amines at low pH, but the relation may be reversed as pH increases.

Further simplification of Equation (17) requires an assumption that will ultimately prove to be invalid: that protons are not involved in the detachment of the surface metal complex once the ligand-metal complex forms at the surface. In an extreme case, one can imagine that proton-and ligand-promoted dissolution pathways are mechanistically nearly identical; the ligand simple enhances the acidity and reactivity of the important M-O bonds. Nevertheless, for the assumptions enumerated above, the rate law is simplified to:

$$Rate = k_L\{\triangleright SL\} \tag{23}$$

which allows us to test the hypothesis.

Table 7. Rates (R_L) and rate coefficients (k_L) for the ligand-promoted dissolution of bunsenite (from Ludwig et al., 1995) along with the stability constant for Ni(II)-ligand binding.

Ligand	R_L(pH=6) [mol m^{-2} s^{-1}]	log k_L(pH=6) [s^{-1}]	R_L(pH=8.5) [mol m^{-2} s^{-1}]	log k_L(pH=8.5)	log K_{NiL}
NTA	87.4($\pm$11)$\times10^{-12}$	-4.58($\pm$0.06)	20.0($\pm$1.8)$\times10^{-11}$	-4.15	11.50
IDA	42.5($\pm$5.3)$\times10^{-12}$	-4.84($\pm$0.06)	10.3($\pm$1.0)$\times10^{-11}$	-4.53	8.30
GLY	18.2($\pm$2.8)$\times10^{-12}$	-5.11($\pm$0.07)	2.8($\pm$0.3)$\times10^{-11}$	-5.17	5.78
AC	6.6($\pm$1.7)$\times10^{-12}$	-5.57($\pm$0.13)	4.9($\pm$2.5)$\times10^{-13}$	-----	0.84

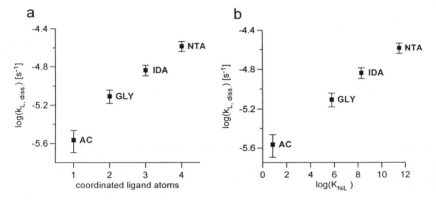

Figure 9. Rate coefficients (k_L) for ligand-promoted dissolution of bunsenite [NiO(s)] at pH = 6.0 (from Ludwig et al., 1995). In (a) the rate coefficients are compared with the number of coordinated ligand atoms (e.g. one in acetic acid = AC; two in glycine = GLY, three in iminodiacetic acid = IDA; and four in nitrilotriacetic acid = NTA). In (b) the same rate coefficients are compared with the logarithm of the stability constants [log(K_{NiL})] for forming the dissolved Ni(II)-ligand complexes.

The rate coefficients for ligand-promoted dissolution of bunsenite are compiled in Table 7 and illustrated in Figure 9. As one can see, the rate coefficients for ligand-promoted dissolution increase substantially with the number of coordinating functional groups. The simplest explanation for this result is that the ligand functional groups displace surface hydroxyl groups or water molecules from around a surface Ni(II) site to form an inner-sphere surface complex at a single metal center. Therefore the rate coefficients for dissolution increase with the number of coordinated ligand atoms in the surface complex, just as is observed for the dissolved metal-ligand complexes (Fig. 3). Furthermore, for many Ni(II)-complexes in solution, the rate coefficients vary with the equilibrium constant for ligand-metal binding (compare Figs. 3b and 9b). This correlation, if verified for surfaces (likely!), promises a means of some predicting rate coefficients for surface reactions from the equilibrium constants.

Such a correlation will not be observed for all ligands. Solution pH is an important complicating factor because adsorbed ligand complexes can protonate differently than complexes in the aqueous solution, leading to desorption (or dechelation) of the ligand. High-molecular weight ligands, particularly those with disparate functional groups, can undergo acid-base reactions on the surface. For these cases, several surface complexes must be identified that change in relative concentration as a function of solution pH. Furthermore, adsorbed protons, separate from the adsorbed ligand, may be involved in the subsequent dissolution reaction. Such a difference in protonation may account for the observed variation in the dissolution rate coefficients for the higher-molecular-weight ligands (NTA, IDA) with pH (Table 7). The larger ligands have more sites to protonate in a surface complex than the smaller ligands. We interpret the variation to result from partial protonation of the surface-ligand complexes at pH = 6, which decreases the reactivity.

The converse may also be true. The rates of exchange of waters around Ni(II) in an outersphere $Ni(OH_2)_6^{2+} \cdot (SO_4^{2-})$(aq) complex is roughly half the rate of exchange around the $Ni(OH_2)_6^{2+}$(aq) complex. The effect is almost an factor of ten for Be(II) complexes (Margerum et al., 1978). One logically predicts that the dissolution of NiO(s) may be slightly lower in some sulfate solutions than in a more inert electrolyte, such as perchlorate. This pathway may complement the existing mechanisms for retarding rates of dissolution, such as the formation of multinuclear surface polymers (e.g. Biber et al., 1994). Even with these potential complexities, rates of ligand-exchange around dissolved complexes (Fig. 3) and the rates of ligand-promoted dissolution (Fig. 9) vary in a remarkably similar fashion, indicating similar processes at work.

When will it fail? Prediction of reactivities becomes difficult:

(1) If the structure of the activated complex for the ligands in solution and at the surface can't be compared. For example, the closure of partially complexed ligands into rings at the surface may be impossible, and distinctly different from the dissolved complexes, because some of the functional groups do not protonate the same in these two settings (see Margerum et al., 1978; Ludwig et al., 1995). Imagine the complexities that arise when ligands have a denticity that approaches the coordination number of the metal (or is even higher). As a further example, the lattice of a mineral can include point defects. These point defects are effectively sites where the metal has a different valence state, which should dramatically affect the dissolution rate. There is, of course, no extended lattice for solutes.

(2) Large ligands can interact with the extensive surface in ways that are impossible for a single metal. Consider the example of a large ligand with alkyl moieties that interact hydrophobically (and indiscriminately) with the surface and ionizable functional groups that can bind to the metal.

(3) The rate controlling step is not hydration and detachment of the formed surface complex. Much evidence (see Casey, 1991; Casey and Westrich, 1992; Ludwig et al., 1995; Ludwig et al., in press; Dove and Czank, 1995) indicates that the rate-controlling step in many dissolution reactions, as well as in many ligand-exchange reactions, involves movement of a water molecule in the surface complex. This movement is sometimes coincident with detachment of the metal-ligand surface complex (see Furrer, 1985), but may not always be so.

(4) Hydroxide ions compete with ligands for metal sites more favorably at surfaces than they do in solution (e.g. Ludwig and Schindler, 1995).

ORTHOSILICATE MINERALS

Orthosilicate minerals are ideal for studying the similarity between ligand exchange and dissolution because these minerals have no polymerized covalent network; they dissolve to release isolated silicate tetrahedra as silicic acid.

Experimental dissolution of orthosilicates

Careful work by many research groups has resulting in a set of data for orthosilicate minerals, generally of the olivine and willemite class, that are more comprehensive data for any other mineral class. Geochemists have determined the surface chemistries (e.g. Schott and Berner, 1985) and pH dependencies of dissolution (e.g. Blum and Lasaga, 1988; Wogelius and Walther, 1991; Westrich et al., 1993; Grandstaff 1980; Casey et al., 1992), rate enhancement attributable to ligands (e.g. Sanemasu et al., 1972; Grandstaff, 1980; Wogelius and Walther, 1991), changes in mineral composition (Westrich et al., 1993; Terry and Monhemius, 1983; Schott and Berner, 1985; Casey and Westrich, 1992), lattice defects (Lin and Shen, 1993a,b), and changes in temperature (Westrich et al., 1993; Casey et al., 1992).

These dissolution experiments provide the most convincing evidence that dissolution resembles a ligand-exchange reaction. In Figure 8 we saw that the dissolution rates of simple orthosilicate minerals scale like the rates of exchange of water molecules around the corresponding hydrated ion. The similarity spans many orders of magnitude at a constant solution composition and temperature. The rate of reaction is controlled solely by the strengths of bonds between the divalent metals and oxygens.

The pH-dependencies of these isostructural minerals are virtually identical, even though these minerals differ dramatically in reactivity. In Figure 10, the pH dependencies of dissolution of Co_2SiO_4, $CaMgSiO_4$ (monticellite) and Mn_2SiO_4 (tephroite) are compared. The pH-dependencies of these rates are indistiguishable within the experimental uncertainties. The pH-dependence of dissolution is directly proportional to the concentration of hyperprotonated surfaces sites (Eqn. 18), which at the conditions of $pH << ZPC?$ is equivalent to the concentration of surface charge. The variation with pH of

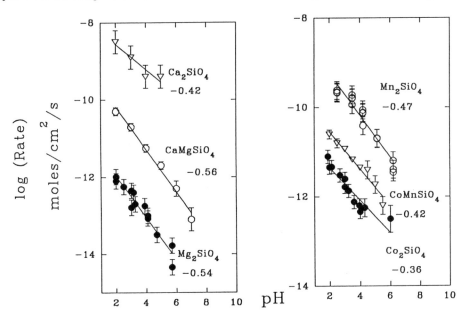

Figure 10. The logarithms of dissolution rate versus solution pH for six different orthosilicate minerals (from Westrich et al., 1993). Although the overall reactivities differ considerably, the rates decrease similarly with pH. The slopes range from -0.4 to -0.6, which are experimentally identical to the variation in surface-charge concentrations (Fig. 11).

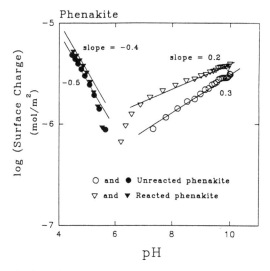

Figure 11. Variation in the logarithm of charge concentration on the surface of phenakite [$Be_2SiO_4(s)$]. The data are from Zemitis (1993). At pH values much less than the mineral ZPC, and at pH values much greater than the mineral ZPC, the concentration of surface charge are virtually equal to the concentrations of $\triangleright SOH_2^+(s)$ and $\triangleright SO^-(s)$, respectively. The surface charge concentrations at pH < 6 (●, Δ) are positive while those at pH > 7 (O, Δ) are negative. The logarithm of the surface-charge concentrations vary in a similar way with solution pH (slope of -0.4 to -0.5) as the variation in the logarithm of dissolution rates with pH values at pH < 6 (Fig. 10).

the logarithm of positive and negative charge on the surface of phenakite [$Be_2SiO_4(s)$] is shown in Figure 11 (from Zemitis, 1993; see also Blum and Lasaga, 1988; Wogelius and Walther, 1991). Please note on this figure that both positive (at pH < 6) and negative surface charge (at pH > 7) are plotted together. As one can see (Fig. 11), the pH-variation in charge is equal to the pH-variation of the logarithms of dissolution rates of the orthosilicate solids (Fig. 10).

Depolymerization of the surface is preceded by the rapid adsorption of protons, consistent our view of ligand-exchange kinetics. Rapid adsorption proceeds to equilibrium and introduces a distinct pH dependence to the subsequent slow reaction. These proton-adsorption reactions contribute a large amount of enthalpy to the apparent activation energy (E_{app}) for dissolution, which is the most common expression of the temperature dependence. Proton-adsorption reactions raise or lower the activation barrier to the reaction by an amount which corresponds to the enthalpy of adsorption: $E_{app} = \Delta H_{ads} + E_{surf}$, where ΔH_{ads} is the heat of adsorption and E_{surf} is the activation energy of the rate-controlling reaction that follows rapid proton adsorption.

The equations to describe the variation in E_{app} values with pH are not complicated but can be obscure. The important geochemistry is illustrated by differentiating Equations (21) and (22) and by application of Van't Hoff's relation (see Casey and Sposito, 1992), yielding:

$$E_{app} = -R\left[\frac{\partial \ln k_{H+}}{\partial 1/T}\right]_{pH} - nR(1 - x_{SOH_2^+})\left[\frac{\partial \ln K'}{\partial 1/T}\right]_{pH} \tag{24}$$

Equation (24) not only includes a variation in the rate coefficient, κ_{H^+}, with temperature, but also includes a temperature variation in the conditional equilibrium constant, K', which varies with pH because more work is required to move a proton to a positively charged surface than to a neutral or negatively charged surface. The key conclusion to draw from Equation (24) is that, just as the protonation state of the surface varies with both pH and temperature, and so too does the derivative parameter, E_{app}, vary.

Correspondingly, the E_{app} values for orthosilicate minerals exhibit a strong pH dependence (Fig. 12), but it is a great surprise to find that they are virtually identical at a

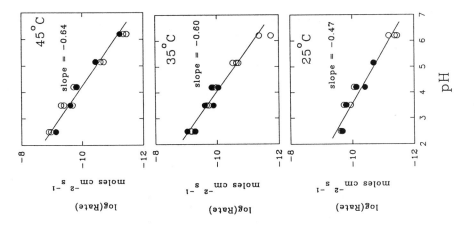

Figure 12 (left). Logarithm of the dissolution rate for orthosilicate minerals varies with both temperature and solution pH. These ortho-silicate minerals represent a wide range of compositions and reactivities, yet the dissolution rates vary similarly with temperature at a given pH (from Westrich et al., 1993).

Figure 13. The pH-dependence of dissolution rates of tephroite [$Mn_2SiO_4(s)$] also varies as a function of temperature (from Casey et al., 1993b).

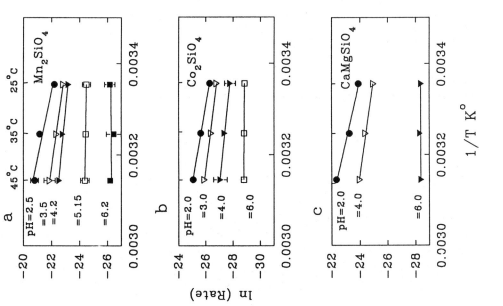

given pH condition (Fig. 13). In general, the enthalpy contribution to E_{app} varies with pH and reaches a minimum near the mineral ZPC where the contribution of adsorption enthalpy to the E_{app} is least. For orthosilicates the E_{app} value near the mineral ZPC is near zero and largely independent of the metal composition. Similarly, the pH-dependence of dissolution rates varies with temperature (Fig. 14) because the adsorption isotherm for protons depends upon temperature. In a sense, this pH-dependence of E_{app} shown in Figures 12 and 13 is misleading and has its origin in the fact that the logarithm of *rates*, and not *rate coefficients*, are typically treated by experimentalists with the Arrhenius relation.

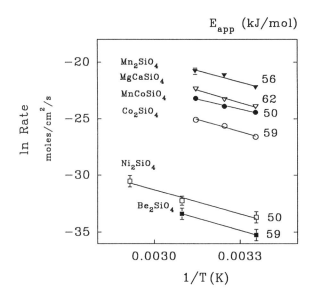

Figure 14. The temperature variations of the logarithm of the dissolution rate for a wide range of orthosilicate minerals are constant at pH = 2.0 (from Westrich et al., 1993).

Nevertheless, the simple connection between the E_{app} values and the ΔH_{ads} values yields an important result: the ΔH_{ads} values can be measured calorimetrically (e.g. Machesky and Anderson, 1986; Casey, 1994). Deviation of a E_{app} value from that predicted from measured enthalpies of proton adsorption indicate that other reactions are involved, such as leaching. Understanding these enthalpy contributions is needed to assess the accuracy of molecular-orbital models of the elementary step reactions, which yield activation energies as an experimentally accessible prediction (e.g. Lee et al., 1994; Casey et al., 1990), but as one can see, the temperature dependencies of dissolution rates of orthosilicate minerals depend upon solution composition and are largely independent of the metals. The rate differences among isostructural minerals are largely accounted for through the pre-exponential term in the Arrhenius rate law and not the activation energy, as one expects from a set of reactions that are driven by differences in the activation *entropy*, not the activation enthalpy (see Casey et al., 1992).

Orthosilicate mineral dissolution in nature

Given the wealth of information on orthosilicate dissolution, we certainly understand how these minerals weather in nature, right? Well, there are some successes and some abysmal failures. First, garnet orthosilicates tend to accumulate Fe(III) and Al(III) in a layer at the mineral surface in oxic soil (Velbel, 1984). In these environments, the rate-controlling step may actually be transport of reactive solutes to the site of dissolution through this armoring layer (Velbel, 1984).

Examination of weathered olivines at a finer scale reveals an even more complicated story. Banfield et al. (1990) employed high-resolution transmission-electron microscopy to show that weathering preferentially proceeds along planar defects and the features produced by subsolidus oxidation of the solid. The olivine grains transform directly to weathering products, including smectite and hematite. The reaction sites are smectite-filled pores that are commonly of a similar size as the hydration sphere of a dissolved metal. The olivine transforms topotactically to the smectite, so that octahedra in olivine and smectites are partly coincident. Weathering is pervasive and distictly different than the products of experimental dissolution (see the chapter by Hochella and Banfield (this volume) for some reasons).

CONCLUSIONS AND PREDICTIONS

Dissolution of oxide and silicate minerals can be understood by drawing an analogy with ligand exchange around dissolved metal-ligand complexes. The reactive species have a similar coordination chemistry in both cases. This similarity allows one to predict reactivity trends and, in some cases, to unravel the reaction mechanisms. The correlation works well for simple oxide and orthosilicate minerals that have no extensive covalent polymer in their structure and where dissociation of simple M-O bonds controls the overall reaction rates. Minerals with extensive polymerized fabrics exhibit a more complicated chemistry because the dissociation of individual M-O bonds is difficult to isolate within the overall process of leaching and dissolution.

Some predictions about surface chemistry are clear:

(1) The detailed spectroscopy of dissolved metal-ligand complexes can be used to infer structural characteristics of the surface complexes and the activated complexes for disequilibrium reactions.

(2) From experience with solution chemistry it is likely that LFER relations exist to correlate rate coefficients and the equilibrium constants (see Brezonik, 1990; Ludwig et al., 1995) or rate coefficients for different metals (see Sverjensky, 1992; Casey, 1991; Casey and Westrich, 1992). Such a simple LFER is shown for proton desorption in Table 5 and they are everywhere. In uncoverng an LFER, greater progress is commonly made through experiments with well-suited, but rare, minerals (like bunsenite) than by direct assault of the rock-forming minerals. The whacko phases provide well-undersood standards.

(3) Coordinated ligands, protons, and hydroxyl ions appear to enhance the rates of dissolution and ligand-exchange around metal-ligand complexes to similar extents. The ligand NTA, for example, which is the most effective of the ligands discussed in this chapter, increases the rates of water exchange in dissolved Ni(II) complexes by a factor of $\approx 10^{1.5}$. This range is approximately equal to that observed for the corresponding ligand-promoted dissolution rate coefficients (compare data in Table 4 and Table 7). Likewise, desorption of a proton from a hydration water around a dissolved metal causes the rates of exchange of the remaining waters to increase by a factor of $\approx 10^2$ or greater. Over a 4 to 6 unit range in pH-PZNPC values, dissolution rates increase by about factor on the same order ($\approx 10^2$) because of proton- or hydroxyl-ion adsorption (compare Blum and Lasaga, 1986; Furrer and Stumm, 1986; and Westrich et al., 1993).

(4) The most important determinant of reactivity is the M-O bond strength. Adsorption of protons or reactive ligands can modify the reactivity of a M-O bond, but the enhancement is small relative to the total range of dissociation rates of different metals.

Finally, considerable attention needs to be paid to the sources of enthalpy that contribute to the activation energies for dissolution. The experimentally measurable parameters include contributions of enthalpy from a wide range of adsorption reactions and are apparently (and surprisingly!) independent of the metal-oxygen bond strength. A good place to start this reconciliation is by modeling ligand-exchange reactions in solution, for which sound and simple experimental data describe the transition states (see Lee et al., 1994).

ACKNOWLEDGMENTS

Support was from the National Science Foundation grant EAR-9302069. The authors thank Paul Ribbe, Sue Brantley and Art White for careful review, and Carrick Eggleston for stimulating conversation.

REFERENCES

Amrhein C, Suarez DL (1988) The use of a surface complexation model to describe the kinetics of ligand-promoted dissolution of anorthite. Geochim Cosmochim Acta 52:2785-2793

Astumian RD, Sasaki M, Yasunaga T, Shelly, ZA (1981) Proton adsorption-desorption kinetics on iron oxides in aqueous suspensions, using the pressure-jump method. J Phys Chem 85:3832-3835

Banfield JF, Veblen DR, Jones BF (1990) Transmission electron microscopy of subsolidus oxidation weathering of olivine. Contrib Mineral Petrol 106:110-123

Bennett PC (1991) Quartz dissolution in organic-rich aqueous systems. Geochim Cosmochim Acta 55:1781-1797

Biber MV, Afonso MD, Stumm W (1994) The coordination chemistry of weathering 4. Inhibition of the dissolution of oxide minerals. Geochim Cosmochim Acta 58:1999-2010

Bish D, Burnham CW (1984) Structure energy calculations on optimal distance model structures: application to the silicate olivines. Am Min 69:1102-1109

Blum A, Lasaga AC (1988) Role of surface speciation in the low-temperature dissolution of minerals. Nature 331:431-433

Boehm HP, Knozinger H (1983) Nature and estimation of functional groups on solids surfaces. In: Catalysis Science and Technology (JR Anderson, M Boudart, eds), Springer-Verlag, New York, p 39-207

Brady PV, Walther JV (1989) Controls on silicate dissolution rates in neutral and basic pH solutions at 25°C. Geochim Cosmochim Acta 53:2823-2830

Brady PV, Walther JV (1993) Surface chemistry and silicate dissolution at elevated temperatures. Am J Sci 53:2822-2830

Brezonik PL (1990) Principles of linear free-energy and structure-activity relationships and their applications to the fate of chemicals in aquatic systems. Ch 4 in (W Stumm, ed) Aquatic Chemical Kinetics. Wiley-Interscience, New York, p 113-143

Budd SM (1961) The mechanisms of chemical reaction between silicate glass and attacking agents. Phys Chem Glasses 2:111-114

Burgess, J (1988) Ions in Solution: Basic Principles of Chemical Interactions. Ellis-Norwood Limited, Chichester, UK

Burgess, J (1990) Metal Ions in Solution. Ellis-Norwood Limited, Chichester, UK

Carroll SA (1989) The dissolution behavior of corundum, kaolinite, and andalusite: A surface complex reaction model for the dissolution of aluminosilicate minerals in diagenetic and weathering environments. PhD dissertation, Northwestern University, Evanston, IL

Carroll-Webb SA, Walther JV (1988) A surface complex reaction model for the pH-dependence of corundum and kaolinite dissolution rates. Geochim Cosmochim Acta 52:2609-2623

Casey WH (1991) On the relative dissolution rates of some oxide and orthosilicate minerals. J Coll Interf Sci 146:586-589

Casey WH, Lasaga AC, Gibbs GV (1990) Mechanisms of silica dissolution as inferred from the kinetic-isotope effect. Geochim Cosmochim Acta 54:3369-3378

Casey WH, Sposito G (1992) On the temperature dependence of mineral dissolution rates. Geochim Cosmochim Acta 56:3825-3830

Casey WH, Hochella MF Jr, Westrich HR (1992) The surface chemistry of manganiferous silicate minerals as inferred from experiments on tephroite (Mn_2SiO_4). Geochim Cosmochim Acta 57:785-793

Casey WH, Westrich HR (1992) Control of dissolution rates of orthosilicate minerals by divalent metal-oxygen bonds. Nature 355:157-159

Casey WH, Banfield JF, Westrich HR, McLaughlin L (1993a) What do dissolution experiments tell us about natural weathering? Chem Geol 105:1-15

Casey WH, Westrich HR, Banfield JF, Ferruzzi G., Arnold GW (1993b) Leaching and reconstruction at the surfaces of dissolving chain-silicate minerals. Nature 366:253-256

Casey WH (1994) Enthalpy changes for Brønsted acid-base reactions on silica. J Coll Interf Sci 163:407-419

Cromack K Jr, Sollins P, Graustein WC, Speidel K, Todd AW, Spycher G, Li CY, Todd RL (1979) Calcium oxalate accumulation and soil weathering in mats of the hypogeous fungus Hysterangium crassum. Soil Biol Biochem 11:463-468

Day VW, Klemperer WG, Malthie DJ (1987) Where are the protons in $H_3V_{10}O_{28}^{-3}$? J Am Chem Soc 109:2991-3002

Dove PM, Czank CA (1995) Crystal chemical controls on the dissolution kinetics of the isostructural sulfates: celestite, anglesite, and barite. Geochim Cosmochim Acta 59:1907-1916

Furrer G (1985) Die Oberflächenkontrollierte Auflösung von Metalloxiden: Ein Koordinationschemischer Ansatz zur Verwitterungskinetik. PhD dissertation, Eidgenössische Tech. Hochschule (ETH), Zürich

Furrer G, Stumm W (1983) The role of surface coordination in the dissolution of δ-Al_2O_3. Chimia 37:338-341

Furrer G, Stumm W (1986) The coordination chemistry of weathering: I. Dissolution kinetics of δ-Al_2O_3 and BeO. Geochim Cosmochim Acta 50:1847-1860

Furrer G, Ludwig C, Schindler PW (1992) On the chemistry of the Keggin Al_{13} polymer I: acid-base properties. J Coll Interf Sci 149:56-67

Furst B (1976) Das coordinationschemische Adsorptionsmodell: Oberflächen-komplexbildung von Cu(II), Cd(II) und Pb(II) and SiO_2 (Aerosil) und TiO_2. PhD dissertation, Universität Bern, 80 p

Graustein WC, Kromack K Jr, Sollins P (1977) Calcium oxalate-occurence in soils and effect on nutrient and geochemical cycles. Science 198:1252-1254

Grandstaff DE (1980) The dissolution rate of forsteritic olivine from Hawaiian Beach Sand. In: Third Int'l Symp on Water-Rock Interaction, Edmonton. Canada, Alberta Research Council.

Griesser R, Sigel H (1970) Ternary complexes in solution VIII: Complex formation between the copper(II)-2,2' bypyridyl 1:1 complex and ligands containing oxygen and/or nitrogen donor atoms. Inorg Chem 9:1238-1243

Hachiya K, Ashida M, Sasaki M, Kan H, Inoue T, Yasunaga T (1979) Study of the kinetics of adsorption-desorption of Pb^{2+} on a γ-Al_2O_3 surface by means of relaxation techniques. J Phys Chem 83:1866-1871

Hachiya K, Sasaki M, Ikeda T, Mikami N, Yasunaga T (1984) Static and kinetic studies of adsorption-desorption of metal ions on a γ-Al_2O_3 surface 2. Kinetic study by means of pressure-jump technique. J Phys Chem 88:27-31

Ikeda T, Sasaki M, Hachiya K, Astumian RD, Yasunaga T, Schelly ZA (1982) Adsorption-desorption kinetics of acetic acid on silica-alumina particles in aqueous suspensions, using the pressure-jump relaxation method. J Phys Chem 86:3861-3866

Jauregui MA, Reisenauer HM (1982) Dissolution of oxides of manganese and iron by root exudate components. Soil Sci Soc Am J 46:314-317

Kanno H (1988) Hydrations of metal ions in aqueous electrolyte solutions: a Raman study. J Phys Chem 92:4232-4236

Kinraide SD, Swaddle TW (1988) Silicon-29 NMR studies of aqueous silicate solutions-1: Chemical shifts and equilibria. Inorg Chem 27:4253-4259

Kinraide SD, Swaddle TW (1988) Silicon-29 NMR studies of aqueous silicate solutions-2: Transverse ^{29}Si relaxation and the kinetics and mechanism of silicate polymerization. Inorg Chem 27:4259-4269

Knight CTG, Kirkpatrick RJ, Oldfield E (1988) Two-dimensional silicon-29 nuclear magnetic resonance spectroscopic study of chemical exchange pathways in potassium silicate solutions. J Magn Reson 78:31-40

Knight CTG, Kirkpatrick RJ, Oldfield E (1989) Silicon-29 multiple quantum filtered NMR spectroscopic evidence for the presence of only six single site silicate anions in a concentrated potassium silicate solution. J Chem Soc Chem Comm 1989:919-921

Koch G (1965) Kinetik und mechanismus der Auflösung von Berylliumoxyd in Säuren. Berichte Bunsengesellshaft 69:141-145

Kummert R, Stumm W (1980) The surface complexation of organic acids on hydrous γ-Al_2O_3. J Coll Interf Sci 75:373-385

Lapeyrie F (1988) Oxalate synthesis from soil bicarbonate by the mycorrhizal fungus Paxillus involutus. Plant and Soil 110:3-8

Lee MA, Winter NW, Casey WH (1994) Investigation of the ligand exchange reaction for the aqueous Be^{2+} ion. J Phys Chem 98:8641-8647.

Lin CC, Shen PY (1993b) Directional dissolution kinetics of willemite. Geochim Cosmochim Acta 57:27-53

Lin CC, Shen PY (1993a) Role of screw axes in dissolution of willemite. Geochim Cosmochim Acta 57:1649-1655

Lopez-Quintela MA, Knoche W, Veith J (1984) Kinetics and thermodynamics of complex formation between aluminum(III) and citric acid in aqueous solutions. J Chem Soc Faraday Trans 80:2313-2321

Ludwig C, Casey WH, Devidal J-L Toward understanding of siderophore geochemistry: The effect of different functional groups on the ligand-promoted dissolution of minerals. Geochim Cosmochim Acta (submitted)

Ludwig C, Casey WH, Rock PA (1995) Prediction of ligand-promoted dissolution rates from the reactivities of aqueous complexes. Nature 375:44-47

Ludwig C, Casey WH (1995) On the mechanisms of dissolution of bunsenite [NiO(s)] and other simple oxide minerals. J Coll Interf Sci (in press)

Ludwig C, Schindler PW (1995) Surface complexation on TiO_2: Ternary surface complexes-coadsorption of Cu(II) and organic ligands (2,2'-bipyridyl, 8-aminoquinoline, and o-phenylenediamine) onto TiO_2 anatase). J Coll Interf Sci 169:291-299

Lundström U, Öhman L-O (1990) Dissolution of feldspars in the presence of natural, organic solutes. J Soil Sci 41:359-369

Lutz B, Wendt H (1970) Fast ionic reactions in solutions VII: Kinetics of the fission and formation of the dimeric isopolybases $(FeOH)_2^{4+}$ and $(VOOH)_2^{2+}$. Berichte Bunsen-Gesellschaft 74:372-380

Machesky ML, Anderson MA (1986) Calorimetric acid-base titrations of aqueous goethite and rutile suspensions. Langmuir 2:582-587.

Malajczuk N, Cromack K Jr (1982) Accumulation of calcium oxalate in the mantle of ectomycorrhizal roots of Pinus radiata and Eucalyptus marginata. New Phytol 92:527-531

Margerum DW, Cayley GR, Weatherburn DC, Pagenkopf GK (1978) Kinetics and mechanisms of complex formation and ligand exchange. Ch 1 in Coordination Chemistry (A Martell, ed) ACS Monograph 174, American Chemical Society, Washington, DC 2:1-220

Mast MA, Drever JI (1987) The effect of oxalate on the dissolution rates of oligoclase and tremolite. Geochim Cosmochim Acta 51:2559-2568

Merbach AE, Akitt JW (1990) High-resolution variable pressure NMR for chemical kinetics. NMR Basic Principles and Progress 24:190-232

Plettinck S, Chou L, Wollast R (1994) Kinetics and mechanisms of dissolution of silica at room temperature and pressure. Mineral Mag 58A:728-729

Pohlmann AA, McColl JG (1986) Kinetics of metal dissolution from forest soils by soluble organic acids. J Environ Qual 14:86-92

Pulfer K, Schindler PW, Westall JC, Grauer R (1984) Kinetics and mechanism of dissolution of bayerite (γ-$Al(OH)_3$) in HNO_3-HF solutions at 298.2 K. J Coll Interf Sci 101:554-564

Richens DT, Helm L, Pittet P-A, Merbach AE, Nicolo F, Chapuis G (1989) Crystal structure of and mechanism of water exchange on $[Mo_3O_4(OH_2)_9]^{+4}$ from X-ray and Oxygen-17 NMR studies. Inorg Chem 28:1394-1402

Rowland TV (1975) Oxygen-17 NMR studies of the rate of water exchange from partially complexed nickel ion. PhD dissertation, University of California Berkeley, 76 p

Sanemasu I, Yoshida M, Ozawa T (1972) The dissolution of olivine in aqueous solutions of inorganic acids. Bull Chem Soc Japan 45:1741-1746.

Schindler PW, Stumm W (1987) The surface chemistry of oxides, hydroxides and oxide minerals. In: Aquatic Surface Chemistry (W Stumm, ed) John Wiley, New York, p 83-107

Schindler PW (1990) Co-adsorption of metal ions and organic ligands: formation of ternary surface complexes. Ch 7 In: Rev Min 23: Mineral-Water Interface Geochemistry

Schott J, Berner, RA (1985) Dissolution mechanisms of pyroxenes and olivines during weathering. In: The Chemistry of Weathering (JI Drever, ed) Reidel, Boston, p 35-55

Sparks DL (1989) Kinetics of Soil Chemical Processes. Academic Press, New York, 210 p

Sposito G (1984) The surface chemistry of soils. Oxford Press, New York, 234 p

Stumm W, Wollast R (1990) Coordination chemistry of weathering: kinetics of surface-controlled dissolution of oxide minerals. Rev Geophys 28:53-69

Sverjensky D (1992) Linear free energy relations for predicting dissolution rates. Nature 358:310-313

Terry B, Monhemius AJ (1983) Acid dissolution of willemite [$(Zn,Mn)_2SiO_4$] and hemimorphite [$Zn_4Si_2O_7(OH)_2 \cdot H_2O$]. Metall Trans 14B:335-346

Velbel MA (1984) Natural weathering mechanisms of almandine garnet. Geology 12:631-634

Wehrli B, Wieland E, Furrer G (1990) Chemical mechanisms in the dissolution kinetics of minerals; the aspect of active sites. Aquatic Sciences 52:3-31

Westrich HR, Cygan RT, Casey WH, Zemitis C, Arnold GW (1993) The dissolution kinetics of mixed cation orthosilicate minerals. Am J Sci 293: 869-893

Wieland E, Wehrli B, Stumm W (1988) The coordination chemistry of weathering-III: A generalization on the dissolution rates of minerals. Geochim Cosmochim Acta 52:1969-1981

Wilkens RG (1974) The Study of Kinetics and Mechanisms of Reactions of Transition Metal Complexes. Allyn and Bacon, Boston

Wilkens RG (1991) Kinetics and Mechanisms of Reactions of Transition Metal Complexes, VCH, 465 p

Wirth GW, Gieskes JM (1979) The initial kinetics of the dissolution of vitreous silica in aqueous media. J Coll Interf Sci 68:492-500

Wogelius RA, Walther, JV (1991) Olivine dissolution at 25°C: Effects of pH, CO_2 and organic acids. Geochim Cosmochim Acta 55:943-954

Zemitis CR (1993) Dissolution kinetics of three beryllate minerals. MS thesis, University of California Davis, 121 p

Zhang PC, Sparks DL (1989a) Kinetics and mechanisms of molybdate adsorption-desorption at the goethite/water interface using pressure-jump relaxation. Soil Sci Soc Am J 53:1028-1034

Zhang PC, Sparks DL (1989b) Kinetics and mechanisms of sulfate adsorption-desorption at the goethite/water interface using pressure-jump relaxation. Soil Sci Soc Am J 54:1266-1273

Zhang PC, Sparks DL (1990) Kinetics and mechanisms of selenate and selenite adsorption-desorption at the goethite/water interface. Environ Sci Tech 24:1848-1856

Zinder B, Furrer G, Stumm W (1986) The coordination chemistry of weathering:II. Dissolution of Fe(III) oxides. Geochim Cosmochim Acta 50:1861-1869

Chapter 4

CHEMICAL WEATHERING RATES OF
PYROXENES AND AMPHIBOLES

S.L. Brantley and Y. Chen

Department of Geosciences
Pennsylvania State University
University Park, PA 16802 U.S.A.

INTRODUCTION

The principal input of Fe and Mg to natural waters derives from the weathering of pyroxenes and amphiboles in almost every type of igneous rock. In particular, because of its variable composition, its relative abundance, and its relative instability under weathering conditions, hornblende comprises the principal source of Fe, Mg, and some trace elements in soils (Huang, 1977). Despite the importance of weathering of these minerals, few studies offering quantitative evaluation of pyroxene or amphibole weathering rates— and in particular weathering rates of Fe-containing inosilicates such as hornblende—have been published. In this chapter we summarize the current state of knowledge concerning the quantification of rates of dissolution of pyroxenes and amphiboles under natural conditions. We also try to integrate laboratory dissolution kinetics with the published studies of hornblende etching rates in soils. Summary of published estimates of pyroxene and amphibole weathering based upon mass-balance studies are reviewed in other chapters (White, this volume; Drever, this volume).

Assessment of the rates of weathering of Ca- and Mg-containing minerals in these silicate families is also important given the role that dissolution of Ca- and Mg-containing silicates plays in the maintenance of balance in the longterm carbon cycle. As many workers have pointed out (see Berner, this volume), weathering of Ca- and Mg-containing silicates withdraws CO_2 from the atmosphere, ultimately precipitating in the oceans as carbonate minerals. This CO_2 flux out of the atmosphere (forward reactions described in Eqns. 1 and 2) balances the CO_2 produced from metamorphism and volcanism on 10^5 to 10^6 y timescales:

$$CaSiO_3 + CO_2 = CaCO_3 + SiO_2 \qquad (1)$$

$$MgSiO_3 + CO_2 = MgCO_3 + SiO_2 \qquad (2)$$

These so-called Urey reactions (Urey, 1952) represent a shorthand descriptor for the processes which occur and which must be in balance to maintain constant levels of atmospheric CO_2. In these equations, Ca- and $MgSiO_3$ represent all Ca- and Mg-containing silicates. To first order, however, the fast weathering Ca-silicates present in large quantity at the earth's surface are predominantly plagioclase feldspars and Ca-pyroxenes and amphiboles, while fast weathering Mg-silicates of importance are Mg-pyroxenes and amphiboles. A better understanding of the rates of dissolution of these minerals is therefore of great importance (Berner, this volume), and greater emphasis must be placed on integration of our understanding of pyroxene and amphibole dissolution at all scales.

MINERALOGY OF COMMON SOIL INOSILICATES

The pyroxenes and amphiboles comprise the inosilicate mineral class, along with a few extremely rare three-chain minerals. In the inosilicates, the silicate tetrahedral groups are linked in single (pyroxene) or double (amphibole) chains, by sharing two to three oxygens per tetrahedron. The Si:O ratio in single chain silicates is 1:3 with an anionic group of $(SiO_3)_n^{2n-}$, where each tetrahedron shares two bridging oxygens. The Si:O ratio in the double chain silicates is 4:11, and the anionic group is $(Si_4O_{11})_n^{6n-}$ where each tetrahedron shares two to three bridging oxygens. Single chain silicates in which the silica tetrahedra of each chain do not comprise a common basal plane (i.e. the chains are twisted), are classified as pyroxenoids. The connectedness, defined as the number of bridging oxygens per tetrahedral unit (Liebau, 1985), is therefore 2 for pyroxenes and pyroxenoids, and 2.5 for amphiboles.

The chains in inosilicates are cross-linked by cations which are defined by their position relative to the apices and the bases of the silica tetrahedra (Liebau, 1985). The apical oxygens of two facing chains are octahedrally coordinated by small cations such as Mg or Fe: these sites are classified as the *M1* site in pyroxenes and the *M1, M2* and *M3* sites in amphiboles. Larger cations such as Ca usually occupy the sites formed by opposed tetrahedral bases where coordination is irregular and depends upon the cation present: for example, the *M2* site in pyroxene and the *M4* site in amphibole. The amphiboles are also characterized by the presence of OH⁻ anions which lie in the center of the ring formed by the 6 apical oxygens.

Inosilicate grains usually appear as prismatic, bladed, or fibrous crystals, with the elongated direction parallel to the axis of the chains (the *c* axis). Because the internal bond strengths between chains are not as strong as within the chains, inosilicates often develop prismatic cleavage parallel to the crystal *c* axis. The cleavage is distinctly different between pyroxenes and amphiboles, exhibiting angles of 87° and 93° for pyroxenes and 124° and 56° for amphiboles.

The general formula for the pyroxene family can be expressed as XYZ_2O_6, where X represents Ca, Na, Mg, Fe^{2+}, or Li in the *M2* site; Y represents Mg, Fe^{2+}, Fe^{3+}, Mn, Cr, Li, Ni, Al, or Ti in the *M1* site; and Z represents Si or Al in the tetrahedral site. The most common pyroxene compositions and solid solutions include the orthopyroxenes,

  enstatite-ferrosilite—$(Mg,Fe)_2Si_2O_6$;

and the clinopyroxenes,

  diopside-hedenbergite—$Ca(Mg,Fe)Si_2O_6$,
  augite— $(Ca, Mg, Fe^{2+}, Al)_2(Si,Al)_2O_6$,
  pigeonite— $(Mg, Fe^{2+} ,Ca)(Mg, Fe^{2+}) Si_2O_6$,
  aegerine-augite— $(Ca, Na)(Mg, Fe^{3+},Fe^{2+}) Si_2O_6$,
  spodumene—$LiAlSi_2O_6$, and
  jadeite—$NaAl Si_2O_6$.

The common pyroxenoids whose dissolution kinetics have been analyzed include

  wollastonite ($CaSiO_3$), and
  rhodonite ($Mn,Ca,Fe)SiO_3$.

The general formula for the amphiboles is $W_{0-1}X_2Y_5Z_8O_{22}(OH,F)_2$, where W represents either a vacancy or Na or K in the site between two tetrahedral rings; X represents Ca, Na, Mg, or Fe^{2+} in the *M4* site; Y represents Mg, Fe^{2+}, Fe^{3+}, or Al in the *M1, M2*, and *M3* sites; and Z represents Si or Al in the tetrahedral site. Major amphibole end-member compositions are:

anthophyllite—$(Mg, Fe^{2+})_7 Si_8 O_{22}(OH)_2$,
cummingtonite-grunerite—$(Mg,Fe,Mn)_7 Si_8 O_{22}(OH)_2$,
tremolite-ferroactinolite—$Ca_2(Mg,Fe^{2+})_5 Si_8 O_{22}(OH)_2$,
hornblende—$(Na,K)_{0-1} Ca_2 (Mg,Fe^{2+},Fe^{3+},Al)_5 Si_{6-7.5} Al_{2-0.5} O_{22}(OH)_2$, and
glaucophane—$Na_2 Mg_3 Al_2 Si_8 O_{22}(OH)_2$.

LABORATORY CHEMICAL REACTORS

Batch reactors

In order to quantify chemical weathering rates for individual minerals, many investigators have run laboratory experiments on mineral separates under controlled conditions of temperature, pH, and solution composition. Experiments have been run in chemical reactors of several types: batch, continuously stirred, plug flow, and fluidized bed (Hill, 1977; Posey-Dowty et al., 1986; Chou and Wollast, 1985; Rimstidt and Dove, 1986). Batch reactors, simple in use and setup, allow experimenters to follow the progress of a closed system reaction in which solution chemistry changes with time due to accumulation of reaction products. Batch reactors are simply agitated or stirred tanks which are set up closed or open to the atmosphere with mineral sample and solution. By monitoring the concentration of dissolution products (c) as a function of time, and correcting for removal of sample during monitoring, the reaction rate can be calculated as dc/dt (Laidler, 1987). The reaction rate can then be expressed as either the release rate of a given component, or as the rate of dissolution of moles of the phase per unit time. Typically, these rates are then normalized per unit surface area (A) or per gram of mineral. The complication of the batch reactor is that the measured rate may change as a function of time as the solution chemistry changes, or as minerals precipitate.

Flow-through reactors

Stirred flow-through reactors are more simple to interpret and allow the experimenter to maintain constant solution chemistry during dissolution. In a continuously stirred tank reactor (CSTR), also known as a mixed flow reactor (Rimstidt and Dove, 1986) a mineral sample is placed in a tank of volume V_o through which fluid is continuously pumped at flow rate Q. Such a reactor is continuously stirred by a propeller or by agitation. The rate of the reaction, R (mol cm^{-2} s^{-1}) is assessed by comparing the inlet concentration (c_i) to the outlet concentration (c_o) of a component derived from dissolution of the mineral:

$$R = \frac{Q(c_o - c_i)}{A_s m} \tag{3}$$

where A_s is the specific surface area of the mineral (cm^2 g^{-1}) and m is its initial mass (g). Reactors are run until outlet concentration reaches a constant value, the so-called steady state value which yields the steady state dissolution rate. Dissolution rates are then reported as observed rate normalized by initial or final surface area, with respect to solution chemistry as measured in the effluent. Thus, if inlet pH does not equal outlet pH, measured rate is reported with respect to the outlet value. If final mass differs significantly from initial mass, then this must also be measured and included in determining the rate.

One form of the CSTR, the fluidized bed reactor, utilizes two flows, one which is single pass, and a second recirculating flow which stirs the mineral powder. The "stirring" flow is operated as a very fast flow rate which maintains particles suspended in the reactor. The "single pass" flow is analyzed for c_i and c_o just as in a CSTR to analyze the chemical reaction rate (see Chou and Wollast, 1985).

A plug-flow reactor, while representing a closer analog to natural systems, is more complicated to model than a CSTR. In an ideal plug-flow reactor, it is assumed that every packet of fluid moves through a packed bed of mineral grains with the same residence time (Hill, 1977). For this reason, these reactors are also called packed bed reactors. The ideal residence, or contact, time is defined as the ratio of the pore volume of the reactor autoclave (V_o) to the fluid flow rate, Q. For the assumptions of no volume change in the reaction, no radial flow, and no pooling of fluid in a column or pipe reactor, the outlet concentration is related to the inlet concentration by the following equation:

$$\frac{1}{1-n}\left[\frac{1}{c_o^{n-1}} - \frac{1}{c_i^{n-1}}\right] = \frac{kA_s m}{Q} \tag{4}$$

for $n \neq 1$ and, for $n = 1$,

$$c_O = c_i \exp\left(\frac{-kA_s m}{Q}\right) \tag{5}$$

where we have assumed the following rate equation:

$$R = kc^n \tag{6}$$

Interpretation of plug-flow reactors are especially difficult due to the changing solution composition as a function of position in the reactor, and the possibility of precipitation of secondary phases (e.g. Gonzalez Bonmati et al., 1985).

Interpreting reactor experiments

Interpretation of mineral dissolution rates using chemical reactors can be equivocal. For example, dissolution experiments are usually interpreted with respect to interface- or transport-control, which may be a function of the stirring rate of the reactor. Whereas a packed bed reactor may easily become transport-controlled, a fluidized bed reactor suspends mineral particles and obviates against the maintenance of thick hydrodynamic boundary layers around mineral grains. However, fluidized reactors have been criticized for causing continuous abrasion and surface area changes during dissolution. On the other hand, mineral grains may not be thoroughly mixed with water in CSTRs, allowing the stagnation of fluid at the grain surface.

Another problem with mineral dissolution experiments in chemical reactors is that effluent concentrations may not reach steady values for many thousands of hours. Furthermore, due to competing effects of changing surface area, dissolving defects, or changing surface properties, concentrations may never reach a true steady state during an experiment. In addition, if secondary minerals precipitate within the reactor, the steady state value will be significantly smaller than that expected from dissolution alone. Where pH changes within the reactor are substantial, some investigators have used buffer solutions; however, with many silicates, the presence of cations and anions other than H^+ or OH^- influence the rate and mechanism of dissolution. In addition, many silicates are observed to dissolve nonstoichiometrically, either because of precipitation of secondary phases or because of accelerated leaching of certain elements out of the structure. Therefore, careful experiments require analysis of several elements in the effluent solution. Nonstoichiometric dissolution can also be related to the fact that, many mineral separates, although hand-picked and carefully cleaned, contain impurity minerals whose dissolution can contribute to the observed chemistry of outlet solutions. Where exsolution has occurred in the phase of interest, preferential dissolution of exsolved phases can also cause apparent nonstoichiometric dissolution experiments. Many such problems are discussed with respect to dissolution experiments of the inosilicates in the subsequent sections.

A particular problem associated with analysis and reporting of inosilicate dissolution rates is related to the assumed stoichiometry of the mineral. Due to fast release of cations such as Ca and Mg from inosilicate dissolution, most researchers have reported dissolution rates calculated from observed Si release. Where the stoichiometry of the mineral is known and the mineral dissolves stoichiometrically, release rates of other cations should be equal to the appropriately corrected Si release. Such mineral dissolution rates can be reported as mol mineral cm^{-2} s^{-1}, as we have reported here. However, such dissolution rates are dependent upon the formula unit chosen: for example, if enstatite is written as $MgSiO_3$, the reported dissolution rate *based upon an observed Si release rate* will be twice the value reported if the formula unit is written as $Mg_2Si_2O_6$. For this reason all mineral formulas, as used to report the rates, are collated here.

Many publications describing dissolution experiments on inosilicates do not provide useful dissolution rate data because they do not report pH or mineral surface area (e.g. Tunn, 1939; McClelland, 1950; Huang and Keller, 1970; Bailey and Reesman, 1971; Sanemasa and Katsura, 1973; Bailey, 1974; Barman et al., 1992), because precipitation reactions occurred during the experiment (Siever and Woodford, 1979; Gonzalez Bonmati et al., 1985), because batch experiments were run without reporting solution chemistry with time (Huang and Keller, 1970), and because very brief reaction periods were utilized (Choi et al., 1974; White and Yee, 1985).

LABORATORY DISSOLUTION RATES

Measurement of true steady state

Several of the first laboratory measurements of area-normalized dissolution using well documented techniques revealed parabolic dissolution kinetics (rate $\propto$ $t^{-1/2}$, Helgeson, 1971): e.g. for all experiments at pH > 1.65 for enstatite (Luce et al., 1972), for all experiments between pH 1 and 4.5 for bronzite pyroxene (Grandstaff, 1977), and for bronzite at pH 6 in the presence of oxygen (Fig. 1, Schott and Berner, 1985). However, Schott et al. (1981) showed that the parabolic kinetics were not observed on enstatite, tremolite, or diopside if the samples were first pre-treated by ultrasonic washing, followed by etching in HCl and 5% HF – 0.1N H_2SO_4. These workers attributed the observation of parabolic kinetics to fast dissolution of ultrafines and reactive surface features. Petrovich (1981) pointed out that nearly all crushed mineral grains have reactive surfaces related to ultrafines, sharp edges, and high energy areas which dissolve preferentially. Most subsequent researchers have concluded that dissolution of Fe-free inosilicates occurs as linear kinetics (independent of time) after an initial transient period. Published measurements of linear kinetics of Fe-free pyroxene and amphibole dissolution normalized to surface area at controlled pH at 25°C are summarized in Figures 2 and 3 and Table 1.

Schott and Berner (1983) reported that, in the absence of oxygen at pH 6 or at low pH (1 to 1.5), dissolution of pre-cleaned bronzite is also linear for reaction periods greater than 600 h. However, Schott and Berner (1983, 1985) reported that at pH 6.0 the rate of Si release from bronzite is not constant under oxygenated conditions up to about 70 d.

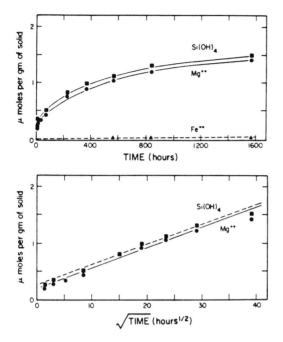

Figure 1. Release of elements to solution during dissolution of Webster bronzite at pH 6 [Used by permission of the editor of *Geochimica Cosmochimica Acta*, from Schott and Berner (1983), Fig. 3, p. 2236]. (a) Concentration of Mg, Fe, and $Si(OH)_4$ released as a function of time for dissolution of bronzite in the presence of oxygen ($P_{O_2} = 0.2$ atm). (b) Data replotted vs. $t^{1/2}$ shows that the dissolution is parabolic. Release is normalized to grams of bronzite.

Zhang (1990) also reported that release of Si from hornblende was not perfectly linear or stoichiometric in the presence of oxygen at pH 4 even after 115 d. Zhang therefore modelled dissolution of hornblende with an exponential rate law, but he noted that rates became more stoichiometric and linear as reaction proceeded to 115 d. In the presence of oxygen, we found no reported linear dissolution rate measurements for Fe-containing inosilicates. Silica release rates under parabolic or exponential kinetics are thus reported for the few Fe-containing minerals investigated in Table 1. These rates are reported as mol Si cm^{-2} s^{-1} as observed at the end of the dissolution experiments.

As mentioned, in the experiments of Schott et al. (1981), the samples were pre-etched with a hydrofluoric-sulfuric acid mixture; however, many subsequent researchers found that preparation of samples by grinding followed by ultrasonic washing in acetone or other organic solvents was adequate to clean the surfaces of ultrafines. Furthermore, Perry et al. (1983) showed that pre-etching of feldspar with HF solutions produced fluoridated surfaces, which changed the reactivity. Similarly, Petit et al. (1987) showed that HF may open channels of transport that allow deeper penetration of hydrogen into diopside. Eggleston et al. (1989) observed that washing diopside in HF-H_2SO_4 decreases the rate of dissolution during the early transient stage (<112 h), although they saw no evidence for introduction of fluorine into the diopside surface. For these reasons, most subsequent researchers used only an ultrasonication, rather than an etching pre-treatment (see however, Brady and Carroll, 1994). Zhang et al. (1993) reported an additional complication: in experiment cycles where hornblende grains were dissolved, dried, and dissolved, they reported observing enhanced dissolution after drying.

Figure 2 (opposite page). Log rate of dissolution of 8 pyroxenes or pyroxenoids vs. pH. Data are in Table 1 and discussed in text. Note that, for consistency, minerals have been written on the basis of 6 oxygens, and all rates are moles inosilicate cm^{-2} s^{-1}. Mineral formulas chosen for this figure may differ from Table 1.

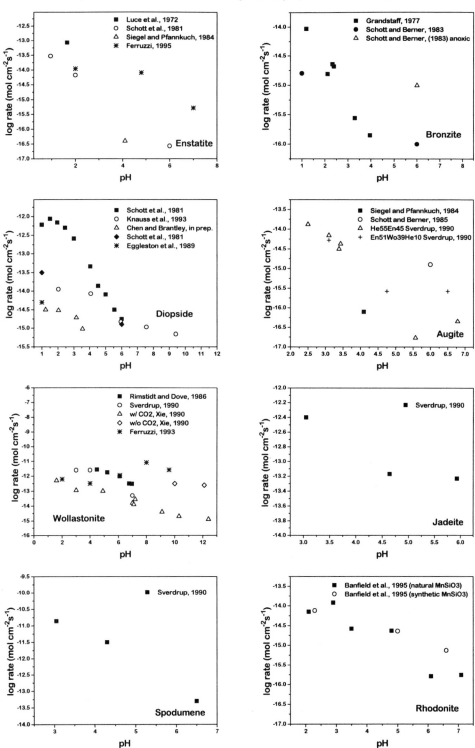

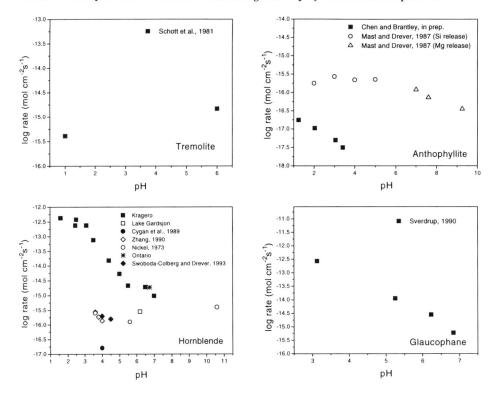

Figure 3. Log rate of dissolution of four amphiboles vs. pH. Data are compiled in Table 1 and discussed in text. Three hornblende compositions from Sverdrup (1990) are included: Kragero, Lake Gardsjon, and Ontario. Anthophyllite dissolution rate from Mast and Drever (1987) are calculated either from Mg or Si release, as noted.

Although early experiments were completed in batch reactors (Luce et al., 1972; Grandstaff, 1977; Schott et al., 1981; Siegel and Pfannkuch, 1984), most of the subsequent experiments were completed in flow-through reactors where solution chemistry remained constant throughout the experiment (see however, Sverdrup, 1990; Zhang et al., 1993; Xie, 1994). Despite the constancy of solution composition in continuously stirred flow-through reactors, dissolution of most ground pyroxenes and amphiboles, like the feldspars, occurs quickly at first, and then slows to a steady state release rate after several days. Knauss et al. (1993) attributed the slow decrease in rate to adjustment of the mineral surface to the solution and/or removal of disturbed material on the ground mineral surface. Ferruzzi (1993) and Casey et al. (1993) argued that repolymerization reactions occurring on the metal-leached silica-rich surface may explain the observed decrease in dissolution rate with time of such phases as $MnSiO_3$ and wollastonite. A thorough investigation of the disordered surface of ground diopside and the early transient stage of dissolution of diopside are reported by Peck et al. (1988) and Eggleston et al. (1989) respectively.

Rate discrepancies

It is clear from Figures 2 and 3 that there are considerable discrepancies among dissolution rate data. Unfortunately, not all workers summarized estimates of

Table 1. Measured dissolution rates for inosilicates under ambient conditions.

All rates reported as moles inosilicate /cm^2/sec, calculated from Si release data unless otherwise noted.

Phase	pH	log R	Reported composition	Reference
enstatite 25 °C	1.6	-13.1[1,m]	$Mg_{1.85}Fe_{0.15}Si_2O_6$	Luce et al., 1972
enstatite[t] 22 °C	1.0	-13.5	$Mg_{2.00}Fe_{0.04}Ti_{0.002}Al_{0.004}Si_{1.97}O_6$	Schott et al., 1981
	2.0	-14.2		
	6.0	-16.5		
72 °C	6.0	-15.1		
enstatite[s] 25 °C	2.0	-13.7	$MgSiO_3$	Ferruzzi, 1993
	4.8	-13.8		
	7.0	-15.0		
enstatite[t] 22 °C	4.1	-16.4	$Na_{0.1}Ca_{0.03}Mg_{1.6}Fe_{0.6}Al_{0.2}Si_{1.6}O_6$ [k]	Siegel & Pfannkuch, 1984
bronzite	1.19	-14.0[p]	$Mg_{1.54}Fe_{.42}Ca_{0.04}Si_2O_6$	Grandstaff, 1977
	2.12	-14.8 [p]		
	2.33	-14.6 [p]		
	2.39	-14.7 [p]		
	3.30	-15.6 [p]		
	3.96	-15.8 [p]		
bronzite	1.0	-14.8[o,l]	$Mg_{1.77}Fe_{0.23}Si_2O_6$	Schott & Berner, 1983
	6.0	-15.0[a,l]		
	6.0	-16.0[o,p]		
diopside[n] 22 °C	1.0	-12.2	$Ca_{1.04}Mg_{1.01}Fe_{0.01}Al_{0.015}Si_{1.96}O_6$ Gouverneur	Schott et al., 1981
	1.5	-12.1		
	2.0	-12.2		
	2.4	-12.3		
	3.0	-12.6		
	4.0	-13.3		
	4.5	-13.9		
	5.0	-14.1		
	5.6	-14.5		
	6.0	-14.8		
diopside 20 °C	1.0	-13.5	$Ca_{1.04}Mg_{0.98}Fe_{0.02}Ti_{0.004}Al_{0.01}Si_{1.95}O_6$ Pitcairn	Schott et al., 1981
	6.0	-14.9		
diopside 25 °C	2.0	-14.0	$Ca_{0.97}Mg_{0.89}Fe_{0.08}Al_{0.04}Si_{1.99}O_6$	Knauss et al., 1993
	4.1	-14.1		
	6.0	-14.8		
	7.6	-15.0		
	9.4	-15.2		
	12.1	-15.2[2]		
diopside 20 °C	1.0	-14.3	$CaMgSi_2O_6$	Eggleston et al., 1989*
diopside[i] 25 °C	1.2	-14.5	$Ca_{0.8}Mg_{0.8}Fe_{0.2}Al_{0.1}Si_2O_6$	Chen & Brantley, in prep.
	2.0	-14.5		
	3.2	-14.7		
	3.6	-15.0		
diopside[f] 25 °C	1.2	-14.7	$Ca_{0.8}Mg_{0.8}Fe_{0.2}Al_{0.1}Si_2O_6$	Chen & Brantley, in prep.
	2.0	-15.0		
	3.2	-15.0		
	3.6	-15.2		
augite 22 °C	4.1	-16.1	$Na_{0.1}Ca_{0.6}Mg_{0.4}Fe_{0.77}Al_{0.05}Si_{1.8}O_6$	Siegel & Pfannkuch, 1984

Phase	pH	log R	Reported composition	Reference
augite 20 °C	6.0	-14.9	$(Ca,Mg,Fe)_2Si_2O_6$**	Schott &
50 °C	6.0	-13.4		Berner, 1985
64 °C	6.0	-13.0		
72 °C	6.0	-12.7		
augite	2.5	-13.9	$Ca_{0.55}Fe_{0.55}Mg_{0.90}Si_2O_6$	Sverdrup, 1990
25 °C	3.1	-14.2		
		-14.4		
	3.4	-14.5		
	5.6	-16.8		
	6.8	-16.4		
augite	3.1	-14.3	$Mg_{1.02}Ca_{0.88}Fe_{0.10}Si_2O_6$	Sverdrup, 1990
25 °C	4.8	-15.6		
	6.5	-15.6		
wollastonite[f]	4.5	-11.3	$CaSiO_3$	Rimstidt &
23.5 °C	5.2	-11.4		Dove, 1986
	6.1	-11.7		
	6.8	-12.2		
	7.0	-12.2		
wollastonite	3.0	-11.6	$Ca_{1.7}Mg_{0.11}Si_{2.1}O_6$	Sverdrup, 1990
25 °C	4.0	-11.6		
	7.0	-13.3		
wollastonite	1.6	-12.0	$CaSiO_3$	Xie, 1994
25 °C	3.0	-12.6		
	4.9	-12.7		
	7.0	-13.5		
	7.1	-13.6		
	7.2	-13.2		
	10	-12.2^{wo}		
	12.1	-12.3^{wo}		
30 °C	9.1	-14.1^{w}		
50 °C	9.1	-13.7^{w}		
80 °C	9.1	-13.6^{w}		
30 °C	10.3	-14.4^{w}		
50 °C	10.3	-14.1^{w}		
30 °C	12.4	-14.6^{w}		
50 °C	12.4	-14.3^{w}		
wollastonite[s]	2.0	-11.9	$CaSiO_3$	Ferruzzi, 1993
25 °C	4.0	-12.2		
	6.1	-11.6		
	8.0	-10.8		
	9.6	-11.3		
jadeite	3.0	-12.4	$NaCa_{0.2}Fe_{0.3}AlSi_2O_6$	Sverdrup, 1990
25 °C	4.6	-13.2		
	5.9	-13.2		
spodumene	3.0	-10.9	$LiAl_{0.86}Fe_{0.07}Si_2O_6$	Sverdrup, 1990
25 °C	4.3	-11.5		
	6.5	-13.3		
$MnSiO_3$[s]	2.3	-13.8	$MnSiO_3$	Banfield et al.,
25 °C	5.0	-14.3		1995
	6.6	-14.8		
$MnSiO_3$[n]	2.1	-13.8	$MnSiO_3$	Banfield et al.,
25 °C	2.9	-13.6		1995
	3.5	-14.3		
	4.8	-14.3		
	6.1	-15.5		
	7.1	-15.5		

Phase	pH	log R	Reported composition	Reference
tremolite	1.0	-15.4	$Ca_{1.91}Fe_{0.03}Mg_{5.09}Si_8O_{22}(OH)_2$	Schott et al.,
20 °C	6.0	-14.8		1981
antho-	2.0	-15.8	$Mg_{1.88}Ca_{0.08}Na_{0.04}(Mg_{4.51}Mn_{0.42}Fe_{0.02}Al_{0.02})Si_8O_{22}(OH)_2$	Mast & Drever,
phyllite	3.0	-15.6		1987
22 °C	4.0	-15.7		
	5.0	-15.6		
	7.0	-15.9[m]		
	7.6	-16.1[m]		
	9.2	-16.4[m]		
antho-	1.2	-16.8	$Mg_{5.7}Fe_{1.0}Al_{0.1}Si_{7.4}O_{22}(OH)_2$	Chen &
phyllite[i]	2.0	-17.0		Brantley, in
25 °C	3.0	-17.3		prep.
	3.4	-17.5		
antho-	1.2	-16.7	$Mg_{5.4}Fe_{0.5}Al_{0.03}Si_8O_{22}(OH)_2$	Chen &
phyllite[f]	2.0	-17.0		Brantley, in
25 °C	3.0	-17.4		prep.
	3.4	-17.5		
hornblende	3.6	-15.6	$Na_{0.4}Ca_{1.8}Mg_{3.7}Fe_{0.9}Al_{0.1}Si_{7.1}Al_{0.9}O_{22}(OH)_2$	Nickel, 1973
	5.6	-15.9		
	10.6	-15.4		
hornblende	4.0	-16.8	$Ca_2Mg_{3.3}Fe_{1.7}Si_{7.5}Al_{0.5}O_{22}(OH)_2$	Cygan et al.,
25 °C				1989
hornblende	3.6	-15.6[e]	$K_{0.11}Na_{.66}Ca_{1.66}(Mg_{3.11}Fe^{2+}_{1.02}Al_{0.79}Fe^{3+}_{0.32})-$	Zhang et al.,
25 °C	3.8	-15.7	$(Al_{1.64}Si_{6.36})O_{22}(OH,F)_2$	1990; Zhang,
	4.0	-15.9		1990
hornblende	1.6	-12.4	$Ca_{1.7}Mg_{3.5}Fe_{1.3}Al_{1.4}Si_{6.9}O_{22}(OH)_2$	Sverdrup, 1990
25 °C	2.4	-12.6	Kragero	
	2.5	-12.4		
	3.1	-12.6		
	3.5	-13.1		
	4.4	-13.8		
	5.0	-14.3		
	5.5	-14.7		
	6.5	-14.7		
	7.0	-15.0		
hornblende	6.2	-15.6	$Na_{0.08}Ca_{2.1}Mg_{4.5}Al_{2.1}Si_7O_{22}(OH)_2$	Sverdrup, 1990
25 °C			Lake Gardsjon	
hornblende	6.8	-14.7	$Na_{0.16}Ca_2Mg_4Al_{0.4}Si_{8.3}O_{22}(OH)_2$	Sverdrup, 1990
25 °C			Ontario	
hornblende	4.0	-15.7	$Na_{0.14}K_{0.09}Ca_2Fe_{1.78}Mg_2Al_1Si_7O_{22}(OH)_2$	Swoboda-
25 °C	4.5	-15.8		Colberg &
				Drever, 1993
glaucophane	3.13	-12.6	$Na_{1.7}Ca_{0.7}Mg_{1.6}Fe_{1.4}Al_{2.3}Si_{7.6}O_{22}(OH)_2$	Sverdrup, 1990
25 °C	5.25	-14.0		
	6.23	-14.6		
	6.83	-15.2		

*as reported by Knauss et al., 1993, **assumed composition (no composition data reported), [2] may have been supersaturated with secondary minerals [a]rate under anoxic conditions, [e] reported as an exponential rate: this rate represents the mean rate observed during the last several days of dissolution, [f]normalized by final surface area, [fe]reported chemical composition contained 21% Fe_2O_3, although authors reported XRD analysis revealed only enstatite + talc + chlorite, [i]normalized by initial surface area, linear dissolution observed at this pH, [m]natural mixture of pyroxmangite and rhodonite, [m] Mg release used to calculate rate instead of Si release, [o]rate under oxic conditions, [p] reported as a parabolic rate: this rate represents the average rate observed at the end of the experiment, [s]synthetic mineral, [t] may have contained talc, [tt]contained 5% tremolite; also, no surface area was reported--we assumed the same surface area as for Pitcairn diopside reported by the same workers, [w] with CO_2, in borate buffer, [wo] without CO_2

experimental uncertainty. Knauss et al. (1993) carefully summarized the propagated errors in their rate calculations, but concluded that these errors may be smaller than the true experimental errors, especially for runs where outlet concentrations were extremely low. They concluded that a conservative interpretation of the rate data yields an error estimate of ±0.25 log units for rate constants. Ferruzzi (1993) also estimated experimental uncertainty in his rate measurements for dissolution of enstatite, wollastonite, and rhodonite on the order of a factor of two. Xie (1994) reported errors on the order of ±0.2 log units for rate measurements on wollastonite.

Run duration. One large source of discrepancy is the lack of consistency in run duration. If we assume that the initial fast transient dissolution period for silicates is always followed by a slow approach to steady state, we can conclude that the short experiments (2 to 40 d) of Schott et al. (1981) may not have reached steady state. Similarly, the wollastonite dissolution experiments of Rimstidt and Dove (1986), run for less than 20 h, yielded faster (possibly non-steady-state) dissolution rates for wollastonite, as compared to the experiments of Xie (1994) run for up to 500 h. Even longer run durations (thousands of hours) are needed for the slow-dissolving inosilicates.

Mineral chemistry or structure. The wide range in rates observed plotted in Figures 2 and 3 may also be due to small differences in mineral chemistry or structure. Knauss et al. (1993) pointed out that the data of Schott et al. (1981) for Pitcairn and Gouverneur diopside showed that these two phases dissolved at similar rates under neutral pH, but at different rates at low pH. At pH 1, the rates differed by an order of magnitude (Fig. 2). Pitcairn diopside was reported to have slightly more Al + Fe than observed in the Gouverneur diopside. Differences in composition or surface area of diopside from Jaipur, Rajastad and from Kangan, Andhra-Pradesh (India) may also account for the observed differences between dissolution rates reported by Knauss et al. (1993) and Chen and Brantley (in prep.). Differences in dissolution rate for pyroxenes from different localities suggests that observed differences in weathering susceptibility of pyroxenes and amphiboles from soil profile to soil profile or from laboratory to laboratory may be related to physical and chemical factors intrinsic to individual minerals as well as external variables. On the other hand, Sverdrup (1990) reported that dissolution rates of hornblendes from three different localities showed only minor differences in dissolution rate (Fig. 3).

Presence of contaminants may also explain discrepant dissolution rates. For example, Siegel and Pfannkuch (1984) reported an extremely slow rate of dissolution of enstatite in long duration (2000 h) experiments (Fig. 2). However, although they reported that X-ray diffraction of the phase indicated enstatite ± chlorite ± talc, their compositional analysis indicated 21 wt % Fe_2O_3. Several other workers noted the presence of contaminants in samples (Table 1).

Aging of powders. Another complication in comparing experimental results of dissolution of ground powders is the observation of effects of aging or preparation on powder surface area and composition before experimental dissolution. Peck et al. (1988) reported the presence of a hydrated glass-like surface phase on dry-ground diopside powder detected using ^1H-^{29}Si cross-polarization magic angle spinning nuclear magnetic resonance (CP MAS-NMR). They argued that the thin surface layer approached a structure with purely random distribution of bridging and nonbridging oxygens. Eggleston et al. (1989) reported that the initial dissolution rate (20°C, pH 1, HCl-H_2O) of diopside powders (<75 μm) decreased with age after grinding, regardless of powder pre-treatment. They argued, therefore, that none of the four pre-treatments tested (including no treatment, water washing, acetone washing, HF-H_2SO_4 pre-etch) removed all artifacts of grinding and that some of the aging may consist of slow relaxation at room temperature of surface structural distortion or slow healing of surficial microcracks. They noted that BET measurements of surface area documented a small decrease in surface area during powder aging (on the laboratory shelf) over 250 d. The effects of powder aging has also been documented for feldspars (Stillings et al., 1995).

Several workers have also documented changes in surface chemistry of mineral samples as a function of aging which would affect dissolution rates measured on powders ground at different times. Schott and Petit (1987) observed that while the H concentration profile of an untreated diopside surface was stable over 1 year of storage, the Ca/Si ratio of the powder washed in water or acetone or HF - H_2SO_4 increased during aging over 7 mos. The Mg/Si ratios did not change significantly during aging. These authors argued that channels of transport into the bulk from the surface opened during dissolution may have allowed Ca mobility during aging. Eggleston et al. (1989) also showed that surface compositions as measured by X-ray photoelectron spectroscopy (XPS) of a leached fracture surface from a single crystal of diopside and of unwashed diopside powder were stable during aging over 7 months, but that the Ca/Si ratio of treated diopside powders increased during that same period. Mg/Si ratios did not change significantly.

Surface area. Several authors reported rates normalized to mass or using assumed surface areas. Indeed, the discrepancies observed by Schott et al. (1981) in the dissolution rates of Pitcairn and Gouverneur diopside may be related to surface area: these workers did not report surface area measurements for Gouverneur diopside, and we have reported rates assuming a surface area based upon the Pitcairn surface area. Therefore, the observed rate differences could be explained by surface area differences. Siegel and Pfannkuch (1984) also reported dissolution rates based upon an assumed rather than measured surface area. Because they noted no systematic dependence upon surface area for dissolution rates up to 60 d, Zhang et al. (1993) reported all of their measurements normalized by mass of hornblende (for consistency, in Table 1 we have renormalized their rates by their reported surface area). Further problems with respect to measurement and normalization by surface area is discussed in a later section.

Chemical affinity. Another source of discrepancy may be differences in the chemical affinity of the reactant solution with respect to the phase of interest, which may cause slow dissolution or even the precipitation of secondary phases. Although not all workers reported their solution chemistry data, evaluation of the existing data reveals that many of the solutions were supersaturated with one or more secondary phases. For example, Zhang et al. (1993) reported their solution chemistry of hornblende dissolution at pH 4 and 25°C for four different size fractions. Using the computer code SOLMINEQ.88 (Kharaka et al., 1988), we calculated the saturation state of their solutions and found that the solutions of the fractions smaller than 0.25 mm were supersaturated with respect to gibbsite; for smaller size fractions, solutions were supersaturated with respect to kaolinite as well. Precipitation of the secondary phases may therefore explain the observed nonlinear dissolution for their 90-day experiment. Evaluation of the data reported by Schott et al. (1981) shows that their diopside (Gouverneur) dissolution was supersaturated with gibbsite, kaolinite, and smectite, assuming stoichiometric dissolution. Their other solutions were also supersaturated with kaolinite (enstatite, pH 6, 50°C), with SiO_2 (untreated enstatite, pH 6, 20°C), and with gibbsite and kaolinite (Pitcairn diopside, pH 6, 20°C), assuming stoichiometric dissolution. Augite dissolution at pH 4 and 25°C by Siegel and Pfannkuch (1984) was also supersaturated with kaolinite assuming stoichiometric dissolution.

Differences in solution chemistry not related to affinity effects may also contribute to discrepancies in measured dissolution rates; these effects are discussed in the next section.

Effects of varying solution composition

Several workers have concluded that solution chemistry, other than pH, has an insignificant effect on pyroxene or amphibole dissolution rates. Sanemasa and Katsura (1973) concluded that Ca and Mg in solution did not affect cation release of diopside glass, but may have slightly inhibited release rates of diopside crystal. They also reported, however, that dissolution in sulfuric acid was pH-dependent even at extremely low pH while dissolution in HCl became pH-independent at low pH. Grandstaff (1977) reported that short term (<55 h) bronzite dissolution at 1 < pH < 4.5 was unaffected by differences in concentration of Na, K, sulfate, nitrate, chloride, citrate, or acetate. However, he observed that the presence of fluoride increased the dissolution rate. Schott et al. (1981) reported no effect of varying ionic strength from 0.01 to 0.30 m (KCl) on the rate of dissolution of enstatite or diopside, and they reported the same pH dependence for dissolution of diopside in KCl-HCl under pH-stat conditions as measured in KH phthalate buffer (0.001 to 0.01 m) between pH 3 and 6. In their work on wollastonite, Rimstidt and Dove (1986) saw very little effect of $Ca_{(aq)}$ on the dissolution rate: they reported the slope of log rate versus log m_{Ca} equals -0.07 ± 0.08. Mast and Drever (1987) concluded that the presence of oxalate in concentrations between 0.5 and 1 mM also had no effect on dissolution of anthophyllite. They pointed out that these organic acid values are much higher than most soil water solutions and concluded that organic acids in nature probably have an insignificant effect on tremolite dissolution. Ferruzzi (1993) reported that use of two different concentrations of Bis-tris propane/LiCl/HCl buffer at pH 6.1 yielded no observed difference in dissolution rate of a rhodonite/pyroxmangite mixture.

On the other hand, Sverdrup (1990) argued that discrepancies between his dissolution rate data for augite without organic acids, and the data of McClelland (1950) and Siever and Woodford (1979) for augite dissolved in the presence of acetic acid argue for a rate enhancement by organic acid. Xie (1994) observed a rate enhancement for wollastonite dissolved in the presence of potassium hydrogen phthalate or succinic acid as compared to HCl-H$_2$O at a given pH. He documented this rate acceleration to be pH-dependent, where the effect of organic ligand increased with decreasing acidity of the solution. He suggested that the effect of KHP and succinic acid on wollastonite dissolution is unimportant at pH values around 2 or lower, but that at neutral pH, the acceleration can be as large as an order of magnitude. Some of the discrepancy between rate measurements of diopside by Chen and Brantley and Knauss et al. (1993), who used HCl-H$_2$O and buffer solutions respectively, may be due to solution effects (Fig. 2).

The effect of varying P_{CO2} on dissolution rates of inosilicates is also not well understood. Bailey (1974) reported a rate enhancement for wollastonite in the presence of 5% CO$_2$ at pH 6 compared to dissolution in the presence of lower amounts of CO$_2$. However, Grandstaff (1977) reported that short term (<55 h) bronzite dissolution at 1 < pH < 4.5 was unaffected by differences in P_{CO2}. Knauss et al. (1993) reported that a tenfold decrease in P_{CO2} below atmospheric had no significant effect on the rate of Si release from diopside at high pH. However, under these alkaline conditions they did observe that release of Ca and Mg from diopside during dissolution was enhanced under conditions of low P_{CO2} relative to atmospheric conditions, and they speculated that perhaps the Mg and Ca sites on the mineral surface were poisoned under alkaline conditions by the attachment of a carbonate species, similar to that observed for olivine

(Wogelius and Walther, 1991). Brady and Carroll (1994) reported no observed P_{CO2} dependence (between $10^{-3.5}$ and 10^0 atm P_{CO2}) for augite dissolution in 10^{-3} M acetic acid adjusted with HCl at pH 4. However, at pH 10 and 12, Xie (1994) observed that the silica release rate of wollastonite was decreased in the presence of ambient CO_2 by about two orders of magnitude compared to dissolution in CO_2-free NaOH solution, and explained this inhibition by postulating a surface Ca-carbonate complex (Fig. 3). The slow dissolution rate measurements reported by Xie (1994) which were attributed to the presence of CO_2 could also have been slower because they were run in batch reactors, rather than in fluidized bed reactors, or because they were run with borate buffer instead of H_2O-NaOH $\pm$ KCl. Clearly, however, the large discrepancies between some wollastonite dissolution rates measured by Ferruzzi (1993) and Xie (1994) argue that some uncontrolled variable, such as buffer composition or CO_2, may be very important (Fig. 2).

Few researchers have reported inhibition by other elements such as Al in solution. Although Sverdrup (1990) reported qualitative data suggesting Al inhibition of hornblende, Zhang (1990) pointed out that those experiments may have been in the range where $Al(OH)_3$ precipitated on the mineral surface. On the other hand, observations of Al-inhibition for feldspar dissolution (Blum and Stillings, this volume) suggest that similar effects may occur for hornblende. For example, Brantley et al. (1993) reported hornblende weathering rates from a soil catena, and concluded that slow weathering of hornblende in a poorly drained as compared to a well drained soil may have resulted from accumulation of dissolution products (such as Al) in the poorly drained soil porewaters. If Al acts as an inhibitor, then experiments run in batch reactors must be clearly distinguished from experiments run in flow reactors.

Mineral surface area

Surface area of starting material. BET-measured surface area (Table 2) of ground and cleaned single chain inosilicates (75 to 150 μm) measured in different laboratories before dissolution vary between 490 cm^2/g (1-year aged diopside, Knauss et al., 1993) and 1900 cm^2/g (wollastonite, Xie and Walther, 1994). Our own measurement of surface area of ultrasonically cleaned diopside of 75 to 150 μm grain size (860 cm^2/g) emphasizes that differences between laboratories or differences between minerals of the same composition but from different localities may be as large as ±50%. Most researchers report errors in replication of BET surface area measurements between 10 to 20% (e.g. Knauss et al., 1993). Such discrepancies in surface area will be reflected in discrepancies in rate reported between laboratories.

Another problem with using surface area to normalize dissolution rate is that surface area changes with time and sample treatment *prior* to dissolution (Table 3). Eggleston et al. (1989) reported that aging of diopside powders (<75 μm grain size) over 250 d caused a small decrease in BET-measured surface areas. Knauss et al. (1993) further observed that a diopside sample (125 to 75 μm) degassed for 15 h under vacuum had a surface area of 610 ± 12 cm^2/g, whereas after 4 h heating at 200°C the same sample had a measured area of 550 ± 15 cm^2/g (all measurements completed using 5-point BET). One year later the sample had an area of 490 ± 14 cm^2/g after again being heated for 4 h at 200°C. Most researchers have not investigated such changes in surface area with time, and simply have reported one BET measurement with Kr or N_2.

Relationships between surface area and dissolution. Almost all workers have implicitly assumed that dissolution rates are directly proportional to the initial surface area of ground minerals (Table 2). In the only study of the rate of dissolution as a

Table 2. Measured BET surface areas of inosilicates.

Phase	Grain size, μm	Prepar- ation	Surface area (cm²/g), BET gas	Reference
enstatite	75 - 150		576 (air perm.)	Luce et al., 1972
enstatite[1]	83 - 124	G, US, E	5500	Schott et al., 1981
enstatite	25 - 75	G,US	3300	Ferruzzi, 1993
bronzite	74 - 149	G	1720	Grandstaff, 1977
bronzite	100 - 200	G, US, E	600	Schott & Berner, 1983
diopside[p]	83 - 124	G,E	600	Schott et al., 1981
diopside	75 - 150	G,US	550±15 (Kr)	Knauss et al., 1993
diopside	75 - 150	G,US	860 (Kr)	Chen & Brantley, in prep.
augite	180 - 420	G,US	1200	White & Yee, 1985
wollastonite	150 - 250	G,US	1330 (N₂)	Rimstidt & Dove, 1986
wollastonite[s]	25 - 75	G,US	4450 (N₂)	Ferruzzi, 1993
wollastonite	74 - 149	G,US	1900 (Kr)	Xie and Walther, 1994
wollastonite	149 - 250	G,US	<3000 (N₂)	Xie and Walther, 1994
MnSiO₃	25 - 75	G,US	5400 (N₂)	Banfield et al., 1995
rhodonite[s]	25 - 75	G,US	1600 (N₂)	Banfield et al., 1995
tremolite	needles	A,E	200	Schott et al., 1981
anthophyllite[2]	75 - 150	US	570	Mast & Drever, 1987
anthophyllite	fibers	G,M,US	12,100 (Kr)	Chen & Brantley, in prep.
amosite	45 - 75	G	37,300 (N₂)	Choi et al., 1974
crocidolite	45 - 75	G	43,000 (N₂)	Choi et al., 1974
hornblende	20 - 35	G,R,E	2370 (N₂)	Nickel,1973
hornblende	180 - 420	G,US	5500	White & Yee, 1985
hornblende[3]	37 - 149	G, US	49,300±1500 (N₂)	Cygan et al., 1989
hornblende	110 - 250	G, US	970 cm²/g	Zhang, 1990
	250 - 500		860 cm²/g (N₂)	

G = ground, US = ultrasonically cleaned, E = etched, R = rinsed repeatedly, M = mixed for 0.5 h in blender, [1]contained talc, [2]contained quartz, [3]unshocked, [p]Pitcairn, [s]synthetic

function of grain size, however, Zhang et al. (1993) reported that the rate of release of Si (after 30 d) was not a linear function of the surface area for hornblende grains of four different grain sizes (silica release rate $\propto$ area$^{0.4}$). They did observe that the rate of release of Al, Fe, and Mg was linear with respect to surface area. Anbeek et al. (1994) analyzed their data and argued that their dissolution rates may be independent of grain diameter (see, however, Bloom et al., 1994). Documenting a similar problem, Banfield et al. (1995) noted that pre-etching of a rhodonite/pyroxmangite mixture with HCl increases the specific surface area from 5400 cm²/g to 12,000 cm²/g, although the Si release rate of this etched material does not similarly increase by a factor of 2 (Ferruzzi, 1993). Rimstidt and Dove (1986) pointed out that etching on the surface of silicates such as wollastonite might cause dissolution to become rate-limited by diffusion in and out of deep etch pits, as observed for some skeletal carbonates. Such an effect would result in dissolution rates not linear with respect to surface area. Many workers have discussed the difference between effective (or reactive) surface area and total, BET surface area (e.g. Helgeson, 1971), but no investigations have focused on defining the effective surface area of dissolving inosilicates. Casey et al. (1993) argued that the leached layer of chain silicates reconstructs and becomes less reactive with time, even as the specific surface area increases.

Table 3. Surface area measurement vs. time.

Phase	Treatment duration	BET surface area (cm^2/g)	Grain size, μm	Run solution	Reference
diopside	G,US aged 1 yr, heated	610 ± 12 490 ± 14	125-75	--	Knauss et al., 1993
diopside	G,US 3400 h	860 1900	75-150	-- pH 2.03 HCl-H₂O	Chen & Brantley, in prep.
wollastonite	G,US 3.8 h 3.78 h 5.92 h 20.67 h	1330 6790 8400 15,040 20,340	150-250	-- pH 3	Rimstidt & Dove, 1986
wollastonite	0 min 10 min	4000 16,000		-- 0.01 M HCl	Casey et al., 1993
wollastonite	G,US 2 wks	10,000 1,560,000	< 74	-- HNO₃, pH 3	Xie & Walther, 1994
MnSiO₃ (natural)	G,US 3523 h	5400 310,600	25-75	-- pH 2.1, buffer solution	Ferruzzi, 1993
anthophyllite	3400 h	12,100 12,800	38-500	-- pH 2.03 HCl-H₂O	Chen & Brantley, in prep.
amosite	G 1 h	37,300 51,000	45-75	-- H₂O	Choi et al., 1974
crocidolite	G 1 h	43,000 95,000	45-75	-- H₂O	Choi et al., 1974
hornblende	0 d 1 d 2 d 5 d 10 d 20 d 40 d	970 1260 1670 1850 2220 2420 2920	110 -250	-- pH 4.0 buffer	Zhang et al., 1993
hornblende (naturally weathered)	40,000 y 250,000 y 600,000 y	3400 7200 6700	250-500		White et al., submitted

G = ground, US = ultrasonically cleaned

For very long duration dissolution experiments where surface area changes during dissolution, laboratory rates should be normalized by final reactive surface area. Larger surface areas at the end of a dissolution experiment necessarily translate into slower area-normalized dissolution rates. In some cases, however, increased surface area after long dissolution is due to a flaky, silica-rich leached layer: use of final measured BET surface areas to normalize dissolution may not be appropriate in those cases, if this layer does not contribute to dissolution. Of the published dissolution rates compiled here, only the dissolution rates of Rimstidt and Dove (1986) for wollastonite were normalized by surface area of the mineral after dissolution. Rimstidt and Dove (1986) observed an increase in surface area of a factor of 15 in 20 h of dissolution of wollastonite at pH 3 at 25°C (Table 3). Ferruzzi (1993) observed that the specific surface area of natural $MnSiO_3$ reacted at pH 2.1 increased by a factor of 60 after about 3500 h of dissolution. Zhang et al. (1993) reported multiple measurements of surface area of hornblende during dissolution: surface area increased rapidly at first and then more slowly as dissolution proceeded (see Table 3 and Fig. 4). Surface area increased by a factor of about 3. In our experiments on diopside, we only observed an increase in surface area of a factor of 2 over 3000 h of dissolution under ambient conditions and subneutral pH. We observed virtually no change in surface area of anthophyllite dissolved under similar conditions (perhaps due to the simultaneous development of etch pits and dissolution of ultrafines present due to the fibrous nature of the mineral). Choi et al. (1974) observed a doubling in surface area of amosite and crocidolite during dissolution in water. Obviously, the increase in surface area during dissolution varies from mineral to mineral, perhaps increasing, for example, from anthophyllite < diopside < hornblende < wollastonite < $MnSiO_3$ (Table 3).

Figure 4 shows that the BET surface area measured for 100 to 200 μm hornblende dissolved in the laboratory at pH 4 for 2 months, (3000 cm²/g, Zhang et al., 1993), is still smaller than the surface area measured for hornblende grains (250 to 500 μm) recovered from soils weathered for 100,000 y (7200 cm²/g) in the Merced chronosequence, California (White et al., subm). The discrepancy between surface areas of mineral grains in the field and laboratory is one of the reasons field rates, when normalized by BET surface areas, are significantly slower than laboratory rates (White and Peterson, 1990).

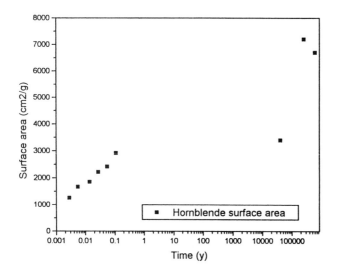

Figure 4. The surface area, measured by BET, for hornblende dissolved for different run durations (time < 1 y; Zhang et al., 1993) and weathered in soils of the Merced sequence (White et al., subm.). All soils developed on similar lithology.

Extremely high BET surface areas are measured in the laboratory only for samples of very small particle size or etched in low pH solution (Tables 2 and 3).

Surface roughness. Assuming a cubic geometry for particles, Zhang et al. (1983) concluded that the surface roughness (= BET surface area /calculated surface area) for hornblende dissolved at pH 4 for 2 months varied between 9 and 32. Sverdrup (1990) reported roughness ratios varying between 1.5 and 71 for 28 silicate minerals. Specifically for inosilicate minerals, he reported roughness values of 4.5 for wollastonite, 48 for hedenbergite, 50 for augite, 19 for jadeite, 9.4 for spodumene, 10 for hornblende, and 19 for glaucophane.

Etch pits. Increased surface area of inosilicate grains during laboratory dissolution or natural weathering is explained by formation of etch pits on the mineral surface (Fig. 5). For example, etch pits parallel to the (100) surface and along fractures on the (100) surface of dissolved diopside formed in the experiments reported by Knauss et al. (1993). Increases in surface area during dissolution of hornblende have been attributed to formation of etch pits along cleavage planes, and deep weathering parallel to silica chains (Zhang et al., 1993). These workers argued that hydration of hornblende promotes release of cations that hold silicate chains together. This dissolution is manifested by deep (sometimes > 1000 Å) etch pits parallel to the c axis of the mineral. Similar etch pits have been observed on highly weathered natural hornblendes (Fig. 6). The characteristic lens shape of pits on hornblendes have also been documented on soil grains of bronzite, hypersthene, diopside, and especially, augite (Berner et al., 1980b; Berner and Schott, 1982). Such lens-shaped pitting can be reproduced in the laboratory with HF + HCl etching.

Little definitive work has been accomplished assessing the nature of the defects causing preferential dissolution. Although dissolution along dislocation cores could create etch pits on the hornblende surface, Cygan et al. (1989) concluded, based upon dissolution measurements of shocked hornblende, that dissolution of dislocation etch pits did not dominate the dissolution flux. Surface area-normalized dissolution rate of hornblende which was explosively shocked was measured in batch reactors at pH 4.0. Shocked hornblende, with dislocation densities several orders of magnitude higher than unshocked, only showed a limited dissolution rate enhancement at pH 4 (factor of ~4), similar to shocked oligoclase and labradorite. Dissolution of shocked hornblende was also observed to be more stoichiometric than unshocked hornblende. The BET surface area of hornblende also changed little during shock loading (less than a factor of 2) in contrast to labradorite which increased 9-fold in surface area. The measurable, but unexpectedly small, acceleration of dissolution rate of hornblende with respect to dislocation density suggested to these authors that the surface dislocations produced by shock loading were not the primary sites of dissolution. This conclusion was substantiated for $MnSiO_3$ by Banfield et al. (1995), who also observed no dissolution rate differences between defective natural $MnSiO_3$ and relatively defect-free synthetic $MnSiO_3$, even though dissolution was accelerated at defect outcrops on the surface. Overall, the relative contribution of different defects toward controlling inosilicate dissolution is poorly known: for example, Banfield and Barker (1994) reported that chain width defects in amphibole increase the resistance of the amphibole to weathering, whereas stacking faults are sites of enhanced dissolution.

Etch pits and the creation of internal porosity have also been analyzed. Zhang et al. (1993) reported that measurement of N_2 adsorption-desorption isotherms shows no hysteresis for unreacted hornblende but hysteresis for material reacted for 2 months at pH 4. These workers concluded that the hysteresis indicates the presence of fine slit-like

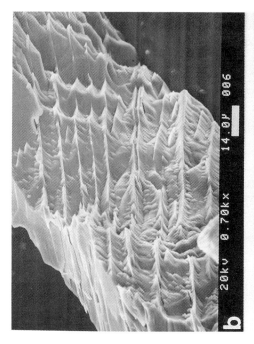

Figure 5 (opposite page). (a) Diopside starting material, before dissolution. Surface area = 860 cm² g⁻¹. (b) Diopside dissolved at pH 2.16 in a continuously stirred tank reactor for 1,800 h at 90°C. (c) Anthophyllite starting material, before dissolution. Surface area = 12,000 cm² g⁻¹. (d) Anthophyllite dissolved at pH 1.07 at 90°C in a continuously stirred tank reactor. Scale bars for each photo are shown in lower right, labelled in microns. Experiments summarized in tables and text (from Chen and Brantley, in prep.).

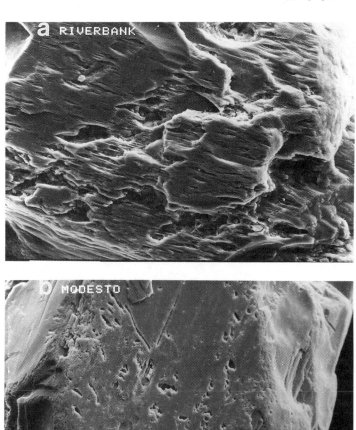

Figure 6. Scanning electron photomicrographs of naturally weathered hornblende grains from soils of the Merced sequence (see White, this volume), showing changes in extent of etching as a function of soil age. (a) Hornblende from the Riverbank soil (250,000 years). (b) Hornblende from the Modesto soil (40,000 years). Scale bars = 40 μm. Soils are described in White et al. (submitted). SEM photomicrographs by Marjorie Schulz.

pores smaller than 20 nm in diameter. This internal porosity was not thought to contribute greatly to measured surface area. However, Anbeek et al. (1994) argued that a reinterpretation of their data suggests that micropores in the natural hornblende starting material may have dominated the surface area, and inferred that microporosity in naturally weathered hornblende grains may be more important in the natural environment than in laboratory etching (however, see Bloom et al., 1994).

Relative elemental release rates

Early observations. Most workers analyzed whether the dissolution reactions of inosilicates release cations to solution in stoichiometric proportion. For example, Barman et al. (1992) reported that Ca was released more rapidly than Mg from hornblende dissolved in four organic acids (oxalic, citric, salicylic, glycine); however, no rates were reported. Luce et al. (1972) reported that enstatite dissolution was nonstoichiometric (Mg > Si) at first, but that dissolution became more stoichiometric with time (up to 100 h), although dissolution remained parabolic. Sanemasa and Katsura (1973) reported that crystalline and glassy diopside dissolved nonstoichiometrically (Ca, Mg > Si), but that crystalline material released Ca preferentially over Mg, while glassy material released Ca and Mg in equivalent proportions. Grandstaff (1977) also observed preferential release of Mg with respect to Si for dissolution of bronzite at subneutral pH (<4.5). In short-term (<550h) batch experiments, he never observed stoichiometric or linear dissolution of this orthopyroxene.

Enstatite, bronzite, diopside, tremolite. In experiments with pre-etched diopside and tremolite, Schott et al. (1981) observed preferential release of Ca and, to a lesser extent, Mg with respect to Si. Preferential release of Fe, present predominantly in the M2 site in bronzite, was reported at pH 1 or pH 6 under anoxic conditions (Schott and Berner, 1983, 1985). However, Schott and Berner (1985) reported that bronzite under oxic conditions at pH 6 loses no Fe to solution, presumably because of ferric oxide precipitation on the mineral surface.

These workers argued that their data was consistent with a greater mobility of the cations in the M2 site over the M1 site in the pyroxene structure (Ca was released more quickly than Mg from diopside) and of the M4 site over the M1, M2, or M3 site in amphibole (Ca was released more quickly than Mg from tremolite). The observation of no difference in release rate between Ca and Mg reported for dissolution of diopside glass (Sanemasa and Katsura, 1973) supports such an observation, because the cations in the glass do not reside in distinct sites. Schott and Berner argued that the absolute value of the Madelung energy is smaller for a cation in the M2 and M4 site as compared to a cation in one of the other sites. A cation with a lower absolute Madelung energy is thus bonded more weakly and is more mobile. This model satisfactorily explains why the apparent activation energy of release of Ca is lower than that of Mg between 35° and 65°C (50 vs. 63 kJ/mol, Sanemasa and Katsura, 1973). Schott and Berner (1985) reported that the small preferential loss of Mg from the enstatite surface (which is enhanced at pH 1 as compared to pH 6) can similarly be explained by assuming that loss of Mg was confined to the M2 sites.

Although Schott et al. (1981) were unable to conclude whether enstatite dissolution eventually became stoichiometric, Ferruzzi (1993) reported congruent dissolution of enstatite observed after 500 h of dissolution at pH 7.0 (particle size of 25 to 75 µm), and close to stoichiometric dissolution at pH 2 and 4.8 (difference in log rate Si release and Mg release on the order of 0.3 log units). Schott and Berner (1983, 1985) reported that initially incongruent dissolution of tremolite and diopside eventually became stoichiometric, and that, at pH 1 or pH 6 under anoxic conditions, bronzite dissolution also became congruent after initial (100 to 300 h) nonstoichiometric dissolution. They interpreted these observations to indicate that a steady-state leached layer formed on the inosilicate surface and they assumed that the thickness of the leached layer was determined by the relative balance between cation leaching rate and silicate network hydrolysis. Calculations of leached layer thickness based upon the evolution of solution chemistry predicted leached layers which were several Å in thickness, and which, they argued, could not inhibit dissolution.

In their experiments, Knauss et al. (1993) reported enhanced release of Ca and Mg and added that days to weeks were necessary in flow-through experiments on diopside to achieve stoichiometric dissolution at temperatures between 25° and 70°C. The difference in duration of the nonstoichiometric dissolution period between the early experiments of Schott et al. (1981) and those of Knauss et al. (1993) may be related to the difference in powder preparation: Schott et al. (1981) acid-etched their starting material while Knauss et al. (1993) only cleaned their samples ultrasonically. In our experiments, with diopside and anthophyllite samples also cleaned ultrasonically, we observed a several week long period of nonstoichiometric dissolution (Chen and Brantley, in prep.).

Wollastonite. While Bailey and Reesman (1971) observed that Ca was released more quickly than Si from dissolved wollastonite, Rimstidt and Dove (1986) reported stoichiometric dissolution for runs up to 20 h. Ferruzzi (1993) reported stoichiometric dissolution of wollastonite (25 to 75 µm) in experiments of 300 to 400 minutes duration at pH 6.1 to 9.6, but nonstoichiometric dissolution (Ca > Si) even after 350 min at pH 2.0 and 4.0 in buffer solutions. He also observed that, while wollastonite dissolves nonstoichiometrically at pH 2, pseudo-wollastonite, composed of calcium ions and silicate anions arranged in three-membered rings rather than chains, dissolves stoichiometrically in 10 min experiments (Ferruzzi, 1993; Casey et al., 1993). They hypothesized that the slow hydrolysis of the silicate chains relative to the fast release of Ca precludes release of stoichiometric ratios of Ca and Si in wollastonite.

Xie and Walther (1994) and Xie (1994) reported that the initial dissolution of alcohol-washed wollastonite is nonstoichiometric (Ca > Si) in HCl-H_2O solution under ambient conditions. During their subneutral pH experiments (both with and without organic ligands present), they never saw stoichiometric dissolution of wollastonite, even after 200 h of dissolution. However, they observed that the enhancement of Ca release over Si decreased at higher pH and that dissolution at pH 7 is almost stoichiometric. Their experiments also showed that, in CO_2-free NaOH solutions, the initial dissolution is nonstoichiometric with Si > Ca release, but that in some experiments in CO_2-free alkaline solutions, dissolution of wollastonite becomes stoichiometric. In general, however, the presence of CO_2 decreases the Si release rate significantly. Xie and Walther (1994) hypothesized that dissolution over longer time periods must become stoichiometric and they cited the observation of Bailey and Reesman (1971) that, in acid solutions, the dissolution of wollastonite becomes stoichiometric over time. More work is needed to sort out the stoichiometry of dissolution of wollastonite with and without CO_2 present, given the discrepancies in the literature, and the different observations in the presence of different buffer solutions and reactor types (Xie, 1994).

Rhodonite. Banfield et al. (1995) reported that dissolution of a rhodonite/pyroxmangite mixture exchanges Mn for protons instantaneously upon placement into water. The faster release of Mn continues for hundreds to thousands of hours in acid solutions, but the discrepancy between Si and Mn flux decreases with time. They did not observe stoichiometric dissolution of the rhodonite/pyroxmangite mixture even after 2000 h at neutral pH. Ferruzzi (1993) reported that the largest degree of incongruence of dissolution of natural $MnSiO_3$ (pyroxmangite-rhodonite mixture) was observed at pH 6.1 and 7.1 where the Mn flux was about 10 times the Si flux. In contrast, he observed the largest degree of incongruence (factor of five) in dissolution of synthetic rhodonite at pH 2.3.

Hornblende. The dissolution of hornblende was also observed to be nonstoichiometric at pH 4, with very fast early release of Ca (Zhang et al., 1993), as expected, because Ca fills the M4 site in hornblende. However, interpretation of the Ca release rate was clouded by the presence of trace calcite in the sample (Bloom, pers. comm.). From 0 to 39 d, pre-weathered hornblende dissolution was characterized by enhanced release of all metals except Al as compared to Si. However, Zhang (1990) reported that preferential release of Al with respect to Si was observed in other longer experiments at pH 3.6 and 4.0. He also reported that preferential release of Al, Mg, and Fe increases systematically with increasing pH from 3.6 to 4.0, and argued that this observation can be explained by a more rapid approach to steady state dissolution at lower pH. Dissolution of four grain size separates (0.045 to 1.0 mm) all maintained nonstoichiometric dissolution although the finest grain size (0.045 to 0.075 mm) was the most nonstoichiometric. Experiments lasting greater than 2 months approached stoichiometric dissolution for all grain sizes (Zhang, 1990), although secondary minerals may have precipitated.

SURFACE ANALYSIS OF Fe-FREE INOSILICATES

Spectroscopy of laboratory-dissolved surfaces

Enstatite. Using solution chemistry from dissolution experiments, Luce et al. (1972) calculated the depth of the Mg-leached layer on enstatite as less than one unit cell thick when dissolved at 25°C under subneutral pH. These workers did not observe a leached layer using X ray diffraction, infrared, or mass spectrometry after ion bombardment of the mineral surface. Schott et al. (1981) were able, however, to measure very shallow Mg-leaching of enstatite reacted at pH 6 for 2 to 40 days using X-ray photoelectron spectroscopy. They estimated this cation-depleted layer to be only a few Å thick. The XPS also showed greater depth of leaching of Mg from the enstatite surface at pH 1 than observed at pH 6, as documented by the solution chemistries during dissolution. Because they observed no shifts in the absorption bands under XPS, they suggested that H^+ had replaced Mg ions in the structure without rearrangement of the silica network bonding. However, Casey et al. (1993) reported no observable penetration of H^+ into the enstatite surface after dissolution at 50°C and pH 2.

Diopside. Schott et al. (1981) and Schott and Berner (1983, 1985) reported small or negligible depletion of Mg as observed by XPS of the surface of diopside dissolved at pH 6 under ambient conditions up to 60 d. Even at 60°C, Schott and Berner (1983, 1985) reported negligible Mg depletion of the diopside surface leached close to neutral pH. Ca depletion from the diopside was much greater than Mg depletion; however, after long durations, the Ca/Si ratio of the surface remained constant indicating congruent dissolution. At pH 1, Mg and (especially) Ca were both released faster than Si from the diopside surface. Eggleston et al. (1989) also reported that leaching of a diopside fracture surface in pH 1.0 HCl - H_2O decreased both the Mg/Si and Ca/Si ratio of the surface as measured by XPS. Assuming a model of linear depletion with depth, or a totally depleted surface layer, Schott and Berner concluded that leached layers were less than 30 Å in depth.

Despite the fact that Ca leaching of the diopside surface depends on pH, Petit et al. (1987), Schott and Petit (1987), and Peck et al. (1988) observed that the extent of hydration of the diopside surface was pH-independent. Indeed, Petit et al. (1987) argued that there was no direct correlation between hydrogen penetration and cation release at the diopside surface, supporting the hypothesis for molecular water diffusion into the crystalline surface, especially along dislocations and other defects. Petit et al. (1987) used

Resonant nuclear reaction analysis (RNRA) and showed that H_2O, H^+, or H_3O^+ penetrates diopside surfaces during dissolution at pH 2 or 6, and that the hydrated surface zone extends up to 700 Å. Undissolved diopside crystal surface shows only limited hydrogen penetration. Heating of the diopside caused a loss of about 25% of the surface H, indicating some of the H is present as water, and some as a species bonded to the crystalline silica network. Petit et al. (1987) also reported that extent of hydration decreases with temperature.

Schott and Petit (1987) used secondary ion mass spectrometry (SIMS) on the same samples and measured a decrease in the signal of all constituent elements—indicating an increase in porosity of the surface of a diopside crystal leached for 5 d at pH 5.5 and 100°C. The discrepancy between the depth of leaching of Ca and Mg as calculated based upon XPS and that estimated based upon SIMS may be explained by the fact that XPS laterally averages a chemical signal over a wide surface area, and the depth of leaching is only inferred from a model calculation (see Schott and Petit, 1987; Mogk and Locke, 1988). In addition, Schott and Petit (1987) pointed out that leaching of the surface may be extremely heterogeneous, varying as a function of defects present at the crystal surface. They summarized their results by postulating the formation of a porous hydrated silicate depleted in cations on the dissolved diopside surface. Casey et al. (1993) corroborated these results when they documented that the surface of diopside became enriched in H to a depth of 1000 Å after reaction for 522 h at 50°C.

Wollastonite. In contrast, the wollastonite surface becomes enriched in H to depths greater than several thousand Å (Casey et al., 1993), when reacted at pH 2 for 522 h at 50°C. The concentration of hydrogen in wollastonite reacted at this temperature was much greater in a wollastonite surface dissolved at pH 2 than one dissolved at pH 5. Ferruzzi (1993) reported that the depth of calcium depletion measured by elastic recoil detection (ERD) is roughly equal to the depth of H enrichment of the wollastonite.

A broad band observed in the Raman spectra of reacted wollastonite at 350 to 500 cm^{-1}, interpreted as breathing or stretching modes of silicate tetrahedra, was not observed in spectra taken from unreacted wollastonite (Fig. 7). Bands characteristic of bending/stretching modes of silicate chains (at 637 and 950 to 970 cm^{-1}) were similar between reacted and unreacted mineral surfaces, and were attributed to chains that still exist in the reacted region or to unreacted areas of the wollastonite surface. Based on such Raman evidence, Casey et al. (1993) argued for the presence of 4-membered silicate rings on the leached wollastonite surface. Changes in the surface of wollastonite during dissolution were also documented by Xie and Walther (1994) who reported that surface leaching of wollastonite is so great that the surface charge properties of acid-reacted wollastonite are similar to those of microporous silica.

Rhodonite. Ferruzzi (1993) reported Rutherford back-scattering evidence for Mn depletion on the surface of leached $MnSiO_3$. He pointed out that the extent of leaching increases with decreasing pH, and that leaching of Mn is less than leaching of Ca in wollastonite. The surface leached layer on $MnSiO_3$ was also documented by Casey et al. (1993) and Banfield et al. (1995) with high-resolution transmission electron microscopy (HRTEM). Banfield et al. (1995) reported that surfaces of grains of a rhodonite/pyroxmangite mixture show a thin (3 to 7 nm) hydrogen- and silica-rich amorphous layer after reaction in a $H_3BO_3/Na_2B_4O_7$ buffer solution at pH 7.1 for more than 2000 h. They suggested that the surface layer could be related to leaching during dissolution, grinding artifacts, synthesis artifacts, or adsorption and polymerization of solutes on the mineral surface during dissolution. Because the thickness of the layer was consistent with a calculated leached layer based upon observed cation fluxes during dissolution, they

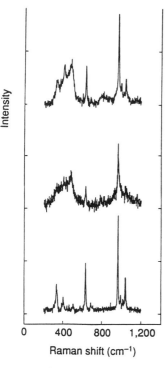

Figure 7. Raman spectra from unreacted wollastonite (bottom) and wollastonite reacted by suspension in large carboys of solution at fixed pH and 0.1 M ionic strength at 50°C for 522 h (after Casey et al., 1993). Middle and top spectra represent wollastonite reacted at pH 2 and 5 respectively. Dissolved silica concentrations were maintained below amorphous silica saturation. The analytical spot size allowed for 1 μm resolution and sampling depth of 5 to 10 μm. Spectra were collected in flowing helium using the 457.9 nm excitation line of an Ar⁺ laser (Casey et al., 1993).

interpreted the layer as a leached layer with density consistent with cristobalite. During the experiment, the bulk solution was never supersaturated with respect to amorphous silica.

Electron photomicrographs of the leached layer of reacted $MnSiO_3$ also revealed dark spheres on the surface which were attributed to silica condensation reactions occurring at the surface (Banfield et al., 1995). Reactions such as the following were inferred to have occurred even at neutral pH where the hydration depth of the mineral surface was only 10 to 100 Å thick:

$$\equiv SiOH + \equiv SiOH ==> \equiv Si\text{-}O\text{-}Si\equiv + H_2O \tag{7}$$

Where this reaction occurs, they argued that the proton content of the surface is lower than that predicted based upon loss of cations. Ferruzzi (1993) pointed out that the observation that specific surface area increases with time during dissolution of $MnSiO_3$ while metal fluxes decrease indicates that the leached layer is reconstructing and becoming less reactive with time. However, the data for surface area vs. time compiled in Ferruzzi (1993) shows that most of the increase in surface area probably happens very early in the experiment (in the first several hours), well below the residence time of the fluid in the chemical reactor, making it difficult to relate large initial surface area changes to solution chemistry. Casey and coworkers reported that the extent of restructuring of the reacted inosilicate surface varies with solution composition (especially pH), mineral structure, and composition. Samples which were dried and stored for up to a year often showed the best evidence for recrystallization features (Banfield et al., 1995).

Tremolite. Schott et al. (1981) also reported XPS evidence showing a slight depletion of Mg and large depletion of Ca with respect to Si for dissolution of tremolite at pH 6 at 20°C. Both elements are more extensively leached at pH 1. They reported that the trends observed at 60°C are similar to those observed at 20°C.

SURFACE ANALYSIS OF NATURALLY WEATHERED MINERALS

Berner and Schott (1982, Schott and Berner, 1983) reported slight depletion in Ca on the upper few tens of Å of surface of bronzite, hypersthene, diopside, and hornblende soil grains after pre-cleaning the grains by ultrasonication. Cation depletion of the natural samples was observed in all cases except augite to be similar but deeper in the naturally weathered samples as compared to laboratory-etched samples. Because their XPS measurements yielded no direct depth measurements, their calculation of the depth of depletion assumed that leaching was either complete over the calculated depth or that the extent of leaching of a cation decreased linearly with depth. The observed depletion of Ca was greater than the depletion observed for Mg, in agreement with the model based upon laboratory measurements whereby preferential loss of cations occurs from the *M4* and *M2* sites.

Berner and Schott (1982) also reported that only naturally weathered diopside showed a surface enrichment of Fe, and that such surface precipitates were easily removed by ultrasonication of the soil grains. For example, surfaces of partly weathered bronzite grains taken from soils where soilwaters were lower than pH 6 exhibited the same Fe/Si, O/Si and Mg/Si ratios as measured with XPS as the underlying unaltered material (Schott and Berner, 1983). They concluded that no armoring surface layer of ferric oxyhydroxide composition, as postulated by Siever and Woodford (1979), formed. However, the observation of binding energy shifts of about 1.2 eV for surface Fe was attributed to the presence of a surface layer of hydrous ferrous silicate of a few tens of Å thickness. Because they documented no thick surface layers on soil grains, they hypothesized that altered surface layers on Fe-containing silicates may break down to ferric oxide and flake off before becoming prohibitively thick in soil systems.

Mogk and Locke (1988) reported that Auger electron spectroscopy (AES) of a naturally weathered hornblende showed depletion in Ca, Fe, and Mg with respect to Al, which was assumed to be conservative. After sputtering the surface, these workers determined that the surface depletion reached to 120 nm. Mogk and Locke (1988) emphasized the point made by Berner and Schott (1982) in explaining the difference in cation depletion depth observed with XPS and AES: no direct depth data was available in the earlier XPS measurements, and the depth of depletion was thus calculated from a model assumption for the change in depletion as a function of depth. However, even assuming uniformly depleted surfaces, Mogk and Locke (1988) calculated a 47 nm depleted zone for Ca and Mg, and a 28 nm Si- and Fe-depleted zone relative to Al—a much thicker depleted zone than the previous estimates. No secondary phase was observed on the surface. These results are in qualitative agreement with the solution chemistry observations of Zhang et al. (1993) for enhanced dissolution of Fe, Mg, and Ca from the hornblende surface dissolved at pH 4 in the laboratory.

COUPLED OXIDATION-REDUCTION OF Fe-CONTAINING INOSILICATES

The presence of metals of variable redox state (especially Fe and Mn) in the inosilicates allows a complex set of oxidation-reduction reactions to occur in parallel to the simple hydrolysis reactions. Siever and Woodford (1979) investigated the effects of oxygen on the rate of weathering of hypersthene and other Fe-containing minerals. They concluded that the dissolution rate of hypersthene at pH 4.5 decreased in the presence of oxygen. They hypothesized that dissolution in the presence of oxygen caused precipitation of iron oxyhydroxides which in turn armored the dissolving mineral grains. In their experiments, which were run in batch mode, they observed an initial, fast increase followed by a slow decrease in {Fe}: from this observation they concluded that the iron was released as Fe^{2+}, and was oxidized in solution and precipitated. Because their solutions contained iron oxyhydroxides which sorbed or coprecipitated other ions, they could not determine accurate dissolution rates.

Schott and Berner (1983,1985) reported that XPS evidence for the bronzite surface dissolved under anoxic conditions at pH 6 or especially under oxic conditions at pH 1 documented incongruent release of Mg, and especially, Fe relative to Si. No oxidation of surface Fe was observed. Under these conditions, they argued that the release of Fe from bronzite is similar to the release of Ca in diopside, and may be accompanied by protonation of the surface. In the presence of oxygen at pH 6, however, Schott and Berner (1983) argued that the observed increase in Fe/Si and O/Si ratios at the surface of bronzite measured by XPS, as well as binding energy shifts observed for Fe and O, is well explained by release of Fe^{2+} to solution, followed by oxidation and precipitation of ferric oxide similar to goethite at the mineral surface. These XPS observations essentially confirmed observations made from solution chemistry.

Schott and Berner (1983,1985) reported several lines of evidence that the increased Fe/Si ratio of the bronzite surface after dissolution at pH 6 under oxic conditions is due to a hydrous ferric oxide precipitate. They reported that ultrasonic cleaning of the mineral surface showed that the precipitate layer was tens of Å thick and was non-adhering. The bronzite surface revealed after removal of the precipitate showed no change in the O_{1s} binding energy but an XPS energy shift of about 1.1 eV indicative of oxidation of Fe(II) to Fe(III), indicating that the inner surface layer consisted of a Fe^{3+}-Mg silicate, perhaps protonated nontronite. These results are consistent with oxidation of Fe occurring not only after release of Fe^{2+} to solution but also at the mineral surface. They argued that this ferric silicate surface should be less reactive than the original ferrous silicate and that its increase in thickness with time could explain the observed parabolic dissolution behavior of bronzite. To maintain charge balance at the surface, Mg was probably released during oxidation, but they reported that their data could not be used to estimate the thickness of this altered layer. However, because no known Fe^{3+}-Mg-silicate is stable at pH 6, they believed that no thick surface protective layer could be maintained under those conditions.

White and Yee (1985) investigated surface oxidation reactions of augite and hornblende at ambient temperature in aqueous solutions between pH 1 and 9. Using XPS, they documented that both augite and hornblende, like the bronzite dissolved by Schott and Berner (1983), show a significantly decreased Fe/Si surface ratio when reacted in solutions at pH below 5 or 6 under closed anoxic conditions. Above this pH value, they observed no Fe depletion in the mineral surface. As observed previously by Schott and Berner (1983, 1985) for bronzite, they concluded that the surface Fe was dominated by ferric ions. While they reported seeing linear release rates of Si, K, Ca, and Mg over month-long periods, they observed that ferrous iron in solution increased rapidly at first and then decreased as iron oxides precipitated (> pH 3.5). The rate of decrease of iron in solution was accelerated in the presence of oxygenated atmospheres. Even at pH values as low as 3.5 they observed $Fe(OH)_3$ saturation in their experiments.

They also reported ample evidence for a coupled surface-solution oxidation-reduction mechanism whereby ferric iron added to solution is reduced by ferrous iron on the mineral surface:

$$[Fe^{2+}, 1/z \; M^{z+}]_{surface} + {}^*Fe^{3+} \Rightarrow [Fe^{3+}]_{surface} + {}^*Fe^{2+} + 1/z \; M^{z+} \qquad (8)$$

where the surface iron is indicated by the "surface" subscript, and the asterisk indicates that atoms are conserved in solution (distinguishing the reaction from an ion exchange process). The oxidation of the surface iron is accompanied by release of a lattice cation, to maintain charge balance. Because the surface iron on the silicate is predominantly oxidized, ferrous iron from several Å to tens of Å depth must be accessed in these reactions.

Hydrolysis of the silicate can also occur either simultaneously or subsequently:

$$[Fe^{2+}, 1/z \; M^{z+}]_{surface} + 3H^+ \Rightarrow [3H^+]_{surface} + 1/z \; M^{z+} + Fe^{2+} \qquad (9)$$

Therefore, cations can be released from the hornblende or augite surface either during hydrolysis reactions (with uptake of protons), or coupled with oxidation of the iron at the surface with no pH change. When Fe^{3+} is released during hydrolysis, it is reduced in solution to Fe^{2+}, regenerating the oxidized Fe surface layer by the coupled oxidation of Fe^{2+} deep in the silicate surface.

White and Yee (1985) observed that release of Ca is accelerated from the augite surface in the presence of ferric chloride in solution, and that this reaction does not change the pH of the solution: they therefore concluded that Reaction 8 was dominating. During this reaction, solution Fe^{3+} decreases and Fe^{2+} increases in concentration in solution. However, after the Fe^{3+} in solution is consumed, Ca release is accompanied by a pH increase, as hydrolysis reactions proceed (Reaction 9). For pre-weathered augite they noticed that surface oxidation (Reaction 8) preceded hydrolysis (Reaction 9), but for freshly ground augite the two reactions occurred in parallel. As shown in Figure 8, their data indicates that while surface oxidation of the Fe^{2+} occurs, hydrolysis is inhibited in the pre-weathered mineral surface. Similar observations were made for hornblende: for this phase they calculated that Fe^{2+} at the surface is a slightly better reducing agent than aqueous Fe^{2+}. The estimated standard half cell potential ranges from +0.33 to +0.52 for surface iron oxidation, bracketed by the ability to spontaneously reduce Fe^{3+} in solution, but the inability to reduce Fe^{3+} in precipitated $Fe(OH)_3$. Although these coupled reactions occurred at pH values above 3 and below 9, precipitation of iron oxyhydroxides also occurred simultaneously at higher pH values.

At least one other surface oxidation reaction also occurred in their experiments, causing a decrease in oxygen fugacity with time:

$$[Fe^{2+}, 1/z \; M^{z+}]_{surface} + 1/4 \; O_2 + H^+ \Rightarrow [Fe^{3+}]_{surface} + 1/z \; M^{z+} + 1/2 \; H_2O \qquad (10)$$

The rate of this reaction for both augite and hornblende was observed to increase with decreasing pH.

The authors concluded that these coupled cation-electron transfer reactions may be important in mediating the reduction of such aqueous species as UO_2^{2+}, VO^{2+}, Cr^{6+} and Cu^{2+}. Reduction of these species as a coupled reaction with the augite or hornblende surface might be especially important in neutral or alkaline pH regimes where little Fe is found in solution.

The results of this study suggest that the presence of oxygen might enhance the weathering rates of Fe-containing silicates. Zhang (1990) also suggested that oxidation of Fe (II) to Fe(III) in hornblende might increase dissolution rates. In agreement with this

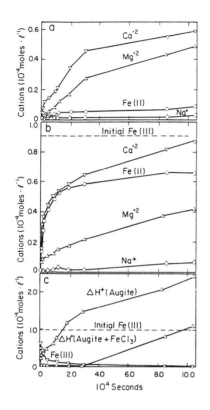

Figure 8. Aqueous chemistry during reaction with weathered augite (after White and Yee, 1985): (a) Major cations in a starting solution of deionized water, (b) major cations in a starting solution of 1×10^4 molar $FeCl_3$ as represented by the horizontal dashed line, (c) hydrogen ion and ferric iron decreases in the respective deionized water and $FeCl_3$ solutions with time. Experiments were completed at pH 3 to eliminate precipitation of ferric hydroxide. Note that during oxidation of the augite in the presence of $FeCl_3$ (Fig. 8b) the rate of Ca release is accelerated (compare to Fig. 8a) and parallels the reduction of iron. The rate of hydrogen ion uptake is essentially zero during the surface oxidation-reduction reaction, and only after all the ferric ion is consumed are the hydrogen uptake rates parallel.

observation, Brantley et al. (1993) documented that etching rates of hornblende grains from a soil catena showed that hornblende etching was slower in poorly drained (and reducing) conditions, than in well drained (and oxygenated) conditions. However, several workers have suggested that the presence of armoring precipitates of Fe oxide might cause slower dissolution rates in the presence of oxygen. As discussed above, Siever and Woodford (1979) observed that the presence of oxygen slowed dissolution of hypersthene, and Schott and Berner (1983) similarly observed that the dissolution of bronzite was lower under oxic as opposed to anoxic conditions at pH 6. However, in their batch experiments without oxygen, these latter workers introduced titanium (III) citrate solution as a reductant to maintain low oxygen. The difference in chemistry between citrate-free (oxic) and citrate-containing (anoxic) experiments may therefore be reflected in the observed rates. Moreover, these same workers have used XPS and SEM to study surface composition of hypersthene, augite, bronzite, and hornblende soil grains, and have concluded that the outermost 20 Å is oxidized to Fe(III), but that there exists no evidence for the presence of an Fe-rich, Si-poor armoring surface layer on the soil grains (Berner and Schott, 1982).

RATE LAWS FOR DISSOLUTION

pH dependence

Although there is some evidence supporting a pH-independent acid regime (pH < ~1) for some inosilicates (e.g. Sanemasa and Katsura, 1973; Schott et al., 1981; Rimstidt and Dove, 1986), many workers have fit dissolution rate data to empirical rate laws such as the following:

$$log\ R = log\ k_H + n\ pH \qquad (11)$$

Values of n reported for dissolution of the inosilicates vary from 0.11 to -0.8, although most are between -0.2 and -0.8 (Table 4).

In the first analysis of the effect of pH on pyroxene dissolution, Luce et al. (1972) concluded that there is little to no pH dependence for the kinetics of dissolution of enstatite. As mentioned, however, their experiments all showed parabolic reaction kinetics for pH > 2. The data of Grandstaff (1977), Schott et al. (1981), and Sverdrup (1990), based upon batch experiments for up to several hundred hours, revealed higher values for n: -0.8 (enstatite), -0.7 (diopside), -0.7 (spodumene, jadeite). Ferruzzi (1993) reported $n = -0.25$ for long duration (1500 to 2000 h) dissolution experiments for enstatite in buffer solutions between pH 2 and 7. Knauss et al. (1993) concluded that $n \approx -0.2$ for diopside dissolution in buffers between pH 2 and 12. They also concluded that the pH-slope was independent of temperature from 25° to 70°C, although the scatter in their data could be interpreted as a higher pH-dependence at acid values and a flattening of the dependence above pH 6. Our 3400 h measurements show a limited pH dependence ($n \approx$ -0.2) for dissolution of diopside below pH 4 in pure HCl-H_2O solutions (Chen and Brantley, in prep.), and an increase in slope at 90°C.

For the pyroxenoids, Rimstidt and Dove (1986) reported a value of $n = -0.40$ for short-term (< 20h) dissolution of wollastonite between pH 4 and 7 in KCl solutions. However, because their data shows that short duration dissolution at low pH became pH-independent, they ignored many of the low pH experiments in order to force a fit to model Equation (11). Ferruzzi (1993) reanalyzed all their data and reported that $n = -0.17$. Ferruzzi (1993) also reported wollastonite dissolution measurements in which he observed an increase in dissolution rate ($n = 0.15$) from pH 2 to 10 for 25 to 75 μm particles in buffer solutions. He reported nonstoichiometric dissolution at pH 2 and 4 in experiments run for 300 to 400 min. For this same phase, the 100 to 500 h experiments of Xie (1994) yielded a value of $n = -0.24$. For the only other pyroxenoid investigated, Banfield et al. (1995) reported $n = -0.33$ (rhodonite) for long-term experiments.

Schott et al. (1981) reported the pH dependence for dissolution of tremolite (based upon two experiments) is close to zero ($n = 0.11$). Mast and Drever (1987) similarly reported a very shallow pH dependence (from pH 2 to 5, $n = 0$) for anthophyllite dissolution for 2 < pH < 9 (note that they reported their mineral composition erroneously as tremolite). Our 3400 h data for dissolution of anthophyllite at 25°C is consistent with a value of $n = -0.37$ (1 < pH < 4, Chen and Brantley, in prep.).

For Fe-containing minerals, there is scarcely any data. Grandstaff (1977) and Schott and Berner (1985) reported a value of $n = -0.5$ to -0.6 for bronzite dissolved in short duration experiments (rates were parabolic) and Sverdrup (1990) reported $n = -0.7$ for augite and glaucophane dissolution. Sverdrup (1990) also presented data for hornblende dissolution between pH 0 and 11 for less than 600 h which fit a value of $n = -0.6$ to -0.7 for the subneutral range. He warned, however, that the variability of aluminum concentration negated separating a true pH-dependence from an Al-dependence.

In another investigation of hornblende dissolution, Zhang (1990) reported that the pH-dependence observed in 1 month experiments was shallower than that observed after 115 d. After 115 d, Si and Al release showed close to a first-order dependence on {H^+};

Table 4. Reported pH dependence for Si release (pH < 7).

Phase	n	Run duration	Solution	Reference
enstatite	0	16-100 h	$HNO_3 + H_2O$	Luce et al., 1972
enstatite	-0.8 (20°C)	2-40 d	K-phthalate ± KCl + NaOH + HCl buffers	Schott et al., 1981; Schott & Berner, 1985
enstatite	-0.25	1500-2000 h	0.1 M buffer solutions	Ferruzzi, 1993
bronzite	-0.5 to -0.6	50-550 h	buffer solutions	Grandstaff, 1977; Schott & Berner, 1985
diopside	-0.7 (20°C) -0.75 (50°C)	2-40 d	K-phthalate ± KCl ± NaOH ± HCl	Schott et al., 1981; Schott & Berner, 1985
diopside [i]	-0.2 (25, 70, 90°C)	40-60 d	buffer solutions pH 2 - 10	Knauss et al., 1993
diopside [i]	-0.20 (25°C)	3400 h	$HCl-H_2O$	Chen & Brantley, in prep.
diopside [f]	-0.23 (25°C) -0.77 (90°C)	3400 h	$HCl-H_2O$	Chen & Brantley, in prep.
augite	-0.7	2-40 d	buffer solutions	Schott and Berner, 1985
augite ($He_{55}En_{45}$, $En_{50}Wo_{39}He_{10}$)	-0.7	48-600h	$H_2SO_4 + H_2O$	Sverdrup, 1990
wollastonite	-0.22 (rate for Ca-release only)	120 - 335 h	buffer solutions	Bailey & Reesman, 1971; as reported by Ferruzzi, 1993
wollastonite	-0.4	<20 h	buffer solutions	Rimstidt & Dove, 1986
wollastonite	0.14	300-350 min	0.1 M buffers pH 2 - 10	Ferruzzi, 1993
wollastonite	-0.24	80 - 220 h	$HCl ± H_2O ±$ KCl	Xie, 1994
rhodonite/pyroxman-gite & syn rhodonite	-0.33 ± 0.06	2000-3500 h	buffer solutions	Banfield et al., 1995
jadeite	-0.7	48-600 h	$H_2SO_4 + H_2O$	Sverdrup, 1990
spodumene	-0.7	48-600 h	$H_2SO_4 + H_2O$	Sverdrup, 1990
anthophyllite [i]	0	1000-2000 h	$HCl + H_2O$	Mast & Drever, 1987
anthophyllite [i]	-0.33 (25°C)	3400 h	$HCl + H_2O$	Chen & Brantley, in prep.
anthophyllite [f]	-0.37 (25°C) -0.64 (90°C)	3400 h	$HCl + H_2O$	Chen & Brantley, in prep.
tremolite	>0	2-40 d	buffer solutions	Schott et al., 1981
hornblende	-0.6 to -0.7	48-600 h	$H_2O-H_4SO_4$	Sverdrup, 1990
hornblende	-0.7	115 d	0.01 M HOAc-LiOAc	Zhang, 1990
glaucophane	-0.7	48-600 h	$H_2O-H_4SO_4$	Sverdrup, 1990

[i] indicates normalized by initial surface area , [f] indicates normalized by final surface area

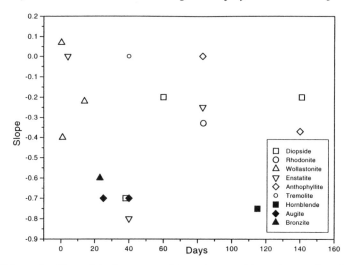

Figure 9. Values of n (slope of the log rate vs. pH curve) reported by various workers for inosilicates, plotted vs. the duration of experimental dissolution. Note that measurements by Sverdrup (1990) for jadeite, spodumene, and hornblende have not been included because they all plotted at approximately the same position on the graph ($n = -0.7$ and 600 h). Minerals which do not contain appreciable Fe or Al are indicated by hollow symbols.

however, the release rates of Fe and Mg showed a lower dependence. Zhang suggested that the true value of n is in the range of -0.8 to -1, and pointed out that the estimates of Sverdrup (1990) may have emphasized short-term nonstoichiometric dissolution, or release of Fe but not Si. He emphasized that the true value of n cannot be determined until stoichiometric dissolution is achieved (>115 d). One complication in the experiments on Fe-containing inosilicates is the possible precipitation of ferric hydroxides. White and Yee (1985) reported precipitation of ferric hydroxide during dissolution of augite and hornblende under oxic conditions even at pH 3.5.

Some of the inconsistencies in observations of n for inosilicates may be related to unconstrained variables such as the duration of dissolution (Fig. 9). Although compositional differences might also explain variability in n, there is a suggestion that long duration experiments yield values of $n = -0.2$ to -0.3 for Fe-free and Al-free inosilicates (open symbols in Fig. 9). There is also evidence for a steeper pH dependence for Fe- and/or Al-containing inosilicates (solid symbols in Fig. 9), although more long-term dissolution data is needed.

Best estimates of empirical rate laws

To estimate dissolution rates consistent with the observation that long duration experiments better exemplify true steady state dissolution, we have calculated a best fit dissolution rate-pH expression using Equation (11) for each mineral studied (Fig. 10, Table 5). Only for diopside, where the data of Knauss et al. (1993) extends from pH 2 to 10, have we extended the rate prediction to high pH. Most aluminosilicate minerals also show an increase in rate at high pH, and two more terms are sometimes added to the empirical rate equation:

$$R = k_H \{H^+\}^{-n} + k_w + k_{OH}\{OH^-\}^m \qquad (12)$$

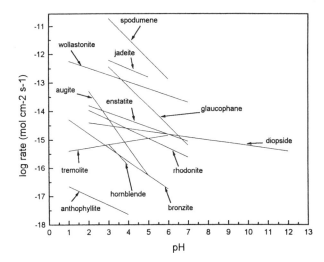

Figure 10. Best estimates of log rate vs. pH for several inosilicates (see Table 5 for rates and references). Note that data for tremolite is based only upon two data points, and could thus be suspect. Rates are expressed as mol inosilicate cm^{-2} s^{-1} where all pyroxenes are expressed as six oxygens per formula unit.

For the data compiled here, only one mineral is reported to show an increase in dissolution at high pH: wollastonite under CO_2-free conditions (Fig. 2). For this mineral under these conditions, $k_{OH} = -15.6 \pm 0.6$ and $m = 0.27 \pm 0.07$, for pH ≥ 7 (Xie, 1994). Because of discrepancies between dissolution rates at high pH between Xie (1994) and Ferruzzi (1993), we have not plotted alkaline dissolution rates for wollastonite.

Where only one dataset was available, we have used it to determine a best fit rate; where there was more than one dataset for a mineral, we have chosen long duration experiments only, as noted in Table 5. For example, for enstatite, we assumed that only the 2000 h experiments of Ferruzzi (1993) represent steady-state. For diopside, where two datasets are available, we have fit all of the data together (Table 5). Note, however, that the two datasets which we have regressed together for diopside may be different due to the inherent difference in the two starting materials, suggesting that the range between these measurements represents a true range and should not necessarily be regressed together. We have also summarized our data for diopside and anthophyllite normalized by both initial and final surface areas. For jadeite, we have regressed all three datapoints to determine n, which explains why our value of this parameter is lower than that reported by Sverdrup (1990). The rate law included for tremolite is based upon only two measured rates reported by Schott et al. (1981), representing one of the more poorly constrained rate measurements in Table 5. For hornblende, we have used the long duration experiments of Zhang (1990) and a slope of $n = -0.75$, based upon his measurements between pH 3.4 and 4.0. Although these three rate measurements do not correspond to a large spread in pH values, the value of n is roughly confirmed by the short duration experiments of Sverdrup (1990).

Trends in dissolution rates. Casey and Westrich (1992) and Casey et al. (1993) have documented a strong case for a systematic trend in orthosilicate dissolution rate at pH 2 as a function of the constituent cation (see Casey, this volume). In particular, these workers have related dissolution rate to the rate of solvent exchange around the metal cation. They predicted that such reactivity trends should exist for other silicate minerals. However, Banfield et al. (1995) argued that the dissolution rates at pH 2 of chain silicates

Table 5. Best estimate of rate laws for dissolution of selected inosilicate phases.

All rates expressed as mol inosilicate $cm^{-2} s^{-1}$, where inosilicate formula units are written with the following number of Si atoms per unit: 2 (pyroxenes and pyroxenoids), 8 (tremolite, anthophyllite), 7.6 (glaucophane), 6.36 (hornblende). Rates are determined from Si release in experiments.

Phase	log k	n	pH range	Data from...
enstatite [i]	-13.3 ± 0.7	-0.25 ± 0.14	2 - 7	Ferruzzi, 1993
bronzite* [i]	-13.8 ± 0.3	-0.49 ± 0.13	< 5	Grandstaff, 1977; Schott and Berner, 1983
diopside [i]	-13.6 ± 0.2	-0.18 ± 0.03	2 - 10	Knauss et al., 1993
diopside [i]	-13.4 ± 0.4	-0.22 ± 0.10	2 - 6	Knauss et al., 1993
diopside [i]	-14.2 ± 0.2	-0.10 ± 0.04	2 - 10	Knauss et al., 1993; Chen & Brantley, in prep
diopside [i]	-14.2 ± 0.2	-0.20 ± 0.07	1 - 4	Chen and Brantley, in prep.
diopside [f]	-14.5 ± 0.2	-0.20 ± 0.06	1 - 4	Chen and Brantley, in prep.
augite* [i]	-11.3 ± 0.6	-0.99 ± 0.14	< 6	Sverdrup, 1990; Siegel and-Pfannkuch, 1984
wollastonite * [i]	-12.0 ± 0.2	-0.24 ± 0.04	≤ 7.2	Xie, 1994
wollastonite* [i]	-15.6 ± 0.6	0.27 ± 0.07	≥ 7	Xie, 1994
rhodonite [i]	-13.3 ± 0.3	-0.33 ± 0.06	2.1 - 7.1	Banfield et al., 1995
spodumene* [i]	-8.6 ± 0.4	-0.72 ± 0.08	3 - 7	Sverdrup, 1990
jadeite* [i]	-11.6 ± 0.6	-0.30 ± 0.12	3 - 6	Sverdrup, 1990
anthophyllite [i]	-16.3 ± 0.1	-0.33 ± 0.03	1 - 4	Chen and Brantley, in prep.
anthophyllite [f]	-16.2 ± 0.007	-0.37 ± 0.003	1 - 4	Chen and Brantley, in prep.
tremolite * [i]	-15.5	0.11	1, 6	Schott et al., 1981
hornblende [i]	-12.9 ± 0.2	-0.75 ± 0.06	3.6 - 4.0	Zhang, 1990
glaucophane* [i]	-10.4 ± 0.3	-0.69 ± 0.05	3 - 7	Sverdrup, 1990

* indicates dissolution measured for less than 1000 h,   [i] indicates normalized by initial surface area,
[f] indicates normalized by final surface area.

exhibit a poor correlation with the first-order rate coefficient describing solvent exchange around the hydrated metal. They note that most of the chain silicate minerals dissolve at approximately 10^{-13} to 10^{-14} mol $cm^{-2} s^{-1}$ at pH 2 under ambient conditions. They further argued that the slow rate of dissolution of rhodonite (manganese inosilicate) compared to tephroite (manganese orthosilicate) or wollastonite (calcium inosilicate) compared to pseudowollastonite (calcium ring silicate) at pH 2 cannot be explained by changes in acid-base properties of the mineral surface, suggesting that correlations between rate and composition may be less predictable for the inosilicates, and may be controlled predominantly by the polymerized chain structure. However, Casey et al. (1993) reported no differences in dissolution rate for two Mn pyroxenes with different silicate chain repeat lengths: rhodonite and pyroxmangite both dissolve at a rate of $10^{-13.5}$ mol $cm^{-2} s^{-1}$ at pH 2 and 25°C. This argument supports the idea that the silicate chain and the number of bridging oxygens dominates the dissolution rate, rather than the exact geometry of the chains.

Banfield et al. (1995) pointed out that the one exception to the generally similar rates of dissolution of inosilicates is the fast weathering rate of wollastonite, a phase which becomes extensively leached of Ca and hydrated at the surface. Sverdrup (1990) pointed out that the dissolution rates of spodumene and jadeite are also fast. It is possible, however, that the large dissolution rate of these phases might be comparable to other

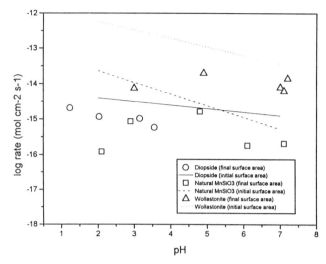

Figure 11. Log rate vs. pH for three minerals, each calculated as rate normalized by either initial (lines) or final surface area (symbols). Data for diopside from Chen and Brantley (in prep.), for $MnSiO_3$ from Ferruzzi (1994), and for wollastonite from Xie (1994). Final surface area for wollastonite was derived from estimates based upon Rimstidt and Dove (1986) for shorter duration dissolution experiments. Increase in measured surface area during dissolution is much larger for the pyroxenoids than for diopside.

pyroxenes if the rates were normalized by *final* rather than *initial* surface area. Figure 11 shows dissolution rates for three pyroxenes where final and initial surface area measurements are available. Diopside dissolution is about a factor of two slower when normalized by final surface area. Dissolution rate of $MnSiO_3$ is comparable to diopside dissolution when normalized by final surface area (data from Ferruzzi, 1993). Although the final surface area was not reported by Xie and Walther (1994), we have assumed that the surface area changes observed by Rimstidt and Dove (1986) could be used to correct the data of Xie (1994) for rough estimates of wollastonite dissolution. Clearly, the rate of dissolution of wollastonite is closer to the rate of dissolution of diopside and $MnSiO_3$ when all rates are normalized by final surface areas. Perhaps some of the discrepancies in plotting log rate vs. log (ligand exchange constant) would disappear if pyroxene dissolution rates were normalized by final rather than initial surface areas. Note that this effect would not be as important for the amphiboles (Table 3).

The dissolution rates of Fe-free amphiboles (anthophyllite—Mast and Drever, 1987; Chen and Brantley, in prep; tremolite—Schott et al., 1981) are slower than dissolution of comparable pyroxenes (i.e. enstatite, diopside) under similar pH (Fig. 10). This is in agreement with the so-called Goldich weathering series for observations from soils (Goldich, 1938). Plotting the rate constant for dissolution of the comparable ortho-, and single-chain and double-chain inosilicates (expressed now as mol Si release cm^{-2} s^{-1}) shows a clear decrease in dissolution rate with increasing polymerization of the silica tetrahedral structure (Fig. 12). The data in Figure 12 defines a slope of -2, and clearly demonstrates that the increase in bridging oxygen per tetrahedral silicate unit decreases the rate of dissolution of comparable endmember phases. The order of the rates at each level of connectedness (= bridging oxygen per tetrahedral unit) generally follows the order of metal-ligand exchange predicted by Casey et al. (1993). We can therefore use Figure 12 to predict dissolution rates for endmember inosilicate compositions, based

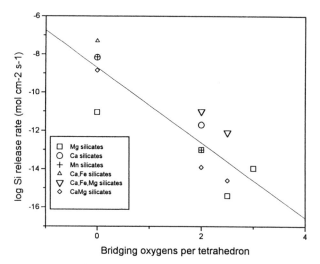

Figure 12. The log of the dissolution rate constant (k_H) plotted vs. the connectedness (number of bridging oxygens per tetrahedral unit) for data from Table 5. Orthosilicate data from Westrich et al. (1993) and Wogelius and Walther (1991); chrysotile data from Bales and Morgan (1985). The slope of the best fit regression line to all the data is -2. For each value of connectedness, the log rate generally increases with the increasing rate of cation-water exchange, as described by Casey (this volume).

upon values of log k_H for the orthosilicates: for example, given a log k_H for CaFeSiO$_4$ from Westrich et al. (1993), the log k_H for hedenbergite and actinolite should be approximately -11.5 and -12.3 respectively. The plot also suggests that the log k_H for enstatite dissolution measured by Ferruzzi (1993) may be about an order of magnitude too high, and for chrysotile dissolution measured by Bales and Morgan (1985) may be two orders of magnitude too high.

TEMPERATURE DEPENDENCE OF DISSOLUTION

Most workers analyzing inosilicate dissolution vs. temperature have used a form of the simple Arrhenius equation to fit rate vs. temperature data :

$$k = k_o \, exp \, (-E_a/RT) \tag{13}$$

where k is the rate constant, k_o is the pre-exponential or frequency factor, and E_a is the apparent activation energy of dissolution. Where the pH dependence of dissolution has been measured, a pH-independent rate constant can be determined using a rate equation such as Equation 11. This pH-independent rate constant can be conceptualized as the rate of dissolution of the phase in a solution at pH 0. Many reported activation energies are thus reported for a nominal solution at pH 0 (Table 6). However, some researchers have measured the E_a for dissolution of a phase at one pH and reported an apparent activation energy for that pH value. If the pH-dependence of dissolution changes as a function of temperature, then the E_a (pH 0) will not equal the E_a reported for any other pH. For this reason, the pH of the reported E_a is noted in Table 6.

The measured activation energies which we have tabulated for inosilicates vary from 38 kJ/mol to 150 kJ/mol (Table 6). Values on the order of 50 to 120 kJ/mol are similar to activation energies reported for albite (Chen and Brantley, subm.), olivine (Wogelius and

Table 6. Activation energies reported for inosilicates

Phase	pH range	pH for reported E_a	Temp. range, °C	Solution chemistry	E_a kJ/mol	Reference
enstatite	6.0	6.0	20-60		49	Schott et al., 1981
enstatite	4.0	4.0	25 - 65		41.2	Bailey, 1977
enstatite		0	25 - 72		~80	Calc'd; see text
bronzite	2.1	2.1	1 - 42	HCl-KCl	44.1	Grandstaff, 1977
diopside glass	0-1	0	5 - 65	HCl-H_2O	67	Sanemasa & Katsura, 1973
diopside	0-1	0	35 - 65	HCl-H_2O	50-63	Sanemasa & Katsura, 1973
diopside	1.7 - 4.0	pH-dependent	20 - 60		50 - 150	Schott et al., 1981
diopside	4 - 6	4-6			38	Schott & Petit, 1987
diopside	2	2			81	Schott & Petit, 1987
diopside	2 - 12	0	25 - 70		40.6	Knauss et al., 1993
diopside	1 - 4	0	25 - 90	HCl-H_2O	96.1	Chen & Brantley, in prep.
augite	6	6	20 - 75		78	Schott & Berner, 1985
augite	4	4	25 - 60	acetic acid	115.4	Brady & Carroll, 1994
wollastonite[1]	7	7	23.5 - 40		79.2	Rimstidt & Dove, 1986
wollastonite[2]	4.0	4.0	26 - 65		54.6	Bailey, 1977
wollastonite[2]	4.0	4.0	26 - 65		72	Bailey, 1977; reinterpreted by Murphy & Helgeson, 1989
anthophyllite	1 - 4	0	25 - 90	HCl-H_2O	77.2	Chen & Brantley, in prep.
forsterite	1.8 - 9.8	0	25 - 65		79. 7	Wogelius & Walther, 1992
albite	1 - 4	0	5 - 300		65.4	Chen & Brantley, submitted

[1] Si release, [2] Ca release

Walther, 1992), and other silicates (Wood and Walther, 1983), and are significantly higher than values measured for transport in solution (~20 kJ mol^{-1}). Furthermore, Brady and Carroll (1994) have cited evidence that the Ca-containing silicates in general have higher activation energies than the alkali silicates. For these reasons, we conclude that it is possible that the value for E_a for diopside measured by Knauss et al. (1993) may be too low, perhaps due to the presence of buffers. The value of this same parameter reported by Sanemasa and Katsura (1973) only reflects apparent E_a for Ca and Mg release during

40-minute experiments; nonetheless, the E_a they report is higher than that reported by Knauss et al. (1993), and closer to the value reported by Schott and Petit (1987) and Chen and Brantley (in prep.).

Measured activation energies summarized for bronzite and enstatite in Table 6 also appear low; however, these values are reported for pH 2.1 and 4.0 to 6.0 respectively. If the absolute value of n for enstatite is lower at 25°C than 60°C, then the reported E_a for enstatite at pH 6.0 (49 kJ/mol, Schott et al., 1981), when recalculated for pH 0 might be similar to the reported values for anthophyllite (77 kJ/mol) and forsterite (80 kJ/mol, see Table 6). For example, if we assume that the rate law summarized in Table 5 describes dissolution of enstatite, and use the rate of dissolution of enstatite measured by Schott and Berner (1985) at 72°C and pH 6 (where short duration experiments may have reached steady state), then we calculate an activation energy of 80 kJ/mol by assuming only a slight increase in the absolute value of n with temperature.

Such an increase in the absolute value of n with temperature would be in agreement with our observations for diopside and anthophyllite (Table 4). Casey and Sposito (1992) have argued that n should increase with increasing temperature for dissolution at pH values less than the point of zero net proton charge (pH_{pznpc})—where the surface charge due to positive sites equals the surface charge due to negative sites and where there are no specifically adsorbing cations. For all the minerals investigated here, E_a was measured for $pH < pH_{pznpc}$ (Tables 6 and 7), suggesting that n should increase with increasing temperature, as observed in our experiments. However, Berner et al. (1980a) and Knauss et al. (1993) reported no such increase in the absolute value of n with temperature from 20° to 50° or 70°C for diopside dissolution.

Brady (1991) pointed out the importance of using measured rates for Ca and Mg silicate dissolution to quantify the weathering term used in long-term global carbon models. In particular, he argued that such models are very dependent on the assumed E_a of dissolution used for these phases, and the dependence of E_a on P_{CO2} and organic acids. Brady (1991) argued that E_a for Ca and Mg silicate weathering fell in the range 40 to 80 kJ/mol, and showed that river chemistry data as a function of mean annual temperature was well explained by this assumption. Brady and Carroll (1994) presented rate data for augite in acetic acid and argued that the presence of organics did not markedly change the value of E_a with respect to inorganic solutions. They observed, however, that E_a of Ca-containing silicates could be much higher than 80 kJ/mol (Table 6). Much data is still needed in order to more accurately assess the temperature dependence of inosilicate dissolution, especially considering the importance of these Ca- and Mg-containing phases toward balancing the long-term carbon cycle.

MECHANISM OF DISSOLUTION

Armoring surface layers

Several models for inosilicate dissolution have been presented. Luce et al. (1972) noted that enstatite dissolution proceeded incongruently, and hypothesized that a layer of non-enstatite composition remained on the surface of the mineral. Several models for this layer have been debated in the literature for Fe-Mg-Ca inosilicate dissolution. For example, Luce et al. (1972) assumed layers consisting of Fe-Mg-Ca-depleted but proton-enriched silicate residual material limited transport at the surface and caused parabolic kinetics. Other workers posited a surface layer of reprecipitated Fe-Mg silicate of a different structure than the host mineral (see Helgeson, 1971; Eggleton and Boland,

1982), or an iron oxide precipitate layer (Siever and Woodford, 1979). For these models, the layer was assumed to be "protective" or "armoring", meaning that it limited transport of products or reactants. Sanemasa and Katsura (1973) argued against a diffusion-limited surface layer on diopside dissolved at pH 0 to 1, although they observed some parabolic dissolution behavior. The work of Schott et al. (1981) and Schott and Berner (1985), in which powders cleaned of ultrafines showed linear dissolution behavior, cast doubt on the observation of parabolic kinetics for Fe-free minerals or for Fe-containing minerals under anoxic conditions, leading to the general acceptance of surface-limited rather than diffusion-limited models for dissolution. However, Schott and Berner (1985) reported that dissolution of bronzite, despite careful sample pre-treatment, remains parabolic under oxic conditions (Fig. 1). Since the work of Grandstaff (1977) and Schott and coworkers, no studies of bronzite dissolution have been accomplished. However, Zhang (1990) reported that hornblende dissolution also showed nonlinear dependence on time for dissolution in the presence of oxygen. Therefore, the dissolution of Fe-containing minerals in the presence of oxygen may show parabolic dissolution behavior.

Interface-controlled models

Steady state leached layers. After the work of Schott et al. (1981), however, most workers concluded that the observation of high activation energies, nonlinear dependence upon H^+ activity, independence of rate with respect to stirring speed and flow rate, and observations of ubiquitous etch pits on the surface of dissolved inosilicates indicates that Fe-free inosilicate dissolution is interface-limited rather than transport-limited in laboratory experiments. The many measurements of nonstoichiometric dissolution and leached surface layers on laboratory-dissolved inosilicates suggests that dissolution is characterized by a leached layer of steady state thickness controlled by the relative rates of cation diffusion across leached layers and the rate of silica hydrolysis.

One assumption implicit to the hypothesis of a steady state leached layer on inosilicates is that dissolution, while nonstoichiometric initially, must eventually become stoichiometric. However, as described earlier, not all inosilicates investigated have been observed to dissolve stoichiometrically. Berner and Schott (1982) argued that the observation of thin leached layers based upon interpretations of XPS measurements on naturally weathered augite, diopside, hypersthene, bronzite, and hornblende indicated that the cation leaching was limited and that stoichiometric dissolution occurred for these minerals. However, Mogk and Locke (1988) used their AES evidence for a thick (120 nm) partially cation-depleted layer on naturally weathered hornblende to argue that the extent of leaching might be time-dependent—in other words, at least for the case of hornblende, leached layers may not have reached steady state thickness even after 100,000 y. They pointed out that a strict surface-controlled mechanism cannot explain such deep leaching, and argued that volume diffusion must also be occurring, at least on naturally weathered hornblende.

Helgeson model. Despite the possible contribution of deep diffusional processes, most researchers have modelled the effect of aqueous species on the rates of dissolution by the assumption of surface complexation models which do not include cation leaching (Grandstaff, 1977; Lasaga, 1981; Schott and Berner, 1985; Stumm et al., 1985; Furrer and Stumm, 1986). For example, Murphy and Helgeson (1987, 1989) argued that experimental data (Grandstaff, 1977; Schott et al., 1981; Schott and Berner, 1983) are consistent with a model where the exchange reaction of H^+ for $M2$ cations on the pyroxene surface goes to completion in solutions which are undersaturated with respect to the dissolved mineral. For such a model, the fraction of exchanged species is close to unity and is independent of pH. These workers argued that the rate of release of the $M2$ cation or Si from dissolving pyroxene is constant during the initial transient period of cation surface exchange, and concluded from this that the activated complex for the detachment reaction formed at both unexchanged and exchanged surface sites to form two activated complexes. They adduced the following rate law:

$$R = k\gamma \left[\frac{a_H K_a}{1 + a_H K_a} \right]^{0.5}$$

(14)

which, at low values of $a_H K_a$, yields $n = -0.5$, or at very low pH yields $n = 0$ if fit to Equation (11). Although this model accurately fit the pyroxene dissolution data of Schott et al. (1981), the more complete dataset collected in experiments of longer duration (Table 4) suggests that most Fe-free pyroxenes exhibit a value of n of about -0.2 to -0.3, contradicting Equation (14). On the other hand, higher values of n have been reported for Fe- and Al-containing silicates, although for short duration experiments.

Rimstidt and Dove model for wollastonite. Rimstidt and Dove (1986) analyzed their data for wollastonite dissolution by suggesting two possible models. The first model for wollastonite dissolution assumed an empirical rate equation such as Equation (11) (Table 4). They argued, however, that a second model better explained all of their experimental data. They suggested that the uptake of H^+ by the reaction,

$$CaSiO_3 + H_2O + H^+ \rightarrow Ca^{2+} + H_4SiO_4$$

(15)

in short-term dissolution experiments for wollastonite (< 20 h) can be modelled as an interaction between a hydrogen ion and a bridging oxygen in the wollastonite structure. They suggested that, if each exposed bridging oxygen represents an adsorption site, and if one hydrogen adsorbs per site, then the reaction can be modelled as directly proportional to the number of occupied sites:

$$R = k_{ads}\theta$$

(16)

where k_{ads} is the rate constant for this adsorption model and θ is the fraction of sites occupied by protons. Expressing θ as a Langmuir adsorption isotherm, and assuming competition between adsorption of the proton and the Ca^{2+} ion, they fit the following model to their dissolution data:

$$R = \frac{k_{ads}K_H[H^+]}{1 + K_H[H^+] + K_{Ca}[Ca^{2+}]}$$

(17)

where K_H and K_{Ca} are the equilibrium constants for proton and Ca adsorption respectively, at the bridging oxygen site. For their model fit, they cited $K_H = 2.08 \times 10^5$ mol^{-1}, $K_{Ca} = 5.52$ mol^{-1}, and $k_{ads} = 9.80 \times 10^{-12}$ mol cm^{-2} s^{-1}.

Stumm's surface protonation model. Perhaps the currently most popular model for silicate dissolution is the model of Furrer and Stumm (1986) developed for simple oxide dissolution. These workers argued that inner sphere surface complexes on an oxide-water interface polarize and thus weaken metal-oxygen bonds, enhancing hydrolysis of the metals. Furrer and Stumm (1986) argued that the rate of dissolution is proportional to the

mth power of the concentration of surface complexes, where m equals the valence of the metal ion (see Casey, this volume). The rationale for the model is that m represents the number of protonated bonds in the detachment reaction. Stumm et al. (1985) argued that the rate-limiting step is the reaction of the protonated surface species with an H_2O molecule.

Wieland et al. (1988) further elaborated the model by proposing the following rate equation for acid- and ligand-promoted dissolution of minerals:

$$R = kx_a P_j S \qquad (18)$$

where k is the rate constant (s^{-1}), x_a is the mole fraction of dissolution active sites (kinks, edges, etc.), P_j is the probability that the site is coordinated as a precursor complex, and S is the surface site density (mol cm^{-2}). The term, $x_a P_j S$, describes the surface site density of the precursor of the activated complex, which is assumed to be in local equilibrium with the activated complex. For example, if the precursor species is a surface hydroxyl group, $\equiv MOH$, which is protonated, then the following equilibrium, described by equilibrium constant $K_{\neq}$, is assumed:

$$\equiv MOH_2^+ \Leftrightarrow \equiv (MOH_2^+)^{\neq} \qquad (19)$$

where $(\equiv MOH_2^+)^{\neq}$ is the activated complex. The dissolution rate is assumed proportional to the activity of the activated complex:

$$R = k'\{\equiv MOH_2^{+\neq}\} = k'K_{\neq}\{\equiv MOH_2^+\} \qquad (20)$$

and k', the rate constant, equals the product of a frequency and an activity coefficient term. For ligand-promoted dissolution, the proton is replaced by a ligand in the appropriate equations.

To determine $\{\equiv MOH_2^+\}$, Wieland et al. (1988) reported the following master equation, based upon a Frumkin-Fowler-Guggenheim isotherm, for concentration of protonated surface sites on a simple oxide:

$$\{\equiv MOH_2^+\} = \frac{K_F}{[H^+]_{pznpc}^m}[H^+]^m \qquad (21)$$

where $pK_F = -6.51$ and $m = 0.21$. Extension of this equation to multiple oxides such as the inosilicates has not been yet attempted.

In a different approach to estimating $\{\equiv MOH_2^+\}$, Furrer and Stumm (1986) argued that a sufficient number j of oxide or hydroxide groups neighboring the central metal ion at the surface must be protonated in order to form the surface precursor for proton-promoted dissolution. Wieland et al. (1988) used lattice statistics to derive the following relationship:

$$P_j = \frac{4!}{j!(4-j)!}x_H^j \qquad (22)$$

where j is the number of nearest neighbor oxygens which are protonated, x_H is the mole fraction of protonated sites and the first term in the expression describes the number of possible geometric arrangements of j protons on four neighboring sites. The proton-promoted rate (Eqn. 18) is expressed:

$$R_H = k\left(\frac{4!}{j!(4-j)!}\right)x_H^j x_a S \tag{23}$$

Stumm et al. (1985) argued that j was a function of the valency of the metal cation: for example, j equals 3 for alumina and 2 for α-FeOOH. This model suggests that for any oxide, the values of j, m and n are therefore related.

Surface charging of inosilicates. According to the model of Furrer and Stumm (1986), dissolution of an oxide should be lowest at its pH_{pznpc} and should increase at lower and higher pH. Brady and Walther (1989, 1992) therefore predicted that the pH-dependence of aluminosilicate mineral dissolution should increase in alkaline solution where dissolution is controlled by adsorption on exposed silica tetrahedra, and increase below pH 5, where adsorption of protons on non-silica oxide components dominates the surface.

However, the prediction of alkaline acceleration of dissolution is specifically for pH > pH_{pznpc}, and very few experimental measurements of this parameter have been made for the inosilicates. Parks (1967) noted that the pH_{pznpc} of a mixed oxide such as an inosilicate can be calculated as a weighted summation of the values of the pH_{pznpc} of the simple oxides: 12.5 (MgO), 12.9 (CaO), 9.3 (Fe_2O_3), 11.8 (FeO), 6.8 ($Al_2^{IV}O_3$), 9.2 ($Al_2^{VI}O_3$), 1.8 (SiO_2). Calculated values of the pH_{pznpc} for each mineral are tabulated in Table 7, along with calculated values from Sverjensky (1994) using a newer technique incorporating the dielectric constant of the bulk.

Table 7. Values of pH_{pznpc} for selected silicates.

Inosilicate mineral	Parks (1967) method	Sverjensky (1994)
enstatite	7.15	
diopside	7.25	7.3
hedenbergite	7.08	5.9
wollastonite	7.25	7.0
spodumene	4.3	
anthophyllite	6.79	6.6
tremolite	6.84	7.0

A complication with respect to using calculated values of the pH_{pznpc} for inosilicates is that placement of inosilicates in water immediately affects the surface composition of the mineral: for example, Xie and Walther (1994) reported that exchange of Ca^{2+} for H^+ on the wollastonite surface occurred in acid solutions within minutes of immersion. These workers argued that the surface protonation properties of the acid-reacted wollastonite are similar to microporous silica, rather than to stoichiometric wollastonite. However, the pH of the point of zero salt effect (pH_{pzse}) of wollastonite was observed to be approximately 4.5 to 5.5 during titration. Although this value is much lower than the theoretical pH_{zpc} for wollastonite (Table 7), it is higher than the value of the pH_{pznpc} predicted for a microporous silica (3.0, Tadros and Lyklema, 1968). Xie and Walther (1994) documented that measurable adsorption of alkali and alkaline earth cations occurs on the acid-reacted wollastonite surface, further complicating the interpretation of the point of zero net proton charge. Repolymerization of the leached silica surface must also affect the surface charging properties. Casey et al. (1993) have documented a strong case for repolymerization of the wollastonite and rhodonite surface during dissolution even at neutral pH.

Alkaline dissolution of inosilicates. As indicated by Table 7, most values of the pH_{pznpc} for inosilicates are within several log units of neutral, suggesting that alkaline enhancement of dissolution should be observed above pH 7 for inosilicates, as argued by Brady and Walther (1989, 1992). Although more rate data and a better understanding of the pH_{pznpc} is needed in order to make unequivocal conclusions, our critical summary of the literature suggests no trough in the rate-pH curve for the Fe- and Al-free inosilicates, such as is characteristic of aluminosilicates such as feldspar and kaolinite (e.g. Brady and Walther, 1992). Rather, for those inosilicates where data exists, the rate has been observed to decrease steadily from pH 2 to 12, with no alkaline acceleration of dissolution (Mast and Drever, 1987; Knauss et al., 1993). The rates of diopside dissolution at all temperatures from 25 to 70°C decrease from pH 1 to 10, and perhaps are constant between pH 10 and 12 (Knauss et al. ,1993; Chen and Brantley, in prep.).

However, Knauss et al. (1993) noted that inhibition of dissolution under alkaline conditions may also occur, due to adsorption of CO_2 as surface complexes. Such a mechanism has been postulated for Mg-silicates to explain CO_2-inhibition of olivine dissolution (Wogelius and Walther, 1991). Xie (1994) reported a significant inhibition of dissolution of wollastonite under alkaline pH in the presence of CO_2 and postulated a bidentate CO_3^{2-} complex with two calcium ions at the mineral surface. We have pointed out, however, that the experiments reported by Xie (1994), while intriguing, are inconclusive, given the other experimental variables which were changed in variation between a CO_2-containing and a CO_2-free environment. Whether inosilicates show an alkaline enhancement of dissolution in CO_2-free solutions is still an open question and more experiments such as those described by Xie (1994) are needed.

Point of minimum dissolution rate. If we define the pH_{pmdr}, the pH at the point of minimum dissolution rate, then we conclude that all of the inosilicate minerals for which there is data dissolve in a manner such that the $pH_{pmdr} > pH_{pznpc}$ (compare Figs. 2 and 3, Table 7). Exactly the same conclusion has been reached for feldspar dissolution (Stillings et al., 1995), where the $pH_{pznpc} << pH_{pmdr} \approx 7$ to 8. Several conclusions might be inferred from these observations: (1) the simple proton-promoted model does not work for mixed oxide silicates; (2) extensive surface leaching, cation readsorption, and surface (de)polymeri-zation reactions change the silicate surface charging properties such that $pH_{pmdr} > pH_{pznpc}$; (3) surface properites of one component oxide of the mixed oxide dominate the pH_{pznpc}. Without more theoretical and experimental work, however, no firm conclusions can be drawn concerning the mechanism of inosilicate dissolution.

NATURAL INOSILICATE WEATHERING RATES

Quantitative evaluation of etching rates

Several workers have attempted quantitative estimates of inosilicate dissolution rates by using mass balance arguments for monitored field settings (see White, this volume; Drever, this volume). In general, dissolution rates estimated in the field are reported to be

10 to 10⁴ times slower than laboratory rates. In this section we summarize another type of dissolution rate estimate for natural weathering based upon etch pits on hornblende.

Mean maximum etch depth. The extent of etching of hornblende has been used as a criterion to assess extent of weathering, because of the increase in etching with soil age (Fig. 6). The use of hornblende is also due to its ease of identification, its ubiquity in soils derived from crystalline rocks, its instability, and its characteristic weathering pattern of "cockscomb" soil grain terminations (Locke, 1986). These etched grain terminations presumably form from coalescence of lens-shaped etch pits parallel to the *c* axis (Berner et al., 1980b; Berner and Schott, 1982). In the first published quantitative study, Locke (1979) concluded that development of cockscomb terminations on hornblende in Arctic soils correlates with relative age and paleoclimate. He examined at least 100 grains per sample and classed each by the mean maximum etching depth (MMED). He observed that the MMED for hornblende from fresh, unweathered till shows a minimum value, while in soils 3200 y or older, the MMED is larger, and the etch extent at any depth increases with age in a nonsmooth function. The *rate* of etching penetration decreases exponentially with time (Fig. 13). In addition, the etching decreases logarithmically with increasing depth (Fig. 13). In a comparison of hornblende etching observed in several soils, Locke (1986) observed differences in etching rates among relatively nearby soils, which he could attribute to differences in the frequency of wetting events.

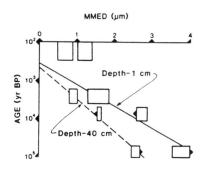

Figure. 13. MMED (maximum mean etching depth (see text)) of hornblende grains plotted vs. soil age for 20 soils of the Cape Dyer area of Cumberland Peninsula, Baffin Island, N.W.T. Canada. [Used by permission of the editor of *Rates of Chemical Weathering of Rocks and Minerals,* from Locke (1986), Fig. 8, p. 139]. Rectanges show statistical uncertainty in etching estimate (1 standard error of the estimate) and qualitative estimate of the dating error. Lines drawn through data for etching measured at 1 and 40 cm depth in the soil represent constant logarithmic trends excluding the late Neoglacial data. Corrections to a constant, modern climate are indicated by arrowhead directions. Note that, although the etching at any given depth in the soil increases with age, the rate of etching decreases with increasing age.

Hall and Martin (1986) reported many of the same observations for hornblende etching in tills from Montana, USA, including evidence for accelerated etching in soils formed in basins with heavier precipitation. Most of their evidence also showed that etching decreases logarithmically with age, occurring most rapidly in the first 7500 y, but continuing through 500,000 y. Hall and Horn (1993) reported that etching rates for four chronosequences in Wyoming and Montana (USA) determined at 12,000 y after deposition showed that etching is slowest in coarse-grained granitic material where mean annual precipitation is low, and significantly faster where precipitation was higher and the parent material more fine-textured. Within individual chronosequences, hornblende deposits at higher elevations show accelerated etching rates which the authors attributed to higher orographic precipitation.

In order to calculate an area-normalized dissolution rate, R (mol cm⁻² s⁻¹), for this etching rate data, a simple normalization can be completed:

$$R_{pit} = EV^{-1} \tag{24}$$

where E is the etching rate (cm s^{-1}) and V is the molar volume of hornblende (cm^3 mol^{-1}). For this transformation, the measured weathering rates of Locke (1986) varied between 2.6×10^{-19} mol cm^{-2} s^{-1} for the dissolution of hornblende between 800 and 75,000 y at 100 cm depth to 9.4×10^{-19} mol cm^{-2} s^{-1} between 330 and 3200 y at 1 cm depth. However, Equation (24) does not take into account the fact that, as the pit gets deeper, more material is released to solution. For example, if we model cockscomb terminations as etch pits with an assumed pit geometry of a right pyramid, then the volume of the pit can be expressed as

$$V_{pit} = \frac{4a^3 \tan^2 \theta}{3}$$ (25)

where a is the measured depth of the pit (MMED) and θ is the angle subtended by the pit wall and the perpendicular to the mineral surface. The assumption of square pit cross-sections *over*estimates etching rate. The dissolution rate can be expressed:

$$R_{pit} = \frac{4a^2 \tan^2 \theta}{V} Ep$$ (26)

where p is the density of pits (number cm^{-2}). If we assume that $\theta = 45°$, and $V = 1.69 \times 10^{-4}$ m^3 mol^{-1} (Brantley et al., 1993), then R is defined if we know p. No pit densities are reported by Locke (1986), so a value for this parameter must be assumed. If we assume etch pit densities comparable to those observed by Cremeens et al. (1992) for etched hornblende from a 12,500 y soil in Illinois (4 to 8×10^7 pits cm^{-2}), then we calculate rates between 3×10^{-19} and 2×10^{-17} mol cm^{-2} s^{-1}, for etch pit rates and values of MMED cited in Locke (1986, his Tables 2 and 3) for the Cape Dyer soils. These rates are still slower than most of the rates measured in the laboratory for hornblende dissolution (Table 1), or even in plot-scale weathering experiments (1.8×10^{-16} mol mineral cm^{-2} s^{-1}, Swoboda-Colberg and Drever, 1993). However, these rates are strictly due to dissolution at etch pits, and do not include the contribution of dissolution at other sites (corners, edges) on the grain.

Pit size distributions. Similar rates were calculated by MacInnis and Brantley (1993) and Brantley et al. (1993), using the etch pit size distribution (PSD) model to analyze the etching rate of hornblende grains weathered in a soil catena in loess (age = 12,500 y; Cremeens et al., 1992). The PSD model assumes a population balance of etch pits on the surface of a dissolving soil grain, maintained by steady state nucleation and annihilation of pits (see White, this volume). In this model, the etching rate of the mineral surface can be determined from the pit growth (or opening) rate, G, and the pit wall slope, θ, the mean lifetime of the pit, τ, and the nucleation density (density of nucleated pits) at time t_o, n^o:

$$R_{pit} = \frac{n^o \tan\theta}{V} G^4 \tau^3$$ (27)

Here, R_{pit} is the contribution from dissolution of etch pits to the overall dissolution rate, assuming a pit geometry of a right pyramid. The model is useful for young soils or where soil grains have experienced minimal etching. The values of G and n^o can be derived from plots of log n vs. W (Fig. 14), where n is the number of etch pits of a given size per unit area of surface, and W is the pit diameter (Brantley, et al., 1993). The value of τ must be assumed: a minimum value of R_{pit} can be determined by assuming that τ equals the age of the soil, as discussed in MacInnis and Brantley (1993). Brantley et al. (1993) reported minimum etching rates for hornblende between 6 and 9×10^{-19} mol cm^{-2} s^{-1}, comparable to rates calculated above for the data of Locke (1986).

hornblende

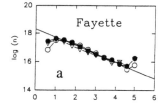

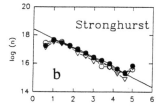

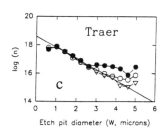

Figure 14. Pit size distribution plots for three horizons of three soils comprising a soil catena (after Brantley et al., 1993). Line in each case represents regression of data for each soil for the top two horizons (some data points are not included in the regression). Abscissa represents etch pit size in microns: ordinate represents log n, where n is the number of etch pits per size range per m^2 area. (a) Fayette (well-drained) soil, (b) Stronghurst (somewhat poorly drained soil), (c) Traer (poorly drained soil). In every diagram, the open circle is the uppermost soil horizon (E or BE), the solid circle is the middle horizon (Bt or Btg), and the triangle is the lowermost horizon (C or Cg) of the sampled horizons (see Brantley et al., 1993).

In the Illinois catena, decreasing soil drainage correlates with an increase in pit nucleation density, $n°$, while decreased drainage correlates with a decrease in pit growth rate, G. Overall etching rates of hornblende were also observed to decrease with decreasing drainage. Brantley et al. (1993) attributed this decrease to the increased solute concentration which might be expected in soil porewaters of poorly drained soils. However, the presence of oxygen in the well drained soils may also enhance the rate of dissolution of hornblende, based on laboratory experiments of White and Yee (1985) discussed earlier. One problem with the use of the PSD model for the poorly drained soil (Fig. 14), however, is the presence of a superabundance of large etch pits. These large etch pits, presumably present due to enhanced coalescence of pits, do not fit a linear PSD model. Because Brantley et al. (1993) simply applied a linear PSD model, they may have underestimated etching in the poorly drained soil where pit coalescence operated more extensively.

Rate control in natural systems

The common observation of highly weathered "boxworks" (angular shells of silicates) as well as etch pits on naturally weathered minerals has been used as an argument for the interface-limited nature of natural dissolution (e.g. Berner et al., 1980b; Berner and Schott, 1982). The presence of etch pits argues against a dissolution mechanism operating in natural systems which is rate-limited by transport across a leached layer. For neutral pH solutions without organic complexing agents, however, dissolution of Fe-, Ti-, and Al-containing inosilicates must be accompanied by nearby precipitation of secondary minerals containing these components. For example, Berner and Schott (1982) reported that microcracks and -caves on the surface of several naturally weathered pyroxenes and amphiboles were filled by clay precipitates. However, XPS analysis of mineral surfaces after ultrasonication, which easily removed the clays, revealed compositions similar to grain interiors, and these authors concluded that weathering did not produce an armoring altered layer. The small depletion in Ca and Mg observed was similar to that observed on laboratory-dissolved samples, and they concluded that prolonged weathering in the field did not produce "appreciably thick..." protective surface layers, suggesting that the high reactivity of such layers limited their development. Although Mogk and Lock (1988) showed that cation-depleted layers on naturally weathered hornblende were much thicker than originally envisioned by Berner and Schott (1982), they agreed that their Auger electron spectroscopic evidence showed no evidence for a layer of any newly nucleated phase on the soil grain surface. The Pilling Bedworth rule from corrosion theory has also been used to argue that the volume of precipitated products is too small, even in the case of formation of ferruginous boxworks, to limit the interface reaction (Velbel, 1993).

Although these workers have concluded that surface precipitates are not rate-limiting for weathering of individual soil grains, other workers have concluded that much secondary alteration and mineral precipitation occurs while minerals are still present in bedrock or saprolite, where transport-limited reaction is possible. For example, Veblen and Buseck (1980, 1981), Nakajima and Ribbe (1980), and Eggleton and Boland (1982) reported evidence that orthopyroxene weathered naturally often shows lattice coherence between pyroxene and layer silicates. Banfield and Barker (1994) noted that topotactically oriented alteration products on amphiboles vary on scales of tens of nanometers, often following parent mineral structures. Eggleton (1986) argued for the intimate relationship between dissolution at initial diffusion avenues which allow access of water into the mineral, followed by precipitation or alteration to secondary products which block diffusion channels and subsequently cause slow weathering. Casey et al. (1993) have even argued that formation of gel-like surfaces on pyroxenes may provide a mechanism for topotactic conversion of inosilicates to phyllosilicates. According to this mechanism, silicic acid would not form as an aqueous component before it became incorporated into an alteration layer on or within the mineral. These workers have argued that the large water-solid ratio in laboratory chemical reactors may preclude topotactic conversion, perhaps explaining the differences in dissolution rates observed between laboratory and field.

A careful investigation of this hypothesis was recently reported by Banfield et al. (1995). These workers concluded that alteration of $MnSiO_3$ in saprolite or bedrock began at grain boundaries and planar defects parallel to the silicate chains. The alteration produced elongate etch pits which were sometimes partially filled by smectite oriented such that the 2:1 layers of the clay were parallel to the chains in the inosilicate. These observations are similar to those of earlier workers for other phases (Eggleton, 1975; Nahon and Colin, 1982; Eggleton and Boland, 1982; Banfield et al., 1991). Smectite contained considerably more Zn than the altering rhodonite/pyroxmangite starting material: the authors concluded that nearby alteration of sphalerite introduced Zn into the system, while Mn was lost. They reported, however, no evidence of leached layers or amorphous products observed on naturally weathered surfaces, in contrast to the incongruent dissolution and restructuring of product layers observed at the surface of dissolving $MnSiO_3$ in the laboratory. Although these workers reported no evidence for the presence of amorphous weathering products, other workers have documented amorphous products associated with inosilicate weathering (Eggleton, 1975; Eggleton and Boland, 1982; Nahon and Colin, 1982; Singh and Gilkes, 1993). The relative importance of topotactic conversion of inosilicates as compared to complete dissolution without local reprecipitation of secondary products in natural systems is still not well understood, making quantitative predictions of natural rates difficult.

CONCLUSIONS

Considering the relative importance of inosilicate weathering toward release of trace metals to natural systems (e.g. Huang, 1977), and toward balancing the long-term carbon cycle (e.g. Brady, 1991), it is surprising that the rates and mechanisms of dissolution of these phases are still so poorly understood. Our understanding of the weathering of redox-sensitive silicates such as augite and hornblende is particularly deficient. However, a few conclusions may be drawn from this critical survey of the literature.

Long-term dissolution (hundreds to thousands of hours) of most silicates yields rates which are constant in time—so-called linear kinetics. Linear kinetics have never been observed, however, for most Fe-containing inosilicates in the presence of oxygen. The slope of the log rate-pH curve for most inosilicates is shallow at subneutral pH (-0.1 to -0.3); however, there is limited evidence that inosilicates containing iron, or, especially, aluminum, show steeper slopes (-0.7). Most workers have concluded that the pH dependence decreases above about pH 7, and little evidence exists to support an alkaline acceleration of dissolution (although the effects of CO_2 on alkaline dissolution need to be further investigated). Most workers have concluded that rates are not strongly dependent on solution chemistry, although intriguing evidence suggests that rates may be affected by some organics and by varying P_{CO2}.

The surface area of some inosilicates—notably the pyroxenoids—increase drastically during dissolution, while that of the pyroxenes and amphiboles apparently increase by at most a factor of 2 to 3. The linear dependence between dissolution rate and BET surface area for inosilicates remains to be unequivocally documented. BET surface areas of hornblende weathered naturally are a factor of 2 larger than those weathered at laboratory time-scales. Increases in surface area during dissolution are attributed to etch pits, micropores, and (in laboratory samples) production of cation-poor silica surface layers.

Many workers have shown that cations in the *M1, M2, M3* and *M 4* sites are selectively removed from the inosilicate surface during laboratory dissolution under subneutral conditions. Extent of cation leaching increases with decreasing pH and the *M2* and *M4* sites are particularly prone to cation loss. Penetration of hydrogen into the surface has also been documented, although the process has been observed to be pH-independent. Pyroxenes and amphiboles weathered naturally also show similar surface leaching, perhaps to greater depths than observed in the laboratory. Intriguing evidence exists to suggest that dissolution of hornblende in natural systems may not reach steady state with respect to surface chemistry even after thousands of years.

No model for the rate and mechanism of inosilicate dissolution has been generally accepted. Most models use surface complexation theory to relate dissolution rates to the surface density of protonated sites. Although deep depletion of cations and penetration of hydrogen into some inosilicate surfaces during dissolution have been documented, no model incorporates both surface complexation and cation depletion.

The relationship between dissolution rate and the rate constant for sovent exchange (k_{KOH}) which has been used successfully to correlate dissolution rates of orthosilicates

(Casey, this volume), is less successful when used to predict rates of dissolution of the more slowly dissolving inosilicates. However, the log rate is a strong function of the connectedness, c, of the silicate ($\log R \propto c^{-2}$). The correlation between R and k_{KOH} and c should eventually yield strong predictive tools for inosilicate dissolution in inorganic solutions. More data is needed from long-term experiments to quantify these relationships.

Quantitative evaluation of weathering in natural systems generally yields rates which are lower than those estimated in the laboratory. Measurement of etch pits on hornblende represent a unique opportunity to quantify natural weathering using a method which is distinct from the normal reliance on assessment of mass balance at the soil or watershed scale (White, this volume; Drever and Clow, this volume). Slow rates estimated for soil grains are probably not related to inorganic precipitates of clays or iron phases, which several workers have argued are non-armoring. However, dissolution in the deeper saprolite and bedrock environment may lead to mechanisms of alteration and topotactic conversion which are completely distinct from dissolution as it occurs in chemical reactors with high water-mineral ratios.

The extreme difficulty of extrapolation from laboratory to field necessitates a better understanding of mineral-water interfacial area and water chemistries in weathering environments, among other variables. The common presence of Fe^{2+} in inosilicate minerals also necessitates an understanding of P_{o2} or other redox parameters on the grain scale in soils and saprolites. Limited experiments on the redox reactions of Fe-containing inosilicates suggest that coupled oxidation-reduction reactions occur at the mineral surface. These coupled reactions oxidize iron deep in the silicate structure, beneath the surface oxidized layer. The ability of organisms to catalyze such redox reactions also suggests that a better understanding of bacteria-inosilicate reactions will be necessary in order to further investigate field settings.

The high measured values of apparent activation energies of dissolution of Mg-silicates (≈ 80 kJ/mol) and Ca-silicates (80 to 120 kJ/mol) suggests a strong temperature dependence for the rate of CO_2 drawdown due to silicate weathering on 10^5 to 10^6 y timescales. A better understanding of this temperature sensitivity is necessary in order to accurately parameterize and model earth's climate throughout the geologic record.

ACKNOWLEDGMENTS

SLB acknowledges the hospitality of Art White and the US Geological Survey office in Menlo Park, where most of this chapter was written. This work was funded by NSF EAR-9305141 and DOE grant DE-FG02-95ER14547.

REFERENCES

Anbeek C, Bloom PR, Nater EA, Zhang H (1994) Comment on "Change in surface area and dissolution rates during hornblende dissolution at pH 4.0" by H. Zhang, P.R. Bloom, and E.A. Nater. Geochim Cosmochim Acta 58:4601-4613

Bailey A, Reesman AL (1971) A survey study of the kinetics of wollastonite dissolution in H_2O-CO_2 and buffered system at 25 degrees C. Am J Sci 271:464-472

Bailey A (1974) Effects of temperature on the reaction of silicates with aqueous solutions in the low temperature range. Proc 2nd Int'l Symp on Water-Rock Interaction, Prague, p 375-380.

Bales RC, Morgan JJ (1985) Dissolution kinetics of chrysotile at pH 7 to 10. Geochim Cosmochim Acta 49:2281-2288

Banfield JF, Jones BF, Veblen DR (1991) An AEM-TEM study of weathering and diagenesis, Abert Lake, Oregon; I, Weathering reactions in the volcanics. Geochim Cosmochim Acta 55:2781-2793

Banfield JF, Barker WW (1994) Direct observation of reactant-product interfaces formed in natural weathering of exsolved, defective amphibole to smectite: Evidence for episodic, isovolumetric reactions involving structural inheritance. Geochim Cosmochim Acta 58:1419-1429

Banfield JF, Ferruzzi GG, Casey WH, Westrich HR (1995) HRTEM study comparing naturally and experimentally weathered pyroxenoids. Geochim Cosmochim Acta 59:19-31

Barman AK, Varadachari C, Ghosh K (1992) Weathering silicate minerals by organic acids: I, Nature of cation solubilisation. Geoderma 53:45-63

Berner RA, Schott J (1982) Mechanism of pyroxene and amphibole weathering; II. Observations of soil grains. Am J Sci 282:1214-1231

Berner, RA, Sjoberg, EL, Schott, J (1980a) Mechanism of pyroxene and amphibole weathering; 1, Experimental studies. Proc 3rd Int'l Symp Water-Rock Interaction, Edmonton, Alberta, A Campbell (ed) p 44-45.

Berner RA, Sjoberg EL, Velbel MA, Krom MD (1980b) Dissolution of pyroxene and amphiboles during weathering. Science 207:1205-1206

Bloom, PR, Nater, EA, Zhang, H (1994) Reply to the Comment by C. Anbeek on "Change in surface area and dissolution rates during hornblende dissolution at pH 4.0". Geochim. Cosmochim. Acta 58:1851.

Brady PV, Walther JV (1989) Controls on silicate dissolution rates in natural and basic pH solutions at 25 degrees C. Geochim Cosmochim Acta 53:2823-2830

Brady PV (1991) The effect of silicate weathering on global temperature and atmospheric CO_2. Journal of Geophysical Research B, Solid Earth and Planets 96:18101-18106

Brady PV, Walther JV (1992) Surface chemistry and silicate dissolution at elevated temperatures. Am J Sci 292:639-658

Brady PV, Carroll SA (1994) Direct effects of CO_2 and temperature on silicate weathering: Possible implications for climate control. Geochim Cosmochim Acta 58:1853-1856

Brantley SL, Blai A, Cremeens DL, MacInnis I, Darmody RG (1993 Natural etching rates of feldspar and hornblende. Aquatic Sciences 55:262-272

Casey WH, Sposito G (1992) On the temperature dependence of mineral dissolution rates. Geochim Cosmochim Acta 56:3825-3830

Casey WH, Westrich HR (1992) Control of dissolution rates of orthosilicate minerals by divalent metal-oxygen bonds. Nature 355:157-159

Casey WH, Westrich HR, Banfield JF, Ferruzzi G, Arnold GW (1993) Leaching and reconstruction at the surface of dissolving chain-silicate minerals. Nature 366:253-256

Chen Y, Brantley SL (1995) Temperature and pH dependence of albite dissolution rates at subneutral pH. Chemical Geology, submitted

Choi I, Malghan SG, Smith RW (1974) The dissolution kinetics of fibrous amphibole minerals in water. Proc Int'l Symp Water-Rock Interaction, Prague, Cadek J, Paces T (eds) p 36

Chou L, Wollast R (1985) Steady-state kinetics and dissolution mechanisms of albite. Am J Sci 185:963-993

Cremeens DL, Darmody RG, Norton LD (1992) Etch-pit size and shape distribution on orthoclase and pyriboles in a loess catena. Geochim Cosmochim Acta 56:3423-3434

Cygan RT, Casey WH, Roslough MB, Westrich HR, Carr MJ, Holdren GR (1989) Dissolution kinetics of experimentally shocked silicate minerals. Chemical Geology 78:229-244

Eggleston CM, Hochella Jr MF, Parks GA (1989) Sample preparation and aging effects on the dissolution rate and surface composition of diopside. Geochim Cosmochim Acta 53:797-804

Eggleton RA (1975) Nontronite topotaxial after hedenbergite. Am Mineral 60:1063-1068

Eggleton RA, Boland JN (1982) Weathering of enstatite to talc through a sequence of transitional phases. Clays Clay Minerals 30:11-20

Eggleton RA (1986) The relation between crystal structure and silicate weathering rates. In Rates of chemical weathering of rocks and minerals, Colman SN, Dethier DP (eds) p 21-40

Ferruzzi GG (1993) The character and rates of dissolution of pyroxenes and pyroxenoids. MS thesis, Univ California, Davis, CA

Furrer G, Stumm W (1986) The coordination chemistry of weathering; I. Dissolution kinetics of delta-Al_2O_3 and BeO. Geochim Cosmochim Acta 50:1847-1860

Goldich SS (1938) A study in rock weathering. J Geol 46:17-58

Gonzalez Bonmati JG, Vera Gomez MP, Garcia Hernandez JE (1985) Kinetic study of the experimental weathering of augite at different temperatures. Volcanic Soils, Catena Supp 7:47-61

Grandstaff DE (1977) Some kinetics of bronzite orthopyroxene dissolution. Geochim Cosmochim Acta 41:1097-1103

Hall RD, Martin RF (1986) The etching of hornblende grains in the matrix of alpine tills and periglacial deposits. In Rates of chemical weathering of rocks and minerals. Colman SM, Dethier DP (eds) p 101-128

Hall RD, Horn LL (1993) Rates of hornblende etching in soils in glacial deposits of the northern Rocky Mountains (Wyoming-Montana, USA); influence of climate and characteristics of the parent material. Chemical Geology 105:17-29

Helgeson HC (1971) Kinetics of mass transfer among silicates and aqueous solutions. Geochim Cosmochim Acta 35:421-469

Hill CG Jr (1977) An Introduction to Chemical Engineering Kinetics and Reactor Design. John Wiley & Sons, New York, 594 p

Huang WH, Keller WD (1970) Dissolution of rock-forming silicate minerals in organic acids; simulated first-stage weathering of fresh mineral surfaces. Am Mineral 55:2076-2094

Huang PM (1977) Feldspars, olivines, pyroxenes, and amphiboles. In: Minerals in Soil Environments. Dixon JB, Weed SR, Kittrick JA, Milford MH, White JL (eds) p 553-602

Kharaka YK, Gunter WD, Aggarwal PK, Perkins EH, DeBraal JD (1988) SOLMINEQ.88: A computer program for geochemical modeling of water-rock interactions. US Geol Surv Water-Resources Invest Report 88-4227

Knauss KG, Nguyen SN, Weed HC (1993) Diopside dissolution kinetics as a function of pH, CO_2, temperature, and time. Geochim Cosmochim Acta 57:285-294

Laidler KJ (1987) Chemical Kinetics (3rd Edn). Harper & Row, New York

Lasaga AC (1981) Rate laws of chemical reactions. In: Kinetics of Geochemical Processes, Lasaga AC, Kirkpatrick RJ (eds) Rev Mineral 8:1-66

Liebau F (1985) Structural Chemistry of Silicates. Springer-Verlag, Berlin

Locke WW III (1979) Etching of hornblende grains in Arctic soils; an indicator of relative age and paleoclimate. Quat Res 11:197-212

Locke WW (1986) Rates of hornblende etching in soils on glacial deposits, Baffin Island, Canada. In: Rates of Chemical Weathering of Rocks and Minerals. Colman SM, Dethier DP (eds) 129-145

Luce RW, Bartlett WB, Parks GA (1972) Dissolution kinetics of magnesium silicates. Geochim Cosmochim Acta 36:35-50

MacInnis IN, Brantley SL (1993) Development of etch pits size distributions on dissolving materials. Chemical Geology 105:31-49

Mast MA, Drever JI (1987) The effect of oxalate on the dissolution rates of oligoclase and tremolite. Geochim Cosmochim Acta 51:2559-2568

McClelland JE (1950) The effect of time, temperature, and particle size on the release of bases from some common soil-forming minerals, Soil Sci Soc Am Proc 15:301-307

Mogk DW, Locke WW III (1988) Application of auger electron spectroscopy (AES) to naturally weathered hornblende. Geochim Cosmochim Acta 52:2537-2542

Murphy WM, Helgeson HC (1987) Thermodynamic and kinetic constraints on reaction rates among minerals and aqueous solutions; III, Activated complexes and the pH-dependence of the rates of feldspar, pyroxene, wollastonite, and olivine hydrolysis. Geochim Cosmochim Acta 51:3137-3153

Murphy WM, Helgeson HC (1989) Thermodynamic and kinetic constraints on reaction rates among minerals and aqueous solutions; IV. Retrieval parameters for the hydrolysis of pyroxene, wollastonite, olivine, andalusite, quartz, and nepheline. Am J Sci 289:17-101

Nahon DB, Colin F (1982) Chemical weathering of orthopyroxenes under lateritic conditions. Am J Sci 282:1232-1243

Nakajima Y, Ribbe PH (1980) Alteration of pyroxenes from Hokkaido, Japan to amphibole, clays, and other biopyriboles. Neues Jahrb Mineral Monatsh 6:258-268

Nickel E (1973) Experimental dissolution of light and heavy minerals in comparison with weathering and intrastratal solution. Contrib Sedimentol, Stability of heavy minerals 1:3-68

Parks GA (1967) Aqueous surface chemistry of oxides and complex oxide minerals. In: Equilibrium Concepts in Natural Water systems. Am Chem Soc Adv Chem Ser 67:121-160

Peck JA, Farnan I, Stebbins JF (1988) Disordering and the progress of hydration at the surface of diopside; a cross-polarisation MAS NMR study. Geochim Cosmochim Acta 52:3017-3021

Perry DL, Tsao L, Gangler KA (1983) Surface study of HF- and HF/H_2SO_4 treated feldspar using Auger electron spectroscopy. Geochim Cosmochim Acta 47:1289-1291

Petit JC, Della MG, Dran JC, Schott J, Berner RA (1987) Mechanism of diopside dissolution from hydrogen depth profiling. Nature 325:705-707

Petrovich R (1981) Kinetics of dissolution of mechanically comminuted rock-forming oxides and silicates - I. Deformation and dissolution of quartz under laboratory conditions. Geochim Cosmochim Acta 45:1665-1674

Posey-Dowty J, Crerar D, Hellmann R, Chang CD (1986) Kinetics of mineral-water reactions; theory, design and application of circulating hydrothermal equipment. Am Mineral 71:85-94

Rimstidt JD, Dove PM (1986) Mineral/solution reaction rates in a mixed flow reactor: wollastonite hydrolysis. Geochim Cosmochim Acta 50:2509-2516

Sanemasa I, Katsura T (1973) The dissolution of $CaMg(SiO_3)_2$ in acid solutions. Bull Chem Soc Japan 46:3416-3422

Schott J, Berner RA, Sjoberg EL (1981) Mechanism of pyroxene and amphibole weathering - I. Experimental studies of iron-free minerals. Geochim Cosmochim Acta 45:2123-2135

Schott J, Berner RA (1983) X-ray photoelectron studies of the mechanism of iron silicate dissolution during weathering. Geochim Cosmochim Acta 47:2233-2240

Schott J, Berner RA (1985) Dissolution mechanisms of pyroxenes and olivines during weathering. In: The Chemistry of Weathering. JI Drever (ed) NATO ASI Series C; Mathematical and Physical Sciences 149:35-53

Schott J, Petit JC (1987) New evidence for the mechanisms of dissolution of silicate minerals. In: Aquatic Surface Chemistry; Chemical Processes at the Particle-Water Interface. Stumm W (ed) Swiss Fed Inst Technol, Zürich, Switzerland, p 293-315

Siegel DI, Pfannkuch HO (1984) Silicate mineral dissolution at pH 4 and near standard temperature and pressure. Geochim Cosmochim Acta 48:197-201

Siever R, Woodford N (1979) Dissolution kinetics and the weathering of mafic minerals. Geochim Cosmochim Acta 43:717-724

Singh B, Gilkes RJ (1993) Weathering of spodumene to smectite in a lateritic environment. Clays Clay Minerals 41:624-630

Stillings LL, Brantley SL, Machesky ML (1995) Proton adsorption at an adularia feldspar surface. Geochim Cosmochim Acta 59:1473-1482

Stumm W, Furrer G, Wieland E, Zinder B (1985) The effects of complex-forming ligands on the dissolution of oxides and aluminosilicates. NATO ASI Series C; Mathematical and Physical Sciences 149:55-74

Sverdrup HU (1990) The Kinetics of Base Cation Release due to Chemical Weathering. Lund Univ Press

Sverjensky DA (1994) Zero-point-of-charge prediction from crystal chemistry and solvation theory. Geochim Cosmochim Acta 58:3123-3129

Swoboda-Colberg NG, Drever JI (1993) Mineral dissolution rates in plot-scale field and laboratory experiments. Chemical Geology 105:51-69

Tadros TF, Lyklema J (1968) Adsorption of potential-determining ions at the silica-aqueous electrolyte interface and the role of some cations. J Electroanalyt Chem 17:267-275

Tunn VW (1939) Untersuchungen über die verwitterung des tremolit. Chemie Erde 12:275-303

Urey HC (1952) The planets, their origin and development. Yale Univ Press, New Haven, CT

Veblen DR, Buseck PR (1980) Microstructures and reaction mechanisms in biopyriboles. Am Mineral 65:599-623

Veblen DR, Buseck PR (1981) Hydrous pyriboles and sheet silicates in pyroxenes and uralites; intergrowth microstructures and reaction mechanisms. Am Mineral 66:1107-1134

Velbel MA (1993) Formation of protective surface layers during silicate-mineral weathering under well leached, oxidizing conditions. Am Mineral 78:405-414

Westrich HR, Cygan RT, Casey WH, Zemitis C, Arnold GW (1993) The dissolution kinetics of mixed-cation orthosilicate minerals. Am J Sci 293: 869-893.

White AF, Yee A (1985) Aqueous oxidation-reduction kinetics associated with coupled electron-cation transfer from iron-containing silicates at 25 degrees C. Geochim Cosmochim Acta 49:1263-1275

White AF, Peterson ML (1990) Role of reactive-surface-area characterization in geochemical kinetic models. In: Chemical Modeling of Aqueous Systems II. Am Chem Soc Symp Series 416:461-475

White AF, Blum AE, Schulz MS, Bullen TD, Harden JW, Peterson ML (1994) Chemical weathering of a soil chronosequence on granite alluvium I. Reaction rates based on changes in soil mineralogy. Geochim Cosmochim Acta (submitted)

Wieland E, Wehrli B, Stumm W. (1988) The coordination chemistry of weathering; III, A generalization on the dissolution rates of minerals. Geochim Cosmochim Acta 52:1969-1981

Wogelius RA, Walther JV (1991) Olivine dissolution at 25°˙C; effects of pH, CO_2, and organic acids. Geochim Cosmochim Acta 55:943-954

Wogelius RA, Walther JV (1992) Olivine dissolution kinetics at near-surface conditions. Chemical Geology 97:101-112

Wood BJ, Walther JV (1983) Rates of hydrothermal reactions. Science 222:413-415

Xie Z (1994) Surface properties of silicates, their solubility and dissolution kinetics. PhD dissertation, Northwestern Univ, Evanston, IL

Xie Z, Walther JV (1994) Dissolution stoichiometry and adsorption of alkali and alkaline earth elements to the acid-reacted wollastonite surface at 25 degrees C. Geochim Cosmochim Acta 58:2587-2598

Zhang H, Bloom PR, Nater EA (1983) Morphology and chemistry of hornblende dissolution products in acid solutions. In: Soil Micromorphology: A Basic and Applied Science. Developments in Soil Science 19:551-556

Zhang H (1990) Factors determining the rate and stoichiometry of hornblende dissolution. PhD dissertation, Univ Minnesota, Minneapolis, MN

Zhang H, Bloom PR, Nater EA (1993) Change in surface area and dissolution rates during hornblende dissolution at pH 4.0. Geochim Cosmochim Acta 57:1681-1689

Chapter 5

DISSOLUTION AND PRECIPITATION KINETICS
OF SHEET SILICATES

K. L. Nagy

Geochemistry Department, MS 0750
Sandia National Laboratories
Albuquerque, NM 87185 U.S.A.

INTRODUCTION

Chemical weathering of sheet silicates encompasses not only dissolution of detrital phases such as muscovite and biotite, but also nucleation and growth of authigenic phases, for example, smectites and kaolinite. Thus, sheet silicates are a source of solutes in surface waters via dissolution and an elemental sink through the processes of adsorption and precipitation. Weathering of sheet silicates (phyllosilicates) and related clay minerals is important on both local and global scales, and occurs in a wide variety of surface and near-surface environments. Rates of dissolution and precipitation are critical for controlling nutrient availability during plant growth, in forming large economic deposits of metals and clays (Samama, 1986), and in global elemental cycling (e.g. "reverse weathering" (Mackenzie and Garrels, 1966)). Dissolution of sheet silicates contributes to the consumption of atmospheric CO_2 and helps neutralize acid rain. The creation of new clay mineral surfaces during soil formation (Dixon and Weed, 1989) plays a significant role in the sorption of toxic metals and nutrients (McBride, 1994; Sposito, 1984). Weathering continues underwater in fresh and alkaline lakes, in evaporative seawater settings, and in seawater (Lerman et al., 1975; Mackin and Swider, 1987). Modeling of these processes requires information on the rates of dissolution and growth of sheet silicates under a wide range of surficial environmental conditions.

Jackson et al. (1948, 1952) first proposed a weathering sequence for clay minerals (Table 1) in which primary phases and sheet silicates containing iron were least stable, followed by mixed layer phyllosilicates, and smectites, and finally by phases of increasingly simpler compositions and structures (e.g. kaolinite, aluminum oxides and hydroxides, Fe-oxides and hydroxides.) Observational data that document weathering of primary igneous or metamorphic minerals, including phyllosilicates, to form kaolinite, smectites, illites and simple oxides and hydroxides are voluminous. As a starting point for those with further interest in the observational data and deduced weathering mechanisms, I refer the reader to recent papers by Banfield and coworkers (Banfield et al., 1995; Banfield and Barker, 1994; Banfield et al., 1991a,b). Also, recent modeling efforts of lateritic weathering are highly informational (Merino et al., 1993; Wang et al., 1995), as are estimates of weathering rates from field data (Velbel, 1984, 1985).

In this chapter, I review what is known about sheet silicate dissolution and growth kinetics at low temperatures, principally from an experimental standpoint. Whereas experimental work to determine rates and mechanisms of sheet silicate weathering has been ongoing for over forty years, within the scope of this chapter I will concentrate on recent work. In addition to true "sheet" silicates, data will be presented for phases such as sepiolite, in which the sheets form edge-linked ribbons, aluminum-oxyhydroxides

Table 1. Mineral weathering sequence for grains < 5 μm

STABLE	anatase, zircon, rutile, ilmenite, leucoxene, CORUNDUM
•	hematite, goethite, limonite
•	GIBBSITE, BOEHMITE, ALLOPHANE
•	KAOLINITE, HALLOYSITE
•	SMECTITES
•	MIXED LAYER MINERALS
•	2:1 LAYER SILICATES WITHOUT INTERLAYER CATIONS
•	MUSCOVITE, ILLITE
•	quartz, cristobalite
•	albite, anorthite, stilbite, microcline, orthoclase
•	BIOTITE, GLAUCONITE, ANTIGORITE, NONTRONITE
•	olivine, hornblende, pyroxene, diopside
•	calcite, dolomite, aragonite, apatite
UNSTABLE	gypsum, halite, Na-nitrate

Note: Sheet silicate, Al oxide, hydroxide and oxyhydroxide phases are capitalized. From Jackson et al. (1948, 1952).

(e.g. gibbsite), and magnesium hydroxide (brucite). The latter two phases represent the structural environment of Al and Mg in the octahedral sheets of phyllosilicates. Fe-rich octahedral sheets also exist in phyllosilicates (e.g. biotite, chlorite); however, Fe-oxyhydroxides will not be considered explicitly because a thorough review would require a separate chapter.

Although the state of knowledge on absolute rates and mechanisms has advanced in the past fifteen years, many gaps remain. Because "sheet silicates" comprise many structurally and compositionally related minerals, and because weathering environments are highly varied, there are numerous areas for new research opportunities. Despite the fact that sheet silicates are generated in large volumes at and near the earth's surface, most quantitative kinetics studies have addressed their dissolution rather than growth. Mainly, this is because it is easier to measure dissolution at low temperatures for a variety of experimental reasons. Growth of sheet silicates has been documented experimentally at low temperatures, but quantification of growth rates and mechanisms is rare, and often growth conditions do not represent those occurring in nature. In addition, in-situ effects of biota need to be quantified under controlled conditions. Combining research on adsorption, a critical step in dissolution and growth, new approaches and data from experimental studies on phyllosilicates at both surface and higher temperatures, and recent progress in surface characterization at the molecular level should rapidly advance our understanding of the kinetics of weathering of sheet silicates.

BONDING IN SHEET SILICATES

In general, sheet silicate dissolution and growth occur at specific sites on two types of surfaces: the basal surface parallel to the layers (in the *ab* plane), and the edge surface (parallel to *c*). Each surface is distinct in the nature of its average electrical charge, metal-oxygen bond characteristics, cation (metal) composition, and degree of hydroxylation in aqueous solutions. The underlying control on these characteristics is exerted by the mineral's crystal structure. Particularly clear explanations of the structure and composition

Tetrahedral Sheet

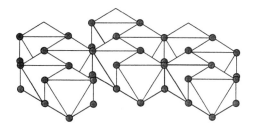

Figure 1. Schematic diagram of tetrahedral (upper) and dioctahedral (lower) sheets in layer silicates. The view of each sheet is perpendicular to the *b*-axis. Modified after Sposito (1984).

Dioctahedral Sheet

of sheet silicates can be found in Bailey (1980, 1984a, 1988a). Here I adopt Bailey's terminology and briefly review the classification of phyllosilicates with emphasis on characteristics important to kinetic reaction mechanisms. For further in-depth information on phyllosilicates and related clay minerals and their chemical properties I refer the reader to two previous Mineralogical Society of America volumes (Bailey, 1984b, 1988b), two Mineralogical Society monographs (Brindley and Brown, 1980; Newman, 1987), Veblen and Wylie (1993), and Bish and Guthrie (1993).

Sheet and layer structures

The fundamental unit of a phyllosilicate is the "sheet." Sheets are composed of linked two-dimensional networks of oxygen tetrahedra or octahedra. The composition of the tetrahedral sheets is T_2O_5. Tetrahedra link by sharing three corner (basal) oxygens. Octahedra are linked in two-dimensions by shared edges (Fig. 1).

Sheets combine in different ways to form types of "layers" (Fig. 2, Table 2). A 1:1 layer is a tetrahedral sheet connected to an octahedral sheet through the nonshared apical oxygen of the tetrahedra. Unshared OH (or F) groups also lie in the common plane between tetrahedral and octahedral sheets and are positioned in the center of the tetrahedral hexagonal rings in the same plane as the oxygens. A 2:1 layer is an octahedral sheet sandwiched between two tetrahedral sheets. Each tetrahedral sheet bonds to the octahedral sheet through the nonshared apical oxygen of the tetrahedra, requiring that one of the tetrahedral sheets be inverted. This differs from 1:1 layer silicates in that structural hydroxyls are exposed only at the edges.

If layers have an electrical charge due to cation substitutions (see below) then cations or molecules form an "interlayer" to neutralize the charge. The layer charge is concentrated in the center of the hexagonal or ditrigonal rings of silica tetrahedra, and the interlayer

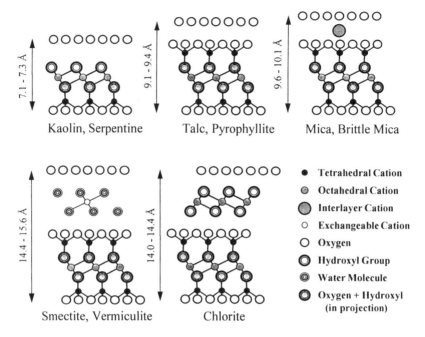

Figure 2. Projection along the b axis of unit structures of the clay groups listed in Table 1. The vertical dimension is the d-spacing parallel to the *c*-axis. Modified after Bailey (1980).

Table 2. Phyllosilicate classification

Layer Type	Group	Charge per formula unit	Interlayer Composition	Octahedral Filling	Minerals
1:1	Kaolin	~ 0	none	dioctahedral	kaolinite
	Serpentine	~ 0	none	trioctahedral	chrysotile, antigorite
2:1	Pyrophyllite	~ 0	none	dioctahedral	pyrophyllite
	Talc	~ 0	none	trioctahedral	talc
	Smectite	~ 0.2 to 0.6	cations	dioctahedral	montmorillonite
				trioctahedral	saponite
	Vermiculite	~ 0.6 to 0.9	cations or water	di- or tri	vermiculite
	Mica	~ 1.0	cations	dioctahedral	muscovite
				trioctahedral	phlogopite, biotite
	Brittle Mica	~ 2.0	hydroxide sheet	dioctahedral	margarite
				trioctahedral	clinonite
	Chlorite	variable	hydroxide sheet	di-, tri-, or di-tri	clinochlore (tri-)
2:1	Sepiolite	variable	none		sepiolite
inverted ribbons	Palygorskite	variable	none		palygorskite

Note: All charges refer to a formula unit for smectite, vermiculite, mica, and brittle mica of $O_{10}(OH)_2$ (modified after Bailey et al., 1980).

cations position themselves to fit in the center of those rings on the two facing tetrahedral surfaces. A 2:1 layer plus an interlayer is a "unit structure." Interlayers may consist of single ions (e.g. K^+ in muscovite), water, other molecules, or hydroxide octahedral sheets (e.g. the brucite interlayer in chlorites).

Internal structure modifications to the layers can occur in response to slight misfits (3° to 5°) between the ideal hexagonal arrangements of octahedral and tetrahedral sheets and to distortions of the octahedra around cations of different sizes and consequent distortions of the bonded tetrahedral sheets. In many sheet silicates the tetrahedral sheet dimensions are larger than those of the octahedral sheet. In order to fit together, alternating tetrahedra rotate inward, reducing the crystal symmetry. In decreasing order of strength the following forces determine the degree of rotation of the tetrahedra: (1) attraction between basal oxygens and surface hydroxyl groups, (2) attraction of basal oxygens by octahedral cations in the same layer, and (3) attraction of basal oxygens by octahedral cations in the underlying layer. If the tetrahedral sheet is smaller than the octahedral sheet, tetrahedra may tilt, resulting in curling of 1:1 layers (e.g. asbestiform chrysotile). Tetrahedra may invert in a periodic pattern to form a "modulated structure" such as sepiolite and palygorskite, which have 2:1 layers that extend as ribbons parallel to the c-axis. Each ribbon connects to four others at corners where the tetrahedra are inverted. In modulated structures, portions of the crystal may have 1:1 layers while others have 2:1 layers. All of these modifications cause lateral changes in dimensions resulting in strained bonds that can affect a mineral's dissolution and growth kinetics.

External structure modifications to layers result in rotations of successive layers to form mica polytypes. Stability of such structures is a function of the degree of repulsion between superposed cations in the tetrahedral and octahedral sheets in 2:1 structures and tetrahedral and octahedral cations in superposed layers in 1:1 structures. The lower the degree of repulsion, the more stable is the mineral.

Composition

Cations. Tetrahedral cations are generally Si^{4+} with Al^{3+} and Fe^{3+} substitutions (Table 3). The octahedral layer composition varies depending on the charge of the tetrahedral sheet and of interlayer cations (if present). The octahedral cations are predominantly Al^{3+} with Mg^{2+}, Fe^{2+}, and Fe^{3+} substitutions (Table 3). Other minor octahedral cations include Li, Ti, V, Cr, Mn, Co, Ni, Cu, and Zn. The nature of the cation composition determines the structural fit of tetrahedral and octahedral sheets. For minerals with nonmodulated structures, three octahedral sites are included in the smallest structural unit. If three or two sites are occupied, the phyllosilicate is termed "trioctahedral" or "dioctahedral," respectively (Tables 2, 3). Gibbsite ($Al(OH)_3$) is dioctahedral and brucite ($Mg(OH)_2$) is trioctahedral.

When cation substitutions occur, the layers may acquire varying amounts of permanent structural electrical charge. 1:1 layers must always be electrically neutral. 2:1 layers can maintain the permanent negative charge which arises from (1) substitution of M^{3+} or M^{2+} for Si^{4+} in tetrahedral sites, (2) substitution of M^{1+} or M^{2+} for M^{2+} or M^{3+} in octahedral sites, (3) vacant octahedral sites, or (4) dehydroxylation of OH to O. To compensate the negative charge in 2:1 layers, an interlayer charge is added in the form of cations, hydrated cations, or larger molecules.

Anions. Anionic substitutions, consisting mainly of F^- or Cl^- for OH^-, also occur. Such substitutions can affect the binding strength of the interlayer cations.

Table 3. Summary of mineral formulas and some recent references
on dissolution of sheet silicates and related minerals

Mineral	Formula	Reference
Brucite	$Mg(OH)_2$	V69, LC81c
Gibbsite	$Al(OH)_3$	B83, BE87, NL92
Bayerite	$Al(OH)_3$	PSWG84
Corundum	Al_2O_3	CW88
δ-Al_2O_3	Al_2O_3	FS86
Kaolinite	$Al_2Si_2O_5(OH)_4$	CW88, CW90 NBL91, WS92 GML95
Antigorite	$Mg_3Si_2O_5(OH)_4$	LC81c
Chrysotile*	$(Mg,Fe)_3Si_{2.2}O_{5.1}(OH)_4$	BM85
Chrysotile	$Mg_3Si_2O_5(OH)_4$	HR92
Talc	$Mg_3Si_4O_{10}(OH)_2$	LC81c
Montmorillonite*	$K_{0.310}(Si_{3.976}Al_{0.024})(Al_{1.509}Fe_{0.205}Mg_{0.286})O_{10}(OH)_2$	FZS93
Smectite*	$K_{0.19}Na_{0.51}Ca_{0.21}Mg_{0.08}(Al_{2.56}Fe_{0.42}Mg_{1.02})(Si_{7.77}Al_{0.23})O_{20}(OH)_4$	CGL94
Muscovite*	$K_{1.84}Na_{0.16}(Al_{3.75}Fe_{0.22}Mg_{0.12}Ti_{0.2})[Si_{6.06}Al_{1.92}O_{20}](OH)_4$	LC81a, KW89
Illite	$KAl_2(Si_3Al)O_{10}(OH)_2$	H66, FS75
Biotite	$K_2(Mg_3Fe_3)Al_2Si_6O_{20}(OH)_4$	AB92
Phlogopite	$KMg_3AlSi_3O_{10}(OH)_2$	LC81ab, KA95
Chlorite*	$(Mg_{1.9}Al_{0.9}Fe^{2+}_{0.1}Fe^{3+}_{0.1})(Si_{3.0}Al_{1.0})(Mg_{3.0})O_{10}(OH)_6$	MASBD95
Sepiolite	$(H_2O)_4(OH)_4Mg_8Si_{12}O_{30}8H_2O$	A-LW69
Palygorskite	$(H_2O)_4(OH)_2Mg_5Si_8O_{20}4H_2O$	A-LW69, CMS90

Note: All formulas are general unless determined from analytical data (*) in a specific study.
2:1 layer formulas are written with either $O_{10}(OH)_2$ or $O_{20}(OH)_4$.
Illites generally are deficient in Si. The illite formula listed is the same as for ideal muscovite.
References are coded by initials of authors last names and abbreviated year of publication.

Interlayer bonding

Where a residual layer charge is absent (e.g. kaolinite, serpentine), attraction between layers occurs primarily by relatively weak hydrogen bonding. This weak bonding controls the ability of phyllosilicates to be cleaved easily perpendicular to *c*. The strength of interlayer bonding in the case of 2:1 layer phyllosilicates is also a function of the electrostatic attraction between permanently-charged tetrahedral and octahedral sheets and the interlayer cations. If permanent charge is located in the tetrahedral sheet, the interlayer bonding strength will be greater than when charge is located in the octahedral sheet (see Giese, 1984).

Surface chemistry, bonding, and reaction pathways

Surface chemical properties are what largely control reaction rates of sheet silicates in aqueous solutions. Like other minerals, when a sheet silicate is immersed in pure water, the surfaces become hydrated; that is, water, protons (or H_3O^+), and/or hydroxyl ions adsorb to the surface (e.g. Sposito, 1984, 1990). In natural waters, this picture becomes more complex as inorganic and organic ions and molecules also sorb to the surface (Schindler and Stumm, 1987; Schindler, 1990) (Fig. 3). As for simple oxides, the nature of and bonding arrangement of oxygens to the metal center determines the strength of hydration, protonation, or hydroxylation at a phyllosilicate surface. Net proton charge (the summation of charge due to adsorbed protons and hydroxyls) and the charge arising from other adsorbed ions vary as a function of solution pH. Sposito (1984) defined the net total charge

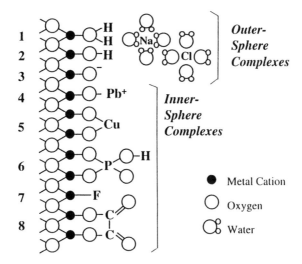

Figure 3. Schematic diagram of adsorption sites on oxide surfaces. When exposed to aqueous solutions, phyllosilicate surfaces form aluminol (>Al-OH) and silanol (>Si-OH) groups. Further protonation and deprotonation of these groups leads to charged sites where adsorption takes place. Outer sphere complexes (e.g., hydrated Na[+] and Cl[-]) interact electrostatically with the surface. These solvated ions will cluster near the surface to neutralize the surface charge. Inner sphere complexes form through ionic and covalent bonding. (1) Protonated surface hydroxyl, (2) Surface hydroxyl, (3) Deprotonated surface hydroxyl, (4) Monodentate inner sphere complex, (5) Bidentate inner sphere complex, (6) Binuclear adsorbed phosphate, (7) Mononuclear adsorbed fluoride, (8) Bidentate mononuclear adsorbed oxalate. Reactivity depends on the solution pH, the types of surface species, and the metal composition of the mineral. Modified after Stumm et al. (1987).

density σ_P at a particle surface as

$$\sigma_P \equiv \sigma_0 + \sigma_H + \sigma_{IS} + \sigma_{OS} \tag{1}$$

where σ_0 is the permanent structural charge, σ_H is the net proton charge, σ_{IS} is the charge from ions adsorbed as inner-sphere complexes, and σ_{OS} is the charge from ions adsorbed as outer-sphere complexes. σ_P will not necessarily equal zero, and must be balanced by ions in solution close to the surface that do not form complexes with surface functional groups.

It is important to remember that a surface-controlled reaction rate is a function of the number of reactive sites at the surface. For phyllosilicates, the most important surface "active" sites are the potentially charged (i.e. protonated or deprotonated) metal sites, such as aluminols (Al-O-H) and silanols (Si-O-H). These hydrated functional groups occur at edges where sheet silicates have broken bonds with unsatisfied charge. Aluminols also occur on the basal surfaces of aluminum phases as well as on the gibbsitic sheet in kaolinite. The siloxane basal surfaces are considered to be hydrophobic unless there is some permanent negative structural charge due to ionic substitution in the tetrahedral sheet (for micas) and/or in the octahedral sheet (for chlorites). If a surface contains permanent negative charge then protons or cations will sorb to the ditrigonal siloxane cavities to balance that charge. In solution, these basal cations can exchange with solution protons or cations.

Net proton surface charge has been shown to control far-from-equilibrium dissolution rates of clay minerals that vary as a function of pH (Pulfer et al., 1984; Bales and Morgan, 1985; Furrer and Stumm, 1986; Carroll-Webb and Walther, 1988; Wieland and Stumm, 1992; Furrer et al., 1993). Surface charge is also thought to play a role in controlling rates of precipitation (Stumm et al., 1983; Fleming, 1986; Brady, 1992); however, more data on the significance of this property during phyllosilicate crystal growth is needed. Adsorption of H^+, OH^-, or other ions to the surfaces is measured in titration experiments and modeled using the surface-complexation approach. In this model, surface complexes are patterned after similar solution complexes. Formation constants for the surface complexes can be treated using equilibrium thermodynamics. Typically, for sheet silicates in pure water or dilute solutions, three surface complex reactions will be important from mildly acidic to slightly above neutral pH solutions typical of weathering solutions (Brady and Walther, 1989; Wieland and Stumm, 1992):

$$>Al\text{-}O\text{-}H + H^+ \; = \; >Al\text{-}O\text{-}H_2^+ \qquad (2)$$

$$>Al\text{-}O\text{-}H \; = \; >Al\text{-}O^- + H^+ \qquad (3)$$

$$>Si\text{-}O\text{-}H \; = \; >Si\text{-}O^- + H^+ \qquad (4)$$

Each of these has an equilibrium constant that is typically obtained by combining acid/base titration data with an iterative numerical fitting routine assuming a particular correction for the Coulombic energy of the charged mineral surface, such as the constant capacitance model (e.g. Wieland and Stumm, 1992). Other metal-oxygen edge sites such as Mg-O in chrysotile (Bales and Morgan, 1985) are modeled similarly. Adsorption of metal ions or organic ligands instead of protons to these sites also can be treated (e.g. Furrer and Stumm, 1986). For example, Kline and Fogler (1981c) modeled the catalytic effects of salts on dissolution rate of kaolinite assuming that adsorption occurred at surface hydroxyls, and Bales and Morgan (1985) modeled the dissolution of chrysotile in the presence of catechol and oxalate.

Because polarization and subsequent breaking of metal oxygen bonds are the primary steps in dissolution of sheet silicates, one must also consider their geometry within the two- and three-dimensional views of the mineral surface. For example, Ganor et al. (1995) emphasized that protonation of the oxygen bonded to both Al and Si at edges is involved in the rate-limiting step for dissolution of kaolinite. They proposed that sequential breaking of Al-O-Si bonds results in an unzipping of the kaolinite layers. Wieland and Stumm (1992), on the other hand, modeled kaolinite dissolution as simultaneous breakage of both Al-O-Si and Al-OH-Al bonds, on the edges and basal surfaces, respectively. Clearly, bond strength depends on the local arrangement of atoms on either basal or edge surfaces as well as the short range electrostatic and long range van der Waals forces arising from the bulk mineral structure. Since the two types of surfaces, basal and edge, have different types and densities of charged sites, their relative amounts play a critical role in controlling the overall reaction rate. This is important when considering surface reaction rates of bulk samples, whether they are experimental powders, soils, or sediments. The charged nature of the two surfaces also contributes to particle/particle interactions such as aggregation, which can reduce the number of reactive sites.

How many reactive sites exist on phyllosilicate surfaces? The number depends on the surface density of protonated sites (a function of pH) and the amount of surface area exposed to solution that contains those sites. Sposito (1984) estimated the density of reactive sites on phyllosilicate edges to be ~8 sites nm^{-2}. However, for the example of kaolinite, he estimated that edge surface area was only 8% of the total, resulting in a total site density of ~0.6 sites nm^{-2}. Xie and Walther (1992) considered site density in terms of the total metal cations on both edge and basal surfaces of kaolinite and estimated that as much as 50% of the kaolinite studied by Carroll-Webb and Walther (1988) was edge surface. This led to a total site density of 20 sites nm^{-2}, still too low to explain Carroll-Webb and Walther's (1988) potentiometric titration results. Therefore, Xie and Walther (1992) suggested that H$^+$ or OH$^-$ can penetrate between layers to react with inner hydroxyls. Wieland and Stumm (1992) estimated that 2.5 and 2 sites nm^{-2} were required for dissolution of the edge and basal gibbsitic surfaces on kaolinite. Their estimate is smaller because they did not count the Si sites and because they estimated the edge surface area at 20% of the total area. White and Zelazny (1988) evaluated the nature and density of charged edge sites on phyllosilicates in solutions of pH 3 to 9 using the crystal growth theory of Hartman (1973). Different charge distribution is predicted on edges perpendicular to different periodic bond chains (an uninterrupted chain of nearest-neighbor particles). They also predicted a decrease in the number of charged edge sites per unit area concomitant with an increase in permanent negative structural charge. With increasing lattice substitution (first tetrahedral, then octahedral) at pH 6.5, for example, edges approach zero surface charge.

Although dissolution occurs preferentially at edge sites, some researchers have demonstrated that basal surfaces are also highly reactive, particularly with respect to water adsorption and subsequent dissolution and precipitation (e.g. Johnsson et al., 1992). Shallow step edges, structural defects, and impurities on the basal surface appear to nucleate dissolution. In many cases we do not know true proportions of edge and basal sites for particles and aggregates in experiments, let alone in nature. Atomic force microscopy has been used to determine clay mineral morphologies (Blum, 1994; Nagy, 1994; Zhou and Maurice, 1995) and the relative proportions of edge and basal surface areas (Nagy et al., 1995) (Fig. 4).

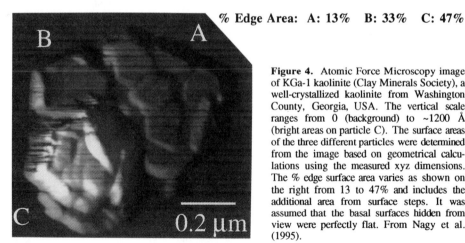

% Edge Area: A: 13%   B: 33%   C: 47%

Figure 4. Atomic Force Microscopy image of KGa-1 kaolinite (Clay Minerals Society), a well-crystallized kaolinite from Washington County, Georgia, USA. The vertical scale ranges from 0 (background) to ~1200 Å (bright areas on particle C). The surface areas of the three different particles were determined from the image based on geometrical calculations using the measured xyz dimensions. The % edge surface area varies as shown on the right from 13 to 47% and includes the additional area from surface steps. It was assumed that the basal surfaces hidden from view were perfectly flat. From Nagy et al. (1995).

DISSOLUTION KINETICS

Introductory comments

The primary goal of experimental weathering studies of phyllosilicates has been to determine dissolution rather than precipitation kinetics. Generally, dissolution rates are measured in far-from-equilibrium solutions where they are relatively fast and easy to quantify. In contrast, when precipitation is measured in far-from-equilibrium solutions, rates are also fast, but complicated by the formation of metastable phases. Both dissolution and precipitation kinetics can be difficult to unravel when determined in solutions closer to equilibrium because of secondary phase formation. Also, dissolution is characterized reasonably well by measuring changes in solution composition only, whereas documenting growth requires additional surface characterization that is more difficult to perform. Dissolution rates of sheet silicates have been measured as a function of solution pH, temperature, organic ligand concentration, Al concentration, and saturation state or reaction affinity. Less attention has been given to electrolyte concentration and composition, and solid properties such as lattice structure, composition, and charge. Some effort has been made to evaluate the effects of adsorption on dissolution rates.

Both surface-coordination complexes and transition state theory are used to model dissolution rate data. Nonetheless, no elementary mechanism controlling dissolution or growth has yet been confirmed by spectroscopic techniques. The surface-coordination complex approach appears to work well for modeling dissolution in undersaturated solutions and should represent the initial dissolution step when rainwater contacts a phyllosilicate mineral. However, in solutions closer to equilibrium and of more complex compositions, it is unclear how well the surface-coordination complex approach accurately represents dissolution rates in nature. Furthermore, few data exist for testing the use of transition state theory in modeling dissolution at any reaction affinity.

The dissolution reaction

The dissolution of a sheet silicate in natural weathering solutions can be described using the example of kaolinite. Dissolution occurs in an overall sense by reacting the solid with H^+ in solution to produce the dissolved species Al^{3+}, H_4SiO_4 and water:

$$Al_2Si_2O_5(OH)_4 + 6\,H^+ = 2\,Al^{3+} + 2\,H_4SiO_4 + H_2O \qquad (5)$$

If one is interested in the reaction in the basic pH range the Al species can be converted to $Al(OH)_4^-$ through the equilibrium hydrolysis constants of aqueous Al. In this case, the reaction could be written

$$Al_2Si_2O_5(OH)_4 + 2\,OH^- + 5\,H_2O = 2\,Al(OH)_4^- + 2\,H_4SiO_4 \qquad (6)$$

Reasons for switching speciation include having better knowledge of the relevant Al speciation in the solutions of interest or for numerical calculations where errors may arise during computations with very small numbers (e.g. Steefel and Van Cappellen, 1990). However, if aqueous speciation is well known, the reaction can be treated using either equation.

Reactions (5) and (6) give no information about reaction mechanism, or the rate-limiting step in the overall reaction. For example, the reaction may not occur stoichiometrically, invalidating the assumption of six simultaneous protonation reactions from (5). Al and Si may not be simultaneously released. It is probable that one cation is detached first and the elementary reaction describing that detachment may or may not be the rate-limiting step. The reaction does not account for reaction intermediates such as catalyst or inhibitor ions that might participate in the dissolution mechanism but do not appear as reactants or products. In the following sections, I review experimental data and surface reaction models that shed light on likely mechanisms that control sheet silicate dissolution. However, it is important to keep in mind that identification and characterization of specific mechanisms are an open topic for future research.

Acid-leaching experiments

Numerous data (not all of which can be cited in this chapter) exist on cation leaching from phyllosilicates in acidic solutions. The purposes of these studies include simulation of weathering, artificial dissolution of reservoir clays during petroleum production, and ore processing. Although the extreme acidities investigated often are outside the range of weathering solutions, the results are useful for identifying common characteristics of phyllosilicate dissolution. Also, release rates of all components of a particular clay generally are not reported, rarely are leach rates normalized to a surface area, and saturation states with respect to secondary phases are not determined.

One reproducible result from early acid-leaching studies showed that octahedral cations are released more readily than tetrahedral cations in 2:1 layer phyllosilicates. For example, Brindley and Youell (1951) leached chlorite in >10% HCl and measured Al, Fe, and Mg as a function of time and temperature. Assuming that octahedral Al was released first, followed by tetrahedral aluminum, they determined a formula very similar to the known structural formula. Osthaus (1954, 1956) performed similar studies on montmorillonite and nontronite and determined that octahedral Al, Fe, and Mg all had approximately the same release rate. Gastuche (1963) and Gilkes et al. (1973b) indicated that tetrahedral Al and Si are released more slowly from biotite than octahedral cations resulting in formation of siliceous rims. Feigenbaum and Shainberg (1975) showed that octahedral Mg^{2+} and Fe^{3+} are released preferentially from illite. Oxidized biotite and vermiculite release octahedral and interlayer cations to maintain charge neutrality (Farmer et al., 1971; Robert and Pedro (1969); Gilkes, 1973b). Gilkes (1973b) showed that the dissolution rate of biotite is inversely proportional to the degree of oxidation. Much of this earlier work is summarized in papers by Ross (1967), Abdul-Latif and Weaver (1969), Gilkes et al. (1973b), and Kline and Fogler (1981a,b).

Reconstitution of acid-leached clays (Mering, 1949; Brindley and Youell, 1951; Abdul-Latif and Weaver, 1969) by treating with $NaOH/MgCl_2$ solutions showed that the intensity of the 10 Å XRD peak increased, indicating the presence of a mica phase. Presumably H^+ ions that had replaced original cations in the octahedral sheets during acid-leaching were exchanged with Mg^{2+}, suggesting that the clay structure is retained to a certain extent. Dissolution of halloysite (a hydrous kaolin mineral) in 42% to 60% sulfuric acid at 125° to 188°C released up to 94% of the aluminum and increased the BET surface area by a factor of four (Chon et al., 1978a,b), suggesting again that tetrahedral sheets resist dissolution to a much greater extent than octahedral sheets.

A uniform conclusion is that sheet silicates preferentially dissolve inward from the edges (e.g. Gastuche, 1963; Ross, 1967; Gilkes, 1973b). Many researchers explained the leaching rates in terms of a diffusion mechanism parallel to the layers, while others posited a surface dissolution mechanism. Kline and Fogler (1981a,b) determined the dissolution rates of a series of phyllosilicates (kaolinite, pyrophyllite, muscovite, illite, Na-montmorillonite, talc, and phlogopite) in HF (pH < 1) solutions at 25°C. Dissolution of all these minerals was congruent. From SEM observations, they concluded that kaolinite dissolves at exposed octahedral sites on edge and basal surfaces, whereas all 2:1 layer clays dissolve predominantly at exposed octahedral sites on edges. However, kaolinite can have a high concentration of structural defects (Giese, 1988) which could cause higher than expected dissolution on basal surfaces. From literature data on muscovite dissolution using rotating disk and fission track etching they determined that dissolution at edges is about four orders of magnitude faster than dissolution on basal surfaces (Kline and Fogler, 1981a).

Parallel reaction pathways were proposed by Kline and Fogler (1981b) who investigated the dissolution of kaolinite in HF/HCl mixtures. They found that adsorbed H^+ and HF catalyzed the dissolution reaction by adsorbing simultaneously to different surface sites, bridging oxygen and hydroxylated metal sites, respectively. They showed that the exchange constant for H^+ on Na-montmorillonite is about 20 times higher than on kaolinite (8.0 vs. 0.39), indicating the large effect of the permanent negative charge on basal surfaces of montmorillonite. Much earlier, Coleman and Craig (1961) showed that H^+ on exchange sites accelerated montmorillonite dissolution.

The activation energy (E_a) for acid dissolution ranges from ~40 to 100 kJ mol^{-1}. Ross (1967) summarized previous work on kaolinite, biotite, glauconite, montmorillonite, and vermiculite as well as his own on clinochlore. He determined an E_a for congruent dissolution of clinochlore at 25° to 60°C in 2 N HCl of 87.9 kJ mol^{-1}. The values of E_a from the previous work obtained under similar conditions ranged from 71 to 92 kJ mol^{-1}. Abdul-Latif and Weaver (1969) investigated the acid dissolution of palygorskite (2.5 and 5.0 N HCl; the latter from 60° to 91°C) and sepiolite (1.0 N HCl; 30.5°C). They determined an E_a of 77 kJ mol^{-1} for dissolution of palygorskite and found that sepiolite dissolves about 2.5 orders of magnitude slower than palygorskite if the palygorskite rates are extrapolated to 30°C. Thomassin et al. (1977) measured chrysotile dissolution in 0.1 N oxalic acid from 22° to 80°C and determined an activation energy of 63 to 84 kJ mol^{-1}. Kline and Fogler (1981a,b) determined that E_a (over a maximum range of 0° to 70°C) for dissolution in HF solutions is a function of Mg content. $E_a = 42$ kJ mol^{-1} for Mg-silicates, and ~54 kJ mol^{-1} for all other aluminosilicates studied except Na-montmorillonite. Its E_a lies in between reflecting 10% substitution of Mg for Al in the octahedral sheet. The lower E_a in HF solutions (Kline and Fogler, 1981a) than those in HCl solutions (e.g. Ross, 1967; Abdul-Latif and Weaver, 1969; Thomassin et al., 1977) indicates that HF is a stronger catalyst to bond-weakening in a precursor step in the overall dissolution reaction.

Dissolution data from solubility experiments

Information on dissolution rates is available from equilibrium solubility studies. Solids initially react quickly in dilute solutions and then more slowly as solutes build up to equilibrium concentrations. Often the data cannot be used to determine the degree of congruency of dissolution, initial rate data are rarely obtained, and secondary or accessory phases can complicate rate interpretations. These characteristics of the dissolution process raise questions about the attainment of true equilibrium for sheet silicates at low temperatures. Nonetheless, many useful qualitative facts related to weathering kinetics of sheet silicates can be extracted from solubility data.

At room temperatures, equilibrium between pure water and aluminosilicate clay mixtures in closed systems is reached in 2 to 4 years (Polzer and Hem, 1965; Reesman, 1974; Mattigod and Kittrick, 1979; Rosenberg et al., 1984; May et al., 1986) (Fig. 5). Sepiolite and kerolite (a hydrous, disordered talc-like mineral) also equilibrate from initially dilute undersaturated solutions in ~3 years (Stoessell, 1988). Commonly, only a small mass of solid is reacted in order to have enough solution to analyze over this long time

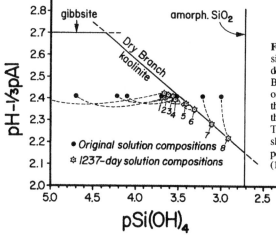

Figure 5. Diagram showing the composition of aqueous fluids in experiments designed to measure solubility of Dry Branch, Georgia, USA kaolinite. Not all of the initial solution compositions reach the equilibrium composition (indicated by the diagonal line) even after 1237 days. This indicates that the reaction kinetics slow considerably with time. Used by permission of the authors, from May et al. (1986).

period. In a soil or sediment closed system, the solid/solution ratio would probably be much greater and equilibrium would be reached in shorter times. However, permeable soil or sediment systems are more open to flowing fluids, leading to generally faster reaction rates further from equilibrium.

For cation-exchangeable 2:1 layer silicates, (i.e. smectites) 3 to 4 years may not be long enough to achieve equilibrium due to the control of Al and Si solubility by other phases (Churchman and Jackson, 1976; May et al., 1986). It is apparent when determining the solubility of kaolinite that a trace silica phase contributes to nonstoichiometric dissolution (Polzer and Hem, 1965; May et al., 1986). This complication means that detailed modeling of the distribution of reactive surface species will be invalid unless the solid phase is well characterized. Rosenberg et al. (1984) suggested that muscovite solubility may be controlled by a mica phase of different composition than the bulk. They based their conclusions on AEM analysis of the samples studied by Mattigod and Kittrick (1979) and suggested that edges enriched in Si and slightly depleted in K and Al controlled the apparent equilibrium solubility.

Ion exchange and early dissolution experiments

Early experiments designed to determine ion-exchange characteristics of micas were undertaken to understand nutrient availability in soils and the process of vermiculite formation (e.g. Mortland, 1958; Mortland and Lawton, 1961; Raussel-Colom et al., 1965; Scott, 1968; Newman, 1969; Gilkes et al., 1973a). For example, it was recognized early on that, during weathering, exchange of interlayer cations for H^+ occurred (e.g. Raussel-Colom et al, 1965). However, it soon became clear that other processes besides diffusion of ions in and out of the interlayer were occurring.

In order to assess these processes, Gaines and Rutkowski (1957) performed one of the earliest sets of experiments on muscovite dissolution in which Al and Si release rates were measured for two days in 0.1 N HCl, dilute HCl (pH = 4.5), 0.1 N KCl (pH ≈ 7), NaHCO$_3$ (pH = 8), and 0.1 N KOH (pH = 13). They observed that the extraction rate was pH-dependent with a minimum at pH near 7. At pH 1, the released Si/Al ranged from 0.14 to 0.37, less than 1/2 of the stoichiometric ratio. At pH 13, Si/Al ranged from 0.6 to 3.6.

Therefore, they concluded that in acid, attack was primarily at edge octahedral layer sites and in basic solutions, dissolution occurred primarily at tetrahedral sheet sites. These two observations agree with the current view that Al- and Si-sites control dissolution rates in acidic and basic solutions, respectively (e.g. Brady and Walther, 1989).

Internal strain resulting from cation exchange in micas during vermiculitization could affect dissolution rate (Von Reichenbach and Rich, 1969). In experiments where K^+ was exchanged for Ba^{2+} in muscovite, they proposed that expansion of edges caused stresses at the boundaries of the exchanged and unexchanged areas. Newman and Brown (1969) demonstrated that the edge exchange zone forms unevenly. Alteration of phlogopite in boiling 0.25 N NaCl showed that a sharply fractured surface formed underwater hardly altered when compared to a less-sharply cut surface prepared in air. Disturbed edges (formed by grinding, dragging cuts with sharp blades, and cutting with abrasive powders) exchanged K^+ rapidly. Excess swelling, especially in thicker particles, produced sufficient strain to cleave the particle. Leonard and Weed (1970) compared K release rates in eight muscovites, biotite, phlogopite, lepidolite, and synthetic fluorphlogopite. Muscovite was about two orders of magnitude more stable (in terms of release rates of K^+) than phlogopite and biotite. Substitution of F^- for OH^- causes lepidolite to be more stable than muscovite by at least a factor of three and fluorphlogopite to be as stable as muscovite. Also, they proposed that the shorter K-O bond length in dioctahedral micas (~2.85 Å) compared to trioctahedral micas (~3 Å) may contribute to the relative stabilities. They summarized previous work showing that the *b*-dimension of muscovite contracts with K^+-removal, whereas that of phlogopite expands, and suggested that such structural adjustments may control the weathering release rates.

Edge modification also was observed by 'T Serstevens et al. (1978) for muscovite and phlogopite dissolution in distilled water and $NaNO_3$ solutions at 60°C. Phlogopite dissolved congruently at pH 2 and the rate of dissolution was lowered by the addition of 0.2 M $NaNO_3$. At neutral pH, both water and $NaNO_3$ solutions caused K^+ exchange and incipient vermiculitization. Grinding was thought to induce initial excess Al release from muscovite; however, dissolution at pH 2 of predissolved muscovite led to congruent dissolution. At neutral pH, K^+/H^+ exchange and Si dissolution occurred, but gibbsite formed on the surface based on aluminum and excess water analyses. This agrees with the results of Von Reichenbach and Rich (1969) who observed boehmite formation during the Ba^{2+}/K^+ exchange. However, these data contrast with those of Rosenberg et al. (1984) and may be due to differences in duration of exposure of muscovite to solution. Results of Al exchange experiments on kaolinite, montmorillonite, muscovite, illite, and biotite, also support the splitting of mica edge surfaces (Cabrera and Talibudeen, 1978). An initial rapid ion exchange in illite occurs in acidic solutions when H^+ is exchanged for interlayer cations (Feigenbaum and Shainberg, 1975), again resulting in splaying of layers at the edges.

Far-from-equilibrium rates and pH-dependence

Recently, the most studied aspect of dissolution of clay minerals relevant to weathering has been the dependence of rates on pH. Dissolution rates for many Al oxide and hydroxide phases, and sheet silicates have been determined, although the pH range and dependencies vary. In general, the data show the standard behavior for aluminosilicate minerals, wherein rates decrease with increasing pH in the acidic range, show a near constancy in value in neutral pH solutions, and increase with increasing pH in the basic range (Tables 4 and 5). Some investigators have observed smoothly varying functions whereas others have seen step functions where the pH-dependence changes over narrow

Table 4. [H^+]-dependence of sheet silicate and related mineral dissolution rates in acidic solutions

Mineral	Solution composition	pH range	Reaction order n	T (°C)	Reference
Brucite	0.1 M KCl	1 - 5	~ 0.5	25	Vermilyea (1969)
Gibbsite	HNO_3	1.5 - 2.1	1.6	25	Bloom (1983)
Gibbsite	"	2.4 - 3.2	$0 \to -0.7$	25	"
Gibbsite	0.1 M KNO_3	1.7 - 2.5	1.0	25	Bloom & Erich (1987)
Gibbsite	"	2.5 - 3.9	$1 \to 0$	25	"
Gibbsite	0.1 M K_2SO_4	1.8 - 2.8	1.0	25	"
Gibbsite	10^{-4} M KH_2PO_4	1.8 - 3.0	0	25	"
Gibbsite	$HClO_4$, HNO_3	3.3 - 3.9	0.29	25	Mogollón et al. (1994)
Bayerite	1.0 M KNO_3/HNO_3	3 - 4	1	25	Pulfer et al. (1984)
δ-Al_2O_3	0.1 M $NaNO_3/HNO_3$	2.5 - 3.5	~ 0	25	Furrer & Stumm (1986)
δ-Al_2O_3	"	3.5 - 6	$0.4 \to 0$	25	"
Corundum	pH buffers; I≤0.05M	1 - 4	~ 0	25	Carroll-Webb & Walther (1988)
Corundum	"	4 - 6	~ 0.3	25	"
Kaolinite	$HClO_4$	3.1 - 4.2	0.5	25	Ganor et al. (1995)
Kaolinite	"	3.2 - 4.2	0.4	50	"
Kaolinite	"	2.1 - 3.1	-0.09	50	"
Kaolinite	"	2.0 - 3.2	0.4	80	"
Kaolinite	0.1 M $NaNO_3/HNO_3$	2.0 - 3.0	-0.02	25	Wieland & Stumm (1992)
Kaolinite	"	3.0 - 4.0	0.34	25	"
Kaolinite	"	4.0 - 6.5	-0.09	25	"
Kaolinite	pH buffers;I=0.05M	0.5 - 6.0	0.09	25	Carroll & Walther (1990)[g]
Kaolinite	"	2.1 - 6.4	0.28	60	"
Kaolinite	"	1.0 - 6.9	0.48	80	"
Kaolinite	H_2SO_4	3 - 6	0.06[d]	25	Heydemann (1966)
Chrysotile	HCl	2.1 - 5.7[a]	0	37	Hume & Rimstidt (1992)
Montmorillonite	0.1 M KCl/HCl	1 - 4	0.38[b], 0.22[c]	25	Furrer et al. (1993)
Montmorillonite	H_2SO_4	3 - 6	0.23[ed], 0.30[fd]	25	Heydemann (1966)
Muscovite	dilute pH buffers	2.1 - 5.0	0.37	70	Knauss & Wolery (1989)
Muscovite	0.1 M $NaNO_3$	3 - 5	0.1	25	Stumm et al. (1987)
Muscovite	HCl, CO_2, NH_3	0.2 - 5.6	0.08	25	Nickel (1973)
Illite	H_2SO_4	3 - 6	0.02[d]	25	Heydemann (1966)
Illite	$HCl/CaCl_2$	3 - 7.5	0.4[d]	25	Feigenbaum & Shainberg (1975)
Biotite	H_2SO_4	3.0 - 5.5	0.34	22	Acker & Bricker (1992)
Biotite	H_2SO_4/H_2O_2	5 - 5.7	0.7	22	Acker & Bricker (1992)
Chlorite	H_2SO_4	3.1 - 5.2	~ 0.5	25	May et al. (1995)

Note: All rates are based on Si release; log R = pH^n. *a:* 3.4 - 7.4 final; *b:* batch experiment; *c:* mixed-flow experiment; *d:* slope based on 2 data points; *e:* Upton, Wyoming; *f:* Polkville, Missouri; *g:* additional data in Carroll-Webb and Walther (1988) at I < 0.05 M.

pH intervals. Certain data fit models describing a direct correspondence to the concentration of protonated surface species.

Overall, most sheet silicates have approximately the same dissolution rates on a mol mineral basis at 25°C and near neutral pH (e.g. pH = 5) ~10^{-13} mol m^{-2} sec^{-1} ±1/2 order of magnitude (Table 6). (One mol refers to a mineral's formula reduced to a single layer, i.e. a 2:1 interlayer phase such as muscovite would include a TOT layer plus one interlayer of potassium.) Gibbsite, which approximates the octahedral sheet in 1:1 and 2:1 layer clays, dissolves at a similar rate. Talc and chrysotile which contain no aluminum dissolve slightly faster. Brucite's dissolution rate is about 7 orders of magnitude faster than any of the other

Table 5. [H⁺]-dependence of sheet silicate and related mineral dissolution rates in basic solutions

Mineral	Solution Composition	pH range	Reaction Order, n	T (°C)	Reference
Corundum	pH buffers; I ≤ 0.05 M	9 - 11	~ -1.0	25	Carroll-Webb & Walther (1988)
Kaolinite	pH buffers; I = 0.05 M	7.5 - 12	-0.25	25	Carroll & Walther (1990)[d]
Kaolinite	"	7.5 - 11	-0.34	60	"
Kaolinite	"	8 - 12	-0.48	80	"
Kaolinite	NH_4Cl/NH_4OH	6 - 10	-0.048	25	Heydemann (1966)
Chrysotile	pH buffers	7 - 10	0.24 (Mg)	25	Bales & Morgan (1985)
Chrysotile	"	7 - 8.3	0.19	"	"
Chrysotile	"	8.3 - 11	-0.23	"	"
Muscovite	dilute pH buffers	7.0 - 12.0	-0.22	70	Knauss & Wolery (1989)
Muscovite	CO_2, NH_3	5.6 - 10.6	-0.10	25	Nickel (1973)
Illite	NH_4Cl/NH_4OH	6 - 10	-0.11	25	Heydemann (1966)
Montmorillonite	NH_4Cl/NH_4OH	6 - 10	-0.13^{ac}, 0.075^{bc}	25	Heydemann (1966)

Note: All rates based on Si release, unless otherwise noted; log R = pH^n. *a:* Upton, Wyoming; *b:* Polkville, Missouri; *c:* slope based on 2 data points; *d:* additional data in Carroll-Webb and Walther (1988) at I < 0.05 M.

Table 6. Selected dissolution rates of sheet silicates and related minerals at 25°C and pH 5

Mineral	Rate (mol m^{-2} sec^{-1})	Reference
Brucite	6 x 10^{-6}	Vermilyea (1969)
Gibbsite*	3 x 10^{-13}	Bloom & Erich (1987), Mogóllon et al. (1994)
Kaolinite	1 x 10^{-13}	Carroll-Webb & Walther (1988)
	4 x 10^{-13}	Wieland & Stumm (1992)
Chrysotile	1 x 10^{-12}	Bales and Morgan (1985)
Talc	1 x 10^{-12}	Lin & Clemency (1981c)
Montmorillonite	4 x 10^{-14}	Furrer et al. (1993)
Muscovite	1 x 10^{-13}	Lin & Clemency (1981b), Stumm et al. (1987)
Biotite	6 x 10^{-13}	Acker & Bricker (1992)
Phlogopite	4 x 10^{-13}	Lin & Clemency (1981a)
Chlorite	3 x 10^{-13}	May et al. (1995)

Note: *Gibbsite rate was extrapolated to pH 5 using both data sets each of which extended up to pH 3.9. 2:1 layer silicate rates are normalized to $O_{10}(OH)_2$ formula unit. All rates except for brucite and gibbsite are based on Si release data.

phases. Below, I review data on the dissolution of Al hydroxide and oxide phases, 1:1, and 2:1 layer silicates in relatively undersaturated solutions as a function of pH and solution composition. This order represents progressively increasing compositional and structural complexity of the minerals (Tables 2, 3).

Aluminum hydroxide phases. Dissolution of gibbsite and bayerite has been investigated in the acidic pH range because of the strong interest in the effects of acid rain on Al mobility and toxicity (Fig. 6). Studies in extreme acidic and basic solutions are conducted to improve processing of bauxite ores in the aluminum industry.

Bloom (1983), Bloom and Erich (1987), and Mogollón et al. (1994) measured gibbsite ($Al(OH)_3$) dissolution rates as a function of pH in acidic solutions. Pulfer et al. (1984) studied bayerite (γ-$Al(OH)_3$) which is polymorphic with gibbsite. The dependence of rate (R_{diss}, a positive number) on H⁺ concentration ([H⁺]) as expressed by the reaction order "n" in the following equation:

Al-Oxide & Al-Hydroxide Dissolution Rates

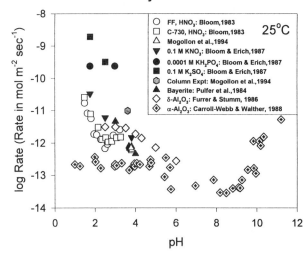

Figure 6. Summary of dissolution data for Al-oxide and Al-hydroxide phases showing rates as a function of pH. FF (open circles) and C-730 (open squares) refer to two different gibbsite materials (Bloom, 1983). Rates are based on release rates of Al. Note that higher rates are obtained for gibbsite in 0.1 M K_2SO_4 and 0.0001 M KH_2PO_4. Little or no dependence on pH is seen for the gibbsite rates in 0.0001 M KH_2PO_4 and HNO_3 solutions from pH 2 to 3, for δ-Al_2O_3 from pH 2 to ~3.5, and for α-Al_2O_3 (corundum) from pH 1 to 5.

$$R = k \, [H^+]^n \tag{7}$$

varies from 0 to 1.7 over the acidic pH range of ~1.7 to 4 for gibbsite (Fig. 6 and Table 4). From pH 1.5 to 2.1, Bloom (1983) observed a reaction order of 1.7; whereas from pH 2.4 to 3.2, the reaction order ranged from 0 to -0.7, depending on the gibbsite source. (Bloom and Erich (1987) later noted that the NO_3 concentration varied in the experiments of Bloom (1983) and this could have caused values of n > 1 or < 0.) Mogollón et al. (1994), on the other hand, observed that n = 0.3 from pH 3.3 to 3.9 in HNO_3 and $HClO_4$ solutions. In 0.1 M KNO_3 solutions Bloom and Erich (1987) found that n = 1.0 from pH 1.7 to 2.5 and decreased gradually to zero at pH 3.9. Pulfer et al. (1984) also observed n = 1 for bayerite dissolution from pH 3 to 4 in 1.0 M KNO_3. In 0.1 M K_2SO_4, n = 1 at pH 1.75 to 2.8 (Bloom and Erich, 1987) and in 0.0001 M KH_2PO_4, n = 0 for pH 1.8 to 3.0.

Gibbsite dissolution in extremely acidic solutions depends on the product $a_{H^+} a_{H_2O}^2$, where the activities represent those of surface adsorbed species. (Packter and Dhillon, 1969). Bloom (1983) also proposed that water participates in the reaction at pH 2.4 to 3.3.

Dissolution results at low temperatures in basic solutions have been determined for base-digestion processes of importance to the aluminum mining industry. In extremely basic solutions (i.e. pH > 13.25), the dissolution rate of gibbsite has a first order dependence on the NaOH concentration (Scotford and Glastonbury, 1972; Packter and Dhillon, 1973, 1974; Peric' et al., 1985).

Al-oxide phases. Two studies of Al-oxide phases have been carried out in part to better characterize the behavior of Al sites in aluminosilicate dissolution (Fig. 6). Furrer and Stumm (1986) investigated the dissolution of δ-Al_2O_3 from pH 2.5 to 6. Carroll-Webb and Walther (1988) measured dissolution rates for α-Al_2O_3 (corundum) from pH 1 to 11.2. Because Al is in octahedral coordination in both phases its chemical environment may represent that in sheet silicates.

Furrer and Stumm (1986) observed a pH dependence in 0.1 m $NaNO_3/HNO_3$ solutions for $\delta-Al_2O_3$ similar to that for gibbsite in KNO_3/HNO_3 solutions (Bloom, 1983). A plateau in the dissolution rate occurred at pH 2.5 to about 3.5. At higher pH the rate steadily decreased to approach asymptotically a value of about 6×10^{-13} mol m^{-2} sec^{-1} at pH 6. From pH 3.5 to 4.5, the reaction order n is ~0.41, similar to the reaction order of 0.3 observed by Mogollón et al. (1994) from pH 3.2 to 3.9.

Although there is more scatter in the data due probably to variations in ionic strength, the data of Carroll-Webb and Walther (1988) for corundum dissolution show a similar pH dependence to that of $\delta-Al_2O_3$ and gibbsite at pH > 4 where n ≈ 0.3. At pH < 4, a flattening with pH is observed. In mildly acidic solutions, the corundum dissolution rates (Carroll-Webb and Walther, 1988) are an order of magnitude less than those for $\delta-Al_2O_3$ (Furrer and Stumm, 1986). Both sets of data were obtained in batch experiments; however, Furrer and Stumm (1986) determined rates after only a few days, whereas Carroll-Webb and Walther (1988) measured rates at ≥200 hours of reaction. The short-term rates of Carroll-Webb and Walther (1988) are faster. The work on corundum by Carroll-Webb and Walther (1988) provides the only data available for an aluminum oxide in basic solutions (buffered with Na borate/NaOH) of geological importance. They observed a pH-dependent reaction order of n ≈ -1 from pH 9 to 11.2. This is similar to the reaction order for gibbsite dissolution in strong NaOH base solutions (Scotford and Glastonbury, 1972; Packter and Dhillon, 1973, 1974).

Both Furrer and Stumm (1986) and Carroll and Walther (1988) performed acid/base surface titrations to determine the surface concentration of adsorbed H^+ or OH^-. For $\delta-Al_2O_3$ the reaction order with respect to surface concentration of H^+ was 3. For $\alpha-Al_2O_3$ the reaction order was 1 with respect to H^+ and 4 with respect to OH^-.

Brucite.   Brucite, $Mg(OH)_2$, forms a trioctahedral sheet. The value of n is ~0.5 from pH 1 to 5 (Vermilyea, 1969). Lin and Clemency (1981c) measured a dissolution rate at pH 8.5 that was similar to that determined by Vermilyea (1969). Wogelius et al. (1995) measured periclase (MgO) dissolution rates at pH 2 to 4 for single crystals that compared well with periclase powder experiments of Vermilyea (1969). The pH-dependency was -0.35 and -0.4, respectively, similar to that of brucite. Vermilyea (1969) proposed, and Wogelius et al. (1995) and Refson et al. (in press) demonstrated that the surface of periclase transforms to brucite in aqueous solutions, explaining why the dissolution rates and pH dependencies are similar.

Kaolinite.   Kaolinite is a 1:1 layer phyllosilicate with an Al-octahedral sheet similar to the gibbsite structure bonded to a Si-tetrahedral sheet. Although it is a secondary mineral that forms during primary weathering reactions, continued leaching of kaolinite in soil profiles results in Al-rich mineral phases that ultimately form bauxitic ores. Therefore, an understanding of its dissolution is directly applicable to soil weathering as well as of interest for understanding dissolution of more complex phyllosilicates. Pure kaolinite has no layer charge, although a small amount of permanent charge may be a result of Al substitution for Si in tetrahedral positions or due to the fact that many kaolinites contain some interstratified smectite (e.g. Lim et al., 1980).

Rates from pH 0.5 to 12 were determined by Carroll-Webb and Walther (1988) and Carroll and Walther (1990) at 25°, 60°, and 80°C (Fig. 7). All the rates were measured in undersaturated solutions (I ≤ 0.05 M); however, solution composition varied due to the use of different pH buffer salts. They evaluated buffer concentration and within the scatter of

Kaolinite Dissolution Rates

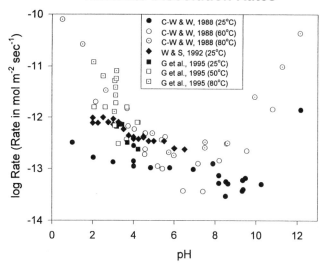

Figure 7. Summary of dissolution rate data for kaolinite as a function of pH and temperature. All rates are based on Si release data. References are abbreviated by first initials of authors' last names.

the experimental data observed no systematic effects. Rate data from Wieland and Stumm (1992) from pH 2 to 6.5 at 25°C (in 0.1 M NaNO₃) and from Ganor et al. (1995) (25°C, pH 3 to 4; 50°C, pH 2 to 4; 80°C, pH 3 in HClO₄ solutions) are also included in Figure 7. At 25°C, rates determined in batch experiments by Carroll-Webb and Walther (1992) are less than those determined by Wieland and Stumm (1992). Again, as for the Al-oxide phases, the difference in rates may be a result of differences in the time period over which the rates were determined. Alternatively, the lower rates when compared to the data from flow-through experiments (Ganor et al., 1995) may reflect effects of saturation state or Al-adsorption. Carroll-Webb and Walther (1988), Xie and Walther (1992), and Wieland and Stumm (1992) fit acid/base titration data to surface complexation models to determine the distribution of protonated surface species with pH (Fig. 8).

A linear dependence on pH in both acidic and basic ranges where the magnitude of n increases with increasing temperature was observed by Carroll-Webb and Walther (1988) and Carroll and Walther (1990) (Fig. 9). In contrast, in acidic solutions, Wieland and Stumm (1992) and Ganor et al. (1995) observed that at 25° and 50°C there is no pH dependence at pH < 3, and that from pH 3 to 4, n ≈ 0.4 (Fig. 10). The latter value is in best agreement with the 80°C data of Carroll and Walther (1990).

Based on slightly incongruent dissolution where Al is released at a slower rate than Si at 25°C, Carroll-Webb and Walther (1988) and Wieland and Stumm (1992) proposed that removal of Al is the rate limiting step in acidic solutions. At pH 4, the dissolution rate of δ-Al₂O₃ (Furrer and Stumm, 1986) is 1/2 to 1 order of magnitude faster than that of kaolinite. Corundum (Carroll-Webb and Walther, 1988) dissolves at about the same rate as kaolinite. Therefore, Al-O bond-breaking in the pure octahedral form appears to be at least equal to if not faster than in the kaolinite structure. Wieland and Stumm (1992) concluded that breaking of the Al-O-Si and Al-OH-Al bonds on the gibbsite-like basal surface and the edge occur simultaneously (Fig. 11). Ganor et al. (1995) hypothesized that the rate-limiting step was dissolution of the Al-O-Si bond only (Fig. 12). Si detachment is proposed as the rate limiting step in basic solutions (Brady and Walther, 1989; Carroll and Walther, 1990).

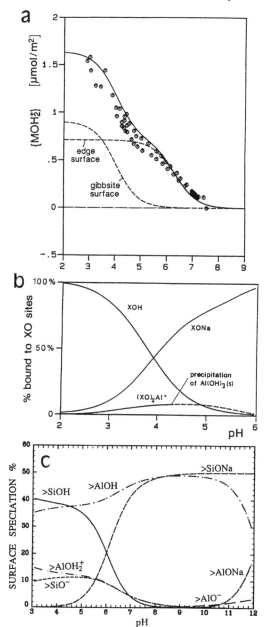

Figure 8. (a) Proton density determined experimentally (data points) and modeled (solid line) expressed as concentration of protonated sites on kaolinite surface as a function of pH. Titration curve has been modeled assuming two sites - on edge and gibbsite-like basal surfaces (dashed lines). Used by permission of the authors, from Wieland and Stumm (1992). (b) Surface speciation at metal (XO) sites on the siloxane layer. Modeled reactions include H^+/Na^+ and H^+/Al^{3+} ion exchange at the silanol sites. Used by permission of the authors, from Wieland and Stumm (1992). (c) Calculated surface speciation at kaolinite surface investigated by Carroll-Webb and Walther (1988). Modeled reactions include the silanol deprotonation reaction, in addition to H^+/Na^+ and H^+/Al^{3+} ion exchange. Used by permission of the authors, from Xie and Walther (1992).

Chrysotile. Chrysotile is a 1:1 layer serpentine mineral. It is trioctahedral with Mg and Fe as major and minor cations, respectively. Chrysotile forms cylindrical grains with an outer sheet of brucite. Under similar pH conditions brucite dissolves about 1000 times faster than chrysotile (Bales and Morgan, 1985) when rates are normalized to 1 mol Mg. This difference may be related to a stronger Mg-O-Si bond in chrysotile or to morphological differences that could result in different relative amounts of exposed surface

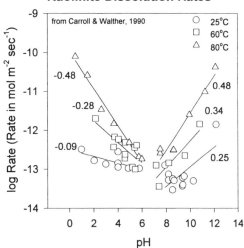

Figure 9. Kaolinite dissolution rates from Carroll and Walther (1990) as a function of temperature. Linear fits to the data in the acidic and basic regions were determined for this chapter. Slopes of lines indicate an increasing dependence on [H^+] concentration in solution with increasing temperature.

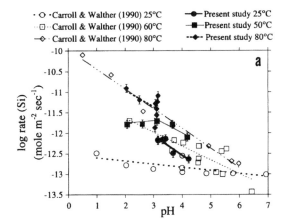

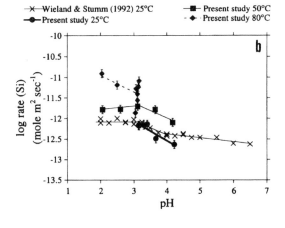

Figure 10.

(a) Kaolinite dissolution rates as a function of pH showing relation between data of Ganor et al. (1995) and those of Carroll and Walther (1990). Note the difference between slopes of both sets of data at 25° and 60°C. "Present study" refers to that of Ganor et al. (1995).

(b) Comparison of data from Ganor et al. (1995) with those of Wieland and Stumm (1992). Note similarity of slopes at 25°C from pH 3 to 4 and at 50°C from pH 2 to 4. Used by permission of the authors, from Ganor et al. (1995).

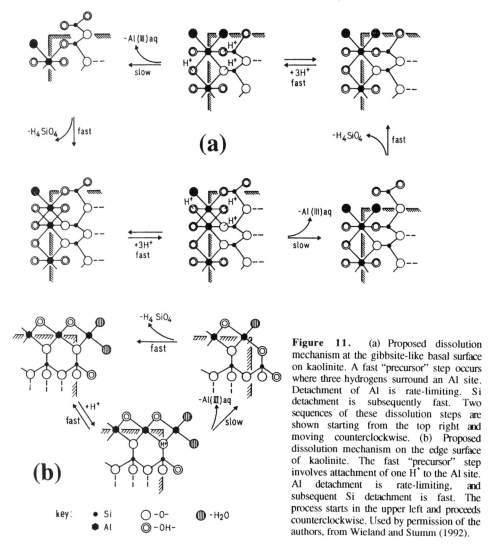

Figure 11. (a) Proposed dissolution mechanism at the gibbsite-like basal surface on kaolinite. A fast "precursor" step occurs where three hydrogens surround an Al site. Detachment of Al is rate-limiting. Si detachment is subsequently fast. Two sequences of these dissolution steps are shown starting from the top right and moving counterclockwise. (b) Proposed dissolution mechanism on the edge surface of kaolinite. The fast "precursor" step involves attachment of one H^+ to the Al site. Al detachment is rate-limiting, and subsequent Si detachment is fast. The process starts in the upper left and proceeds counterclockwise. Used by permission of the authors, from Wieland and Stumm (1992).

active sites. Bales and Morgan (1985) measured chrysotile dissolution rates in batch experiments at 25°C from pH 7 to 10 (Fig. 13). They observed that Mg was preferentially removed over a period of up to 5 days, and that the initial rate during the first few hours was much higher. For Mg dissolution, the reaction order n = 0.24 which the authors suggested may indicate a dissolution mechanism involving both H^+ and H_2O adsorption. The Si dissolution rates showed a slightly positive or no dependence on $[H^+]$ from pH 7 to ~8.5 and a negative dependence of -0.19 from pH 8.5 to 10 (Fig. 13). Surface speciation calculations showed that $>Mg-OH_2^+$ is the dominant charged surface site (Fig. 14). The reaction order with respect to the concentration of this site is 0.75. Hume and Rimstidt (1992) investigated chrysotile dissolution in solutions representing lung fluids (final pH of 3.4 to 7.4, Si and Mg in the range of 10^{-6} to 10^{-3} m) at 37°C. They observed no pH-dependence over this range. The average rate calculated after only three hours of

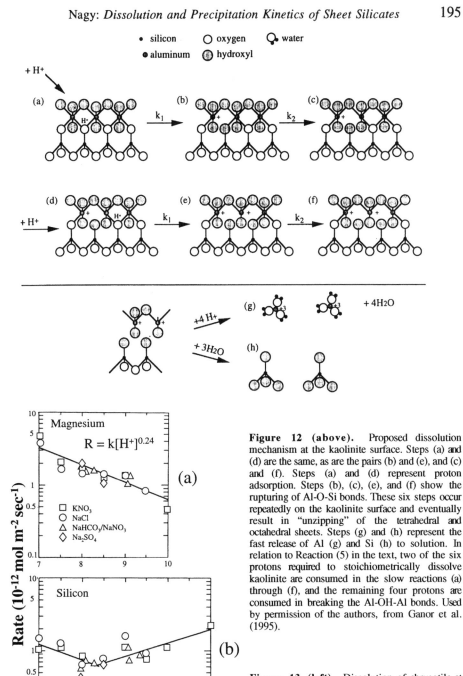

Figure 12 (above). Proposed dissolution mechanism at the kaolinite surface. Steps (a) and (d) are the same, as are the pairs (b) and (e), and (c) and (f). Steps (a) and (d) represent proton adsorption. Steps (b), (c), (e), and (f) show the rupturing of Al-O-Si bonds. These six steps occur repeatedly on the kaolinite surface and eventually result in "unzipping" of the tetrahedral and octahedral sheets. Steps (g) and (h) represent the fast release of Al (g) and Si (h) to solution. In relation to Reaction (5) in the text, two of the six protons required to stoichiometrically dissolve kaolinite are consumed in the slow reactions (a) through (f), and the remaining four protons are consumed in breaking the Al-OH-Al bonds. Used by permission of the authors, from Ganor et al. (1995).

Figure 13 (left). Dissolution of chrysotile at 25°C. (a) Rates based on Mg release. Decrease in rate with increasing pH may be partly a function of increasing saturation of the solution. (b) Rates based on Si release. Modified after Bales and Morgan (1985).

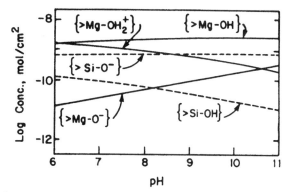

Figure 14. Distribution of surface species on chrysotile. Used by permission of the authors, from Bales and Morgan (1985).

dissolution was 6×10^{-10} mol m^{-2}s^{-1}, about 2.5 orders of magnitude higher than observed by Bales and Morgan (1985).

Antigorite. Lin and Clemency (1981c) investigated the dissolution rate of antigorite at 25°C in CO_2-buffered near-neutral solutions. Antigorite is also a 1:1 layer serpentine that is polymorphic with chrysotile, but with a much larger a dimension. Its rate of dissolution at pH = 6.5 is similar to the lowest rates of dissolution measured for chrysotile at pH 8 by Bales and Morgan (1985).

Talc. Talc is a 2:1 dioctahedral layer silicate with essentially no permanent charge, and hence no interlayer cations. Lin and Clemency (1981c) investigated the relative rates of dissolution of brucite, antigorite, talc, and phlogopite, a series of Mg-layer phases that increase in structural and compositional complexity. Batch "pH-drift" experiments with dry-ground solids were conducted at 25°C, 1 atm P_{CO_2}, and a final pH of ~5. They found decreasing rates of Mg release normalized to surface area in the order brucite > antigorite > phlogopite > talc. However, based on Si release data the dissolution rate is calculated to be 1.3×10^{-12} mol m^{-2}s^{-1}, faster than the phlogopite dissolution rate based on Si release (see below). Mg was always released in excess of its stoichiometric value.

Muscovite. Muscovite is also a 2:1 dioctahedral layer silicate, but with a permanent structural charge of ~1, which is satisfied by the presence of interlayer K. A summary of dissolution rates as a function of pH is shown in Figure 15. All rates are relative to a unit formula of $KAl_2(Si_3Al)O_{10}(OH)_2$.

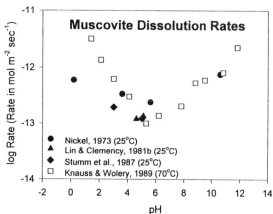

Figure 15. Muscovite dissolution rates as a function of pH and temperature.

At 25°C, Nickel (1973) determined muscovite dissolution rates at at pH 0.2, 3.6, 5.6, and 10.6. When normalized to the initial BET surface area, the three rates in order of increasing pH are 6, 3.4, 2.4, and 7.5 10^{-13} mol m^{-2} s^{-1}. Using previously dissolved muscovite, he repeated the pH 10.6 experiment and obtained a dissolution rate about 1/2 the original rate. Lin and Clemency (1981b) measured dry-ground muscovite dissolution in a CO_2 buffered (pH = 4.6 to 5.1) experiment at 25°C. Dissolution was incongruent, and Si had the slowest release rate. Two different muscovites had release rates an order of magnitude apart: 1.2×10^{-14} and 1.4×10^{-13} mol m^{-2} s^{-1}. K^+ and Na^+ rapidly exchanged for H^+ during the initial stages of the experiment, during which no change in surface area was observed. Mg release was used to follow the dissolution of the octahedral layers. They pointed out that in previous studies muscovite dissolution was both congruent and incongruent and that this may be because the muscovite had been wet ground or sieved leading to different extents of preferential initial leaching. Dry-grinding may have adverse affects as well and may cause preferential weakening of octahedral over tetrahedral bonds at surfaces and edges. Stumm et al. (1987) reported muscovite dissolution rates at pH 3 and 5; details of these experiments are found in Wieland (1988). Their rates compare to the faster rate of Lin and Clemency (1981b). Nickel (1973) reported ~1% mineralogic impurities in his muscovite sample which Knauss and Wolery (1989) suggested could contribute to the slightly faster rates.

Boles and Johnson (1983) performed surface potentiometric titrations of muscovite (and biotite) at 25°C in 0.01, 0.1, and 0.5 M NaCl solutions using HCl and NaOH as titrants. They determined that muscovite adsorbed H^+ only below pH 5.5. At higher pH, Na was preferentially adsorbed and H^+ was released. (Biotite adsorbed H^+ from pH 4 to 10.) Increasing the concentration of NaCl decreased the amount of proton adsorption, suggesting a competing adsorption reaction. However, neither muscovite nor biotite surface complex speciation has been modeled in relation to dissolution rate data.

Knauss and Wolery (1989) determined muscovite dissolution rates as a function of pH at 70°C in dilute pH-buffered solutions ranging in ionic strength from 0.002 to 0.13 m. The primary difference in this and the previous two studies is the pretreatment of the muscovite. Knauss and Wolery (1989) used larger (100 μm) particles and removed fines by repeated settling and ultrasonic cleaning in isopropanol. They also used single-pass flow-through experiments vs. the batch dissolution experiments used in prior studies. Experimental duration was similar in all three studies. The dissolution rate at neutral pH is 1/8 the rate determined by Lin and Clemency (1981b) at 25°C. Knauss and Wolery (1989) suggested that the absence of fines in their study may account for their lower rates. They observed no evidence for dissolution such as etch pit formation on the basal surface using SEM. The low dependence of log rate on pH seen in the data of Nickel (1973) is surprising in that the rates measured at extreme pH's are lower than those of Knauss and Wolery (1989) (at 70°C) whereas the rate at near-neutral pH is higher (Fig. 15). These slope differences cannot be explained solely by temperature or the presence of impurities.

Illite. Illite is a "nonexpanding, dioctahedral, aluminous, potassium mica-like mineral which occurs in the < 4 μm fraction" and is naturally-occurring (Srodon and Eberl, 1984). Heydemann (1966) measured illite (Fithian, Illinois, USA) dissolution at pH 3, 6, and 10. She monitored solution composition in small batch reactors which had their solution replaced periodically. I have interpreted her data by applying the numerical analysis one would use for a single-pass flow-through reactor to her experimental approach. Based on rates of Si release, a low pH-dependence in both acidic and basic solutions (-0.02 and 0.11, respectively) is obtained and rates are on the order of 10^{-15} to

10^{-14} mol m^{-2} sec^{-1}, at least a factor of 10 lower than for any other phyllosilicate. Feigenbaum and Shainberg (1975) also measured dissolution rates of Fithian illite at 5° and 25°C in pH ~7.5 and 2.9 (0.001 M CaCl$_2$ and 0.001 N HCl, respectively) using ion exchange resins. At 25°C, the rate calculated from K release between 144 and 840 hours of reaction time using the same surface area as reported by Heydemann (1966) is three orders of magnitude higher than the rate at pH 6 from Heydemann (1966), ~4×10^{-12} vs. 6×10^{-15} mol m^{-2} s^{-1}. At pH 2.9, 25°C, the calculated rate is 2.4 x 10^{-10} mol m^{-2} s^{-1}. The reaction order with respect to [H$^+$] is 0.39, in line with that determined more recently for kaolinite, muscovite and biotite over a similar pH range. Feigenbaum and Shainberg (1975) also measured Mg, Al, and Fe release rates. They found that Al and K release were in stoichiometric proportions, but Al and Fe release were not. They proposed that the larger Mg^{2+} and Fe^{3+} ions strained the crystal lattice resulting in weaker metal-oxygen bonds, and that breakage of an Al-O bond in the Al-O-Si linkage requires more energy than breaking an Mg-O or Fe-O bond in the Mg-O-Si or Mg-O-Al linkages.

Montmorillonite. Montmorillonite is a dioctahedral smectite mineral. Its permanent surface charge ranges from 0.2 to 0.6, and thus, like biotite, it can have complex cation compositions in both the tetrahedral and octahedral sheets. Heydemann (1966) measured montmorillonite dissolution rates (~10^{-16} to 10^{-15} mol m^{-2} sec^{-1}) as a function of pH at 25°C. Recently, Furrer et al. (1993) measured montmorillonite (SWy-1) dissolution rates of 8.6×10^{-13} to 4.3×10^{-14} mol m^{-2} sec^{-1} at pH 1 to 5. Their rates are reported in units of mol g^{-1} hr^{-1}, and can be converted to units of mol m^{-2} s^{-1} for comparison with data collected for other sheet silicates using the surface area (32 m^2 g^{-1}) for SWy-1 montmorillonite in Van Olphen and Fripiat (1979). The montmorillonites studied by Heydemann (1966) had [H$^+$]-dependencies from pH 3 to 6 of 0.23 and 0.30. Furrer et al. (1993) obtained 0.38 and 0.22 in batch and mixed-flow experiments, respectively.

Biotite. Biotite is a trioctahedral mica which can have complex octahedral and interlayer cations. During weathering, it can alter to vermiculite, hydrobiotite (regularly interstratified biotite-vermiculite), and kaolinite. Acker and Bricker (1992) investigated biotite dissolution in acidic solutions at 25°C (Fig. 16). Their data cover a pH range of 3 to 5.5; oxidizing and reducing solutions, and saturated vs. unsaturated flow conditions. When compared to other 2:1 phyllosilicate dissolution rates, the pH-dependence is similar to that

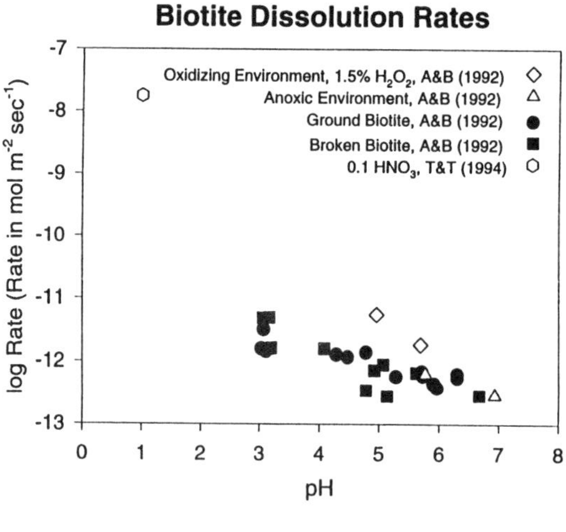

Biotite Dissolution Rates

Legend:
Oxidizing Environment, 1.5% H$_2$O$_2$, A&B (1992) ◇
Anoxic Environment, A&B (1992) △
Ground Biotite, A&B (1992) ●
Broken Biotite, A&B (1992) ■
0.1 HNO$_3$, T&T (1994) ◇

y-axis: log Rate (Rate in mol m^{-2} sec^{-1})
x-axis: pH

Figure 16. Biotite dissolution rates as a function of method of solid preparation and oxidizing level of solutions (A&B = Acker and Bricker; T&T = Turpault and Trotignon). Ground biotite was prepared by grinding small broken flakes of biotite with a ceramic ball mill. Data are presented from three size fractions: >40 mesh, 40-100 mesh, and < 100 mesh. Broken biotite was prepared by hand. Data are presented from 40-100 and 45-100 size fractions. No consistent trends were observed between the two methods of preparation or in the size fraction results. Anoxic conditions were obtained by continuous bubbling with nitrogen. Single cut crystals of biotite were used in experiments of Turpault and Trotignon (1994).

observed by Knauss and Wolery (1989) at 70°C for muscovite and to that for montmorillonite at 25°C (Furrer et al., 1993) (Table 4). Also, the magnitude of the dissolution rate is higher than for muscovite by about approximately a factor of 5. Like previous workers, they confirmed that dissolution occurs from the edges inward, based on SEM examination of dissolved grains. They proposed that edges alter to vermiculite, particularly in the saturated column experiments. The alterations appear to cause the biotite dissolution reaction to slow down. Trotignon and Turpault (1992) and Turpault and Trotignon (1994) investigated biotite dissolution in HNO_3 solutions. The rate at pH 1 in Turpault and Trotignon (1994) is about an order of magnitude slower than reported in their earlier paper (Fig. 16). They determined that the dissolution rate was a strong function of the edge surface area and only a weak function of the basal surface area. Thus, they proposed that the rates in Acker and Bricker may be underestimated because they were normalized to total BET-determined surface area. They also proposed that the dissolution rate of biotite was limited by Si release and that the edges develop an alteration front of amorphous Si. This is in agreement with results of many early acid-leaching studies.

Phlogopite. Lin and Clemency (1981a) measured dry-ground phlogopite dissolution in a CO_2 buffered (pH = 5.34) closed system at 25°C. Assuming linear reaction kinetics with time, the dissolution rate based on Si release was 3.8×10^{-13} mol m^{-2} sec^{-1}. K^+/H^+ exchange occurred early in the experiment and equaled the measured cation exchange capacity. Surface area remained constant during initial ion exchange indicating that 1/3 of the cations come from basal surfaces and 2/3 from deeper within the basal surface or from edges. They suggested that precipitated Al and Fe hydroxide phases account for an increased surface area from 3.8 to 5.8 m^2 g^{-1} after 1010 hours. Mg release was about 1.5 times greater than Si release. F$^-$ release was constant with time suggesting that F$^-$ was controlled by exchange equilibrium at octahedral edge sites. Using an ion-exchange resin to adsorb dissolved cations and to provide a source of H^+ to maintain an acidic pH, Clemency and Lin (1981) measured a congruent dissolution rate at pH 3.3 equal to 2.0×10^{-10} mol m^{-2} sec^{-1} during the first 200 hours of dissolution. However, an estimate based on Si release using the data for change in weight percent of phlogopite gives a slower rate that can be compared with the rate at pH 5.3 from Lin and Clemency (1981a). Al and Fe immediately precipitated and were not used to calculate rate. At pH 3.3 and 5.3 the rate of Si release is 4.4×10^{-11} and 3.8×10^{-13} mol m^{-2} sec^{-1}, respectively. The calculated value of n is 1.0. Although higher by a factor of 2 than any other reported [H^+]-dependence for phyllosilicates, this value is similar to that for gibbsite (Bloom and Erich, 1987) and bayerite (Pulfer et al., 1984). The value of n for kaolinite dissolution ranges up to 0.5 from 25° to 80°C (Carroll and Walther, 1990; Ganor et al., 1995). Biotite in H_2O_2 solutions has n = 0.66 from pH 5 to 5.7 (Acker and Bricker, 1992).

Chlorite. From pH 3 to 5.2, n is ~0.5 (May et al., 1995). At pH 5, the dissolution rate is 3×10^{-13} mol m^{-2} s^{-1} which they report is one to three orders of magnitude slower than other rates for chlorites with higher iron contents. This rate is similar to the phlogopite dissolution rates cited above and slightly slower than the biotite dissolution rate (Acker and Bricker, 1992). It is also similar to chrysotile if the rate at pH 7 (Bales and Morgan, 1985) is extrapolated to pH 5 assuming n = 0 (Hume and Rimstidt, 1992).

Effect of temperature

Dissolution rates of phyllosilicates and related minerals are modeled using the Arrhenius rate law ($R_{diss} = k \exp(-E_a/RT)$) where k is a rate coefficient, E_a is the activation energy for dissolution, R is the gas constant and T is temperature in K. Summaries of activation energies are given in Tables 7 and 8 including data from acid-leaching studies.

Gibbsite. Activation energies for gibbsite dissolution have been determined for a variety of conditions (Table 7). In acidic dilute solutions, the activation energy is 45 to 68 kJ mol^{-1}. In extremely basic solutions, E_a is ~70 to 130 kJ mol^{-1}.

Brucite. From 25° to 75°C, Vermilyea (1969) determined an activation energy for brucite dissolution of ~42 kJ mol^{-1}.

Table 7. Activation energies for dissolution kinetics of gibbsite

T (°C)	E_a (kJ mol^{-1})	pH range	Solution Composition	Reference
35 - 65	45	-0.8 - -0.5	HCl, HClO$_4$	Packter & Dhillon (1969)
"	60	-0.5	H$_2$SO$_4$	"
"	52	-0.8 - -0.5	HClO$_4$	"
25 - 55	68	1.75	HNO$_3$, KNO$_3$	Bloom (1983)
10 - 40	60 - 67	1.75 - 2.19	HNO$_3$, H$_2$SO$_4$,	Bloom & Erich (1987)
"	60	2.27 - 2.93	H$_3$PO$_4$	"
"	64	2.31		"
25 - 80	56	3	HCl/HClO$_4$	Nagy & Lasaga (1992); Mogollón et al. (1995)[a]
50 - 90	73 - 77	14.4 - 14.6	NaOH	Peric´ et al. (1985)
50 - 90	97	14.7	NaOH	Peric´ et al. (1985)
35 - 65	92 - 97	13.8 - 15	NaOH	Packter & Dhillon (1973)
20 - 65	76 - 83	13.3 - 15	NaOH	Packter & Dhillon (1974)
25 - 100	129, 107[b]	14.2	NaOH	Scotford & Glastonbury (1971)

Note: a: Data from the two studies were combined; b: bauxite ore

Table 8. Activation energies for dissolution kinetics of sheet silicates and brucite

Mineral	T (°C)	E_a (kJ mol^{-1})	pH range	Solution Composition	Reference
Brucite	25 - 75	~ 42	1 - 3	0.1 M KCl	Vermilyea (1969)
Chrysotile	22 - 80	63 - 84	1	0.1 N oxalic acid	Thomassin et al. (1977)
Kaolinite	25 - 80	67 → 7.1	1 - 7	pH buffers, I = 0.05 M	Carroll & Walther (1990)
Kaolinite	25 - 80	14.2 → 41	8 - 12	"	"
Kaolinite	25 - 80	29.3	3 - 4	HClO$_4$	Ganor et al. (1995)
Kaolinite	0 - 50	~ 54	-0.7 - 1	HF	Kline and Fogler (1981a)
Illite	0 - 45	"	"	"	
Pyrophyllite	25 - 70	"	"	"	
Muscovite	38 - 70	"	"	"	
Muscovite*	25 - 70	22	3	HCl, dilute buffers, 0.1 M NaNO$_3$	Nickel (1973), Stumm et al. (1987), Knauss & Wolery (1989)
Talc, Phlogopite	25 - 60	~ 42	-0.7 - 1	HF	Kline and Fogler (1981a)
Phlogopite	50 - 120	29 - 42	2	0.01 N HCl/ 0.1 M NaCl	Kuwahara & Aoki (1995)
Na-Montmorillonite	0 - 25	~ 48	-0.7 - 1	HF	Kline and Fogler (1981a)
Clinochlore	20 - 60	88	-0.3	HCl	Ross (1967)
Palygorskite	60 - 91	77	-0.7 - -0.3	HCl	Abdul-Latif & Weaver (1969)

Note: Muscovite activation energy was calculated using data from all three studies listed.

Table 9. Kaolinite dissolution E_a

E_a (kJ mol^{-1}) ($\pm$ 2.5 kJ mol^{-1})	pH
66.9	1
55.6	2
43.1	3
32.2	4
20.5	5
9.6	6
7.1	7
14.2	8
22.2	9
20.3	10
35.6	11
41.1	12

Note: From Carroll & Walther (1990).

Kaolinite. Carroll and Walther (1990) and Ganor et al. (1995) determined E_a for kaolinite dissolution from 25° to 80°C (Table 8). The former observed that E_a varied with pH (Table 9), with a minimum at neutral pH. Ganor et al. (1995), on the other hand, observed that over the limited pH range of 3 to 4, activation energy was constant from 25° to 80°C at a value of 29 kJ mol^{-1}. From pH 3 to 4, Ganor et al. (1995) constrained their value of E_a primarily using 25° and 50°C data. Their 80°C rates were obtained at one pH only (~3.2) and show almost an order of magnitude variation. At pH 2 to 3, the data of Ganor et al. (1995) at 50° and 80°C suggest that E_a is higher in more acidic solutions.

Muscovite. Data for muscovite dissolution as a function of Si release rate exists at two temperatures: 25°C (Nickel, 1973; Lin and Clemency, 1981b, Stumm et al., 1987) and 70°C (Knauss and Wolery, 1989) (Fig 15). They show a similarity to the kaolinite dissolution data in that the slopes of the dissolution rates as a function of pH increase with temperature. At nearly neutral pH the 70°C rates are similar to if not lower than the 25°C rates. Casey and Sposito (1992) proposed that negative E_a values are possible if interpreted as a sum of enthalpies of similar magnitude but opposite sign. These enthalpies include that of the activated complex equilibrium and that of proton adsorption/desorption. The proton adsorption enthalpy affects E_a through the concentration of surface charged sites. If the absolute rates reported by Nickel (1973), which appear to be on the high side possibly due to sample impurities, are ignored, an E_a = 22 kJ mol^{-1} is obtained at pH 3. This is about 50% of the E_a at pH 3 for kaolinite (Carroll and Walther, 1990) and about 75% of the value from Ganor et al. (1995).

Phlogopite. Kuwahara and Aoki (1995) studied phlogopite dissolution in 0.01 N HCl/0.1 M NaCl solutions at pH 2 from 50° to 120°C. They observed nearly congruent dissolution at 50° and 80°C. Only at 120°C was dissolution incongruent due to the precipitation of Al and Fe oxyhydroxides. They observed that elemental release rates decreased in the order K > Fe > Mg, Al > Si. At 50°C, about 40% of the initial phlogopite was transformed to vermiculite after about 30 days in 100 ml of solution as determined by XRD. In contrast, at 120°C, the same percentage of transformation occurred after 10 days in 20 ml of solution. Calculated E_a's of 29 to 42 kJ mol^{-1} were obtained between 50° and 120°C. This range of E_a is within that for kaolinite dissolution from 25° to 80°C (Carroll and Walther, 1990).

Chrysotile. Dissolution rates exist at more than one temperature (25° and 37°C) for chrysotile (Bales and Morgan, 1985; Hume and Rimstidt, 1992). At pH = 7, the rates from these two studies overlap. Using an E_a of 30 kJ mol^{-1} and the rate at pH 7 from Bales and Morgan (1985), the rates reported in Hume and Rimstidt (1992) are about 400 times higher than estimated. Hume and Rimstidt (1992) measured rates after 3 hours in solutions simulating lung fluids. These rates may be fast compared to rates measured under longer-term weathering conditions.

Stoichiometry versus nonstoichiometry of dissolution

The question of stoichiometric or congruent dissolution versus nonstoichiometric or incongruent dissolution has been raised for many sheet silicates in many types of solutions. The general observation for many 2:1 minerals in acidic to neutral solutions is that the rate of release is greatest for Mg-octahedral sheets and presumably Fe, although this tends to precipitate out precluding its use for determining accurate rates (e.g. Acker and Bricker, 1992), intermediate for Al-octahedral sheets, and slowest for Si-tetrahedral sheets. This contrasts with the observations on kaolinite, a 1:1 layer mineral, where under acidic conditions at 25°C, Al release rates are observed to be slower than Si release rates.

All of the dissolution rates reported above were based on release of Si (except where otherwise noted) because octahedral cations such as Al or Mg can rapidly reprecipitate or adsorb to the clay surfaces in circumneutral solutions. In general, nonstoichiometric dissolution almost always occurs under initial dissolution conditions, but with increasing duration of an experiment, at T > 25°C, or by pretreating solids, dissolution approaches congruency rapidly when undersaturation with respect to Al or Mg phases is maintained.

Aside from acid-leaching studies in which it was shown that Al and Mg can be rapidly dissolved relative to Si from clay edges, relative rates of dissolution of Si, Al, and Mg from sheet silicates have been considered in two other studies at higher pH. Lin and Clemency (1981a,b,c) compared dissolution rates of brucite, talc, antigorite, and phlogopite at near-neutral pH. Although there were some slight complications because solution conditions were not always exactly the same, they were able to show that the cation dissolution rate decreased in the order Mg > Al > Si. May et al. (1995) demonstrated the same order of incongruency by looking carefully at the ratios of cations released from dissolving chlorite in a fluidized bed reactor at pH 3 to 5 and 25°C. Their data support the interpretation that release rate is correlated to metal-oxygen bond strength. The relatively short experiments on chrysotile dissolution by Bales and Morgan (1985) at pH 7 to 10 also showed that Mg release was in excess of its stoichiometric value.

Acker and Bricker (1992) observed that biotite dissolved incongruently from pH 3 to 7 with preferential release of Mg and Fe from the octahedral sheet. In column experiments under saturated flow conditions at pH 4, the incongruency was consistent with vermiculite formation. Turpault and Trotignon (1994) noted that initial fast ion exchange of H^+ for K^+ occurs, in agreement with early results of Newman (1969). This is followed by release of Al and Fe, then Mg and Ti, and finally Si. They suggested that octahedral cation release rates depend on the structure of the octahedral sites as well as ligand-promoted dissolution in which the ligands also dissolve from the biotite. In contrast, Furrer et al. (1993) observed nearly congruent dissolution of pretreated montmorillonite from pH 1 to 4 at 25°C and said this supported dominant dissolution at edge sites. Cama et al. (1994) observed congruent dissolution of pretreated smectite based on release rates of tetrahedral and octahedral cations at pH 8.8 and 80°C. For solutions with input Si concentrations <100 μM, the release of Al, Si, and Mg was congruent. Fe, Ca, and K, on the other hand, dissolved incongruently. Fe was suspected to precipitate as $Fe(OH)_3$, and Ca and K in the interlayers may have rapidly exchanged with Na from the borax buffer despite two months of pretreatment in the buffer solution.

One might think that such data could be extended to explain the incongruency of dissolution rates for more simple sheet silicates such as kaolinite. For example, Carroll-

Webb and Walther (1988) showed that during initial dissolution of kaolinite (Washington Co., Georgia) Al was released at a faster rate than Si. On the other hand, Wieland and Stumm (1992) examined dissolution rates of St. Austell kaolinite from Cornwall, England, and observed incongruency in the opposite sense (Fig. 17). Nagy et al. (1991) used XPS to examine dissolved Twiggs Co., Georgia kaolinite and found no significant variation in the Si/Al ratio (0.96 ± 0.03) in the near-surface region of the solid for either basal or edge surfaces. Carroll and Walther (1990) and Nagy et al. (1991) both reported congruent dissolution of kaolinite at $T > 25°C$, essentially from the beginning of batch and stirred-flow reactor experiments, respectively. In all of these studies, the kaolinite was pretreated, although by different procedures.

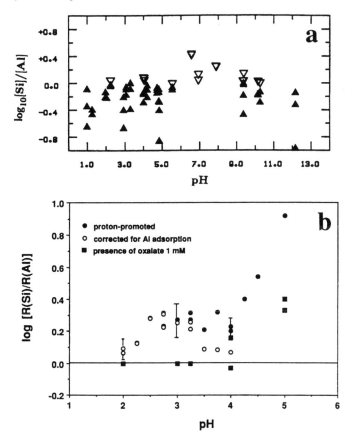

Figure 17. Stoichiometry of dissolution of kaolinite. (a) Long-term rate data showing deficiency of Si release (closed upright triangles). Excess Si release (open inverted triangles) is observed at circumneutral pH due to back precipitation or adsorption of aluminum. Used by permission of the authors, from Carroll-Webb and Walther (1988). (b) Rate data showing excess Si release in acidic solutions. When data are corrected for Al adsorption the ratio of rates of release of the two elements approaches 1. In the presence of 1mM oxalate the release data become stoichiometric from pH 2 to 4. Used by permission of the authors, from Wieland and Stumm (1992).

How can we explain the wide range in incongruency observed for kaolinite dissolution? Three explanations are probable: (1) the composition of the reacting solids was different, (2) temperature and previous dissolution events control the extent of incongruency which is only a transient phenomenon, and (3) aluminum and silicon dissolve congruently, but aluminum immediately readsorbs causing apparent incongruency. Kaolinite is a phyllosilicate that can form under low to high temperature conditions. Georgia (USA) kaolinites for the most part form by weathering of igneous rocks and are redeposited as sedimentary units a short distance away. Some continued authigenic growth may occur (Fripiat and Van Olphen, 1979) resulting in the presence of additional aluminous or siliceous phases such as gibbsite, quartz, or amorphous silica or alumina. Such phases are difficult to remove in pretreatment steps and may account for some of the observed incongruency. These phases will dissolve quickly at higher temperatures, thus making the incongruent dissolution period appear shorter or perhaps go unnoticed depending on sample spacing. Cornwall kaolinite is of hydrothermal origin and formed as veins in areas of metal ore deposits. It could also be contaminated with faster reacting phases. However, Wieland and Stumm (1992) conducted a more extreme series of pretreatment steps in order to prepare their kaolinite for dissolution experiments. This could explain the fact that the incongruency they observed was inverted relative to that observed by Carroll-Webb and Walther (1988). Wieland and Stumm (1992) also explained their incongruent data as being caused by Al adsorption and precipitation of an aluminous phase on the dissolving kaolinite surface. In the presence of Al-complexing ligands, their dissolution data were congruent. Nagy et al. (1991) proposed that incongruency of kaolinite dissolution near equilibrium resulted from precipitation of an aluminous phase, although Al adsorption could have been occurring as well.

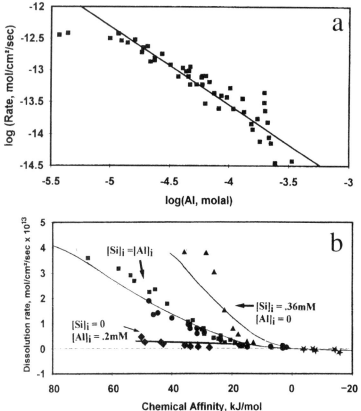

Figure 18. Dissolution rate of kaolinite at 150°C, pH 2. (a) Log of the dissolution rate is a linear function of Al concentration except in extremely undersaturated solutions. (b) Dissolution rate (all data except stars) as a function of chemical affinity. Positive affinity values represent undersaturation; negative values represent supersaturation. The curves were obtained from Reaction (14). Inlet solution concentrations of Al and Si are given for each curve. Used by permission of the authors, from Oelkers et al. (1994).

Rate catalysts and inhibitors

Rate catalysts and inhibitors are dissolved ions or molecules that may accelerate the rate-limiting elementary reaction or prevent it from occurring. They may play a chemical role in the rate-limiting dissolution mechanism or in the case of an inhibitor may physically block reactive sites. Below are some data that have been obtained defining the effects of catalysts and inhibitors during Al-oxide, Al-hydroxide, and phyllosilicate dissolution.

Aluminum and silicon. Devidal et al. (1992) and Oelkers et al. (1994) modeled the dissolution of kaolinite at 150°C, pH 2 and 7.8 as a function of reaction affinity and Al concentration. They proposed that an Al-deficient precursor complex on the kaolinite surface controls the dissolution rate over a significant range of Al concentration (Fig. 18). The reaction describing the equilibrium between the precursor and solution is

$$\text{M-}n\text{Al} + 3n\text{H}^+ + \sum_i \nu_i A_i \Leftrightarrow P^{\bullet} + n\text{Al}^{3+} \tag{8}$$

where M-nAl is an Al surface site, n is the number of Al ions in each potential precursor site, A_i is the ith aqueous species involved in the precursor formation, ν_i are stoichiometric coefficients, and $P^{\bullet}$ is the surface precursor complex. Under far-from-equilibrium conditions, the reaction order with respect to m_{Al} is 1.

Furrer et al. (1993) also observed that montmorillonite dissolution was retarded by a factor of ~2 to 3 in the presence of dissolved Al (up to 100 μmol), but only in pH 4 solutions. Addition of dissolved Al from 10 to 100 μmol in single-pass flow experiments at pH 1 to 3 did not change the steady-state dissolution rates. At pH > 2.5 they proposed that adsorption of dissolved Al may cause the slightly higher release rates of Si. They also indicated that additional Al may enhance particle aggregation, thereby reducing the amount of reactive sites exposed to solution. Walker et al. (1988) showed that Al adsorbed preferentially over H$^+$ to the surfaces of kaolinite, vermiculite, and montmorillonite from 10° to 25°C at pH 3.0 to 4.1. They showed that the rate of adsorption followed a first order dependence on initial aluminum concentration, and that the formation constant for an Al^{3+} surface complex decreased in the order kaolinite > montmorillonite > vermiculite. The fraction of complexed Al^{3+} on montmorillonite increases by a little less than a factor of 2 going from 10 to 100 μmol in solution, approximately consistent with the observed decrease in dissolution rate observed by Furrer et al. (1993). Jardine et al. (1985) also studied Al adsorption on phyllosilicates.

Siever and Woodford (1973) determined the sorptive capacity of kaolinite, montmorillonite, illite, and gibbsite for Si at pH values from 4.2 to 9.1. They observed that below a certain initial Si value that varied with pH and mineral phase, all these clays dissolved, whereas above the same initial Si value, Si was adsorbed. In particular, gibbsite was particularly efficient at adsorbing Si. On one hand, Bloom and Erich (1987) did not observe a change in dissolution rate of gibbsite in acidic solutions (< pH 3) containing 100 μm Si. Jepson et al. (1976), on the other hand, showed that Si adsorption on gibbsite at pH 4 and 9 does inhibit Al dissolution. The data of Devidal et al. (1992) indicate that Si adsorption does not appear to inhibit kaolinite dissolution in acidic to neutral solutions.

Other metals and anions. No difference was observed in gibbsite dissolution rates with acid composition (HCl or HClO$_4$) at pH 3, 80°C (Nagy and Lasaga, 1992) and at pH 3.3 to 3.9, 25°C (Mogollón et al., 1994). In contrast, in extremely acidic solutions (1 to 6 molal), rates of gibbsite dissolution are about 3 times faster in HCl than in HClO$_4$ and in H_2SO_4, rates are 15 to 30 times greater than in $HClO_4$ for equi-molar concentrations (Packter and Dhillon, 1969). An increase of 30 times also was observed at pH 1.75 to 2.8 in K_2SO_4 over KNO_3 solutions (Bloom and Erich, 1987).

Bloom and Erich (1987) observed that gibbsite dissolution rates varied with acid anion concentration as expressed by

$$R = k\,[anion]^n \tag{9}$$

The reaction order with respect to nitrate (pH = 1.74), sulfate (pH = 2.16), and phosphate (pH = 2.38) concentrations are 0.56, 0.36, and 0.88, respectively. A much lower concentration of phosphate (0.0001 M) yielded no pH dependence and rates that were both less than (at pH < 2.7) and greater than rates in 0.1 M K_2SO_4 solutions (Fig. 6).

Pulfer et al. (1984) observed an approximately linear ($n \approx 1$ to 1.5) variation in bayerite dissolution rate with F^- concentration at high $[F^-]$. At low concentrations, rate is independent of fluoride. Based on these two studies, the increase in the absolute value of n with respect to acid anion concentration is $SO_4^{2-} < NO_3^- < PO_4^{3-} < F^-$. However, relative dissolution rates also depend on pH (Pulfer et al., 1984; Bloom and Erich, 1987). Therefore, surface complexation models used to explain the pH dependence must consider complexes involving both these acid anions and OH^- (e.g. Pulfer et al., 1984). Zutic´ and Stumm (1984) determined the effect of fluoride concentration (5×10^{-6} to 10^{-3} M) on dissolution of hydrous alumina using a rotating disc aluminum electrode coated with an amorphous Al-oxide layer. They observed that even a small amount of F^- greatly increased the dissolution rate over the pH range 3 to 6. Additional data on surface chemistry of aluminum phases that can be related to dissolution rates are found in the review by Davis and Hem (1989).

Neither Mg nor Si release rates from chrysotile are affected by increasing SO_4^{2-}, or NO_3^- from 0.01 to 0.1 mol L^{-1} at pH 8 to 8.5 (Bales and Morgan, 1985). Short term dissolution rates of chrysotile are enhanced in the presence of F^- and PO_4^{3-} (Gupta and Smith, 1975). At pH 1 to -3, no effect of acid anion (Br^-, I^-, Cl^-, NO_3^-) composition on kaolinite dissolution was observed (Kline and Fogler, 1981b,c). However, enhanced rates were observed in HF solutions. They related the cation catalysis to the total charge at the kaolinite surface arising from adsorbed ions at surface hydroxyls.

Rates of dissolution of gibbsite in $Ca(NO_3)_2$ and $Mg(NO_3)_2$ are similar to those in KNO_3 (Bloom and Erich, 1987). In contrast, in very basic solutions at temperatures of 20° to 50°C, the dissolution rate of gibbsite slows down in the presence of Ba, Ca, and Sr (Packter and Panesar, 1986), presumably due to the precipitation of metal hydroaluminate hydrate phases of these metal cations on the gibbsite surface.

Increasing KCl concentration from 0.1 to 1.0 M doubled the dissolution rate of montmorillonite at pH 2.5 (Furrer et al., 1993). They suggested that modification of the electric double layer at the mineral surface caused an improved mobility of hydrogen ions to the surface. Carroll-Webb and Walther (1988) did not observe systematic effects on kaolinite or corundum dissolution rates at 25°C from changing buffer salt concentrations in I = 0.005 to 0.05 M solutions. Kline and Fogler (1981b,c) found that Na^+, Li^+, and NH_4^+ cations catalyzed the dissolution rate of kaolinite in HF solutions at 25°C over the range of about $pH \approx 0$ to 3.

Organic ions and molecules. Organic ions and molecules are produced by living organisms and by decomposition of organic materials. Because of the intimate association of plants, animals, and minerals in soils and sediments, the potential effects of the organic components on phyllosilicate and related mineral weathering rates are great. The most important effect of organic acids on weathering of sheet silicates is that of cation-complexation. Complexation takes place on the surface of the mineral to form surface chelates by ligand exchange with surface hydroxyl groups (Kummert and Stumm, 1980; Bales and Morgan, 1985; Furrer and Stumm, 1986). Organic ligand-promoted dissolution is assumed to be a parallel reaction to proton-promoted dissolution (Furrer and Stumm, 1986). The surface chelates weaken the metal-oxygen bonds within the mineral structure, leading to metal detachment. Cation complexation can also occur in solution, altering the driving force for cation adsorption and precipitation.

Clay mineral dissolution rates increase in the presence of organic acids (acetic, aspartic, salicylic, tartaric) over pure H_2O or CO_2-charged solutions (Huang and Keller, 1970, 1971, 1973). Salicylic and tartaric acids have the greatest effect. Tan (1975) used IR to investigate complexation of Si and Al from kaolinite and bentonite dissolution with humic and fulvic acids. After reaction at pH 7 (humic) and pH 3, 5, 7, and 9 (fulvic), the IR spectra contained new peaks for Si-O and Al-OH bonds. The degree of complexation increases with pH for fulvic acid. Solutions yielded significantly higher Si and Al in the presence of humic acids than in water, indicating that the acids may participate in the dissolution mechanism. Kodama and Schnitzer (1973) investigated the effect of fulvic acids on chlorite dissolution. Schnitzer and Kodama (1976) found that phlogopite, biotite, and muscovite dissolve congruently in the presence of fulvic acids, whereas Arshad et al. (1972) observed incongruent dissolution of trioctahedral micas in organic solutions.

Bidentate ligands that form mononuclear complexes enhance dissolution rates of δ-Al_2O_3 at 25°C. Chelates that form five- and six-membered rings when bound to the mineral surface (oxalate, catechol, malonate, and salicylate) enhance rates more than those that form seven-membered rings (phthalate, succinate) (Fig. 19, Furrer and Stumm, 1986). Mono-dentate ligands (benzoate) adsorb, but inhibit dissolution by displacing bidentate ligands. Organic acid anions compete with fluoride for adsorption sites on hydrous alumina and slow the rate at pH 4 under selected diffusion-controlled experimental conditions (Zutié and Stumm, 1984). The ligands bind to the surface with increasing strength in the order: formate $\approx$ chloride $\approx$ carbonate < acetate < sulphate < salicylate < fumarate < maleate < malonate << oxalate $\approx$ fluoride $\leq$ citrate.

From pH 2 to 6, 0.001 M oxalate increases the dissolution rate of kaolinite by a factor of 1.7 to 2.6. At pH 4 the dissolution rate increases by a factor of 7 in 0.05 M oxalate and by less than a factor of 2 in 0.0005 M oxalate. Salicylate enhances the dissolution rate only at pH 4 to 6 (Wieland and Stumm, 1992). At lower pH, the surface-bound salicylate chelate is protonated, resulting in an opening of the structure and a diminishing of its ability to bind to surface Al sites. Ganor and Lasaga (1994) observed a greater effect of oxalate at 80°C, pH 3; the far-from-equilibrium dissolution rate increases by a factor of 3 when only 117 µm of oxalate is added. They proposed that an additional effect on the reaction affinity occurs under near-equilibrium conditions, where increasing oxalate concentration in solution results in solution complexes of Al and effectively increases the degree of undersaturation. Carroll-Webb and Walther (1988) varied the concentration of buffer solutions containing organic-complexing ligands (KH phthalate, succinic acid, and tris(hydroxymethyl) aminomethane; pH 3 to 9) in order to assess their effect on dissolution rates of kaolinite and corundum. They did not observe a significant enhancement of rates

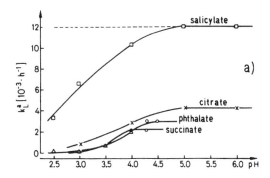

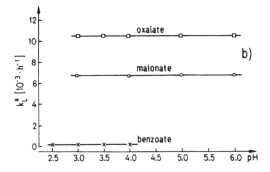

Figure 19. Dissolution rate constants, k_L^a, of δ-Al_2O_3 at 25°C in the presence of organic acids as a function of pH. The organic ligands in (a) become protonated at lower pH and change from a chelate to a monodentate surface complex. The monodentate complex is less active in promoting dissolution. In (b) the ligands oxalate and malonate maintain their bidenticity over the entire range of pH. Benzoate always forms a monodentate inner sphere complex. Used by permission of the authors, from Furrer and Stumm (1986).

when buffer solution ionic strength was raised from 0.005 M to 0.05 M. At pH 5 to 9, the presence of Al-complexers malonic acid and 2,4 pentanedione did prevent Al precipitation, but did not enhance the dissolution rates.

Chin and Mills (1991) examined effects of low molecular weight organic ligands (oxalate, malonate, salicylate, o-phthalate), soil humic acid, and stream-water dissolved organic matter on kaolinite dissolution. At pH 4.5 they observed an increase in dissolution rate compatible with the order of increase observed by Furrer and Stumm (1986) (oxalate > malonate > salicylate > phthalate). Stream water dissolved organic matter (40 mg L^{-1}) caused some variation in the release of Al with time, but did not change the overall dissolution rate. Soil humic acid (50 mg L^{-1}), on the other hand, appeared to inhibit the dissolution of Al. They proposed that large humic acid molecules bind to multiple surface Al sites both preventing adsorption of H^+ and small organic ligands and necessitating breaking of many Al-O bonds to result in dissolution.

Mg release from chrysotile appeared to show no significant enhancement in the presence of catechol and oxalate up to 10 mmol L^{-1} (Bales and Morgan, 1985), beyond a slight increase during the initial 12 to 24 hours of dissolution. Catechol slightly increased Si release at pH 8. Oxalate inhibited Si release at pH 8. Thomassin et al. (1977) dissolved chrysotile in 0.1 N oxalic acid in order to complex Mg and prevent its precipitation. They observed no difference in rates when chrysotile dissolved in 6 N HCl and concluded either that Si dissolution was the rate-limiting step or diffusion of Mg through a Si-gel layer was rate-limiting.

Role of biota

It is unclear to what extent plants affect phyllosilicate dissolution rates (e.g. Weed et al., 1969; Berthelin and Leyval, 1982; Robert and Berthelin, 1986; Hinsinger et al., 1992). Although K and Mg are plant nutrients, dissolution rates of micas in the vicinity of plant roots may or may not be related to metal uptake by the plants. Microbial activity, on the other hand, may play a role in mica alteration. Bhatti et al. (1994) observed that bacterial oxidation of pyrrhotite and pyrite in ground schists at pH $\geq$ 2 produced Fe^{3+} that was the driving force for jarosite precipitation and cation exchange in phlogopite (Mg^{2+} for K^+) which resulted in the formation of vermiculite. Simultaneously Si and Al dissolution from phlogopite was thought to occur by proton attack. Berthelin (1983) reviewed microbial weathering of minerals including phyllosilicates.

Dependence on reaction affinity or Gibbs free energy of reaction

Reaction affinity and ΔG_r (Gibbs free energy of reaction) are two ways to express the same variable in a chemical system. Both characterize the deviation from the equilibrium state. Reaction affinity (A) is defined by (Denbigh, 1971)

$$A = \left(\frac{\partial G}{\partial \xi} \right)_{T,P} = -\sum v_i \mu_i \tag{10}$$

where G is the Gibbs Free Energy of the system, ξ is the extent of reaction, v_i are stoichiometric coefficients and μ_i are chemical potentials of reactants and products. The Gibbs free energy of reaction, ΔG_r, is given by

$$\Delta G_r = -RT \ln \left(\frac{Q}{K_{eq}} \right) \tag{11}$$

where Q is the ion activity product or quotient in the nonequilibrium solution; K_{eq} is the same product at equilibrium; R is the gas constant, and T is temperature in K. For a dissolution reaction written from left to right, these parameters have the same absolute value, but are opposite in sign. Reaction affinity will be positive; ΔG_r will be negative.

Recently, a few studies have been conducted to assess the effect of reaction affinity or ΔG_r on the dissolution rates of clays. In general, the results confirm that dissolution rates slow down as equilibrium is approached, although the exact functionality of the rate change with reaction affinity varies from phase to phase and is a function of solution composition. Nagy and Lasaga (1992) observed that gibbsite dissolution showed a marked decrease in reaction rate at a ΔG_r value (-0.2 to -0.5 kcal mol^{-1}) that happened to be equal to that calculated for dissolution at a dislocation on the gibbsite basal surface. Under more undersaturated conditions (to -1.14 kcal mol^{-1}), the rate appeared to be constant. This was confirmed and extended to even more undersaturated conditions (-3.5 kcal mol^{-1}) by Mogollón et al. (1994) and Soler et al. (1994) for gibbsite dissolution rates measured in a column experiment at 25°C. The equation for gibbsite dissolution rate (mol m^{-2} sec^{-1}) given in Nagy and Lasaga (1992) is

$$R_{diss} = -(4.72 \pm 0.28) \times 10^{-10} \left[1 - \exp\left\{ (-8.12 \pm 1.02)(\Delta G_r / RT)^{3.01 \pm 0.05} \right\} \right] \tag{12}$$

They pointed out that using a simple Transition State Theory approach where the overall reaction represents the rate-limiting elementary reaction does not appear to hold (Fig. 20).

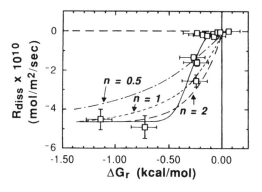

Figure 20. Dissolution rates of gibbsite at 80°C, pH 3 as a function of Gibbs free energy of reaction, ΔG_r (Nagy and Lasaga, 1992). Negative values represent undersaturation; ΔG_r = zero represents equilibrium. Data show a sigmoidal variation with ΔG_r. Reaction (12) describes the solid curve. A reaction dependence based on Transition State Theory, where $R = k (1 - \exp(\Delta G_r))^n$, and in which k is the rate constant and n = 0.5 to 2, does not fit the data.

Nagy et al. (1990; 1991) reported on a similar study for Twiggs County, Georgia kaolinite. At pH 3, 80°C, they observed the following dependence on saturation state

$$R_{diss} = -1.2 \pm 0.12 \times 10^{-12} [1 - \exp\{(0.85 \pm 0.24)(\Delta G_r/RT)\}] \tag{13}$$

In contrast to the more complex dependence seen for gibbsite, a compositionally simpler mineral, the kaolinite data could be explained using a simple Transition State Theory approach; however, the density of data collected was less for kaolinite. Soong (1993) confirmed this form of a rate equation for congruent dissolution of kaolinite (Dry Branch, Georgia, USA) at pH 4.2 and 7.3 from 130° to 230°C (Fig. 21). Values of the rate coefficient increased with increasing temperature.

Devidal et al. (1992) examined the dependence of kaolinite dissolution rate on reaction affinity at pH 2 and 7.8 at 150°C (Fig. 18). When no Si and Al were added to the initial solutions or when the initial ratio of Si to Al was 1:1, the dependence of rate on **A** appears to be similar to that observed by Nagy et al. (1991) and Soong (1993) for saturations over a similar range. However, Devidal et al. (1992) conducted their experiments to much greater undersaturations than did Nagy et al. (1991) and Soong (1993). Devidal et al. (1992) also examined initially nonstoichiometric solution compositions, with either high concentrations of Si or Al. These data were analyzed by Oelkers et

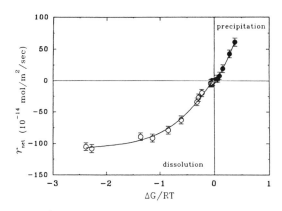

Figure 21. Dissolution and precipitation rates of kaolinite at 170°C, pH 7.3 shown as a function of normalized Gibbs free energy of reaction $\Delta G_r/RT$. Stoichiometry of reaction was observed under all saturation conditions. The reaction order with respect to $(1 - \exp(\Delta G_r/RT))$ is ~1, in agreement with the results of Nagy et al. (1991) at 80°C, pH 3. Used by permission of the author, from Soong (1993).

al. (1994) who furthered the explanation of the observed rate dependence. Their rate (mol m^{-2} sec^{-1}) equation is given by

$$R_{diss} = k^{\bullet} \left(\frac{\left(\dfrac{a_{H^+}^{3n}}{a_{Al^{3+}}^n}\right)\prod_i a_{A_i}^{v_i}}{1 + K^{\bullet}\left(\dfrac{a_{H^+}^{3n}}{a_{Al^{3+}}^n}\right)\prod_i a_{A_i}^{v_i}}\right)\left(1 - \exp\left(\frac{-A}{n_{Si}RT}\right)\right) \tag{14}$$

where A_i is the ith aqueous species involved in the formation of a precursor complex at the surface, v_i are stoichiometric coefficients, and n_{Si} is the stoichiometric number of moles of Si per mole of kaolinite. The effective rate coefficient is $k^{\bullet} = 1.87 \times 10^{-9}$ mol m^{-2} sec^{-1}, and the equilibrium constant for the formation of an Al-deficient precursor complex at the kaolinite surface is $K^{\bullet} = 0.445$. Their model equation assumes that each Si in kaolinite forms a precursor complex that is Al-deficient, and that one equation describes both dissolution and precipitation. Far-from-equilibrium (i.e. at $\Delta G_r \cong -16.5$ kcal mol^{-1}), the reaction is independent of saturation state. Over a significant region of undersaturation there is a slowing effect due to the increasing concentration of Al in solution. Through the law of mass action for the supposed precursor surface complex this results in a decreasing number of reactive sites. Very near equilibrium, the chemical affinity term again dominates and the rate slows even further.

In contrast, smectite appears to dissolve as a linear function of ΔG_r to undersaturations of ~-22 kcal mol^{-1} at 80°C and pH 8.8 (Cama et al., 1994). However, the absolute value of ΔG_r depends on the choice of mineral formula (e.g. 10 vs. 20 structural oxygens for 2:1 layer phyllosilicates), so ΔG_r values cannot be directly compared.

Oelkers et al. (1994) proposed that rates of kaolinite dissolution are limited by decomposition of a surface complex containing Si-O bonds. This contrasts with the interpretation by Carroll-Webb and Walther (1988) and Wieland and Stumm (1992), who argued that in acidic solutions, the breaking of Al-O bonds was rate limiting. Early leaching studies showed that Al can be readsorbed on phyllosilicate surfaces during dissolution. The determination of an exact surface species that controls dissolution rate is an area of current controversy and merits more research on the rate-limiting mechanism(s). The work carried out so far on the effect of the ΔG_r or reaction affinity of the overall dissolution reaction also demonstrates that the application of transition state theory for a simple elementary reaction to an overall reaction is an oversimplification of sheet silicate dissolution. Keep in mind that these studies have all been conducted at temperatures above those of the weathering environment but may be appropriate to subsurface diagenetic processes. It is probable that inhibitory mechanisms may play a more important role at lower temperatures. More experimental work, spectroscopic information and molecular modeling results may provide important data to address this issue in the near future.

Surface defects

Surface defects consist, in part, of steps and kinks that result in a degree of surface roughening that can be measured using techniques such as atomic force microscopy. The steps and kinks provide sites of different reactivity because they present different bonding arrangements to adsorbing protons (or other ions). Surface defects may also consist of sites with higher energies arising from structural defects within the bulk clay. Dislocation outcrops provide points of localized excess strain which are unstable relative to the bulk and will have a higher reactivity. Compositional impurities at mineral surfaces may also be sites of local high strain energies (Blum and Lasaga, 1987). Dislocations probably are more important under near-equilibrium conditions where a relatively low density of highly active sites can control the overall dissolution rate. Under far-from-equilibrium conditions, unnaturally high concentrations of dislocations are necessary to strongly affect rates (e.g. Blum et al., 1990).

If they occur in high enough concentration, surface defects may control the dissolution rates of phyllosilicates. Johnsson et al. (1992) used AFM to observe that shallow (< 10 Å) etch pits formed on muscovite basal surfaces in pH 5.7 solutions at 22°C. Adding XPS and LEED data they indicated that island networks of a fibrous, amorphous, Al-hydroxide phase appeared to form simultaneously on the surface. They postulated that compositional impurities might control pit nucleation because the depths were shallow. Pits formed at dislocations should be deeper. Nagy and Lasaga (1992) observed that gibbsite dissolution rates sharply decreased in magnitude over a short range of reaction affinity when moving from an undersaturated solution toward an equilibrium state at 80°C, pH 3. The range over which the transition occurred, -0.5 to -0.2 kcal mol^{-1} corresponded well with the ΔG_r required to overcome the lattice strain built up at a dislocation outcrop on the basal face of gibbsite and nucleate an etch pit. This correspondence in solution chemistry and calculated critical ΔG_r necessary to form an etch pit appeared to be confirmed by SEM observation of the gibbsite powders. Turpault and Trotignon (1994) suggested that dissolution on the biotite basal surface is focused at structural defects.

Veblen and Wylie (1993) discussed potential effects arising from structural surface differences between spiral and cylindrical chrysotile. The spiral structure has a permanent ledge which could provide step sites that accelerate dissolution.

Oxidation of octahedral Fe

During conversion to vermiculite, biotite Fe must be oxidized to compensate for the charge lost by removing interlayer K. At pH 4, under anoxic and mildly oxidizing conditions for short time periods, at least some Fe is released from biotite in the ferrous state (Acker and Bricker, 1992). This agrees with the results of White and Yee (1985) who showed that Fe^{2+} was released from biotite in deionized water and $FeCl_3$ solutions. The dissolution rate of biotite based on Si release was 3 times faster at pH 4 in solutions under strongly oxidizing conditions (H_2O_2) vs. solutions at atmospheric PO_2 (Acker and Bricker, 1992) (Fig. 16). They suggested that Mg dissolution and partial oxidation of Fe may occur simultaneously, resulting in creation of an octahedral vacancy to balance the charge. The results from H_2O_2 solutions also show that dissolution of the tetrahedral layer is strongly linked to dissolution of the octahedral layer.

Experimental approaches and caveats

Type of experiment.   Three types of experiments are used to measure clay dissolution rates: batch, single-pass flow, and column experiments. Batch experiments involve reacting a fixed amount of solid with a solution in a closed system container, and analyzing extracted solution and/or solid as a function of time. In mixed-flow and stirred-flow reactors the solution is passed over the clay sample only once, whereas in fluidized bed reactors, the reacting fluid is recirculated for a short time to suspend the solids. With single-pass flow reactors, one can measure a steady-state rate as a function of controlled solution composition. Column experiments simulate groundwater percolation through soils in which case flow may be saturated or unsaturated. Although not as precise for determining absolute rates, column experiments are useful for direct comparison with weathering rates determined in the field.

In general, rates from batch and single-pass flow reactors are comparable. For example, kaolinite dissolution in batch reactors at pH 3, 80°C (Carroll and Walther, 1990) and in stirred-flow reactors (Nagy et al., 1991) are the same within a factor of 2. At 25°C, rates are within 1/2 order of magnitude (Carroll-Webb and Walther, 1988 (batch); Wieland and Stumm, 1992 (batch); Ganor et al., 1995 (stirred-flow)) at pH 4 to 6. Dissolution rates of montmorillonite in batch experiments are about 5 times faster at pH 1 and approach mixed-flow reactor rates as pH increases to 4 (Furrer et al., 1993). They suggested that gel-like particle aggregation in the mixed-flow reactors increased with time and decreasing pH because of increased edge-to-face aggregation in acidic solutions where edge charge is positive due to proton saturation and face charge is negative due to permanent structural charge. Such a process would also affect any interpretation of a surface-controlled reaction because diffusion through the aggregate could play a role.

Dissolution rates of biotite at 25°C in fluidized-bed reactors were about 1/2 order of magnitude higher than those in saturated column experiments at pH 4 to 5 (Acker and Bricker, 1992). Dissolution rates of a bauxitic ore rich in gibbsite (HNO_3 or $HClO_4$, 25°C, pH 3.2 to 4) in a packed column experiment are also about 1/2 order of magnitude smaller than gibbsite dissolution rates (Bloom and Erich, 1987) in batch experiments (KNO_3 solutions, pH 3.5) (Mogollón et al., 1994).

In unsaturated flow columns biotite was periodically wetted and dried over 7 months (Acker and Bricker, 1992). Individual rates were faster because they were determined over short time intervals or because drying locally increased H^+ concentration and consequently, the dissolution rate. The result is that surface amorphous phases may form and be easily removed during the subsequent wetting cycle. Cumulative dissolution was less than in the saturated columns and fluidized-bed reactor experiments. However, when normalized to the total amount of wetted time, the biotite in the unsaturated column dissolved twice as fast than in the saturated column using Si and Al release rates, and four times as fast using Fe and Mg release rates.

Solution composition.   Two aspects of solution composition require some thought when designing an experiment. The first is the method used to adjust pH. pH can be adjusted with buffers, although some buffers can complex clay cations such as Al which may affect the measured reaction rates. Knauss and Wolery (1989) used dilute buffers to minimize such effects. Carroll-Webb and Walther (1988) also discussed potential negative buffer effects. pH can also be set using strong acids and bases such as HCl, $HClO_4$, H_2SO_4, and NaOH. However, Bloom and Erich (1987) showed that gibbsite dissolution is increased by 30 times when the acid anion is SO_4^{2-} vs. NO_3^-. Nagy and Lasaga (1992) and Mogollón et al. (1994) saw no difference between gibbsite dissolution rates measured in HCl and $HClO_4$ solutions.

The second concern is finding a method to control Al (Mg, or Fe) solubility in the solution. Carroll-Webb and Walther (1988), for example, used an Al-complexing agent at near-neutral pH to prevent precipitation or adsorption of Al on the clays. Feigenbaum and Shainberg (1975) and Clemency and Lin (1981) used ion-exchange resins to control solution composition (Al and pH) during dissolution of illite and phlogopite, respectively.

Solids pretreatment. The manner in which clay mineral powders are pretreated prior to a dissolution measurement can have a drastic effect on the rates that are measured. It is difficult to find pure natural starting clay minerals. Often other phases, in particular amorphous surface coatings, can be present. One must remove extraneous phases or ignore them at the risk of damaging the primary mineral surface. Grinding is also used to obtain a consistent grain size fraction. Even ultrasonication can have an effect on surface area by cleaving particles (Blum, 1994). However, both mechanical and chemical pretreatments may affect the stoichiometry of dissolution as well as the initial rates by disturbing the composition and structure of the surfaces. This has been shown by Bloom (1983), Carroll and Walther (1988), Nagy et al. (1991), Wieland and Stumm (1992), and Furrer et al. (1993), among others. This can be important in any kind of experiment and will be most important during the initial dissolution stages. Typically, an effective pretreatment consists of physical removal of obvious mineral contaminants, size fractionation by settling in a solvent (Knauss and Wolery, 1989) or pretreatment solution, and chemical "annealing" of the surface by conditioning the solid in the solution in which the dissolution rate will be measured (Carroll and Walther, 1988; Furrer et al., 1993) or in a near-equilibrium solution at the same pH and temperature as that of the dissolution experiment (Nagy et al., 1991).

NUCLEATION KINETICS

Introductory comments

In contrast to dissolution, less attention has been paid to systematic and mechanistic quantification of nucleation and growth kinetics of sheet silicates. This is probably because it is presumed to be a slower process to measure in the laboratory. Also, until recently, characterization of growth surfaces was difficult. However, it is clear that in nature, sheet silicates grow quite rapidly and in a variety of weathering environments at the earth's surface. In fact, many investigators have demonstrated that it is possible if not easy to synthesize sheet silicates in the laboratory at low temperatures. These experiments generally were conducted using gels as starting materials that were aged and/or dehydrated to form various clay mixtures. The problems in taking such data and extracting kinetic information are numerous because of the complexity of the system composition and number of phases that form. Therefore, recently there have been attempts to quantify sheet silicate growth rates and mechanisms using solid/solution experiments under controlled conditions, application of surface-complexation modeling, and high intensity X-ray spectroscopy.

Nucleation

Sheet silicates may form by homogeneous nucleation from solution in certain geologic environments (e.g. sepiolite from seawater), but in most cases it is thought that their nucleation occurs in the presence of another phase, and hence is heterogeneous. Banfield and Barker (1994) and Banfield et al. (1995) have advanced the hypothesis supported by TEM evidence that many clays that form secondarily during weathering of primary igneous or metamorphic minerals, nucleate via a topotactic mechanism. Topotactic nucleation occurs when the dissolving mineral is leached of some cations, but retains partial structural integrity involving linked SiO_4 tetrahedra, that provides the basic building blocks of clays. In apparent contradiction, the bulk solution chemistry can indicate undersaturation with respect to the new phase (Casey et al., 1993). Epitaxial nucleation also occurs in many surficial environments. For example, the formation of smectite in alkaline waters such as tidal lagoons or lakes, is considered to occur by nucleation on detrital clays (Banfield et al.,

1991a,b) or on diatom frustrules (Badaut and Risacher, 1983). This type of nucleation step must be limited, in part, by adsorption of metals to the nucleating surface and may be catalyzed by surface amorphous Si layers, whatever their origin. Below I review the few experimental data that exist supporting either topotaxial or epitaxial nucleation mechanisms and the role of adsorption.

Topotaxy. Casey et al. (1993) recently showed that leaching of pyroxene surfaces results in a Si-rich surface layer with 4-membered Si-O rings based on the similarity of Raman spectra of reacted wollastonite to those published for amorphous silica produced by a sol-gel method. In the unreacted wollastonite structure, Si-O linkages form chains. During leaching, Ca ions are exchanged for protons, and the silicate anions restructure to form isolated rings. One can imagine that in the presence of aqueous solutions containing other metal ions such as Mg and or Al, the surfaces could transform to clay structures such as kerolite, talc, or even Mg-chlorite. However, TEM examination of dissolved wollastonite and pyroxene structures do not show new clay structures, suggesting that the abundance of fluid in the typical batch or single-pass flow dissolution experiments does not mimic the environment of the transformations observed in natural samples. Casey et al. (1993), therefore, suggested that natural clay formation or intergrowth at the surfaces of minerals unstable in weathering environments may occur by dehydration of a gel phase at these minerals' surfaces. This is consistent with the observations on reconstitution of acid-leached phyllosilicates (Mering, 1949; Brindley and Youell, 1951; Abdul-Latif and Weaver, 1969).

Recently, Refson et al. (in press) and Wogelius et al. (1995) performed an interesting study of periclase dissolution. They observed that in pH 2 and 4 aqueous solutions, MgO forms a hydroxylated surface that reconstructs itself to become brucite. Elastic Recoil Detection Analysis (ERDA) was used to determine that at pH 2, a surface layer about 900 Å thick had an H/Mg ratio of about 2, consistent with the brucite formula $Mg(OH)_2$. They stated that precipitation from solution could not explain their data because their solutions were 16 orders of magnitude below brucite solubility. Using quantum mechanical *ab initio* models of water adsorption to the (001) and (111) faces of MgO, they proposed that transformation to brucite stabilized the (111) face over the (001) face generally formed during MgO synthesis at high temperatures. When a layer of Mg ions is replaced by a layer of H (two H replace every Mg), the d-spacing of brucite is obtained to within 2% (Fig. 22). The authors suggested that surface coverage is uneven under these extremely acidic conditions, but that at neutral pH, this step would occur faster than Mg detachment, and surface coverage could be more uniform. Hence, the authors propose an explanation for why brucite and periclase dissolution rates are the same (Vermilyea, 1969). Their data also provide evidence for the type of topotactic relations observed by Banfield in her TEM studies and by Casey et al. (1993) in their experimental pyroxene dissolution data.

Epitaxy. Epitaxial nucleation is invoked when observational evidence points to clay interstratification. In such cases, nucleation occurs through a chemical and structural match of the new clay layer onto the templating surface. Banfield et al. (1991a,b) suggested that epitaxial nucleation occurs when Mg-smectites form in alkaline lakes. Epitaxial nucleation is also invoked in subsurface diagenesis, for example when kaolinite forms on expanded detrital mica surfaces (e.g. Crowley, 1991; Pevear et al., 1991; Jiang and Peacor, 1991). Hydroxy Al-interlayers in smectite can adsorb Si to form a proto-imogolite structure on the the interlayer surfaces and a slightly expanded lattice parallel to the *c*-axis (Lou and Huang, 1993). Also, replacement reactions in which gibbsite replaces kaolinite in bauxitic soil formation (Merino et al., 1993), may initiate by an epitaxial nucleation step.

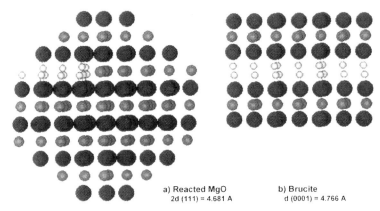

a) Reacted MgO
2d (111) = 4.681 A

b) Brucite
d (0001) = 4.766 A

Figure 22. Proposed transformation of periclase surface to brucite via hydroxylation of structural Mg. Large dark gray spheres are oxygen, intermediate gray spheres are magnesium, and small light gray spheres are hydrogen. (a) Atomic representation of periclase structure showing replacement of Mg ions at corner by hydrogen ions in 2 to 1 ratio. Two times the d-spacing of the (111) planes = 4.681 Å. Hydroxylation of (111) is determined to be thermodynamically more stable than hydroxylation of (001), the stable face of periclase during crystal growth based on ab initio DFT/LDA calculations. (b) Brucite structure showing a correspondence between the arrangement of ions and (0001) d-spacing to those shown in (a). Used by permission of the authors, from Wogelius et al. (1995).

However, despite the abundant observational evidence for epitaxial nucleation, the mechanism(s) by which epitaxial growth is initiated is unknown. For example, in the case of kaolinite growth on muscovite or biotite, does a kaolinite layer nucleate on the mica or does the mica growth terminate with an Al octahedral sheet? Tsipursky et al. (1992) provided some TEM evidence that 1M illite particles are terminated by either a kaolinite layer, a single tetrahedral sheet, or a single octahedral sheet. Brown and Hem (1975) showed that Al polymeric species nucleate and precipitate on phyllosilicate surfaces. This is consistent with the observations of Johnsson et al. (1992) on the reactivity of the muscovite basal surface. Using atomic force microscopy, XPS, and LEED analyses, Johnsson et al. (1992) showed that a freshly-cleaved muscovite basal surface reacts quickly in aqueous solution and even in the presence of humid air to produce islands of an apparent Al-rich material. Given that sheet silicates react more quickly at their edges, it is possible that the majority of the Al comes from the grain edges or, more likely, from edges of steps and 10 Å deep dissolution pits also observed on the basal surface by Johnsson et al. (1992). These results agree with acid-leaching data for muscovite and with the very early stages of dissolution in aqueous experiments (e.g. Rosenberg et al., 1984) in which edges are enriched in Si. Nucleation of a kaolinite layer or aluminum octahedral sheet on illite is also required to explain the exact position of the ~10 Å XRD peak (Schuette and Pevear, 1993). Such an observation would explain the chemical composition of an illite, i.e. would account for the excess water (due to excess hydroxyls on the terminated Al-octahedral surface) and excess aluminum (Srodon and Eberl, 1984).

Epitaxial nucleation and growth of Al-oxide (corundum) on muscovite have been obtained by vapor deposition using the "van der Waals epitaxy (VDWE)" method (Steinberg et al., 1993). The mechanism of nucleation is driven by weak van der Waals forces between the oxide film and the cleaved tetrahedral surface, even though the lattice mismatch observed between the two phases was 9%. Although outside the realm of weathering, it is clear that mica surfaces have an affinity for adhering to phases of similar composition or similar structure.

Role of adsorption. Sposito (1986) clearly pointed out the need to distinguish adsorption from precipitation when deriving kinetic mechanistic interpretations based on solution chemistry data alone. He stressed that spectroscopic techniques are the only methods suitable for determining bonding arrangements on mineral surfaces. Such techniques can distinguish two-dimensional structures (i.e. adsorbed layers) from three-dimensional structures (i.e. nucleated layers). Recently, XAS (X-ray Adsorption Spectroscopy) has been used to identify bonding environments of adsorbed species on mineral surfaces (e.g. see reviews by Charlet and Manceau, 1993; Brown et al., 1995; Greaves, 1995). Although interpretations of XAS data are still somewhat debatable when discussing adsorption versus precipitation, recent work by Charlet and Manceau (1994) argues for interpreting some data as evidence for sheet silicate nucleation.

Charlet and Manceau (1994) reinterpreted published Co(II) and Ni(II) sorption data, including those in which XPS, Auger and EXAFS (extended X-ray absorption fine structure) spectroscopic techniques were used, to indicate the formation of new hydrous silicates that precipitated on the sorbent mineral surfaces. They proposed that average metal-metal and metal-Si distances for absorption onto silicates and silica in the presence of Si-containing solutions are consistent with those distances in the metal-bearing sheet-silicate phases, thus indicating the formation of clay-like local structures (Fig. 23). O'Day et al. (1994a) used EXAFS to show that at low surface coverage of Co on kaolinite, Co-Co distances are consistent with absorption of mononuclear Co at Al-O-Si edge sites or at Al-O-Si and Al-OH edge sites. At higher surface coverage, O'Day et al. (1994a,b) interpreted their data to indicate that multinuclear clusters of $Co(OH)_2$ formed either at the edges or on the basal aluminol surface. Continuous monolayer coverage was not expected due to slight misregistry of the Co-hydroxide layer on the aluminol sheet of kaolinite. They also suggested that a metastable hydroxide phase may form initially, later transforming by the Ostwald step rule to the stable phase. It is apparent that there is considerable potential for understanding the crossover between adsorption and heterogeneous nucleation from such spectroscopic data, especially when they are combined with other experimental evidence, such as solution saturation states or cation exchange data.

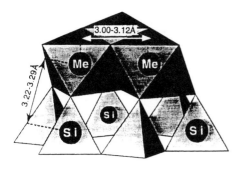

Figure 23. Schematic diagram showing metal-metal and metal-Si bonding distances in trioctahedral layered silicates. EXAFS data can be interpreted to obtain metal-metal distances as well as the number of silicon atoms surrounding the metal cation. If the metal-metal distance is 3.00 to 3.12 Å and the number of neighboring silicons is three, then it can be argued that a stoichiometric $Si_4Me_3O_{10}(OH)_2$ structure has formed as opposed to a $Me(OH)_2$ layer structure. Used by permission of the authors, from Charlet and Manceau, (1994).

Gel recrystallization. Experimental data confirm that formation of clays is possible in a day to a few months at low temperatures given certain compositional conditions and initial gel compositions similar to those of the final clays (e.g. De Kimpe et al., 1961; De Kimpe, 1964, Decarreau, 1980, 1981; Decarreau and Bonnin, 1986; Decarreau et al., 1987, 1989; Harder, 1971, 1974, 1976, 1977, 1978; Flehmig, 1992).

The general experimental method involves preparation of oxy-hydroxide gels of the appropriate metal cation(s); aging of the gels in solutions of varying composition; and final characterization of the crystallized solids by X-ray Diffraction (XRD) and Scanning Electron Microscopy (SEM). Given suitable conditions and starting materials, smectites, illite, paragonite, kaolinite, or almost any clay can be synthesized in relatively short times at room temperature (1 day to a few months, depending on the mineral and solution composition). However, precise identification has not been always possible due to the poor quality of the XRD patterns obtained, and the inability to make identification based on morphology in SEM images. Some general observations were (1) clays formed when the concentration of Si in solution was below amorphous silica saturation, (2) the gel mixtures must have a composition similar to the clay that forms, (3) high pH accelerates the clay formation rate, and (4) greater concentrations of Si in solution formed clays of increasing expandability.

Organic templates. It is known that sulfides, carbonates (Mann et al., 1993), mesoporous silica molecular sieves (Monnier et al., 1993; Tanev and Pinnavaia, 1995), and ceramics (Messersmith and Stupp, 1992) nucleate on organic substrates. Cairns-Smith and Hartman (1986) argued that clays provided the templates on which pre-life organic molecules nucleated. In reverse, it has been shown that organic molecules decrease the nucleation time and increase the yield of kaolinite in synthesis experiments (Hem and Lind, 1974, Linares and Huertas, 1971). La Iglesia and Van Oosterwyck-Gastuche (1978) emphasized that fulvic acids are large sheetlike molecules which may induce epitaxial nucleation of kaolinite. Although no systematic investigations have been undertaken to determine nucleation kinetics of phyllosilicates on organic substrates, the potential for molecular recognition (Weissbuch et al., 1991) processes to occur in nature is great. This is a subject for future research.

PRECIPITATION (GROWTH) KINETICS

Introductory comments

An example of a precipitation reaction is obtained by reading Reaction (5) in reverse. However, the same discussion on kinetics holds, that is, mechanisms and driving forces cannot be determined from such an equation. In addition to the data discussed below, transformation reactions in which dissolution of a metastable phase and growth of a more stable phase occur concurrently (e.g. Tsuzuki and Kawabe, 1983; Steefel and Van Cappellen, 1990) are important when determining the rate-limiting step for growth in some systems. Below is a summary of data describing kinetic information that has been obtained for Al-hydroxides and oxyhydroxides, kaolinite, and Mg-silicates.

Precipitation data

Al phases. The volume of work on formation of Al-hydroxides and oxyhydrox- ides at low temperatures is great. Extensive review of earlier work is found in the papers by Hem and Roberson (1967), Schoen and Roberson (1970), and Smith and Hem (1972) who investigated precipitation of Al phases under weathering conditions. A more recent summary is found in Hem and Roberson (1990). Schoen and Roberson (1970) suggested that polarization of hydroxyl ions in aqueous aluminum complexes controls the $Al(OH)_3$ polymorph that precipitates from a solution. In particular, bayerite and nordstrandite are rare in nature, but precipitate easily at pH > 5.8. Schoen and Roberson (1970) also showed that gibbsite grows faster at its edges based on crystallite morphology. Homogeneous precipitation of Al phases can occur over the entire pH range of natural solutions. Aging of the solutions containing Al polymeric colloids produces stable crystals of gibbsite, bayerite or nordstrandite with increasing pH. The smallest polymeric structure is a six-membered ring of aluminum ions, in which each aluminum is bonded to a neighboring Al through shared pairs of hydroxyl ions (Smith and Hem, 1972).

Formation of Al_{13} polymer, $Al_{13}O_4(OH)_{24}(H_2O)_{12}^{7+}$, occurs when solutions at pH 3 to 4 containing dissolved Al are neutralized to pH 5.3 to 6 (Furrer et al., 1992). The presence of gibbsite surfaces does not inhibit the polymer formation. Mechanisms for the transformation of Al_{13} to gibbsite have been proposed by Hsu (1988) and Bottero et al. (1987).

Stol et al. (1976) studied the homogeneous precipitation of $Al(OH)_3$ from acidified aluminum nitrate and chloride solutions at 25° to 90°C. They interpreted their data to show that up to 2.5 OH^- per Al ion formed bridges between Al^{3+} ions. In the presence of anions such as NO_3^- or Cl^-, the charge is neutralized and polymer aggregates form. In a series of papers, Van Straten and coworkers investigated the precipitation of $Al(OH)_3$ as a function of pH, alkali ions, and temperature (Van Straten and De Bruyn, 1984; Van Straten et al., 1984, 1985a,b). Bayerite precipitation was found to be second order in Al concentration and in H^+ concentration. Growth of bayerite was proportional to the surface area and the square of the supersaturation, and has an activation energy of ~60 to 80 kJ mol^{-1} from 25° to 90°C.

Kaolinite.   Kaolinite has been synthesized from aqueous solutions at low temperatures using a wide range of experimental designs. La Iglesia and Serna (1974), La Iglesia and Martín-Vivaldi (1975), La Iglesia et al. (1976), La Iglesia and Van Oosterwyck-Gastuche (1978), and Van Oosterwyck-Gastuche and La Iglesia (1978) summarized the early work and pointed out a number of conditions necessary for kaolinite growth at low temperatures. There must be (1) Al in six-fold coordination, (2) low solution concentrations of Al and Si, (3) acidic conditions, (4) dehydration of ions and gels, and (5) epitaxial heteronucleation.

La Iglesia and Martín-Vivaldi (1975) found that release of H^+ from carboxylic resins produces better crystalline kaolinite than sulfonic resins. Siffert (1962) successfully grew kaolinite by letting Al-oxalate (and citric, tartaric and salicylic) complexes slowly decompose in solution. Additional apparent catalysts to kaolinite yields include NaCl and ions obtained from hydrolysis of phases such as feldspar rather than from direct precipitation from solution (indicating possible topotaxial relations).

Kittrick (1970) grew kaolinite (and/or halloysite) on montmorillonite seeds in solutions supersaturated with respect to kaolinite at 25°C after 3 to 4 years. Linares and Huertas (1971) precipitated kaolinite after 1 month of aging at 25°C in solutions containing silicic acid and where Al was complexed by fulvic acid with Si/Al ratios of 0.05 to 5 at pH of 4 to 9. Other phases formed included gibbsite, bayerite, boehmite, and amorphous silicoalumina gels. Hem and Lind (1974) used an organic flavone, quercetin ($C_{15}H_{10}O_7$) in aqueous solutions of Si and Al at pH of 6.5 to 8.5. After 6 to 16 months of aging, they obtained up to 5% well-formed kaolinite plates. They observed that the presence of quercetin inhibited the formation of Al-polymers, and microcrystalline gibbsite. They proposed that fulvic acid catalyzed kaolinite formation by providing an Al-O bonding arrangement for forming an Al-O-Si bond in the kaolinite lattice. Al-polymers require double OH bridging bonds, not Al-O bonds. In inorganic solutions (Hem et al., 1973), the kaolinite did not form, even after two years of aging.

Sepiolite, palygorskite, and brucite. Wollast et al. (1968) investigated sepiolite growth from seawater solutions. They suggested that in earlier work brucite nucleation and growth in high pH experimental solutions may have catalyzed the formation of sepiolite. Their product had the same Mg/Si ratio as sepiolite but more water. By raising the solution pH to 9.4 they precipitated brucite + sepiolite. They also showed semi-quantitatively that the rate of precipitation depended on the concentration of OH^-. La Iglesia (1978) precipitated sepiolite at pH = 6 to 9 using periclase as a Mg source in solutions with 20 ppm Si and in some cases 20 ppm Al. In the systems without Al they formed brucite + sepiolite as determined by XRD. In systems with Al, at pH ≈ 6, hydrotalcite and brucite formed. Sepiolite formed in addition at pH > 6, and possibly palygorskite formed at pH > 7. Kent and Kastner (1985) also formed sepiolite in synthetic seawater solutions seeded with amorphous silica. They determined that precipitation rate increased with increasing supersaturation and that there appeared to be a pH dependence. Above pH ≈ 8.6, precipitation was significant.

La Iglesia (1977) also precipitated palygorskite by using dissolving forsterite as a Mg source at 25°C in solutions of pH 6 to 9 containing 20 ppm Al + 20 ppm Si or 20 ppm Al. In both precipitation studies, La Iglesia (1977, 1978) suggested that precipitation occurred homogeneously from solution.

pH-dependence

In addition to bayerite precipitation as a function of pH (Van Straten et al., 1984), the only other phase for which the pH-dependence of growth rates has been determined is sepiolite (Kent and Kastner, 1985; Brady, 1992). Brady (1992) determined seeded growth rates in 0.0125 M sodium borate solutions adjusted to pH's of 8.9 to 9.5 using HCl or NaOH (Fig. 24). The buffers were effective in maintaining constant pH during the growth reaction. The removal of Mg and Si from solution occurred at nearly the stoichiometric ratio of 2:3 Mg:Si. He also performed surface adorption titrations of Mg and Si onto the sepiolite seed. Combining both sets of data led to the following equation for the precipitation rate (R_{ppt} in mol m^{-2} sec^{-1}).

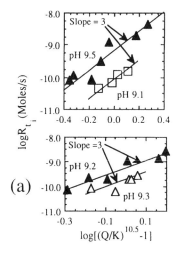

(a)

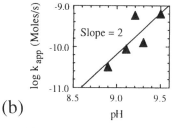

(b)

Figure 24. Precipitation rates of sepiolite based on finite difference derivatives of the measured Si concentrations as a function of time in batch experiments. (a) Precipitation rates as a function of saturation state and at constant pH. At pH 9.1 to 9.5, the reaction order with respect to the given function of saturation state is 3. (b) Calculated rate coefficient (K_{app}) as a function of pH. Used by permission of the author, from Brady (1992).

$$R_{ppt} = k[> Mg]^2 \left(\left(\frac{Q}{K_{eq}} \right)^{10.5} - 1 \right)^3 \tag{15}$$

where k is an apparent rate coefficient equal to 2.5×10^3 over the experimental conditions, $[>Mg]^2$ is the concentration of adsorbed Mg at the sepiolite surface which varies as a function of pH, and the last term expresses the saturation state dependence of the precipitation reaction. The power of 10.5 is derived by summing the number of reactant species according to methods outlined in Van Cappellen and Berner (1991). Brady (1992) hypothesized that the rate-limiting step was dehydration of Mg adsorbed onto the crystal surface. The solid was not characterized after the experiment to confirm formation of crystalline sepiolite.

Rate catalysts and inhibitors

Inorganic ions. Inhibition and catalysis of crystal growth is an important mechanism that controls crystal morphologies as well as growth rates. However, essentially no work has been conducted with respect to the growth of sheet silicates at low temperatures. If we knew something about solution controls on morphology we could infer information on paleosolution compositions. For example, Sun and Baronnet (1989a,b) investigated the effects of solution concentration of Ti and Cr on the morphology and growth mechanism of phlogopite at 522° to 535°C and P of 1 kbar in KOH (6.5 m) solutions. They found that the presence of Ti and Cr caused two-dimensional nucleation or one-dimensional roughening of all lateral faces, and therefore, slower growth kinetics, over the supersaturation range investigated. Although these studies may not directly apply to the weathering environment, a brief description of them may serve to instigate investigations in this direction at low temperatures, especially since two-dimensional nucleation and not spiral dislocations appears to control phyllosilicate growth at low temperature (Nagy and Lasaga, 1992; Nagy, 1994).

Hem et al. (1973) found that Al-hydroxide crystallization rates from aqueous solutions at 25°C were slowed considerably in the presence of Si as low as 2 mg L^{-1}. Al phases formed in a few days to weeks of aging time in the absence of Si. Two months or more were needed in the presence of Si. Schoen and Roberson (1970) proposed that the presence of Si in solution may prohibit the formation of bayerite and nordstrandite in nature. Ion impurities (chloride, nitrate, sulfate, phosphate, fluoride) are found to the slow the rate of formation. Na$^+$ and K$^+$ do not affect the growth rate of bayerite; however, Li$^+$ dramatically accelerates the growth rate, suggesting that a Li-aluminate precursor acts to nucleate bayerite (Van Straten et al., 1985a).

Veesler and Boistelle (1994) suggested that growth of gibbsite in high pH NaOH solutions was inhibited at low supersaturations by contaminant Fe. They conducted their growth experiments in single-pass flow reactors with 30 μm gibbsite seed particles. Therefore, the expected dependence on the square of the supersaturation term (Q/K - 1) as predicted by Burton et al. (1952) must be modified by replacing "1" with the critical supersaturation which must be exceeded to grow the crystals. They also determined an activation energy for growth of 121 kJ mol^{-1} from 55° to 80°C. Adu-Wusu and Wilcox (1990) also reported that Si adsorption inhibits gibbsite crystallization in NaOH solutions.

Organic ions. In many weathering environments, particularly in soil formation, organic components may be the key to rapid clay formation. Salicylate and pthalate appear to catalyze and enhance Al_{13} polymer formation slightly, primarily by shifting its formation constant to lower pH (Furrer et al., 1992). Violante and Violante (1980), Kodama and Schnitzer (1980), and Van Straten et al. (1985b) showed that organic chelates accelerated Al-hydroxide and Al-oxyhydroxide formation (see also, Lind and Hem, 1975). Organic compounds have been shown to enhance kaolinite formation (Linares and Huertas, 1971; Hem and Lind, 1974). Hem et al. (1973) prepared aluminosilicate gels according to the method of DeKimpe (1969) by mixing equilmolar quantities of aluminum isopropoxide ($Al[OCH(CH_3)_2]_3$) and ethyl orthosilicate ($Si(OCH_2CH_3)_4$) and then placed these in aqueous $NaClO_4$ solutions at pH 4.7 and 8.8. Tubular halloysite was formed. They proposed that the cation-oxygen-carbon bonding structure promotes transformation to aluminum-oxygen-silicon bonding necessary to form a kaolinite-type layer. They pointed out that in dilute aqueous solutions it is difficult to form a sheet silicate mineral structure because bonding between Al and Si or Si and Si requires removal of an OH^- and H^+ at each bonding site.

Dependence on reaction affinity or Gibbs free energy of reaction

A primary control on the precipitation or growth rate of a crystal is the solution saturation state. If nucleation occurs topotactically from a gel-mediated phase, the rate of absorption of metal ions into the gel may determine the growth rate. If nucleation occurs epitaxially on detrital sheet silicates, growth must be limited by the supply of ions from solution. Assuming growth is a surface-controlled process, then the saturation state of the solution will affect the growth rate by determining the number of ions in solution and, therefore, the probability of their attachment to the surface.

Few studies have been conducted to determine the effect of saturation state on growth rates of sheet silicates at low temperatures (Nagy et al., 1991; Nagy and Lasaga, 1992; 1993). In the gel experiments described above, clays always formed relatively rapidly (in a day to a few months) at 25°C. However, the quantity formed and the formation mechanisms were never determined. In contrast, phyllosilicate growth rates and mechanisms in supersaturated hydrothermal solutions have been quantified (e.g. Baronnet, 1972, 1973, 1982, 1984; Devidal et al., 1992; Soong, 1993). Although temperature and pressure conditions are drastically different from those that occur during weathering, some of the observations may bear on mechanisms that are important at low temperatures. Below, data are presented in order of increasing mineral complexity.

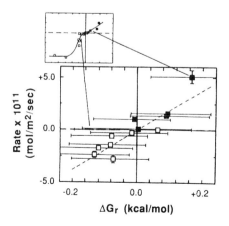

Figure 25. Near-equilibrium dissolution (negative values) and precipitation (positive values) rates of gibbsite at 80°C, pH 3. Precipitation rates are linear with ΔG_r. The equilibrium solubility obtained agreed well with results from Palmer and Wesolowski (1992). From Nagy and Lasaga (1992).

Gibbsite. Nagy and Lasaga (1992) measured gibbsite growth rates at 80°C and in pH 3 solutions from near equilibrium to a supersaturation of ~0.5 kcal mol^{-1}. They used stirred-flow reactors seeded with gibbsite powder. The rates obtained were an approximately linear function of ΔG_r (Fig. 25). Although they could not determine that the precipitate (at most 2% of the total seed mass) was truly gibbsite, they did obtain closure on the equilibrium solubility value (as determined from their dissolution rate determinations as a function of solution saturation state), that compared well with solubility measurements (Palmer and Wesolowski, 1992). This indicated that formation of another phase such as boehmite was unlikely. They also obtained XPS analyses of the reacted gibbsite surface and observed an Al/O atomic ratio expected for gibbsite as opposed to boehmite.

Kaolinite. Nagy et al. (1990, 1991) obtained precipitation rates for kaolinite (Twiggs Co., Georgia, USA) as a function of the degree of supersaturation in similar experiments to those described above for gibbsite. They found that it was more difficult to document stoichiometric precipitation and were unable to design experiments in which the saturation state could be maintained at values above ΔG_r ~0.5 kcal mol^{-1}. Nonetheless, their data show that precipitation rates for kaolinite also appear to be a linear function of supersaturation, at least over the limited range of saturation studied (Fig. 26a). Nagy and Lasaga (1993) extended the range of supersaturation slightly by raising the saturation state of the inlet fluids with respect to kaolinite by entering the supersaturated field of gibbsite. Knowledge of the gibbsite precipitation rates under the same conditions enabled correction of the measured Al and Si precipitation for gibbsite growth, and by difference, determination of stoichiometric kaolinite precipitation rates. These new rates confirmed the approximately linear dependence of kaolinite growth rates on solution saturation state (Fig. 26a). In fact, a fit of the data not constrained to pass through the previously determined solubility from Nagy et al. (1991) suggests that the equilibrium solubility for this newly-formed kaolinite may be lower than determined from the undersaturated side. This could be possible if the dissolving kaolinite contained defects or impurities and the growing kaolinite were more perfect. One deficiency in the mixed-seed study is that growth was determined from changes in solution composition only. The identification of the growth phase was inferred, but not verified directly, and may account for the variability observed in calculated gibbsite growth rates (Fig. 26b).

Devidal et al. (1992) and Soong (1993) also both determined kaolinite growth rates using the same hydrothermal single-pass experimental apparatus as for their dissolution work. In both studies, kaolinite growth rates were determined to be linear functions of supersaturation to ~4 and 0.5 kcal mol^{-1}, respectively. It is surprising that at the higher temperatures of their experiments (130° to 230°C), a nonlinear function of supersaturation was not observed. Typically, for electrolyte salt crystals, growth rate is a function of the square of saturation state (Nielsen, 1984; Nielson and Toft, 1984). This has been explained classically as control by a spiral growth mechanism (Burton et al., 1952), and has been documented for phyllosilicate growth under hydrothermal conditions (Baronnet, 1973; Pandey et al., 1982). However, apparently at temperatures extending through upper diagenetic conditions, the expected spiral growth mechanism does not take place. This is documented by AFM studies of both platy and fibrous illite morphologies from sedimentary basins (Blum, 1994; Nagy, 1994). Only hydrothermal illites exhibit evidence for spiral steps on their basal surfaces (Eberl and Blum, 1993; Blum, 1994).

Surface defects

Devouard and Baronnet (1994) and Devouard (1995) proposed that the crystal growth of chrysotile is determined by the elastic energy of curvature stored in the fiber

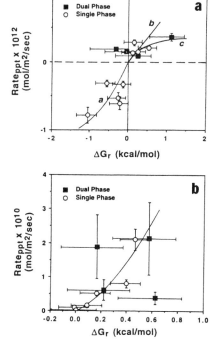

Figure 26. (a) Precipitation rates of kaolinite in single phase (Nagy et al., 1991) and dual phase (Nagy and Lasaga, 1993) experiments. Data are in reasonable agreement as a function of ΔG_r. Equations for fitted curves are given in Nagy and Lasaga (1993). (b) Precipitation rates of gibbsite in single phase (Nagy and Lasaga, 1992) and dual phase (Nagy and Lasaga, 1993) experiments. Error bars are based on analytical uncertainties. Two data points fall off the fitted curve (from Nagy and Lasaga, 1992) indicating either experimental uncertainty or incomplete characterization of reactive sites.

layers which balances the volume energy. They suggested that the nature of the fibers, whether cylindrical, conical, spiral, or helical, represents equilibrium states with varying degrees of saturation of the solutions in which the fibers grew. Structural and impurity defects in micas may nucleate growth of Al-hydroxides (Johnsson et al., 1992). Linearity of growth rates of gibbsite and kaolinite suggests two-dimensional nucleation and growth on rough or impure surfaces (Nagy and Lasaga, 1992; Nagy et al., 1990; Sun and Baronnet, 1988, 1989a,b).

SUMMARY REMARKS

Although dissolution and precipitation of sheet silicate and related clay minerals have been studied for decades, there are still many areas for new research opportunities. My intent in this chapter was to review the literature by consolidating research on a variety of related phases according to broadly-defined kinetic mechanisms. Although many data are semi-qualitative in that combinations of environmental variables such as solution composition, temperature, and pH indicate relative rates of dissolution or growth, we still lack measurements of actual rates under controlled conditions where the effects of each variable can be isolated.

Areas for new research include coupling the numerous existing data for sorption of aqueous components other than H^+, OH^-, and water, onto phyllosilicates to measurements of rates in the presence of these components. Al and Si adsorption have certain effects on dissolution and precipitation, but some reported results are in conflict. We also need to better characterize samples with small grain sizes in terms of the density of reactive sites on edge and basal surfaces as well as the percentages of edge and basal surfaces exposed to solution in experiments and in nature. Effects of compositional impurities and structural defects need to be quantified. Regarding nucleation and growth, high intensity X-ray techniques should be applied more frequently to the products of kinetic experiments to understand the mechanisms of attachment to the growth surface. Organic templates need to be investigated to a greater extent, as much experimental and observational data point to these as potentially common growth nucleation sites and "catalysts" in the weathering environment. Gel dehydration experiments in which phyllosilicates crystallize could be revisited with the goal of quantifying the nucleation and growth process using new experimental and analytical techniques.

Although a broad topic from the mineralogical and geochemical points of view, the dissolution and precipitation kinetics of sheet silicates and related minerals is exciting because of the wide range of disciplines in which kinetic data are important. Applications to weathering have been emphasized in this chapter. However, even on this topic, much remains to be discovered.

ACKNOWLEDGMENTS

I thank Susan Carroll, Jim Krumhansl, Henry Westrich, and Art White for reviewing this chapter. Their comments and time are much appreciated. This work was supported by the U.S. Nuclear Regulatory Commission, the U.S. Department of Energy Office of Basic Energy Sciences/Geoscience, and Laboratory Directed Research and Development under contract DE-AC04-94AL85000 to Sandia National Laboratories.

REFERENCES

Abdul-Latif N, Weaver CE (1969) Kinetics of acid-dissolution of palygorskite (attapulgite) and sepiolite. Clays Clay Minerals 17:169-178

Acker JG, Bricker OP (1992) The influence of pH on biotite dissolution and alteration kinetics at low temperature. Geochim Cosmochim Acta 56:3-73-3092

Adu-Wusu K, Wilcox WR (1991) Kinetics of silicate reaction with gibbsite. J Coll Interface Sci 143:127-138

Arshad MA, St Arnaud RJ, Huang PM (1972) Dissolution of trioctahedral layer silicates by ammonium oxalate, sodium dithionite-citrate-bicarbonate, and potassium pyrophosphate. Can J Soil Sci 52:19-25

Badaut D, Risacher F (1983) Authigenic smectite on diatom frustules in Bolivian saline lakes. Geochim Cosmochim Acta 47:363-375

Bailey SW (1980) Structures of layer silicates. In: Brindley GW, Brown G (eds) Crystal Structures of Clay Minerals and their X-ray Identification. Mineral Soc Monograph 5:1-124

Bailey SW (1984a) Classification and structures of the micas. In: Bailey SW (ed) Micas. Rev Mineral 13:1-12

Bailey SW (1984b) (ed) Micas. Rev Mineral 13. Mineral Soc Am, Washington, DC, 584 p

Bailey SW (1988a) Introduction. In: Bailey SW (ed) Hydrous Phyllosilicates (exclusive of micas). Rev Mineral 19:1-8

Bailey SW (1988b) (ed) Hydrous Phyllosilicates (exclusive of micas). Rev Mineral 19. Mineral Soc Am, Washington, DC, 725 p

Bales RC, Morgan JJ (1985) Dissolution kinetics of chrysotile at pH 7 to 10. Geochim Cosmochim Acta 49:2281-2288

Banfield JF, Ferruzzi GG, Casey WH, Westrich HR (1995) HRTEM study comparing naturally and experimentally weathered pyroxenoids. Geochim Cosmochim Acta 59:19-31

Banfield JF, Barker WW (1994) Direct observation of reactant-product interfaces formed in natural weathering of exsolved, defective amphibole to smectite: Evidence for episodic, isovolumetric reactions involving structural inheritance. Geochim Cosmochim Acta 58:1419-1429

Banfield JF, Jones BF, Veblen DR (1991a) An AEM-TEM study of weathering and diagenesis, Abert Lake, Oregon: I. Weathering reactions in the volcanics. Geochim Cosmochim Acta 55:2781-2793

Banfield JF, Jones BF, Veblen DR (1991a) An AEM-TEM study of weathering and diagenesis, Abert Lake, Oregon: II. Diagenetic modification of the sedimentary assemblage. Geochim Cosmochim Acta 55:2795-2810

Baronnet A (1972) Growth mechanisms and polytypism in synthetic hydroxyl-bearing phlogopite. Am Mineral 57:1272-1293

Baronnet A (1973) Sur les origines des dislocations vis et des spirales de croissance dans les micas. J Crystal Growth 19:193-198

Baronnet A (1982) Ostwald ripening in solution. The case of calcite and mica. Estud Geol 38:185-198

Baronnet A (1984) Growth kinetics of the silicates. A review of basic concepts. Fortschr Mineral 62:187-232

Berthelin J (1983) Microbial weathering processes. In: Krumbein WE (ed) Microbial Geochemistry, p 223-262. Blackwell Scientific Publications, Oxford

Berthelin J, Leyval C (1982) Ability of symbiotic and nonsymbiotic rizospheric microflora of maize (Zea mays) to weather micas and to promote plant growth and plant nutrition. Plant Soil 68:369-377

Bhatti TM, Bigham JM, Vuorinen A, Tuovinen OH (1994) Alteration of mica and feldspar associated with the microbiological oxidation of pyrrhotite and pyrite. In: Alpers CN, Blowes DW (eds) Environmental Geochemistry of Sulfide Oxidation. Am Chem Soc Symp Ser 550:90-105

Bish DL, Guthrie GD Jr (1993) Mineralogy of clay and zeolite dusts (exclusive of 1:1 layer silicates). In: Guthrie GD Jr, Mossman BT (eds) Health Effects of Mineral Dusts, Rev Mineral 28:139-184

Bloom PR (1983) The kinetics of gibbsite dissolution in nitric acid. Soil Sci Soc Am J 47:164-168

Bloom PR, Erich MS (1987) Effect of solution composition on the rate and mechanism of gibbsite dissolution in acid solutions. Soil Sci Soc Am J 51:1131-1136

Blum AE (1994) Determination of illite/smectite particle morphology using scanning force microscopy. In: Nagy KL, Blum AE (eds) Scanning Probe Microscopy of Clay Minerals, Clay Minerals Soc Workshop Lectures 7:171-202

Blum AE, Lasaga AC (1987) Monte Carlo simulations of surface reaction rate laws. In: Stumm W (ed) Aquatic Surface Chemistry: Chemical Processes at the Particle-Water Interface. John Wiley & Sons, New York, p 255-292

Blum AE, Yund RA, Lasaga AC (1990) The effect of dislocation density on the dissolution rate of quartz. Geochim Cosmochim Acta 54:283-297

Boles JR, Johnson KS (1983) Influence of mica surfaces on pore-water pH. Chem Geol 43:303-317

Bottero JY, Axelos M, Tschoubar D, Cases JM, Fripiat JJ, Fiessinger F (1987) Mechanism of formation of aluminum trihydroxide from Keggin Al_{13} polymers. J Coll Interface Sci 117:47-57

Brady PV, Walther JV (1989) Controls on silicate dissolution rates in neutral and basic pH solutions at 25°C. Geochim Cosmochim Acta 53:2823-2830

Brady PV (1992) Surface complexation and mineral growth: Sepiolite. In: Kharaka YK, Maest AS (eds) Water Rock Interactions. AA. Balkema, Rotterdam, p 85-88

Brindley GW, Youell RF (1951) A chemical determination of the tetrahedral and octahedral aluminums in a silicate. Acta Crystallogr 4:495-497

Brindley GW, Brown G (1980) (eds) Crystal Structures of Clay Minerals and their X-ray Identification. Mineral Soc Monograph No. 5, Mineral Soc, London, 495 p

Brown GE Jr, Parks GA, O'Day PA (1995) Sorption at mineral-water interfaces: macroscopic and microscopic perspectives. In: Vaughan DJ, Pattrick RAD (eds) Mineral Surfaces, Mineral Soc Ser 5:129-183

Brown DW, Hem JD (1975) Reactions of aqueous aluminum species at mineral surfaces. US Geol Surv Water-Supply Paper 1827-F, 48 p

Burton WK, Cabrera N, Frank FC (1952) The growth of crystals and the equilibrium structure of their surfaces. Phil Trans Roy Soc London A243:299-358

Cabrera F., Talibudeen O. (1978) The release of aluminum from aluminosilicate minerals. I. Kinetics. Clays Clay Minerals 26:434-440

Cairns-Smith AG, Hartman H (1986) Clay Minerals and the Origin of Life. Cambridge Univ Press, Cambridge, UK, 193 p

Cama J, Ganor J, Lasaga AC (1994) The kinetics of smectite dissolution. Mineral Mag 58A:140-141

Carroll SA, Walther JV (1990) Kaolinite dissolution at 25°, 60°, and 80°C. Am J Sci 290:797-810

Carroll-Webb SA, Walther JV (1988) A surface complex reaction model for the pH-dependence of corundum and kaolinite dissolution rates. Geochim Cosmochim Acta 52:2609-2623

Casey WH, Sposito G (1992) On the temperature dependence of mineral dissolution rates. Geochim Cosmochim Acta 56:3825-3830

Casey WH, Westrich HR, Banfield JF, Ferruzzi G, Arnold GW (1993) Leaching and reconstruction at the surfaces of dissolving chain-silicate minerals. Nature 366:253-256

Charlet L, Manceau A. (1993) Structure, formation, and reactivity of hydrous oxide particles: Insights from X-ray absorption spectroscopy. In: Buffle J, van Leeuwen HP (eds) Environmental Particles, Environmental Analytical and Physical Chemistry Ser, p 117-164. CRC Press, Boca Raton, LA

Charlet L, Manceau A. (1994) Evidence for the neoformation of clays upon sorption of Co(II) and Ni(II) on silicates. Geochim Cosmochim Acta 58:2577-2582

Chin P-FK, Mills GL (1991) Kinetics and mechanisms of kaolinite dissolution: effects of organic ligands. Chem Geol 90:307-317

Chon MC, Tsuru T, Takahashi H (1978a) Changes in pore structure of kaolin mineral by sulfuric acid treatment. Clay Sci 5:155-162

Chon MC, Tsuru T, Takahashi H (1978b) Changes in surface properties of kaolin mineral in dealumination process by sulfuric acid treatment. Clay Sci 5:163-169

Churchman GJ, Jackson ML (1976) Reaction of montmorillonite with acid aqueous solutions: solute activity control by a secondary phase. Geochim Cosmochim Acta 40:1251-1259

Clemency CV, Lin F-C (1981) Dissolution kinetics of phlogopite. II. Open system using an ion-exchange resin. Clays Clay Minerals 29:107-112

Coleman NT, Craig D (1961) The spontaneous alteration of hydrogen clay. Soil Sci 91:14-18

Crowley SF (1991) Diagenetic modification of detrital muscovite: an example from the Great Limestone Cyclothem (Carboniferous) of Co. Durham, U. K. Clay Minerals 26:91-103

Davis JA, Hem JD (1989) The surface chemistry of aluminum oxides and hydroxides. In: Sposito G (ed) The Environmental Chemistry of Aluminum. CRC Press, Boca Raton, LA, p 185-219

Decarreau A (1980) Cristallogène expérimentale des smectites magnésiennes: hectorite, stévensite. Bull Minéral 103:579-590

Decarreau A (1981) Cristallogenèse à basse température de smectites trioctaédriques par vieillissement de coprécipités silicométalliques de formule $(Si_{4-x})M_3{}^{2+}O_{11} \cdot nH_2O$, où x varie de 0 à 1 et où M^{2+} = Mg-Ni-Co-Zn-Fe-Cu-Mn. C R Acad Sci Paris 292:61-64

Decarreau A, Bonnin D (1986) Synthesis and crystallogenesis at low temperature of Fe(III)-smectites by evolution of coprecipitated gels: experiments in partially reducing conditions. Clay Minerals 21:861-877

Decarreau A, Bonnin D, Badaut-Trauth D, Couty R, Kaiser P (1987) Synthesis and crystallogenesis of ferric smectite by evolution of Si-Fe coprecipitates in oxidizing conditions. Clay Minerals 22:207-223

Decarreau A, Mondesir H, Besson G (1989) Synthèse et stabilité des stévensites kérolites et talcs, magnésiens et nickelifères, entre 80 et 240°C. C R Acad Sci Paris 308:301-306

De Kimpe C (1969) Crystallization of kaolinite at low temperature from an alumino-silicic gel. Clays Clay Minerals 17:37-38

De Kimpe C, Gastuche MC, Brindley GW (1961) Ionic coordination in alumino-silicic gels in relation to clay mineral formation. Am Mineral 46:1370-1381

Denbigh K (1971) The Principles of Chemical Equilibrium, Cambridge University Press, New York, 494 p

Devidal JL, Dandurand JL, Schott J (1992) Dissolution and precipitation kinetics of kaolinite as a function of chemical affinity (T = 150°C, pH = 2 and 7.8). In: Kharaka YK, Maest AS (eds) Water Rock Interactions, p 93-96. AA Balkema, Rotterdam

Devouard B (1995) Structure and crystal growth of chrysotile and polygonal serpentines. PhD dissertation, University of Aix-Marseille III, 182 p

Devouard B, Baronnet A (1994) Synthetic chrysotile fibers: from cylindrically to conically wrapped structures. Implications for the growth mechanisms of chrysotile. Int'l Mineral Assoc 16th Gen Meeting Abstr:95

Dixon JB, Weed SB (1989) (eds) Minerals in Soil Environments, 2nd Edn. Soil Sci Soc Am, Madison, Wisconsin, 1244 p

Eberl DD, Blum AE (1993) Illite crystal thickness by X-ray diffraction. In: Reynolds RC Jr, Walker Jr (eds) Computer Applications to X-Ray Powder Diffraction Analysis of Clay Minerals. Clay Minerals Soc Workshop Lectures 5:123-153

Feigenbaum S, Shainberg I (1975) Dissolution of illite - a possible mechanism of potassium release. Soil Sci Soc Am Proc 39:985-990

Farmer VC, Russel JD, McHardy WJ, Newman ACD, Ahlrichs JLA, Rimsaite JL (1971) Evidence for loss of protons and octahedral iron from oxidized biotites and vermiculites. Mineral Mag 38:121-138

Flehmig W (1992) The synthesis of 2M1-illitic micas at 20°C. N Jb Miner Mh H.11:507-512

Fleming BA (1986) Kinetics of reaction between silicic acid and amorphous silica surfaces in NaCl solutions. J Coll Interface Sci 110:40-64

Furrer G, Stumm W (1986) The coordination chemistry of weathering: I. Dissolution kinetics of δ-Al_2O_3 and BeO. Geochim Cosmochim Acta 50:1847-1860

Furrer G, Trusch B, Müller C (1992) The formation of polynuclear Al_{13} under simulated natural conditions. Geochim Cosmochim Acta 56:3831-3838

Furrer G, Zysset M, Schindler PW (1993) Weathering kinetics of montmorillonite: Investigations in batch and mixed-flow reactors. In: Manning DAC, Hall, PL, Hughes, CR (eds) Geochemistry of Clay-Pore Fluid Interactions. Chapman & Hall, London, p 243-262

Gaines GL Jr, Rutkowski CP (1957) The extraction of aluminum and silicon from muscovite mica by aqueous solutions. J Phys Chem 61:1439-1441

Ganor J, Lasaga AC (1994) The effects of oxalic acid on kaolinite dissolution rate. Mineral Mag 58A:315-316

Ganor J, Mogollón JL, Lasaga AC (1995) The effect of pH on kaolinite dissolution rates and on activation energy. Geochim Cosmochim Acta 59:1037-1052

Gastuche MC (1963) Kinetics of acid dissolution of biotite. Proc Int'l Clay Conf Stockholm 1:67-83

Giese RF Jr (1984) Electrostatic energy models of micas. In: Bailey SW (ed) Micas. Rev Mineral 13:105-144

Giese RF Jr (1988) Kaolin minerals: structures and stabilities. In: Bailey SW (ed) Hydrous Phyllosilicates (exclusive of micas). Rev Mineral 19:29-66

Gilkes RJ, Young RC, Quirk JP (1973a) Artificial weathering of oxidized biotite: I. Potassium removal by sodium chloride and sodium tetraphenylboron solutions. Soil Sci Soc Am Proc 37:25-28

Gilkes RJ, Young RC, Quirk JP (1973b) Artificial weathering of oxidized biotite: II. Rates of dissolution in 0.1, 0.01, 0.001 M HCl. Soil Sci Soc Am Proc 37:29-33

Greaves GN (1995) New X-ray techniques and approaches to surface mineralogy. In: Vaughan DJ, Pattrick RAD (eds) Mineral Surfaces. Mineral Soc Ser 5:87-128

Gupta AK, Smith RW (1975) Kinetic study of the reaction of acids with asbestos minerals. Proc Int'l Symp Water-Rock Interaction. Czechoslovakia, p 417-424

Harder H (1971) The role of magnesium in the formation of smectite minerals. Chem Geol 10:31-39

Harder H (1974) Illite mineral synthesis at surface temperatures. Chem Geol 14:241-253

Harder H (1976) Nontronite synthesis at low temperatures. Chem Geol 18:169-180

Harder H (1977) Clay mineral formation under lateritic weathering conditions. Clay Minerals 12:281-288

Harder H (1978) Synthesis of iron layer silicate minerals under natural conditions. Clays Clay Minerals 26:65-72

Hartman P (1973) Structure and morphology. In: Hartman P (ed) Crystal Growth: An Introduction. North-Holland Publishers, Amsterdam, p 367-402

Hem JD, Lind CJ (1974) Kaolinite synthesis at 25°C. Science 184:1171-1173

Hem JD, Roberson CE (1967) Form and stability of aluminum hydroxide complexes in dilute solution. US Geol Survey Water-Supply Paper 1827-A, 55 p

Hem JD, Roberson CE (1990) Aluminum hydrolysis reactions and products in mildly acidic aqueous systems. In: Melchior DC, Bassett RL (eds) Chemical Modeling of Aqueous Systems II, Am Chem Soc Symp Ser 416:429-446

Hem JD, Roberson CE, Lind CJ, Polzer WL (1973) Chemical interactions of aluminum with aqueous silica at 25C. US Geol Survey Water-Supply Paper 1827-E, 57 p

Heydemann A (1966) Über die chemische Verwitterung von Tonmineralen (Experimentelle Untersuchungen). Geochim Cosmochim Acta 30:995-1035

Hinsinger P, Jaillard B, Dufey JE (1992) Rapid weathering of a trioctahedral mica by the roots of ryegrass. Soil Sci Soc Am J 56:977-982

Hsu PH (1988) Mechanisms of gibbsite crystallization from partially neutralized aluminum chloride solutions. Clays Clay Minerals 36:25-30

Huang WH, Keller WD (1970) Dissolution of rock-forming silicate minerals in organic acids: simulated first-stage weathering of fresh mineral surfaces. Am Mineral 55:2076-2094

Huang WH, Keller WD (1971) Dissolution of clay minerals in dilute organic acids at room temperature. Am Mineral 56:1082-1095

Huang WH, Keller WD (1973) Kinetics and mechanisms of dissolution of Fithian illite in two complexing organic acids. Proc Int'l Clay Conf, Madrid, 321-331

Hume LA, Rimstidt JD (1992) The biodurability of chrysotile asbestos. Am Mineral 77:1125-1128

Jackson ML, Tyler SA, Willis AL, Bourbeau GA, Pennington RP (1948) Weathering sequence of clay-size minerals in soils and sediments. I. Fundamental generalizations. J Phys Coll Chem 52:1237-1260

Jackson ML, Hseung Y, Corey RB, Evans EJ, Vanden Heuvel RC (1952) Weathering sequence of clay-size minerals in soils and sediments. II. Chemical weathering of layer silicates. Soil Sci Soc Am Proc 16:3-6

Jardine PM, Zelazny LW, Parker JC (1985) Mechanisms of aluminum adsorption on clay minerals and peat. Soil Sci Soc Am J 49:862-872

Jepson WB, Jeffs DG, Ferris AP (1976) The adsorption of silica on gibbsite and its relevance to the kaolinite surface. J Coll Interface Sci 55:454-461

Jiang W-T, Peacor DR (1991) Transmission electron microscopic study of the kaolinitization of muscovite. Clays Clay Minerals 39:1-13

Johnsson PA, Hochella MF Jr, Parks GA, Blum AE, Sposito G (1992) Direct observation of muscovite basal-plane dissolution and secondary phase formation: An XPS, LEED, and SFM study. In: Kharaka YK, Maest AS (eds) Water Rock Interactions. AA Balkema, Rotterdam, p 159-162

Kent DB, Kastner M (1985) Mg^{2+} removal in the system Mg^{2+}-amorphous SiO_2-H_2O by adsorption and Mg-hydroxysilicate precipitation. Geochim Cosmochim Acta 49:1123-1136

Kittrick JA (1970) Precipitation of kaolinite at 25°C and 1 atm. Clays Clay Minerals 18:261-267

Kline WE, Fogler HS (1981a) Dissolution kinetics: The nature of the particle attack of layered silicate in HF. Chem Eng Sci 36:871-884

Kline WE, Fogler HS (1981b) Dissolution kinetics: Catalysis by strong acids. J Coll Interface Sci 82:93-102

Kline WE, Fogler HS (1981c) Dissolution kinetics: Catalysis by salts. J Coll Interface Sci 82:103-115

Knauss KG, Wolery TJ (1989) Muscovite dissolution kinetics as a function of pH and time at 70°C. Geochim Cosmochim Acta 53:1493-1502

Kodama H, Schnitzer M (1973) Dissolution of chlorite minerals by fulvic acid. Can J Soil Sci 53:240-243

Kodama H, Schnitzer M (1980) Effect of fulvic acid on the crystallization of aluminum hydroxides. Geoderma 24:195-205

Kummert R, Stumm W (1980) The surface complexation of organic acids on hydrous γ-Al_2O_3. J Coll Interface Sci 75:373-385

Kuwahara Y, Aoki Y (1995) Dissolution process of phlogopite in acid solutions. Clays Clay Minerals 43:39-50

La Iglesia A (1977) Precipitación por disolución homogénea de silicatos de aluminio y magnesio a temperatura ambiente. Síntesis de la paligorskita. Estudios Geol 33:535-544

La Iglesia A (1978) Síntesis de la sepiolita a temperatura ambiente por precipitación homogénea. Boletín Geológico Minero T. LXXXIX:258-265

La Iglesia A, Martín-Vivaldi JL (1975) Synthesis of kaolinite by homogeneous precipitation at room temperature I. Use of anionic resins in (OH) form. Clay Minerals 10:399-405

La Iglesia A, Serna J (1974) Cristalización de caolinita por precipitación homogenea. Parte II. Empleo de resinas catiónicas en fase H^+. Estud Geol Madrid 30:281-287

La Iglesia A, Martín-Vivaldi JL Jr, López Aguayo F (1976) Kaolinite crystallization at room temperature by homogeneous precipitation-III: Hydrolysis of feldspars. Clays Clay Minerals 24:36-42

La Iglesia A, Van Oosterwyck-Gastuche MC (1978) Kaolinite synthesis. I. Crystallization conditions at low temperatures and calculation of thermodynamic equilibria. Application to laboratory and field observations. Clays Clay Minerals 26:397-408

Leonard RA, Weed SB (1970) Mica weathering rates as related to mica type and composition. Clays Clay Minerals 18:187-195

Lerman A, Mackenzie FT, Bricker OP (1975) Rates of dissolution of aluminosilicates in seawater. Earth Planet Sci Lett 25:82-88

Lim CH, Jackson ML, Koons RD, Helmke PA (1980) Kaolins: Sources of differences in cation-exchange capacities and cesium retention. Clays Clay Minerals 28:223-229

Lin F-C, Clemency CV (1981a) Dissolution kinetics of phlogopite. I. Closed system. Clays Clay Minerals 29:101-106

Lin F-C, Clemency CV (1981b) The kinetics of dissolution of muscovites at 25°C and 1 atm CO_2 partial pressure. Geochim Cosmochim Acta 45:571-576

Lin F-C, Clemency CV (1981c) The dissolution kinetics of brucite, antigorite, talc, and phlogopite at room temperature and pressure. Am Mineral 66:801-806

Lind CJ, Hem JD (1975) Effects of organic solutes on chemical reactions of aluminum. US Geol Surv Water-Supply Paper 1827-G, 83 p

Linares J, Huertas F (1971) Kaolinite synthesis at room temperature. Science 171:896-897

Lou G, Huang PM (1993) Silication of hydroxy-Al interlayers in smectite. Clays Clay Minerals 41:38-44

Mackin JE, Swider KT (1987) Modeling the dissolution behavior of standard clays in seawater. Geochim Cosmochim Acta 51:2947-2964

Mackenzie FT, Garrels RM (1966) Chemical mass balance between rivers and oceans. Am J Sci 264:507-525

Mann S, Archibald DD, Didymus JM, Douglas T, Heywood BR, Meldrum FC, Reeves NJ (1993) Crystallization at inorganic-organic interfaces: Biominerals and biomimetic synthesis. Science 261:1286-1292

Mattigod SV, Kittrick JA (1979) Aqueous solubility studies of muscovite: Apparent nonstoichiometric solute activities at equilibrium. Soil Sci Soc Am J 43:180-187

May HM, Kinniburgh DG, Helmke PA, Jackson ML (1986) Aqueous dissolution, solubilities and thermodynamic stabilities of common aluminosilicate clay minerals: Kaolinite and smectites. Geochim Cosmochim Acta 50:1667-1677

May HM, Acker JG, Smyth JR, Bricker OP, Dyar MD (1995) Aqueous dissolution of low-iron chlorite in dilute acid solutions at 25°C. Clay Minerals Soc Prog Abstr 32:88

McBride MB (1994) Environmental Chemistry of Soils. Oxford University Press, New York, 406 p

Mering J (1949) Les reactions de la montmorillonite. Bull Soc Chim France D218-D223

Merino E, Nahon D, Wang Y (1993) Kinetics and mass transfer of pseudomorphic replacement: application to replacement of parent minerals and kaolinite by Al, Fe, and Mn oxides during weathering. Am J Sci 293:135-155

Messersmith PB, Stupp SI (1992) Synthesis of nanocomposites: Organoceramics. J Mater Res 7:2599-2611

Mogollón JL, Perez DA, Lo Monaco S, Ganor J, Lasaga AC (1994) The effect of pH, $HClO_4$, HNO_3 and ΔG_r on the dissolution rate of natural gibbsite using column experiments. Mineral Mag 58A:619-620

Monnier A, Schüth F, Huo Q, Kumar D, Margolese D, Maxwell RS, Stucky GD, Krishnamurty M, Petroff P, Firouzi A, Janicke M, Chmelka BF (1993) Cooperative formation of inorganic-organic interfaces in the synthesis of silicate mesostructures. Science 261:1299-1303

Mortland MM (1958) Kinetics of potassium release from biotite. Soil Sci Soc Am Proc 22:503-508

Mortland MM, Lawton K (1961) Relationships between particle size and potassium release from biotite and its analogues. Soil Sci Soc Am Proc 25:473-476

Nagy KL (1994) Application of morphological data obtained using scanning force microscopy to quantification of fibrous illite growth rates. In: Nagy KL, Blum AE (eds) Scanning Probe Microscopy of Clay Minerals. Clay Minerals Soc Workshop Lectures 7:203-239

Nagy KL, Lasaga AC (1993) Simultaneous precipitation kinetics of kaolinite and gibbsite at 80°C and pH 3. Geochim Cosmochim Acta 57:4329-4335

Nagy KL, Lasaga AC (1992) Dissolution and precipitation kinetics of gibbsite at 80°C and pH 3: The dependence on solution saturation state. Geochim Cosmochim Acta 56:3093-3111

Nagy KL, Blum AE, Lasaga AC (1991) Dissolution and precipitation kinetics of kaolinite at 80°C and pH 3: The dependence on solution saturation state. Am J Sci 291:649-686

Nagy KL, Cygan RT, Brady PV (1995) Kaolinite morphology: AFM observations, model predictions, and adsorption site density. Clay Minerals Soc Prog Abstr 32:95

Nagy KL, Steefel CI, Blum AE, Lasaga AC (1990) Dissolution and precipitation kinetics of kaolinite: initial results at 80C with application to porosity evolution in a sandstone. In: Meshri ID and Ortoleva PJ (eds) Prediction of Reservoir Quality through Chemical Modeling. Am Assoc Petrol Geol Memoir 49:85-101

Newman ACD (1969) Cation exchange properties of micas. J Soil Sci 20:298-373

Newman ACD (1987) (ed) Chemistry of Clays and Clay Minerals. Mineral Soc Monograph 6, Wiley Interscience, New York, 480 p

Newman ACD, Brown G (1969) Delayed exchange of potassium from some edges of mica flakes. Nature 223:175-176

Nickel E (1973) Experimental dissolution of light and heavy minerals in comparison with weathering and intrastratal solution. Contrib Sedimentology 1:1-68

Nielsen AE (1984) Electrolyte crystal growth mechanisms. J Crystal Growth 67:289-310

Nielsen AE, Toft JM (1984) Electrolyte crystal growth kinetics. J Crystal Growth 67:278-288

O'Day PA, Parks GA, Brown GE Jr (1994b) Molecular structure and binding of cobalt(II) surface complexes on kaolinite from X-ray absorption spectroscopy. Clays Clay Minerals 42:337-355

O'Day PA, Brown GE Jr, Parks GA (1994b) X-ray absorption spectroscopy of cobalt(II) multinuclear surface complexes and surface precipitates on kaolinite. J Coll Interface Sci 165:269-289

Oelkers EH, Schott J, Devidal JL (1994) The effect of aluminum, pH, and chemical affinity on the rates of aluminosilicate dissolution reactions. Geochim Cosmochim Acta 58:2011-2024

Osthaus BB (1954) Chemical determination of tetrahedral ions in nontronite and montmorillonite. Clays Clay Minerals 2:404-417

Osthaus BB (1956) Kinetic studies on montmorillonite and nontronite by the acid-dissolution technique. Clays Clay Minerals 4:301-321

Packter A, Dhillon HS (1969) The heterogeneous reaction of gibbsite powder with aqueous inorganic acid solutions; Kinetics and mechanism. J Chem Soc A:2588-2592

Packter A, Dhillon HS (1973) The kinetics and mechanism of the heterogeneous reactions of crystallized gibbsite powders with aqueous sodium hydroxide solutions. J Phys Chem 77:2942-2947

Packter A, Dhillon HS (1974) Studies on recrystallized aluminium hydroxide precipitates: Kinetics and mechanism of dissolution by sodium hydroxide solutions. Coll Polymer Sci 252:249-256

Packter A, Panesar KS (1986) The heterogeneous reactions of recrystallised aluminium trihydroxide precipitates with alkaline-earth metal hydroxide solution: kinetics and mechanism. Z Phys Chemie, Leipzig 267:9-14

Palmer DA, Wesolowski DJ (1992) Aluminum speciation and equilibria in aqueous solution: Part 2. The solubility of gibbsite in acidic sodium chloride solutions from 30 to 70°C. Geochim Cosmochim Acta 56:1093-1111

Pandey D, Baronnet A, Krishna P (1982) Influence of stacking faults on the spiral growth of polytype structures in micas. Phys Chem Minerals 8:268-278

Peric J, Krstulovic R, Feric T (1985) Kinetic aspects of the gibbsite digestion process in aqueous solution of sodium hydroxide. Croat Chem Acta 58:255-264

Pevear DR, Klimentidis RE, Robinson GA (1991) Petrogenetic significance of kaolinite nucleation and growth on pre-existing mica in sandstone and shales. Clay Minerals Soc Prog Abstr 28:125

Polzer WL, Hem JD (1965) The dissolution of kaolinite. J Geophys Res 70:6233-6240

Pulfer K, Schindler PW, Westall JC, Grauer R (1984) Kinetics and mechanism of dissolution of bayerite in HNO_3-HF solutions at 298.2K. J Coll Interface Sci 101:554-564

Raussel-Colom JA, Sweatman TR, Wells CB, Norrish K (1965) Studies in the artificial weathering of mica. Experimental pedology. Proc Univ Nottingham 11th Easter Sch Agr Sci: 40-72

Reesman AL (1974) Aqueous dissolution studies of illite under ambient conditions. Clays Clay Minerals 22:443-454

Refson K, Wogelius RA, Fraser DG, Payne MC, Lee MH, Milman V (1995) Water chemisorption and reconstruction of the MgO surface. Phys Rev B (in press)

Robert M, Berthelin J (1986) Role of biological and biochemical factors in soil mineral weathering. In: Huang PM, Schnitzer M (eds) Interactions of Soil Minerals with Natural Organics and Microbes. Soil Sci Soc Am Spec Pub 17:453-495

Robert M, Pedro G (1969) Etude des relations entre les phenomenes d'oxydation et l'aptitude à l'ouverture dans les micas trioctaedriques. Proc Int Clay Conf Japan 1:455-473

Rosenberg PE, Kittrick JH, Alldredge JR (1984) Composition of the controlling phase in muscovite equilibrium solubility. Clays Clay Minerals 32:480-482

Ross GJ (1967) Kinetics of acid dissolution of an orthochlorite mineral. Can J Chem 45:3031-3034

Samama J-C (1986) Ore Fields and Continental Weathering, Van Nostrand Reinhold Company, New York, 326 p

Schindler PW, Stumm W (1987) The surface chemistry of oxides, hydroxides, and oxide minerals. In: Stumm W (ed) Aquatic Surface Chemistry. John Wiley, New York, p 83-110

Schindler PW (1990) Co-adsorption of metal ions and organic ligands: formation of ternary surface complexes. In: Hochella MF Jr, White AF (eds), Mineral-Water Interface Geochemistry. Rev Mineral 23:281-307

Schnitzer M, Kodama H (1976) The dissolution of mica by fulvic acid. Geoderma 15:381-391

Schoen R, Roberson CE (1970) Structures of aluminum hydroxide and geochemical implications. Am Mineral 55:43-77

Schuette JF, Pevear DR (1993) Inverting the NEWMOD© X-ray diffraction forward model for clay minerals using genetic algorithms. In: Reynolds RC Jr, Walker JR (eds) Computer Applications to X-ray Powder Diffraction Analysis of Clay Minerals, Clay Minerals Soc Workshop Lectures 5:19-41

Scotford RF, Glastonbury JR (1971) Dissolution of gibbsite and boehmite. Can J Chem Eng 49:611-616

Scotford RF, Glastonbury JR (1972) The effect of concentration on the rates of dissolution of gibbsite and boehmite. Can J Chem Eng 50:754-758

Scott AD (1968) Effect of particle size on interlayer potassium exchange in micas. Trans 9th Int'l Congr Soil Sci 3:649-660

Siever R, Woodward N (1973) Sorption of silica by clay minerals. Geochim Cosmochim Acta 37:1851-1880

Siffert B (1962) Quelques reactions de la silice en solution: la formation des argiles. Memoires du Service de al carte geologique d'Alsace et de Lorràine 21:86 p

Smith RW, Hem JD (1972) Effect of aging on aluminum hydroxide complexes in dilute aqueous solutions. US Geol Survey Water-Supply Paper 1827-D, 51 p

Soler JM, Mogollón JL, Lasaga AC (1994) Coupled fluid flow and chemical reaction: a steady state model of the dissolution of gibbsite in a column. Boletín Sociedad Española Mineralogía 17:17-28

Soong C (1993) Hydrothermal kinetics of kaolinite-water interaction at pH 4.2 and 7.3, 130°C to 230°C, PhD dissertation, The Pennsylvania State University, State College, PA, 106 p

Sposito G (1984) The Surface Chemistry of Soils. Oxford University Press, New York, 234 p

Sposito G (1986) Distinguishing adsorption from surface precipitation. In: Davis JA, Hayes KF (eds) Geochemical Processes at Mineral Surfaces. Am Chem Soc Symp Ser 323, p 217-228. Am Chem Soc, Washington, DC

Sposito G (1990) Molecular models of ion adsorption on mineral surfaces. In: Hochella MF Jr, White AF (eds), Mineral-Water Interface Geochemistry. Rev Mineral 23:261-279

Srodon J, Eberl DD (1984) Illite. In: Bailey SW (ed) Micas. Rev Mineral 13:495-544

Steefel CI, Van Cappellen P (1990) A new kinetic approach to modeling water-rock interaction: The role of nucleation, precursors, and Ostwald ripening. Geochim Cosmochim Acta 54:2657-2677

Steinberg S, Ducker W, Vigil G, Hyukjin C, Frank C, Tseng MZ, Clarke DR, Israelachvili JN (1993) Van der Waals epitaxial growth of α-alumina nanocrystals on mica. Science 260:656-659

Stoessell RK (1988) 25°C and 1 atm dissolution experiments of sepiolite and kerolite. Geochim Cosmochim Acta 52:365-374

Stol RJ, Van Helden AK, De Bruyn PL (1976) Hydrolysis-precipitation studies of aluminum (III) solutions 2. A kinetic study and model. J Coll Interface Sci 57:115-131

Stumm W, Furrer G, Kunz B (1983) The role of surface coordination in precipitation and dissolution of mineral phases. Croatica Chem Acta 56:593-611

Stumm W, Wehrli B, Wieland E (1987) Surface complexation and its impact on geochemical kinetics. Croatica Chemica Acta 60:429-456

Sun BN, Baronnet A (1989a) Hydrothermal growth of OH-phlogopite single crystals. I. Undoped growth medium. J Crystal Growth 96:265-276

Sun BN, Baronnet A (1989b) Hydrothermal growth of OH-phlogopite single crystals II. Role of Cr and Ti adsorption on crystal growth rates. Chem Geol 78:301-314

Tan KH (1975) The catalytic decomposition of clay minerals by complex reaction with humic and fulvic acid. Soil Sci 120:188-194

Tanev PT, Pinnavaia TJ (1995) A neutral templating route to mesoporous molecular sieves. Science 267:865-867

Thomassin JH, Goni J, Baillif P, Touray JC, Jaurand MC (1977) An XPS study of the dissolution kinetics of chrysotile in 0.1 N oxalic acid at different temperatures. Phys Chem Minerals 1:385-398

Trotignon L, Turpault M-P (1992) The dissolution kinetics of biotite in dilute HNO_3 at 24°C. In: Kharaka YK, Maest AS (eds) Water Rock Interactions. AA Balkema, Rotterdam, p 123-125

'T Serstevens A, Rouxhet PG, Herbillon AJ (1978) Alteration of mica surfaces by water and solutions. Clay Minerals 13:401-410

Tsipursky SJ, Eberl DD, Buseck PR (1992) Unusual tops (bottoms?) of particles of 1M illite from the Silverton caldera. Clay Minerals Soc Prog Abstr 29:381

Tsuzuki Y, Kawabe I (1983) Polymorphic transformations of kaolin minerals in aqueous solutions. Geochim Cosmochim Acta 47:59-66

Turpault M-P, Trotignon L (1994) The dissolution of biotite single crystals in dilute HNO_3 at 24°C: Evidence of an anisotropic corrosion process of micas in acidic solutions. Geochim Cosmochim Acta 58:2761-2775

Van Cappellen P, Berner RA (1991) Fluorapatite crystal growth from modified seawater solutions. Geochim Cosmochim Acta 55:1219-1234

Van Olphen H, Fripiat JJ (1979) Data Handbook for Clay Materials and other Non-Metallic Minerals. Pergamon Press, New York, 346 p

Van Oosterwyck-Gastuche MC, La Iglesia A (1978) Kaolinite synthesis. II. A review and discussion of the factors influencing the rate process. Clays Clay Minerals 26:409-417

Van Straten HA, De Bruyn PL (1984) Precipitation from supersaturated aluminate solutions II. Role of temperature. J Coll Interface Sci 102:260-277

Van Straten, HA, Holtkamp BTW, De Bruyn PL (1984) Precipitation from supersaturated aluminate solutions I. Nucleation and growth of solid phases at room temperature. J Coll Interface Sci 98:342-362

Van Straten HA, Schoonen MAA, De Bruyn PL (1985a) Precipitation from supersaturated aluminate solutions III. Influence of alkali ions with special reference to Li^+. J Coll Interface Sci 103:493-507

Van Straten HA, Schoonen MAA, Verheul RCS, De Bruyn PL (1985b) Precipitation from supersaturated aluminate solutions. IV. Influence of citrate ions. J Coll Interface Sci 106:175-185

Veblen DR, Wylie AG (1993) Mineralogy of amphiboles and 1:1 layer silicates. In: Guthrie GD Jr, Mossman BT (eds) Health Effects of Mineral Dusts. Rev Mineral 28:61-137

Veesler S, Boistelle R (1994) Growth kinetics of hydrargillite $Al(OH)_3$ from caustic soda solutions. J Crystal Growth 142:177-183

Velbel MA (1984) Weathering processes of rock-forming minerals. In: Fleet ME (ed) Short Course in Environmental Geochemistry. Mineral Assoc Canada, p 67-111

Velbel MA (1985) Geochemical mass balances and weathering rates in forested watersheds of the southern Blue Ridge. Am J Sci 285:904-930

Vermilyea DA (1969) The dissolution of MgO and $Mg(OH)_2$ in aqueous solutions. J Electrochem Soc 116:1179-1183

Violante A, Violante P (1980) Influence of pH, concentration and chelating power of organic anions on the synthesis of aluminium hydroxides and oxy-hydroxides. Clays Clay Minerals 28:425-434

Von Reichenbach HG, Rich CI (1969) Potassium release from muscovite as influenced by particle size. Clays Clay Minerals 17:23-29

Walker WJ, Cronan CS, Patterson HH (1988) A kinetic study of aluminum adsorption by aluminosilicate clay minerals. Geochim Cosmochim Acta 52:55-62

Wang Y, Wang Y, Merino E (1995) Dynamic weathering model: Constraints required by coupled dissolution and pseudomorphic replacement. Geochim Cosmochim Acta 59:1559-1570

Weed SB, Davey CB, Cook MG (1969) Weathering of mica by fungi. Soil Sci Soc Am Proc 33:702-706

Weissbuch I, Addadi L, Lahav M, Leiserowitz L (1991) Molecular recognition at crystal interfaces. Science 253:637-645

White AF, Yee A (1985) Aqueous oxidation-reduction kinetics associated with coupled electron-cation transfer from iron-containing silicates at 25°C. Geochim Cosmochim Acta 49:1263-1275

White GN, Zelazny LW (1988) Analysis and implications of the edge structure of dioctahedral phyllosilicates. Clays Clay Minerals 36:141-146

Wieland E (1988) Die Verwitterung schwerlöslicher Mineralien - ein koordinationschemischer Ansatz zur Beschreibung der Auflösungskinetik. PhD dissertation, Eidgenössischen Technischen Hochschule Zürich, 251 p

Wieland E, Stumm W (1992) Dissolution kinetics of kaolinite in acidic aqueous solutions at 25°C. Geochim Cosmochim Acta 56:3339-3355

Wogelius RA, Refson K, Fraser DG, Grime GW, Goff JP (1995) Periclase surface hydroxylation during dissolution. Geochim Cosmochim Acta 59:1875-1881

Wollast R, Mackenzie FT, Bricker OP (1968) Experimental precipitation and genesis of sepiolite at Earth-surface conditions. Am Mineral 53:1645-1662

Xie Z, Walther JV (1992) Incongruent dissolution and surface area of kaolinite. Geochim Cosmochim Acta 56:3357-3363

Zhou Q, Maurice PA (1995) The microtopography of well and poorly crystallized kaolinite standards imaged by atomic force microscopy. VM Goldschmidt Conf Prog Abstr p 102

Zutic V, Stumm W (1984) Effect of organic acids and fluoride on the dissolution kinetics of hydrous alumina. A model study using the rotating disc electrode. Geochim Cosmochim Acta 48:1493-1503

Chapter 6

KINETIC AND THERMODYNAMIC CONTROLS ON SILICA REACTIVITY IN WEATHERING ENVIRONMENTS

P. M. Dove

School of Earth and Atmospheric Sciences
Georgia Institute of Technology
Atlanta, GA 30332 U.S.A.

"The behavior of silica in water solutions at low temperatures remains one of the most stubborn problems of geochemistry." K.B. Krauskopf (1956)

INTRODUCTION

Since the 'decay of quartzyte' observations by Dana (1884), earth scientists have pondered the nature of terrestrial weathering. Throughout these investigations, quartz and other silica polymorphs have played unique roles. First, silica minerals and materials have a near-ubiquitous occurrence in natural systems. The exposed crust of the Earth is estimated to be 20% quartz by volume (Nesbitt and Young, 1984) and this resistate mineral is a rate-limiting component in rock weathering to detritus and the subsequent degradation of soil and sediment constituents (Goldich, 1938; Harriss and Adams, 1966). Second, the silica polymorphs are unique as the end-member oxides to the immense classes of rock-forming silicate minerals and silicate glasses. As such, the fundamental controls of crystal structure and crystallinity on the reactivity of these relatively simple Si–O bonded compounds are an important baseline to understanding the more complex aqueous, amorphous, and crystalline silicate minerals, gels, and other materials found in surficial weathering environments.

Biogeochemistry of silica in terrestrial weathering environments

The unique character of the silica polymorphs in weathering systems can be illustrated as a simple model of silica reservoirs and processes which control mobility. Figure 1 shows the eventual transformation of SiO_2 from a rock constituent to aqueous or particulate marine silica involves a number of chemical and physical processes. Of the chemical processes, the translocation and deposition of silica within both primary and secondary weathering reservoirs is in large part controlled by dissolution and precipitation reactions. Soil environments are a central reservoir in these weathering reactions where silica is recycled between numerous inorganic and biological sinks. This recycling suggests a special property of the silica polymorphs as both primary reactants and secondary products in weathering phenomena.

Implicit in Figure 1 is the concept that silica weathering necessarily involves processes which occur over a wide range of scales. From regional-scale rock degradation and transport of clastic or aqueous silica to biogenic silica dissolution in microbial environments, silica translocation occurs over length and time scales ranging from macroscopic to microscopic. Since the relevant scale is dependent upon both the physical characteristics of an individual earth system and the transient nature of individual and system phenomena, each new scientific question must be metered accordingly. The intertwined concepts of length and time scales are considered throughout this chapter, but

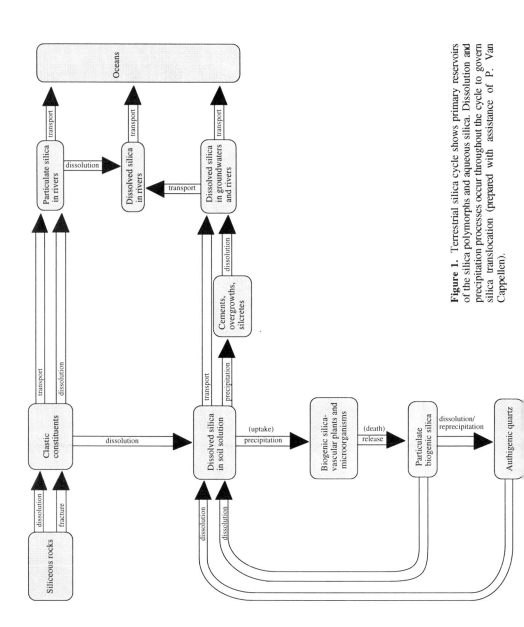

Figure 1. Terrestrial silica cycle shows primary reservoirs of the silica polymorphs and aqueous silica. Dissolution and precipitation processes occur throughout the cycle to govern silica translocation (prepared with assistance of P. Van Cappellen).

my focus upon kinetics of chemical weathering constrains most discussion to shorter time frames where reactions occur over the small distances associated with particular microenvironments.

Intent of this review

This review explores kinetic and thermodynamic controls on the partitioning of silica between reservoirs in terrestrial weathering environments. The term, silica, is defined broadly in reference to aqueous forms of silicon as well as the crystalline and amorphous SiO_2 polymorphs (e.g. Krauskopf, 1956). Though silica behavior is intimately tied to reactions with silicate minerals, these processes are addressed elsewhere in this volume. Owing to the nature of this volume, the review is largely limited to low temperature studies. For a discussion of silica-water interactions in both low temperature and hydrothermal environments, the reader is referred to Dove and Rimstidt (1994).

The chapter begins with the nature and occurrences of silica polymorphs in terrestrial weathering environments. Following a brief description of mineral and solvent properties at the SiO_2-water interface, the surface controls upon the kinetics of dissolution, precipitation and growth of the silica polymorphs are reviewed. Particular emphasis is placed upon the catalyzing or inhibiting roles of sorbed and near-surface aqueous ions and compounds found in weathering environments. In the last part, a new model quantifying the chemical controls upon the kinetics of siliceous rock fracture is briefly examined. This type of silica terrestrial weathering may be considered a special case of early-stage rock degradation that is governed by both chemical environment and stress regime.

SILICA POLYMORPHS IN WEATHERING ENVIRONMENTS

Occurrence and forms

At least eight different silica polymorphs are found in the silica reservoirs represented in Figure 1. These occur as quartz, cristobalite, tridymite, coesite, stishovite, lechatelerite (silica glass), opal and inorganic amorphous silica (Jones and Segnit, 1971; Drees et al., 1989). Another distinct SiO_2 polymorph, moganite, is now known to have widespread occurrence in association with microcrystalline quartz (Heaney and Post, 1992).

Recent reviews of silica polymorph mineral chemistry and structures are presented in Graetsch (1994), Heaney (1994), and Knauth (1994). Drees et al. (1989) reviews silica polymorph occurrences in soil environments. All of these discussions show that quartz is the most abundant silica polymorph in weathering environments as it comprises 12% of the Earth's crust by volume and is the most resistant to chemical and physical weathering. Most quartz occurs as the familiar coarse variety, but microcrystalline phases are also found in diverse sedimentary environments with the fibrous form of chalcedony or the granular forms of flint and chert (e.g. Klein and Hurlbut, 1993; Knauth, 1994). As quartz is mobilized from its host rock source, it is primarily a clastic constituent. However, quartz may also form as a secondary precipitate in soil weathering environments (Mackenzie and Gees, 1971; Robinson, 1980; Milnes et al., (1991), sediments (Merino, 1975; Williams et al., 1985) and through the recrystallization of biogenic silica (Williams and Crerar, 1985). While Figure 2 shows that quartz is a stable mineral phase at surficial temperatures and pressures, the high temperature or high pressure polymorphs are sometimes found in metastable association with weathering siliceous volcanic rocks or at high pressure events such as at the sites of meteorite impacts (e.g. Heaney, 1994). Lechatelerite is found in isolated occurrences as the result of a lightning strike of sandy soils or sediments.

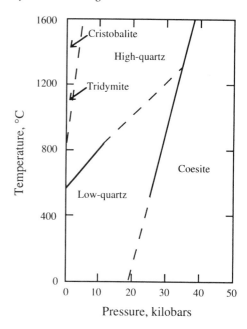

Figure 2. Stability of the silica polymorphs
(modified from Klein and Hurlbut, 1993).

Opaline silica occurs extensively in weathering environments. Most opal has a biogenic origin and exists in the form of opal-A (e.g. Drees et al., 1989; Heaney, 1994). Some of these biogenic forms are produced when certain grass species biomineralize aqueous silica to form these phytoliths in their tissues. Subsequent degradation of this biomass results in extensive accumulation of opaline phytoliths in some environments at concentrations ranging from <1 to 30 g kg^{-1} of soil. Figure 3 shows the morphological complexity of phytoliths found within the epithelium of a leaf. Although most of this material is less than 5 microns in diameter (Drees et al., 1989), biogenic opal can persist in soils for many years. For example, Beavers and Stephen (1958) noted that phytoliths produced by native prairie grasses in the Midwestern United States persists in the A horizons of soils over time scales sufficient to use as an index mineral for the identification

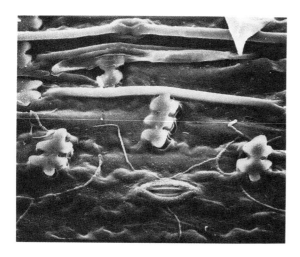

Figure 3. A SEM micrograph of silica bodies in the abaxial epidermis of *Olyra Latifolia Linnaeus* ×867 (used with permission from Simkiss and Wilbur, 1989).

of paleosols. Biogenic opal may also be transported to soils from nearby marine environments (Smithson, 1956; Smithson, 1958; Smithson, 1959). Opaline silica may also contribute to mesoscale silica cycling through atmospheric transport as it can comprise a significant fraction of airborne dusts. Folger et al. (1967) found that 25% of the silt fraction collected off the West African coast was composed of phytoliths. Opaline silica undergoes a transition to quartz via dissolution-reprecipitation reactions.

Macroscopic to microscopic: Grain textures and surfaces in weathering

Many field and laboratory studies have examined the decomposition of quartz-bearing source rocks and the resultant detritus in efforts to correlate grain sizes and textures with weathering and paleoclimate histories. Figure 1 suggests the complex combination of mechanical abrasion and crushing processes as well as the chemical dissolution and precipitation reactions that can affect mineral textures at every scale.

Fragmentation. Since Griggs (1936) found that temperatures changes alone have an insignificant affect on the degradation of quartz-rich host rocks, investigations have focused on the role of water as an agent in physical and chemical degradation. Field experiments by Roth (1964) in the Mojave Desert showed the importance of water content as "an important reagent" in chemical weathering. Temperature cycling over the range of -4.5° to 13°C in the presence of water caused significant breakage of quartz-rich granitic detritus in experimental studies of Moss et al. (1981). Grain fragmentation was attributed to the combined action of ice and adsorbed water, implying a combined physical and chemical effects on degradation. The presence of (1) adsorbed water and (2) adsorbed water plus ice showed similar effects on grain size distributions of grains with the production of large proportions of silt-sized materials (20 to 60 μm). These findings suggested that 'static' breakage during soil weathering occurs by similar mechanisms of grain splitting along major weaknesses. Studies of quartz silt formation in a humid tropical environment by Pye (1983) and Pye and Sperling (1983) also found that quartz grain fragmentation occurs by silica solution along microfractures and dislocation defects. These findings confirm their hypothesis that weathering processes are a primary mechanism in the formation of silt-sized particles and may contribute to the formation of loess and siltstone deposits. Asumadu et al. (1988) suggest that these fragmentation processes are so extensive as to invalidate the use of quartz as a stable index mineral in soil pedogenesis studies.

Grain surface shapes. Morey et al. (1962) noted that quartz dissolution must continue indefinitely in silica undersaturated solutions. Thus, dissolution must govern the persistence of mineral components in soils. Petrographic studies of Crook (1968) supported this hypothesis with observations that quartz sand grains in soils, silcretes and certain quartz-rich sedimentary rocks exhibit textures which are the result of in situ dissolution. These observations lead to Crook's classification of soil mineral grains upon the shapes and rounding of 're-entrants' and 'appendages' formed upon quartz grains which have undergone pedogenesis. Crook (1968) and Cleary and Conolly (1972) proposed that weathering intensity, climate, and soil maturity can be qualitatively measured by the degree and amount of roundness and the embayment of grain surfaces.

Other studies of chemical controls on the development of quartz grain shape and surface textures indicate the importance of solution composition in affecting weathering rates. Field observations show that inorganic and organic solutes act as catalyzing agents in the surface corrosion of quartz mineral grains in soil environments. Citing other studies, Crook (1968) postulated that organic molecules play a significant catalyzing role in mobilizing silica from quartz to form smoothly rounded surfaces. This hypothesis is supported in recent work (see later discussion). Jones et al. (1981) noted the contribution of biological products to weathering rates through observations of corroded quartz grain surfaces found beneath growing lichens. These textures were attributed to chelating effects of oxalic acid released by lichens at the lichen-mineral interface. Inorganic solutes are also implicated in the surface weathering of quartz grains. Young (1987) found that sodium-rich salts promote the weathering of quartzose sandstones in Western Australia and New South Wales. Experimental studies of crushed Brazilian quartz by Magee et al. (1988) supported salt-enhanced degradation of quartz. Their SEM observations of quartz grain surfaces suggest that various salt solutions promote the development of pits and precipitation/solution textures within 50 to 140 hours of exposure. Their observations are consistent with recent kinetic studies (see later discussion), showing that rates were fastest in basic pH solutions containing Na_2CO_3.

Etch pits. At length scales of tens of microns, weathered quartz grains may have etch pits on their surfaces. Brantley et al. (1986a,b) examined quartz grains from a soil profile that had developed in situ on the Parguaza granite in Venezuela and found that grains showed a gradual change from triangular pitted surfaces to rounded surfaces at just above bedrock (see Fig. 4). They proposed that pit development is governed by the aqueous silica concentration such that pits cannot nucleate above a critical saturation state (see later discussion). Investigations of etch pit densities on naturally weathered surfaces suggest that dissolution from etch pits becomes increasingly important with long exposure times to weathering conditions (Anbeek et al., 1994).

Correlation of grain textures and weathering history. After 60 years of research, the complexity of grain shapes and surface textures produced by weathering processes continues to limit quantitative correlation of textures with histories of soils and sediments. The general model of Margolis and Krinsley (1974) related physical and chemical factors that produce microfeatures on quartz grains in different sedimentary environments to quartz crystallographic properties. Though fracture and cleavage features were found to be ubiquitous across environments, specific categories of minute features were related to grain size and transport. Darmody (1985) noted that textures exhibited by weathered quartz grains are size dependent and thus, comparisons of grains across size fractions may have an inherent bias. He proposed a semi-quantitative approach to eliminate size dependent variability in surface textures by using only one grain size fraction . These grains were characterized with SEM imaging by a qualitative scoring system based upon the presence or absence of ten surface features. Qualitative features ranged from coarse shape to fine scale surface topography. More recently, Pye and Mazzullo (1994) developed a Fourier shape analysis technique to describe the progress of post-depositional weathering of quartz grains in a humid tropical climate. Using SEM images collected for a 180 to 250 μm grain size fraction, the shape was partitioned into a series of 23 standard shape components. Lower order harmonics represented gross shape or form while higher order harmonics represent components of fine shape or roundness. Comparing grains from Queensland dune sands of two ages, they found that the older sands were significantly more angular and quantitative differences were most pronounced in the higher harmonics region. They suggest this type of image analysis approach offers future promise as a quantitative method for evaluating weathering histories.

Silica polymorphs as coatings on other mineral constituents

The formation of silica coatings on other mineral surfaces as a secondary product of weathering is widely recognized in soils and sediments (e.g. Crook, 1968; Pollard and Weaver, 1973; Langford-Smith, 1978; Milnes et al., 1991). The terminology is diverse,

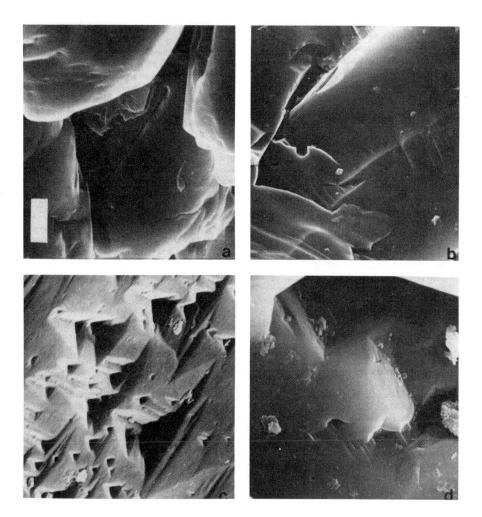

Figure 4. SEM micrographs of surface features of sand grains from a soil profile developed on the Parguaza granite, Venezuela. Sample from (a) 2 cm just above bedrock, (b) 90 cm deep, (c) 60 cm deep, (d) surface (from Brantley et al., 1986b).

but these (usually) amorphous silica coatings occur as silcretes, cements, or other secondary precipitates. Silcretes constitute an extreme form of cementation where a highly indurated, quartz-rich material is cemented by the precipitation of mobilized silica into crystalline and amorphous silica polymorphs (Birkeland, 1984). Factors leading to the development of these features are of particular interest in parts of Australia where silcrete development is so extensive as to affect the evolution of some landscapes. Information contained in these indurated layers may also aid in reconstructing paleo landscapes and environments (e.g. Birkeland, 1984).

Amorphous silica cements also occur in soil environments with less extensive precipitation. Nahon (1991) discusses the characteristics of this accumulated material and illustrates the formation of coatings on mineral surfaces in porous environments. While quantitative descriptions of silica cementation processes are not forthcoming, Nahon (1991) uses an extensive descriptive terminology has arisen to characterize the types and environments of "cutanic" accumulations which result by "plasmic concentrations". Silica cementation in soils appears to be a significant mechanistic pathway for the conversion of soluble silica to solid polymorphic forms within/between silica reservoirs. Yet the structure and reactivities of these translocated silica phases are not well understood.

REACTIVITY IN SATURATED ENVIRONMENTS: SILICA SOLUBILITY

In saturated soil weathering environments, aqueous solutions may approach saturation with respect to the silica polymorphs. Though the non-crystalline and micro-crystalline forms materials are minor constituents in these environments (Drees et al., 1989), their solubilities may markedly influence weathering chemistry.

The dissolution reaction

The silica polymorphs dissolve in water by the reaction

$$SiO_2(s) + 2 H_2O = H_4SiO_4(aq) \qquad (1)$$

and the equilibrium constant for this reaction is

$$K = \frac{a_{H_4SiO_4}}{a_{SiO_2} a_{H_2O}^2} \qquad (2)$$

Although Equation (1) explains most of the solubility behavior of silica, there are reasons to believe that the ratio of SiO_2:H_2O in the aqueous species may not be exactly 1:2 (Walther and Helgeson, 1977) so that some prefer to generalize Equation (1) to

$$SiO_2(s) + n H_2O = SiO_2 \cdot n H_2O(aq) \qquad (3)$$

The SiO_2:H_2O ratio is less than 1:2 if hydrogen bonded waters of hydration are counted into the stoichiometry of the aqueous species. On the other hand, the formation of polynuclear species eliminates water molecules and thus increases the SiO_2:H_2O ratio. NMR (Cary et al,, 1982), Raman (Alvarez and Sparks, 1985) and diffusion rate (Applin, 1987) studies of dilute silica solutions all provided evidence for the presence of silica dimers formed by the reaction

$$2 H_4SiO_4(aq) = H_6Si_2O_7(aq) + H_2O \qquad (4)$$

Figure 5, based on $K = 330$ for Reaction (4) (Applin, 1987), suggests that this dimer contributes little to the solubility of quartz but accounts for about 40% of dissolved silica in equilibrium with amorphous silica. Thus, representing dissolved silica as H_4SiO_4 is similar to representing dissolved CO_2 as H_2CO_3 even though there is clear evidence that CO_2 solutions contain both $CO_2(aq)$ and $H_2CO_3(aq)$ species. This representation of dissolved silica creates no problems in thermodynamic calculations but is inadequate for understanding the kinetics of solvent-silica interactions.

Solubility of silica phases

The solubility of silica polymorphs increases with changes in crystal structure to less ordered and less dense structures (e.g. Stöber, 1967; Iler, 1979). By convention, the solubility of the polymorphs is given by defining an activity of one for each crystalline solid and the activity of the pure solvent as one. Thus, $a_{SiO_2} = 1$ and $a_{H_2O} = 1$ so that

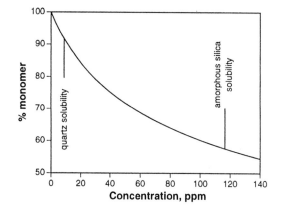

Figure 5. Percent of dissolved silica in the monomer form (H_4SiO_4) as a function of total dissolved silica concentration. The balance of the dissolved silica occurs as dimers ($H_6Si_2O_7$). This diagram is based on the association constant of 330 at 25°C reported by Applin (1987).

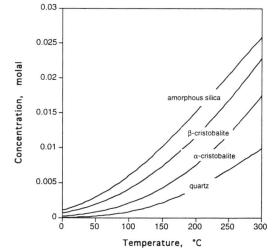

Figure 6. Solubility of quartz, cristobalite, and amorphous silica along the three phase curve, $SiO_2(s)$–$H_2O(l)$–$H_2O(v)$ based on equations from Rimstidt and Barnes (1980).

Equation (2) simplifies to

$$K = a_{H_4SiO_4} .$$ (5)

The effect on solubility of varying the ΔG_f° of the silica phase is illustrated in Figure 6 which shows quartz, cristobalite, and amorphous silica solubility along the three phase curve $SiO_2(s)$–$H_2O(l)$–$H_2O(v)$. The solubility of silica phases increases from quartz to α-cristobalite to β-cristobalite to amorphous silica: $\Delta G_f^\circ(\text{quartz}) < \Delta G_f^\circ(\alpha\text{-cristobalite})$ $< \Delta G_f^\circ(\beta\text{-cristobalite}) < \Delta G_f^\circ(\text{amorphous silica})$. This variation in the ΔG_f° of silica phases is conventionally reflected by assigning successively higher values of K for their dissolution equilibria. It is instructive, however, to change the convention so that K is always constant and the variation of solubility is expressed by changing a_{SiO_2} (Table 1). This table shows that at 25°C, the activity of cristobalite is twice that of quartz and amorphous silica is ten times that of quartz. Inserting these values into Equation (2) and using the value of K for quartz, the solubility of these other phases can be calculated from

$$a_{H_4SiO_4} = a_{SiO_2} K_{qz} .$$ (6)

The solubility of these phases can be found by multiplying quartz solubility by a_{SiO_2} for

Table 1. Comparison of the solubility of silica phases at 25°C based on data from Robie et al. (1978). The conventional values of the equilibrium constant K assume that $K = 1.11 \times 10^{-4}$, the solubility constant for quartz, for all dissolution reactions.

Phase	ΔG_f°, kJ mol^{-1}	K	a_{SiO2}
quartz	-856.288	1.11×10^{-4}	1.00
cristobalite	-854.512	2.28×10^{-4}	2.05
tridimite	-853.812	3.02×10^{-4}	2.71
coesite	-850.85	9.98×10^{-4}	8.97
glass	-850.559	1.12×10^{-3}	10.08

these polymorphs. On this diagram, the phase with the lowest activity is the most stable, and therefore, the least soluble. Note that although this convention is a convenient way to show the relative solubilities of silica phases, it creates certain computational difficulties, and therefore, it is not used for most thermodynamic calculations.

Recent investigations of silica solubility are considering the solubility of other polymorphs not previously considered as a reactant or product in weathering processes. Gíslason et al. (1995) showed that the solubility of a quartz/moganite mixtures are greater than that of quartz alone. They suggest that the low temperature solubility of this mixture brackets the silica concentration of most rivers in the world.

Effect of pH. The solubility of crystalline and amorphous silica phases are essentially constant up to pH 8.5. Near pH 9, the weakly acidic H_4SiO_4 dissociates appreciably such that the first dissociation reaction is written

$$H_4SiO_4 = H_3SiO_4^- + H^+ \tag{7}$$

Reaction (7) is an important natural buffer for some natural systems where free silica is present. If a reaction consumes enough hydrogen ions to increase the pH above the pK for Reaction (7), then silica dissolves by the reaction

$$SiO_2(s) + 2\ H_2O = H_3SiO_4^- + H^+ \tag{8}$$

and supplies one mole of hydrogen ions per mole of silica that dissolves. As a result only highly unusual, low-silica environments like carbonatites and serpentinites can attain a pH above the first pK of silica. This buffering effect insures that the second and higher dissociation reactions

$$H_3SiO_4^- = H_2SiO_4^{2-} + H^+ \tag{9}$$

$$H_2SiO_4^{2-} = HSiO_4^{3-} + H^+ \tag{10}$$

and
$$HSiO_4^{3-} = SiO_4^{4-} + H^+ \tag{11}$$

are of little importance in most natural environments. Note that when a solution is in equilibrium with amorphous silica, ionized polymers such as $H_6Si_4O_7^{2-}$ also occur at significant concentrations as shown on Figure 7. The solubility of quartz follows the same pattern shown in Figure 7 but the total solubility is about one-tenth that of amorphous silica and only small concentrations of polymers form at these lower silica concentrations. The solid solubility and first dissociation constant of silica changes with increasing temperature

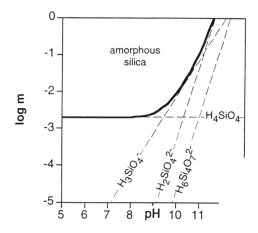

Figure 7. Solubility of amorphous silica as a function of pH (solid line). The dashed lines show the concentration of aqueous species. The shaded area indicates the approximate conditions where polynuclear species make up a significant fraction of the aqueous species. Diagram based on equilibrium constants from Stumm and Morgan (1981) for solutions with $I = 0.5$.

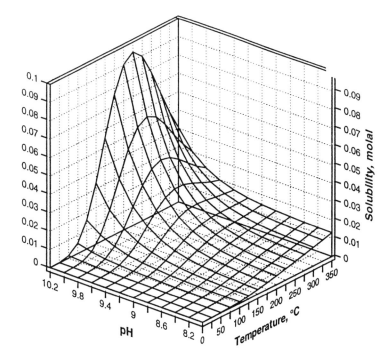

Figure 8. Solubility of quartz as a function of pH and temperature for conditions along the three phase curve (quartz-solution-vapor) based on data from Fleming and Crerar (1982).

to produce a fairly complex behavior for quartz solubility as a function of pH and temperature (Fig. 8) for conditions along the three phase curve $SiO_2(s)$–solution–$H_2O(v)$.

Effect of solution species. The presence of dissolved constituents can affect silica solubility in two ways. Some solutes react with H_4SiO_4 to form complexes and others interact with water molecules to change the hydration energy of H_4SiO_4. The activity ($\approx$ concentration) of H_4SiO_4 is fixed by Reaction (1) so that if some of the H_4SiO_4 reacts

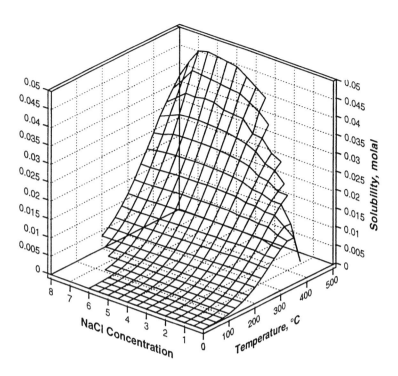

Figure 9. Solubility of quartz in sodium chloride solutions along the three phase curve quartz-solution-vapor based on the model of Fournier (1983). The solubility surface is terminated along the upper right edge by the critical point of the solution and on the lower left corner by the four phase field quartz-halite-solution-vapor. This diagram shows that silica is salted into solution.

with a solute species to form a complex, more solid dissolves to replenish the reacted H_4SiO_4. Because the total amount of dissolved silica is the sum of all of the silica-containing aqueous species, these complex-forming reactions increase the solubility of silica. There is evidence that aqueous silica interacts with a variety of organic and inorganic species to form complexes. Reported values of equilibrium constants show the following decreasing affinity for reaction with $H_3SiO_4^-$: $H^+ > Fe^{3+} > UO_2^+ >> Mg^{2+} > Ca^{2+} >>$ Na^+ (Porter and Webber, 1971; Olsen and O'Melia, 1973). In addition, F^- interacts strongly with silica displacing OH groups to form SiF_4 and SiF_6^{2-} (Roberson and Barnes, 1978). Finally, several silica-organic complexes (Bennett, 1991) may be quite important in weathering reactions (see later discussion). Other solutes, such as sodium, can increase silica solubility by changing the solvent properties of the solution. Quartz solubility is greater in NaCl solutions (Fournier, 1982; Xie and Walther, 1993) than in pure water presumably because Na^+ and/or Cl^- ions change the structure of water in such a way to increase its hydrogen bonding with H_4SiO_4. This effect is minimal at the low temperatures of weathering environments as illustrated in Figure 9.

Apparent solubility.   Coatings of organic and inorganic compounds may reduce apparent solubilities of the silica polymorphs through chemisorption onto silica. Bartoli and Wilding (1980) observed that opaline phytolith solubilities are affected by hydration state, age, rate of organic matter biodegration of the encasing plant tissues, and aluminum content. Sorbed aluminum is widely recognized to reduce apparent solubilities of quartz, phytoliths and other biogenic silicas (Lewin, 1961; Iler, 1973; Bartoli, 1985), silica gel (Krauskopf, 1959). This topic is discussed later (see Fig. 30 below).

Solubility as a function of particle size

Relationships between solubility, surface area and particle size contribute to the precipitation of metastable forms of silica. This phenomena is widely recognized in diagenetic processes (Hurd and Birdwhistell, 1983; Williams et al. 1985; Williams and Crerar, 1985). When a silica particle is broken to form smaller fragments, work must be done to create the new surface. The amount of work required to create a unit area of new surface is called the surface free energy (σ, erg cm^{-2}). The growth of silica particles produces new surface area so that the free energy, ΔG_f (particle), required to form a particle with a surface area of A cm^2 containing n moles of substance is

$$\Delta G_f(\text{particle}) = n\, \Delta G_f(\text{bulk solid}) + A\, \sigma \tag{12}$$

this relationship can be recast in terms of the radius of the particle (r, cm) (assuming that it is spherical)

$$\Delta G_f(\text{particle}) = \frac{-4\pi r^3 \Delta G_f^\circ}{3\, V_m} + 4 \times 10^{10} \pi r^2 \sigma \tag{13}$$

where V_m is the molar volume of the solid (cm^3 mol^{-1}). This relationship can be further modified to give the solubility of the particle (c, arbitrary units) relative to the bulk solubility of the substance (c$^\circ$, arbitrary units)

$$\ln\left(\frac{c}{c^\circ}\right) = \frac{\frac{2}{3}\, 10^{-7}\, \sigma\, V\, B}{r\, R\, T} \tag{14}$$

As these groups dissociate, hydronium ions are produced which diffuse from the surface to develop a pH-dependent surface charge and potential. This surface charge, in turn, attracts a diffuse cloud of counterions to preserve electroneutrality. The resulting surface-solution interface that exists at virtually all wetted mineral surfaces (Fig. 11.IV, below) is called the where B is a geometric factor (16.8 for spheres) and R is the Universal gas constant. The surface free energy of amorphous silica in contact with solution is around 45 erg cm^{-2} (Alexander, 1957) and the surface free energy of quartz in contact with solution is around 120 erg cm^{-2} (e.g. Rimstidt and Cole, 1983; Parks, 1984). Figure 10 shows the solubility of quartz and amorphous silica as a function of the radius of curvature of the silica surface. The solubility of convex surfaces (positive radius of curvature) increases with decreasing particle size so that for radii less than about 0.1 μm the particles have a measurably increased solubility. This causes small silica particles to dissolve while large particles increase in size. This physical result of the Kelvin equation for particle-water interactions (Atkins, 1994) is widely known as Ostwald ripening. On the other hand, Equation (14) predicts that the solubility of silica at concave surfaces (negative radius of curvature) decreases with decreasing radius of curvature. This causes silica to precipitate at or near the tips of cracks or in environments where small menisci lead to a very small radius of curvature. These processes thermodynamically favor large particles with smooth faces or spherical shape.

THE SILICA-WATER INTERFACE

Processes governing the kinetics of silica-water interactions in weathering environments originate at the interface between solution and solid. Recent advances in quantifying the reactivity of silica and perhaps all mineral phases emphasize the role of

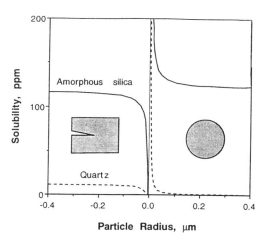

Figure 10. The solubility of quartz and amorphous silica at 25°C as a function of the radius of curvature of the surface of the solid. Note that the solubility of particles with convex surfaces increases with decreasing radius of curvature and the solubility of concave regions decreases with decreasing radius of curvature. This causes small particles to dissolve and small fractures to be filled (from Dove and Rimstidt, 1994).

I. FRESH FRACTURE

II. HYDROXYLATED

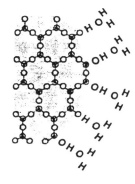

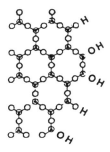

III. ADSORBED MOLECULAR WATER

IV. WET SURFACE

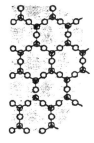

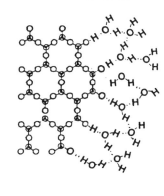

Figure 11. Schematic diagram illustrating the stepwise water-surface interactions with freshly cleaved or fractured quartz (after Parks, 1984).

surface structure and chemistry. Silica surface geochemistry is extensively discussed elsewhere and the reader is referred to Iler (1979), Parks (1990), and references therein. For the purposes of this review, a brief discussion introduces terminology and concepts useful to subsequent discussion of rates and mechanisms of silica-water interactions.

One means of understanding silica-water interactions is to examine the nature of the quartz-water interface. Parks (1984) shows the stepwise development of this interface as a freshly cleaved quartz surface is hydrated to a fully wetted surface. When quartz is fractured in vacuum, the freshly exposed initial surface (Fig. 11.I) is composed of 'dangling' silicon and oxygen bonds. These 'sites' and the corresponding surface structures are unstable and hydroxylate easily with available water within seconds, even at ultra high vacuum conditions. This hydroxylated surface (Fig. 11.II) is dominated by >SiOH or silanol groups. As this adsorbed water film increases beyond approximately three monolayers (Fig. 11.III), its properties become more like bulk water (Parks, 1984). In the presence of molecular water, the silanol groups ionize, producing mobile protons that associate/dissociate with the surface to impart an electrical conductivity to the surface. electrical double layer (e.g. Yates et al., 1974; Davis et al., 1978). Discussion of the mechanisms that develop the interfacial field gradients found at silica-water interfaces are found elsewhere (see Li and DeBruyn, 1966; James and Healy, 1972; Iler, 1979; Hayes and Leckie, 1987).

Extensive work shows that adsorbed water adjacent to the surface is oriented and has properties (e.g. entropy, mobility, dielectric constant, dissociation constant, viscosity, rates of water exchange, proton transfer) which are different from those of bulk water (Fripiat et al., 1965; Anderson and Parks, 1968; Peschel and Aldfinger, 1970; Tait and Franks, 1971; Churaev et al., 1972; Drost-Hansen, 1977; Davies and Rideal, 1963; Roberts and Zundel, 1979; Sermon, 1980; Sposito, 1989; Zhu and Robinson, 1991; Israelachvili, 1992). Dove (1994) suggested that these high interfacial charge gradients affect quartz reactivity by promoting reorientation of water in the interfacial environment to affect local solvent dissociation or rates of near-surface solvent exchange. This reorientation model was recently confirmed by experimental Infrared-Sum Frequency Generation (SFG-IR) spectroscopic methods showing that water molecules at quartz-water interfaces have a strong orientational and bond ordering dependence upon solution pH and sodium concentration (Du et al., 1994). Figure 12a shows that a quartz-water interface in an acidic solution has an ice-like structure where interfacial water exists as a time-averaged symmetrical arrangement of molecules in tetrahedral coordination (Du et al., 1994). In the transitional range of pH 3 to 8, surface structures are partially ionized. Figure 12b,c,d show an unbuffered condition where the interfacial solvent structure is readily modified by changes in pH or solution composition. Water assumes a disordered structure intermediate between the low and high pH 'ordered' cases. This intermediate structure varies significantly over this pH range, suggesting that the average solvent molecular arrangement is readily perturbed by small changes in surface ionization. At high solution pH, surface silanol groups become ionized ($>SiO^-$) to produce a strong electric field. This affects near-surface solvent structure by polarizing and reorienting interfacial water so that protons are directed towards the silica surface (Fig. 12e). Correlations of these interfacial solvent properties and silica reactivity suggest that the properties of near-surface water may directly control kinetics of mineral-water reactions (see later discussion).

Most of what is known about silica surfaces comes from studies of amorphous silica and quartz. Iler (1979) suggests that these findings are largely applicable to all silica polymorphs as surface chemistry investigations show that crystalline and amorphous forms share important similarities in the sorption of water and development of charged interfacial

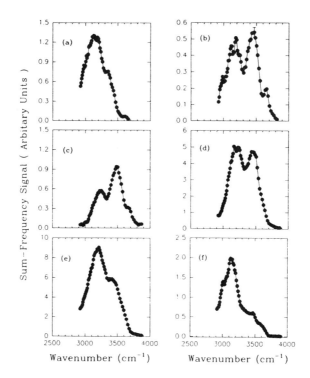

Figure 12. Infrared-Sum Frequency Generation spectra of a quartz-water interface reveal the pH-dependent nature of near-surface water interactions with silanol groups. (a) In a pH 1.5 solution, the solvent has an ice-like structure where the interfacial water exists as a time-averaged symmetrical arrangement of molecules in tetrahedral coordination. (b,c,d) In the transitional pH range between acidic and basic conditions, surface structures are partially ionized. This intermediate structure varies significantly over the pH range of 3.8 to 8.0, suggesting that the average solvent molecular arrangement is readily perturbed by small changes in surface ionization. (e) At high solution pH, surface silanol groups become ionized ($>SiO^-$) to produce a strong electric field. This affects near-surface solvent structure by polarizing and reorienting interfacial water so that protons are directed towards the silica surface. This alignment causes of near-surface hydroxyls a large degree, but different type, of solvent order (reproduced with permission of Du et al., 1994).

environments. However, there are important exceptions. Iler (1979) also notes that comparisons between polymorph structures should account for density differences between the phases and the resultant effect on silanol site densities on the surfaces. For fine grained or porous forms of silica, particle size or pore radius of curvature may also significantly perturb the sorptive properties per the Kelvin effect (e.g. Atkins, 1994; Kenny and Sing, 1994; Unger, 1994).

Surface structures and properties

Electrical double layer of the silica-water interface. Recent advances in surface complexation theory formalize the chemical and physical structure of the electrical double layer into a model that assumes surface reactions mimic aqueous complexation (e.g. Davis and Kent, 1990 and references therein). In one such model, a silica-water interface is described as three electrostatically charged regions. Figure 13a shows that each of these regions can be defined as the o, β, and d planes where each is associated with an electric potential and surface charge. Comparing Figures 13a and 13b, idealized planar surfaces correspond to the decreasing charge distribution and electrical potential that occurs with

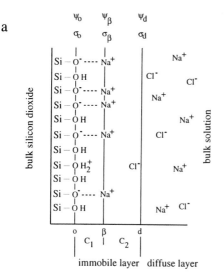

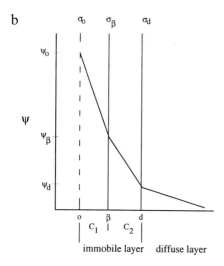

Figure 13. Schematic illustration of the triple layer model (modified from Hayes and Leckie, 1987) of an idealized planar quartz-water-sodium chloride interface. (a) Each layer has an associated interfacial potential, ψ_i (volts), and charge density, σ_i (Coulomb m^{-2}), that determine the inner (C_1) and outer (C_2) layer capacitances (Faraday m^{-2}), by the relationship

$$C_j = \Delta \sigma_i \left(\Delta \psi_i \right)^{-1}.$$

The magnitude of these parameters decreases with increasing distance from the mineral surface into the solution side of the interface and finally to the bulk solution. In this model, the bulk uncharged solution is beyond the diffuse, d, layer. Hydrogen ions bind with unsaturated oxygens in the innermost o layer, whereas, weakly sorbed ions such as sodium interact only from distances associated with the β layer at low temperatures. (b) Schematic representation of the corresponding charge distribution and the potential decay away from the surface (from Dove and Elston, 1992).

increasing distance from the surface. Hydrogen ions coordinate with the unsaturated sites of the interface at the innermost o layer (as an 'inner-sphere complex'). Sodium and other weakly bound hydrated cations are positioned at the β layer (as an 'outer-sphere complex') or the d layer (near the bulk solution). Low temperature surface complexation models do not permit sodium to specifically interact with the surface. This is because potentiometric titration data and coordination theory suggest sodium is prevented from specifically binding to the surface because of shielding by its own solvation sphere. However, sodium may exist in the β layer in an 'ion pair' coordination with the surface that promotes its partial dehydration and water dissociation within the charged interface (Conway, 1981; Fokkink et al., 1990). Although hydration spheres are implicitly ignored in this representation, the same relative surface-metal distances are predicted for the surface coordination of H$^+$ and Na$^+$ at o and β planes, respectively.

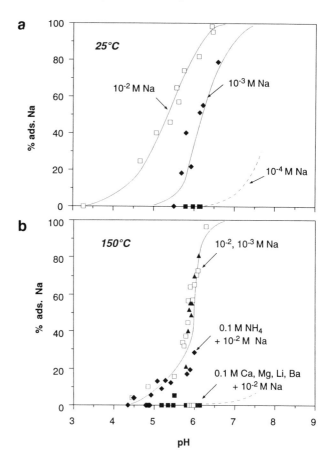

Figure 14. Adsorption isotherms of sodium and lead on amorphous silica powder at (a) 25° and (b) 150°C from nitrate-bearing solutions of variable ionic strength (from Berger et al., 1994).

The Berger et al. (1994) experiments revealed a that sodium sorption behavior on silica changes with increasing temperature. Figure 14a shows that at 25°C, sodium sorption edges shift with increasing concentration suggesting that sodium interacts only weakly with the surface and is easily displaced by either an ionic strength effect or competitive displacement by other electrolytes or protons. This behavior is indicative of 'outer sphere' sorption of the metal, meaning that the cation interacts only weakly with the surface through its solvation water. In contrast at 150°C, Figure 14b shows that the sodium sorption edge is independent of ionic strength. This implies an important change in the sodium-silica surface speciation with temperature and suggests that the hydration energy is lower and the average solvation sphere may be 'thinned' or lost depending upon system conditions.

Investigations of SiO_2 surface chemistry have produced a generally accepted representation for site distributions on quartz surfaces in aqueous solutions (e.g. Iler, 1979). Briefly, quartz in contact with a solution composition containing only an alkali salt (i.e. quartz-water-sodium chloride system) has surface structures that can be described by three 'complexes' that are denoted by a '>' symbol and have a population balance of

$$\theta_{>SiOH} + \theta_{>SiO^-} + \theta_{>SiO-Na^+} = 1.0 \tag{15}$$

where

$\theta_{>SiOH}$ = fraction of total sites as >SiOH species

$\theta_{>SiO^-}$ = fraction of total sites as >SiO⁻ species

$\theta_{>SiO-Na^+}$ = fraction of total sites as >SiO-Na⁺ species.

Note that these 'species' are model constructs describing surface titration data and the surface charge relationships of oxides. Of the three complexes listed, only >SiOH has been directly observed using spectroscopic methods (Anderson and Wickersheim, 1964; Gallei and Parks, 1972; Morrow and Cody, 1976a,b). The other two complexes may or may not have physical meaning as they describe the time-averaged degree of surface ionization. For the purpose of later relating surface complex distributions to dissolution kinetic data, the >SiO⁻ and >SiO-Na⁺ are co-dependent upon changes in solution pH and sodium concentration and cannot be evaluated independently. These terms are added and referred to as >SiO⁻$_{tot}$.

Ionization and surface charge. Surface complexes describing silica-solution interfaces are not static, but are rather calculated distributions reflecting an average electronic state resulting from proton, cation, and hydroxyl ion interactions with the undersaturated oxygens at the mineral surface (Prigogine and Fripiat, 1974). The resulting population balance is largely controlled by the relative magnitudes of association constants for the surface reactions listed in Table 2. Titration studies show the pK_a for ionization of silanol groups to >SiO⁻$_{tot}$ complexes (see Table 2) is about 6.8, indicating that the surface is only weakly acidic. Recent investigations of surface complexation equilibria by Sverjensky and Sahai (in press) show that surface protonation is correlated with dielectric constant and metal-OH Pauling bond strength for several oxides, including quartz. Their findings suggest the importance of bulk crystal structure in governing the bonding and equilibria of surface protonated species.

The resulting surface equilibria from these titration and theoretical studies lead to the pH and sodium dependence of average >SiO⁻$_{tot}$ distributions as represented in Figure 15. In general, net negative charge increases with increasing solution pH and/or alkali cation concentration until about pH 10 or 11. Above this pH, further pH increases or addition of alkali have smaller effects on net negative charge.

Metals other than the alkali cations also interact with silica surfaces, some quite strongly. Interaction mechanisms appear to range from simple ion exchange into the β plane

Table 2. Acid-base quartz surface complex reactions and corresponding association constants at 25°C.

	K_a	Reference
>SiOH = >SiO⁻ + H⁺	$10^{-6.8}$	a
>SiOH + Na⁺ = >SiO-Na⁺ + H⁺	$10^{-7.1}$	b
>SiOH₂⁺ = >SiOH + H⁺	$10^{-2.3}$	c
>SiOH + H⁺ + Cl⁻ = >SiOH₂Cl	$10^{-6.4}$	d

a) Schindler and Kamber,1968; b) Kent et al., 1988; Dugger et al., 1964;
c) Schindler and Stumm, 1987; d) Kent et al., 1988.

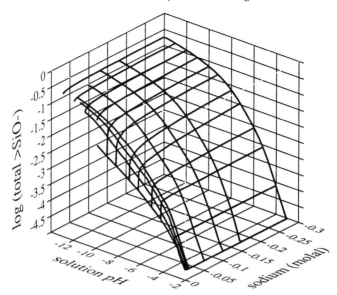

Figure 15. The three-dimensional dependence of fractional surface charge upon bulk solution pH and sodium concentration for the quartz-water-sodium chloride system at 25°C. The fractional sum of calculated >SiO⁻ and >SiO-Na⁺ populations is plotted as a function of solution pH and sodium concentration. Three general environments are recognized. At low pH, negative surface charge is low and independent of sodium concentration. At near-neutral pH, surface charge is sensitive to the addition of small concentrations of sodium. At high pH, sodium has only a small effect on net surface charge as the basic solution inherently contains sodium as sodium hydroxide and surface ionization nears saturation (from Dove, 1994).

to cation-surface specific binding at the innermost o layer. For example, surface association constants, K_a, vary from $10^{-7.8}$ for lithium (Dugger et al., 1964) to $10^{-1.8}$ for ferric iron (Schindler et al., 1976) at 25°C. Later discussion will show that metal-surface interactions significantly enhance or inhibit silica reactivity. The direction and degree of these effects are partially related to interaction strengths although the specifics are considerably more complex. It is significant to note that silica surfaces can 'scavenge' strongly sorbed ions such as ferric iron and aluminum from low concentration solutions. This process of 'super equivalent sorption' is so extensive as to exceed the amount required to satisfy silica surface charge such that the aluminum reverses charge from negative to positive (Weise et al., 1976). This chemisorption occurs even at low pH where net negative charge is zero (Stanton and Maatman, 1963). Figure 16 shows the effect of aluminum hydrous oxide coatings on the surface charge of quartz (Hendershot and Lavkulich, 1983).

REACTIVITY IN NON-EQUILIBRIUM ENVIRONMENTS: KINETICS

Basic principles

Aqueous diffusion. The rates of reactions between minerals and solutions are either limited by diffusion of aqueous species to or from the surface or by rates of bond breaking and formation at the mineral surface. At the temperatures of terrestrial weathering, bond breaking and formation is clearly the rate limiting step for silica dissolution or precipitation. The best evidence for this is the relatively high activation energies for these reactions. As a rule-of-thumb, the activation energy for a bond-breaking reaction is about 20% of the enthalpy of formation of that bond. The $\Delta H_f°$ of quartz is -910.700 kJ mol⁻¹

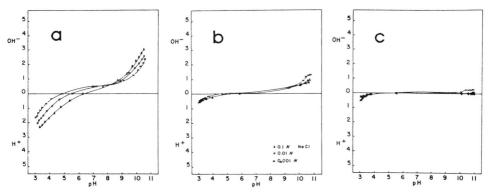

Figure 16. Potentiometric titration curves for quartz that has been (a) Fe-coated; (b) Al-coated; (c) uncoated. Y axis units are given as H^+ or OH^- adsorbed (cmol kg^{-1}) (from Hendershot and Lavkulich, 1983).

(Robie et al., 1978) and each mole of quartz contains four moles of Si–O bonds so the $\Delta H_f°$ (Si–O) is -227.675 kJ mol^{-1}. Twenty percent value of this is 45.5 kJ mol^{-1} which is low, but consistent with experimental measurements of the activation energy for quartz dissolution and precipitation (see Table 3). The activation energy for diffusion of aqueous species is on the order of 20 kJ mol^{-1}. However, because the activation energy for the bond-breaking reaction is higher than the activation energy for the diffusion reaction, some temperature exists where these two rates are equal as illustrated in Figure 17. This temperature varies between systems depending on the geometry- and flow-controlled diffusion process, but diffusion controlled transport occurs well above the temperatures of weathering environments (e.g. Casey, 1987).

Dissolution and precipitation. A fundamental reference point in describing the kinetics of silica dissolution and precipitation is the relationship to equilibrium solubility between the mineral phase and the solution. Considering the simplest scenario for opposing forward and reverse reactions, one can write the expression

$$SiO_2 + 2 H_2O(l) = H_4SiO_4 \tag{16}$$

where dissolution and precipitation, respectively, proceed by

$$SiO_2 + 2 H_2O = (*)^{\ddagger} \tag{17}$$

$$(SiO_2 \cdot 2 H_2O)^{\ddagger} = H_4SiO_4 \tag{18}$$

and

$$H_4SiO_4 = (*)^{\ddagger} \tag{19}$$

$$(*)^{\ddagger} = SiO_2 + 2 H_2O \tag{20}$$

The $(*)^{\ddagger}$ indicates an activated intermediate species whose stoichiometry is unknown. At equilibrium, Reactions (18) and (20) proceed at equal rates. From this basis, Rimstidt and Barnes (1980) derived an integrated rate equation which accounted for both reaction directions to determine that since

$$r_+ = (dn_{H_4SiO_4}/dt)_+ = Ak_+ a_{SiO_2} a^2_{H_2O} \tag{21}$$

and

$$r_- = (dn_{H_4SiO_4}/dt)_- = Ak_- a_{H_4SiO_4} \tag{22}$$

where $n_{H_4SiO_4}$ is the number of moles of H_4SiO_4, A is the interfacial area (m^2) and k_+ and

Table 3. Summary of some activation energies reported for the precipitation and dissolution kinetics of the silica polymorphs.

Reference	Solution Composition	Temperature °C	E_a kJ mol^{-1}
Precipitation-Quartz			
Bird et al. (1986)	deionized water	121-255	51-55
Precipitation-Cristobalite			
Renders et al. (1995)	deionized water	150-300	53.7
Precipitation-all silica polymorphs			
Rimstidt and Barnes (1980)	deionized water	25-300	49.8
Dissolution-Quartz			
Rimstidt and Barnes (1980)	deionized water	25-300	67-4-76.6
Gratz et al. (1990)	pH 10 to 13	148-236	86.4[1]-90.2[2]
Casey et al. (1990)	pH 3, 11	20-70	36[3], 53[4]
Brady and Walther (1990)	pH 2-11.7	25, 60	54[5], 96[6]
Dove and Crerar (1990)	pH 5.7, Na$^+$ 0 to 0.2	150-300	
Gratz and Bird (1993)	pH 10 to 13	148-236	78.6
Tester et al. (1994)	deionized water	25-625	89±5
Dove (1994)	pH 2 to 12, Na$^+$ 0 to 0.3	25-300	66.0[7]- 82.7[8]
House and Hickenbotham (1992)	pH 10	5-35	83.2
Dissolution-Cristobalite			
Rimstidt and Barnes (1980)	deionized water	25-300	68.7[9], 65.7[10]
Gu (1994)	Na$^+$ 0 to 0.10	25-75	41.7[11]
Dissolution-Amorphous silica			
Rimstidt and Barnes (1980)	deionized water	25-300	60.9-64.9
Fleming (1986)	deionized water	25-100	54.8±3.8
Liang and Readey (1987)	hydrofluoric acid	24-70	33, 30, 26[12]
Mazer and Walther (1994)	deionized water	40-85	93.7±15.5[13]
Gu (1994)	Na$^+$ 0 to 0.10	25-75	37.5[13]

[1]prism average; [2]rhomb average; [3]pH 3; [4]pH 11; [5]pH 3; [6]pH 11; [7]low pH range with transition to [8]high pH range; [9]a-cristobalite; [10]b-cristobalite; [11]opal-CT; [12]0%HF, 25%HF, 49% HF, respectively; [13]fused quartz.

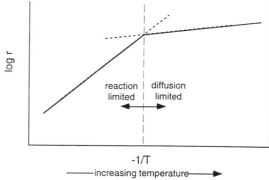

Figure 17. Schematic diagram illustrating the relationship between the surface reaction limited rate and the rate of diffusion of H$_4$SiO$_4$ away from the mineral surface as a function of temperature. At low temperatures the rate limiting step for silica dissolution is bond breaking at the surface but this reaction has a high activation energy so that its rate increases rapidly with increasing temperature. At some temperature, depending on the geometry of the system and the relative fluid velocity, the reaction limited rate releases silica to the solution faster than it can diffuse away from the surface. The rate of dissolution is limited by diffusion at this and higher temperatures (from Dove and Rimstidt, 1994).

k_- are the dissolution and precipitation rate constants, respectively. The net rate is the sum of (25) and (26) such that

$$r = (dn_{H_4SiO_4}/dt) = A (k_+ a_{SiO_2} a^2_{H_2O} - k_- a_{H_4SiO_4}).$$ (23)

Since $n_{H_4SiO_4}$ can be recast as $a_{H_4SiO_4}$ for

$$n_{H_4SiO_4} = m_{H_4SiO_4} (M)$$ (24)

where M is the mass of water. Assuming that the mass of water in the system is constant, this allows the reaction rate to be expressed relative to the rate in a system containing one kg of water such that (23) is written

$$r = (dm_{H_4SiO_4}/dt) = (A/M) (k_+ a_{SiO_2} a^2_{H_2O} - k_- a_{H_4SiO_4}).$$ (25)

Rimstidt and Barnes (1980) designed the A/M term to quantify the extent of the system, that is, the ratio of the relative surface area to the relative mass of water present. Normalizing the reaction rate to one kilogram of water and one square meter of interfacial surface area, they defined apparent rate constants, k'_+ , -, that included A/M such that

$$k'_+ = (A/M) \gamma_{H_4SiO_4} a_{SiO_2} a^2_{H_2O} k_+$$ (26)

$$k'_- = (A/M) \gamma_{H_4SiO_4} a_{SiO_2} a^2_{H_2O} k_-$$ (27)

Fundamental rate constants that can be compared must separate this A/M contribution from the measured apparent constants. Substituting Equation (26) and (27) into (23) converts the rate equation to the simple form

$$r = k'_+ - k'_- a_{H_4SiO_4}.$$ (28)

Since

$$K = k_+ / k_-$$ (29)

then (28) is rearranged to

$$r = k'_+ (1 - Q/K)$$ (30)

where Q is the activity quotient of the system such that

$$Q = (a_{H_4SiO_4}) / (a_{SiO_2}) (a_{H_2O})^2$$ (31)

Equation (30) gives the dissolution rate of quartz as a first order expression written in terms of silicic acid release. Numerous studies have assumed first order behavior of quartz dissolution since Rimstidt and Barnes (1980) presented Equation (30). However, this has been confirmed experimentally only in hydrothermal studies. Berger et al. (1994) measured quartz dissolution rates in deionized water containing 0 to 10 mmoles of silica at 300°C and found the reaction rate follows a first order rate law in silicic acid concentration over the tested range of reaction affinity as shown in Figure 18a. In contrast, Figure 18b and c show that the rate law deviates significantly from first order behavior in the presence of low sodium or lead concentrations. This difference in behavior is explained in terms of competitive adsorption between dissolved cations and H_4SiO_4 at the silica surface.

While the rate equations developed by Rimstidt and Barnes (1980) are consistent with thermodynamics, mineral-water reaction rates are perhaps better evaluated in terms of processes occurring at the mineral-solution interface. This means removing emphasis on the behavior of the bulk solution and refocusing on the reactants. That is, the surface site densities and interfacial solvent population and behavior.

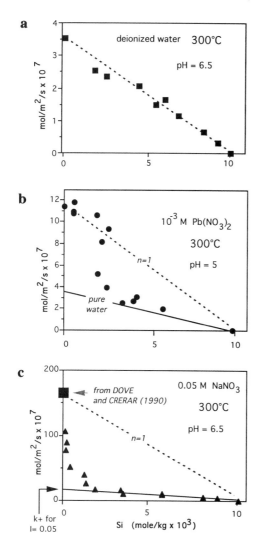

Figure 18. Experimental quartz dissolution rates showing the dependence upon reaction affinity at 300°C. The dotted lines represent calculated rates. (a) deionized water; (b) 10^{-3} molal $Pb(NO_3)_2$; (c) 0.05 molal $NaNO_3$ (from Berger et al., 1994).

Transition state theory (TST) describes the energetics of elementary reactions on a molecular level and recent reviews of this subject have shown its direct application to describing mineral-water reaction rates (Lasaga, 1981; Lasaga, 1990). TST is briefly applied here to show how silica-dissolution and precipitation reaction rates can be reformulated from Equation (25) to a kinetic expression based on the surface and solvent reactant concentrations. For simple rate-limiting dissolution reactions of the forms,

$$A + B = AB^{\ddagger} = C + D \tag{32a}$$

$$A = A^{\ddagger} = P, \tag{32b}$$

the absolute identity of A, B and the activated intermediates, $AB^{\ddagger}$ or $A^{\ddagger}$, are not usually known, but can be assumed to have one of these stoichiometries with reasonable certainty. Dove and Crerar (1990) proposed that the rate-determining step for quartz dissolution involves the attack of silica surfaces by H_2O to break Si—O bridging bonds by the simple reaction

$$>Si\text{-}O\text{-}Si< + H_2O = (>Si\text{-}O\text{-}Si< \cdot H_2O)^{\ddagger} \rightarrow 2(>Si\text{-}O\text{-}H). \tag{33}$$

which is analogous to Equation (32a). The TST formulation of the rate constant, k_+, describing the frequency that an adsorbed water molecule binds to a silica surface to form an activated complex is conventionally written as

$$k'_+ = \left(\frac{kT}{h}\right)\left(\frac{\gamma_{ads}}{\gamma^{\ddagger}}\right)\left(\exp\frac{\Delta S^{\ddagger}}{R}\right)\left(\exp\frac{-\Delta H^{\ddagger}}{RT}\right) \tag{34}$$

where k'_+ = apparent rate constant (sec^{-1})

$\frac{kT}{h}$ = frequency factor $(9.85 \times 10^{12}\ sec^{-1}$ at $200°C)$

$\frac{\gamma_{ads}}{\gamma^{\ddagger}}$ = activity coefficients for adsorbed H_2O and intermediate species

$\Delta S^{\ddagger}$ = standard activation entropy $(J\ mol^{-1}\ K^{-1})$

$\Delta H^{\ddagger}$ = standard activation enthalpy $(kJ\ mol^{-1})$.

The k'_+ in Equation (24) has units of (sec^{-1}) because it represents the rate that an individual adsorbed water molecule attacks the quartz structure. However, when BET or geometric mineral surface areas are known, experimental dissolution rate data give the reaction rate, $r_{H_4SiO_4}$, for a population of water molecules that reacts per unit area of mineral surface. The parameters, k'_+ and $r_{H_4SiO_4}$, are linked by recasting the net reaction rate in terms of the number of moles of water molecules per reactive area $(mol\ m^{-2})$ such that the reaction rate, $r_{H_4SiO_4}$, is equal to the number of adsorbed H_2O molecules multiplied by the apparent rate, k'_+. This gives

number moles of adsorbed H_2O per square meter $= X_{H_2O}\ N_{t,ads}$ \hfill (35)

where X_{H_2O} = mole fraction of sites accessible to water molecules, and $N_{t,ads}$ = moles of reactive sites on the mineral surface $(mol\ m^{-2})$
so that the forward rate is expressed

$$r = X_{H_2O}\ N_{t,ads}\ k'_+. \tag{36}$$

The forward reaction rate in Equation (17) is given by

$$r = k_+ (a_{SiO_2})(a_{H_2O})^2 . \tag{37}$$

Equating Equations (36) and (37) and solving for k_+ gives

$$k_+ = \frac{X_{H_2O}\ N_{t,ads}}{(a_{SiO_2})(a_{H_2O})^2}\ k'_+. \tag{38}$$

Substituting (34) for k'_+, Dove and Crerar (1990) obtain the rate constant for reaction of water molecules with the quartz surface

$$k_+ = \frac{X_{H_2O}\ N_{t,ads}}{(a_{SiO_2})(a_{H_2O})^2}\left(\frac{kT}{h}\right)\left(\frac{\gamma_{ads}}{\gamma^{\ddagger}}\right)\left(\exp\frac{\Delta S^{\ddagger}}{R}\right)\left(\exp\frac{-\Delta H^{\ddagger}}{RT}\right) \tag{39}$$

where k_+, the fundament rate constant has units of $(mol\ m^{-2}\ sec^{-1})$. This description of the rate constant has a number of advantages over the *A/M* convention used in Equation (25).

First, the standard state for the activated reaction becomes one mole of sites per square meter. The units of this standard state are physically consistent with the activated complex concentration that exists within the two-dimensional space of the interface. Second, it produces a system which describes the thermodynamic variables, and thus, r and k_+ with internally consistent units (e.g. Lasaga, 1981). Another advantage of using this convention is that general rate equations can be written with multiple rate *constants* to describe the kinetics of silica-water dissolution over a range of conditions, such as solution pH, where multiple rate-limiting reactions are operative. Thus, a net rate equation can be described by a sum of processes as shown later in Equations (53 to 56).

Although the previous discussion focuses on dissolution, Renders et al. (1995) suggests that the reverse reaction rate of silica precipitation can be also described with this convention for the rate constant as shown in Figure 19. Briefly, one can rewrite Equations (25) and (26) as

$$r_+ = (dn_{H_4SiO_4}/dt)_+ = k_+ a_{SiO_2} \, a^2 \, H_2O \tag{40}$$

and

$$r_+ = (dn_{H_4SiO_4}/dt)_- = k_- a_{H_4SiO_4} \tag{41}$$

where k_+ and k_- are described by Equation (25) for a single forward and reverse rate-limiting reaction, respectively.

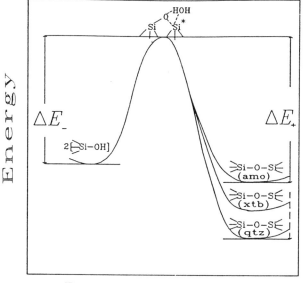

Figure 19. Schematic illustration showing the similarity between forward and reverse reaction path for microscopically reversible silica dissolution and precipitation. Activation energy, $E_{a,-}$ is thought to be a constant for the precipitation of all silica polymorphs, but $E_{a,+}$ varies with the silica polymorph. (from Renders et al., 1995).

Nucleation. The precipitation of the first silica particles from a homogeneous supersaturated solution is inhibited by the high solubility of very small silica particles. This high solubility, which is the result of the work required to form silica surfaces (see discussion of Eqn. 12), can be visualized as an energy barrier that must be crossed via an activated complex which is called the critical nucleus as illustrated in Figure 20. The rate of formation of critical nuclei has been modeled by Nielsen (1959) who showed that the flux of aqueous species to the growing nuclei, J, is given by:

$$J = A \exp\left(\frac{-\Delta G^*}{kT}\right) \tag{42}$$

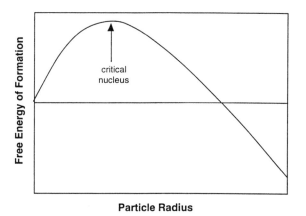

Figure 20. Schematic illustration of the change in free energy of formation of a growing particle as a function of size. Particles smaller than the size of the critical nucleus will tend to dissolve while those larger than the critical nucleus will tend to grow to macroscopic size (from Dove and Rimstidt, 1994).

where A is a factor relating to the collision rate of the aqueous species involved in the formation of the nuclei (on the order of 10^{30} for most systems), k is the Boltzman constant, T is temperature in degrees Kelvin, and ΔG^* is the free energy of formation of a critical nucleus as given by the following expression:

$$\Delta G^* = \frac{16\,\pi\,\sigma^3\,V^2}{3\left[k\,T\,\ln\left(\frac{c}{c_o}\right)\right]^2} \tag{43}$$

where σ is the surface free energy of the solid, V is its molar volume, c is the concentration of aqueous species, and c_o is the equilibrium concentration of the aqueous species. This equation shows that high values of σ produce slow rates of nucleation. Because quartz has a high surface free energy, it is much more difficult to nucleate than amorphous silica. As a result, amorphous silica often precipitates from solutions instead of quartz. This effect is illustrated in Figure 21 which shows the flux of silica onto growing nuclei with decreasing temperature.   This explains the formation of opal and amorphous silica phases in systems

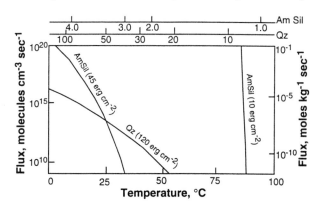

Figure 21. The flux of dissolved silica (J) to form nuclei of quartz (Qz) or amorphous silica (Am Sil) by homogeneous nucleation is essentially zero until the solution reaches 25°C (rates below 1×10^{-7} mol kg^{-1} sec^{-1} precipitate less than 10% of the dissolved silica per day). Observations of siliceous sinter formation shows that amorphous silica forms near 100°C suggesting that heterogeneous nucleation occurs as illustrated by the line on the right. This probably happens on iron and manganese oxyhydroxide substrates. The axes at the top of the diagram indicate the degree of saturation of quartz and amorphous silica (from Dove and Rimstidt, 1994).

that are substantially supersaturated with respect to quartz. The more soluble phases have a lower surface free energy than less soluble phases so that in general the most soluble polymorph of a material forms first from a supersaturated solution and this polymorph transforms stepwise to less soluble polymorphs until the most stable, least soluble polymorph finally forms via Ostwald's step rule (Stumm, 1992). Thermodynamically, opal-A often precipitates first from solution; it then transforms to opal-CT which has a lower solubility; and finally, opal-CT transforms to quartz (e.g. Williams et al., 1985). However, in low temperature weathering environments, metastable forms of silica persist for long times (see earlier discussion). For example, Kastner et al. (1977) and Kastner and Gieskes (1983) find that rates of opal-A to opal-CT transformation become very slow below 50°C.

Etch pit development

Observations of etch pits on naturally weathered quartz surfaces show that dissolution under sufficiently undersaturated conditions results in the formation of etch pits on the mineral surface (see earlier discussion). Brantley et al. (1986a) proposed that when the silica mineral is unstrained, the free energy of formation of an etch pit at a mineral surface, ΔG_p, is the sum of the free energy driving the dissolution, ΔG_v, and the work needed to create the new surface area of the etch pit, ΔG_s such that:

$$\Delta G_p = \Delta G_v + \Delta G_s = \frac{\pi r^2 a\, RT \ln\left(\frac{c}{c_o}\right)}{V} + 2\,\pi\, r\, a\, \sigma \qquad (44)$$

where r is the radius of the pit, a is the depth of the pit, c is the concentration of silica in solution, c_0 is the equilibrium solubility of the silica mineral, V is the molar volume of the silica mineral, R is the gas constant, T is temperature in Kelvins, and s is the surface free energy of the mineral (Lasaga and Blum, 1986; Brantley et al., 1986a,b). According to this equation, there is always a free energy barrier to the growth of etch pits in perfect silica minerals but the magnitude of this barrier decreases as the solution becomes more and more undersaturated (Brantley et al., 1986a,b). There is an additional source of free energy, ΔG_d, that is present where dislocations intersect crystal surfaces so that equation (44) acquires an additional term:

$$\Delta G_p = \Delta G_v + \Delta G_s + \Delta G_d = \frac{\pi r^2 a\, RT \ln\left(\frac{c}{c_o}\right)}{V} + 2\,\pi\, r\, a\, \sigma\ - \frac{a\, K\, b^2 \ln\left(\frac{r}{r_o}\right)}{4} \qquad (45)$$

where K is an energy factor, b is the Burgers vector of the dislocation, and r_0 is the radius of the dislocation core. Lasaga and Blum (1986) discuss ways to model the ΔG_d term which has the effect of removing the barrier to nucleation of etch pits for low values of c/c_0 so that etch pits easily nucleate and spontaneously grow. Brantley et al. (1986b) determined that for quartz in water at 300°C, etch pits would nucleate and spontaneously grow when $c/c_0 < 0.75\pm0.15$. Thus, silica phases in contact with highly undersaturated aqueous solutions will develop etch pits at places where dislocations outcrop on the crystal surface. Above some critical degree of undersaturation, $c/c_0 > 0.75$, there is little or no etch pit formation.

Recent work on the kinetics of etch pit development on quartz surfaces led to the *negative crystal method* (Gratz et al., 1990) for isolating the rates of dissolution of specific crystallographic faces. Figure 22 shows four types of etch pit features that are produced when quartz is reacted in alkaline solutions. Gratz et al. (1990) estimated dissolution rates of individual negative crystals by measuring the size of individual crystals during a sequence of dissolution steps. This method has powerful advantages over other

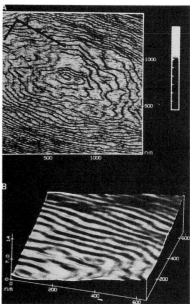

Figure 22 (left). Features produced by etching prismatic faces of quartz in alkaline solution. (a) large negative crystal- a flat bottomed pit with crystallographic walls; (b) two jumbo pits- large depressions with pointed centers and curved walls; (c) etch tunnel undercutting the surface and d.) abundant small pits superimposed on negative crystals. All figures are scaled to the 100 micron bar shown in (c) (from Gratz et al., 1990).

Figure 23 (right). SFM images of a microscopic quartz etch pit formed at 148°C, then scanned in ethanol with <10^{-8} Newton of applied force (from Gratz et al., 1991). (A) Map view of an etch pit that is image-enhanced to show the edges between topographic levels. Image area is 1490 nm by 1490 nm with a height difference of 20 nm between the bottom of the pit and the highest edge of the surface. (B) Surface plot of one quadrant of the etch pit. Image size is 650 by 650 by 14 nm (used with permission of authors).

studies which measure rates by changes in bulk solution composition in that it acknowledges the large variations in dissolution rate with crystal orientation and cracks.

With developments in Scanning Force Microscopy (SFM) Gratz et al. (1991) was the first to reveal the microscopic nature of etch pit development on quartz surfaces. Figure 23 shows the 7 Å monolayer steps which comprise the floors and walls of etch pits such as those in Figure 22. Gratz and Bird (1993a,b) proposed a ledge motion rate law that described the dissolution kinetics of all single crystals and powders in terms of surface roughening and the mathematics of monolayer step motion.

Controls of temperature and solution composition on reactivity.

Numerous studies have qualitatively and quantitatively investigated factors affecting silica-water reaction rates such as saturation state, temperature, solution pH, salinity, dissolved metals, and organics. These factors are interrelated, but grouped somewhat arbitrarily to review some of the most important results. Findings from higher temperature studies are considered in this discussion of silica reactivity in weathering environments. This is because they constrain findings of low temperature behavior and provide

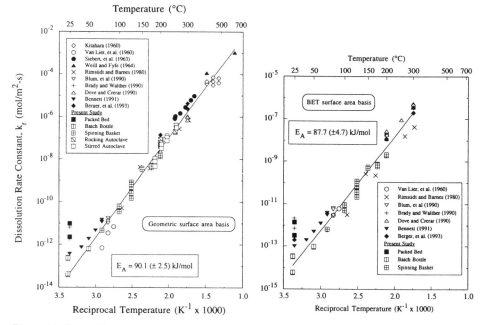

Figure 24. Quartz dissolution kinetics in pure water from 25° to 300°C using (a) a geometric surface area basis and Arrhenius coordinates; (b) BET-determined surface area basis and Arrhenius coordinates (from Tester et al., 1994).

information on the temperature dependent behavior of the silica polymorphs in paleoclimate and predictive studies.

Temperature dependence in deionized water. The dissolution and precipitation kinetics of the silica polymorphs in deionized water have been investigated in several studies over the temperature range of 25° to 300°C. In perhaps the first investigation of silica reactivity in hydrothermal systems, Rimstidt and Barnes (1980) determined the dissolution rate of four silica polymorphs as empirical functions of temperature. They found that for dissolution in deionized water, activation energies of these polymorphs were similar (within the error of their experiments) with values ranging from 60 to 76 kJ mol[-1]. Table 3 shows that these activation energies are consistent with more recent findings. Tester et al. (1994) combined new data with existing rates published in ten different studies to determine an updated temperature function (spanning 25° to 625°C) for the dissolution kinetics of quartz in deionized water. Their experimental design carefully considered the potential rate-enhancing or inhibiting effects of dissolved components from fluids or bomb walls on the reaction rate. One finding of this study was that both geometric and BET-derived surface areas were successfully used to normalize dissolution rates and determine a simple Arrhenius expression for quartz dissolution rate in deionized water. Figures 24a and b illustrate the fit of both temperature functions to the dissolution rate data.

There are few quantitative measurements of silica precipitation at the low temperatures of weathering environments. Most early studies of silica precipitation kinetics in deionized water investigated the formation of high purity quartz crystals for materials applications at hydrothermal temperatures (e.g. Laudise, 1959; Hosaka and Taki, 1981a,b; Hosaka and Taki, 1986; Martin and Armington, 1983; Armington and Larkin, 1985). Others considered rates of silica precipitation in reservoir systems (Bohlmann et al., 1980; Thornton, 1988; Thornton and Radke, 1988).

In one quantitative investigation, Rimstidt and Barnes (1980) found that a single rate constant expression with an activation energy of 40 kJ mol[-1] described the precipitation of quartz, α-cristobalite, β-cristobalite and amorphous silica. Renders et al. (1995) studied the precipitation kinetics of synthetic cristobalite to investigated the reactivity of this polymorph which often forms preferentially over other thermodynamically-favored forms of silica (see *Nucleation* section). They report that cristobalite precipitates only when precursor cristobalite nuclei are present and when silica concentrations are above cristobalite thermodynamic saturation. Like Rimstidt and Barnes (1980), they found an activation energy for precipitation which is similar to values determined for quartz and amorphous silica (see Table 3). Thus, they conclude that all silica polymorphs precipitate by the same rate-limiting elementary reaction. Renders et al. (1995) also show that the presence of cristobalite catalyzes the metastable precipitation of hydrothermal cristobalite.

To my knowledge, few studies report the kinetics of SiO_2 precipitation in low temperature environments except for investigations of particle ripening (see earlier discussion). However, the importance of silica sorption in soil environments is widely recognized since the findings of Beckwith and Reeve (1963, 1964). They suggest that sesquioxide minerals are responsible for the capacity of soils to remove H_4SiO_4 from solution. It is recognized that chemisorption of silica with metallic ions results in the precipitation of amorphous and crystalline silicate precipitates (Drees et al., 1989; Karathanasis, 1989; Nahon, 1991), yet little quantitative information exists.

Catalysis by alkali cations. Numerous studies have qualitatively shown that the presence of alkali cations markedly increases the dissolution rate of quartz and amorphous silica at the conditions of weathering environments (Diénert and Wandenbulcke, 1923; Van Lier et al., 1960; Weigel, 1964; Kamiya and Shimokata, 1976; Fleming, 1986). Alkali cations enhance quartz and amorphous silica dissolution rates in near-neutral pH solutions at ambient and hydrothermal temperatures. Bennett (1991) found that sodium and potassium chloride increase rates at 25° to 70°C by a factor of 5 to 8. Likewise, Dove and Elston (1992) and Berger et al. (1994) determined that electrolytes cause a similar rate increase (factor of 6) for quartz dissolution pH 6.5 to 7 and 25°C . House (1994) reports that $CaCl_2$ enhances rates of Fontainbleu sand at ionic strengths less than 0.1 molal. ^{29}Si-NMR studies of silica gel dissolution in basic alkali cation-bearing solutions shows that rates increase in the order (LiOH $\approx$ CsOH) < (RbOH $\approx$ NaOH) < KOH (Wijnen et al., 1989; Wijnen et al., 1990). At higher temperatures, the catalyzing effect of salt on quartz dissolution increases to a factor of 33 with the addition of as little as 0.05 molal sodium or potassium chloride to deionized water at 100° to 300°C (Dove and Crerar, 1990). The magnitude of this rate increase is cation-dependent where the effect of Na $\approx$ K > Li > Mg.

Little work has been done to investigate the effect of electrolyte solutions on the precipitation kinetics of silica polymorphs at low temperatures. However, it has been shown that quartz precipitation rates increase as salinity increases at hydrothermal conditions (Corwin and Swinnerton, 1951; Laudise, 1959). Hosaka and Taki (1981a,b) found that quartz growth rates are faster in NaCl- and KCl-bearing solutions than in deionized water at 350° to 500°C. At more saline electrolyte concentrations, Bohlmann et al. (1980) examined rates of silica deposition in reactors packed with several forms of silica (α-quartz, polycrystalline α-quartz, porous Vycor) in NaCl brine solutions at pH 5 to 8 and temperatures of 60° to 120°C. Kinetic interpretation of the data is complicated since much of the silica removal appeared to be dominated by nucleation onto the silica surfaces, but the final analysis yield a silica deposition rate that predicts rapid plugging of underground geothermal systems.

Although few data exist for precipitation kinetics at conditions relevant to weathering environments, the available rates suggest that the contribution of simple dissolved components can have catalyzing effects. This has profound implications for the accuracy of geochemical models as past numerical codes have incorporated kinetic constants from dissolution experiments conducted in deionized water. Recent work quantifying the effect of simple salts suggests that predicted rates of silica dissolution from these models are likely slower than what actually occurs in the natural surface waters or geothermal fluids which contain significant concentrations of Ca, Mg, Na and K. The rate-enhancing effect of alkali is also important for interpreting experimental studies. Cation-bearing solutions are often used in rate experiments to examine ligand effects (e.g. Bennett et al., 1988), control ionic strength or to buffer solution pH (e.g. Thornton and Radke, 1988; Knauss and Wolery, 1988; Schwartzentruber et al., 1987). Bennett (1991) illustrates the careful experimental design required to clearly separate alkali effects and organic ligand effects.

Solution pH. Numerous studies have investigated both the general pH-dependent behavior of silica dissolution rates and specifics of behavior at particular ranges of solution pH. In general, the silica polymorphs appear to have a dissolution rate minimum near the zero point of net proton charge, ZPNPC, of pH 2 for silica. In both low temperatures (Liang and Readey, 1987; Knauss and Wolery, 1988; Wollast and Chou, 1988; Bennett, 1991) and hydrothermal temperatures (Dove and Crerar, 1991; Berger et al., 1994) dissolution rates increase in solutions that are more acidic or basic on either sides of this pH. This discussion focuses on rates measured in the pH range of about 2 to 12, which cover the solution pH values observed in natural systems.

Baumann (1955) made the first observations of the pH-dependence of silica dissolution with a qualitative study of a commercial silica gel (Aerosil) at 30°C. Figures 25a and b show the general pH-dependence of dissolution where rates have a minimum near the ZPNPC of pH 2 for silica and increases exponentially until leveling off near pH 10. Studies of quartz dissolution rates versus pH have since shown similar trends in reactivity (Henderson et al., 1970; Knauss and Wolery, 1988; Wollast and Chou, 1988; Brady and Walther, 1990; House and Orr, 1992).

Silica dissolution behavior exhibits a complex rate dependence upon both variable pH and electrolyte concentration. Wirth and Gieskes (1979) noted the dependence of both parameters in measurements of vitreous silica dissolution rates. Figure 26a shows that rates of amorphous silica dissolution plotted versus solution pH show considerable scatter with variable sodium concentration. In contrast, these same data form the single trend shown in Figure 26b when rates are replotted in terms of calculated negative surface charge at the interface. A later analysis of the pH-dependence of quartz dissolution rate by Dove and Elston (1992) showed that a direct comparison of experimental values reported by seven laboratories showed a disconcerting amount of scatter between investigators. Recasting the reported rate data in terms of the influence of both solution pH and electrolyte concentration on the predominant negative surface complex distributions, most discrepancies between investigators could be explained by the pH-dependent influence of alkali cations contained in reagents used to adjust pH and ionic strength. Thus, the data reported by seven laboratories using different sources of quartz and experimental methods show consistent behavior. Figure 27 illustrates the complex pH- and alkali cation- rate dependence of quartz dissolution resulting from this study.

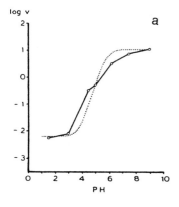

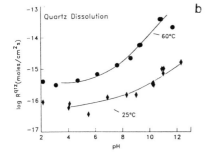

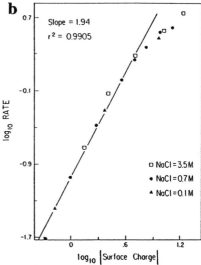

Figure 25. (a) Early work showing the pH dependence of Aerosil dissolution rate at 23°C (from Baumann, 1955); (b) pH dependence of quartz dissolution rate vs. pH at 25° and 60°C corrected to ionic strength of 10^{-3} (from Brady and Walther, 1990).

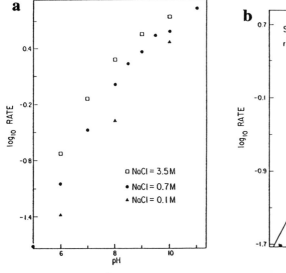

Figure 26. Dissolution rate (mol cm^{-2} s^{-1}) of amorphous silica in three concentrations of sodium chloride plotted as (a) a function of bulk solution pH and (b) a logarithm of absolute surface charge (from Wirth and Gieskes, 1979).

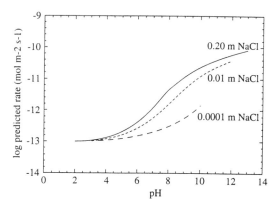

Figure 27. The predicted dissolution rates of quartz at 25°C as a function of pH and contoured by sodium concentration. The reaction rate is independent of cation concentration below approximately pH four (from Dove and Elston, 1992).

The pH and electrolyte dependence of dissolution kinetics is also observed for other silica polymorphs. Investigations of vitreous silica glass by Mazer and Walther (1994) show a similar solution pH dependence to that observed for quartz such that dlog k/dpH is 0.5 at pH >7. Also, Mekonnen (1995) finds that the dissolution rate of amorphous silica has a strong (OH-) dependence at pH >2.5. Ionic strength has only a small catalyzing effect on rate at pH 2.5, but strongly affects rates at pH 3.9. Similar findings are reported by Gu (1995) in studies of amorphous silica and opal-CT dissolution rates.

Combined temperature, electrolyte, and pH relations. The kinetic studies of pH, temperature and electrolyte dependence have been combined to develop a comprehensive model of quartz reactivity in variable pH and simple salt solutions. Dove (1994) uses low temperature and new hydrothermal data with reported rate measurements to determine a general rate equation that quantifies reaction kinetics from 25° to 300°C in solutions having an pH_T (in situ pH at temperature of experiment) of 2 to 12, and 0 to 0.3 molal sodium. The rate equation gives a good fit to experimental data that span reaction rates over a range of 10^{11} (see Eqn. 56). The expression has several properties suggesting robustness: It describes the pH dependence of dissolution rates, cation-specific effects observed at near-neutral pH, the diminishing effects of cations at higher solution pH, the increasing rate-enhancing effects of alkali with increasing temperature and has a reaction order near one, implying first order behavior (see discussion of **Mechanistic Controls**). Figure 28 shows the behavior of this predictive expression for two temperatures.

Modifiers: Effects of sorbates and coatings on reactivity

Organic acids. High concentrations of dissolved silica are often found in association with organic rich soil and sediment pore waters. Organic acids present in these environments are believed to increase silica reactivity through ligand exchange complexation reactions. In a study of a glacio-fluvial sand aquifer that had been contaminated by crude petroleum, Bennett and Siegel (1987) and Bennett et al. (1987) report evidence for organo-enhanced complexing and reactivity of silica. They observe that quartz sand grains from uncontaminated and contaminated regions of the aquifer show different etching patterns. While uncontaminated quartz grains showed a complete lack of chemical etch pits, contaminated grains revealed the extensive development of triangular pits and a general roughening of the surface. These observations lead to experimental studies (Bennett et al., 1988) which quantified the rates of quartz dissolution in dilute solutions of salicylic, oxalic and humic acids. Citrate was found to increase dissolution rates by a factor of 8 to 10. Bennett (1991) later resolved individual contributions of these competing mixed organic-

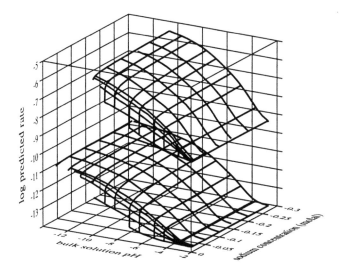

Figure 28. The dissolution rate of quartz in solutions of variable pH$_T$ and sodium concentration at two temperatures. The predicted rates at 25°C (lower surface) and 200°C (upper surface) show the dissolution behavior described by Equation (57). Little pH-dependence of dissolution rate is observed in pure water at 25°C but increases with increasing sodium concentration or temperature (from Dove, 1994).

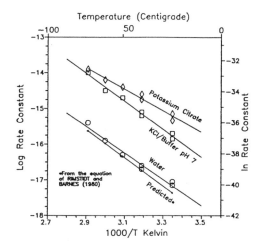

Figure 29. The effect of potassium citrate, KCl buffer and deionized water on the dissolution rate of quartz between 25° and 70°C (from Bennett, 1991).

inorganic electrolyte solutions to silica reactivity as seen in Figure 29. The organic acids appear to increase dissolution rates by a factor of four by decreasing the activation energy 20% and increasing apparent quartz solubility.

Aluminum. It is qualitatively well-documented that minute concentrations of sorbed aluminum as well as other bi- and trivalent metals can affect the surface properties (see Fig. 16), apparent solubility and dissolution kinetics of the silica polymorphs. Table 4 lists several ions qualitatively found to decrease rates of dissolution for many of the SiO_2 polymorphs. Of these, aluminum and ferric iron (see following section) occur the most extensively in natural environments, albeit in low concentrations for the pH range of most weathering environments.

Table 4. Experimental investigations reporting ions qualitatively observed to inhibit the dissolution kinetics of SiO_2 polymorphs.

Reference	Polymorph, silica-rich phase
Al^{3+}	
Lewin (1961)	diatoms
Jones and Handreck (1963)	monosilicic acid, natural soil
Wiegel (1961)	silica glass
Iler (1973)	amorphous silica
Hudson and Bacon (1958)	sodic glass (71% silica)
Ballou et al. (1973)	silica glass (96% silica)
Fe^{3+}	
Lewin (1961)	diatoms
Jones and Handreck (1963)	monosilicic acid, natural soil
Wiegel (1961)	silica glass
Hudson and Bacon (1958)	sodic glass (71% silica)
Ballou et al. (1973)	silica glass (96% silica)
Zn^{2+}	
Wiegel (1961)	silica glass
Hudson and Bacon (1958)	sodic glass (71% silica)
Cu^{2+}	
Wiegel (1961)	silica glass
Hudson and Bacon (1958)	sodic glass (71% silica)
Be^{2+}	
Lewin (1961)	diatoms
Hudson and Bacon (1958)	sodic glass (71% silica)
Ga^{3+}, Gd, Sc, Ti^{4+}, Y, Yb	
Lewin (1961)	diatoms

The phenomena of aluminum-silica interactions has led to extensive studies of colloid reactivity the natural waters (Matijevic et al., 1971; Allen and Matijevic, 1971; Hem et al., 1973; Brown and Hem, 1975), biominerals in biological systems (Bartoli, 1985; Birchall, 1994), phytoliths and particles in soil weathering environments (Bartoli and Wilding, 1980; Jardine, 1985a,b) and diatoms of marine sediments (Lewin, 1961; Hurd, 1973; Van Bennekom, 1981; Van Bennekom et al., 1991; Van Cappellen and Qui, 1995a,b). Qualitative estimates of the reduction in silica dissolution rates by aluminum range from three to eight orders of magnitude (Hurd, 1973). However, quantified relationships between dissolution rate, the concentration of aluminum in solutions and processes occurring at the SiO_2-solution interface are not presently known. For example, Figure 30a shows the extent of our quantitative understanding of aluminum effects on diatom dissolution.

Aluminum is believed to reduce silica solubility and dissolution by two or more processes that are determined by aluminum concentration (saturation with respect to its oxyhydroxides), aging time, and solution pH (Iler, 1979). Iler (1973) proposed that SiO_2 surfaces exposed to solutions containing aluminum formed aluminosilicate surface complexes by reactions with the hypothetical stoichiometry:

$$>SiOH + Al(OH)_3^+ \leftrightarrow >SiOAl(OH)_2 + H^+. \tag{46}$$

This expression for the binding of a surface silanol group with aqueous aluminum suggests a pH dependence of sorption that is confirmed in the laboratory. The surface interaction mechanism in Reaction (1) is supported by recent NMR studies of silica gel by Stone et al. (1993) showing that aluminum is incorporated into the silica tetrahedral framework. Iler (1973) also showed that low aluminum concentrations (in pH 8 solutions) give significant

a Effect of Al on biogenic silica dissolution
(from Lewin, 1961)

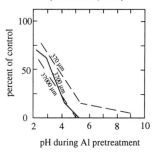

Figure 30. (a) The effect of aluminum concentration and pH on the dissolution of acid-cleaned silica valves. Samples were pretreated for three hours with solutions of the compositions shown, then silicon concentrations were measured at the end of ten days and plotted as percent of the control silicon concentration (modified from Lewin, 1961). These qualitative data suggest that aluminum has little effect on dissolution at low pH and the reduction in silica reactivity is strongly pH-dependent. (b) Aluminum concentration affects the solubility of colloidal silica (modified from Iler, 1973).

b SiO_2(am) solubility in Al solutions at 25°C
(from Iler, 1973)

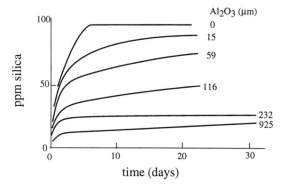

reductions in both rate of solution and solubility (Fig. 30b) of silica gel. Van Cappellen and Qui (1995a) propose that apparent silica solubility is strongly dependent upon pore water aluminum concentrations in marine sediments.

It is not clear from existing studies that aluminum reduces silica reactivity if present at concentrations below the solubility of its oxyhydroxides. 'Small amounts' of sorbed aluminum could have little effect on reactivity since rates of water exchange with aluminum are similar to silicon. Therefore, minute aluminum sorption may not necessarily slow the kinetics of silica dissolution (e.g. Dove, 1994). It is significant to note that silica surfaces can 'scavenge' aluminum from low concentration solutions (see earlier discussion).

At higher concentrations, aluminum sorbs/precipitates onto SiO_2 surfaces as aluminum-oxyhydroxides (Brown and Hem, 1975). Aluminum sorption likely begins at steps and structural defects and continues with progressive armoring of the entire surface. With exposure to higher dissolved aluminum concentrations, electron micrographic evidence suggests aluminum continues to sorb/precipitate(?) to form surface coatings of amorphous aluminum oxyhydroxides and/or crystalline aluminum oxides (Brown and Hem, 1975). This continued sorption of aluminum results in armored coatings extensive enough to reverse net SiO_2 (or TiO_2) surface charge from negative to positive (Matijevic et al., 1971; Jednacak and others, 1974; James et al., 1977).

These discussions of aluminum-SiO_2 *surface* interactions are necessarily descriptive because previous work in this area is largely qualitative. Until recently, the available techniques necessitated investigations of bulk solution chemistry and *ex situ* examination of the initial reactants and final products. Findings from these two approaches provide our current picture of aluminum-silica interactions and describe many of the complexities in the aluminum-silica speciation, sorption and aging processes. However, most of this understanding is from indirect inference of the processes occurring at the mineral surface, without direct spectroscopic or microscopic evidence.

New work is integrating techniques in geochemical kinetics and surface analysis to characterize the influence of sorbed aluminum on silica reactivity (e.g. Dove et al., 1994; Dixit et al., 1995). Presently the controls on the chemistry of these coatings and their effect on silica reactivity is not well-understood. Investigations of Al-treated quartz surfaces by Atomic Force Microscopy illustrates the extent of these coatings as a result of exposing a fresh (100) quartz surface to a pH 4.05 solution of 0.02 m AlCl₃ for 48 hours (solution prepared per method of Hsu, 1988). Aluminum-free controls show that the starting etched surface is covered with 'bathtub' pits like those observed by Gratz (1990) while the Al-exposed surface is overlain by a thick coating of sorbate/precipitates.

Ferrous-ferric iron.   Field and laboratory observations of iron-stained quartz support titration studies showing that iron binds strongly with silica surfaces. Table 4 suggests that sorbed or precipitated iron also reduces silica reactivity. This phenomena could be important to understanding paleoclimate during development of Banded Iron Formations. Some studies suggest that the formation of these deposits was controlled by the rate of quartz removal in surficial weathering processes. Morris and Fletcher (1987) conducted experiments to test this hypothesis by measuring quartz solubility at 60°C in a series of iron and sodium-bearing solutions over a pH range of 4 to 8.6. Figure 31a shows that fine grained quartz (A_{sp} = 0.22 m² g⁻¹) exposed to starting solutions containing ferrous iron for 1 to 9 days became significantly more soluble following an 'oxidizing period'. In this period, samples were aerated continuously with bubbling oxygen and without significant abrasion of the quartz grains. The observed increases in solubility were further enhanced by delaying the oxidizing period to later times. Although the results are somewhat difficult to interpret, the data clearly suggest a significant effect on apparent solubility or dissolution rates.

Morris and Fletcher (1987) suggest that these effects on quartz reactivity are caused by surface reactions of iron with silica. Figure 31b shows their proposed multistep reaction of ferrous iron with the quartz surface, exposure to oxidizing conditions, then re-exposure to reducing conditions. If the effects are real, they could also be important to understanding silica mobility in a range of modern iron-rich settings where soil and groundwaters undergo alternating reducing and oxidizing conditions.

These qualitative data suggest that iron sorption and armoring processes have profound effects on the reactivity and behavior of all silica phases. They may exert an important influence on microbiological processes in weathering environments. For example, iron coatings may influence the attachment of bacterial colonies to SiO₂ surfaces (Scholl, 1989; Mills et al., 1994) and the weathering and development of secondary porosity in some iron-cemented sandstones. Biological contributions to the dissimilatory reduction of Fe(III) compounds likely complicate the prediction of weathering rates of iron coatings and cements beyond the abilities of kinetic and thermodynamically-based models alone. Recent investigations of live bacterial adhesion to silica surfaces using in situ Fluid Tapping Atomic Force Microscopy shows that *S. putrifaciens* (a closely related member of the *Pseudomonas* genus) attach strongly to iron-coated silica surfaces compared to the uncoated silica controls (Grantham and Dove, 1995). These findings suggest the primary

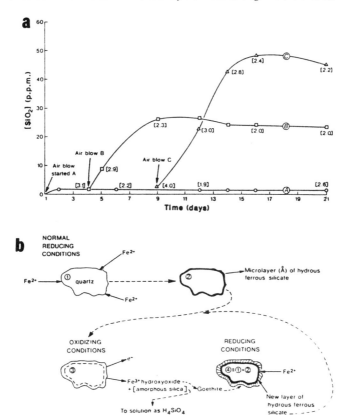

Figure 31. Graph shows silica accumulation during quartz dissolution at 60°C in solutions containing ferrous iron and an initial pH of 6.6. The 'A' indicates aeration at the start of the experiment, B, aeration after 4 days of reaction and C, aeration after 9 days of reaction. pH values are shown in square brackets. Lower figure illustrates hypothesis for the reaction of iron with quartz surfaces: (1) quartz reacts with ferrous iron in ground water to form a hydrous ferrous iron/silica surface layer, (2) layer breaks down under oxidizing conditions to a goethite precursor, (3) release of silica to solution (from Morris and Fletcher, 1987, used with permission).

role that coatings must play in governing bacterial adhesion and transport through silica-rich groundwater and soil weathering environments.

Coatings on silica surfaces in natural settings. Organic and inorganic coatings may form on silica surfaces in weathering environments. The organic coatings may be forms of natural organic matter sorbed from solution or biological tissues surrounding biomineralized particles. Earlier discussions in this review showed that organic acids can catalyze dissolution through ligand complexation reactions with surface groups (e.g. Jones et al., 1981). However, amorphous or crystalline 'rinds' of sorbed or precipitated coatings may also indirectly retard silica reactivity by isolating silica surfaces from the aqueous environment. Thick coatings of natural organic matter are widespread on mineral surfaces in natural environments (Hunter and Liss, 1979). Anbeek (1993) suggests that these organics may contribute to the slower rates of dissolution observed in field settings compared to the laboratory. In studies of organic carbon accumulation in continental shelf sediments, Mayer (1994) shows evidence that organic matter sorption is sufficiently pervasive to be sorbed into small pores with diameters <10nm. Such extensive coatings may limit the available reactive surface area of amorphous or crystalline silica phases. However, dissolution rate measurements of H_2O_2-washed biogenic silica show that the formation of protective organic coatings do not explain observed reductions in biogenic silica reactivity (Van Cappellen and Qui, 1995b). These findings suggest that inorganic controls on dissolution such as aluminum sorption are more significant.

Figure 3 shows that biological tissues of silica biominerals is another type of organic matter coating. Bartoli and Wilding (1980) suggested that organic matter degradation of encasing grass vegetative material must retard the early stage dissolution rate of opaline phytoliths.

Inorganic coatings are also widespread in soils and sediments. Heald and Larese (1974) noted that coatings of chlorite, illite, hematite, or carbonate often form on quartz grains in soils and sediments. Iron coatings may also form on silica surfaces. Wang et al. (1993) used improved separation methods, to find that iron-rich layers on some soil mineral particles may be comprised of microcrystalline particles of ferrihydrite, lepidocrocite and goethite. It is also possible for near-surface silica to react with pore water constituents to precipitate a surface layer. In studies of saline waters, amorphous silica combined with magnesium-rich solutions forms a Mg-hydroxysilicate resembling sepiolite (Kent and Kastner, 1985).

Surface and mechanistic controls on reactivity

Numerous experimental and theoretical studies of silica polymorph reactivity in aqueous solutions have lead to some consensus on the mechanism of silica nucleation, growth and dissolution. This discussion of mechanisms draws from the earlier description of the silica-water interface to emphasize the roles of mineral surface charge and coordination chemistry in reactivity, surface-solvent interactions and reaction processes.

Development of surface complexation models of dissolution. The work of Wieland et al. (1988) established surface coordination chemistry as a basis for correlating the surface-controlled dissolution kinetics of many oxides and silicate minerals and materials with observed pH-dependence of dissolution (also see discussion in Sparks, 1989). They proposed that surface-controlled dissolution can be represented by the simplified multistep reactions

$$>M + L(aq) = >ML \tag{47}$$

$$>ML = >ML^{\ddagger} \tag{48}$$

$$>ML^{\ddagger} = ML(aq) \tag{49}$$

and the dissolution rate, r, is generally proportional to the surface concentration of the precursor site, [>ML], such that

$$r = k\,[>ML]. \tag{50}$$

This approach produces a more satisfactory description of the pH dependence of mineral dissolution than rate laws based on proton activities in solution such that

$$r = k\,[H^+]^n \tag{51}$$

where r and k are the experimental dissolution rate (mol m^{-2} s^{-1}) and rate constant, respectively. The rate constant typically exhibits a fractional dependence upon reaction order, n. Wieland et al. (1988) predicted that surface complexation models would "prove to be essential tools for the quantitative description of reactive surface species in dissolution". While equilibrium surface complexation models are finding success in describing the solution-composition dependence of silica (and other mineral) dissolution kinetics, it is important to keep in mind that these rate equations employ simple correlations between equilibrium surface complexes and do not represent actual stoichiometries.

The first applications of surface complexation concepts to describing silica reactivity predates Wieland et al. (1988). A study by Wirth and Gieskes (1979) found that the logarithm of the dissolution rate of vitreous silica in solutions having a pH range of 5 to 11 correlated linearly with the logarithm of the net negative surface excess charge (as [$>OH^-$]) with a slope of 1.94, suggesting a second order dependence of reaction. They also showed that different dissolution rates measured for solutions containing different sodium concentrations, but similar pH, could be condensed into the single trend shown in Figure 26b. Similar findings were reported by Fleming (1986) who measured the pH and salt dependence of amorphous silica (Ludox) polymerization kinetics. Rates of aqueous silica condensation were described in terms of the surface concentration of negatively charged $>SiO^-$ complexes:

$$r \propto k_o A [>SiO^-]C \tag{52}$$

where r is given by the rate constant, k_o, in mol sec^{-1}, surface area of reactant, A, surface concentration of ionized surface sites, [$>SiO^-$] and reaction affinity, A.

Brady and Walther (1990) showed that quartz dissolution rates also correlate with negative surface charge. They related surface speciation at the quartz-solution interface to the kinetics of dissolution as a function of solution pH and salt concentration to write a rate equation as a sum of near-neutral pH (assumed proportional to the detachment rate of $>SiOH$) and basic pH mechanisms:

$$r = k_n [>SiOH] + k_b [>SiO^-] \tag{53}$$

where k_i is the dissolution rate constant and surface concentrations are given as the number of species per unit area. Like Wirth and Gieskes (1979) and Bohlmann et al. (1980), the data suggest that the reaction order approaches two at higher solution pH.

Similar rate expression based on multiple activated complex (MUSIC and MAC) models include a variety of activated complexes (Hiemstra and van Riemsdijk, 1989a,b; Hiemstra and van Riemsdijk, 1990). These complexes are hypothesized as silica species that were singly, doubly and triply bound to the surface. Using their approach, only two protonation Reactions (16) and (18) are required to describe reactions occurring over the pH 0 to 14 range. This gives the rate equation

$$r = k_{r1} \{[>SiO^-]^n + [>SiOH_2^+]^n\} + k_{r2} [>SiOH^0]^n. \tag{54}$$

By setting the reaction order, n, to assume first or second order behavior, they fitted the reaction rate constant. They found that different reaction orders for [$>SiO^-$] were required to fit rate data from different investigators. For example, the Knauss and Wolery (1988) and Wollast and Chou (1988) quartz dissolution rate data fit Equation (54) if they assumed second or first order behavior, respectively. Dove and Elston (1992) also found that a single reaction order could not describe the reported rates of multiple investigators. They use a surface reaction model similar to Hiemstra and van Riemsdijk (1990) to suggest that dissolution rates have both a first and second order dependence upon $\theta_{>SiOtot}$, the fractional sum of $>SiO^-$ and $>SiO-Na^+$ complexes.

These studies investigated dissolution rate behavior and surface charge dependence over numerous small ranges of solution composition and/or temperature. A single comprehensive model describing all of these effects (Dove, 1994) combined the rate data from these individual studies with new rate data and supporting evidence from the surface chemistry literature to extend these surface reaction models (see earlier section) to obtain:

$$r = k_{>SiOH} \left(\theta_{>SiOH}\right)^{n_{>SiOH}} + k_{>SiO^-_{tot}} \left(\theta_{>SiO^-_{tot}}\right)^{n_{>SiO^-_{tot}}} \tag{55}$$

where

k_i = rate constant for the reaction of surface species i,

n_i = reaction order of surface species i.

$\theta_{>SiOH}$ = fraction of total surface sites occupied by hydrogen ion as >SiOH,

$\theta_{>SiO^-_{tot}}$ = sum of the fractions of total sites existing as deprotonated >SiO⁻ site and as a complex with sodium ion interaction as >SiO-Na⁺.

The temperature dependence of Equation (55) is introduced by expanding the rate constants, k_i, into their components per Equation (39) for mineral-solution reactions. It is not straightforward to determine the temperature dependence of the hydrothermal kinetic data because the mineral surface complex distributions (i.e. $\theta_{>SiOH}$) cannot be directly measured. Existing surface complexation models have been developed from titration experiments conducted at 25°C and so the behavior of these complexes is still largely unknown at hydrothermal temperatures. Therefore, Dove (1994) estimated surface site complex distributions at higher temperatures by assuming that:

- total surface site number remains is independent of temperature,
- distributions of surface complexes are primarily controlled by the relative magnitudes of the surface association constants,
- these distributions are, in turn, primarily controlled by changes in the dissociation constant of water with temperature (Lyklema, 1987; Fokkink et al., 1989; Blesa et al., 1990; Brady, 1992),
- therefore, site *distributions* at in situ pH (pH_T) remain approximately constant.

Thus, equilibrium surface complex distributions can be estimated for the temperature of each experiment by using pH_T and the association constants for Equations (16) and (17).

The regression fit of rate data and the corresponding site distributions for each experiment listed in Dove (1994) obtained estimates of five parameters for the expression

$$r = \exp^{-10.7 \pm 1.1} T \, \exp\left(\frac{-66.0 \pm 2.3}{10^{-3}RT}\right)\left(\theta_{>SiOH}\right)^1$$
$$+ \exp^{4.7 \pm 0.8} T \, \exp\left(\frac{-82.7 \pm 2.1}{10^{-3}RT}\right)\left(\theta_{>SiO^-_{tot}}\right)^{1.1 \pm 0.06} \tag{56}$$

where the dissolution rate in mol m⁻² s⁻¹. The '±' values give two standard errors of each estimate. The reaction order associated with $\theta_{>SiOH}$, was indeterminate from the fitting procedure but inferred to equal one from investigations of quartz dissolution rates and probable reaction mechanisms (Lasaga and Gibbs, 1990; Berger et al., 1994). Equation (56) describes quartz dissolution rates from 25° to 300°C for pH 2 to 12, variable ionic strength, and sodium concentrations from 0 to as high as 0.5 molal. The first and second terms in Equation (56) contributes to total reaction rate at the solution composition where the corresponding surface complexes are predominant. That is, the first term largely describes rates at acidic solution pH near the ZPNPC. The second term describes dissolution rates in higher pH solutions and the contribution of alkali cations to reactivity.

Although Equation (56) is an empirical description of quartz dissolution kinetics, it has properties suggesting robustness. First, a comparison of measured and predicted rates for seven temperatures illustrates the general good quality of this expression over reaction rates spanning a factor of 10^{11} (Dove, 1994). The model predicts dissolution rates in variable solution compositions and temperatures including the numerous data at 25°C and extends to give dissolution rates comparable to estimates at 385° to 430°C by Murphy and Helgeson (1989). Second, the reaction order associated with the >SiO$^-_{tot}$ term is near one implying first order behavior. This near-integer value suggests physical meaning. Third, estimates of apparent enthalpy and entropy associated with the higher pH conditions (where the >SiO$^-_{tot}$ term dominates rates) give good agreement with other investigations (see later discussion). Finally, the good quality of fit is also manifested by the model's ability to predict large increases in dissolution rate with the introduction of sodium ion to solutions having near-neutral pH and the smaller effect at higher pH. Figure 28 illustrates the relationship between sodium chloride concentration and the dissolution rate of quartz at near-neutral pH. Plotting the rate predicted by Equation (56) for two temperatures as surfaces, the ability of sodium to increase dissolution rates at near-neutral pH is temperature dependent. This agrees with observations that sodium has a small effect on quartz dissolution rate at 25°C compared to hydrothermal temperatures (see earlier section).

The terms in Equation (56) describe the *combined* reactions leading to and including the formation of an activated complex and so, is necessarily empirical. It gives no indication of reaction stoichiometry, water properties, attachment geometry or specific site chemistry for individual rate-limiting reactions. These processes must be evaluated with evidence external to the dissolution rate data. While this method is still an indirect approach to describing the detailed reactivity of surface bonds, it is an improvement over rate models which correlate reaction rates with bulk solution composition parameters (i.e. pH) or bulk surface charge. Ideally, quartz reactivity could be related to detailed changes in electronic structure of mineral surfaces. However, this is not yet possible. As a beginning, this representation relates reaction rates to modeled surface complexes. The end result quantifies silica dissolution rates in solutions of variable pH and sodium concentration and provides insight on mechanisms of solvent-surface controls on reactivity .

Solvent-surface controls on reactivity. Silica polymorph dissolution and precipitation rates are determined by the mutual interactions of solvent and mineral surface in hydration-dehydration reactions. Surface charge models of reactivity appear to be constructs that describe the time-averaged state of near-surface solvent and its interaction with the solid-state mineral structures. These changes result in the widely observed correlation of solution pH and solute compositions with quartz (and other mineral) dissolution rates. Three lines of evidence from dissolution studies suggest that silica reactivity is controlled by solvent-surface interactions:

(1) Dissolution rates mimic net surface charge and solvent behavior at interface. At the slow reaction rates associated with a low solution pH (see Fig. 15), surface charge and potential are small and thus have little influence on the structure of water near the surface compared to bulk water (Conway, 1981). Under these conditions, hydrogen bonding among the water molecules themselves and between water molecules and >SiOH groups at the surface control structure (Eisenberg and Kauzmann, 1969; Tait and Franks, 1971; Klier and Zettlemoyer, 1977). This view of the interface is supported by dielectric constant measurements of water at low field strengths that are approximately equal to the bulk value of 78 (Davies and Rideal, 1963). Extensive intermolecular hydrogen bonding results in a relatively large proportion of molecular water adjacent to the surface (Bérubé and deBruyn, 1968; Zhu and Robinson, 1991) and is the primary species available for

reaction. The weak nucleophilic properties of molecular water are manifested by the relatively slow dissolution rate of quartz in low pH solutions (Casey et al., 1990). Dissolution in this environment is associated with a low entropy change as indicated by the low value of the pre-exponential in the first term of Equation (56) and a corresponding low $\Delta S_{xp,>SiOH}$ of -244 J mol^{-1} K^{-1}. Low values of ΔS_{xp} are also reported by House and Hickenbotham (1992) and Gratz and Bird (1993). Lasaga and Gibbs (1990) proposed that in this environment, the electronegative oxygen of water is the primary reactant in a mechanism where the interaction of oxygen with a surface silicon atom forms a silicon-oxygen complex having five-fold coordination with oxygen. At very acidic (positive surface charge) and ZPNPC (net neutral surface charge) solution pH, this mechanism may dominate by reaction of the electronegative oxygen of the water dipole with the quartz surface (see Fig. 12). If true, this may also explain the lower observed $\Delta H_{xp,>SiOH}$ of 66 kJ mol^{-1} (Dove, 1994) since little additional energy of water dissociation is involved.

At higher solution pH, the accumulated negative charge (see Fig. 12) results in increased polarization of the solvent. The resulting net-negative charge affects near-surface water structure to induce a locally different pH at the mineral-solution interface (e.g. Dove, 1994). This increased ionization of water is confirmed by dielectric measurements of water at high field strengths showing a decreased dielectric constant to values near eight (Davies and Rideal, 1963; also see Toney et al., 1994). This environment is also suggested by the larger pre-exponential term, A, shown in Equation (56). The increase in preexponential from the first to second terms is indicative of a change in the reaction mechanism and/or the properties of the surface-solution interface. Equation (39) shows that increases in the pre-exponential must be derived from increases in $\Pi\gamma_i$, C_S, X_{H_2O}, or ΔS. Although individual contributions of these terms cannot be evaluated from dissolution rate data, it appears that only ΔS can make significant contributions to the larger $A_{>SiO_{tot}^-}$. This change in ΔS or $A_{>SiO_{tot}^-}$ may reflect the effect of high interfacial charge gradients in promoting reorientation/dissociation of water in the interfacial environment or by affecting rates of near-surface solvent exchange (e.g. discussion of Figs. 12 and 15).

(2) Diminishing cation-specific effects observed at higher solution pH. At near-neutral pH, cation-specific effects on dissolution rate were observed by Dove and Crerar (1990) where rates increased by a factor of 6 to 25 from solutions containing lithium or magnesium to sodium or potassium. In contrast, rates measured by Gratz et al. (1990) conducted at pH 12 found that dissolution rates in solutions containing lithium, sodium, or potassium are similar within a factor of two. This difference is consistent with a solvent-surface interaction model since at conditions of higher surface charge, or where solution pH is greater than near-neutral, transition to faster dissolution mechanism is nearly complete, and the second term of Equation (56) controls dissolution rates at these conditions (Dove, 1994). Thus, cation specific effects are expected to diminish at higher solution pH because differences between the hydration properties of cations impose little additional effect on water striction and polarization to promote further the hydroxyl dominated reaction (e.g. discussion of Fig. 12 and 15).

One proposed mechanism for a hydroxide-surface interaction is the oriented sorption of the proton of a hydroxyl group onto the bridging oxygen of a >Si-O-Si< surface group. Experimental evidence for the reactivity of a bridging oxygen at silica surfaces has been reported (Morrow and Cody, 1975; Morrow and Cody, 1976a,b; Foley, 1986; Gallei and Parks, 1972). Molecular orbital and *ab initio* calculations of silicon-oxygen bonded analogs estimate that a bridging oxygen bears a net negative charge ranging from -0.7 to -0.9 esu (deJong and Brown, 1980; Geerlings et al., 1984; Foley, 1986) and is sensitive to adsorbing molecules (Mortier et al., 1984). Susceptibility of negative charge at the bridging oxygen to a hydrogen ion in proximity to the surface at a charged interface also suggests that hydrolysis could involve proton or oriented hydroxyl sorption at the bridging oxygen. *Ab initio* calculations indicate this process is accompanied by a net flow of charge from the bridging oxygen to the proton (Ugliengo et al., 1990). Following this step, redistribution of electronic charge may permit reaction of oxygen (as hydroxide?) with the adjacent silicon atom to form a five-coordinated transition state complex (see Lasaga, this volume). Casey et al. (1990) also called for early transfer of hydrogen to a bridging oxygen, followed by nucleophilic attack of silicon by hydroxyl ion. These mechanisms may involve higher net energies as water dissociation is folded into the apparent enthalpy (and entropy). This is consistent with the larger observed $\Delta H_{>SiO^-tot}$ discussed earlier.

(3) Dissolution rates correlate with $\Delta G_{hy, \, cation}$. Alkali cations appear to indirectly enhancing quartz dissolution rates by affecting solvent properties to promote a faster reaction mechanism involving hydroxyl ion. Fokkink et al. (1990) suggested that cadmium ions (having similar behavior to alkali cations) are 'stuck' in the interfacial region in a passive way by their waters of hydration and weak attraction for the surface. This 'ion-pair' surface-interaction results in partial release and polarization of low-entropy waters of hydration to affect water dissociation to a degree proportional to the free energy of hydration, ΔG_{hy}, of each alkali cation (see earlier discussion and Fig. 12). Alkali cations added to solutions at near-neutral pH may behave similarly as hydrated cations in the negatively charged interface by their weak association with the surface. At pH conditions where interfacial charge is sensitive to the addition of alkali (see Figs. 12 and 15) dissolution rates should be sensitive to the hydration properties of individual alkali ions. This relationship is again observed for cation-specific quartz dissolution rates. Figure 32 shows that rates of quartz dissolution in solutions containing a variety of alkali cations

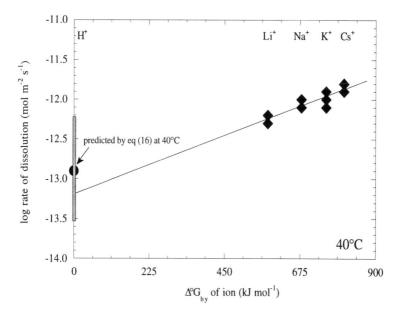

Figure 32. Measured rates of quartz dissolution in near-neutral pH solutions containing alkali metal cations correlate with the standard free energy of hydration of the ion. This trend extrapolates to the dissolution rate of quartz in the absence of cations by predicting a rate of $10^{-13.2}$ mol m^{-2} s^{-1} for pure water. Shaded area indicates the 95% confidence interval for predicted rate in deionized water (from Dove, 1994).

correlate with ΔG_{hy} of the corresponding cation. Extrapolating this trend to a zero ΔG_{hy} (standard state for proton) obtains $10^{-13.2}$ mol m^{-2} s^{-1}. This value is close to the predicted rate of $10^{-12.9\pm0.7}$ mol m^{-2} s^{-1} at 40°C for pure water (Dove, 1994) and also near to 40°C rates measured by Bennett (1991), Tester et al. (1994) and predicted by Rimstidt and Barnes (1980) and Dove and Elston (1992).

Theoretical mechanistic models. This topic is reviewed elsewhere and the reader is referred to Lasaga (this volume) as well as treatments in Lasaga and Gibbs (1990) and Lasaga (1990).

REGIONAL-SCALE WEATHERING: SILICEOUS FRACTURE KINETICS

Another aspect of silica weathering kinetics is the water-promoted degradation of quartzose host rocks by tensile fracture. While this might be viewed as a highly specific form of terrestrial weathering, Figure 1 shows that fracture is an early stage contributor to the decomposition of host rock along the pathway of converting siliceous rocks to clastic constituents. At regional scales, fractures can be an important control on the development of structural landforms (e.g. Scheidegger, 1991) over the long term of weathering processes. This phenomena has important environmental implications since long-range groundwater transport patterns can be dominated by the structure and arrangement of local and regional-scale joint sets (e.g. Segall and Pollard, 1983; Evans and Nicholson, 1987; Pollard and Aydin, 1993; Olson, 1994).

At tensile stresses far below critical values, cracks may form and grow in crystalline and glassy materials. These fractures propagate at velocities ranging from 10^{-10} to 10^{-1} m s^{-1} (Anderson and Grew, 1977; Atkinson, 1984; Segall, 1984; Kirby, 1984) in a processes called subcritical fracture growth (e.g. Lawn and Wilshaw, 1975; Atkinson and Meredith, 1981; Scholz, 1990; Lawn, 1993). For materials subjected to saturated and/or aggressive fluid environments over the long-term loadings found in the Earth's crust, the dominant mechanism for fracture growth may be chemical stress corrosion or dissolution (e.g. Renshaw and Pollard, 1994; Kronenberg, 1994). Many studies have documented that the kinetics of fracture growth are influenced by the chemical composition of fluids in contact with quartz (Atkinson and Meredith, 1981) and silica glass (White et al., 1987; Weiderhorn and Johnson, 1973). Atkinson (1984) suggested that processes that govern subcritical crack growth of quartz include stress corrosion, diffusion, ion exchange, microplasticity and dissolution. Of these, subcritical fracture processes approximate dissolution through the relatively slow rupture of chemical bonds (e.g. Atkinson and Meredith, 1981) at rates that are not limited by the flow of fluid from the material into the fracture void space (Renshaw and Pollard, 1994).

To better understand the interactions of water with quartz in stress environments, a kinetic model links physical and chemical controls on the subcritical fracture kinetics of quartz (Dove, in press). It is based upon the assumption that molecular level reactions governing fracture and dissolution proceed by similar pathways. The model formulation combines fracture theory with a mechanistically-based description of chemical and thermal controls on mineral dissolution kinetics to quantify the effect of chemical, thermal, and tensile stresses on reactivity in aqueous environments. The fracture model states that water, as a vapor or liquid, promotes rupture of Si–O bonds by endmember processes: (1) reaction of a protonated surface with molecular water and (2) reaction of hydroxyl ions at an ionized surface. In humid environments, reaction frequency is determined by water accessibility to the crack tip. In wetted environments, the relative contributions of these mechanisms are determined by bulk solution composition which affects surface ionization and solvent-surface interactions. Macroscopic fracture rate, r_{Si-O} (m sec^{-1}), is given by the fractional sum of these endmember reaction mechanisms per the first order equation:

$$r_{Si-O} = 10^{-3.95}T \exp\left(\frac{-10^{-4.82}}{RT}\right) \exp\left(10^{-4.57}K_I\right)\left(\theta_{Si-O}^{H_2O}\right)^1$$

$$+ 10^{3.40}T \exp\left(\frac{-10^{4.92}}{RT}\right) \exp\left(10^{-3.74}K_I\right)\left(\theta_{Si-O}^{OH^-}\right)^1 \tag{57}$$

from temperature (T, K), stress intensity (K_I, Nm$^{-3/2}$), and fractional contribution (θ_{Si-O}^i) of endmember reactions. The agreement of this empirical rate expression with reported measurements of quartz fracture rates suggests the model is robust. Equation (57) was obtained by fitting the first and second terms of the rate model to fracture rate data shown in Figure 33a for experiments in acidic and alkaline solutions, respectively. Figure 33b shows that Equation (57) gives a good fit to fracture rates in deionzed water over six orders of magnitude. The model also explains the (1) pH-dependence of fracture rates measured in aqueous solutions, (2) absence of cation-specific rates at high pH, (3) dependence of fracture rate upon crystallographic direction, and (4) thermal-dependence of rate over 20° to 80°C. Results of the model indicate that common chemical constituents in groundwater may affect silica reactivity in near-surface and deeper earth environments, but this has not been investigated in experimental studies. This quartz fracture rate model may lead to a better understanding of the time-dependent degradation of rock and glassy materials in contact with the fluids of near-surface weathering environments.

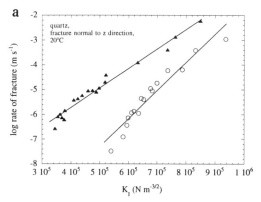

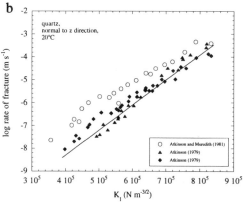

Figure 33. (a) Quartz fracture rate data for experiments at 20°C in 2 M HCl and 2 M NaOH show that steepest slopes are associated with the lowest pH solution (data from Atkinson and Meredith, 1981). Solid lines are obtained by fitting rate measurements in low and high pH solutions to the first (reaction by molecular water) and second (reaction by hydroxyl ion) terms to Equation (57), respectively. (b) The resulting rate expression gives good agreement (solid line) with data from Atkinson (1979) for fracture rates in deionized water at 20°C. It is not known if quartz fracture rates of neutral pH solutions are further modified by the presence of dissolved salts inherent to natural hydrologic fluids per the effects shown in Figure 27 (from Dove, in press).

CONCLUDING REMARKS

The translocation of silica between reservoirs in terrestrial weathering environments is influenced by complex inorganic and organic controls on reactivity of SiO_2 polymorphs. While progress has been made in quantifying processes of dissolution and precipitation, much remains to be understood. There is a continuing and urgent need for quantitative experimental and field studies as quantified relationships for all but the most simple mineral-fluid systems are still unknown. Figure 1 showed that an understanding of precipitation phenomena is essential to estimating silica movement between reservoirs. However, the kinetics of silica nucleation and growth are poorly quantified for virtually all of the crystalline and amorphous silica polymorphs. This is particularly true at the low temperatures of most weathering environments. Studies quantifying the effect of temperature, solution pH, and simple salts on dissolution rates of quartz suggest that pH and cation-dependent dissolution trends hold for all of the silica polymorphs. Yet absolute rate constants remain undetermined for the non-quartz polymorphs, even though they occur extensively in weathering environments and have greater reactivities than quartz. Finally, there is much need for understanding the controls of sorbed organic and inorganic constituents on silica reactivity. These and other gaps in our quantitative understanding will continue to limit our ability to compare laboratory and field measured mineral reactivities (see Sverdrup and Warfvinge, this volume). Quantitative research in these and other areas is necessary to fully characterize silica translocation and better understand mineralogical and geochemical factors governing the reactivity of their silicate relatives in weathering environments.

ACKNOWLEDGMENTS

I thank Don Rimstidt for allowing me to reproduce some of the materials developed for our recent review, *Silica-Water Interactions*, and thank Philippe Van Cappellen for helpful discussion of the terrestrial silica cycle. Susan Brantley and Art White provided comments on the final draft. Whitfield Roberts greatly assisted in the literature search and figure preparation. This work was supported by National Science Foundation grant EAR-9405362.

REFERENCES

Alexander GB (1957) The effect of particle size on the solubility of amorphous silica in water. J Phys Chem 61:1563-1564
Allen LH, Matijevic E (1971) Stability of colloidal silica. III. Effect of hydrolyzable cations. J Coll Interf Sci 35:66-76
Alvarez R, Sparks DL (1985) Polymerization of silicate anions in solutions at low concentrations. Nature 318:649-651
Anbeek C (1993) The effect of natural weathering on dissolution rates. Geochim Cosmochim Acta 57:4963-4975.
Anbeek C, van Breemen N, Meijer EL, van der Plas L (1994) The dissolution of naturally weathered feldspar and quartz. Geochim Cosmochim Acta 58:4601-4613
Anderson OL, Grew PC (1977) Stress corrosion theory of crack propagation with applications to geophysics. Rev Geophys Space Phys 15:77-103
Anderson JH, Wickersheim KA (1964) Near infrared characterization of water and hydroxyl groups on silica surfaces. Surface Sci 2:252-260
Anderson JH, Parks GA (1968) The electrical conductivity of silica gel in the presence of adsorbed water. J Phys Chem 72:3662-3668
Applin KR (1987) The diffusion of dissolved silica in dilute aqueous solutions. Geochim Cosmochim Acta 51:2147-2151
Armington AF, Larkin JJ (1985) The growth of high purity, low dislocation quartz. J Cryst Growth 71:799-802
Asumadu K, Gilkes RJ, Armitage TM, Churchward HM (1988) The effects of chemical weathering on the

morphology and strength of quartz grains- an example from S.W. Australia. J Soil Sci 39:375-383

Atkins PW (1994) Physical Chemistry. Freeman, New York

Atkinson BK (1984) Subcritical crack growth in geological materials. J Geophys Res 89:4077-4114

Atkinson BK, Meredith PG (1981) Stress corrosion cracking of quartz: a note on the influence of chemical environment. Tectonophysics 77:T1-T11

Ballou EV, Leban MI, Wydeven T (1973) Solute rejection by porous glass membranes III. Reduced silica dissolution and prolonged hyperfiltration service with feed additives. J Appl Chem Biotech 23:119-130

Bartoli F, Wilding LP (1980) Dissolution of biogenic opal as a function of its physical and chemical properties. Soil Sci Soc Am J 44:873-878

Bartoli F (1985) Crystallochemistry and surface properties of biogenic opal. J Soil Sci 36:335-350

Baumann H. (1955) Über die auflösung von SiO_2 in wasser. Beiträge zur Silikose Forsch 37:47-71

Beavers AH, Stephen I (1958) Some features of the distribution of plant-opal in Illinois soils. Soil Sci 86:1-5

Beckwith RS, Reeve R (1963a) Studies on soluble silica in soils I. The sorption of silicic acid by soils and minerals. Aust J Soil Res 1:157-168

Beckwith RS, Reeve R (1963b) Studies on soluble silica in soils II. The release of monsilicic acid from soils. Aust J Soil Res 2:33-45.

Bennett PC, Melcer ME, Siegel DI, Hassett JP (1988) The dissolution of quartz in dilute aqueous solutions of organic acids at 25°C. Geochim Cosmochim Acta 52:1521-1530.

Bennett PC (1991) Quartz dissolution in organic-rich aqueous systems. Geochim Cosmochim Acta 55: 1781-1797

Bennett PC, Siegel DI (1987) Increased solubility of quartz in water due to complexing by organic compounds. Nature 6114:684-686

Bennett PC, Siegel DI, Hill BM, Glaser PH (1987) Fate of silicate minerals in a peat bog. Geology 19:328-331

Berger G, Cadore E, Schott J,Dove, PM (1994) Dissolution rate of quartz in lead and sodium electrolyte solutions between 25 and 300°C: Effect of the nature of surface complexes and reaction affinity. Geochim Cosmochim Acta 58:541-551

Bérubé YG, deBruyn PL (1968) Adsorption at the rutile-solution interface. J Coll Interf Sci 28:92-105

Birchall JD (1994) Silicon-aluminum interactions and biology. In: Bergna HE (ed) The Colloid Chemistry of Silica. Adv Chem Series 234:601-616, Am Chem Soc, Washington, DC

Bird G, Boon J, Stone T (1986) Silica transport during steam injection into oil sands 1. Dissolution and precipitation kinetics of quartz: New results and review of existing data. Chem Geol 54:69-80

Birkeland PW (1984) Soils and Geomorphology. Oxford Univ Press, New York

Blesa MA. Maroto AJG, Regazzoni AE (1990) Surface acidity of metal oxides immersed in water: A critical analysis of thermodynamic data. J Coll Interf Sci 140:287-290

Bohlmann EG, Mesmer RE, Berlinski P (1980) Kinetics of silica deposition from simulated geothermal brines. Soc Pet Eng J 20:239-248

Brady PV, Walther JV (1990) Kinetics of quartz dissolution at low temperatures. Chem Geol 82:253-264

Brady PV (1992) Silica surface chemistry at elevated temperatures. Geochim Cosmochim Acta 56:2941-2946

Brantley SL, Crane SR, Crerar DA, Hellman R, Stallard R (1986a) Dislocation etch pits in quartz. In: Davis JA, Hayes KF (eds) Geochemical Processes at Mineral Surfaces, Vol 323:635-649, Am Chem Soc, Washington, DC

Brantley SL, Crane SR, Crerar DA, Hellman R, Stallard R (1986b) Dissolution at dislocation etch pits in quartz. Geochim Cosmochim Acta 50:2349-2361

Brown DW, Hem, JD (1975) Reactions of aqueous aluminum species at mineral surfaces. U.S. Geological Survey Water Supply Paper 1827-F

Cary, L. W., deJong, B. H. W. S. and Dibble Jr., W. E. (1982) A 29Si NMR study of silica species in dilute aqueous solutions. Geochim Cosmochim Acta, 46, 1317-1320

Casey, W. H. (1987) Heterogeneous Kinetics and Diffusion Boundary Layers: The Example of Reaction in a Fracture. J Geophysical Res, 92, 8007-8013

Casey, W. H., Lasaga, A. C. and Gibbs, G. V. (1990) Mechanisms of silica dissolution as inferred from the kinetic isotope effect. Geochim Cosmochim Acta, 54, 3369-3378.

Churaev NV, Sobolev VD, Zheleznyi BV (1972) Thermal expansion of water in thin quartz capillaries. Russian J Phys Chem 46:1320-1322

Cleary WJ, Conolly JR (1972) Embayed quartz grains in soils and their significance. J Sed Petrology 42:899-904

Conway, B. E. (1981) Ionic Hydration in Chemistry and Biophysics. Elsevier Scientific, Amsterdam, 773 p

Corwin, J. F. and Swinnerton, A. C. (1951) The growth of quartz in alkali halide solutions. J Am Chem Soc 73:3598-3601

Crook KAW (1968) Weathering and roundness of quartz sand grains. Sedimentology 11:171-182

Dana JD (1884) On the decay of quartzyte, and the formation of sand, kaolin and crystallized quartz. Am J Sci 28:448-452

Darmody RG (1985) Weathering assessment of quartz grains: A semiquantitative approach. Soil Sci Soc Am J 49:1322-1324

Davies JT, Rideal EK (1963) Interfacial Phenomena. Academic Press, New York

Davis, J. A., James, R. O. and Leckie, J. O. (1978) Surface ionization and complexation at the oxide/water interface. J Coll Interface Sci 63:480-499

Davis JA, Kent DB (1990) Surface complexation modeling in aqueous geochemistry. In: Hochella, Jr. MF, White, AF (eds) Mineral-water Interface Geochemistry. Rev Mineral 23:177-250

DeJong, BHWS, Brown Jr., GE (1980) Polymerization of silicate and aluminate tetrahedra in glasses, melts and aqueous solutons- II. The network modifying effects of Mg^{2+}, K^+, Na^+, Li^+, H^+, OH^-, F^-, Cl^-, H_2O, CO_2 and H_3O^+ on silicate polymers. Geochim Cosmochim Acta 44:1627-1642

Diénert F (1923) Sur le dosage de la silice dans les eaux. Comptes rendus des seances de l'academié des sciences 176:1478-1480

Dixit S, Dove PM, Roberts TW (1995) Controls on the surface reactivity of silica: Investigations of aluminum sorbates in weathering environments. Geol Soc Am Abstracts, New Orleans, in press

Dove PM (1994) The dissolution kinetics of quartz in sodium chloride solutions at 25° to 300°C. Am J Sci 294:665-712

Dove PM (1995) Geochemical controls on the kinetics of quartz fracture at subcritical tensile stresses. J Geophys Res, in press

Dove PM, Crerar DA (1990) Kinetics of quartz dissolution in electrolyte solutions using a hydrothermal mixed flow reactor. Geochim Cosmochim Acta 54:955-969

Dove PM, Elston SF (1992) Dissolution kinetics of quartz in sodium chloride solutions: Analysis of existing data and a rate model for 25°C. Geochim Cosmochim Acta 56:4147-4156

Dove PM, Krishnan V, Murovitz JR (1994) Controls on the surface reactivity of silica: The effect of aluminum on the kinetics of quartz dissolution. Geol Soc Am Abstr 26:A-110

Dove PM, Rimstidt JD (1994) Silica-water interactions. In: Heaney PJ, Prewitt CT, Gibbs GV (eds) Silica: Physical Behavior, Geochemistry, and Materials Applications. Rev Mineral 29:259-308

Drees LR, Wilding LP, Smeck NE and Senkayi AL (1989) Silica in soils: Quartz and disordered silica polymorphs. In: Dixon JB, Weed SB (eds) Minerals in Soil Environments, p 913-974, Soil Sci Soc Am, Madison, WI

Drost-Hansen W (1977) Effects of vicinal water on colloidal stability and sedimentation processes. J Coll Interf Sci 58:251-262

Du Q, Freysz E, Shen YR (1994) Vibrational spectra of water molecules at quartz/water interfaces. Phys Review Lett 72:238-241

Dugger DL, Stanton JH, Irby BN, McConnell BL, Cummings WW, Maatman, RW (1964) The exchange of twenty metal ions with the weakly acidic silanol group of silica gel. J Phys Chem 68:757-760

Eisenberg D Kauzmann W (1969) The Structure and Properties of Water. OxfordUniv Press, New York

Evans DD, Nicholson TJ (1987) Flow and transport through unsaturated fractured rock: An overview. In: Evans DD, Nicholson TJ (eds) Flow and Transport Through Unsaturated Fractured Rock. Monograph 42:1-10, Am Geophys Union, Washington, DC

Fleming BA, Crerar DA (1982) Silicic acid ionization and calculation of silica solubility at elevated temperature and pH. Geothermics 11:15-29

Fleming BA (1986) Kinetics of reaction between silicic acid and amorphous silica surfaces in NaCl solutions. J Coll Interf Sci 110:40-64

Fokkink LGJ, De Keizer A, Lyklema J (1989) Temperature dependence of the electrical double layer on oxides: rutile and hematite. J Coll Interf Sci 127:116-131

Fokkink LGJ, De Keizer, A, Lyklema J (1990) Temperature dependence of cadmium adsorption on oxides. J Coll Interf Sci 135:118-131

Foley JA (1986) Hydronium ion and water interactions with SiOSi, SiOAl, and AlOAl tetrahedral linkages. MS thesis, Virginia Polytechnic Institute & State University, Blacksburg

Folger DW, Burckle LH, Heezen BC (1967) Opal phytoliths in a North Atlantic dust fall. Science 155:1243-1244

Fournier RO, Rosenbauer RJ, Bischoff JL (1982) The solubility of quartz in aqueous sodium chloride solution at 350°C and 180 to 500 bars. Geochim Cosmochim Acta 46:1975-1978

Fournier RO (1983) A method of calculating quartz solubilities in aqueous sodium chloride solutions. Geochim Cosmochim Acta 47:579-586

Fripiat JJ, Jelli A, Poncelet G, Andre J (1965) Thermodynamic properties of adsorbed water molecules and electrical conduction in montmorillonites and silicas. J Phys Chem 69:2185-2197.

Gallei E, Parks GA (1972) Evidence for surface hydroxyl groups in attenuated total reflectance spectra of crystalline quartz. J Coll Interf Sci 38:650-651

Geerlings P, Tariel N, Botrel A, Lissilour R, Mortier WJ (1984) Interaction of surface hyroxyls with adsorbed molecules. A quantum-chemical study. J Phys Chem 88:5752-5759

Gíslason SR, Heaney PJ, Oelkers EH, Schott, J (1995) Quartz and quartz/moganite mixtures: Temperature dependence of dissolution rates and solubilities. VM Goldschmidt Conf Abstracts, p 49, State College, Pennsylvania

Goldich SS (1938) A study in rock weathering. J Geol 46:17-58

Graetsch H (1994) Structural characteristics of opaline and microcrystalline silica minerals. In: Heaney PJ, Prewitt CT and Gibbs GV (eds) Silica: Physical Behavior, Geochemistry, and Materials Applications. Rev Mineral 29:209-232

Grantham MC, Dove PM (1995) Biogeochemistry of quartz in weathering environments: Influence of iron oxyhydroxide coatings and cements on bacterial adhesion. Geol Soc Am Abstracts, New Orleans, in press

Gratz AJ, Bird P, Quiro GB (1990) Dissolution of quartz in aqueous basic solution, 106-236°C: Surface kinetics of "perfect" crystallographic faces. Geochim Cosmochim Acta 54:2911-2922

Gratz AJ, Manne S, Hansma PK (1991) Atomic force microscopy of atomic-scale ledges and etch pits formed during dissolution of quartz. Science 251:1343-1346.

Gratz AJ, Bird P (1993a) Quartz dissolution: Negative crystal experiments and a rate law. Geochim Cosmochim Acta 57:965-976

Gratz AJ, Bird P (1993b) Quartz dissolution: Theory of rough and smooth surfaces. Geochim Cosmochim Acta 57:977-989

Griggs DT (1936) The factor of fatigue in rock exfoliation. J Geol 44:783-796.

Gu J (1994) The dissolution rates of amorphous silica and opal-CT. MS thesis, Virginia Polytechnic Inst & State Univ, Blacksburg, VA

Hayes KF, Leckie JO (1987) Modeling ionic strength effects on cation adsorption at hydrous oxide/solution interfaces. J Coll Interf Sci 115:564-572

Harriss RC, Adams JS (1966) Geochemical and mineralogical studies on the weathering of granitic rocks. Am J Sci 264:146-173

Heald MT, Larese RE (1974) Influence of coatings on quartz cementation. J Sed Petrology 44:1269-1274

Heaney PJ, Post JE (1992) The widespread distribution of a novel silica polymorph in microcrystalline quartz varieties. Science 255: 441-443

Heaney PJ (1994) Structure and chemistry of the low-pressure silica polymorphs. In: Heaney PJ, Prewitt CT and Gibbs GV (eds) Silica: Physical Behavior, Geochemistry, and Materials Applications. Rev Mineral 29:1-40

Hem JD, Roberson CE, Lind CJ, Polzer WL (1973) Chemical interactions of aluminum with aqueous silica at 25°C. U.S. Geol Surv Water Supply Paper 1827-E

Hendershot WH and Lavkulich LM (1983) Effect of sesquioxide coatings on surface charge of standard mineral and soil samples. Soil Sci Soc Am J 47:1252-1260

Henderson JH, Syers JK, Jackson ML (1970) Quartz dissolution as influenced by pH and the presence of a disturbed surface layer. Israel J Chem 8:357-372

Hiemstra T, Van Riemsdijk WH, Bolt GH (1989a) Multisite proton adsorption modeling at the solid/solution interface of (hydr)oxides: A new approach I. Model description and evaluation of intrinsic reaction constants. J Coll Interf Sci 133:91-104

Hiemstra T, Van Riemsdijk WH, Bolt GH (1989b) Multisite Proton adsorption modeling at the solid/solution interface of (hydr)oxides: A new approach II. Application to various important (hydr)oxides. J Coll Interf Sci 133:105-117

Hiemstra T, Van Riemsdijk WH (1990) Multiple activated complex dissolution of metal(hydr)oxides: A thermodynamic approach to quartz. J Coll Interf Sci 136:132-150

Hosaka M, Taki S (1981a) Hydrothermal growth of quartz crystals in NaCl solutions. J Crystal Growth 52:837-842

Hosaka M, Taki S (1981b) Hydrothermal growth of quartz crystals in KCl solutions. J Crystal Growth 53:542-546

Hosaka M, Taki S (1986) Hydrothermal growth of quartz crystals at low fillings in NaCl and KCl solutions. J Crystal Growth 78:413-417

House, W. A. and Orr, D. R. (1992a) Investigation of the pH dependence of the kinetics of quartz dissolution at 25°C. J Chem. Soc. Faraday Trans., 88, no. 2, 233-241

House WA, Hickinbotham LA (1992b) Dissolution kinetics of silica between 5 and 35°C. J Chem Soc Faraday Trans 88:2021-2026

House WA (1994) The role of surface complexation in the dissolution kinetics of silica: Effects of monovalent and divalent ions at 25°C. J Coll Int Sci 163:379-390

Hsu PH (1988) Mechanisms of gibbsite crystallization from partially neutralized aluminum chloride solutions. Clays Clay Min 36:25-30

Hudson GA, Bacon FR (1958) Inhibition of alkaline attack on soda-lime glass. Ceram Bull 37:185-188

Hunter KA, Liss PS (1979) The surface charge of suspended particles in estuarine and coastal waters. Nature

282:823-825

Hurd DC (1972) Factors affecting the solution rate of biogenic opal in seawater. Earth Planet Sci Lett 15:411-417

Hurd DC (1973) Interactions of biogenic opal, sediment and seawater in the Central Equatorial Pacific. Geochim Cosmochim Acta 37:2257-2282

Hurd DC, Birdwhistell S (1983) On producing a more general model for biogenic silica dissolution. Am J Sci 283:1-28

Iler RK (1973) Effect of adsorbed alumina on the solubility of amorphous silica in water. J Coll Interf Sc 43:399-408

Iler R (1979) The Chemistry of Silica, John Wiley & Sons, New York

Israelachvili J (1992) Interfacial and Surface Forces, Academic Press, New York

James RO, Healy TW (1972) Adsorption of hydrolyzable metal ions at the oxide-water interface. I. Co(II) adsorption on SiO_2 and TiO_2 as model systems. J Coll Interf Sci 40:42-52

James RO, Wiese GR, Healy TW (1977) Charge reversal coagulation of colloidal dispersions by hydrolysable metal ions. J Coll Interf Sci 59:381-385

Jardine PM, Zelazny LW, Parker JC (1985a) Mechanisms of aluminum adsorption on clay minerals and peat. Soil Sci Soc Am J 49: 862-867

Jardine PM, Zelazny LW, Parker JC (1985b) Kinetics and mechanisms of aluminum adsorption on kaolinite using a two-site nonequilibrium transport model. Soil Sci Soc Am J 49:862-867

Jednacak J, Pravdic V (1974) The electrokinetic potential of glasses in aqueous electrolyte solutions. J Coll Interf Sci 49:16-23

Jones LHP, Handreck KA (1963) Effects of iron and aluminum oxides on silica in solution in soils. Nature 198:852-853

Jones JB, Segnit ER (1971) The nature of opal. I. Nomenclature and constituent phases. J Geol Soc Australia 18:419-422

Jones D, Wilson MJ, McHardy WJ (1981) Lichen weathering of rock-forming minerals: Application of scanning electron microscopy and microprobe analyis. J Microscopy 124:95-104

Kamiya H, Shimokata K (1976) The role of salts in the dissolution of powdered quartz. In: Cadek J, Paces T (eds) Proceedings of International Symposium of Water-rock Interaction, Czechoslovakian Geological Survey, Prague

Karathasis AD (1989) Solution chemistry of fragipans- Thermodynamic approach to understanding fragipan formation. In: Fragipans: Occurrence, Classification and Genesis, v 24, p 113-140, Soil Sci Soc Am, Madison, WI

Kastner M, Gieskes JM (1983) Opal-A to opal-CT transformation: A kinetic study. In: Iijima A, Hein JR, Siever R (eds) Siliceous Deposits in the Pacific Region, p 211-228, Elsevier, Amsterdam

Kastner M, Keene JB, Gieskes JM (1977) Diagenesis of siliceous oozes- I. Chemical controls on the rate of opal-A to opal-CT transformation- an expermimental study. Geochim Cosmochim Acta 41:1041-1059

Kent DB and M Kastner (1985) Mg^{2+} removal in the system Mg^{2+}-amorphous SiO_2-H_2O by adsorption and Mg-hydroxysilicate precipitation. Geochim Cosmochim Acta 49:1123-1136

Kent DB, Tripathi VS, Ball NB, Leckie JO, Siegel MD (1988) Surface-complexation modeling of radionuclide adsorption in subsurface environments: U.S. Nuclear Regulatory Comm Report NUREG/CR-4807:113, Sandia National Laboratories, Albuquerque, NM

Kenny MB, Sing KSW(1994) Adsoptive properties of porous silica. In: The Colloid Chemistry of Silica, Adv Chem Series 234:505-516, Am Chem Soc, Washington, DC

Kirby SH (1984) Introduction and digest for the special issue on chemical effect of water on the deformation and strengths of rocks. J Geophys Res 89:3991-3995

Klein C, Hurlbut Jr. CS (1993) Manual of Mineralogy. John Wiley, New York

Klier K, Zettlemoyer AC (1977) Water at interfaces: Molecular structure and dynamics. In: Kerker M, Zettlemoyer AC, Rowell RL (eds) Coll Interf Sci 1:231-244

Knauth LP (1994) Petrogenesis of chert. In: Heaney PJ, Prewitt CT, Gibbs GV (eds) Silica: Physical Behavior, Geochemistry, and Materials Applications. Rev Mineral 29:233-258

Knauss KG, Wolery TJ (1988) The dissolution kinetics of quartz as a function of pH and time at 70°C. Geochim Cosmochim Acta 52:43-53

Krauskopf KB (1956) Dissolution and precipitation of silica at low temperatures. Geochim Cosmochim Acta 10:1-26

Krauskopf KB (1959) The geochemistry of silica in sedimentary environments. In: Silica in Sediments. Spec Pub 7:4-19, Soc Econ Paleon Mineral, George Banta Co, Menasha, Wisconsin

Kronenberg AK (1994) Hydrogen speciation and chemical weakening of quartz. In: Heaney PJ, Prewitt CT, Gibbs GV (eds) Silica: Physical Behavior, Geochemistry, and Materials Applications. Rev Mineral 29:123-176

Langford-Smith T (1978) A select review of silcrete research in Australia. In: Langford-Smith (ed) Silcrete in Australia, p 1-11, Univ New England, Armidale, Australia

Lasaga AC (1981) Transition state theory. In: Lasaga AC, Kirkpatrick RJ (eds) Kinetics of Geochemical Processes. Rev Mineral 8:135-139

Lasaga AC, Blum AE (1986) Surface chemistry, etch pits and mineral-water reactions. Geochim Cosmochim Acta 50:2363-2379

Lasaga AC, Gibbs GV (1990) Ab-initio quantum mechanical calculations of water-rock interactions: Adsorption and hydrolysis reactions. Am J Sci 290:263-295

Lasaga AC (1990) Atomic treatment of mineral-water surface reactions. In: Hochella, Jr. MF, White, AF (eds) Mineral-Water Interface Geochemistry. Rev Mineral 23:17-85

Laudise RA (1959) Kinetics of hydrothermal quartz crystallization. J Am Chem Soc 81:562-566

Lawn B (1993) Fracture of Brittle Solids. Cambridge Univ Press, New York

Lawn BR, Wilshaw TR (1975) Fracture of Brittle Solids. Cambridge Univ Press, New York

Lewin JC (1961) The dissolution of silica from diatom walls. Geochim Cosmochim Acta 21:182-198

Li HC, de Bruyn,PL (1966) Electrokinetic and adsorption studies on quartz. Surface Sci 5:203-220

Liang D, Readey DW (1987) Dissolution kinetics of crystalline and amorphous silica in hydrofluoric-hydrochloric acid mixtures. J Am Ceram Soc 70:570-577

Lyklema J (1987) Electric double layers on oxides: Disparate observations and unifying principles. Chem Indust 2:741-747

Mackenzie FT, Gees R (1971) Quartz: Synthesis at Earth surface conditions. Science 173:533-535

Magee AW, Bull PA, Goudie AS (1988) Chemical textures on quartz grains: An experimental approach using salts. Earth Surf Proc Landforms 13:665-676

Margolis SV, Krinsley DH (1974) Processes of formation and environmental occurrence of microfeatures on detrital quartz grains. Am J Sci 274:449-464

Martin JJ, Armington AF (1983) Effect of growth rate on quartz defects. J Crystal Growth 62:203-206

Matijevic E, Mangravite Jr. FJ, Cassell EA (1971) Stability of colloidal silica. IV. The silica-alumina system. J Coll Interf Sci 35:560-568

Mayer LM (1994) Surface area control of organic carbon accumulation in continental shelf sediments. Geochim Cosmochim Acta 58:1271-1284

Mazer JJ, Walther JV (1994) Dissolution kinetics of silica glass as a function of pH between 40 and 85°C. J Non-Cryst Solids 170:32-45

Mekonnen EJ (1995) Composition and Dissolution Rates of Silica, Kaolinite, and Plagioclase Feldspars. PhD dissertation, University of Minnesota, Minneapolis, MN

Merino E (1975) Diagenesis in tertiary sandstones from Kettleman North Dome, California. I. Diagenetic Mineralogy. J Sed Petrology 45:320-336

Mills AL, Herman JS, Hornberger GM, DeJesus TH (1994) Effect of solution ionic strength and iron coatings on mineral grains on the sorption of bacterial cells to quartz sand. Applied Environ Microbiology 60:3300-3306

Milnes AR, Wright MJ, Thiry M (1991) Silica accumulations in saprolites and soils in South Australia. In: Occurrence, Characterization and Genesis of Carbonate, Gypsum, Silica Accumulation in Soils, 26:121-149 Soil Sci Soc Am, Madison, WI

Morey GW, Fournier RO, Rowe JJ (1962) The solubility of quartz in water in the temperature interval from 25 to 300°C. Geochim Cosmochim Acta 22:1029-1043

Morris RC and Fletcher AB (1987) Increased solubility of quartz following ferrous-ferric iron reactions. Nature 330:558-561

Morrow BA, Cody IA (1975) An infrared study of some reactions with reactive sites on dehydroxylated silica. J Phys Chem 79:761-762

Morrow BA, Cody I (1976a) Infrared studies of reactions on oxide surfaces. 5. Lewis acid sites on dehydroxylated silica. J Phys Chem 80:1995-1998

Morrow BA, Cody IA (1976b) Infrared studies of reactions on oxide surfaces. 6. Active sites on dehydroxylated silica for the chemisorption of ammonia and water. J Phys Chem 80:1998-2004

Mortier WJ., Sauer J, Lercher JA, Noller H (1984) Bridging and terminal hydroxyls. A structural and chemical quantum chemical discussion. J Phys Chem 88:905-912

Moss AJ, Green P, Hutka J (1981) Static breakage of granitic detritus by ice and water in comparison with breakage by flowing water. Sedimentology 28:261-272

Murphy WM, Helgeson HC (1989) Thermodynamic and kinetic constraints on reaction rates among minerals and aqueous solutions. IV. Retrieval of rate constants and activation parameters for the hydrolysis constants of pyroxene, wollastonite, olivine, andalusite, quartz, and nepheline. Am J Sci 289:17-101

Nahon DB (1991) Introduction to the Petrology of Soils and Chemical Weathering. John Wiley & Sons, New York

Nesbitt HW, Young GM (1984) Prediction of some weathering trends of plutonic and volcanic rocks based on thermodynamic and kinetic considerations. Geochim Cosmochim Acta 48:1523-1534

Nielsen AE (1959) The kinetics of crystal growth in barium sulfate precipitation. II. Temperature

Dependence and Mechanism. Acta Chem. Scand 13:784-802

Olsen LL, O'Melia CR (1973) Interactions of Fe(III) with Si(OH)$_4$. J Inorgan Nuclear Chem 35:1977-1985

Olson JE (1994) Joint pattern development: Effects of subcritical crack growth and mechanical crack interaction. J Geophys Res 98:12,251-12,265

Parks GA (1984) Surface and interfacial free energies of quartz. J Geophys Res 89:3997-4008

Parks GA. (1990) Surface energy and adsorption at mineral-water interfaces: An introduction. In: Hochella MF, White AF (eds) Mineral-Water Interface Geochemistry. Rev Mineral 23:133-175

Peschel G, Aldfinger KH (1970) Viscosity anomalies in liquid surface zones. IV. Apparent viscosity of water in thin layers adjacent to hydroxylated fused silica surfaces. J Coll Interf Sci 34:505-510

Pollard DD, Aydin A (1993) Progress in understanding jointing over the past century. Geol Soc Am. Bull 100:1181-1204

Pollard CO, Weaver CE (1973) Opaline spheres: Loosely packed aggregates from silica nodule in diatomaceous Miocene fuller's earth. J Sed Petrology 43:1072-1076

Porter RA, Weber Jr. WJ (1971) The interaction of silicic acid with iron (III) and uranyl ions in dilute aqueous solution. J Inorg Nuclear Chem 33:2443-2449

Prigogine M, Fripiat JJ (1974) A possible mechanism of the interaction of adsorbed water molecules with a silica surface. Bull Soc Royale Sciences Liego 43e:449-458

Pye K (1983) Formation of quartz silt during humid tropical weathering of dune sands. Sed Geol 34:267-282

Pye K, Mazzullo J (1994) Effects of tropical weathering on quartz grain shape: An example from Northeastern Australia. J Sed Petrology A64:500-507

Pye K, Sperling CHB (1983) Experimental investigation of silt formation by static breakage processes: the effect of temperature, moisture and salt on quartz dune sand and granitic regolith. Sediment 30:49-62

Renders PJN, Gammons C, Barnes HL (1995) Precipitation and dissolution rate constants for cristobalite at 150 to 300°C. Geochim Cosmochim Acta 59:77-85

Renshaw CE, Pollard DD (1994) Numerical simulation of fracture set formation: A fracture mechanics model consistent with experimental observations. J Geophys Res 99:9359-9372

Rimstidt JD, Barnes HL (1980) The kinetics of silica-water reactions. Geochim Cosmochim Acta 44:1683-1699

Rimstidt JD, Cole DR (1983) Geothermal mineralization I: The mechanism of formation of the Beowawe, Nevada, siliceous sinter deposit. Am J Sci 283:861-875

Roberson CE, Barnes RB (1978) Stability of fluoride complex with silica and its distribution in natural water systems. Chem Geol 21:239-256

Roberts NK and Zundel G (1979) IR studies of long-range surface effects-excess proton mobility in water in quartz pores. Nature 278:726-728

Robie RA, Hemingway BS, Fisher JS (1978) Thermodynamic Properties of Minerals and Related Substances at 298.15 K and 1 Bar (105 Pascals) Pressure and at Higher Temperatures, 456 p, U S Geological Survey, Washington DC

Robinson Jr GD (1980) Possible quartz synthesis during weathering of quartz-free mafic rock, Jasper County, Georgia. J Sed Petrology 50:193-203

Roth ES (1964) Temperature and water content as factors in desert weathering. J Geol, 73:454-468

Scheidegger AE (1991) Theoretical Geomorphology. Springer-Verlag, New York

Schindler PW, Furst B, Dick R, Wolf PU (1976) Ligand properties of surface silanol groups I. Surface complex formation with Fe^{3+}, Cu^{2+}, Cd^{2+} and Pb^{2+}. J Coll Interf Sci 55:469-475

Schindler P, Kamber HR (1968) Die aciditat von silanolgruppen. Helvetica Chim Acta 51:1781-1786

Schindler P, Stumm W (1987) The surface chemistry of oxides, hydroxides and oxide minerals. In: Stumm W (ed) Aquatic Surface Chemistry, p 83-110, John Wiley, New York

Schwartzentruber J, Furst W, Renon H (1987) Dissolution of quartz into dilute alkaline solutions at 90°C: A kinetic study. Geochim Cosmochim Acta 51:1867-1874

Segall P, Pollard DD (1983) Joint formation in granitic rock of the Sierra Nevada. Geol Soc Am Bull 94:563-575

Segall P (1984) Rate-dependent extensional deformation resulting from crack growth in rock. J Geophys Res 89:4185-4195

Scholz CH (1990) The mechanics of earthquakes and faulting. Cambridge Press, New York

Scholl MA (1989) Mineralogical and hydrological influences on bacterial attachment to representative aquifer materials. MS thesis, University of Virginia, Charlottesville

Sermon PA (1980) Interaction of water with some silicas. J Chem Soc Faraday I 76:885-888

Simkiss K, Wilbur KM (1989) Biomineralization. Academic Press, San Diego

Smithson F (1956) Silica particles in some British soils. J Soil Sci 7:122-129

Smithson F (1958) Grass opal in British soils. J Soil Sci 9:148-155

Smithson F (1959) Opal sponge spicules in soils. J Soil Sci 10:105-109

Sparks DL (1989) Kinetics of Soil Chemical Processes. Academic Press, San Diego

Sposito G. (1989) The Chemistry of Soils, Oxford University Press, New York

Stanton J, Maatman RW (1963) The reaction between aqueous uranyl ion and the surface of silica gel. J Coll Sci 18:132-146

Stöber W (1967) Formation of silicic acid in aqueous suspensions of different silica modifications. In: Equilibrium Concepts in Natural Water Systems. Adv Chem Series 67:161-181, Am Chem Soc, Washington, DC

Stone WEE, Shafei GMS, Sanz J, Selim SA (1993) Association of soluble aluminum ionic species with a silica-gel surface. A solid-state NMR study. J Phys Chem 97:10,127-10,132

Stumm W, Morgan JJ (1981) Aquatic Chemistry. John Wiley, New York

Stumm W (1992) Chemistry of the Solid-Water Interface. John Wiley, New York

Sverjensky DA, Sahai N (1995) Theoretical prediction of single-site surface-protonation equilbrium constants for oxides and silicates in water. Geochim Cosmochim Acta, in press

Tait MJ, Franks F (1971) Water in biological systems. Nature 230:91-94

Tester JW, Worley WG, Robinson BA, Grigsby CO, Feerer JL (1994) Correlating quartz dissolution kinetics in pure water from 25 to 625°C. Geochim Cosmochim Acta 58:2407-2420

Thornton SD, Radke CJ (1988) Dissolution and condensation kinetics of silica in alkaline solution. SPE Reservoir Eng 3:743-752

Thornton SD (1987) Role of silicate and aluminate ions in the reaction of sodium hydroxide with reservoir minerals. SPE Reserv Eng, Richardson, Texas, SPE 16277:369-380

Toney MF, Howard JN, Richer J, Borges GL, Gordon JG, Melroy OR, Wiesler DG, Yee D, Sorenson LB (1994) Voltage-dependent ordering of water molecules at an electrode-electrolyte interface. Nature 368:444-446

Ugliengo P, Saunders V, Garrone E (1990) Silanol as a model for the free hydroxyl of amorphous silica: Ab initio calculations of the interaction of water. J Phys Chem 94:2260-2267

Unger KK (1994) Surface structure of amorphous and crystalline porous silicas: Status and prospects. In: The Colloid Chemistry of Silica, Adv Chem Series 234:147-164, Am Chem Soc, Washington, DC

Van Bennekom AJ (1981) On the role of aluminum in the dissolution kinetics of diatom frustrules. In: Ross R (ed) 6th Diatom Symposium, p 445-456, Otto Koeltz, Koenigstein

Van Bennekom AJ, Buma AGJ, Nolting RF (1991) Dissolved aluminum in the Weddell-Scotia confluence and effect of Al on the dissolution kinetics of biogenic silica. Marine Chem 35:423-434

Van Cappellen P, Linqing Q (1995a) Biogenic silica dissolution in sediments of the Southern ocean. I. Solubility. In: Gaillard JF, Tréguer P (eds) The Indian Sector of the Southern Ocean: Hydrology and Benthic Processes Antares 1 France-JGOFS

Van Cappellen P, Linqing Q (1995b) Biogenic silica dissolution in sediments of the Southern ocean. II. Kinetics. In: Gaillard JF, Tréguer P (eds) The Indian Sector of the Southern Ocean: Hydrology and Benthic Processes Antares 1 France-JGOFS

Van Lier JA, de Bruyn PL, Overbeek JTG (1960) The solubility of quartz. J Phys Chem 64:1675-1682

Walther JV, Helgeson HC (1977) Calculation of the thermodynamic properties of aqueous silica and the solubility of quartz and its polymorphs at high pressures and temperatures. Am J Sci 277:1315-1351

Wang HD, White GN, Turner FT, Dixon JB (1993) Ferrihydrite, lepidocrocite, and goethite in coatings from East Texas vertic soils. Soil Sci Soc Am J 57:1381-1386

Weiderhorn SM, Johnson H (1973) Effect of electrolyte pH on crack propagation in glass. J Am Ceram Soc 56:192-197

Weigel VE (1964) Über die Heißauslaugung von Silikätglasern durch Neutralsalzlösungen. Glastechnische Berichte 37:141-147

White G S, Freiman SW, Wiederhorn SM, Coyle TD (1987) Effects of counterions on crack growth in vitreous silica. J Am Ceram Soc 70:891-895

Wiese GR, James RO, Yates DE, Healy TW (1976) Electrochemistry of the Colloid-Water Interface. In: Int'l Reviews Science, Physical Chem 6:53-85, Butterworth, New York

Wieland, E., Wehrli, B. and Stumm, W. (1988) The coordination chemistry of weathering: III. A generalization on the dissolution rates of minerals. Geochim Cosmochim Acta 52:1969-1981

Wijnen PWJG, Beelen TPM, de Haan JW, Rummens CPJ, van de Ven LJM, van Santen RA (1989) Silica gel dissolution in aquous alkali metal hydroxides studied by ^{29}Si-NMR. J Non-Cryst Solids 109:85-94

Wijnen PWJG, Beelen TPM, de Haan JW, van de Ven LJM, van Santen RA (1990) The structure directing effect of cations in aqueous silicate solutions. A [29]Si-nmr study. Colloids Surfaces 45:255-268

Williams LA, Parks GA, Crerar DA (1985) Silica diagenesis. I. Solubility controls. J Sed Petrology 55:301-311

Williams LA, Crerar DA (1985) Silica diagenesis. II. General mechanisms. J Sed Petrology 55:312-321

Wirth GS, Gieskes JM (1979) The inital kinetics of dissolution of vitreous silica in aqueous media. J Coll Interf Sci 68:492-500

Wollast R, Chou L (1988) Rate control of weathering of silicate minerals at room temperature and pressure. In: Lerman A, Meybeck M (eds) Physical and Chemical Weathering in Geochemical Cycles. NATO

ASI 251:11-31, Reidel, Dordrecht, Netherlands

Xie Z, Walther JV (1993) Quartz solubilities in NaCl solutions with and without wollastonite at elevated temperatures and pressures. Geochim Cosmochim Acta 57:1947-1955

Yates DE, Levine S, Healy TW (1974) Site-binding model of the electrical double layer at the oxide/water interface. J Chem Soc Farad Trans 70:1807-1818

Young ARM (1987) Salt as an agent in the development of cavernous weathering. Geology 15:962-966

Zhu SB, Robinson GW (1991) Structure and dynamics of liquid water between plates. J Phys Chem 94:1403-1410

Chapter 7

FELDSPAR DISSOLUTION KINETICS

Alex E. Blum

U. S. Geological Survey
3215 Marine Street
Boulder, CO 80303 U.S.A.

Lisa L. Stillings

Department of Geology and Geophysics
University of Wyoming
Laramie, WY 82071 U.S.A.

INTRODUCTION

Feldspars have the most intensely-studied dissolution kinetics of any of the silicate minerals. This is partly because they are the most abundant silicate mineral in crustal rocks, and because they have a fairly well constrained chemistry and structure, making them amenable to experimental studies. The dissolution kinetics of feldspars have been increasingly studied for many years, especially over the last decade. Some reasons for this interest are: (1) understanding the long-term capacity of soils to neutralize anthropogenic acidic rainfall (e.g. Reuss and Johnson, 1986), and to provide nutrients such as K and Ca to ecosystems after exchangeable reservoirs in soils are exhausted by acid rain and/or deforestation (McClelland, 1950); (2) understanding the importance of silicate weathering in regulating atmospheric CO_2 concentrations over geologic time, and consequently in controlling global climatic change (e.g. Walker et al., 1981; Berner et al. 1983; Lasaga et al., 1985; Volk, 1987, Brady and Carroll, 1994; White and Blum; 1995); (3) incorporation of chemical kinetics into reactive transport models, which requires rate laws for the dissolution and precipitation of major minerals; and (4) advances in surface science and heterogeneous chemical kinetics have provided new analytical and theoretical techniques for solving complex problems of heterogeneous kinetics such as feldspar dissolution.

The overall feldspar weathering process can be simplified as the alteration of feldspar to common clay minerals, such as kaolinite. For example:

$$2\,NaAlSi_3O_8 + 2\,H^+ + 9\,H_2O = Al_2Si_2O_5(OH)_4 + 4\,H_4SiO_{4(aq)} + 2\,Na^+ \qquad (1)$$

It is now widely recognized that the alteration of feldspars generally involves two independent processes: (1) the initial dissolution of feldspar into solution, and (2) the subsequent precipitation of kaolinite and other secondary phases from solution. These two reactions are linked by the effects of the solution composition on their relative rates, and in a closed system will reach steady-state. Because the dissolution kinetics of feldspar are very slow, and most near-surface environments are open systems, the dissolution rate of feldspar should control the rate of the overall feldspar alteration process. Consequently, feldspar dissolution is the focus of much research on mineral alteration and weathering.

A major advance in our understanding of feldspar dissolution kinetics is the recognition of the central importance of the mineral-solution interface. Neither the chemistry of the bulk solution or the feldspar controls the rate of the dissolution reaction; rather, reactions which occur at the mineral/water interface are the crucial concerns. In this chapter we summarize: (1) the surface chemistry of feldspars during dissolution, (2) the experimental data for feldspar dissolution kinetics, including the effects of pH, Al, cations, organics, and temperature, and (3) theoretical approaches which explain the experimental results.

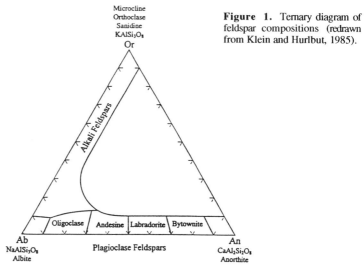

Figure 1. Ternary diagram of feldspar compositions (redrawn from Klein and Hurlbut, 1985).

FELDSPAR MINERALOGY

This brief summary of feldspar chemistry and structure does not begin to reflect the complexity of this mineral group. Feldspars are framework silicates which have a general formula $(Na,K,1/2\,Ca)_n Al_n Si_{(4-n)} O_8$, where n varies from 1 to 2. Corner sharing AlO_4 and SiO_4 tetrahedra are linked in a three-dimensional lattice, and cations with an effective radius of >1Å fill large, irregular cavities in the tetrahedral framework to compensate for the charge deficiency caused by the tetrahedral Al. The feldspars form two continuous solid-solution series during crystallization from a liquid magma: (1) the alkali feldspars from orthoclase (Or, $KAlSi_3O_8$) to albite (Ab, $NaAlSi_3O_8$), and (2) the plagioclase feldspars from albite to anorthite (An, $CaAl_2Si_2O_8$) (Fig. 1). The solubility of the An component in the alkali feldspars and the Or component in the plagioclase feldspars is generally <10 mol %. The exsolution of alkali feldspars during cooling requires only the diffusion of cations, a process which has a relatively low activation energy (~165 to 250 kJ/mol, Yund, 1983). Most igneous alkali feldspars have exsolved during cooling into Or and Ab enriched regions, often with macroscopic lamella several millimeters across. Plagioclase exsolution requires diffusion of tetrahedral Al and Si with a much higher activation energy (~270 to 355 kJ/mol, Yund, 1983), and occurs only on the scale of tens of nanometers in a few of the intermediate plagioclase compositions. Ribbe (1983a) and Parsons (1994) contain excellent summaries of feldspar chemistry and structure. Most (but not all) of the feldspar specimens used in experimental studies have been chosen to have near end-member compositions, and contain a minimal number of structural and compostional complications.

FELDSPAR SURFACE CHEMISTRY DURING DISSOLUTION

Protons (H^+) and hydroxyls (OH^-) are major "potential-determining ions" for oxide surfaces, meaning they chemically adsorb to specific sites located directly on the mineral surface, and they impart charge to that surface (Schindler, 1981). Feldspars are "multi-oxides", with both SiO_2 and Al_2O_3 components; therefore, most studies of the feldspar-water interface begin by determining the concentration H^+ and OH^- adsorbed to the surface as a function of pH. The correlation between the concentration of adsorbed H^+, OH^-, and albite dissolution rates, first noted by Blum and Lasaga (1988), has led to rate expressions of the form

$$rate = k_a[H^+_{ads}]^n \qquad (2)$$

$$rate = k_b[OH^-_{ads}]^m \qquad (3)$$

where k_a and k_b are the rate constants for acid and base pH regions, $[i_{ads}]$ is the concentration of H^+ or OH^- adsorbed at the surface, and n and m are the reaction orders, which have been predicted to range from 1 to 4. Additional studies of feldspar surface speciation (Amrhein and Suarez, 1988; Schott, 1990; Blum and Lasaga, 1991; Unwin and Bard, 1992; Oxburgh et al., 1994; Stillings et al., 1995a) have continued to refine the above model, and to suggest mechanisms for feldspar dissolution.

Dissolution of the feldspar surface should result in the stoichiometric release of feldspar components (i.e. the elements are released into solution in the same ratio as their abundance in the original mineral). However, feldspar dissolution is usually non-stoichiometric during the initial phase of dissolution, which may range from the first few minutes of exposure to aqueous solution to as long as 500 h (Stillings and Brantley, 1995). The non-stoichiometry during the first few minutes is the result of exchange and adsorption/desorption reactions which occur rapidly when fresh feldspar surfaces are first placed in solution. Sustained non-stoichiometric dissolution over longer periods is due to: (1) the preferential removal of one component from the structure, such as Al leaching at low pH, leaving a "leached layer" on the surface, (2) the precipitation of a secondary phase from solution, and/or (3) the preferential dissolution of compositionally distinct regions of the crystal, such as impurities or exsolution lamellae, resulting in elemental release ratios different than those in the bulk crystal (Inskeep et al., 1991; Stillings and Brantley, 1995).

An important issue is the definition of the "surface". We will consider the "surface" to be the portion of the crystal which reacts readily on a molecular level with solutes in solution. The surface may be a fairly distinct plane containing the uppermost few atomic layers, or it may be a near-surface region of the crystal with high solute mobility, or an extensive portion of the bulk crystal accessible to solution via microporosity, microfractures, surface damage, or defects. The nature of the mineral surface is a major issue in the kinetics of silicate mineral dissolution, and will be examined in more detail later in this chapter, as well as elsewhere in this volume.

H^+/OH^- sorption and H^+/cation exchange reactions on the feldspar surface

The concentrations of H^+ and OH^- which react with the feldspar surface are usually determined by acid-base titrations of the feldspar. These experiments last for ~1 to 2 h. During the titrations H^+ and OH^- are assumed to react with the surface in three ways: (1) exchange reactions involving the cation sites, (2) sorption reactions at dangling oxygens, and (3) sorption reactions at bridging oxygens. The available data suggests these reactions reach equilibrium between the surface and solution within minutes and are governed by classical exchange and partitioning relationships.

Surface titrations. Surface titrations of feldspars are conducted by adding acid, base and other solutes to aqueous suspensions of the ground mineral, and monitoring the composition of the solution with either ion-specific electrodes or chemical analysis of solution aliquots. The amount of a species adsorbed or desorbed from the surface is determined by the difference between the amount of species added to the system, and the concentration of species in solution (e.g. Stumm and Morgan, 1981; Huang, 1981). The rapid release and consumption of elements during the first few minutes of a dissolution experiment yields similar information for a single solution composition.

Several cautions in the interpretation of surface titrations are warranted. Because we interpret surface chemistry from changes in solution chemistry, we have no direct information on the actual surface species involved in the reactions. We can only guess at the appropriate surface complexes based upon the observed stoichiometry of species entering and leaving solution. Furthermore, if a single aqueous species is involved in several different reactions, we may not be able to differentiate between the reactions.

Surface titrations also give no information on heterogeneity. They usually average the behavior of all exposed faces, yet we know that different crystallographic faces may have dramatically different chemical behaviors. For example, the (0001) face of quartz dissolves 10^3 times faster than the prismatic faces in HF solution (Ernsberger, 1960), such that a surface titration would be heavily biased toward "unreactive" surface area. We currently have no information on the relative dissolution rates of feldspar surfaces, but the (001) and (010) faces of feldspar grains tend to be less heavily corroded than the irregular faces. Ultramicroelectrode titrations (Unwin and Bard, 1992), which titrate a portion of the surface covered by a single drop of solution, provide information on the behavior of different cleavage surfaces. However, even using ultramicroelectrode measurements, much of the protonation reaction may still occur at high energy "reactive" sites such as dislocations and impurities, which are only a small fraction of the total surface.

The titration behavior of feldspars is dependent on both the aging of the feldspar surfaces after grinding the specimen and the reaction conditions of pretreatment. Freshly cleaved silicate surfaces behave differently than aged surfaces (Petrovic, 1981a,b; Eggleston et al., 1989). For example, Stillings et al., (1995a) observed the immersion pH (the equilibrium pH of a feldspar-water suspension before a titration) of a ground K-feldspar decreased from ~9.2 to ~8.2 over three years of aging in air. In addition, the BET surface area (calculated from the Brunauer-Emmett-Teller adsorption isotherm; Lowell, 1979) decreased from 4.5 to 1.5 m^2/g during this same time period, a change they attributed to surface relaxation and healing of microporosity over time. Pretreatment conditions are important because they may change both composition and structural state of the surface. Schott (1990) observed that ground albite which had been (a) pre-reacted at pH 2, or (b) annealed for 80 h at 1000°C, adsorbed 2 to 3 times less H^+ than freshly ground powders.

Finally, surface titrations are highly susceptible to experimental errors, especially precipitation of secondary phases, errors in electrode measurements, and inaccurate compensation for mineral dissolution and aqueous speciation of solutes. These complications have been poorly investigated for silicates, and discretion must be used in extrapolating the results of surface titrations to explain dissolution kinetics. Nevertheless, surface titrations are probably our best source of information on the stoichiometry and the magnitude of sorption reactions at feldspar surfaces.

Exchange and/or sorption reactions involving the interstitial cations. The cations in the feldspar structure, primarily K^+, Na^+, and Ca^{2+}, are located in irregular cavities within the tetrahedral framework. They are closely bound to the oxygens which bridge the Al and Si atoms in order to compensate for the net negative charge of the tetrahedral framework resulting from the presence of tetrahedral Al (e.g. see Smith, 1974a, 1974b; Ribbe, 1983b). Surface cations are released to solution through exchange with H^+ or other solution cations. Examples of these reactions include:

$$\equiv SO - Na \longleftrightarrow \equiv SO^- + Na^+ \tag{4}$$

$$\equiv SO - Na + H^+ \longleftrightarrow \equiv SO - H + Na^+ \tag{5}$$

$$(\equiv SO)_2 - Ca + 2H^+ \longleftrightarrow 2 \equiv SO - H + Ca^{2+} \tag{6}$$

where $\equiv SO$ represents a cation site accessible to solution, and $\equiv S$ is a tetrahedrally-coordinated Si or Al. These exchange reactions release cations and increase solution pH when feldspars are first placed in solution (e.g. Tamm, 1930; Nash and Marshall, 1956a,b; Garrels and Howard, 1959). Table 1 shows the depth of the cation exchange layer for albite, anorthite, and potassium feldspar formed during surface titrations. Depths are calculated from BET surface areas, assuming a uniform exchange depth. The depth of the exchange layer is several unit cells in albite, and slightly deeper in the K-feldspars and anorthite. This magnitude of cation exchange is greater than can be explained by the cation density at a surface terminating the bulk structure, and suggests a high H^+ concentration and cation mobility at the feldspar surface. Possible explanations include relaxation/disruption of the surface, which may broadly enhance cation mobility in the near surface region, and defects intersecting the surface which may create localized regions of enhanced cation mobility. Both phenomena may also allow penetration of hydrogen species (H^+, H_3O^+, and/or H_2O; Petit et al., 1987; Dran et al., 1988); however, the cause and distribution of the high cation exchange site densities in feldspars are not well understood.

Table 1. Depth of the cation exchange layer.

Reference	Mineral (cation)	depth (A)
Wollast and Chou (1992)[1]	albite (Na)	21
Blum and Lasaga (1991)[1]	albite (Na)	15
Amrhein and Suarez (1988)[1]	anorthite (Ca)	25-30
Schweda (1989)[2]	microcline (K)	37
Schweda (1989)[2]	sanidine (K)	85

[1] determined by surface titration: total number of exchange sites.
[2] determined from non-stoichiometry of initial dissolution at ~30 minutes at pH 3. Assumed to represent total exchange.

In solutions of pH < 8 and low Na concentrations, albite exchanges essentially all its Na^+ for H^+ in the outermost unit cells (i.e. Eqn. 5 goes completely to the right). There has been some controversy about the reversibility of the H^+/Na^+ exchange reaction in albite (Chou and Wollast, 1989; Murphy and Helgeson, 1989). Blum and Lasaga (1988, 1991) did not detect any Na^+ uptake during surface titrations of albite at low Na^+ concentrations; however, Wollast and Chou (1992) presented convincing evidence that the exchange of H^+ for Na^+ in albite is reversible, with significant Na^+ occupancy of the exchange sites at pH > 8, and moderate Na^+ concentrations. Garrels and Howard (1959) studied K^+-H^+ exchange in adularia-water suspensions, and concluded that an activity ratio, a_{K^+}/a_{H^+}, of $10^{9\text{-}10}$ was needed to reverse the reaction. Stillings et al. (1995a) observed an ionic strength dependence in the K^+-H^+ exchange on an adularia surface, which they attributed to competition between H^+ and solution cations for the feldspar exchange site. Thus, in saline environments, the exchange sites on the feldspar surface may have significant alkali-cation occupancy. This is consistent with the observed precipitation of secondary albite and K-feldspar in saline diagenetic environments.

Adsorption and desorption reactions at dangling oxygens at the surface. Dangling oxygens on feldspar surfaces rapidly hydrate with atmospheric or aqueous water to form surface hydroxyl groups (Schindler, 1981), analogous to the behavior of SiO_2 (Iler, 1979), Al_2O_3 (Huang, 1981), and metallic oxides (Stumm and Morgan, 1981). Sorption reactions, governed by equilibrium expressions, form charged surface species at these sites:

$$\equiv S-OH + H_s^+ \xleftrightarrow{\;fast\;} \equiv S-OH_2^+ \qquad K_1 = \frac{a_{[\equiv S-OH_2^+]}}{a_{[\equiv S-OH]}\, a_{[H_s^+]}} \qquad (7)$$

$$\equiv S-O^- + H_s^+ \xleftrightarrow{\;fast\;} \equiv S-OH \qquad K_2 = \frac{a_{[\equiv S-OH]}}{a_{[\equiv S-O^-]}\, a_{[H_s^+]}} \qquad (8)$$

$$\equiv S-OH + M_s^{n+} \xleftrightarrow{\;fast\;} \equiv S-OM^{(n-1)+} + H_s \qquad K_M = \frac{a_{[\equiv S-OM^{(n-1)+}]}\, a_{[H_s^+]}}{a_{[\equiv S-OH]}\, a_{[M_s^+]}} \qquad (9)$$

where $\equiv S$ is a tetrahedral Al or Si on the surface, $\equiv S\text{-}O^-$ is a dangling oxygen on the surface, and M^{n+} is any adsorbing metal with charge n, and a_i is the activity of species i. The subscript s represents a species located at the mineral surface.

Blum and Lasaga (1991) performed surface titrations on albite and found they could describe the results using a Langmuir adsorption isotherm with Equations (7) and (8), where $K_1 = 2.7 \times 10^4$ and $K_2 = 9.4 \times 10^4$. In the acid region, they found a site density of 35 $\mu mol/m^2$, and a density of of 27 $\mu mol/m^2$ in the basic region. Blum and Lasaga (1991) interpreted the charged surface species as protonation and deprotonation of dangling Al-OH surface groups to $\equiv$Al-OH$_2^+$ and $\equiv$Al-O$^-$, respectively. Unwin and Bard (1992) titrated the (010) face of a single albite crystal, and also used a Langmuir model to estimate a K_1 of 4.3 $\times 10^3$ (Eqn. 7) and a site density of 71 $\mu mol/m^2$ in the acid region. Wollast and Chou (1992) examined the albite surface only in the basic pH region, and found a negative site density of 21.3 $\mu mol/m^2$, in good agreement with Blum and Lasaga (1991). However, Wollast and Chou (1992) attribute the negative charge density to the removal of Na^+ from cation exchange sites via Equation (4). Schott (1990) used site densities and intrinsic pKs from SiO_2 and Al_2O_3 surfaces, together with a diffuse layer electrostatic model, to estimate H^+ adsorption on ground albite surfaces from pH 1 to pH 12.5. With this approach, he was able to accurately model adsorption at the surface of a "restored" albite (annealed for 80 h at 1000°C), but his model greatly underestimated H^+ adsorption on freshly ground surfaces at pH < 6. Amrhein and Suarez (1988) performed a surface titration of anorthite and found a total site density of 120 $\mu mol/m^2$.

Site densities reported here are much larger than predicted using a simple geometric calculation of average Al density on the surface, which implies that a simple conceptual model of a planar surface cannot be strictly correct. Limitations of this model include (1) the BET surface area may not be accurate measure of feldspar reactive surface area, (2) protonation reactions may occur at depth within the surface as a result of surface relaxation or defects, analogous to the Na exchange described earlier, or (3) that sites other than simple Al-OH dangling bonds react.

Adsorption at bridging oxygens at the surface

Examples of adsorption at the bridging oxygen between two tetrahedra include the adsorption of H⁺ or Na⁺:

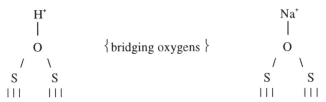

Blum and Lasaga (1991) suggest that only a small proportion of bridging oxygens on feldspar surfaces protonate at any specific time. However, recent work by Oxburgh et al. (1994) and Xiao and Lasaga (1994) suggests that proton adsorption at bridging oxygens is both quantitatively significant, and an important step in the feldspar dissolution mechanism.

Evidence in support of H⁺ adsorption at Si-O-Al bridging oxygens comes from the 2:1 phyllosilicates, which have only bridging oxygens between Si- and Al-tetrahedra exposed on the (001) surfaces. The 2:1 phyllosilicate (001) surfaces act as Lewis bases with the Lewis base parameter (γ) generally increasing with increasing tetrahedral Al content (Giese and van Oss, 1993).

To date only one surface titration study has attempted to differentiate between the fraction of protons which exchange for cations at a feldspar surface and the fraction which adsorb at dangling and bridging oxygens. Stillings et al. (1995a) measured the proton concentration on the surface of a ground adularia feldspar, $[H_s^+]$, as a function of pH and concentrations of Na⁺, K⁺, and $(CH_3)_4N^+$ in solution: $[H_s^+]$ decreased with increasing concentrations of Na⁺ and K⁺ but was not affected by $(CH_3)_4N^+$, a cation with a much larger radius (Fig. 2). Assuming that the exchange and adsorption reactions described by Equations (7-9) occur simultaneously on the feldspar surface, Stillings et al. (1995a) calculated a mass balance for $[H_s^+]$ by separating the fraction of protons consumed in: (1) the exchange reaction, $[H_{ex}^+]$, (2) adsorption to dangling surface hydroxyl site, $[H_{ads}^+]$, and (3) dissolution of the adularia solid during the titration, $[H_{dis}^+]$. Moreover, they identified the exchange site to be the (Al,Si)-bridging oxygen because it is the oxygen in the feldspar structure most closely bound to the exchangeable cation.

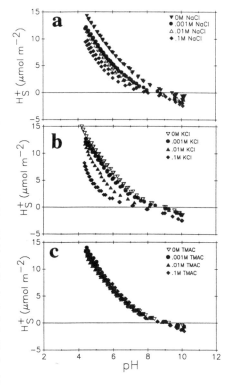

Figure 2. Concentration of protons adsorbed at an adularia (K-feldspar) surface, as a function of (a) NaCl, (b) KCl, and (c) $(CH_3)_4NCl$ concentration. 0 M refers to titrations without added salt.

$[H_s^+]$ was calculated assuming a 1:1 exchange of H_{aq}^+ for $K_{feldspar}^+$. Therefore $[H_{ex}^+]$ equaled the concentration of K^+ produced in solution during the titration minus the $[K^+]$ contributed by stoichiometric dissolution of the adularia solid. The concentration of protons adsorbed to dangling surface hydroxyl sites, $[H_{ads}^+]$ (Eqns. 7 and 8), was then calculated as the difference between $[H_s^+]$ and $[H_{ex}^+]$. Figure 3 illustrates the H^+ mass balance calculated for titrations in 0-0.1 M NaCl solutions. From this analysis the ionic strength dependence of the total proton uptake by the surface, $([H_s^+]$, Fig. 3a), is due to the ionic strength dependence of the proton adsorption at the exchange site, $([H_{ex}^+]$, Fig. 3b), rather than at the surface hydroxyl site, $([H_{ads}^+]$, Fig. 3c). The ionic strength dependence of the exchange reaction results from competition between H_{aq}^+ and Na_{aq}^+ for the $K_{feldspar}^+$ exchange site:

$$KAlSi_3O_8 + xNa_{aq}^+ + (1-x)H_{aq}^+ \longleftrightarrow Na_xH_{(1-x)}AlSi_3O_8 + K_{aq}^+ \qquad (10)$$
$$\text{adularia} \qquad\qquad\qquad\qquad\qquad (Na,H)\text{ - exchanged adularia}$$

The $K_{feldspar}^+ : H_{aq}^+$ exchange was not affected by the presence of $(CH_3)_4N^+$, which is too large to compete effectively for the exchange sites.

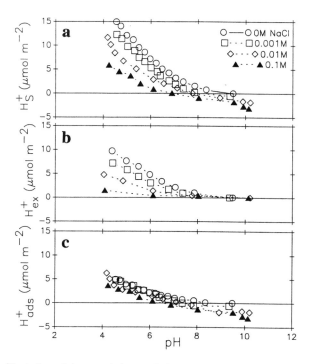

Figure 3. An illustration of the mass balance calculation for H^+ adsorption at an adularia surface, as a function of NaCl concentration. (a) Total concentration of H^+ adsorbed to the surface (see Fig. 2a). (b) The concentration of H^+ adsorbed to the bridging (Al,Si)-O, exchange site. (c) The concentration of H+ adsorbed to the dangling, surface hydroxyl sites. From this analysis, the ionic strength effect observed in the total concentration of H^+ adsorbed to the surface (plot a), is due to competitive adsorption at the exchange site (plot b).

Because feldspar dissolution rates in acid solution are dependent on cation concentration (Schweda, 1990; Stillings and Brantley, 1995a), and the concentration of *exchanged* protons also depends on cation concentration, Stillings et al. (1995a) suggest that Equation (2) be amended to:

$$rate = k_a [H_{ex}^+]^n \tag{11}$$

where $[H_{ex}^+]$ is the concentration of H^+ adsorbed at the exchange site, which is defined as the (Al,Si)-bridging oxygen site in the feldspar structure.

Alteration of the Al-Si tetrahedral framework

There is considerable evidence for the development of Al and cation depleted, Si-rich layers on the feldspar surface after dissolution in the acid region (pH < 5), both from direct measurement by surface analytical techniques (Petrovic et al., 1976; Berner and Holdren, 1979; Casey et al., 1988, 1989; Nesbitt and Muir, 1988; Goossens et al., 1989; Hellmann et al., 1989, 1990; Muir et al.,1989; Sjöberg, 1989; Nesbitt et al., 1991; Muir and Nesbitt, 1991,1992) and from non-stoichiometric dissolution (Chou and Wollast, 1984; Holdren and Speyer, 1985; Mast and Drever, 1987; Schweda, 1990; Hellmann, 1994, 1995; Stillings and Brantley, 1995; and others). The thickness of the Si-rich layer shows a positive correlation with the depth of penetration of hydrogen species (H^+, H_3O^+, and/or H_2O) into the surface. Both the thickness of the layer and the depth of penetration increase with decreasing pH, decreasing ionic strength, increasing temperature, and increasing anorthite mole fraction (Chou and Wollast, 1985; Casey et al., 1988, 1989; Petit et al., 1989; Schweda, 1990; Nesbitt et al., 1991; Stillings and Brantley, 1995). At pH 1 to 2 and 25°C, the Si-rich layer developed on labradorite reaches a thickness of several thousand Ångströms, but when the sample is removed from solution and dried, the Si-rich layer cracks and spalls (Casey et al., 1989). The silica-rich layer formed at low pH is highly-porous, as indicated by transmission electron microscopy (TEM) images and nitrogen porosimetry measurements on labradorite (Casey et al., 1989), and ion beam measurements on albite (Petit et al., 1990), alkali feldspar (Petit et al., 1989), and labradorite (Casey et al., 1989) as well as diopside (Petit et al., 1987; Schott and Petit, 1987). This layer has a structure similar to amorphous, hydrated silica gel. Casey et al. (1989) showed a sharp and distinct interface between the labradorite structure and the amorphous silica layer formed at pH 2, as indicated by the abrupt termination of the feldspar TEM lattice fringe image at the interface with the amorphous Si-rich layer.

Diffusion rates through porous silica layers on glass and zeolites are very similar to aqueous diffusion rates (Bunker et al., 1984, 1986, 1988). Westrich et al. (1989) demonstrated ^{18}O will diffuse within hours into a 1000 Å leached layer formed on labradorite at pH 2. Casey et al. (1988), Westrich et al. (1989), Hellmann et al. (1990), Rose (1991) and Amrhein and Suarez (1992) all conclude that the dissolution rate of feldspars in acid solutions is not limited by diffusion through the Si-rich leached layer, and the formation of the leached layer does not indicate a change in the rate controlling mechanism of dissolution. A Si-rich leached layer may, however, provide a favorable environment for the crystallization of secondary phases, such as clays. Kawano and Tomita (1994) observed smectite formation in a leached layer formed on albite during dissolution at 150° to 225°C.

There is less agreement about the presence and nature of the leached layer above pH 3. Most studies agree that the cation and Al depleted layers reach thicknesses <20 Å between pH 5 and 8 for all the feldspars. This is comparable to the depth of rapid cation exchange measured by Chou and Wollast (1984) and Blum and Lasaga (1991) for albite, and by Amrhein and Suarez (1988) for anorthite, and there is a general consensus that no thick depleted layer forms during feldspar dissolution in the neutral pH region. There is some evidence that the leached layer persists but grows rapidly thinner as pH increases from 3 to 5 (Sjöberg, 1989; Nesbitt et al., 1991; Hellmann et al. 1990, 1994, 1995).

TEM observations of labradorite by Inskeep et al. (1991) showed preferential dissolution of Ca-rich exsolution lamellae with a width of 700 Å to a depth of 1350 Å. They suggested that much of the evidence for non-stoichiometric dissolution of plagioclase based upon surface analytical techniques and dissolution stoichiometries in the pH 3 to 5 region can be explained by the preferential dissolution of calcium-rich exsolution phases. Wilson and McHardy (1980) and Gardner (1983) observed similar preferential dissolution of exsolution lamellae in microcline, although at much larger scales of several μm. Apparent non-stoichiometric dissolution may also be caused by the presence of impurity phases. Stillings and Brantley (1995) observed non-stoichiometry in the overall dissolution reaction due to preferential dissolution of fluorite (found in Amelia albite) and of zoisite and muscovite (found in Mitchell Co. oligoclase).

There is less information about non-stoichiometric feldspar dissolution at basic pH. Casey et al. (1988), used Rutherford backscattering analysis (RBS) and elastic recoil detection (ERD) to examine a labradorite surface after reaction for 2000 h at pH 12 and 25°C. Their analyses revealed no detectable Al-leached layers and very low hydrogen inventories, suggesting the lack of a leached layer at basic pH. Holdren and Speyer (1985) and Schweda (1990) concluded that K-feldspar dissolution remained stoichiometric (i.e. no leached layer formation) up to at least pH 10.9 at 25°C. Brady and Walther (1989) suggest that at high pH silicon detachment controls the overall dissolution rate, and Si-rich layers will not occur. However in contrast, the study of Hellmann et al. (1990) observed a Si enriched layer ~500 Å thick on albite at pH 10.8 and 200°C.

A conceptual model for the pH-dependence of leached layer formation is that during dissolution at low pH, Al sites (including the (Al,Si)-bridging oxygen) are more readily protonated, leading to destabilization of Al-O bonds and the eventual release of Al from tetrahedral sites in the structure. This leavesncompletely coordinated fragments of Si tetrahedra are left on the surface. If the rate of feldspar dissolution is slower than the rate of dissolution of silica, then the silica tetrahedra will detach into solution and no silica-rich layer will form. This is the case above pH ~3 to 5, where feldspar dissolution rates are slow. As feldspar dissolution rates increase below pH 3, silica fragments accumulate near the surface, where they repolymerize to form a continuous porous, hydrous, amorphous silica layer. The dissolution rates of amorphous SiO_2 at 25°C (Rimstidt and Barnes, 1980), normalized on a molar basis to the silica content of the feldspar, are shown on Figures 4-6. The dissolution rate of the alkali feldspars exceeds that of amorphous SiO_2 at ~pH 3, and the intermediate feldspars at ~pH 5. This generally corresponds to the pH conditions under which a Si-rich layer forms on the feldspar surface during dissolution. There is evidence for the repolymerization of silanol groups within the silica layer, based upon hydrogen inventories measured by elastic recoil detection (ERD) by Casey et al., (1988), hydrogen inventories determined by ion beam analysis (Petit et al., 1990), and incorporation of [18]O into the silica-rich layer by Westrich et al. (1989).

In contrast, dissolution of bytownite (An_{76}) and anorthite (An_{90-100}) is nearly stoichiometric, even at pH 2 to 3 and elevated temperatures, despite having much higher dissolution rates than either the alkali feldspars or amorphous SiO_2 (Fig. 6 and Amrhein and Suarez, 1988; Stillings and Brantley, 1995; Oelkers and Schott, in press, b). The change from a non-stoichiometric to a stoichiometric dissolution mechanism may correspond to the dramatic increase in dissolution rates which occurs at compositions greater than An_{70-80} (Fig. 6). The absence of bridging Si-O-Si bonds in anorthite means that

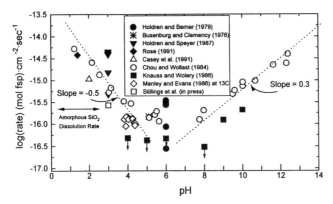

Figure 4. Compilation of albite dissolution rates as a function of solution pH. The arrow indicates the dissolution rates of amorphous silica at 25°C from Rimstidt and Barnes (1980).

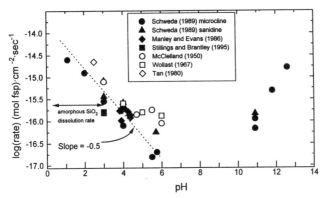

Figure 5. Compilation of K-feldspar dissolution rates as a function of solution pH. The arrow indicates the dissolution kinetics of amorphous silica at 25°C from Rimstidt and Barnes (1980).

there is no fragmented Si network remaining after the removal of Al. Rather, Si tetrahedra are isolated and released into solution along with Al. Thus, even though the dissolution rate of anorthite in acid solution greatly exceeds that of amorphous silica, no Si-rich surface layer develops.

While laboratory dissolution experiments have a duration of several hundred to several thousand hours, natural weathering of feldspars occurs over hundreds of thousands of years. The very slow feldspar dissolution rates at neutral pH mean that laboratory experiments dissolve only a small proportion of the starting material, and there is doubt whether thick leached layers form on natural feldspar surfaces given sufficient time.

X-ray photoelectron spectroscopy (XPS) has been used to look for leached layers on natural samples. The first of these studies (Berner and Holdren, 1979) found no evidence of a leached layer >20 Å thick on naturally weathered plagioclase with XPS. A more recent XPS study (Blum et al., 1992) examined soils which ranged in age from 10,000 to 800,000 years old, with soil pH of 5 to 7. They also observed no cation depletion >50 Å for Na and Ca, or >30 Å for K on the (001) and (010) faces of naturally weathered alkali and oligoclase feldspars. A slight Al enrichment was observed on the very near surface, but there were no non-stoichiometric Al/Si ratios at depths greater than a few Ångströms.

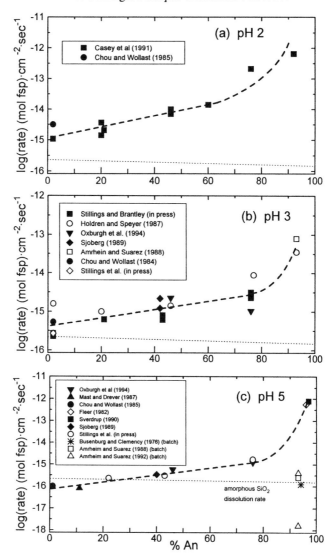

Figure 6. Plagioclase dissolution rates at (a) pH 2, (b) pH 3 and (c) pH 5 as a function of anorthite content (%An). The thick dashed lines are an interpolation of the data. From An_{0-60} the dissolution rate increases gradually with the proportion of An content $(R \propto X_{An}^{n})$ with n » 1.7, 1.1 and 1.6 at pH 2, 3, and 5, respectively. Above An_{-70}, the dependence of the dissolution rate on An component increases dramatically to with n » 6.0, 8.3 and 12.7 at pH 2, 3, and 5, respectively. The thin dotted line is the dissolution kinetics of amorphous silica at 25°C from Rimstidt and Barnes (1980), normalized to the silica content of the feldspar, which causes the gentle slope with An content.

These results agree with experimental observations in near neutral pH solutions; cations are removed by exchange to depths of several tens of Ångströms, and silica and aluminum dissolve stoichiometrically. This indicates that experimental results may mirror the natural weathering processes. However Nesbitt and Muir (1988) used secondary ion mass spectrometry (SIMS) to measure surface compositions of nine oligoclase grains from a

10,000-year-old till in an acidic watershed in Ontario, Canada. They observed consistent Al enrichment of the surface to depths of ~500 Å, which contrasts with experimental and XPS studies discussed above. The reasons for the disparity are unclear, but suggested explanations include high concentrations of Al^{3+} and other solutes in acidic soil solutions (Muir and Nesbitt, 1991; Nesbitt et al., 1991) and failure to remove all secondary precipitates from the surface.

EXPERIMENTAL DETERMINATION
OF FELDSPAR DISSOLUTION KINETICS

Experimental techniques

Feldspar dissolution rates are usually measured using either batch or flow-through reactors (see Brantley and Chen, this volume). Batch experiments are closed systems in which the feldspar sample and starting solution are sealed in a vessel, usually a polyethylene or teflon bottle. Solutes released or consumed by reaction of the feldspar are allowed to accumulate in solution (free drift), resulting in a changing solution composition with time. Buffers may be used to minimize fluctuations in pH and/or ionic strength, and solutions may be changed periodically to prevent the composition from drifting over too large a range. The variable solution compositions, particularly in pH and Al concentration, associated with most batch dissolution experiments often confound interpretation of the results because several variables change simultaneously. Precipitation of secondary phases can be a serious problem in batch systems, particularly at moderate pH, where the system almost always oversaturates with respect to gibbsite and/or kaolinite, rendering the results only a minimum estimate of the dissolution rate.

Flow-through systems continuously add and remove solution from a well-mixed reactor (e.g. Chou and Wollast, 1984). The dissolution rate is calculated from the difference in solute concentrations of the input and output solutions. Continuous flow of solution through the reactor allows the experiment to remain undersaturated with respect to secondary phases, and/or allows the manipulation of solute concentration in a systematic fashion (e.g. Burch et al., 1993). Because the reactor is well mixed, the composition of the output solution is identical to the solution in contact with the mineral. After an initial transitory period, the flow-through system reaches a steady-state at which the output composition is constant with time. Thus, the feldspar dissolution rate can be determined under constant chemical conditions.

There are many factors to consider in determining and evaluating feldspar dissolution rates. Most emphasis in experimental studies has been on the effects of solution pH, temperature, and feldspar composition on the dissolution rate. Other factors to consider are: (1) the catalytic or inhibiting effect of dissolved inorganic and organic solutes (including pH buffers), (2) the effect of the saturation state of the solution, (3) the type and concentration of both point and line defects within the feldspar, and (4) compositional variations within the feldspar crystals. In addition, different studies treat starting material differently, determine mineral surface areas differently, use mineral names inconsistent with the actual composition, and do not always evaluate the possibility of precipitation of secondary phases. These variations mean that the comparison of results from different studies must be done carefully and critically.

Feldspar surface areas

Feldspar dissolution rates are expressed in units of moles (or mass) per unit area per unit time (mol m^{-2} s^{-1}). The mass of material released by dissolution is determined during the experiments, but the surface area of the dissolving mineral is also required to calculate a dissolution rate, and must be determined independently. Surface areas (SA) can be estimated geometrically, based on the size and shape of particles. However, this geometrical estimate will not include surface roughness or any internal surface area. The surface areas used for determining dissolution rates are most often determined by the BET method (Brunauer et al., 1938; Gregg and Sing, 1982) which measures the adsorption of a gas, usually N$_2$ or Kr, on the surface at 77 K. At a low and fixed N$_2$ or Kr partial pressure, the surface area is proportional to the amount of gas absorbed. Because the diameter of the N$_2$ and Kr molecules is ~4 to 5 Å, surface roughness, porosity, and fractures with dimensions >4 Å will be included in a BET surface area measurement. The 4 Å scale is very similar to the effective size of a water molecule (~3 Å), and probably represents a reasonable estimate of the wetted surface area.

The ratio of the BET to geometrical surface areas is called the surface roughness (SR) factor, following the convention used in surface science. SR also includes any internal surface area such as etch pits, fractures or any other void and connected microstructures. Blum (1994) compiled BET surface areas of freshly ground and washed feldspar particles used in dissolution experiments. The SR of the feldspars averaged 9±6, and thus, the BET surface area of crushed feldspar specimens can be estimated within a factor of ~3 from the grain size.

The informal convention in the silicate dissolution literature is to use the surface area of the starting material for computing the dissolution rate, and surface areas after reaction are only occasionally reported. During most feldspar dissolution experiments only a small fraction of a percent of the starting material is dissolved. Yet surface areas of the starting material may increase by as much as a factor of 7 (Stillings and Brantley, 1995) presumably as a result of surface roughening. Weathered feldspars from soils typically have a SR of several hundreds, with the SR increasing systematically with the increasing extent of weathering from <100 to >1000 (White and Peterson, 1990; Blum, 1994; White et al., in press; White, this volume). The interpretation of the large discrepancy between the surface areas of freshly ground experimental and naturally weathered feldspar grains introduces a large uncertainty in the comparison of natural and experimental feldspar dissolution rates.

Rapid initial dissolution rates

During feldspar dissolution experiments, rapid surface adsorption reactions are followed by a period of rapid dissolution with high rates of release of all the feldspar components. The dissolution rate then decreases with time until a nearly constant dissolution rate is reached after several tens to many hundreds (or even thousands) of hours. This long-term steady-state dissolution rate is of the greatest relevance to most geological systems.

The cause of the rapid initial dissolution is not well understood. Holdren and Berner (1979) suggested two possible explanations: (1) the small particles adhering to the surface after sample grinding have a higher surface free energy, and should dissolve more rapidly than large grains, and (2) disruption of the surface during grinding produces a disordered layer which dissolves more rapidly. Subsequent experiments in which most fine particles were removed prior to dissolution did not completely reduce the high initial rates. Petrovic (1981a,b) supported the idea that a disturbed surface layer produced during grinding causes the rapid initial dissolution rates. This is still probably the most widely accepted explanation.

However, Chou and Wollast (1984, 1985) and several subsequent investigators have also observed a rapid period of dissolution immediately after changing the solution pH during continuous flow-through dissolution experiments. Mast and Drever (1987) also observed a high transient Al release after introducing oxalic acid into oligoclase dissolution experiments. They suggested that the rapid release of elements may be a consequence of the presence of a thin, steady-state surface layer with a non-stoichiometric composition. Changing the solution composition may establish a new steady-state surface composition. As the surface reequilibrates, there may be a transient release of those elements which have a lower concentration in the new surface layer. With this interpretation, the rapid dissolution rates observed at the start of the experiments may result from more rapid dissolution before a non-stoichiometric, steady-state surface composition can evolve.

Steady-state feldspar dissolution rates

Albite dissolution kinetics. Most studies of feldspar dissolution kinetics have used albite, and its dissolution behavior is typical in most respects of feldspars of all compositions. Solution pH is the major solution variable controlling the dissolution rate. This is because both H^+ and OH^- are important participants in feldspar dissolution mechanisms, and because H^+ and OH^- have wider fluctuations in commonly occurring natural concentrations than any other solute. The dissolution rate has a minimum at pH 6 to 8, and increases in both the acid and basic regions (Fig. 4), a pattern common to all feldspars, and many other silicates (e.g. Helgeson et al., 1984; Lasaga, 1984; Murphy and Helgeson, 1987). This behavior strongly suggests at least two different dissolution mechanisms for feldspars: a proton-promoted mechanism in the acid region, and a hydroxyl-promoted mechanism in the basic region.

Many workers (e.g. Murphy and Helgeson, 1987; Chou and Wollast, 1985; Knauss and Wolery, 1986; Mast and Drever, 1987; Sverdrup, 1990; Drever, 1994) have suggested that feldspar dissolution rates are independent of pH in the neutral pH region from ~5 to 8. They have interpreted this as a third dissolution mechanism for feldspars which dominates in the neutral pH region and is pH independent. Other workers (e.g. Blum and Lasaga, 1991; Brady and Walther, 1989; Schweda, 1990) proposed only two pH dependent mechanisms.

The determination of very slow feldspar dissolution rates at near neutral pH is problematic. Slow reaction rates mean that solutes may be at or below analytical detection limits, thus affecting the accuracy of the experiments. Also, very long reaction times are needed to dissolve high-energy surface sites and to reach steady-state dissolution of the bulk mineral. Therefore it is not certain whether experiments at neutral pH reflect a true steady-state rate. The neutral pH region includes many natural environments including most soils, and surface and ground waters. A pH independent mechanism in the neutral region would greatly simplify the practical application of feldspar dissolution kinetics.

Potassium-feldspar dissolution kinetics. Figure 5 compiles K-feldspar dissolution rates as a function of solution pH at ~25°C. At pH < 6 the dissolution rates (R) of K-feldspars are indistinguishable from albite, both in their pH dependence (R $\propto$ $[H^+]^{-0.5}$) and in their absolute rate. The major difference between the structures of well ordered albite and K-feldspar is the substitution of K^+ for Na^+ in the cation site, and the slight distortion of the tetrahedral lattice to accommodate the smaller Na^+ cation (Ribbe, 1983b). In acidic solutions both the albite and K-feldspar surfaces undergo rapid exchange of cations with H^+ to form a hydrogen-feldspar surface layer several unit cells thick. This implies that the structure and composition of albite and K-feldspar surfaces at the mineral-solution interface are nearly identical, which is consistent with the similar dissolution kinetics of albite and K-feldspar in the acid region. The presence of a hydrogen feldspar surface layer also implies that the nature of the tetrahedral framework and hydrolysis of bridging oxygen bonds at tetrahedral sites control the dissolution rate of feldspars under acidic conditions, whereas the nature of the charge-balancing cation is of secondary importance.

In most soils, including acidic soils, it is commonly observed that K-feldspar weathers more slowly than sodic plagioclase. This contrasts with the experimental results in Figures 4 and 5, which indicate similar dissolution rates for albite and K-feldspar. The discrepancy may be due to soil porewater chemistry, which tends to be closer to saturation with respect to potassium feldspar than albite. Fritz et al. (1993) have suggested that porewaters quickly reach saturation with respect to K-feldspar during the first days and months of weathering. Likewise, Brantley et al. (in review) analyzed >100 published measurements of soil water solutions from 45 soils, and found that soil waters were always closer to saturation with respect to potassium feldspar than albite. Kinetic models (discussed later) suggest that at this chemical affinity, dissolution rates should decrease with respect to rates measured far from equilibrium (i.e. most laboratory rate measurements).

In the basic pH region, the dissolution rate data for K-feldspar are sparse. However, it appears thatK-feldspar has a slower absolute dissolution rate than albite by a factor of ~10, which may imply that cations have an important influence on feldspar dissolution kinetics in basic solutions. Readsorption of K^+ at cation sites may stabilize the K-feldspar surface more strongly than readsorption of Na^+ on albite surfaces (Wollast and Chou, 1992). K-feldspar dissolution rates may also have a greater pH dependence than albite in basic solutions, although the slope is poorly defined. Schweda (1990) measured the slope of microcline dissolution rates over pH 10.9 to 12.6 and observed it to be highly dependent on the concentration of cations in solution. In his first set of experiments, he used LiOH to adjust pH (i.e. the concentration of Li^+ varied with pH), and found a dependence of $R \propto [OH^-]^{0.73}$. However, when he repeated the experiments with constant Li^+ concentrations, he observed a dependence of $R \propto [H^+]^{0.45}$. The latter is more consistent with previous observations of $R \propto [H^+]^{0.3}$ (Chou and Wollast, 1985) and $R \propto [H^+]^{0.4}$ (Helgeson et al., 1984).

Plagioclase dissolution kinetics. Feldspar weathering rates in soils vary as calcic plagioclase > sodic plagioclase (Goldich, 1938). Abundant morphologic evidence supports this relationship. A common example is the preferential weathering of the calcic cores from zoned plagioclase crystals before the weathering of the more sodic rims. Figures 6a-c compile recent dissolution rate data for plagioclase as a function of feldspar composition (% An) at pH 2, 3, and 5. Casey et al. (1991), Stillings and Brantley (1995), and Stillings et al. (in review) have compiled similar plots. At all three pH values, we observe a linear increase in the logarithm of the rate of feldspar dissolution with increasing anorthite content from An_0 to $\sim An_{80}$. The total increase in dissolution rate from albite to anorthite ranges from a factor of $10^{2.5}$ to 10^4, and the mole fraction of the anorthite component is a strong control on plagioclase dissolution rates. There appears to be a discontinuity in the dependence of plagioclase dissolution rate on composition at $\sim An_{80}$, with anorthite dissolution rates increasing considerably faster than the trend predicted from the less-calcic plagioclase compositions.

The anorthite measurements at pH 5 demonstrate wide disagreements, with differences of almost 10^6 in the measured dissolution rates. The higher values are from Fleer (1982) and Sverdrup (1990), although Fleer (1982) used the geometric surface area of a single large

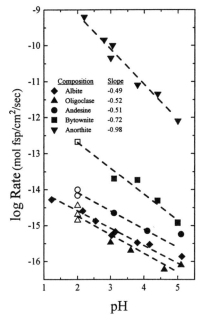

Figure 7. Dissolution rate as a function of solution pH for plagioclase feldspars of different compositions. ▼ anorthite, Sverdrup (1990); ■ bytownite, Oxburgh and et al. (1994); □ bytownite, Casey et al. (1991); ● andesine, Oxburgh et al. (1994); O andesine, Casey et al. (1991); ▲ oligoclase, Oxburgh and et al. (1994); Δ oligoclase, Casey et al. (1991); ◆ albite, Chou and Wollast (1985) (after Oxburgh et al., 1994).

crystal, and may have severely underestimated the BET surface area. The data from Busenburg and Clemency (1976) were collected from batch experiments at 1 atm CO_2, and Busenburg (1978) reported that these experiments precipitated secondary phases. The data from Amrhein and Suarez (1988, 1992) are also batch experiments which were run for up to 4.5 years, and in which significant secondary precipitates were reported. Thus, these low anorthite dissolution rates are probably the result of secondary precipitation, or the high solute concentrations allowed to accumulate during the experiment which reduced the reaction rate. In either case these experiments are not comparable with the other rate data in Figure 6c collected in dilute solutions far from equilibrium, and they have not been included in the indicated trends or in the interpretation. However, discrepancies of a factor of 10^6 illustrate the difficulties in the comparison and interpretation of feldspar dissolution experiments.

Figure 7 shows the dissolution rate for a series of plagioclase compositions as a function of solution pH in the acid region. The plagioclase series from albite to $\sim An_{70}$ (as well as K-feldspar) has a dependence of R $\propto$ $[H^+]^{-0.5}$, but the exponent increases to 0.75 for An_{76} (Oxburgh et al., 1994), and 1.0 for anorthite (Sverdrup, 1990). Data from Welch and Ullman (in review) also suggest that the $[H^+]$ dependence in the acid region increases with increasing An content. The transition from a dependence of ~0.5 appears to occur at $\sim An_{70}$, and corresponds to the observed change in dissolution behavior with mol % An observed at pH 2, 3 and 5 (Fig. 6). There appears to be a critical Al/Si ratio at An_{70-80} which dramatically affects plagioclase dissolution kinetics. This may be related to a decrease in the average size of linked clusters of Si tetrahedra after preferential attack and removal of tetrahedral Al. In albite with an Si/Al ratio of 3, preferential attack and removal of Al will still leave a fragmented but partially linked framework of tetrahedral Si. Removal of the Si requires the hydrolysis of Si-O-Si bonds, although the Si centers will probably have several hydroxyl groups. In anorthite (An_{100}) the Si/Al ratio equals 1, and the ordering imposed by the aluminum avoidance principle means that removal of Al leaves completely detached Si tetrahedra, meaning that hydrolysis of Si-O-Si bonds is not required to decompose the anorthite structure. There may be a critical Al/Si ratio at which the hydrolysis of Si-O-Si is no longer a control on the plagioclase dissolution rate, which corresponds to a change in the overall dissolution mechanism.

Reproducibility of feldspar dissolution rates. The reproducibility of feldspar rate determinations by the same laboratory are generally $<\pm 50\%$ and almost always within a factor of 2. For example, Holdren and Speyer (1987) found a reproducibility of individual experiments of $<\pm 15\%$. The agreement between feldspar rate determinations conducted in different laboratories under similar conditions, such as in dilute solutions with no secondary precipitates, is generally within a factor of 5, maybe a little better at low pH. The differences may reflect subtle differences in composition and microstructure of the mineral specimens, grain size, length of experiment, solution saturation state, cation concentrations, surface area determinations, and the concentrations of other possible catalysts and inhibitors such as organic buffers and Al^{3+}.

Holdren and Speyer (1987) measured the dissolution rates of different grain size fractions from nine different feldspar compositions at pH 3, and one sample each at pH 2 and 6. They found a systematic decrease in dissolution rate with decreasing grain size from >600 mm to <37 mm, with the rates decreasing by a factor from 2.8 to 28. Their data for albite are shown on Figure 4 at pH 3 (Holdren and Speyer, 1987) and 6 (Holdren and Berner, 1979), and the observed change in rate with grain size is greater than the variability between the rates of most other workers. A similar decrease in dissolution rate with decreasing grain size is reported by Amrhein and Suarez (1992) for anorthite. Holdren and Speyer (1987) proposed that defect structures at twin boundaries and between exsolution domains are preferential sites for dissolution. Grinding of feldspar starting material preferentially cleaves the crystals along twin and exsolution boundaries, effectively destroying them. Thus, the density of these high energy boundaries exposed on the grain surfaces decreases with decreasing grain size. This continues until the grain size drops below the average spacing of the defect. However, Murphy and Drever (1993) measured the dissolution rates of seven albite fractions from 0.5 mm to 150 mm at pH 4 and found the dissolution rate did not vary with grain size, in contrast to the data of Holdren and Speyer (1985, 1987). Inskeep et al. (1991) observed differential weathering of the more calcic exsolution lamellae in experimentally dissolved labradorite using TEM, but they did not observe preferential dissolution along the lamellae boundaries. Thus there remains uncertainty in the magnitude and/or mechanism by which grain size affects experimental dissolution rates.

Effects of aluminum concentration and saturation state

The concentrations of feldspar components in solution may affect the dissolution rates, either by the adsorption of a specific component to the surface which inhibits the kinetics of the molecular reactions, or by inducing a back reaction (i.e. precipitation) of feldspar on the surface, thereby reducing the overall dissolution rate. The first mechanism, specific inhibition, is sensitive to the concentration of the individual solution species which forms the inhibiting surface complex. The second mechanism, back precipitation, requires the stoichiometric deposition of all components of the crystal on the surface, and is more sensitive to the solution saturation state (Q/K, where Q is the activity product and K is the equilibrium constant) which is a function of the concentrations of all components in solution, and is a measure of the difference in chemical potential of the components in solution versus in the crystal structure.

Chou and Wollast (1985) examined the effects of Al, Si and Na concentrations on albite dissolution rates at pH 3 by varying the aqueous concentrations of these elements from 10^{-5} to 10^{-3} M. They found that aqueous concentrations of Si and Na had only a weak retarding effect on the albite dissolution rate (R $\propto$ [Si]$^{-0.15}$ and [Na]$^{-0.12}$). The dependence of the dissolution rate on Al concentration was also weak at Al concentrations >10^{-5} M (R $\propto$ [Al]$^{-0.08}$). At Al concentrations <10^{-5} M, Chou and Wollast (1985) found a much stronger dependence of the dissolution rate on Al concentration (R $\propto$ [Al]$^{-0.38}$), indicating a strong inhibition of albite dissolution by aqueous Al at very low concentrations. At higher Al concentrations, the effect of all the elements was involved, suggesting a saturation state effect. However, because the concentrations of all the elements varied simultaneously, it is difficult to differentiate the effects of the individual solute concentrations on the dissolution rate. While Na concentrations did not have a strong effect on albite dissolution rates at pH 3, Chou and Wollast (1985) compared albite dissolution in 0.1 M NaCl and 0.1 M LiCl over a wide pH range. Dissolution rates in NaCl slow above ~pH 9, indicating that Na does have a retarding effect in the basic region. This retardation is consistent with the exchange of Na$^+$ for H$^+$ in the surface layer retarding the dissolution rate (Wollast and Chou, 1992).

Burch et al. (1993) conducted dissolution experiments of albite at 80°C and pH 8.8 at a constant NaCl concentration of 0.1 M. They systematically varied the Al and Si concentrations, and observed a decrease in the albite dissolution rate with increasing solution saturation state (i.e. as they approached equilibrium, indicated by a Gibbs free energy of reaction, $\Delta G_r = 0$ in Fig. 8a). They observed a dramatic decrease in the dissolution rate at a $\Delta G_r \sim -35$ kJ/mol (Fig. 8a). In dilute solutions ($\Delta G_r < -38$ kJ/mol) they found the albite dissolution rate was independent of solution saturation state, reaching a "dissolution plateau." Burch et al. (1993) used an isotach diagram to distinguish whether retardation of dissolution rate is a specific inhibition by Al and/or Si or is a saturation effect dominated by

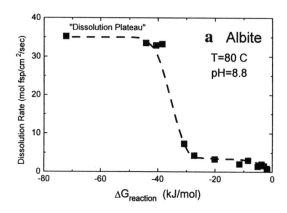

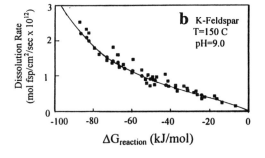

Figure 8. Dissolution rates of (a) albite (after Burch et al., 1993), and (b) K-feldspar (after Gautier et al; 1994) as a function of Gibbs free energy of reaction (ΔG_r). Squares and circles represent data from Al-rich and Si/Al-free input solutions. In both plots, $\Delta G_r = 0$ at equilibrium, and ΔG_r becomes more negative as the solution becomes more undersaturated (to the left), which corresponds to increasing dissolution rates.

ΔG_r. They concluded that the retardation corresponds to a saturation state effect. However, the shape of the rate versus ΔG_r plot (Fig. 8a) does not agree with the predictions of transition state theory (TST) for the effects of saturation state. Burch et al. (1993) propose that the dissolution rate increases dramatically at solution compositions with $\Delta G_r < -35$ kJ/mol as a consequence of the opening of etch pits at dislocations, resulting in accelerated dissolution at the sites of these defects.

Gautier et al. (1994) and Oelkers et al. (1994) conducted dissolution experiments of K-feldspar and albite, respectively, at 150°C and pH 9 while independently varying the Al and Si aqueous concentrations. Their results also demonstrate a dependence of the dissolution rate on the saturation state of the solution (Fig. 8b). The results of Gautier et al. (1994) and Oelkers et al. (1994) differ from those of Burch et al. (1993) in several respects. Their data extend to more dilute solutions (lower ΔG_r) yet they continue to observe a retardation of feldspar dissolution rates as a function of the solution composition, and never see the dissolution plateau observed by Burch et al. (1993). The continued retardation of the dissolution rate suggests that solution composition is a critical factor in feldspar dissolution kinetics, even in extremely dilute solutions typical of those found in many near-surface weathering environments. Secondly, neither Gautier et al. (1994) nor Oelkers et al. (1994) observed the dramatic break in the dissolution rate as a function of ΔG_r observed by Burch et al. (1993). Finally, Gautier et al. (1994) and Oelkers et al. (1994) conclude, based on two lines of evidence, that at least in dilute solutions, Al specifically inhibits K-feldspar dissolution. First, the dissolution rates of both albite and K-feldspar do not show any decrease of the dissolution rate as a function of Si concentration in solutions with constant Al concentrations. Secondly, Gautier et al. (1994) used an isotach diagram to distinguish the specific inhibitory effects of Al and Si on K-feldspar dissolution. They concluded that Al has a specific inhibitory effect on K-feldspar dissolution far from equilibrium, although it appears that the saturation state may dominate the retardation of dissolution in more concentrated solutions (near equilibrium). A transition from a specific Al inhibition in dilute solutions to a saturation effect nearer to equilibrium is consistent with the observations of Chou and Wollast (1985).

Oelkers and Schott (in press) studied anorthite (An_{96}) dissolution at 45° to 95°C and a pH range of 2.4 to 3.2. They found no dependence of the anorthite dissolution rate on either the dissolved Si or Al concentrations. This contrasts with the alkali feldspars for which Al is an effective inhibitor. Once again, this suggests a fundamentally different dissolution mechanism for the Ca-rich plagioclases.

Effects of ionic strength on feldspar dissolution

Past studies suggest that ionic strength does not affect silicate dissolution rates in undersaturated conditions (Fleer, 1982; Schott et al., 1981; Rimstidt and Dove, 1986; Tole et al., 1986). However more recent studies show that dissolution rate is a function of cation identity in both acidic and basic pH.

Acidic pH solutions. Schweda (1990) studied the effect of 10^{-2} M solutions of LiCl, NaCl, NH_4Cl, and $SrCl_2$ on microcline and sanidine dissolution at 25°C and pH 3. He observed that the addition of cations inhibited release of Si and Al components from both microcline and sanidine, and that the degree of inhibition increased in the order $NH_4^+ > Sr^{2+} > Na^+ > Li^+$. Moreover, the inhibiting effect of Na^+ on sanidine dissolution appeared to increase with decreasing pH; rates were proportional to $a_{H^+}^{0.41}$ in 0.0 M NaCl, but dropped to $a_{H^+}^{0.29}$ in 10^{-2} M NaCl. Schweda (1990) hypothesized the presence of a porous layer on the dissolving feldspar surface and suggested that cations from solution migrate below the solid-solution interface and exchange with Na^+ or K^+ in the feldspar structure. This exchange inhibits the transport of H^+ (or H_3O^+) into the surface layer, reducing the rate of Si and Al release. The inhibition of the dissolution rate is related to strength of the H_2O dipoles in the cation's hydration shell. Li^+, with the strongest hydration of the cations studied ($\Delta H_{hydration}$ = -499 kJ/mol), has the smallest influence on dissolution rates because it is least likely to lose waters from its hydration shell, and therefore cannot fit into pores of the K-feldspar surface.

Stillings and Brantley (1995) conducted a similar study of the effect of NaCl and $(CH_3)_4NCl$ on feldspar dissolution rates at pH 3 and 25°C with compositions ranging from An_0 to An_{76}. Because surface titration experiments revealed a decrease in feldspar surface protonation in NaCl but not in $(CH_3)_4NCl$ solutions (Fig. 2), they hypothesized that feldspar dissolution rates should decrease in NaCl but not in $(CH_3)_4NCl$. Experimental results show that dissolution rates do decrease by a factor of 1 to 6 (depending upon feldspar composition) as NaCl increases from 0 to 0.1 M. Dissolution rates in NaCl were modeled using Equation (11), and a 2-term Langmuir isotherm which incorporates the competition between Na^+ and H^+ for adsorption for feldspar cation exchange sites ($[H^+_{ex}]$ in Eqn. 11):

$$[H^+_{ex}] = N_s \left[\frac{K_H\{H^+\}}{1+K_H\{H^+\}+K_{Na}\{Na^+\}} \right] \tag{12}$$

where K_i is the equilibrium constant for the adsorption of species i (H^+ or Na^+), $\{i\}$ is the activity of species i, and N_s is the surface concentration of exchange sites. In the absence of Na^+, or if Na^+ does not adsorb (K_{Na} = 0), then Equation (12) reduces to the Langmuir expression used by Blum and Lasaga (1991) to describe the concentration of H^+ adsorbed at an albite surface.

Substituting Equation (12) into the rate model (Eqn. 11), produces a dissolution rate expression:

$$rate = k_a N_s \left[\frac{K_H\{H^+\}}{1+K_H\{H^+\}+K_{Na}\{Na^+\}} \right]^{0.5} \tag{13}$$

where $k_a(s^{-1})$ is the rate constant, and n = 0.5 was theoretically derived by Brantley and Stillings (in press). Equation (13) was fit to dissolution rate data (Fig. 9) with the parameters tabulated Stillings and Brantley (1995).

Although Stillings and Brantley (1995) predicted that dissolution rates will not decrease in the presence of $(CH_3)_4NCl$, the $(CH_3)_4N^+$ cation does appear to have an inhibiting effect on dissolution rates (Fig. 9). They propose two possible explanations: (1) although $(CH_3)_4N^+$ has a larger hydrated diameter than Na^+, (5.8 vs. 4.0 to 4.5 Å), it may be able to penetrate some pores in the leached layer, which on glass surfaces have effective diameters of ≥ 10 Å (Casey and Bunker, 1990). Therefore $(CH_3)_4N^+$ might penetrate the surface to a limited extent, and compete with H^+ for exchange sites. (2) $(CH_3)_4N^+$ adsorbs to siloxane sites, $\equiv Si\text{-}O\text{-}Si\equiv$ (van der Donck et al., 1993), which may inhibit the rate of network hydrolysis and breakdown of the feldspar structure.

Basic pH solutions. In basic pH solutions, cations have the opposite effect on feldspar dissolution as in acidic solutions, enhancing the dissolution rate rather than inhibiting it. Schweda (1990) measured K-feldspar dissolution at pH 10.9 to 12.6 in the presence of 10^{-2} M solutions of Li_2CO_3, Na_2CO_3, and K_2CO_3. Dissolution rates of microcline increased with increasing Li^+ concentration, and although microcline and

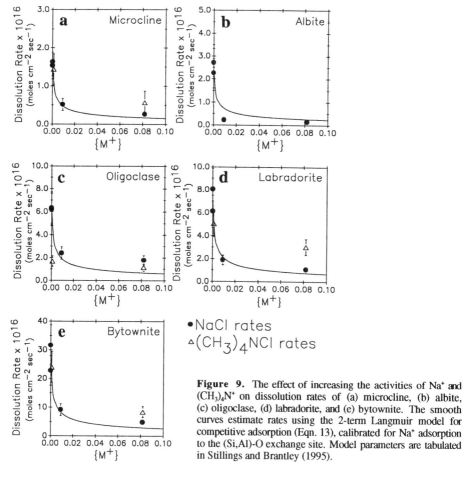

Figure 9. The effect of increasing the activities of Na^+ and $(CH_3)_4N^+$ on dissolution rates of (a) microcline, (b) albite, (c) oligoclase, (d) labradorite, and (e) bytownite. The smooth curves estimate rates using the 2-term Langmuir model for competitive adsorption (Eqn. 13), calibrated for Na^+ adsorption to the (Si,Al)-O exchange site. Model parameters are tabulated in Stillings and Brantley (1995).

sanidine both dissolved most rapidly in the presence of Li^+, Na^+ also enhanced the dissolution rate of both minerals. The effect of K^+ was unclear because it enhanced the dissolution rate of sanidine while inhibiting the rate of microcline.

At basic pH, feldspar dissolution may be controlled by detachment of Si, because Si surface sites are deprotonated and carry negative charge (Brady and Walther, 1989). If feldspar dissolution at this pH is enhanced by Li^+ and Na^+, then these cations must enhance the release of Si from the feldspar structure. It has been suggested that adsorption of Na^+ to surface silanols in quartz accelerates dissolution by providing water a greater access to ≡Si-O-Si≡ linkages (Dove and Crerar, 1990). A similar mechanism may operate to enhance feldspar dissolution at basic pH in the presence of cations.

Temperature dependence of feldspar dissolution

Temperature has a significant influence on the dissolution rate of feldspars, and its effect is typically expressed with the Arrhenius equation:

$$k_+ = A\,e^{\frac{-E_{app}}{RT}} \tag{14}$$

where k_+ is the forward rate constant, A is the pre-exponential frequency factor, E_{app} is the apparent activation energy, R is the gas constant, and T is temperature (K). Lasaga (1994) suggests using the label "apparent" to describe the activation energy because it is usually calculated for the overall, net reaction, and not the elementary reaction described by classical transition state theory.

A number of studies have measured E_{app} (Table 2), of feldspars with results ranging from 14 to 117 kJ/mol. Many of these studies measured E_{app} by varying the temperature

Table 2. Published activation energies for feldspar dissolution.

Feldspar	E_{app} (kJ/mol)	pH	Temp range, °C	References
albite	84	acid	25 - 200	Helgeson et al., 1984
	58.6	pH 3	6.6 - 55.0	Chou, 1985
	117	< 3	25 - 70	Knauss & Wolery, 1986
	54.4	neutral	"	Knauss & Wolery, 1986
	32.2	basic	"	Knauss & Wolery, 1986
	64.3	acid	n.r.	Sverdrup, 1990
	50.7	neutral	"	Sverdrup, 1990
	59.3	basic	"	Sverdrup, 1990
	71.4	pH 1.4	25 - 90	Rose, 1991
	62.8	acid	5 - 300	Chen, 1994
	88.9	acid	100 - 300	Hellmann, 1994
	68.8	neutral	"	Hellmann, 1994
	85.2	basic	"	Hellmann, 1994
	44.0[1]	pH 3	5 - 90	Stillings et al., 1995b
	60.0	acid	5 - 300	Figure 10a
	67.7	neutral	"	Figure 10b
	50.1	basic	"	Figure 10c
K-feldspar	38	neutral	25 - 200	Helgeson et al., 1984
	14.4 - 57.7	1 - 3.6	70 - 95	Bevan & Savage, 1989
	52 - 60	3 - 4	5 - 70	Schweda, 1990
	63 - 70	10.9	"	Schweda, 1990"
	53 - 78.3	acid	n.r.	Sverdrup, 1990
	35 - 37	neutral	"	Sverdrup, 1990
	51.7	acid	5 - 100	Figure 11a
	57.8	basic	"	Figure 11b
Oligoclase	80.3	acid	n.r.	Sverdrup, 1990
	46.1	neutral	"	Sverdrup, 1990
Labradorite	65	1 - 2	15 - 70	Sjöberg, 1989
	48.1[2]	4	21 - 60	Brady & Carroll, 1994
	66.4	acid	5 - 70	Figure 12
Anorthite	33	2	25 - 70	Fleer, 1982
	35	acid	n.r.	Sverdrup, 1990
	107	neutral	"	Sverdrup, 1990
	18.4	2.4 - 3.2	45 - 95	Oelkers, in press
	80.7	acid	25 - 95	Figure 13

[1] experiments conducted in 10^{-3} M oxalic acid

[2] experments conducted in 10^{-3} M acetic acid

n.r. = not reported

and comparing dissolution rates under otherwise similar reaction conditions (primarily constant pH). Lasaga (1984) notes the observed range of activation energy values falls between activation energies for transport in solution (~21 kJ/mol) and the range of activation energies expected for the breaking of bonds in crystals (160 to 400 kJ/mol). He suggests that catalytic effects of adsorption on the surface and/or the role of surface defects are responsible for reducing activation energies to this intermediate range.

To colculate acrivation energies, we have extracted rate constants from dissolution data in the literature, following a method used by Hellmann (1994). Depending upon pH, one of three equations was used. For rates measured in the acidic pH range,

$$k_+ = rate \cdot (a_{H^+})^n \tag{15}$$

In the neutral range,

$$k_+ = rate \tag{16}$$

For rates measured in the basic pH range,

$$k_+ = rate \cdot (a_{OH^-})^{-n} \tag{17}$$

which can be rewritten in terms of pH and pKw (Hellmann, 1994),

$$\log(k_+) = \log(rate) + n(pK_w) - n(pH) \tag{18}$$

Where possible, the rate constants (k_+) were calculated from dissolution rates measured over a range of pH values. However, in some cases rate constants were extrapolated from a single measurement (Tables 3 to 6).

Activation energies can be calculated using a linearized version of Equation (14):

$$\ln k_+ = \frac{-E_{app}}{RT} + \ln A \tag{19}$$

Therefore E_{app} can be calculated from the slope of a plot of $\ln(k_+)$ versus $1/T$. Plots of rate constants from Tables 3 to 5 as a function of $1/T$, pH, and feldspar composition, reveal an apparent activation energy of ~60 kJ/mol for dissolution of albite, K-feldspar, and labradorite (An_{50}-An_{70}) in acidic pH solutions (Figs. 10 to 12), which is similar to that predicted by Lasaga (1984) for the dissolution of silicate minerals. With an E_{app} of ~60 kJ/mol, dissolution rates will increase by approximately two orders of magnitude between 25° and 95°C, a temperature representative of dissolution in sedimentary basin environments.

The activation energy calculated for anorthite dissolution, 80.7 kJ/mol (Fig. 13), is higher than estimates calculated from each individual dataset for anorthite dissolution, which ranged from 18 to 35 kJ/mol (Table 6). This discrepancy may be caused by the relative lack of dissolution data for the An_{90-100} feldspar composition, different experimental techniques and specimens used in the studies, or perhaps to different dissolution mechanisms which operate between 25° and 100°C.

Indeed, the studies cited in Figure 13 and Table 6 are not in agreement with respect to *n*, the order of the elementary reaction, which is a measure of the number of reactant molecules. Because the activated complex has not yet been identified, most dissolution studies report *n* for the net or overall reaction. Amrhein and Suarez (1988) and Sverdrup

(1990) estimate that at 25°C, n = 0.95 and 1.1, for anorthite dissolution. Oelkers and Schott (in press) average their anorthite data across a temperature range of 45° to 95°C, and suggest that n = 1.5. These three studies of anorthite dissolution suggest that n may increase from ~1 to 1.5 as temperature increases from 25 to 95°C. Brady and Walther (1992), Casey and Sposito (1992), and Casey et al. (1993) all predict an increase in n with increasing temperature for silicate dissolution. Hellmann (1994) also observed n to rise with increasing temperature for albite dissolution in acidic pH, from 0.2 at 100°C, and 0.4 at 200°C, to 0.6 at 300°C.

Less data are available for the temperature dependence of dissolution rates in the neutral and basic pH regions, and therefore the activation energies reported here are poorly

Table 3. Calculated rate constants for albite.

log k_+	$\mid n \mid$	T, °C	pH	Source
-10.04	0.5	6.6	3	Chou, 1985
-9.66	0.5	25	3	"
-9.64	0.5	26.2	3	"
-9.4	0.5	35	3	"
-9.37	0.5	39	3	"
-9.19	0.5	45	3	"
-9.11	0.5	50	3	"
-9.00	0.5	55	3	"
-9.69	0.49	25	≤5	Wollast & Chou, 1985
-8.18	0.97	70	≤3	Knauss & Wolery, 1986
-9.5	0.5	25	acid	Sverdrup, 1990
-10.2	0.5	8	acid	"
-8.5	0.2	100	≤5	Hellmann, 1994
-5.9	0.4	200	≤5	"
-4.1	0.6	300	≤5	"
-8.8	0.26	90	3-5	Voigt et al., unpub. data
-12.15	0	25	5-8	Wollast & Chou, 1985
-11.14	0	70	4-8.8	Knauss & Wolery, 1988
-11.8	0	25	neutral	Sverdrup, 1990
-12.4	0	8	neutral	"
-9.5	0	100	5-8.6	Hellmann, 1994
-7.7	0	200	5-8.6	"
-6.2	0	300	5-8.6	"
-9.95	0.3	25	>8	Wollast & Chou, 1985
-9.25	0.48	70	>8.8	Knauss & wolery, 1988
-9.9	0.3	25	basic	Sverdrup, 1990
-10.3	0.3	8	basic	"
-8.3	0.3	100	≥8.6	Hellmann, 1994
-6.3	0.4	200	≥8.6	"
-4.5	0.6	300	≥8.6	"
-8.3[a]	0.3	150	9	Oelkers et al., 1994
-8.71[a]	0.3	150	9	"
-9.16[a]	0.3	150	9	"

[a]k_+ estimates vary with chemical affinity.

Table 4. Calculated rate constants for K-feldspar dissolution.

log k_+	$\lvert n \rvert$	T °C	pH	Source
-9.45	0.4	25	acid	Holdren & Speyer, 1985; 1987
-9.07	0.68	70	1-3.6	Bevan & Savage, 1989
-8.28	0.85	95	1-3.6	"
-10.45	0.5	5	acid	Schweda, 1990
-10.15	0.5	15	acid	"
-9.93	0.5	25	1-5.7	"
-9.55	0..5	35	acid	"
-9.12	0.5	50	acid	"
-8.55	0.5	70	acid	"
-9.6	0.5	25	acid	Sverdrup, 1990
-10.3	0.5	8	acid	Sverdrup, 1990
-10.43[a]	0.45	25	basic	Schweda, 1990
-10.2[a]	0.45	25	basic	"
-9.8[a]	0.73	25	basic	"
-9.2	0.3	25	basic	Sverdrup, 1990
-9.6	0.3	8	basic	"
-6.15[b]	0.3	150	9	Gautier et al., 1994
-6.48[b]	0.3	150	9	"
-7.33[b]	0.3	150	9	"

[a]k_+ and n depend upon Li concentration
[b]k_+ estimates vary with chemical affinity

Table 5. Calculated rate constants for labradorite dissolution.

log k_+	$\lvert n \rvert$	T °C	pH	Source
-8.86	0.5	25	1-5	Sjöberg, 1989
-7.37[a]	0.5	70	1-2	"
-9.20[a]	0.5	15	1-2	"
-8.93	0.5	25	acid	Sverdrup, 1990
-9.67	0.5	8	acid	"
-8.29	0.41	25	3.1-5.3	Welch & Ullman, 1993
-9.33	0.4	25	2-5.1	Oxburgh et al., 1994

[a]k_+ calculated from E_{app} provided by the author.

constrained. The E_{app} calculated for K-feldspar at basic pH, in particular, is highly influenced by the data of Gautier et al. (1994), which, in turn, appear to be function of the concentration of Al in solution. Nonetheless, these estimates still fall within the expected range for surface-reaction mechanism of dissolution.

Effects of organic acids upon feldspar dissolution

Feldspar dissolution in the presence of naturally-occurring organic acids has been extensively investigated in studies of soil formation and fertility, and in diagenetic studies where feldspar dissolution and Al transport influence the formation of secondary porosity

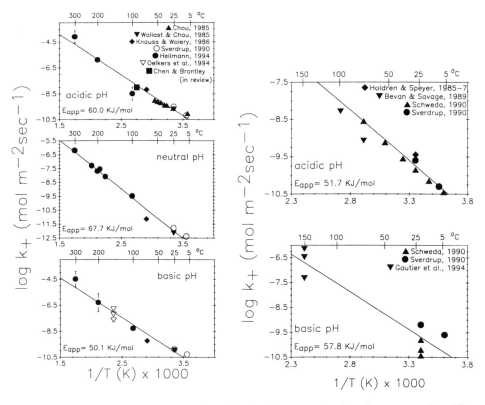

Figure 10 (left). Rate constants (k_+) for albite dissolution as a function of temperature for acidic, neutral, and basic pH solutions. Data plotted here were calculated with Equations (15) to (18), and is summarized in Table 3.

Figure 11 (right). Rate constants (k_+) for K-feldspar dissolution as a function of temperature for acidic and basic pH solutions. Data plotted in these figures are summarized in Table 4.

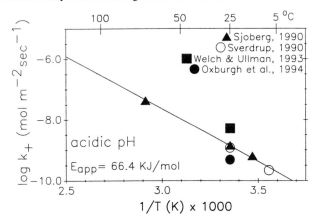

Figure 12. Rate constants (k_+) for labradorite (An_{50-70}) dissolution as a function of temperature for acidic pH solutions. These data are also in Table 5.

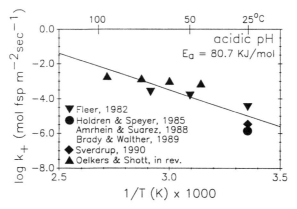

Figure 13. Rate constants (k_+) for anorthite (An$_{90\text{-}100}$) dissolution as a function of temperature for acidic pH solutions. These data are tabulated in Table 6.

Table 6. Calculated rate constants for anorthite dissolution.

log k_+	$\vert n \vert$	T °C	pH	Source
-4.47	1.12	25	2	Fleer, 1982
-3.82	1.12	50	2	"
-3.62	1.14	70	2	"
-5.87	1.12	25	≤5	Brady & Walther, 1989;
				Holdren & Speyer, 1985;
				Amrhein & Suarez, 1988
-5.48	0.95	25	2.2-5	Sverdrup, 1990
-3.13	1.5	45	2.4-3.1	Oelkers et al., in press
-2.98	1.5	60	"	"
-2.84	1.5	75	"	"
-2.72	1.5	95	"	"

in petroleum reservoirs. Most experiments have been performed in batch reactors, where concentrations of Al, Si, Ca, Na, and K released from feldspar powders are measured in a closed system as a function of the type of organic ligand. Results reveal that multi-functional organic ligands (e.g. citrate, oxalate, EDTA) which form strong complexes with polyvalent cations (e.g. Ca^{2+} and Al^{3+}) are the most effective in promoting the release of feldspar components to solution. Bennett and Casey (1994) provide an excellent discussion of mechanisms of silicate dissolution in the presence of organic acids. A summary of feldspar dissolution studies with various organic acids is shown in Appendix I. Many of the studies in Appendix I were designed only to compare concentrations of feldspar components released in various organic acids; they were not designed to quantify rates of dissolution (moles fsp cm^{-2} sec^{-1}) as a function of pH and organic ligand concentration. Therefore, Appendix II compiles data currently available for dissolution rates as a function of pH and ligand concentration.

Huang and Kiang (1972) concisely summarized the effect of organic acids on dissolution of albite, oligoclase, labradorite, bytownite, anorthite, and K-feldspar at 25°C by ranking the effectiveness of organic acids (0.01 M solutions) as citric > salicylic > aspartic > acetic, for each mineral. They concluded that organic acid solutions are more effective in promoting feldspar dissolution than distilled water because of their ability to complex cations such as aluminum. However, because pH was not measured in their experiments, it is difficult to separate the effect of the organic ligand from the effect of pH upon dissolution. Manley and Evans (1986) conducted similar experiments on microcline, albite, and labradorite dissolution at 13°C, and ranked the effectiveness of 10^{-4} M organic acids as citric $\approx$ oxalic > salicylic > protocatechuic $\approx$ gallic > p-hydroxybenzoic > vanillic $\approx$ caffeic. The calculated pH of these solutions ranged from 3.8 to 4.4, and the acid solutions which promoted the highest dissolution rates also had the lowest pH. The authors concluded the effectiveness of organic acids in promoting feldspar dissolution was due more to the strength of the acid than to its ability to complex aluminum.

Surdam et al. (1984) were among the first to investigate the effect of carboxylic acids on feldspar dissolution at temperatures representative of diagenetic environments (75° to 200°C). They concluded that carboxylic acids enhance feldspar dissolution because they increase the mobility of Al by facilitating Al transport as an organic complex. More recent work (Stoessell and Pittman, 1990) suggests that although Surdam et al. (1984) may have overestimated the ability of carboxylic acids to mobilize and transport Al, oxalate and malonate ligands do form significant complexes with Al under reservoir conditions, although acetate and propionate complexes may be insignificant.

Acetic acid (CH_3COOH) is often used in studies of feldspar dissolution for a number of reasons. It is quite commonly found in brines of sedimentary basins (in concentrations of 1 to 8 mM (Carothers and Kharaka, 1978)), in soil solutions (usually < 1mM, but higher in the rhizosphere; Drever and Vance, 1994), and it can be used as a pH buffer (Murphy, 1993; Franklin et al., 1994). Although it is important to determine whether acetate has a significant effect on dissolution rates, a comparison of data between laboratories leads to conflicting results. Welch and Ullman (1993) studied bytownite and labradorite dissolution at 25°C in 0, 0.1, and 1 mM acetic acid solutions, and their data show no enhancement of dissolution rates in the presence of acetic acid (Fig. 14). In contrast, Manning et al. (1991a) found that albite dissolution increased by up to 2 orders of magnitude in acidic acid solutions of 0 to 2.5 M at 150°C and 500 bars (Appendix II). The discrepancy between these two studies may be due to: (1) the higher temperature and acetic acid concentrations used by Manning et al. (1991a), (2) differences in feldspar composition (albite has less Al than labradorite and bytownite), or (3) experimental and procedural differences between laboratories.

The most commonly studied organic acid is oxalic acid (HOOC-COOH), because of its abundance in natural environments and its ability to form strong complexes with polyvalent cations such as Al^{3+} and Ca^{2+} (Drever, 1994). Oxalic acid is commonly found in soils associated with forest litter, roots, fungi (Stevenson, 1967; Graustein et al., 1977; Drever and Vance, 1994); lichens (Wilson et al., 1980), often observed as Ca and Mg oxalate crystals; and in sedimentary brines (McGowen and Surdam, 1988). There has been controversy regarding the prevalence of oxalic acid in sedimentary basins because naturally occurring concentrations of oxalate are limited by highly insoluble Ca-oxalate salts. Data for the effect of oxalic acid on low temperature (13° to 25°C) dissolution rates are varied and contradictory. Manley and Evans (1986) observed that feldspar dissolution rates increase in the presence of 0.1 M oxalic acid, yet they attribute the increase to the acid strength rather than a complexing effect. Mast and Drever (1987) showed that 0.5 and 1 mM oxalate solutions had little effect on the steady-state release of Si from oligoclase from pH 4 to 9. However, Oxburgh (1991) showed that 1 mM oxalate solutions increased

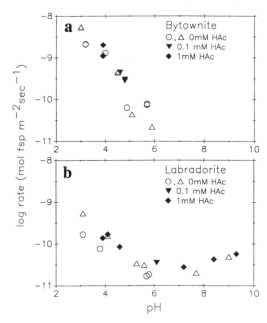

Figure 14. Dissolution rates for (a) bytownite and (b) labradorite in solutions of 0 to 1 mM acetic acid (HAc). These rate data (Welch and Ullman, 1993) show little dependence upon the concentration of acetic acid in solution.

labradorite and bytownite dissolution by a factor of 1 to 3 at pH 3 to 5. Amrhein and Suarez (1988), Welch and Ullman (1993), and Stillings et al. (in review) all report that oxalic acid solutions increase plagioclase dissolution rates by a factor of 2 to 15, depending upon pH and the plagioclase composition.

These apparent contradictions persist at higher temperatures as well. Bevan and Savage (1989) measured the rates of orthoclase dissolution at 70° and 95°C at pH 1, 3.6 and 9, both with and without 0.02 M oxalic acid. Dissolution rates increased in oxalic acid solutions at pH 3.6 and 9 at both temperatures, but *decreased* at pH 1. Stoessell and Pittman (1990) observed enhanced dissolution of microcline in the presence of 0.075 M oxalic acid at 100°C, but questioned whether the effect was due to ligand complexation or the low pH (1 to 3) of the solutions. Franklin et al. (1994) found dissolution rates of albite to increase by ~2 to 4 times in the presence of 5 mM oxalic acid at pH 3 to 5 and 100°C.

One reason for conflicting results in the literature may be the experimental and procedural differences between laboratories. **Figure 15** presents data from 25°C dissolution studies which measured plagioclase dissolution rates as a function of pH and oxalic acid concentrations of 0 and 1 mM. While there is obviously scatter *between* the datasets, *within* each dataset dissolution rates are faster in the presence of 1 mM oxalic acid. The only exception is the oligoclase (An_{13}) data from Mast and Drever (1987) and Oxburgh (1991), which show no rate increase in the presence of 1 mM oxalic acid. The slope of the best-fit lines for data with and without oxalic acid vary from 0.3 to 0.6, increasing with Al-content in the plagioclase. Welch and Ullman (in review) also observed the pH dependence of feldspar dissolution to increase with increasing Al-content.

Another reason for apparent inconsistencies in the literature may be due to speciation of the organic ligand as a function of pH. Oxalate ($C_2O_4^{2-}$) is not the predominant acid anion in solution until the solution pH exceeds the second dissociation constant for oxalic acid (pK_2), which is 4.19 for oxalic acid at 25°C (Linde, 1990). A number of authors

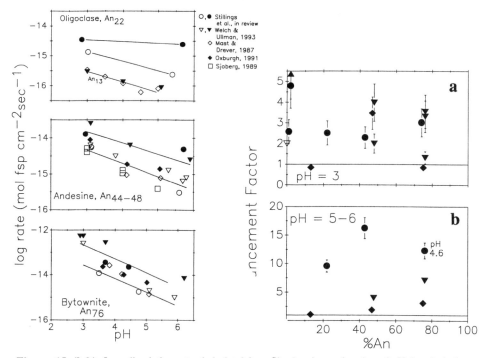

Figure 15 (left). Log dissolution rates (calculated from Si release) as a function of pH for plagioclase dissolution at 25°C. Solid symbols represent dissolution rates in solutions of 1mM oxalic acid, while open symbols represent dissolution in solutions without oxalic acid.

Figure 16 (right). Enhancement of dissolution rates (Eqn. 20) in 1mM oxalic acid solutions. Symbols are defined in Figure 15. There does not appear to be an effect of feldspar composition (as indicated by mol % anorthite component) at pH 3, but individual datasets suggest enhancement may increase with increasing Al-content at pH 5-6.

suggest that oxalic acid should have the greatest effect on dissolution rates at pH's greater than this value. To test this suggestion, data in Figure 15 have been used to calculate a dissolution rate enhancement factor, EF (Amrhein and Suarez, 1988),

$$EF = \frac{R_t}{R_H} \qquad (20)$$

where R_t is the dissolution rate measured in the presence of 1 mM oxalic acid, and R_H is the rate measured without oxalic acid. Figure 16a shows that at pH 3, oxalic acid enhances dissolution rates by a factor of 2 to 5, and the enhancement does not increase with Al-content (anorthite mole fraction) of the plagioclase. At pH 5 to 6, however, the enhancement factor ranges from 2 to 15 times, and the *EF* of individual datasets may increase with increasing anorthite content of the solid (Fig. 16b), as suggested by Amrhein and Suarez (1988).

Rates of dissolution increase with increasing concentrations of oxalic acid. Stillings et al. (in review) present data for andesine dissolution (An$_{47}$) at pH 3 to 5 with 0 to 8 mM concentrations of oxalic acid. Using the model,

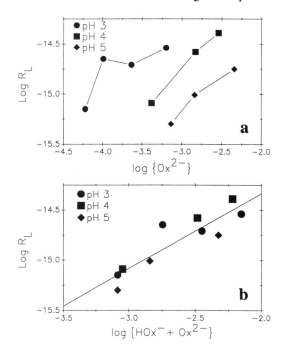

Figure 17. Logarithm of the ligand-promoted rate, R_L, as a function of (a) log oxalate activity, and (b) log (bioxalate + oxalate) activity; $r^2 = 0.8162$ for (b).

$$R_{total} = R_H + R_L \tag{21}$$

where R_{total} represents the observed dissolution rate in the presence of oxalic acid, R_H represents the proton-promoted rate and R_L represents the ligand-promoted rate, they calculated

$$R_L = R_{total} - R_H \tag{22}$$

and plotted R_L as a function of oxalate $(C_2O_4^{2-})$ activity (Fig. 17a). From this plot it is apparent that oxalate alone cannot explain all the variation in R_L. However, by plotting R_L as a function of both oxalate and bioxalate, $HC_2O_4^-$ (Fig. 17b), the data become much more linear, suggesting the bioxalate species also promotes dissolution. From this correlation, Stillings et al. (in review) estimate

$$R_L = 10^{-12.8}(a_{HOx^-} + a_{Ox^{2-}})^{0.75} \tag{23}$$

If both the $HC_2O_4^-$ and $C_2O_4^{2-}$ ligands contribute to the overall dissolution reaction, then Equation (23) should strictly include rate constants (k_{HOx} and k_{Ox}) for each organic ligand. However, it is not certain whether the $HC_2O_4^-$ ligand adsorbs to an individual surface site, or whether, as suggested by Welch and Ullman (1992), $HC_2O_4^-$ dissociates to H^+ and $C_2O_4^{2-}$, leading to a simultaneous attack by both H^+ and $C_2O_4^{2-}$ at a surface site. Welch and Ullman (1992) propose:

$$R = 10^{-7.92}a_{H^+}^{0.71} - 10^{-14.16}pH[ox] + 10^{-12.96}[ox] \tag{24}$$

where R is the rate of Si release (moles•m^{-2}•sec^{-1}); a_{H^+} and $[ox]$ represent the activity of H^+ and the total concentration of oxalate (μM), respectively; and the constants have been evaluated for bytownite (An$_{75}$) dissolution at 25°C, pH 3 to 10, and 0 to 1 mM oxalate.

The middle term defines the contribution of a ligand/proton attack to the overall rate.

Higher temperature studies confirm that dissolution rates increase with increasing concentrations of organic ligands. Franklin et al. (1994) demonstrated that albite dissolution rates at 100°C increase with oxalic acid concentrations of 0.5 to 10 mM. In their study, the ligand-promoted rate term (R_L) is modeled with a Langmuir-type expression:

$$R_{Ox} = 6.7x10^{-10} \left(\frac{2.3x10^{-4} m_{Ox}}{1.0 + 2.3x10^{-4} m_{Ox}} \right) \tag{25}$$

While research to date has primarily focused on the effects of oxalate and acetate on dissolution, there are some data available for the effects of other organic acids, as well. The study of Welch and Ullman (1993) compared the effects of many naturally-occurring organic acids upon plagioclase dissolution. They concluded that the polyfunctional acids oxalate, citrate, succinate, pyruvate, and 2-ketoglutarate, are more effective in promoting dissolution than the monofunctional acids such as acetate and propionate.

Malonic acid also promotes feldspar dissolution. Knauss and Copenhaver (1995) dissolved albite and microcline at 70°C in solutions of pH 4 to 10 with 1 to 100 mM concentrations of malonic acid. Their results indicate that rates increase with increasing concentrations, and that malonate has the greatest effect on dissolution at mildly acidic pH (~5.5), where dissolution rates increased by factors of 2.9 for albite and 6.9 for microcline. At basic pH the effect of the malonate anion was insignificant compared with simple hydrolysis.

A few studies have examined the effect of humic and fulvic acids on dissolution rates. Tan (1980) performed batch experiments of microcline dissolution in humic and fulvic acids at pH 2.5 and 7. The acids were extracted from a Davidson surface soil, a member of the kaolinite-rich, thermic family of Rhodic Paleudult. At pH 7, he observed greater dissolution in humic acid than in fulvic acid. Al was preferentially released in the humic acid experiments, while release was stoichiometric with fulvic acid. At pH 2.5, fulvic acid caused a slightly greater release of Si and Al from microcline than solutions of HCl. Another study conducted by Lundström and Öhman (1990) compared dissolution rates for microcline and oligoclase between solutions of distilled water, stream water, and peat and mor extracts at pH 5.1. They observed the greatest dissolution rates in solutions of streamwater, followed by the peat and mor extracts, then distilled water. Interestingly, when they inoculated their experiments with microorganisms, the dissolution rates dropped to equal those in distilled water, which suggested that microbes consumed some of the organic compounds in these solutions.

Mechanisms by which microorganisms enhance feldspar dissolution are largely unknown. Hiebert and Bennett (1992) and Bennett et al. (in review) have observed accelerated weathering and deep etching of feldspar surfaces under colonies of adhering microbes. Direct contact between bacteria and mineral surfaces is not always necessary, however, and Vandevivere et al. (1994) demonstrated that some bacteria enhance bytownite dissolution by excreting gluconate into their environment. Bacterial secretions may also inhibit dissolution rates, as shown in studies with alginate and high-molecular-weight polyaspartate, or the secretions may have no effect if they are neutral sugars such as cellulose, starch, polysucrose, or gum xanthan (Welch and Vandevivere, 1994). Welch and Vandevivere (1994) concluded that bacterial secretions with negatively charged functional groups have the largest effect on feldspar dissolution, because they bind directly to the mineral surface. In contrast, polymers which form weak bonds with the surface have little or no effect on dissolution kinetics. It is a point of debate whether microbes exert their influence beyond the scale of their microenvironment and active research on this topic may show that microbes have a larger influence than previously assumed.

MODELS OF FELDSPAR DISSOLUTION KINETICS

Lasaga (this volume) describes theoretical approaches to modeling mineral dissolution kinetics, focusing on the strengths and weaknesses in the application of different theoretical approaches to feldspar dissolution kinetics. Any comprehensive model of feldspar dissolution kinetics should be consistent with six basic observations: (1) the pH dependence of dissolution rates, (2) the retarding effects of Al in solution, (3) the pH-dependent retarding/catalyzing effects of increasing concentrations of alkali and alkaline earth cations in solution, (4) the observations of non-stoichiometric dissolution and formation of non-stoichiometric surface layers, at least at acidic pH, (5) the temperature dependence and apparent activation energies of the dissolution reaction, and (6) the ubiquitous presence of high energy surface sites which form etch pits, and their role in the overall dissolution kinetics. Some theoretical approaches are more appropriate than others in addressing different aspects of these observations, and a successful understanding of feldspar dissolution requires integration of several different approaches. However, there is currently no consensus on the overall mechanism of the feldspar dissolution reaction, and geochemists are still in the process of selecting the most appropriate conceptual models. We have, somewhat arbitrarily, divided the proposed models into four categories based upon the conceptual scale: leached-layer/diffusion models, atomistic models, surface speciation models, and macroscopic models.

Leached-layer / diffusion models

Several early studies of silicate dissolution (Wollast, 1967; Luce et al., 1972; Paces, 1973) observed initial non-stoichiometric dissolution along with "parabolic kinetics" in which the rate of dissolution decreased with the square-root of time ($R \propto t^{-1/2}$) during the first few days of reaction. These observations were consistent with dissolution kinetics controlled by the solid-state diffusion of reactants through a protective surface layer, and led to the acceptance of a leached/diffusion layer model for dissolution kinetics.

In a diffusion-controlled reaction model, the rate-limiting step is the transport of products or reactants through a layer at the mineral surface. When a fresh surface is exposed to solution, reaction between the solid and solution occurs at a rapid rate. However, if the products from this reaction form a continuous and tightly-adhering layer on the surface, then reactants and products must diffuse through this surface layer for the reaction to continue. As the reaction proceeds, the layer of reaction product grows thicker, requiring a longer diffusion path of the reactants, thus decreasing the overall reaction rate. Eventually, the dissolution rate of the surface layer approaches the overall reaction rate of the underlying material, and both the reaction rate and the surface layer thickness reach steady-state values.

There is an empirical observation in metallurgy that when the volume of the reaction products is greater than the volume of reactants, surface layers tend to form effective diffusional barriers. Indeed, solid-state diffusion through protective surface layers controls the oxidation kinetics of many metals (e.g. $Fe^\circ \Rightarrow Fe_2O_3$ and $Al^\circ \Rightarrow Al_2O_3$). However, feldspar dissolution does not appear to meet this criterion for diffusionally controlled kinetics, because the volume ratio of products to reactants is <1 (Velbel, 1993). The appearance of etch pits also suggests that protective layers do not form on the feldspar surface.

The leached layer model was questioned when Petrovic et al. (1976) and Berner and Holdren (1979) attributed the early observations of parabolic kinetics to artifacts of sample preparation, including the generation of fine particles which dissolve at rapid rates, and surface damage due to grinding and fracturing of the sample. In addition, these authors conducted XPS measurements which revealed no evidence for surface layers >20 Å on either naturally or experimentally dissolved feldspars at near neutral pH. However other evidence, namely solution data collected during dissolution experiments, suggested that Al- and cation-leached layers on the order of 20 to 30 Å did form on the dissolving feldspar surface (Chou and Wollast, 1984). These layers were thickest at acidic pH where the dissolution rates are most rapid, and thinner at neutral and basic pH where dissolution rates are slower. Rapid dissolution rates in the presence of thick surface layers contradicts the expectations of diffusion-controlled kinetics, because thick surface layers should decrease reaction rates.

All of this evidence led to the general conclusion that feldspar dissolution kinetics are surface-reaction controlled; the kinetics of the hydrolysis reactions which break framework (Al,Si)-O bonds in the feldspar structure control the overall dissolution rate, not diffusion through a surface layer of reaction products. As indicated by Chou and Wollast (1984), leached layers do form on the feldspar surface, at least at acidic pH. The question is whether these leached layers form an diffusional barrier. Several recent studies have proposed modified forms of a diffusion-controlled mechanism, and emphasize the contribution of these thin, non-stoichiometric surface layers to the feldspar dissolution kinetics (Oelkers et al., 1994; Hellmann, 1995; Brantley and Stillings, in press; Oelkers and Schott, in press a, b). These studies are described later in this section.

Atomistic models

Atomistic models of dissolution look at the detailed configuration and energetics of interactions between atoms on the mineral surface and molecules in solution. The most widely applied atomistic approach to kinetics is transition state theory (TST) (Eyring, 1935; Glasstone et al., 1951) which has been applied to many studies of mineral dissolution kinetics (Lasaga, 1981; Aagaard and Helgeson, 1982; Helgeson et al., 1984; Chou and Wollast, 1985; Hellmann et al., 1990, 1994, 1995; Oelkers et al., 1994; Oelkers and Schott, in press, b). Geochemists commonly write overall reactions, which reflect the overall stoichiometry of a chemical process. In contrast, TST applies only to elementary reactions which reflect the exact molecular species actually involved in the reaction at the atomic level, and there may be a large system of elementary reactions involved in a single overall reaction. TST calculates the reaction rate by assuming equilibrium between the reactants and a very short lived ($\sim 10^{-14}$ sec) activated complex, by calculating the concentration of the activated complex, and by estimating its decomposition rate. Both of these quantities are a function of the energetics of the molecular species, and can be calculated from their partition functions. Lasaga (1981) gives a simple and concise summary of TST and its application to geochemistry.

Application of TST to a specific reaction requires knowledge of the elementary reactions of interest and the structure of the activated complex. Rigorous application of TST requires knowing the activity coefficient of the activated complex, which can generally be determined only by quantum mechanical calculations. We do not yet have the detailed knowledge of the feldspar reaction mechanisms necessary for the rigorous application of TST. Previous application of TST to silicate dissolution postulated the reactants in the rate limiting step of the reaction mechanism, but did not calculate rates. However, TST has contributed to our understanding of dissolution mechanisms by focusing attention on the reactions at the mineral-solution interface at the molecular and atomic level. TST provides the conceptual framework in which feldspar dissolution mechanisms are most commonly discussed.

Rapid advances in ab-initio quantum mechanical calculations allow the calculation of the properties of proposed activated complexes and testing the energetics of the proposed reaction mechanisms (Lasaga, this volume). Lasaga and Gibbs (1990a, b) used ab-initio calculations to investigate the hydrolysis of Si-O-Si bonds. They calculated the reaction trajectory and energetics of the transition state for a proposed mechanism, and applied the results to quartz dissolution kinetics. Xiao and Lasaga (1994) used ab-initio calculations to investigate the effect of protonation of terminal hydroxyls and bridging Al-O-Si and Si-O-Si linkages on hydrolysis reaction energetics. They found that protonation of the Al-O-Si linkage catalyzes the hydrolysis reaction with an activation energy of 67 kJ/mol, in good agreement with experimentally measured activation energies of feldspars in acidic solutions (Table 2 and Figs. 10 to 13). This is an example of how quantum mechanical calculations are used to test proposed dissolution mechanisms, and facilitate the quantitative application of TST and atomistic descriptions of feldspar dissolution kinetics.

All of the atomistic approaches share a common weakness. It is difficult to apply atomistically based models to reactions where larger scale characteristics such as defect structures (e.g. surface step structures, point defects and dislocations), or compositional heterogeneities (e.g. exsolution features, impurities and compostional gradients in leached layers) influence the energetics of the reaction. In these cases, there may be a number of slightly different activation complexes at different sites, with a wide range of geometries and energetics, and the overall reaction kinetics is the summation of all these subtly different pathways. The complexity of the overall reaction may be both conceptually and numerically difficult to quantify on the atomic scale. Many of the most successful approaches to nucleation and growth in the material sciences, such as spiral growth theory (Christian, 1975; Bennema, 1984) are controlled by heterogeneities within the system. These growth theories simplify or ignore the processes at the atomic scale. Atomistic models are the most rigorous theoretical and conceptual approach to reaction kinetics, and are clearly extremely useful in clarifying the reactions and molecular species involved in the feldspar dissolution mechanism. However, it remains to be seen whether atomistic models can encompass the complexity of heterogeneous systems and really are a practical method for quantifying the overall dissolution rates of the feldspars. The greatest potential for modeling feldspar dissolution kinetics may involve atomistic models to provide the basic parameters for constraining and applying more general, macroscopic models.

Surface speciation models

Surface speciation models are also based on a molecular scale conceptual framework, but are neither as rigorous nor as detailed as TST models. Surface speciation models attempt to relate an observed dissolution rate to the concentration of specific chemical complexes on the surface. With these models, the dissolution rate is a consequence of a series of elementary reactions.

$$[\text{surface site}] + [\text{aqueous species}] \xleftrightarrow{\text{fast and reversible}} [\text{surface complex}] \qquad (26)$$

$$[\text{surface complex}] + [\text{reactant}] \xleftrightarrow{\text{slow and irreversible}} [\text{dissolved species}] \qquad (27)$$

Early steps in the reaction sequence (Eqn. 26) may be rapid and reversible adsorption/exchange reactions, as described earlier. The configuration of the resulting surface complex lowers the activation energy for the next step in the reaction sequence, a slow and irreversible reaction which breaks the bonds within the structure, and leads to the removal of components from the crystal. In the case of feldspars, the slow step is probably the hydrolysis of bridging tetrahedral oxygen bonds. If the rate-limiting reaction occurs with equal probability at each occurrence of a surface complex and only at sites with this specific surface complex, then

$$Rate = k_+ S_c \qquad (28)$$

so that the dissolution rate (mol cm^{-2} s^{-1}) is directly proportional to the number of surface complexes (S_c, mol m^{-2}), where k_+ is the rate constant (s^{-1}). In terms of TST, the surface complex provides a reaction pathway with a lower activation energy, and is a stable intermediate in the dissolution reaction mechanism.

Identification of the surface complex gives only general insight into the reaction mechanism and rate-limiting step. The actual "activated complex" for the rate-limiting step is still an unknown molecular configuration in a subsequent slow reaction, and *perhaps* the surface complex is one of the reactants. Nevertheless, surface complexation models are extremely powerful. If we identify the surface complex involved in the dissolution reaction: (1) we can predict dissolution rates from only a knowledge of the rate constant (k_+, Tables 3 to 6) and the adsorption isotherms of the solutes involved in the surface complex. Determining adsorption isotherms is a much simpler matter than explicitly measuring the kinetics of slow reactions like feldspar dissolution over the full range of possible solution compositions. (2) The activity of both catalysts and inhibitors may result from competition for the adsorption sites, which are then either blocked or enhanced by the adsorbed species. These effects may also be quantified by surface adsorption measurements, and intelligent guesses as to the effects of solutes may be made even without measurements.

Applications of surface speciation models to feldspar dissolution kinetics. The adsorption characteristics of simple oxides have been studied extensively, and surface complexation models have been developed to estimate the protonation reactions at oxide surfaces, as well as the adsorption of other ligands (Stumm and Wieland, 1990; Davis and Kent, 1990; Wehrli et al., 1990).

Furrer and Stumm (1986) and Guy and Schott (1989) found a correlation between the dissolution rate of simple oxides and their surface charge with a reaction order equal to the formal charge of the metal center:

$$\text{Dissolution rate} \propto c^n \qquad (29)$$

where c is the surface charge and n is the formal charge of the metal center (i.e. $n = 2$ for BeO, 3 for Al_2O_3 and Fe_2O_3, and 4 for SiO_2). This model suggests that dissolution occurs at metal centers at which n of the oxygens coordinated around that metal center are simultaneously protonated. Wieland et al. (1988) present a model for this dependence. They show statistically that if each metal center has n possible adsorption sites, which are filled in a completely random fashion, then the concentration of metal centers with all n sites filled (F) will vary as $F \propto (X_c)^n$ where X_c is the mole fraction of the surface sites which are filled (also see Stumm and Wollast, 1990).

Blum and Lasaga (1988, 1991) measured the change in protonation of an albite surface as a function of pH by surface titration. They observed that the inflection point in the H$^+$ adsorption isotherm occurs at pH 6.8, which Parks (1967) estimated to be the zero point of charge for the surface of a tetrahedrally-coordinated Al oxide. Combining this observation with the calculated depth of the exchange layer, they concluded that protonation of Al sites controls H^+ adsorption at albite surfaces, as a function of pH.

Blum and Lasaga (1991) also observed a strong correlation between surface protonation of albite and its dissolution rate measured by Chou and Wollast (1985). They concluded that albite dissolution is directly proportional to H^+ adsorption in the acid region (pH <6.8), and to OH^- adsorption in the basic region, and that adsorption of both species occurs at Al sites. The rate laws are then:

$$Dissolution\ rate\ =\ k_a c_{\equiv S+}^{1.0} \qquad (pH \leq 6) \qquad (30)$$

$$Dissolution\ rate\ =\ k_b c_{\equiv S-}^{1.0} \qquad (pH \geq 7.5) \qquad (31)$$

where $c_{\equiv S+}$ and $c_{\equiv S-}$ are the surface concentrations of positive and negative charge, the reaction order is 1.0, and the reaction constants for the dissolution mechanisms in acid and basic regions are $k_a = 10^{-6.5}\ s^{-1}$ and $k_b = 10^{-6.1}\ s^{-1}$, respectively. Near the inflection point at pH 6.8, the dissolution rate is a more complex function of both $c_{\equiv S+}$ and $c_{\equiv S-}$. Blum and Lasaga (1988, 1991) did not observe Na^+ uptake during the titrations, and assumed that the protonation and deprotonation of Al dangling hydroxyls were the major reactions controlling the change in surface charge. They concluded that the hydrolysis rate of bridging Al-O-Si bonds adjacent to the protonated aluminum hydroxyls is the rate limiting step in the feldspar dissolution process.

Wollast and Chou (1992) found a similar linear relation between negative surface charge and the dissolution rate of albite in the basic region (R μ $c^{\sim 1.0}$). Schott (1990) conducted similar surface titrations of albite and found a correlation between the dissolution rate and surface charge of albite in the basic region, but with a reaction order of 3.3 (R $\propto c^{\sim 3.3}$). Amrhein and Suarez (1988) found a correlation between the dissolution rate and the surface charge of anorthite with a reaction order of 4 (R $\propto c^{\sim 4.0}$). Reaction orders of 3 to 4 are consistent with the theory of Wieland et al. (1988) (Eqn. 29) for control of feldspar dissolution by surface species at Al centers. This theory contrasts with the reaction order of 1 found by Blum and Lasaga (1991) and Wollast and Chou (1992). Brady and Walther (1989) examined the dissolution versus surface charge behavior over a range of oxides and silicates. They concluded that Si hydroxyls, and not Al, are deprotonated at basic pH, and the detachment of Si controls the dissolution kinetics of feldspars in the basic region.

Oxburgh et al. (1994) conducted surface titrations and dissolution experiments with three plagioclase feldspars, and found that reaction orders in the acid region are strongly dependent on the mole fraction of anorthite component in the solid: $n = 0.46$ for oligoclase (An_{11}), 1.2 for andesine (An_{46}) and 2.0 for bytownite (An_{76}). In their model, proton absorption at (Al,Si)-bridging oxygen sites dominates the dissolution kinetics, and the reaction order is related to the average number of these sites coordinated around each Si center. By assuming all Si on the surface are bound to three bridging oxygens, and that the dangling surface hydroxyl was originally bound to an Al atom, Oxburgh et al. (1994) estimated the average number of (Al,Si)-bridging oxygens around each surface Si to be 0.33 for albite and 3.0 for anorthite. For plagioclase of compositions An_{11}, An_{46} and $An_{76,}$ surface Si atoms are bound to an average of 0.56, 1.3 and 2.1 (Al,Si)-bridging oxygens, respectively. These averages agree with the reaction orders cited above for each composition, thus leading to a model where dissolution rates are dependent upon the protonation of (Al,Si)-bridging oxygens.

One criticism of the above model is that Oxburgh et al. (1994) do not justify the assumption that all dangling surface hydroxyls were originally bound to an Al atom. If surface hydroxyls were originally bound to either a Si or an Al atom with the same distribution as in the bulk crystal, the average number of (Al,Si)-bridging oxygens around each surface Si atom (bound to 3 bridging oxygens) will vary from 1.0 for albite to 3.0 for anorthite, and the predicted reaction orders will also vary from 1 to 3 with increasing An content. Despite this slight difference in calculating (Al,Si)-bridging oxygens, the model of Oxburgh et al. (1994) is still a reasonable explanation for the dependence of the reaction order on the mole fraction of anorthite in plagioclase feldspars. In summary, Oxburgh et al. (1994) suggest that the proposed mechanism of Wieland et al. (1988) is applicable to silicates, but that n is not directly related to the formal charge of the metal center. Instead, n equals the number of nearest neighbor oxygens which can be protonated, which in the case of feldspars, is the average number (Al,Si)-bridging oxygens bound to each surface Si atom.

Gautier et al. (1994), Oelkers et al. (1994) and Oelkers and Schott (in press, a,b) integrated surface speciation, preferential hydrolysis of Al-O-Si bonds, and Si-rich surface layers to explain the feldspar dissolution mechanism. They suggest that destruction of tetrahedrally coordinated Al-O bonds is relatively rapid compared to Si-O bonds. For the alkali feldspars, with a high Si/Al ratio, the rapid and reversible exchange of 3 H^+ for a tetrahedral Al involves breakage of Al-O-Si bonds. This produces an Al-deficient surface "precursor complex," which is a network of partially detached silica tetrahedra remaining on the surface:

$$3(Si - O) \equiv Al + 3\,H^+ \quad \xleftrightarrow{\text{rapid and reversible}} \quad [\; 3(Si - OH) \;]^\dagger + Al^{3+} \tag{32}$$

$$K_{eq} = \frac{[Al^{3+}][3(Si - OH)]^\dagger}{[H^+]^3[3(Si - O) \equiv Al]} \tag{33}$$

where $3(Si - O) \equiv Al$ is the Al-filled surface site, $[3(Si\text{-}OH)]^\dagger$ is the precursor complex, and $[i]$ is the activity of species i. We simplify the notation of Oelkers et al. (1994) by assuming that no other aqueous species is involved in the formation of the precursor. The authors suggest that hydrolysis of Si-O-Si bonds within the residual Si-rich surface layer is the slow step in the feldspar dissolution mechanism, and the exchange of $3H^+$ for an Al^{3+} promotes the hydrolysis of the residual Si-rich layer. Therefore the dissolution rate is proportional to the activity of precursor complexes, $[3(Si\text{-}OH)]^\dagger$, so that:

$$R = k^* [3(Si\text{-}OH)]^\dagger \tag{34}$$

where k^* is the reaction rate per mole of precursor sites.

This model is attractive because it explains the strong dependence of feldspar dissolution on both pH and $[Al^{3+}]$ via the exchange reaction (Eqn. 32), which controls the surface concentration of active precursor sites. It is also reasonable that the presence of Na^+ and other species in solution could retard the dissolution rate by stabilizing Al in these exchange sites, either by occupying a larger proportion of the interstitial cation sites, or by directly stabilizing Al by adsorption to the Si-O-Al oxygen, thus inhibiting acid-catalyzed hydrolysis. However, this model also has several weaknesses: (1) The Al-O-Si bonds are very energetic. As previously discussed, ab-initio models (Xiao and Lasaga, 1994) have calculated activation energies of 109-113 and 70 kJ/mol for hydrolysis and H_3O^+ catalysis, respectively. It seems unlikely that such energetic bonds could participate in rapid and reversible reactions at equilibrium with solution, and such rapid, reversible Al-O-Si hydrolysis would be contrary to previous theories on the importance of Al in the feldspar dissolution mechanism. However, the possibility of an undetermined low activation energy mechanism for Al-O-Si hydrolysis cannot be conclusively eliminated. (2) The thick Si-rich layers found after dissolution at low pH are amorphous and porous, do not seem to form a diffusional barrier, have a sharp boundary with the intact feldspar lattice, and do not seem to retard the feldspar dissolution rate. The evidence we have from the Si-rich layer at very low pH (where it is thick enough to image and study by bulk methods) does not suggest a dissolution mechanism where the dissolution of the Si-rich layer is the rate limiting factor.

Brantley and Stillings (in press) proposed a dissolution mechanism for feldspars in acidic solutions (pH < ~5) which combines a surface to speciation model with diffusion of Al, M^{b+} (the charge to balancing cation in the structure), and H^+ through a leached layer. They assume that the rate-controlling step in the dissolution mechanism is the hydrolysis of protonated Al-O-Si bonds, again as suggested by Oxburgh et al. (1994) and Xiao and Lasaga (1994). However, Brantley and Stillings (in press) propose that the exchange of H^+ (or H_3O^+) for M^{b+}, the interstitial cation in the structure, protonates the Al-O-Si surface site, thus weakening the Al-O-Si bridging bonds for subsequent hydrolysis (Xiao and Lasaga, 1994). Therefore, they propose that the extent of the H^+ for M^{b+} exchange controls the number of potential reaction sites.

They further suggest that the depth and extent of exchange increases as pH and cation concentration decrease (Casey et al., 1988; Muir and Nesbitt, 1991), and that the depth is controlled by counter-diffusion of H^+ and M^{b+} through a leached layer, a mechanism commonly observed during dissolution of silicate glasses (Doremus, 1975; Lanford et al., 1979; Smets and Tholen, 1985). Thus, thicker leached layers at acidic pH's have more protonated sites at which hydrolysis of Al-O-Si may occur, and therefore, more rapid dissolution rates. Also, the pH dependence of dissolution arises from the exchange of H^+ for M^{b+}, with higher H^+ activity driving the exchange by increasing the gradient of H^+ across the leached layer. The observation that increasing Na^+ concentration decreases feldspar dissolution rates at pH < 5 (Schweda, 1990; Stillings and Brantley, 1995a) is a logical consequence of Na^+ competing with H^+ for Al-O-Si adsorption sites. This model explains several observations which have been difficult to reconcile previously: (1) feldspar surfaces are leached of M^{b+} and Al^{3+} at low pH, (2) dissolution rates increase with decreasing pH and salt concentration, (3) surface hydration increases with decreasing pH and salt concentration, (4) proton adsorption at the Al-O-Si site increases with decreasing pH and salt concentration, and (5) as pH decreases from 6 to 3, feldspar dissolution rates increase while dissolution rates for SiO_2 remain approximately constant.

The model of Brantley and Stillings (in press) is attractive in that is integrates the observations from many studies of feldspar dissolution and surface chemistry into a comprehensive model consistent with established models for glass dissolution. However, there two aspects of the model which must be resolved. First, for H^+ penetration to increase with decreasing pH, the concentration gradient of H^+ across the thickness of the leached layer must increase. This appears to require a concentration of H^+ at the surface which is greater than the available density of adsorption sites. This apparent inconsistency is not unique to the model of Stillings and Brantley (1995); many studies have observed that adsorption densities on silica surfaces appear to be far greater than the apparent site density (Tadros and Lyklema, 1968; Abendroth, 1970; Unwin and Bard, 1992). Applications of electrical double layer models for *porous* oxides (Lyklema, 1968, 1971; Perram et al., 1974; Kleijn, 1990) might help to explain high H^+ concentrations at feldspar surfaces. Second, although Brantley and Stillings (in press) explain why dissolution rates increase with increasing thickness of a leached layer, the question remains as to whether or not a leached layer is a diffusional barrier.

The evidence of Westrich et al. (1989) and Casey et al. (1989) suggests that the thick, Si-rich layers, which can reach several microns at very acidic pH, have an amorphous structure and are not particularly important in the dissolution kinetics. This led to surface speciation and atomistic models which envision a surface reaction of the intact tetrahedral framework directly with the solution. Yet there may be thinner, nanometer-scale, non-stoichiometric surface layers on feldspars which may be important in the dissolution kinetics. However, the nature of these layers is poorly constrained both experimentally and theoretically. Discussions of feldspar dissolution must be quite specific about which conceptual model is used to describe the nature of the feldspar surface, particularly the clear definition of any "leached layer".

Macroscopic models

Macroscopic models are based on the configuration of the molecular-scale surface. These approaches are most useful for dealing with features of the system which are defined only at scales larger than molecular clusters, such as the saturation state of the solution, and the effects of dislocations, impurities, and other heterogeneities in the solid.

Surface nucleation models. Surface nucleation models were first proposed for growth (Nielsen, 1964; Ohara and Reed, 1973), but apply equally well to dissolution (Blum and Lasaga, 1987). Surface nucleation models take a basic thermodynamic approach in which there are two competing energy terms: the chemical potential and the surface free energy of the system. When a unit volume of a crystal with a flat surface is dissolved into an undersaturated solution: (1) energy is released by the change in chemical potential (Δm) due to dissolution. The chemical potential is related to the solution composition by $Dm = kT \ln(Q/K)$ where k is Boltzmann's constant, Q is the activity product and K is the equilibrium constant for the dissolution reaction. (2) The surface free energy (σ, erg/cm^2) of the system increases due to creation of dissolution pits increases the crystal surface area.

As a dissolution pit forms on a surface, the surface free energy term initially dominates the increase in energy of the system until the pit reaches a critical radius (r_c). At this point the change in chemical potential term dominates, and further dissolution is energetically favored. At solution compositions common in near-surface environments, Lasaga and Blum (1986) estimated that r_c for feldspars and other silicates is on the order of several nanometers. Thus, there is an energy barrier which must be overcome to nucleate mineral dissolution at a perfect surface, and nucleation models assume that this energy barrier controls the dissolution kinetics. Nucleation theory generally predicts far too strong a dependence of the dissolution rate on the solution saturation state, and greatly underestimates dissolution and growth rates close to equilibrium. However, the concept of the critical radius (r_c) is used in many of the common dissolution theories.

Dissolution at dislocations. Extensive crystallographically-controlled etch pits are a ubiquitous feature in both naturally weathered (Berner and Holdren, 1979; Velbel, 1986; Banfield and Eggleton, 1990) and experimentally dissolved feldspars (Wilson and McHardy, 1980; Holdren and Speyer, 1985; Knauss and Wolery, 1986; Gautier et al., 1994). The morphology of the etch pits is similar to those which form at the intersection of dislocations with the surface. There are several ways in which dislocations may potentially affect feldspar dissolution rates. First, the intersection of screw dislocations with the surface provides a continuous source of steps which circumvents the need to nucleate pits on the surface. This is the basis for the Burton-Cabrera-Frank (BCF) theory for spiral dissolution and growth (Burton et al., 1951; Ohara and Reed, 1973; Christian, 1975; Bennema, 1984). In BCF theory, the step produced at a screw dislocation is a preferential site for dissolution and growth. The geometry imposed by the step being fixed at one end causes the propagating step to wind into a large spiral, either as an etch pit during dissolution or a growth hillock during precipitation. BCF theory has been very successful in describing precipitation in many simple synthetic systems. Growth spirals have also been observed in natural kaolinite, beryl, quartz, illites, and other minerals.

Dislocations may affect dissolution in a second way. In the vicinity of dislocations, the crystal is distorted, introducing strain energy into the structure. The magnitude of this strain energy is too small to effect the overall thermodynamic properties of the mineral (Helgeson et al., 1984; Wintsch and Dunning, 1985). However, the concentration of strain energy in the vicinity of the dislocation provides a localized area of increased solubility. This provides an additional mechanism for forming etch pits which exceed r_c, and are thermodynamically favored to continue dissolution. Thus, strain energy around dislocations provides an alternative mechanism to surface nucleation or BCF for mineral dissolution (Cabrera and Levine, 1956; Van der Hoek et al., 1982; Lasaga, 1983). Note that strain energy accelerates dissolution, but will marginally inhibit growth. Blum and Lasaga (1987) used Monte Carlo simulations to demonstrate that strain energy is probably the dominant mechanism of etch pit formation at dislocations in quartz and feldspars (Fig. 18). They found that far from equilibrium, the strain energy mechanism was ~8 times faster than the BCF mechanism, with the ratio increasing toward equilibrium. An important prediction of this theory is that there is a critical solution saturation state (ΔG_{crit}) below which macroscopic etch pits will not form and above which the dislocation cores will "open" to form etch pits. Brantley et al. (1986) and Lasaga and Blum (1986) suggest that the presence or absence of etch pits is an indicator of solution saturation state in soils and ground water. This relation has been observed experimentally for albite at pH 8 (Burch et al., 1993) and for quartz, potassium feldspar, and hornblende in soils (Brantley et al., 1986, 1993).

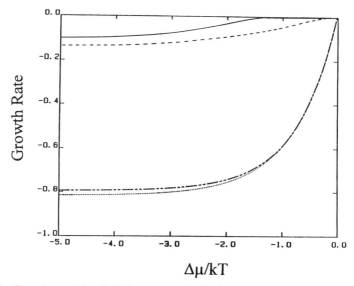

Figure 18. Dependence of the dissolution rate on the dissolution mechanism as determined by Monte Carlo simulations. The X-axis is the saturation state, with 0 at equilibrium and becoming more undersaturated with more negative values of Dm/kT, and more rapid dissolution rates along the Y-axis from 0 to -1. Solid curve = defect free surface, dashed curve = step at screw dislocation only, long and short dashes = strain field around dislocation only, dotted curve = step and strain field around dislocation (from Blum and Lasaga, 1987).

The observation of extreme etch pitting in virtually all feldspars which have undergone extensive dissolution, either experimentally or in natural environments, suggests that a large amount of the total dissolution occurs at etch pits. However, Blum and Lasaga (1987) suggested that dislocation density would have a minimal effect on the overall dissolution rate of quartz and other silicates based upon Monte Carlo model results. Murphy (1989) studied the dissolution rate of sanidine (K-feldspar) with low and high dislocation densities, and found no difference in the dissolution rates. Similar results were found for quartz (Blum et al., 1990), rutile (Casey et al., 1988) and calcite (Schott et al., 1989). All these experiments were conducted far from equilibrium.

Burch et al. (1993) conducted albite dissolution experiments at pH 8 and 80°C, varying the solution composition over a wide range of saturation states. They observed a dramatic decrease in the dissolution rate of albite as equilibrium approached, with an abrupt decrease in the dissolution rate at DG_r = ~-35 kJ/mol, which corresponds to $Q/K = 10^{-4.4}$. Thus, the solution saturation state can significantly depress the dissolution kinetics of albite, even in very dilute solutions. They point out that the shape of this curve is inconsistent with the classical TST formulation of the rate law. However, the abrupt transition in dissolution rate occurs approximately at the ΔG_{crit} predicted for the opening of etch pits at dislocations, and examination of the experimental material with SEM reveals etch pits only in the most undersaturated experiments. A similar dependence of the gibbsite dissolution rate on solution saturation state was observed by Nagy and Lasaga (1992), and also attributed to the opening of etch pits at dislocations. This supports the prediction of Blum et al. (1990) that dislocations should have a larger effect on the overall dissolution rate close to equilibrium.

The overall effect of dislocations on weathering rates in natural systems is still uncertain. Experimental evidence from highly undersaturated solutions does not show a significant effect of dislocation densities on dissolution rates, but the prevalence of etch pits in naturally weathered feldspars suggests highly accelerated rates at these locations. Weathering rates of soil minerals also suggest decreasing reactivity with time (White et al., in press; Taylor and Blum, in press) which can be explained by the preferential dissolution of reactive material at etch pits. It may be that the composition of natural weathering solutions, which are closer to saturation and have higher levels of potential inhibitors, accentuates the contrast between dissolution rates at dislocations versus defect-free portions of the crystal. The effects of defects on dissolution rates in natural systems are discussed in more detail in other chapters of this volume (see chapters by Lasaga, Hochella and Banfield, and White). While surface speciation and atomistic models are the most commonly applied in the feldspar dissolution literature, it is not clear how to incorporate the role of defects into such models, and the description of geological systems may eventually require the development of more comprehensive approaches.

Limitations to modeling natural systems with experimentally-derived rates

The primary motivation for the study of feldspar dissolution is application to natural systems. There is tremendous uncertainty in how to apply experimentally measured dissolution rates to natural systems. Most attempts to compare rates between laboratory and natural systems have found natural feldspar weathering rates to be 10^1 to 10^4 times slower than experimental rates (see chapters by Sverdrup and Warfvinge, Drever and Clow, and White). Especially interesting is the study of Swoboda-Colberg and Drever (1993), which compared dissolution rates for identical mineral material in both field and laboratory experiments. These authors observed dissolution rates in the field to be ~200 to 400 times slower than laboratory rates. The following is a list of factors which might be possible explanations for this discrepancy, and are issues which need to be addressed in the design of future laboratory studies.

(1) Time.   Most feldspars from natural environments have been weathered from hundreds to millions of years. This time frame which cannot be duplicated in the laboratory. Several workers have suggested that dissolution rates continue to decrease with time, even after several thousand hours at 25°C. In addition, the few field studies which have addressed the issue report decreasing mineral weathering rates with time. There is still the considerable question of whether dissolution experiments which dissolve only a small proportion of the sample reflect dissolution of only the most reactive material, and may not represent the reactivity of the bulk of the crystal or the long-term weathering rate.

(2) Surface areas.   Mineral surface areas are a critical quantity in calculating specific dissolution rates (mol m^{-2} sec^{-1}). BET surface areas have been found to vary by a factor of ~2 to 7 between the beginning and end of 3000 hour laboratory experiments (Stillings and Brantley, 1995), and by several orders of magnitude during the evolution of a soil (White, this volume; White et al., in press). The best way to quantify mineral surface areas in natural materials is a complex issue rarely addressed (White and Peterson, 1990; Hochella and Banfield, this volume; White, this volume; White et al., in press), yet it must be resolved in order to apply experimental rates of feldspar dissolution to field situations. Errors in estimates of surface area may impact calculated weathering rates by several orders of magnitude.

(3) Defects.   The manner by which structural and compositional impurities in the crystals, which may be spatially variable, affect the dissolution kinetics is quite uncertain. The issue of defects is central to that of time and surface area; they are heterogeneities which affect both reactivity and the development of porosity and surface roughening.

(4) Unsaturated hydrology.   The complexity of soil hydrology impacts mineral dissolution in several ways. Extreme heterogeneities of permeability lead to preferential flow paths in soils, and most mineral grains may very rarely, if ever, contact the bulk of the soil solutions commonly sampled in field studies. The result is that some grains may not be in contact with solution at all, and a large proportion will be in isolated pores where solutions may accumulate solutes and evolve chemistries very different from the bulk solution. Evaluations of the effects of soil hydrology on weathering, and comparisons between pore water chemistry and laboratory solutions are severely needed.

(5) Effects of solution chemistry: saturation state and local pH. There is good experimental evidence that feldspar dissolution rates are retarded by low concentrations of Al in solution, perhaps as either an inhibitor, or as a saturation state effect. We do not completely understand the effects of solutes, particularly of Al and Fe, which have a ubiquitous presence in natural environments and are often at significant concentrations. This issue becomes particularly acute when considering the effects of isolated pores in soils, which may allow solutes to accumulate to higher concentrations than measured in bulk soil solutions.

(6) Surface coatings.   There are abundant Al and Fe oxide phases in most soils which may form coatings on mineral grains. In addition, organics and clays form well developed films on mineral surfaces. The effects of such coatings on feldspar weathering rates is largely speculative.

(7) Biological effects.   There is little consensus on the effect of plants and microbes on mineral weathering. Mechanisms for biological acceleration of weathering are

unclear, whether they involve direct physical and chemical attack, exudates which alter the local chemical environment (through changes in pH and organic ligand concentrations), or organisms which exploit nutrients as they are released by inorganic processes. There are considerable logical and circumstantial arguments that organisms should exploit feldspars for nutrients such as Ca and K, and evolve mechanisms to accomplish this by dissolving the feldspar structure. However, two factors point to biological effects not being a dominant factor: (1) natural weathering rates are slower than predicted laboratory rates determined in abiotic systems, so that any accelerating effects of organisms may be modest, and (2) environments with lower biological activity, such as arctic and alpine environments, do not appear to have dramatically lower weathering rates (White and Blum, 1995). The role of organisms in feldspar weathering is an area of active research which is critical to both geochemical and ecological applications.

(8) Effects of wetting and drying cycles. Most soils have highly variable soil moisture during and between precipitation events, often undergoing periodic wetting and drying cycles. Drying affects weathering in several ways. Dehydration of the surface destabilizes some surface species, and tends to induce condensation reactions between surface groups, which may make the surface less reactive. The evaporation of solutions will also concentrate solutes, alter the pH, and tend to precipitate secondary phases including salts and amorphous oxides. For example, the cyclic wetting and drying of smectitic clays in the presence of microcline has been shown to form illitic clays, sequestering K, and facilitating the weathering process (Eberl et al., 1986). The effects on the overall weathering rates of dramatic and cyclic changes in the local chemical environment, as is typical of many soil environments, is almost completely unknown.

REFERENCES

Aagaard P, Helgeson HC (1982) Thermodynamic and kinetic constrains on reaction rates among minerals and aqueous solution. I. Theoretical consideration. Am J Sci 282:237-285

Abendroth RP (1970) Behavior of a pyrogenic silica in simple electrolytes. J Coll Int Sci 34:591-596

Amrhein C, Suarez DL (1988) The use of a surface complexation model to describe the kinetics of ligand-promoted dissolution of anorthite. Geochim Cosmochim Acta 52:2785-2793

Amrhein C, Suarez DL (1992) Some factors affecting the dissolution kinetics of anorthite at 25°C. Geochim Cosmochim Acta 56:1815-1826

Banfield JF, Eggleton RA (1990) Analytical transmission electron microscope studies of plagioclase, muscovite, and K-feldspar weathering. Clays Clay Minerals 38:77-89

Bennema P (1984) Spiral growth and surface roughening: Developments since Burton, Cabrera and Frank. J Cryst Growth 69:182-197

Bennett PC, Casey W (1994) Chemistry and mechanisms of low-temperature dissolution of silicates by organic acids. In: Pittman ED, Lewan MD (eds) Organic Acids in Geological Processes. Springer-Verlag, NY, p 162-200

Bennett PC, Hiebert FK, Choi WJ (in review) Rates of microbial weathering of silicates in ground water. Submitted to Chemical Geology

Berner RA, Holdren GR (1979) Mechanism of feldspar weathering - II. Observations of feldspars from soils. Geochim Cosmochim Acta 43:1173-1186

Berner RA, Lasaga AC, Garrels RM (1983) The carbonate-silicate geochemical cycle and its effect on atmospheric carbon dioxide over the past 100 million years. Am J Sci 283:641-683

Bevan J, Savage D (1989) The effect of organic acids on the dissolution of K-feldspar under conditions relevant to burial diagenesis. Mineral Mag 53:415-425

Blake RE, Walter LM (in review) Effects of organic acids on K-feldspar dissolution at 80°C and pH 6. submitted to Chemical Geology

Blum AE (1994) Feldspars in weathering. In: Parsons I (ed) Feldspars and Their Reactions. NATO Advanced Study Inst, Kluwer Academic Pub, Netherlands, p 595-629

Blum AE, Lasaga AC (1987) Monte Carlo simulations of surface reaction rate laws. In: Stumm W (ed) Aquatic Surface Chemistry, J Wiley & Sons, New York, p 255-292

Blum AE, Lasaga AC (1988) Role of surface speciation in the low-temperature dissolution of minerals. Nature 4:431-433

Blum AE, Lasaga AC (1991) The role of surface speciation in the dissolution of albite. Geochim Cosmochim Acta 55:2193-2201

Blum AE, White AF, Hochella MF (1992) The surface chemistry and topography of weathered feldspar surfaces. Trans Am Geophys Union Abstr 73:602

Blum AE, Yund RA, Lasaga AC (1990) The effect of dislocation density on the dissolution rate of quartz. Geochim Cosmochim Acta 54:283-297

Brady PV, Carroll SA (1994) Direct effects of CO_2 and temperature on silicate weathering: Possible implications for climate control. Geochim Cosmochim Acta 58:1853-1856

Brady PV, Walther JV (1989) Controls on silicate dissolution in neutral and basic pH solutions at 25°C. Geochim Cosmochim Acta 53:2823-2830

Brady PV, Walther JV (1992) Surface chemistry and silicate dissolution at elevated temperatures. Am J Sci 292: 639-658

Brantley SL, Stillings LL (in press) Feldspar dissolution at 25°C and low pH. Am J Sci

Brantley SL, Blai AC, Cremeens DL, MacInnis I, Darmody RG (1993) Natural etching rates of feldspar and hornblende. Aquatic Sciences 55: 262-272

Brantley SL, Crane R, Crerar DA, Hellman R, Stallard R (1986) Dislocation etch pits in quartz. In: Davis J, Hayes K (eds) Geochemical Processes at Mineral Surfaces. Am Chem Soc, p 639-649

Brantley SL, Richards P, Murphy S (in review) Soil porewater chemistry and the relative rate of weathering of feldspar and quartz. Submitted to Geochim Cosmochim Acta

Brunauer S, Emmett PH, Teller E (1938) Adsorption of gases in multimolecular layers. J Am Chem Soc 60:309-319

Bunker BC, Arnold GW, Day DE, Bray PJ (1986) The effect of molecular structure on borosilicate glass leaching. J Non-Cryst Solids 87:226-253

Bunker BC, Headley TJ, Douglas SC (1984) Gel structures in leached alkali silicate glass. Mater Res Soc Symp 32:226-253

Bunker BC, Tallant DR, Headley TJ, Turner GL, Kirkpatrick RJ (1988) The structure of leached sodium borosilicate glass. Phys Chem Glasses 29:106-120

Burch TE, Nagy KL, Lasaga AC (1993) Free energy dependence of albite dissolution kinetics at 80°C and pH 8.8. Chem Geol 105:137-162

Burton WK, Cabrera N, Frank FC (1951) The growth of crystals and the equilibrium structure of their surfaces. Phil Trans Roy Soc London A243:299-368

Busenberg E (1978) The products of the interaction of feldspars with aqueous solutions at 25°C. Geochim Cosmochim Acta 42:1629-1686

Busenberg E, Clemency CV (1976) The dissolution kinetics of feldspars at 25°C and 1 atm CO_2 partial pressure. Geochim Cosmochim Acta 40:41-50

Cabrera N, Levine MM (1956) On the dislocation theory of evaporation of crystals. Phil Mag 1:450-458

Carothers WW, Kharaka YF (1978) Aliphatic acid anions in oil-field waters--implications for origin of natural gas. Am Assoc Petrol Geol Bull 62:2441-2453

Casey WH, Bunker B (1990) Leaching of mineral and glass surfaces during dissolution. In: Hochella MF, White AF (eds) Mineral-Water Interface Geochemistry. Rev Mineral 23:397-426

Casey WH, Sposito G (1992) On the temperature dependence of mineral dissolution rates. Geochim Cosmochim Acta 56:3825-3830

Casey WH, Carr MJ, Graham RA (1988) Crystal defects and the dissolution kinetics of rutile. Geochim Cosmochim Acta 52:1545-1556

Casey WH, Hochella MF, Westrich HR (1993) The surface chemistry of manganiferous silicate minerals as inferred from experiments on tephroite (Mn_2SiO_4). Geochim Cosmochim Acta 57:785-793

Casey WH, Westrich HR, Holdren GR (1991) Dissolution rates of plagioclase at pH = 2 and 3. Am Mineral 76:211-217

Casey WH, Westrich HR, Massis T, Banfield, JF, Arnold GW (1989) The surface of labradorite feldspar after acid hydrolysis. Chem Geol 78:205-218

Chen Y, Brantley SL (1994) Steady-state dissolution of albite at temperatures between 5 and 300°C. Geol Soc Am Program Abstracts, A-287

Chen Y, Brantley SL (in review) Temperature and pH dependence of albite dissolution rate at sub-neutral pH. Submitted to Chemical Geology

Chou L (1985) Study of the kinetics and mechanisms of dissolution of albite at room temperature and pressure. PhD dissertation. Northwestern Univ, Evanston, IL

Chou L, Wollast R (1984) Study of the weathering of albite at room temperature and pressure with a fluidized bed reactor. Geochim Cosmochim Acta 48:2205-2218

Chou L, Wollast R (1985) Steady-state kinetics and dissolution mechanisms of albite. Am J Sci 285:963-993

Chou L, Wollast R (1989) Is the exchange of alkali feldspars reversible? Geochim Cosmochim Acta 53:557-558

Christian JW (1975) The theory of transformation in metals and alloys, Part I, 2nd Edition. Pergamon Press.

Davis JA, Kent DB (1990) Surface complexation modeling in aqueous geochemistry. In: Hochella MF, White AF (eds) Mineral-Water Interface Geochemistry. Rev Mineral 23:177-260

Doremus RH (1975) Interdiffusion of hydrogen and alkali ions in a glass surface. J Non-cryst Solids 19:137-144

Dove PA, Crerar DA (1990) Kinetics of quartz dissolution in electrolyte solutions using a hydrothermal mixed flow reactor. Geochim Cosmochim Acta 54:955-969

Dran JC, Della Mea G, Paccagnella A, Petit JC, Trotignon L (1988) The aqueous dissolution of alkali silicate glasses: reappraisal of mechanisms by H and Na depth profiling with high energy ion beams. Phys Chem Glasses 29:249-255

Drever JI (1994) The effect of land plants on weathering rates of silicate minerals. Geochem Cosmochim Acta 58:2325-2332

Drever JI, Vance GF (1994) Role of soil organic acids in mineral weathering processes. In: Pittman ED, Lewan MD (eds) Organic Acids in Geological Processes. Springer-Verlag, NY, p 138-161

Eberl DD, Srodon J, Northrop HR (1986) Potassium fixation in smectite by wetting and drying. In: Davis JA, Hayes KF (eds) Geochemical Processes at Mineral Surfaces. ACS Symp Ser 323:296-326, Am Chem Soc, Washington, DC

Eggleston CM, Hochella MF, Parks GA (1989) Sample preparation and aging effects on the dissolution rate and surface composition of diopside. Geochim Cosmochim Acta 53:797-805

Ernsberger FM (1960) Structural effects in the chemical reactivity of silica and silicates. J Phys Chem Solids 13:347-351

Eyring H (1935) The activated complex and the absolute rate of chemical reactions. Chem Rev 17:65-82

Fleer VN (1982) The dissolution kinetics of anorthite ($CaAl_2Si_2O_8$) and synthetic strontium feldspar ($SrAl_2Si_2O_8$) in aqueous solutions at temperatures below 100°C: Applications to the geological disposal of radioactive nuclear wastes. PhD dissertation, Penn State Univ, University Park, PA

Franklin SP, Hajash A, Dewers TA, Tieh TT (1994) The role of carboxylic acids in albite and quartz dissolution: An experimental study under diagenetic conditions. Geochim Cosmochim Acta 58:4259-4279

Fritz B, Probst A, Richard L (1993) Geochemical modelling of strontium behavior during weathering processes in a small granitic watershed. In: Cerny J (ed) Abstracts, Biogeomon and Workshop on Intergrated Monitoring. Czech Geological Survey, Prague, p 88-89

Furrer G, Stumm W (1986) The coordination chemistry of weathering. I. Dissolution kinetics of δ-Al_2O_3 and BeO. Geochim Cosmochim Acta 50:1847-1860

Gardner LR (1983) Mechanics and kinetics of incongruent feldspar dissolution. Geology 11:418-421

Garrels RM, Howard P (1959) Reactions of feldspar and mica with water at low temperature and pressure. Proc 6th Nat'l Conf on Clays and Clay Minerals, p 68-88

Gautier J -M, Oelkers EH, Schott J (1994) Experimental study of K-feldspar dissolution rates as a function of chemical affinity at 150°C and pH 9. Geochim Cosmochim Acta 58:4549-4560

Gestsdottir K, Manning DAC (1992) An experimental study of dissolution of albite in the presence of organic acids. In: Kharaka YK, Maest AA (eds) Water-Rock Interaction. AA Balkema, Rotterdam, p 315-318

Giese RF, van Oss CJ (1993) The surface thermodynamic properties of silicates and their interactions with biological materials. In: Guthrie GD, Mossman BT (eds) Health Effects of Mineral Dusts. Rev Mineral 28:327-346

Glasstone S, Laidler K, Eyring H (1951) The Theory of Rate Processes. McGraw-Hill, New York, 600 p

Goldich SS (1938) A study in rock weathering. J Geol 46:17-58

Goossens DA, Pijpers AP, Philipparerts JG, Van Tendeloo S, Althaus E, Gijbels R (1989) A SIMS, XPS, SEM, TEM and FTIR study of feldspar surfaces after reacting with acid solutions. In: Miles DL (ed) Proc 6th Int'l Symp on Water-Rock Interaction. AA Balkema, Rotterdam. p 267-270

Graustein WC, Cromack K, Sollins P (1977) Calcium oxalate: occurrence in soils and effect on nutrient and geochemical cycles. Science 198:1252-1254

Gregg SJ, Sing KS (1982) Adsorption, surface area and porosity. Academic Press, New York, 303 p

Guy C, Schott J (1989) Multi-site surface reaction versus transport control during the hydrolysis of a complex oxide. Chem Geol 78:181-204

Helgeson HC, Murphy WM, Aagaard P (1984) Thermodynamic and kinetic constraints on reaction rates among minerals and aqueous solution. II. Rate constants, effective surface area, and the hydrolysis of feldspar. Geochim Cosmochim Acta 48:2405-2432

Hellmann R (1994) The albite-water system: Part I. The kinetics of dissolution as a function of pH at 100, 200, and 300°C. Geochim Cosmochim Acta 58:595-611

Hellmann R (1995) The albite-water system: Part II. The time-evolution of the stoichiometry of dissolution as a function of pH at 100, 200 and 300°C. Geochim Cosmochim Acta 59:1669-1697

Hellmann R, Eggleston CM, Hochella MF, Crerar DA (1989) Altered layers on dissolving albite - 1. Results. In: Miles DL (ed) Proc 6th Int'l Symp on Water-Rock Interaction. AA Balkema, Rotterdam, p 293-269

Hellmann R, Eggleston CM, Hochella MF, Crerar DA (1990) The formation of leached layers on albite surfaces during dissolution under hydrothermal conditions. Geochim Cosmochim Acta 54:1267-1281

Hiebert FK, Bennett PC (1992) Microbial control of silicate weathering in organic-rich ground water. Science 258:278-281

Holdren GJ, Berner RA (1979) Mechanism of feldspar weathering - I. Experimental studies. Geochim Cosmochim Acta 43:1161-1171

Holdren GR, Speyer PM (1985) pH dependent change in the rates and stoichiometry of dissolution of an alkali feldspar at room temperature. Am J Sci 285:994-1026

Holdren GR, Speyer PM (1987) Reaction rate-surface area relationships during the early stages of weathering. II. Data on eight additional feldspars. Geochim Cosmochim Acta 51:2311-2318

Huang CP (1981) The surface acidity of hydrous solids. In: Anderson MA, Rubin AJ (eds) Adsorption of Inorganics at Solid-Liquid Interfaces. Ann Arbor Sci Pub, Ann Arbor, MI, p 183-218

Huang W-L, Longo JM (1992) The effect of organics on feldspar dissolution and the development of secondary porosity. Chem Geol 98:271-292

Huang WH, Keller WD (1970) Dissolution of rock-forming silicate minerals in organic acids: simulated first-stage weathering of fresh mineral surfaces. Am Mineral 55:2076-2094

Huang WH, Kiang WC (1972) Laboratory dissolution of plagioclase feldspars in water and organic acids at room temperature. Am Mineral 57:1849-1859

Iler RK (1979) The Chemistry of Silica. John Wiley & Sons, New York

Inskeep WP, Nater EA, Bloom PR, Vandervoort DS, Erich MS (1991) Characterization of laboratory weathered labradorite surfaces using X-ray photoelectron spectroscopy and transmission electron microscopy. Geochim Cosmochim Acta 55:787-800

Jørgensen SS (1976) Dissolution kinetics of silicate minerals in aqueous catechol solutions. J Soil Sci 27:183-195

Kawano M, Tomita K (1994) Growth of smectite from leached layer during experimental alteration of albite. Clays Clay Mineral 42:7-17

Kleijn JM (1990) The electrical double layer on oxides: Site-binding in the porous double layer model. Colloids Surfaces. 51:371-388

Knauss KG, Copenhaver SA (1995) The effect of malonate on the dissolution kinetics of albite, quartz, and microcline as a function of pH at 70°C. Appl Geochem 10:17-33

Knauss KG, Wolery TJ (1986) Dependence of albite dissolution kinetics on pH and time at 25°C and 70°C. Geochim Cosmochim Acta 50:2481-2497

Lanford WA, Davis K, Lamarche P, Laursen T, Groleau R (1979) Hydration of soda-lime glass. J Non-cryst Solids 33:249-266

Lasaga AC (1981) Transition state theory. In: Lasaga AC, Kirkpatrick RJ (eds) Kinetics of Geochemical Processes. Rev Mineral 8:135-170

Lasaga AC (1983) Kinetics of silicate dissolution. In: 4th Int'l Symp on Water-Rock Interaction. Misasa, Japan, p 269-274

Lasaga AC (1984) Chemical kinetics of water-rock interactions. J Geophy Res 89:4009-4025

Lasaga AC, Blum AE (1986) Surface chemistry, etch pits and mineral-water reactions. Geochim Cosmochim Acta 50:2363-2379

Lasaga AC, Gibbs GV (1990a) Ab initio quantum mechanical calculations of surface reaction - A new era? In: Stumm W (ed) Aquatic Chemical Kinetics. Wiley Interscience, New York, p 259-289

Lasaga AC, Gibbs GV (1990b) Ab-initio quantum mechanical calculation of water-rock interactions: adsorption and hydrolysis reactions. Am J Sci 290:263-295

Lasaga AC, Berner RA, Garrels RM (1985) An improved geochemical model of atmospheric CO_2 over the past 100 million years. In: Sundquist ET, Broecker WS (eds) The Carbon Cycle and Atmospheric CO_2: Natural Variations Archean to Present. Geophys Monogr Ser 32, Am Geophys Union, Washington DC, p 397-411

Lasaga AC, Soler JM, Ganor J, Burch TE, Nagy KL. (1994) Chemical weathering rate laws and global geochemical cycles. Geochim Cosmochim Acta 58:2361-2386

Linde DR (1990) CRC Handbook of Chemistry and Physics. CRC Press, Boca Raton, Florida

Lowell S (1979) Introduction to Powder Surface Areas. Wiley-Interscience, New York

Luce RW, Bartlett RW, Parks GA (1972) Dissolution kinetics of magnesium silicates. Geochim Cosmochim Acta 36:35-50

Lundström U, Öhman L-O (1990) Dissolution of feldspars in the presence of natural, organic solutes. J Soil Sci 41:359-369

Lyklema J (1968) The structure of the electrical double layer on porous surfaces. J Electroanalyt Chem. 18:341-348.

Lyklema J (1971) The electrical double layer on oxides. Croatica Chemica Acta. 43:249-260

Manley P, Evans LJ (1986) Dissolution of feldspars by low-molecular weight aliphatic and aromatic acids. Soil Sci 141:106-112

Manning DC, Gestsdóttir K, Rae,EIC (1991a) Feldspar dissolution in the presence of organic acid anions under diagenetic conditions: an experimental study. Org Geochem 19:483-492

Manning DC, Rae EIC, Small JS (1991b) An exploratory study of acetate decomposition and dissolution of quartz and Pb-rich potassium feldspar at 150°C, 50MPa (500 bars). Mineral Mag 55:183-195

Mast MA, Drever JI (1987) The effect of oxalate on the dissolution rates of oligoclase and tremolite. Geochim Cosmochim Acta 51:2559-2568

McClelland JE (1950) The effect of time, temperature, and particle size on the release of bases from some common soil-forming minerals of different crystal structure. Soil Sci Soc Proc, p 301-307

McGowen DB, Surdam RC (1988) Difunctional carboxylic acid anions in oil field waters. Org Geochem 12:245-259

Muir IJ, Nesbitt HW (1991) Effects of aqueous cations on the dissolution of labradorite feldspar. Geochim Cosmochim Acta 55:3181-3189

Muir IJ, Nesbitt HW (1992) Controls on differential leaching of calcium and aluminum from labradorite in dilute electrolyte solutions. Geochim Cosmochim Acta 56:3979-3985

Muir IJ, Bancroft GM, Nesbitt HW (1989) Characteristics of altered labradorite surfaces by SIMS and XPS. Geochim Cosmochim Acta 53:1235-1241

Murphy K, Drever JI (1993) Dissolution of albite as a function of grain size. Trans Am Geophys Union 74:329

Murphy KM (1993) Kinetics of albite dissolution: The effect of grain size. MS thesis, Univ Wyoming, Laramie, WY

Murphy WM (1989) Dislocations and feldspar dissolution. Eur J Mineral 70:163

Murphy WM, Helgeson HC (1987) Thermodynamic and kinetic constraints on reaction rates among minerals and aqueous solution. III. Activated complexes and the pH-dependence of rates of feldspar, pyroxene, wollastonite, and olivine hydrolysis. Geochim Cosmochim Acta 51:3137-3153

Murphy WM, Helgeson HC (1989) Surface exchange and the hydrolysis of feldspar. Geochim Cosmochim Acta 53:559

Nagy KL, Lasaga AC (1992) Dissolution and precipitation kinetics of gibbsite at 80°C and pH 3: The dependence on solution saturation state. Geochim Cosmochim Acta 56:3093-3112

Nash VE, Marshall CM (1956a) The surface reactions of silicate minerals, Part 1; Reactions of feldspar surfaces with acid solutions. Univ Missouri, College of Agriculture, Research Bull 613

Nash VE, Marshall CM (1956b) The surface reactions of silicate minerals, Part 2; Reactions of feldspar surfaces with salt solutions. Univ Missouri College of Agriculture, Res Bull 614

Nesbitt HW, Muir IJ (1988) SIMS depth profiles of weathered plagioclase and processes affecting dissolved Al and Si in some acidic soil solutions. Nature 334:336-338

Nesbitt HW, Macrae ND, Shotyk W (1991) Congruent and incongruent dissolution of labradorite in dilute, acidic, salt solutions. J Geol 99:429-442

Nielsen JW (1964) Kinetics of Precipitation. Macmillan, New York

Oelkers EH, Schott J (in press, a) The dependence of silicate dissolution rates on their stucture and composition In: Proc 8th Int'l Symp Water-Rock Interaction, Vladivostok, Russia

Oelkers EH, Schott J (in press, b) Experimental study of anorthite dissolution and the relative mechanism of feldspar hydrolysis. Geochim Cosmochim Acta

Oelkers EH, Schott J, Devidal JL (1994) The effect of aluminum, pH and chemical affinity on the rates of aluminosilicate dissolution reactions. Geochim Cosmochim Acta 58:2011-2024

Ohara M, Reed RC (1973) Modeling crystal growth rates from solution. Prentice-Hall, New York

Oxburgh R (1991) The effect of pH, oxalate ion and mineral composition on the dissolution rates of plagioclase feldspars. MS thesis,Univ Wyoming, Laramie, WY

Oxburgh R, Drever, JI, Sun YT (1994) Mechanism of plagioclase dissolution in acid solution at 25°C. Geochim Cosmochim Acta 58:661-669

Paces T (1973) Steady-state kinetics and equilibrium between ground water and granitic rock. Geochim Cosmochim Acta 37:2641-2663

Parks GA (1967) Aqueous surface chemistry of oxides and complex oxide minerals. Equilibrium concepts in natural water systems, ACS Adv Chem Ser 67:121-160

Parsons I (ed) (1994) Feldspars and Their Reactions. NATO Advanced Study Inst, Series C, Kluwer, 629 p

Perram JW, Hunter RJ, Wright HJL (1974) The oxide-solution interface. Aust J Chem. 27: 461-475

Petit JC, Della Mea G, Dran JC, Schott J, Berner RA (1987) Mechanism of diopside dissolution from hydrogen depth profiling. Nature 325:705-707

Petit JC, Dran JC, Della Mea G (1990) Energetic ion beam analysis in the earth sciences. Nature 344:621-626

Petit JC, Dran JC, Paccagenella A, Della Mea G (1989) Structural dependence of crystalline silicate hydration during aqueous dissolution. Earth Planet Sci Letters 93:292-298

Petrovic R (1981a) Kinetics of dissolution of mechanically comminuted rock-forming oxides and silicates. I. Deformation and dissolution of quartz under laboratory conditions. Geochim Cosmochim Acta 45:1665-1674

Petrovic R (1981b) Kinetics of dissolution of mechanically comminuted rock-forming oxides and silicates. II. Deformation and dissolution of oxides and silicates in the laboratory and at the earth's surface. Geochim Cosmochim Acta 45:1675-1686

Petrovic R, Berner RA, Goldhaber MB (1976) Rate control in dissolution of alkali feldspars I. Study of residual grains by X-ray photoelectron spectroscopy. Geochim Cosmochim Acta 40:537-548

Reuss JO, Johnson DW (1986) Acid Deposition and the Acidification of Soils and Waters. Springer-Verlag, New York, 303 p

Ribbe PH (1983a) Feldspar Mineralogy (2nd Edn). Rev Mineral 2, Mineral Soc Am, Washington, DC, 362 p

Ribbe PH (1983b) The chemistry, structure, and nomenclature of feldspars. In: Ribbe PH (ed) Feldspar Mineralogy. Rev Mineral 2:1-19

Rimstidt JD, Barnes HL (1980) The kinetics of silica-water reactions. Geochim Cosmochim Acta 44:1683-1699

Rimstidt JD, Dove PM (1986) Mineral/solution reaction rates in a mixed flow reactor: wollastonite hydrolysis. Geochim Cosmochim Acta 50:2509-2516

Rose NM (1991) Dissolution rates of prehnite, epidote, and albite. Geochim Cosmochim Acta 55:3273-3286

Schindler PW (1981) Surface complexes at oxide-water interfaces. In: Anderson MA, Rubin AJ (eds) Adsorption of Inorganics at Solid-Liquid Interfaces. Ann Arbor Science, Ann Arbor, MI, p 1-50

Schott J (1990) Modeling of the dissolution of strained and unstrained multiple oxides: The surface speciation approach. In: Stumm W (ed) Aquatic Chemical Kinetics. Wiley Interscience, NY, p 337-366

Schott J, Petit JC (1987) New evidence for the mechanisms of dissolution of silicate minerals. In: Stumm W (ed) Aquatic Surface Chemistry. Wiley Interscience, NY, p 293-315

Schott J, Berner RA, Sjoberg EL (1981) Mechanism of pyroxene and amphibole weathering, I. Experimental studies of iron-free minerals. Geochim Cosmochim Acta 45:2123-2135

Schott J, Brantley S, Crerar D, Guy C, Borcsik M, Willaime C (1989) Dissolution kinetics of strained calcite. Geochim Cosmochim Acta 53:373-382

Schweda P (1989) Kinetics of alkali feldspar dissolution at low temperature. In: Miles DL (ed) Proc of the 6th Int Symp on Water-Rock Interaction. AA Balkema, Rotterdam, p 609-612

Schweda P (1990) Kinetics and mechanisms of alkali feldspar dissolution at low temperatures. PhD dissertation, Stockholm University

Sjöberg L (1989) Kinetics and non-stoichiometry of labradorite dissolution. In: Miles DL (ed) Proc 6th Int'l Symp on Water-Rock Interaction. AA Balkema, Rotterdam, p 639-642

Smets BMJ, Tholen MGW (1985) The pH dependence of the aqueous corrosion of glass. Phys Chem Glasses. 26:60-63

Smith JV (1974a) Feldspar minerals: I. Crystal Structure and Physical Properties. Springer-Verlag, Heidelberg , 627 p

Smith JV (1974b) Feldspar minerals: II. Chemical and Textural Properties. Springer-Verlag, Heidelberg, 690 p

Song SK, Huang PM (1988) Dynamics of potassium release from potassium-bearing minerals as influenced by oxalic and citric acids. Soil Sci Soc Am J 52:383-390

Stevenson FJ (1967) Organic acids in soils. In: McLaren AD, Petersen GH (eds) Soil Biochemistry. Academic Press, NY, p 110-146

Stillings LL, Brantley SL (1995) Feldspar dissolution at 25°C and pH 3: Reaction stoichiometry and the effect of cations. Geochim Cosmochim Acta 59:1483-1496

Stillings LL, Brantley SL, Machesky ML (1995a) Proton adsorption at the adularia feldspar surface. Geochim Cosmochim Acta 59:1473-1482

Stillings LL, Brantley SL, Drever JI, Sun Y (1995b) Does oxalic acid increase dissolution rates for feldspar minerals? Book of Abstracts, 209th ACS National Meeting, p. GEOC 055

Stillings LL, Drever JI, Brantley SL, Sun Y, Oxburgh R (in review) Rates of feldspar dissolution at pH 3-5 with 0-8 mM oxalic acid. submitted to Chemical Geology

Stoessell RK, Pittman ED (1990) Secondary porosity revisited: The chemistry of feldspar dissolution by carboxylic acids and anions. Am Assoc Petrol Geol Bull 74:1795-1805

Stumm W, Morgan JJ (1981) Aquatic Chemistry. J Wiley & Sons, New York, 780 p

Stumm W, Wieland E (1990) Dissolution of oxide and silicate minerals: Rates depend on surface speciation. In: Stumm W (ed) Aquatic Chemical Kinetics. Wiley Interscience, New York, p 367-400

Stumm W, Wollast R (1990) Coordination chemistry of weathering: Kinetics of the surface-controlled dissolution of oxide minerals. Reviews Geophys 28:53-69

Sun Y (1994) The effect of pH and oxalate ion on the dissolution rates of plagioclase feldspar. MS thesis, University of Wyoming, Laramie, WY

Surdam RC, Boese SW, Crossey LJ (1984) The chemistry of secondary porosity. In: MacDonald DA, RC Surdam (eds) Clastic Diagenesis. Am Assoc Petrol Geol 37:127-149, Tulsa, OK

Sverdrup HU (1990) The Kinetics of Base Cation Release Due to Chemical Weathering. Lund Univ Press, Lund, Sweden, 246 p

Swoboda-Colberg NG, Drever JI (1993) Mineral dissolution rates in plot-scale field and laboratory experiments. Chem Geol 105:51-69

Tadros ThF, Lyklema J (1968) Adsorption of potential-determining ions at the silica-aqueous electrolyte interface and the role of some cations. J Electroanalyt Chem 17: 267-275

Tamm O (1930) Experimental Studied Umber die Verwitterung und Tonbildung von Feldspäten. Chemie Erde 4:420-430

Tan KH (1980) The release of silicon, aluminum, and potassium during decomposition of soil minerals by humic acid. Soil Sci 129:5-11

Taylor A, Blum JD (in press) Relation between soil age and silicate weathering rates determined from the chemical evolution of a glacial chronosequence. Geology.

Tole MP, Lasaga AC, Pantano C, White WB (1986) The kinetics of dissolution of nepheline ($NaAlSiO_4$). Geochim Cosmochim Acta 50:379-392

Unwin PR, Bard AJ (1992) Ultramicroelectrode voltammetry in a drop of solution: a new approach to the measurement of adsorption isotherms at the solid-liquid interface. Analyt Chem 64: 113-119

van der Donck JCJ, Vaessen GEJ, Stein HN (1993) Adsorption of short-chain tetraalkylammonium bromide on silica. Langmuir 9:3553-3557

Van der Hoek B, Van der Eerden JP, Bennema P (1982) Thermodynamical stability conditions for the occurrence of hollow cores caused by stress of line and planar defects. J Crystal Growth 56:621-632

Vandevivere P, Welch SA, Ullman WJ, Kirchman DL (1994) Enhanced dissolution of silicate minerals by bacteria at near-neutral pH. Microb Ecol 27:241-251

Velbel MA (1986) Influence of surface area, surface characteristics, and solution composition on feldspar weathering rates. In: Davis JA, Hayes KF (eds) Geochemical Processes at Mineral Surfaces. Am Chem Soc Symp Ser 323, Washington, DC, p 615-634

Velbel MA (1993) Formation of protective surface layers during silicate-mineral weathering under well-leached, oxidizing conditions. Am Mineral 78:405-414

Voigt DE, Stillings LL, Brantley SL (unpublished) Effect of oxalic acid and pH on rate of albite dissolution at 90°C.

Volk T (1987) Feedbacks between weathering and atmospheric CO_2 over the last 100 million years. Am J Sci 287:763-779

Walker JCG, Hays PB, Kasting JF (1981) A negative feedback mechanism for the long term stabilization of Earth's surface temperatures. J Geophys Res 86:9776-9782

Wehrli B, Wieland E, Furrer G (1990) Chemical mechanisms in the dissolution kinetics of minerals; the aspect of active sites. Aquatic Sci 52:3-31

Welch SA, Ullman WJ (1992) Dissolution of feldspars in oxalic acid solutions. In: Kharaka YK, Maest AS (eds) Water-Rock Interaction–7, Vol 1, AA Balkema, Rotterdam, p 127-130

Welch SA, Ullman WJ (1993) The effect of organic acids on plagioclase dissolution rates and stoichiometry. Geochim Cosmochim Acta 57:2725-2736

Welch SA, Ullman WJ (in review) Feldspar dissolution in acidic and organic solutions: compositional and pH dependence of dissolution rate. Submitted to Geochim Cosmochim Acta

Welch SA, Vandevivere (1994) Effect of microbial and other naturally occurring polymers on mineral dissolution. Geomicrobiology J 12:227-238

Westich HR, Casey WH, Arnold GW (1989) Oxygen isotope exchange in the leached layer of labradorite feldspar. Geochim Cosmochim Acta 53:1681-1685

White AF, Blum AE (1995) Effects of climate on chemical weathering in watersheds. Geochem Cosmochim Acta 59:1729-1748

White AF, Peterson ML (1990) Role of reactive-surface-area characterization in geochemical kinetic models. Am Chem Soc Symp Ser 416:461-475

White AF, Blum AE, Schulz MS, Bullen TD, Harden JW, Peterson ML (in press) Chemical Weathering of a soil chronosequence on granite alluvium: I. Reaction rates based on changes in soil mineralogy. Geochim Cosmochim Acta

Wieland E, Wehrli B, Stumm W (1988) The coordination chemistry of weathering. III Potential generalization on dissolution rates of minerals. Geochim Cosmochim Acta 52:1969-1981

Wilson MJ, McHardy WJ (1980) Experimental etching of a microcline perthite and implications regarding natural weathering. J Microscopy 120:291-302

Wilson MJ, Jones D, Russell JD (1980) Glushinskite, a naturally occurring magnesium oxalate. Mineral Mag 43:837-840

Wintsch RP, Dunning J (1985) The effect of dislocation density on the aqueous solubility of quartz and some geologic implications: A theoretical approach. J Geophys Res 90: 3649-3657

Wollast R (1967) Kinetics of the alteration of K-feldspar in buffered solutions at low temperature. Geochim Cosmochim Acta 31:635-648

Wollast R, Chou L (1992) Surface reactions during the early stages of weathering of albite. Geochim Cosmochim Acta 56:3113-3122

Xiao Y, Lasaga AC (1994) Ab initio quantum mechanical studies of the kinetics and mechanisms of silicate dissolution: $H^+(H_3O^+)$ catalysis. Geochim Cosmochim Acta 58:5379-5400

Yund RA (1983) Diffusion in feldspars. In: Ribbe PH (ed) Feldspar Mineralogy (2nd Edn) Rev Mineral 2:203-222

APPENDIX I

Dissolution studies of feldspar minerals in the presence of organic acid solutions. Acids are ranked by number of carboxyl groups.

Abbreviations are as follows: or=orthoclase, mic=microcline, san=sanidine, ksp=K-feldspar, alb=albite, olig=oligoclase, and=andesine, lab=labradorite, byt=bytownite, an=anorthite, n.r.= data not reported.

Feldspars	Temp	Initial Acid conc.	pH range (at 25°C)	Reactor Type	Source
		CATECHOL, HOC$_6$H$_4$OH			
mic,	25°C	0.1 M	9.4 - 9.7	batch	Jørgensen, 1976
an	100°C	0.05, 0.1 m	8.2	batch	Huang & Longo, 1992
		ACETIC ACID, CH$_3$COOH			
mic, lab	25°C	0.01 M	n.r.	batch	Huang & Keller, 1970
ksp, alb, olig, lab, byt, an	25°C	0.01 M	n.r.	batch	Huang & Kiang, 1972
mic, alb, and, lab	100°C	10,000 ppm	2.90 - 3.40	batch	Surdam et al., 1984
mic	100°C	0.15 m	3.19, 6.80	batch	Stoessell & Pittman, 1990
alb	150°C	0.1 - 2.5 M	6.00	batch	Manning et al., 1991a
Pb-or	150°C	0.1 - 2.5 M	5.9 - 9.4	batch	Manning et al., 1991b
ksp, olig, an	100°C	0.143 - 0.164 m	2.88 - 7.0	batch	Huang & Longo, 1992
lab, byt	20-22°C	0.1, 1.0 mM	3.9 - 9.3	flow-through	Welch & Ullman, 1993
lab	25, 35, 60°C	1.0 mM	3.92 - 4.02	flow-through	Brady & Carroll, 1994; pers. comm.
alb	100°C	0.07 m	3.4 - 5.3	flow-through	Franklin et al., 1994
		PROPIONIC ACID, CH$_3$CH$_2$COOH			
mic	100°C	0.075, 0.15 m	6.90-7.06	batch	Stoessell & Pittman, 1990
byt	25°C	0.1, 1.0 mM	3.9 - 4.7	flow-through	Welch & Ullman, 1993
		PYRUVIC ACID, CH$_3$COCOOH			
lab, byt	25°C	0.1, 1.0 mM	6.0, 3.7	flow-through	Welch & Ullman, 1993
		SALICYLIC ACID, C$_6$H$_4$(OH)COOH			
mic, lab	25°C	0.01 M	n.r.	batch	Huang & Keller, 1970
ksp, alb, olig, lab, byt, an	25°C	0.01 M	n.r.	batch	Huang & Kiang, 1972
mic, alb, lab	13°C	0.1 mM	4.03 (initial pH)	batch	Manley & Evans, 1986
		PROTOCATECHUIC ACID, (HO)$_2$C$_6$H$_3$COOH			
mic, alb, lab	13°C	0.1mM	4.25 (initial pH)	batch	Manley & Evans, 1986
		GALLIC ACID, (HO)$_3$C$_6$H$_2$COOH			
mic, alb, lab	13°C	0.1mM	4.25 (initial pH)	batch	Manley & Evans, 1986

Appendix I (continued)

Feldspars	Temp	Initial Acid conc.	pH range (at 25°C)	Reactor Type	Source
		p-HYDROXYBENZOIC ACID, C₆H₄(OH)COOH			
mic, alb, lab	13°C	0.1mM	4.32 (initial pH)	batch	Manley & Evans, 1986
		VANILLIC ACID, CH₃O(HO)C₆H₃-COOH			
mic, alb, lab	13°C	0.1mM	4.37 (initial pH)	batch	Manley & Evans, 1986
		CAFFEIC ACID, (HO)₂C₆H₃CH=CHCOOH			
mic, alb, lab	13°C	0.1mM	4.37 (initial pH)	batch	Manley & Evans, 1986
		OXALIC ACID, HOOC-COOH			
alb	25°C	0.001-1mM	2.93 - 5.19	flow-through	Chou, 1985
mic, alb, lab	13°C	0.1 mM	n.r.	batch	Manley & Evans, 1986
olig	25°C	0.1, 0.5, 1.0mM	4 - 9	flow-through	Mast & Drever, 1987
an	25°C	2.5 mM	5.2 - 6.2	batch	Amrhein & Suarez, 1988
mic, or	25°C	0.01 M	2.1 (initial pH)	batch	Song & Huang, 1988
or	70, 95°C	20 mM	1, 3.6, 9.0	batch, flow-through	Bevan & Savage, 1989
mic	100°C	0.075 m	3.0 - 8.3	batch	Stoessell & Pittman, 1990
alb	150°C	n.r.	7.5	batch	Manning et al., 1991a
olig, and, byt	25°C	1 mM	3 - 7	flow-through	Oxburgh, 1991
alb	150°C	0.1 - 2.5 M	6.75 - 8	batch	Gestsdóttir & Manning, 1992
ksp, olig, an	95°C	0.0011 - 0.096 m	5 - 8.5	batch	Huang & Longo, 1992
alb, lab, byt	25°C	0.1, 1.0 mM	3 - 10	flow-through	Welch & Ullman, 1992, 1993
alb	100°C	0.0005 - 0.01 m	3.3 - 5.3	flow-through	Franklin et al., 1994
olig, and, byt	25°C	2 mM - 8 mM	3 - 8	flow-through	Sun, 1994
or	80°C	0.5, 3, 10 mM	6	batch	Blake and Walter, in review
mic, alb, olig, and, byt	25°C	1mM	3 - 5	flow-through	Stillings et al., in review
		MALONIC ACID, HOOCCH₂COOH			
mic	100°C	0.075 m	7.61	batch	Stoessell & Pittman, 1990
alb, mic	70°C	0.001 - 0.1 m	4.1 - 9.9	flow-through	Knauss & Copenhaver, 1995
		SUCCINIC ACID, HOOCCH₂CH₂COOH			
byt	25°C	0.1, 1.0	3.6, 4.4	flow-through	Welch & Ullman, 1993
		TARTARIC ACID, HOOCCHOH-CHOHCOOH			
mic, lab	25°C	0.01M	n.r.	batch	Huang & Keller, 1970

Appendix I (concluded)

Feldspars	Temp	Initial Acid conc.	pH range (at 25°C)	Reactor	Source
ASPARTIC ACID, HOOCCH₂CH(NH₂)-COOH					

ASPARTIC ACID, $HOOCCH_2CH(NH_2)\text{-}COOH$

Feldspars	Temp	Initial Acid conc.	pH range (at 25°C)	Reactor	Source
mic, lab	25°C	0.01 M	n.r.	batch	Huang & Keller, 1970
ksp, alb, olig, lab, byt, an	25°C	0.01 M	n.r.	batch	Huang & Kiang, 1972

a-KETOGLUTARIC ACID, $HOOCCH_2CH_2\text{-}COCOOH$

Feldspars	Temp	Initial Acid conc.	pH range (at 25°C)	Reactor	Source
lab, byt	25°C	0.1, 1.0 mM	3.1 - 9.3	flow-through	Welch & Ullman, 1993

CITRIC ACID, $HOC(CH_2CO_2H)_2CO_2H$

Feldspars	Temp	Initial Acid conc.	pH range (at 25°C)	Reactor	Source
ksp, alb, olig, lab, byt, an	25°C	0.01 M	n.r.	batch	Huang & Kiang, 1972
mic, alb, lab	13°C	0.1 mM	3.92 (initial pH)	batch	Manley & Evans, 1986
mic, or	25°C	0.01M	2.6 (initial pH)	batch	Song & Huang, 1988
ksp, olig	20°C	5.55 mM	5.1 (initial pH)	batch	Lundström & Öhman, 1990
mic, san	25°C	0.1 - 1 mM	4, 5.7	flow-through	Schweda, 1990
alb	150°C	n.r.	3.0	batch	Manning et al., 1991a
byt	25°C	0.01 - 10 mM	2.7 - 5.9	flow-through	Welch & Ullman, 1993

EDTA, $[(HOOCCH_2)_2\text{-}NCH_2\text{-}]_2$

Feldspars	Temp	Initial Acid conc.	pH range (at 25°C)	Reactor	Source
an	95°C	0.025, 0.1 m	8.5	batch	Huang & Longo, 1992

HUMIC AND FULVIC ACIDS

Feldspars	Temp	Initial Acid conc.	pH range (at 25°C)	Reactor	Source
mic	25°C	100 mg/l	2.5, 7.0	batch	Tan, 1980

MOR EXTRACT (from a coniferous forest)

Feldspars	Temp	Initial Acid conc.	pH range (at 25°C)	Reactor	Source
ksp, olig	20°C	filtered suspension of 45g mor/l	5.1	batch	Lundström & Öhman, 1990

Appendix II

Dissolution rates of feldspars in organic acid solutions.
Symbols are defined at the bottom of the table.

Phase (reference)	Temp °C	Final pH$_{25°}$	Organic Acid Concentration	log Rate$_{SI}$ (mol fsp m^{-2} sec^{-1})
		ALBITE		
Acetic Acid				
Ab$_{99}$Or$_1$ (7)	150	6.0	0 M	-8.00*
		6.0	0.1 M	-6.74*
		6.0	1.0 M	-6.48*
		6.0	2.5 M	-6.28*
Ab$_{97}$An$_2$Or$_1$ (12)	100	3.4	0.07 m	-9.22
		4.75	0.07 m	-10.01
		5.3	0.07 m	-10.84
Oxalic Acid				
Ab$_{98}$An$_2$ (1)	25	2.93	1 mM	-10.4
		3.20	1 mM	-10.5
		3.87	0.1 mM	-10.8
		4.21	0.1 mM	-11.0
		5.19	0.01 mM	-11.3
Ab$_{97}$An$_2$Or$_1$ (14)	25	3.00	0 mM	-11.57
		3.32	1 mM	-10.67
Ab$_{97}$An$_2$Or$_1$ (16)	90	3.13	0 mM	-9.72[2]
		3.22	0.1 mM	-9.66
		3.40	0.5 mM	-9.80
		3.30	1 mM	-9.44
		3.26	10 mM	-10.06
		5.72	0 mM	-10.30
		6.52	0.005 mM	-10.22
		5.62	1 mM	-10.08
Ab$_{97}$An$_2$Or$_1$ (11)†	100	3.3	0.005 m	-8.90
		4.8	0.000 5m	-10.10
		4.6	0.001 m	-9.92
		4.6	0.005 m	-9.32[3]
		4.5	0.01 m	-9.46
		5.1	0.005 m	-9.45
		5.72	0 mM	-10.30
		6.52	0.005 mM	-10.22
		5.62	1 mM	-10.08
Malonic Acid				
Ab$_{99}$An$_1$ (13)	70	3.88	0 m	-11.2
		4.04	0.001 m	-11.1
		4.56	0.01 m	-11.1
		4.55	0.1 m	-10.8
		5.40	0 m	-11.2
		6.81	0.01 m	-11.2
		6.43		-10.8
		7.72	0 m	-11.2
		8.23	0.01 m	-11.3
		7.89	0.1 m	-11.0
		9.85	0 m	-10.9
		9.69	0.001 m	-11.0
		9.81	0.01 m	-10.5
		10.02	0.1 m	-10.5
		9.85	0 m	-10.9
		9.69	0.001 m	-11.0
		9.81	0.01 m	-10.5
		10.02	0.1 m	-10.5

Appendix II (continued)

Phase (reference)	Temp °C	Final pH$_{25°}$	Organic Acid Concentration	log Rate$_{SI}$ (mol fsp m^{-2} sec^{-1})
OLIGOCLASE				
Acetic Acid				
Ab$_{69}$An$_{28}$Or$_3$ (8)	100	3.4	0.164 m	-10.66
Oxalic Acid				
Ab$_{87}$An$_{12}$ (2)	25	4.0	0 mM	-11.72
		4.0	0.5 mM	
		4.0	1mM	-11.81
		5.0	0 mM	-11.85
		5.0	0.5 mM	-12.06
		5.0	1 mM	-12.06
		7.0	0 mM	-11.89
		7.0	1 mM	-11.83
		9.0	0 mM	-11.95
		9.0	1 mM	-11.92
Ab$_{87}$An$_{12}$Or$_1$ (7)	25	3.00	0 mM	-11.45
		2.98	1 mM	-11.51
		5.09	0 mM	-12.10
		5.29	1 mM	-12.34
Ab$_{70}$An$_{23}$Or$_7$ (12)		2.95	2 mM	-11.51
		4.09	2 mM	-12.05
		5.45	2 mM	-12.42
		7.79	2 mM	-12.06
Ab$_{73}$An$_{22}$Or$_5$ (14)	25	3.03	0 mM	-11.21
		2.85	1 mM	-10.46
		5.40	0 mM	-11.63
		5.68	1 mM	-10.65
Citric Acid				
Ab$_{78}$An$_{18}$Or$_4$ (5)	20	5.1	5.5 mM	-12.78
Mor extract				
Ab$_{78}$An$_{18}$Or$_4$ (5)	20	5.1	45 g /l	-12.26
ANDESINE				
Acetic Acid				
Ab$_{47}$An$_{48}$Or$_5$ (9)	25		0 mM	-9.81
		3.9	1 mM	-9.86
Ab$_{46}$An$_{48}$Or$_6$		4.1	1 mM	-9.77
		4.6	1 mM	-10.07
		5.75	0 mM	-10.76[2]
		6.1	0.1 mM	-10.45
		7.7	0 mM	-10.70
		7.2	1 mM	-10.56
		8.4	1 mM	-10.37
		9.0	0 mM	-10.31
		9.3	1 mM	-10.24
Pyruvic Acid				
Ab$_{47}$An$_{48}$Or$_5$ (9)	25	5.75	0 mM	-10.76[2]
		6.0	0.1 mM	-10.45

Appendix II (continued)

Phase (reference)	Temp °C	Final pH$_{25°}$	Organic Acid Concentration	log Rate$_{si}$ (mol fsp m^{-2} sec^{-1})
ANDESINE				
Oxalic Acid				
Ab$_{46}$An$_{48}$Or$_6$ (9)	25	3.1	0 mM	-10.2
		3.1	1 mM	-9.61
Ab$_{47}$An$_{48}$Or$_5$		3.8	0 mM	-10.52
		3.7	0.1 mM	-10.15
		5.8	0 mM	-11.16
		5.9	1 mM	-10.66
		7.7	0 mM	-11.11
		7.3	1 mM	-11.08
		9.0	0 mM	-10.72
		9.3	1 mM	-10.08
Ab$_{51}$An$_{47}$Or$_2$	25	3.13	0 mM	-10.64a
(7 & 12)		3.06		-10.51a
		2.97	2 mM	-10.35b
		3.04	4 mM	-10.33b
		3.18	8 mM	-10.29b
		4.11	0 mM	-11.15a
		4.14	1 mM	-10.82a
		5.23	2 mM	-10.81b
		4.11	4 mM	-10.48b
		4.14	8 mM	-10.32b
		5.06	0 mM	-11.24a
		5.11	1 mM	-10.97a
		6.82	2 mM	-11.69b
		6.60	4 mM	-11.49b
		5.54	8 mM	-10.84b
		7.24	0 mM	-11.23a
		7.22	1 mM	-10.95a
		7.69	2 mM	-11.69b
		7.61	4 mM	-11.42b
		7.48	8 mM	-11.30b
Ab$_{50}$An$_{43}$Or$_7$ (14)	25	3.13	0 mM	-11.09
		2.96	1 mM	-9.90
		5.61	0 mM	-10.31
		5.72	1 mM	-11.52
a-Ketoglutaric Acid				
Ab$_{47}$An$_{48}$Or$_5$ (9)	25	3.1	0 mM	-9.78
		3.8	0.1 mM	-9.80
Ab$_{46}$An$_{48}$Or$_6$ (9)		3.1	0 mM	-9.27
		3.1	1 mM	-9.18
		4.1	0 mM	-9.81
		4.0	1 mM	-9.71
		5.6	0 mM	-10.51
		6.0	1 mM	-10.25
		7.7	0 mM	-10.70
		7.1	1 mM	-10.55
		9.0	0 mM	-10.31
		8.7	1 mM	-10.48
		9.3	1 mM	-10.37

Appendix II (continued)

Phase (reference)	Temp °C	Final pH$_{25°}$	Organic Acid Concentration	log Rate$_{si}$ (mol fsp m^{-2} sec^{-1})
LABRADORITE				
Acetic Acid				
An$_{60}$ (10)	21	3.92	1 mM	-11.91
	35	3.97	1 mM	-11.49
	60	4.02	1 mM	-10.92
BYTOWNITE				
Acetic Acid				
Ab$_{23}$An$_{76}$Or$_1$ (9)	25	4.0	0 mM	-8.89[3]
		3.9	1 mM	-8.85[3]
		4.9	0 mM	-10.20
		4.7	0.1 mM	-9.46[3]
Propionic Acid				
Ab$_{23}$An$_{76}$Or$_1$ (9)	25	4.0	0 mM	-8.89[3]
		3.9	1 mM	-8.64
		4.9	0 mM	-10.20
		4.7	0.1 mM	-9.33
Pyruvic Acid				
Ab$_{23}$An$_{76}$Or$_1$ (9)	25	4.0	0 mM	-8.89[3]
		3.7	1 mM	-8.20
Oxalic Acid				
Ab$_{25}$An$_{75}$	25	3.8	0 mM	-9.62a
(7 & 12)		3.8	1 mM	-9.68a
		3.05	2 mM	-9.79b
		4.36	0 mM	-9.97a
		4.30	1 mM	-10.02a
		4.18	2 mM	-10.68b
		5.08	0 mM	-10.9a
		5.02	1 mM	-10.4a
		6.34	2 mM	-10.32b
		7.25	0 mM	-10.86a
		7.12	1 mM	-10.64a
		7.75	2 mM	-11.15b
Ab$_{25}$An$_{75}$ (9)	25	3.1	0 mM	-8.78[4]
		3.0	1 mM	-8.46[6]
		4.0	0 mM	-9.24 [6]
		4.0	0.1 mM	-8.34[3]
		3.7	1 mM	-8.57[1]
		5.9	0 mM	-11.01[1]
		6.2	1 mM	-10.15[1]
Ab$_{24}$An$_{76}$ (14)	25	3.51	0 mM	-10.64
		3.70	1 mM	-9.45
		4.75	0 mM	-9.66
		4.43	1 mM	-10.76
Succinic Acid				
Ab$_{23}$An$_{76}$Or$_1$ (9)	25	4.0	0 mM	-8.89[3]
		3.6	1.0 mM	-8.25
		4.5	0 mM	-9.35[2]
		4.4	0.1 mM	-9.04
a-Ketoglutaric Acid				
Ab$_{23}$An$_{76}$Or$_1$ (9)	25	4.0	0 mM	-8.89[3]
		3.8	1 mM	8.10[2]

Appendix II (continued)

Phase (reference)	Temp °C	Final pH$_{25°}$	Organic Acid Concentration	log Rate$_{SI}$ (mol fsp m^{-2} sec^{-1})
BYTOWNITE				
Citric Acid				
Ab$_{23}$An$_{76}$Or$_1$ (9)	25	3.0	0 mM	-8.26[3]
		3.0	1 mM	-7.97
		3.1	1 mM	-8.33
		2.7	10 mM	-7.89
		4.0	0 mM	-8.89[3]
		3.7	0.1 mM	-8.40
		4.2	0.1 mM	-8.33
		5.7	0 mM	-10.12[2]
		5.9	0.01 mM	-9.06
ANORTHITE				
Acetic Acid				
Ab$_2$An$_{98}$ (8)	100	3.4	0.164 m	-10.39
Oxalic Acid				
Ab$_6$An$_{94}$ (3)	25	5.2	0 mM	-11.75
		5.2	2.5 mM	-10.62
		6.2	0 mM	-12.00
		6.2	2.5 mM	-11.04
		6.6	0 mM	-12.00
		6.6	2.5 mM	-11.22
K-FELDSPAR				
Acetic Acid				
Microcline,	100	6.0	0.143 m	-11.59
Ab$_{26}$An$_1$Or$_{73}$ (8)		3.9	0.164 m	-11.20
		3.4	0.164 m	-10.96
		3.4	0.164 m	-10.66
		3.4	0.164 m	-10.37
Oxalic Acid				
Microcline,	25	3.02	0 mM	-11.79
Ab$_{22}$Or$_{78}$ (14)		3.12	1 mM	-11.08
Orthoclase,	95	1.0	0 mM	-9.75
Ab$_{17}$Or$_{83}$ (4)		1.0	20 mM	-9.83
		3.6	0 mM	-11.33
		3.6	20 mM	-11.73
		9.0	20 mM	-11.62
	70	1.0	0 mM	-9.75
		1.0	20 mM	-9.83
		3.6	0 mM	-11.51
		3.6	20 mM	-11.31
		9.0	0 mM	-12.03
		9.0	20 mM	-11.84

Appendix II (continued)

Phase (reference)	Temp °C	Final pH$_{25°}$	Organic Acid Concentration	log Rate$_{Si}$ (mol fsp m^{-2} sec^{-1})
K-FELDSPAR				
Malonic Acid				
Microcline	70°C	3.88	0 m	-11.1
Ab$_6$Or$_{94}$ (13)		4.03	0.001 m	-10.8
		4.54	0.01 m	-11.0
		4.56	0.1 m	-10.7
		5.37	0 m	-11.5
		5.68	0.001 m	-11.2
		6.85	0.01 m	-11.1
		6.44	0.1 m	-10.7
		7.68	0 m	-11.4
		7.16	0.001 m	-11.4
		8.22	0.01 m	-11.2
		7.90	0.1 m	-11.0
		9.82	0 m	-10.7
		9.71	0.001 m	-11.0
		10.04	0.1 m	-10.4
Citric Acid				
Ab$_{27}$Or$_{71}$Cn$_2$	25	5.7	0 mM	-12.2
			0.01 mM	-12.0
			0.05 mM	-11.8
			0.1 mM	-10.8
			0.5 mM	-11.7
Ab$_{36}$Or$_{64}$ (5)	20	5.1	0 mM	-12.52
		5.1	5.5 mM	-12.08
Mor extract				
Ab$_{36}$Or$_{64}$ (5)	20	5.1	0 g/l	-12.52
		5.1	45 g/l	-12.64

REFERENCES

(1) Chou (1985)
(2) Mast and Drever (1987)
(3) Amrhein and Suarez (1988)
(4) Bevan and Savage (1989)
(5) Lundström and Öhman (1990)
(6) Schweda (1990)
(7) Manning et al. (1991a)
(8) Oxburgh (1991)
(9) Huang and Longo (1992)
(10) Welch and Ullman (1993)
(11) Brady and Carroll (1994)
(12) Franklin et al. (1994)
(13) Sun (1994)
(14) Knauss and Copenhaver (1995)
(15) Stillings et al. (in review)
(16) Voigt et al. (unpublished ms)

Chapter 8

CHEMICAL WEATHERING OF SILICATES IN NATURE: A MICROSCOPIC PERSPECTIVE WITH THEORETICAL CONSIDERATIONS

M.F. Hochella, Jr.

Department of Geological Sciences
Virginia Polytechnic Institute and State University
Blacksburg, VA 24061 U.S.A.

J.F. Banfield

Department of Geology and Geophysics
University of Wisconsin
Madison, WI 53706 U.S.A.

"It is the mark of an instructed mind to rest satisfied with the degree of precision which the nature of the subject permits and not to seek an exactness where only an approximation of the truth is possible."   Aristotle

INTRODUCTION

Attempting to obtain the chemical weathering kinetics of silicate minerals is a noble goal within the earth sciences, especially when one considers the truly fundamental importance of this process to the global geochemical cycling of the elements. Yet there may be even more compelling reasons to study weathering rates. These rates are key to understanding such things as soil development and the distribution of nutrients within these soils, groundwater composition and quality, and the transport of toxic materials in groundwater systems in both a physical and chemical sense.

In general, silicate weathering rates have been studied in two ways. One way is through laboratory experimentation. This is equivalent to an *in vitro* biological study (a study removed from a living system) and is subject to similar advantages and disadvantages. Geoscientists have been dissolving minerals in laboratories for a long time (e.g. Ebelmen, 1847; Daubrée, 1867), and this type of experimentation remains very popular today as shown in the first several chapters of this book. The common theme involves taking a sample, exposing it to an aqueous solution, waiting, and thereafter analyzing the compositional change of the solution and/or the mineral surface. Possible variations involved in this type of experimentation include the starting material (natural or synthetic, the history of the sample, as well as exact crystallography, composition, and texture of the material), sample preparation before the experiment begins (both mechanical, e.g. grinding, and chemical, e.g. pre-leaching of ultrafines), the composition (including pH) of the reacting solution, the type of laboratory weathering reactor used (e.g. batch or flow-through, stirred or unstirred, etc.), the water to rock ratio, the temperature and pressure of the experiment, the length of the experiment, the means by which the solutions and/or minerals are analyzed, and the timing of analysis (at points during the experiment and/or after it). Despite the number of variations, the advantage to this type of study is that the actual experiment can be relatively well-defined. The main underlying disadvantage of this technique is that no laboratory design can ever have total relevance to a field situation. In reality, for better or worse, most experimental designs have very little similarity to the field environment.

The other way to study mineral dissolution is via field studies (e.g. Velbel, 1985, 1992; White, this volume; Drever and Clow, this volume, and many references therein). This is equivalent to an *in vivo* biological study (in a living system). Again, like the experimental methods outlined above, there are a myriad of ways to obtain mineral dissolution rates in nature. Underlying most of these methods is the need to define a natural drainage system (e.g. a watershed), to get various data on that system, and to make a few critical estimates. These estimates include, but are not necessarily limited to, the relative amounts of all weatherable minerals, the surface area of the minerals that water comes in contact with in the drainage system, the stoichiometry of all pertinent weathering reactions, the volume of earth that the discharged water has been in contact with, and the affects of fauna and flora on the entire process (e.g. the net removal of Mg, Ca, and K by an aggrading forest in the watershed area). Next, one measures the chemistry of the water input to and discharged from the system, and together with what is stipulated above, one determines the mineral dissolution rate. This method is again subject to both advantages and disadvantages. Its tremendous advantage is that this rate determination method should result in a realistic natural dissolution rate. The disadvantages include the fact that this rate is only strictly applicable to the setting in which it is measured, and one gathers no information about the actual weathering mechanism. Further, the rates are only as good as the estimates made.

It is obvious that the two basic methods for determining mineral dissolution rates are grossly different. At first it might be assumed that a difference of 2 to 3 orders of magnitude in mineral dissolution rate obtained from the two methods may just be due to the considerable uncertainties and potential variations of both methods. However, one nearly always finds that the laboratory method gives rates that are 1 to 3 orders of magnitude faster than the field-derived rates (e.g. Velbel, 1993b). Such a systematic difference is telling, and has encouraged researchers in this field to continue to explore mineral weathering processes and rates both in the laboratory and field. The reason for this discrepency is touched upon near the end of this chapter, and in fact is a common subject throughout this book.

The goal of this chapter is not to refine methods for obtaining laboratory and field-based dissolution rates. It is, however, our goal to highlight aspects of silicate weathering that are not generally considered in attempting to understand the mechanisms and rates of weathering in either laboratory or field settings. To do this, we will look very carefully at what nature is trying to tell us about how minerals break down, and how and where secondary minerals form. This is best done, in our opinion, by probing naturally weathered samples with scanning and transmission electron microscopy (SEM and TEM, respectively), as well as by utilizing the chemical analyses that these techniques can provide. High-resolution TEM (HRTEM) is definitely the key to this approach, as this technique will afford us a critical look at weathering fronts and defects in minerals with sub-nanometer resolution, and chemical analyses from areas not much bigger than that. This high resolution ability turns out to be critical, as microtextures and weathering fronts (the interface between primary and secondary minerals) have fascinating and important detail on the nanometer scale. Along with this, we will explore what we believe may be some of the most critical controls of mineral weathering rates in nature, particularly in its early stages. Our approach challenges some long-standing thoughts on silicate weathering processes.

It is our hope that readers of this chapter will be better equipped to understand how minerals weather in nature, and how to more realistically model this process. It is also our hope that through this, the reason for gaps between laboratory and field rates will be more clear, and more reliable mineral dissolution rates will ultimately be obtained.

PRECURSORS TO WEATHERING: MINERAL SURFACES

We will assume in our discussion below that the chemical weathering process must occur on a solid surface or surfaces which are separated from a fluid environment by an interface across which reactants and products must pass. The mineral surface is defined here as the topmost layer of atoms of the mineral down to and including regions of the mineral that are structurally and/or chemically different from the bulk average mineral. This is half of the interface. The other half is the fluid 'surface', and like the mineral surface, it includes that portion which is structurally and/or chemically different from the bulk average fluid. The perturbations in both the mineral and fluid adjoining the interface are due to two factors, first their abrupt termination, and second the effects of the presence of one on the other. In this section, we will be primarily concerned with the mineral side of the interface; later parts of the chapter will be primarily concerned with the solution side. It should be made clear at the outset that the atomic perturbations found on the surface of minerals do not have a direct analog to other well-known mineral phase transformations such as reconstructive polymorphic and order-disorder transitions, as well as exsolution. These latter transformations are for the most part solid-state processes.

General characteristics of mineral surfaces

The nature of mineral surfaces and their associated interfaces can be remarkably variable and dynamic (e.g. Hochella and White, 1990; Hochella, 1990, 1994; Smith, 1994; Vaughan and Pattrick, 1995). It has been shown recently, both in the field of surface science in general, and more recently within mineral surface science research, that unexpected processes often occur at surfaces and interfaces. This is a result of our familiarity with bulk, homogeneous chemistry. By far the majority of our experiences and measurements in science are in this three-dimensional realm. Chemistry in two dimensions is, in general, much further from our experience, and one must become accustomed to chemistry which occurs only within a massive defect array. This is precisely what surfaces and interfaces are in their entirety. Further, in bulk chemistry, we are use to the principle that fundamental physical properties of a material are dependent only on the atomic structure of the material and its chemical composition. The properties, that is the chemical reactivities, of surfaces are likewise dependent on the atomic structure of the surface and its chemical composition, but they are also dependent on the shape (topography) of the surface. This relationship (between reactivity and shape) has no direct analog in bulk chemistry. Because we are generally interested in the topography of a surface in the atomic to molecular size range (i.e. the dimensional order of what is reacting on the surface), the term most often used is *microtopography.*

Although the dependence of surface reactivity on atomic structure and composition should be fairly obvious (due to direct analogies to the understanding of bulk chemistry), such is not the case for surface microtopography. However, the importance of surface microtopography cannot be overstated. To illustrate and emphasize this, a few examples are given here.

The first example involves carbon monoxide dissociation on single crystal platinum surfaces with and without the presence of surface steps (Iwasawa et al., 1976). This process has no direct connection with mineral dissolution, but it is one of the most elegant examples of the importance of surface shape in all of surface chemistry and is worth a brief look. In the Iwasawa et al. study, CO gas was introduced into an ultra-high vacuum chamber and was allowed to react with atomically smooth Pt (111) surfaces and stepped Pt (111) surfaces. The steps on the latter surfaces were of the monoatomic type, that is approximately the diameter of a platinum atom in height. For the flat, unstepped surface, X-ray photoelectron spectroscopy (XPS) of the carbon 1s region clearly showed the presence of carbonyl ligands on the surface (that is Pt-CO bonding). On the stepped surface, XPS showed the presence of carbides (that is Pt-C) bonding. Therefore, this work presents a case where steps on a surface can dissociate a molecule into its component parts, whereas the same surface without monoatomic steps is passive in regards to this dissociation reaction. This type of reaction is particularly important in heterogeneous catalysis, where the dissociation of a reacting species is often a critical (sometimes rate-limiting) step.

A second example concerns mineral surfaces, but in this case adsorption and precipitation instead of dissolution. However, adsorption and precipitation reactions can be critical in dissolution reactions. Sorbed species may inhibit or accelerate dissolution (for example sorption of Al inhibiting feldspar dissolution; e.g. Wollast and Chou, 1985, Casey et al., 1988b, Amrhein and Suarez, 1992, and Oelkers et al., 1994). Also, surface precipitation can block dissolution sites on a substrate. In the study of Junta and Hochella (1994), the oxidative sorption of $Mn(II)_{aq}$ was studied on both albite and hematite surfaces. Before this work, it was thought that Mn(II) in solution adsorbed on various reactive sites on these mineral surfaces and were oxidized to Mn(III) from dissolved oxygen in solution. It was shown in Junta and Hochella (1994), using XPS, scanning force microscopy (SFM), and scanning Auger microscopy (SAM), that the initial uptake of Mn(II) is fully dependent on precipitation reactions that occur only at steps on these surfaces. The precipitates are Mn(III) oxyhydroxides. In short, terraces on these surfaces will not participate in these precipitation reactions, but steps provide the configurational entropy for such reactions to proceed. In nature, these reactions eventually produce Mn(III) and Mn(IV) coatings on mineral grains in soils. Although complete armoring is never observed or suggested, portions of the substrate surfaces (in fact, the step sites which are very reactive in dissolution processes) are not available to dissolving agents after such reactions.

In a thermodynamic sense, the importance of surface shape to dissolution reactions has been known for some time. It was first noticed that fine particles have a seemingly higher solubility than coarse particles of the same material (see, e.g. Enustun and Turkevich, 1960, and references therein). This phenomena has been described in a modified version of the Kelvin equation which relates the solubility of a material to its surface free energy and grain size (e.g. Petrovich, 1981; Adamson, 1982):

$$\frac{S}{S_0} = \exp\left[\frac{2\gamma\overline{V}}{RTr}\right] \tag{1}$$

where S is the solubility (in mol/kg H_2O) of grains with inscribed radius r in meters, S_0 is the solubility of the bulk material, γ is the surface free energy in mJ/m^2 (see, e.g. Parks, 1984, 1990, for discussions of surface free energy), $\overline{V}$ is the molar volume in m^3/mol, R is the gas constant in $mJ/mol \cdot K$, and T is the temperature in K. This equation implies that as the mean radius of a particle decreases, the solubility of that particle will go up exponentially relative to the measured solubility of a very large grain (the bulk solubility, where the grain size has no appreciable affect on solubility). Considering the surface free energy of silicates (typically in the range of 100 to several hundred mJ/m^2) and their molar volumes, the solubility only begins to change appreciably as the radius of a particle gets very small, on the order of tens of nanometers. In this size range (and down to the nanometer size range), one could also apply Equation (1) to individual features on a surface by using inscribed radii. In this way, the apparent solubility of an irregular portion of a surface can be roughly compared to a flat terrace. The latter would have an inscribed radius of infinity, resulting in $S = S_0$.

Overall, more and more evidence is being provided which emphasizes the importance of microtopography in mineral dissolution, whether from a theoretical perspective (e.g. Dibble and Tiller, 1981; Lasaga and Blum, 1986) or an experimental one (e.g. Holdren and Speyer, 1985, 1987; Brantley et al., 1986; Schott et al. 1989).

Mineral surface areas

In any complete study of the reactivity of a mineral surface, one must know how much mineral surface there is to react. Dissolution rates are usually reported in terms of moles of component released per surface area per time (i.e. mol m^{-2} s^{-1}). Obviously, mineral surface areas are needed to normalize reaction rates that are measured in the laboratory as well as in the field. Estimating the mineral surface area, or mis-measuring it, can be the largest source of error in a reported reaction rate. In this section, we will take a brief look at how surface areas can be measured and what the measurements mean. Note that in this discussion we are concerned with physical surface area measured in terms of m^2/g. The concept of reactive surface area will be examined in the next section.

Measuring surface areas. There are many ways to measure (or approximate) a surface area. The easiest method makes the assumption that all grains can be adequately described by a simple geometric shape (usually a sphere) so that a rough dimension of a particle will provide its approximate surface area. A surface area obtained in this way is called a *geometric surface area*. Obviously the geometric surface area does not include surface roughness, but it can be adjusted to do so with a *surface roughness factor*. This factor is equal to the ratio of the BET surface area (see below) divided by the area of a sphere with an equivalent diameter (i.e. the geometric surface area). White and Peterson (1990) found that there is a mean roughness factor of about 7 for a wide assortment of natural mineral particles over a considerable size range, although the roughness factor can vary widely for natural grains and can reach values of several hundred (e.g. White et al., 1995).

The Brunauer-Emmett-Teller (BET) method for surface area measurement (Brunauer et al., 1938; Lowell and Shields, 1991) is the most popular quantitative method used today. This method utilizes the physisorption of a known gas onto a surface at different partial pressures of the gas in a vacuum. If one knows or designs the experiment so that only one monolayer of the gas will physisorb, and if one knows the cross-sectional area of the gas on a substrate (that is how much area each molecule of the gas will occupy), it is possible to measure the surface area of a sample. The surface area is obtained by determining how much gas is released when the sample is brought up to room temperature after physisorption at cryogenic temperatures. The gas of choice is generally N$_2$ which will physisorb in a highly predictable and consistent way at its boiling point of 77 K (Kr sorption is also commonly used at the same temperature).

In the discussion that we will present below, it is critical to know the smallest features that can be analyzed by the BET method. This is because weathering front internal spaces in minerals can be exceptionally narrow, on the order of a nanometer or less. The approximate diameter of the N_2 molecule is 3.5 Å. If we expect monolayer coverage of this diatomic molecule on, for example, both walls of a reaction front in order to appropriately measure the surface area within this space, we must expect the space to have an opening of at least twice the molecular diameter of N_2, or 7 Å. If the walls are between 3.5 and 7 Å apart, we might expect physisorption of a single monolayer within the space, with the method then recording half of the surface area available. Below 3.5 Å, the feature may not be available to N_2-BET measurement at all.

Even if the above assessment is correct, there is another unknown that is just as critical. Before gas sorption takes place in a BET measurement, the sample is outgassed at temperature (often 100°C) in vacuum. This provides a relatively clean surface for the measurement. It is not at all clear whether reaction front spaces, in the nanometer size range, will "degas" under these conditions. If they do not, this space will certainly not be available for N_2 penetration and measurement.

By using two surface area measurement techniques, it is possible to obtain surface areas exclusively from features of a certain size. For example, by subtracting the BET-N_2 surface area from that obtained by the mercury porosimetry method (the latter cannot access spaces smaller than about 18 Å in radius), one obtains the surface area of micropores with radius between 3.5 and 18 Å only (Shields and Lowell, 1983).

One further note on measuring surface area is worth mentioning here. Although rarely studied, time is a factor in area analysis. For example, Eggleston et al. (1989) showed that the surface area of a ground diopside powder systematically decreased by approximately 20% over the 250 day experiment as measured by Kr-BET (Fig. 1). Because the same sample was repeatedly used over the duration of the experiment, it was considered a possibility that the surface area pretreatment of outgassing at 100°C for 10 hours may be responsible for the reduction in surface area. Therefore, the original ground material was divided into thirds, and two portions of this were stored in air and measured for the first time at 54 and 243 days, respectively. The results were the same, within error, of the experiment of repeated measurement on the same sample. Why does aging have an

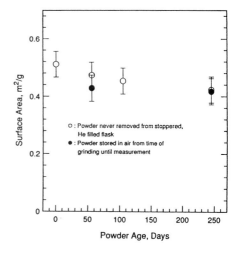

Figure 1. BET-measured surface area versus powder age for ground diopside. The open symbols represent remeasurement of the same sample over time. The filled symbols represent measurement of different previously unmeasured portions of the same powder after 53 and 243 days of aging in air. See text for further details. From Eggleston et al. (1989).

effect on surface area, and why does the surface area decrease? Although far from proven, the answer may depend on the healing of microcracks caused by grinding. When a crack forms, for example in a framework silicate, dangling tetrahedral bonds (designated T-O) are quickly hydroxylated in the presence of water via the reaction (e.g. Parks, 1984, 1990):

$$\equiv Si - O - Si \equiv \; + \; H_2O \; = \; 2 \equiv SiOH \tag{2}$$

If this reaction goes from left to right upon creating virgin surfaces within microcracks by grinding, the reverse reaction may occur with aging. This mechanism was neither proven nor disproven in the work of Michalske and Fuller (1985) who studied the energy required to re-open cracks in silicate glasses. These authors also suggested that, starting with the right side of Equation (2), cracks could reheal with hydrogen bonding across the former crack, or in fact by other cation linkages that may be possible. Stavrinidis and Holloway (1983) determined that crack repropagation energies in silicate glasses were consistent with having to rebreak T-O-T linkages (siloxane linkages if T = Si).

Taken together, the studies referenced in the preceding paragraph strongly imply that grinding of silicates disturbs surfaces. Whatever the true nature of this disturbance, the effects are at least partially reversible. The extent to which this affects laboratory dissolution studies of silicates is not well-known. At least in the case of diopside, it has been shown by Eggleston et al. (1989) that the rates for the initial stages of dissolution my vary by a factor of approximately one order of magnitude depending on how long the sample has been allowed to age from the time of preparation to the beginning of a dissolution experiment (Fig. 2).

Figure 2. Reduction of initial dissolution rates of ground diopside with increasing age of the powder. The grain size for these experiments was <75μm. Dissolution experiments were run at 20°C in stirred polyethylene bottles in DI water adjusted to pH 1.0 with HCl. Silica release data in the first 25 h of the experiments could be divided into 3 stages. Stage I, from 0 to 3.5 hours, was nonlinear. The remainder of the Si-release data, to the end of these experiments (25 h), was divided into two linear regions, designated stage II (from ~4 to 10-13 h), and stage III (from 10-13 h to 25 h). Open symbols represent stage II, closed symbols stage III data. Stage I data are not shown. Data include results from various powder pre-treatments, as shown. The acid-etched powder was exposed to 5% HF, 0.1N H_2SO_4 for 15 min. Error bars are 95% confidence intervals of the slope (rate) estimated from least-squares regression. From Eggleston et al. (1989).

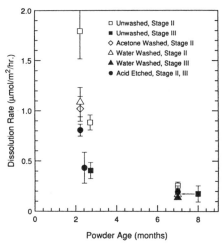

Working definitions of external versus internal surfaces. In our discussion below, it is important to clearly distinguish between external and internal surface areas. First, one might attempt to define these terms conceptually. External surface area could be thought of as the portion of the surface of a mineral grain that a microscope can 'see'. This obviously includes the convex portions of the outer surface, as well as the concave portions such as pits. The internal surface would consist of all other surfaces interior to the grain that are 'out of sight'. However, this definition breaks down when a grain becomes weathered in a skeletal fashion with large internal voids.

For the purposes of modeling and out of necessity, definitions of external and internal surfaces remain mathematically precise, but physically approximate (e.g. Helgeson et al., 1984). For example, surface roughness factors (defined above) can be approximated by the ratio of surface area obtained by the BET technique over the surface area approximated by a macroscopic technique, such as light scattering or sieving. When multiplied by the geometric surface area, an approximation of the external surface area is obtained (e.g. White et al., 1995). This, subtracted from the total physical surface area as measured by the BET technique, gives an approximation of the internal surface area.

A more precise working definition of external versus internal surface area, at least for the purposes of this chapter, utilizes fluid interaction with various parts of the grain. A grain surface over which a free-flowing fluid may pass is defined as an *external surface area*. The flow is the result of a normal hydraulic head. In this case, the movement at any one time is unidirectional, and the rate of transport is a function of the head and the permeability of the inter- and intragrain system in the flow path. Note that the flow over such an external surface does not have to be sustained, as in the vadose zone where flow may be intermittent. Also, 'external' in this sense does not necessarily mean the exterior surface of a grain with the exclusion of intergrain surfaces. Interior surfaces of grains are part of the external surface area if fluid flow can occur over them as described above. On the other hand, a grain surface over which only stagnant fluid resides is defined as *internal surface area*. The channels over these surfaces are generally branches off the flow path.

The main characteristic of this space is that the transport mechanism switches from fluid flow to diffusion. The transport rate is generally slow, controlled not by the hydraulic head and permeability, but by diffusion coefficients, chemical gradients, and the geometry of the fluid space. Note that in this case the transport mode switches from unidirectional to bi-directional, with the transport direction and rate dependent on concentration gradients. Internal surfaces generally define narrow (often microns to nanometers in width), contorted, dead-end passages in grain interiors. Note that dead-end passages do not necessarily qualify as internal surface space. For example, wide, relatively shallow dead-end pores may support turbulent flow sustained by linear flow in a neighboring flow path. Further, portions of grain exteriors may qualify as internal surfaces if, due to clogging of intragranular regions by clays, etc., fluids remain stagnant over them. Finally, there must be transitional zones between moving and truly stagnant fluid where fluid flow and diffusion may be important. These transitional zones will not be specifically addressed in this chapter.

Figure 3 shows examples of external and internal surface area from naturally weathered augite crystals (from Berner and Schott, 1982). These grains, collected from soil developed on a diabase in the north-central Appalachians, were sonically cleaned to remove adhering clay particles. It is interesting to attempt to qualitatively separate the external from internal surfaces area in these photomicrographs. In Figure 3a, numerous elongated etch pits can be seen with submicron openings, with some going down into the tenth micron (100 nm) range or below. If it is assumed that the pit 'bottoms out' within the grain, it is likely that the fluid within the pit was stagnant even if it was not obstructed by clay mineral grains. In Figure 3b, similar pits are seen. However where they have coalesced, a continuous groove several microns wide and running horizontally across the center of the photomicrograph may contribute to the external surface area, even though it is internal to the grain. Whether the interior surface of this groove did contribute to the external surface area depends on the direction of fluid flow over the groove, the extent and depth of the groove, and how flow within it was obstructed by clay minerals.

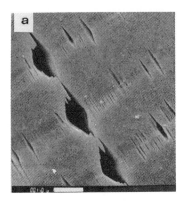 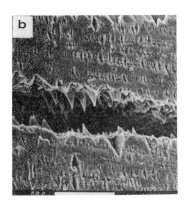

Figure 3. Naturally weathered augite grains from soil associated with the Coffman Hill diabase, Bucks County, Pennsylvania, USA. These grains have been ultrasonically cleaned to remove clay grains and other loosely adhering particles. (a) Crystallographically controlled etch pits. The c-axis is in the direction of the etch pit elongation. Some of the pit openings approach the nanometer range. In the soil, especially with clay grains partially filling these pits, fluid flow within most or all of these pits was probably negligible. The scale bar is 1 μm. (b) Augite face with a very high density of elongated etch pits, many of which have coalesced to form a deep eroded channel several microns from top to bottom. If fluids were moving through this channel while this grain was in place in the soil, the channel walls would be considered external surfaces (according to our definition) even though they are internal to the grain. The scale bar is 10 μm. [Used by permission of the American Association for the Advancement of Science, from Berner and Schott (1982), Plates 1D and 3C.]

Based on the definitions just given which we will use often in the remainder of this chapter, it is traditional in our understanding and study of mineral weathering to consider primarily, and perhaps only, external surface area. This is to say that most (not all) studies, whether laboratory or field based, assume (usually implicitly) that only external surfaces of the grain are reacting with neighboring fluids, and that all weathering processes are associated with these surfaces only. Much of this chapter attempts to demonstrate that it is vital to also consider internal surface area in the weathering process.

Evidence for the importance of internal surface area. Although the existence of internal surface area within mineral grains (see definition above) seems intuitively obvious, defining exactly how much there is in a grain, and exactly where it is located, is a considerable and extensive undertaking. Some of the first indirect evidence of the importance of interior, if not internal, surface area of minerals comes from Folk (1955). He suggested that the turbidity often observed in alkali feldspars in igneous rocks is due to minute (often not resolvable by optical microscopes) vesicles filled with fluid. Since then, a number of studies, among them Montgomery and Brace (1975), Parsons (1978, 1980), Ferry (1985), Worden et al. (1990), Guthrie and Veblen (1991), Waldron and Parons (1992), Fitz Gerald and Harrison (1993), Lee et al. (1995), and Walker et al. (1995) have all supported these findings and studied this phenomenon in various ways. It has been shown that both alkali and plagioclase feldspars (but especially perthitic alkali feldspars) develop this turbidity via interaction with magmatic and/or hydrothermal fluids (e.g. Parsons, 1980), and that dramatic internal microtextural changes are possible, if not likely, in this sort of interaction. Before going further, it is important to realize that these microtextural components (described more fully below) exist before the feldspar enters the weathering regime.

Eggleton and Buseck (1980) were the first to begin to study the details of the microtextures of turbid feldspar with TEM. Wordon et al. (1990) were the first to study the microtextures of turbid feldspars with TEM and SEM in detail. They clearly show that the microporosity of *unweathered* feldspars can be extensive but patchy, resulting in an estimated porosity of up to 4 to 5% in the most affected areas of what presumably was originally a pristine crystal. The porosity development seems to be associated with coarsening of microperthite and development of sub-grain boundaries via interaction with an aqueous fluid. The dimensional coarsening of the perthite in these disturbed areas has been observed to be up to three orders of magnitude. The final results are feldspar crystals which are highly modified consisting of complex mosaic structures. The microporosity most often is in the form of minute tubules which can go down to the nanometer size range.

Worden et al.(1990) suggested that these altered crystals just described should be more reactive in a weathering environment than the original pristine (as-crystallized) rocks. That this must be the case was shown by Walker (1990). In this study, the same feldspars studied in the Worden et al. work were subjected to $H_2^{18}O$-enriched fluids at 700°C and low confining pressure for 75 hours. The point of the study was to determine if the microporosity observed and described by the Worden et al. study was interconnected, providing pathways throughout the interior portions of the grains from the exterior. If so, the ^{18}O should penetrate throughout the grains and exchange with ^{16}O on the internal surfaces. Volume diffusion to the center of these grains would require several thousand years at the temperature of these experiments. After the experiments were complete, an ion microprobe was used to determine the distribution of the ^{18}O within several grains. It was found in this way that ^{18}O had penetrated entire grains, proving that the microporosity observed in the Worden et al. (1990) study did in fact represent a connected network. Importantly, it was also shown that the experiment itself did not change the micro-texture of the interior of the grains.

Although the microporosity studies discussed above have only been performed on a limited number of feldspar samples from a few locations, the connection between turbid feldspars and these interior micro-networks is now clearly established. Given the fact that feldspar in plutonic rocks are nearly always turbid to some extent, the importance of these micro-networks is presumably widespread.

The Walker (1990) study strongly suggests that such microporosity networks in feldspars are a factor in water-rock interactions, yet they do not directly address whether the microporosity is actually micropermiability. So the question remains, do these minute features in feldspars (as well as other minerals; see, e.g. Fig. 11, below) participate in fluid flow through the grain and rock, in which case they would be classified as external surfaces? Or do the surfaces associated with this microporosity generally host stagnant fluids in dead-end passages? As we discuss next, we may not be able to answer these questions directly at this time, but there is indirect evidence that suggests that at least portions of these microporosity networks do not support flow. This indirect evidence comes from the fact that rock aggregates, including soils, have a considerable amount of 'dead' space with respect to fluid flow. This may be intuitively obvious when considering the vadose zone where the rock or soil is not fully saturated. Indeed, much experimental, theoretical, and field evidence exists to show this (see Velbel, 1993b, and the latter sections of this chapter). However, there is also evidence for stagnant fluid under saturated flow conditions (e.g. Nkedi-Kizza et al., 1983, and references therein) and in this case, internal surface areas may form some or most of the boundaries of this dead space. The presence of stagnant fluid under saturated flow conditions is shown by experiments where a marker element (a tracer) is put into the flowstream of a fluid before passing through a packed column (in the lab) or an aquifer (in the field). The input of the tracer is in the form of a pulse, or spike, and the tracer component is measured (concentration versus time) as it appears at some defined point 'downstream' in the column or aquifer. The concentration versus time curve that one measures is called a *break-through curve* (BTC). Assuming that the tracer is not sorbed by the media through which it passes, the BTC might be expected to follow a function such as (e.g. Lapidus and Amundson, 1952; van Genuchten and Wierenga, 1976):

$$\frac{\partial C}{\partial t} = D\frac{\partial^2 C}{\partial z^2} - v_o\frac{\partial C}{\partial z} \tag{3}$$

where C is the concentration, t is the time, D is the dispersion coefficient, z is the distance, and v_o is the fluid velocity. Except under extreme conditions, Equation (3) predicts that BTC's should be approximately sigmoidal or symmetrical. However, BTC's are very commonly asymmetrical with a long tail on the high time side of the curve even when sorption is not a factor. Several models have been proposed to explain this phenomenon. Nearly all of them have a very important characteristic relevant to the discussion above. They divide the water in the system into two regions, one mobile (flowing), the other immobile (stagnant). Differences among these models come from the way the model handles the stagnant water in the system, in particular its location and its exact form within the porous media being tested. This may be quite variable depending on whether air-space is present in the model (saturated versus unsaturated zones). Other differences come from the way the tracer exchange between the flowing and stagnant part of the system is handled. Nevertheless, these models have been very successful at fitting asymmetric BTC data (e.g. Gaudet et al., 1977; DeSmedt and Wierenga, 1979a,b; Nkedi-Kizza et al., 1983). This work, collectively, provides a great deal of evidence in support of important amounts of internal surface area in soils and rocks.

Implications for laboratory-based dissolution studies. One can easily surmise from the previous section that feldspars, seemingly unweathered, which are mechanically ground for laboratory-based dissolution studies will have a much larger proportion of previously exposed and reacted surfaces than expected. This is precisely what was predicted and utilized in the work by Anbeek (1992, 1993). Specifically, Anbeek (1992) re-interpreted the feldspar dissolution experiments of Holdren and Speyer (1985, 1987) based on this assumption. To explain this work, we define new feldspar surfaces exposed by grinding as 'virgin', and internal feldspar surfaces (discussed in connection with Worden et al., 1990, etc. in the previous section) exposed by grinding as 'non-virgin'. [In his papers, Anbeek refers to non-virgin surfaces as 'weathered' which is technically not correct; as explain above, these surfaces are generated under magmatic and/or hydrothermal conditions and may not have seen a weathering environment before grinding for lab work. Nevertheless, the points of his arguments still stand.] According to Anbeek's model, dissolution rates of the virgin surfaces created by grinding decreased with decreasing grain size. Also the ratio of dissolution rates of virgin surfaces to non-virgin surfaces decreased with decreasing grain diameter. He also predicted that at small grain sizes (on the order of microns), non-virgin surfaces would dissolve faster than fresh surfaces, while for larger grain sizes (several millimeters), virgin surfaces would dissolve up to two orders of magnitude faster than non-virgin surfaces. These observations can be explained if the density of surface sites at which dissolution occurs are concentrated on surfaces which are preferentially exposed early in the grinding process. Therefore, as grinding proceeds, these most reactive dissolution sites become 'diluted', resulting in the observations noted above. Whether Anbeek's (1992) modeling accurately reflects reality is not as important as his realization that significant amounts of fresh and previously exposed surfaces are present in most, if not all, feldspar dissolution experiments. Because these surfaces can have very different dissolution rates, comparing measured dissolution rates among laboratories, especially when different starting grain sizes are used, becomes tenuous. Also, the fact that samples used in laboratory dissolution experiments are ground at all will introduce a certain (perhaps large) discrepancy when the results are compared to field results.

In Anbeek (1993), observations made during dissolution experiments of alkali feldspars recovered from a coastal sand supported the findings and suggestions described above from Anbeek (1992). In the 1993 study, dissolution rates were measured at pH 3 and 5 for the unground, naturally weathered grains, as well as from four samples of smaller and smaller grain diameters. The results in these particular experiments indicated that the dissolution rates of non-virgin surfaces were about one order of magnitude slower than fresh surfaces exposed by grinding. If this modeling of dissolution rates of feldspars is correct, it at least explains a portion of the difference between rates determined in the field versus those found in the lab. Additional factors, discussed below and elsewhere in this book, must also be considered.

Reactive surface areas. Before leaving the subject of surface area, the meaning and use of the term *reactive surface area* is briefly explored here. Quite simply, portions of the surface which are 'reactive' will participate in the dissolution process, while the rest of the surface will not participate. In many mineral dissolution rate studies, reactive surface area is assumed to be the same as, or at least proportional to, the analytically measured surface area, for example with the BET technique. This is certainly a simple, commonly used solution to a complex problem. In other studies, some portion of the measured surface area will be considered reactive, while the remaining fraction will be assumed to be passive in the dissolution reaction being studied. Partitioning between reactive and unreactive surface areas or sites may even be arbitrarily assigned, or used as a fitting parameter in a dissolution model. Certainly, it is well known that models which are successful in fitting experimental data cannot be reliably used by themselves to determine reaction mechanisms (e.g. Westall and Hohl, 1980; Sposito, 1986). Further, what sites will or will not participate in a dissolution reaction or any other surface reaction is a subject of intense research, and one of considerable controversy (see, e.g. Blum and Lasaga, 1988, Blum et al., 1990, Hochella, 1990, and Blum and Stillings, this volume).

Some of the more recent studies which closely address that of reactive surface area are those of Junta and Hochella (1994) and Junta-Rosso (1995). These studies deal with mineral surface sorption (in this case precipitation) rather than dissolution, but it is the direct identification of reactive surface area that is important here. As mentioned briefly above, in these studies, hematite and albite substrates were reacted with air-equilibrated aqueous solutions containing $Mn(II)_{aq}$ in the low ppm range at pH around 8. Mn sorbs to these substrates under these conditions, oxidizing to Mn(III), and β-MnOOH precipitates. The various sites at which this reaction occurs on these substrates were followed directly at the mineral-water interface, in real-time and down to the nanometer level, with scanning force microscopy (SFM). For hematite, the precipitation reaction always starts at the base of a step, proceeding away from the step across the surface. However, at the same time, β-MnOOH protocrystallites became thicker as they spread across the surface, meaning that Mn_{aq} also precipitates on surface-bound Mn. In fact, what the SFM directly shows is that both the β-MnOOH/solution interface as well as the β-MnOOH/hematite/solution interface are reactive surface sites for this precipitation reaction. However, based on the growth rate at both interfaces as observed with SFM, the β-MnOOH/solution interface is 70 times less reactive for this sorption reaction than the β-MnOOH/hematite/solution interface. This seems to contradict interpretations drawn from solution experiments under similar conditions by Matsui (1973). These indicate that the rate of Mn(II) oxidation on Mn-oxides is up to 25 times higher than on hematite. Although the surfaces used by Matsui were not (could not) be characterized as those in the Junta and Hochella (1994) study, his results indicate that the Mn-oxides in his study had considerably more reactive surface area on them than ours, presumably in the form of steps. Therefore, without knowing the specific microtopography of the surfaces being studied, the relative reactivities between different minerals, or in fact two samples of the same mineral, can be misinterpreted. Certainly, the same could be said for dissolution experiments.

A schematic drawing is shown in Figure 4 to emphasize the problem concerning reactive surface area assessment. In the figure, both blocks have the same total surface area. However, one has twice the number of steps as the other. If reactive surface sites are only associated with these steps (as was shown in the Mn oxyhydroxide precipitation experiments described above), the lower block will appear to promote the reaction at twice the rate of the upper one as measured by a macroscopic, indirect observable such as solution composition. This could be a problem when two samples of the same mineral are prepared in different ways, possibly creating a difference in microtopography between the two. However, when samples of different minerals are compared, what is shown in Figure 4 may always be a problem. To complicate the situation even more, as the reaction proceeds, the reactive surface area is constantly changing as the surface microtopography constantly changes. Depending on the exact nature of the reaction and the original surface, reactive surface area may increase or decrease relative to the total surface area.

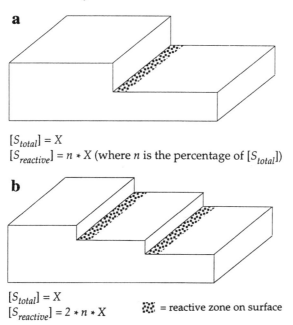

a

$[S_{total}] = X$
$[S_{reactive}] = n * X$ (where n is the percentage of $[S_{total}]$)

b

$[S_{total}] = X$
$[S_{reactive}] = 2 * n * X$ ▓▓ = reactive zone on surface

Figure 4. Schematic of two grains of the *same* mineral, both with the same total surface area (S_T), but one with twice the number of steps. The reactive surface area ($S_{reactive}$) for some hypothetical reaction (actually determined for initial Mn(II)$_{aq}$ attachment on hematite; Junta and Hochella, 1994) is designated by the stippled area. The initial reaction will occur at half the rate on A relative to B. From Junta-Rosso (1995).

The reactive surface area in dissolution/weathering processes is subject to the same uncertainties and complications as in the precipitation reactions just described. We will incorporate these thoughts and findings in our discussion of *internal weathering fronts* in minerals at the end of this chapter. A description of these fronts follows in the next section.

MICROSCOPIC OBSERVATIONS:
RATIONALE, TECHNIQUES, AND RESULTS

In this section, we will scrutinize chemical weathering processes with both SEM and HRTEM. SEM is immensely useful in this type of study, but HRTEM is the unique key that unlocks what we believe to be some of the most fundamental aspects of chemical weathering. In the first two subsections below, we discuss the microscopic approach to weathering studies in general, and then the specific rationale and techniques for HRTEM research in this area. This is followed by results of weathering studies which utilize HRTEM.

The need for microscopic characterization of reacted surfaces

The use of modern electron microscopes (both imaging and compositional analysis) is one of the most important parts of mineral weathering studies. There are many reasons for this. First, weathering rates and mechanisms are often inferred from changes in solution chemistry. But these observations can be misleading. For example, congruent release of constituents could be interpreted as congruent release of ions from mineral surfaces. Alternatively, diffusion modified surface layers created by incongruent dissolution could result in stoichiometric concentrations of dissolved species in solution. These possibilities can be distinguished if the extent of change of surface chemistry and structure can be quantified (Fig. 5; Banfield et al., 1995a).

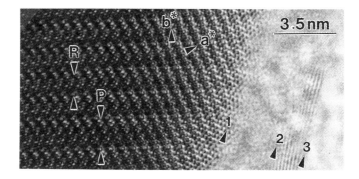

Figure 5. HRTEM image of experimentally weathered intergrown rhodonite (R) - pyroxmangite (P) showing modification of the mineral surface (pH 6.1, room temperature, approximately 2500 h; for other experimental details, see Banfield et al., 1995). The outermost few nanometers have been converted to an amorphous Si rich layer (between '1' and '3'). The outer portion of this layer shows signs of recrystallization (lattice fringes between '2' and '3') after about a year.

Second, mineral dissolution experiments are typically carried out under conditions that differ from those prevailing during natural weathering. Although measured rates can be adjusted to incorporate the effects of variables such as pH and temperature, it is necessary to assume the mechanism for which the rate has been measured is the same in the laboratory and in nature. A second important presumption is that physiochemical conditions at the reaction site are known. Both of these assumptions can only be tested by reference to naturally reacted materials, preferably via microscopic means. Ideally, this would involve direct measurements with atomic-scale resolution. Even in the absence of such measurements, details of reaction mechanisms and conditions at reaction sites can be inferred from physical evidence and chemical characteristics of the solid phases using modern electron microscopes.

Third, in addition to insights of relevance to reaction rate measurements, direct microscopic observation of naturally reacted mineral surfaces can place vital constraints on the process of secondary mineral formation. Solid products of silicate mineral weathering (clay minerals, oxides, oxyhydroxides) constitute essential components of soils and sediments. Controls on the rate of growth, chemistry, and structure of clay minerals formed by mineral weathering are complex, depending on many parameters in addition to the rate of destruction of primary minerals. Microscopic examination of clay mineral-primary mineral interfaces helps clarify the sequence of steps involved in nucleation and growth of secondary minerals after elements are removed from primary minerals.

The HRTEM approach

Characterization of the structure and distribution of reactive surfaces relevant to weathering studies requires methods of investigation capable of providing crystallographic, chemical, and textural data from intact mineral-product and mineral-organic interfaces at the unit cell scale. Further, secondary mineral phases and organic polymers (from biochemical activity) can be in such intimate association with primary mineral surfaces that this can greatly impact the physiochemical nature of mineral-solution interface zones. Therefore, the nature of this intimate association must be characterized. While other techniques can probe various aspects of silicate mineral surfaces with high spatial resolution (e.g. scanning force microscopy and scanning Auger spectroscopy/microscopy), HRTEM has generally proven to be the most versatile and informative in this regard.

The capabilities of HRTEM are adequately described elsewhere (e.g. Buseck, 1993), and will not be repeated here. We will, however, critique the HRTEM method in its role in weathering studies. This approach makes some fundamental assumptions and has some limitations, and these should be stated explicitly. Firstly, it is assumed that interfaces between primary and secondary minerals represent snapshots of a reaction in progress. It is not necessary to assume that the reaction involved growth of a product at the expense of a retreating surface, as chemical, structural, or textural evidence may indicate otherwise. Secondly, it is assumed that less than one millimeter-sized regions selected for examination are relatively representative of the bulk. Sample preparation always begins in the field at the outcrop scale, with selection of fresh and partly reacted samples. These samples are thin sectioned and examined using conventional optical microscopy, X-ray diffraction, and other techniques (e.g. scanning electron microscopy, electron microprobe, etc.). Three millimeter-diameter regions are identified and prepared for electron imaging by polishing and ion milling or microtoming (e.g. Buseck, 1993, and references therein). Only areas around holes created by ion milling are electron transparent and suitable for TEM characterization. Preliminary imaging is done at magnifications of about 30,000 to 60,000× to select regions of interest. Although an inevitable consequence of this approach is that relatively small volumes of material are examined with relatively high resolution, perspective is provided through linking these observations to the field via other lower resolution microscopies.

Artifacts are unavoidable in sample preparation and as a consequence of the microscope vacuum. However, these are rather well understood. Firstly, ion milling creates a thin, readily identifiable amorphous layer (a few nanometers thick) at the specimen surface. The volume of this layer is small compared to the volume of electron transparent material. Secondly, hydrous minerals and organic materials are normally dehydrated in the high vacuum. This can be minimized through various impregnation and fixation protocols, which result in substitution of organic molecules in clay interlayers and cell structures (biological embedment procedures may themselves induce artifacts). Because TEM is a high vacuum technique, experiments cannot be conducted *in situ*. Isotopic data and trace element chemical information are not available. These limitations indicate the need to combine HRTEM data with those obtained from other sources.

Although considerable care is taken to record atomic and unit-cell scale phenomena that represent the overall process of mineral weathering, the question as to how broadly applicable the results may be remains. The overall weathering process depends on local factors (e.g. climate, ground water chemistry, involvement of organic compounds, etc.). However, incipient stages of alteration should be less affected by these variables. The ideal way to test the general relevance of observed phenomena is to compare results of different studies. However, it is rare to find more than one or two microscopic studies describing weathering of a single mineral.

One example of a mineral that has been studied several times is olivine. Although the three studies (Eggleton, 1984; Smith et al., 1987; Banfield et al., 1990) involved samples from very different climatic environments, the basic phenomena observed are remarkably consistent. The only observed variation is the hydration state of secondary Fe-oxides that form. Nanometer-wide goethite crystals are associated with smectite from wetter environments (Eggleton, 1984; Smith et al., 1987), while hematite develops with smectite in olivine weathered under water-poor (desert) conditions (Banfield et al., 1990).

Evidence of microtextural and defect controls of silicate dissolution and weathering

Specific reactions that occur at the mineral-water interface must ultimately dictate overall dissolution kinetics. However, there is a critical caveat here. In many cases, it is the bulk microtextural and/or defect characteristics of the dissolving mineral that are the most important underlying factors in determining what surface reactions will come into play. For example, Wilson and McHardy (1980), in their investigation of a perthitic alkali feldspar, were one of the first to show microscopic evidence that exsolution lamellae must have a significant influence on mineral dissolution rate. Holdren and Speyer (1985, 1987) followed with more specific ideas about this relationship. In their work, which involved the measurement of dissolution rates of alkali feldspars as a function of grain size, they speculated that defects associated with exsolution lamellae were responsible for the lack of complete correlation between dissolution rate and BET-measured surface areas. They observed that dissolution rates correlated well with the measured surface areas for larger grain sizes of their starting material, but that this correlation broke down for the smallest grain sizes. They speculated that, for larger grains, the spacing between bulk defects intersecting the surface is much smaller than grain dimensions, but that for smaller grains, their size begins to approach surface defect spacing on the surface resulting in a loss of correlation between surface area and dissolution rate.

Since these studies, a number of workers have looked for correlation between bulk dislocation densities and dissolution rate. The studies include those of rutile by Casey et al. (1988a), calcite by Schott et al. (1989), sanidine by Murphy (1989), and quartz by

Blum et al. (1990). Although dislocation densities varied in these studies by up to 5 orders of magnitude, dissolution rates varied by a factor of two or less. This apparent paradox may possibly be explained by the fact that all of the studies just mentioned were performed far from equilibrium. Blum et al. (1990), Nagy and Lasaga (1992), and Burch et al. (1993) have all suggested, based on both theory and experiment, that dissolution at dislocations may be much more important to overall rates near equilibrium.

The apparent confusion surrounding the role of bulk defects on the dissolution rates of minerals may primarily reflect our tendency to oversimplify very complex processes. This point can be amplified in part as follows. Because dislocation cores are sites of excess energy, the mineral dissolves much more readily immediately adjacent to these defects that intersect the surface than from defect-free terraces. A small difference in a measured dissolution rate (e.g. a factor of 2) can create very obvious surface topography for this reason. When the driving force for dissolution is small—close to equilibrium in the case of calcite, for example, or close to solution saturation in the case of minerals that cannot achieve equilibrium with solution at room temperature—the only sites that can contribute ions to solution are those with excess energy. These sites may be steps, or bulk defects intersecting the surface, or some other reactive site. As the driving force for dissolution is increased, dissolving flux is generated from more and more kinds of surface sites. This does not mean that defects are not preferentially etched but that this effect may be swamped out by the large flux from other sources, perhaps less reactive but much more abundant than a single defect type intersecting the surface. A final point that should be made is that it is important to understand the nature of defects in a sample because all defects do not affect mineral weathering rates in the same way. For example, specific planar defects in olivine (Banfield et al., 1990) and amphibole (Banfield and Barker, 1994) clearly significantly *lower* the reactivity of the mineral. This effect must be attributed to the low coherencey strain associated with some planar defects and to other factors, such as modification of the oxidation state of Fe in defective material.

Certain recent studies have made it abundantly clear that the microtexture of minerals (including the number and distribution of dislocations) are undoubtedly an important influence on mineral weathering in nature. Confidence in this conclusion comes from the assessment, using TEM and SEM, of the interrelationship between microtexture and defects in minerals and their observed natural dissolution patterns. The beauty of this approach is that it is direct. Experimental work and theoretical assessment, which may have hidden flaws, are not needed. Particularly illustrative examples of this approach concern plagioclase feldspars (Inskeep et al., 1991) and alkali feldspars (Lee and Parsons, 1995). These studies are highlighted below.

Perthitic alkali feldspars. Lee and Parsons (1995) studied perthetic alkali feldspars from the Devonian age Shap granite of north-west England. They describe how the microtexture and defect distribution of this feldspar influence its early stages of dissolution in the natural weathering environment. Their main thesis is that many aspects of these exceedingly complex minerals, including details of their microtexture, defects, size, shape, and bulk and local chemical composition must be understood in order to fully appreciate (understand) their dissolution characteristics both in terms of mechanism and rate. This seems particularly pertinent when one considers that different laboratories have obtained dissolution rates for K-feldspar that are different by well over an order of magnitude (e.g. see data compiled by Blum, 1994, and Blum and Stillings, this volume). The same is true for albite. Although portions of these discrepancies are undoubtedly from different lab protocols and experimental error, the most important factors are most likely what studies like Lee and Parsons (1995) have explored.

Perthitic alkali feldspar from the Shap granite is similar to the turbid perthites described earlier in this chapter studied by Worden et al. (1990) and Walker (1990) (see section entitled 'Evidence of the importance of internal surface area'). Within single phenocrysts of the Shap perthite, there are small amounts of pristine cryptoperthite, 50 to 60% (by volume) pristine lamellar microperthite, and approximately 40% patch perthite. The cryptoperthite consists of less than 75 nm wide albite platelets that are fully coherent with the neighboring K-feldspar (orthoclase). The microperthite consists of 75 to 700 nm albite platelets that are semi-coherent, resulting in relatively abundant edge dislocations along their interface with the neighboring orthoclase (Fig. 6a,b). The overall dislocation density in the microperthite is 2 to $3/\mu m^2$. As shown in Figure 6, these dislocations occur in pairs directly across an albite platelet from each other. The patch perthites are the commonly observed turbid area of feldspars, and consist of microporous, irregular and relatively course incoherent to semi-coherent intergrowths of albite and microcline (Fig. 7a,b). The dislocation density is approximately half of that found in the microperthite, and there are numerous micron to submicron pores within the microcline that occupy 1 to 2 vol % of these regions. As already mentioned above, this intricate microtexture is the result of interaction with magmatic and/or hydrothermal solutions and is fully developed before these feldspars reach the weathering regime.

When these grains are exposed to the weathering environment, the dislocations described above act as nucleation points for etch pits. This is seen best in the SEM photomicrographs of (001) cleavage surfaces (Fig. 8a) which show etch pit arrays that match the distribution of edge dislocations seen by TEM in Figure 6. The pits themselves are almost entirely contained within the albite fraction. These pits, starting on opposite sides of an albite lamellae, eventually coalesce across the lamellae resulting in etch pits with rectangular cross-sections (Fig 8b). In patch perthite regions, etch pits like those shown in the microperthite portions are more irregularly distributed as expected (Fig. 8d). These pits have in places combined with pre-existing micropores to form a skeletal texture (just right of center in the upper half of Fig. 8d). Resin casts of these etch pits (Fig. 9) show that these pits can penetrate as much as 10 to 20 μm from the cleavage surface and can form a complex interlocking tunneling network within the grain. Finally, close scrutiny of the relief of albite to K-feldspar lamellae show that even between etch pits, albite dissolves uniformly but faster than K-feldspar.

– – – – – – – – – – – – –

Figure 6 (top, next page). Bright-field TEM images of a microperthitic region of an unweathered feldspar phenocryst from the Shap granite. Lighter regions are orthoclase, darker regions are albite. (a) The pairs of discontinuities along the albite are edge dislocations (one pair is labeled). In the weathering environment, etch pits nucleate where these dislocations intersect the surface. (b) The albite shown in this image is twinned in the upper right and lower left. A pair of edge dislocations (labeled D) are also seen along the untwinned portion of the albite, with a former micropore associated with one of the dislocations filled with a secondary feldspar (S) during magmatic/hydrothermal alteration. [Used by permission of the editor of *Geochimica et Cosmochimica Acta*, from Lee and Parons (1995), Fig. 2.]

– – – – – – – – – – – – –

Figure 7 (bottom, next page). (a) Back-scattered electron (BSE) image of an unweathered alkali feldspar phenocryst from the Shap granite. The left portion of the image is dominated by microperthite; the albite is darker grey, the orthoclase is lighter grey. The center and right portions of the image are dominated by patch perthite. The course, irregular black veins are albite. The grey regions in between are microcline. Within the microcline are numerous micropores (irregular black dots) resulting in up to a few percent microporosity by volume. (b) SEM image of a (001) surface of the same unweathered alkali feldspar showing the character of the micropore openings. Several of the pores contain secondary feldspars, characterised by euhedral faces pointing into the pore (e.g. at S). [Used by permission of the editor of *Geochimica et Cosmochimica Acta*, from Lee and Parons (1995), Fig. 1.]

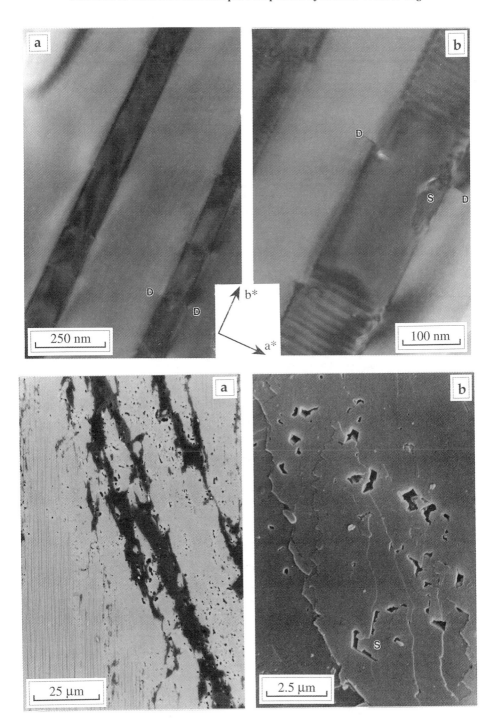

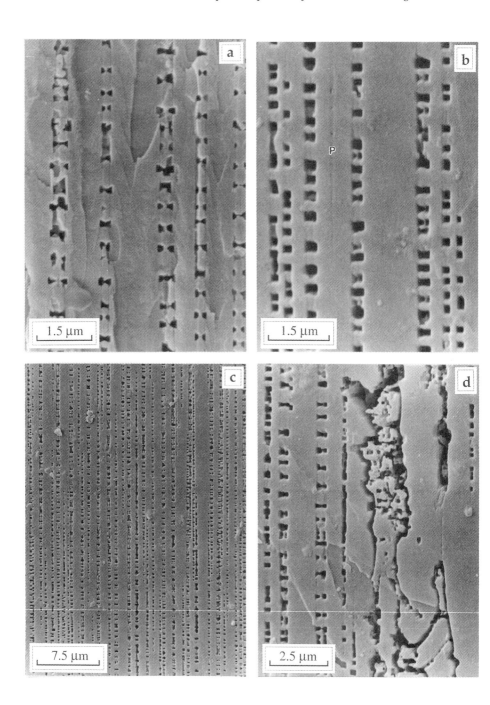

Figure 8 (previous page). SEM images of (001) cleavage surfaces of alkali feldspars from gravels weathered from the Shap granite. The grains have been untreated, except for ultrasonic cleaning to remove adhering clays and organic material. In all images (except the right half of d), albite/orthoclase microperthitic lamellae are vertical, with the etch pits originating at the albite/orthoclase interface at the point of an edge dislocation. Albite is preferentially removed as the pit deepens. (a) At relatively early stages of weathering, pairs of pits across an albite lamellae have not coalesced. (b) At a slightly more advanced stage of weathering, pits have coalesced across albite lamellae, resulting in rectangular shapes. Only a few pits have formed across a particularly fine lamellae, labeled P, but it is slightly recessed and has clearly weathered faster than the surrounding orthoclase. (c) Lower magnification of b) showing the uniformity of pitting. (d) Pitted microperthite on the left, pitted patch perthite on the right. The latter is characterized by coalesced pits which now form channels. Numerous micropores, perhaps partially modified by weathering, can be seen in the upper-right-center of the photomicrograph. Micropores before weathering can be seen in Figure 7. [Used by permission of the editor of *Geochimica et Cosmochimica Acta*, from Lee and Parons (1995), Fig. 8.]

Studies like that of Lee and Parsons (1995), and Inskeep et al. (1991) presented below, should be critically considered by everyone involved in both laboratory and field-based mineral dissolution work. Using the perthitic feldspars as an example, unweathered samples that are ground for laboratory dissolution work will have exposed, highly reactive edge dislocation cores which will have no surface expression as viewed even by high resolution SEM (one should also consider the potentially significant increase in defect density due strictly to the sample grinding). These exposed defects will dissolve quickly at the beginning of laboratory dissolution runs. In the worst case, if no microtexture is considered at all, solution analyses might suggest that sodium is being leached to considerable depths relative to potassium when in fact both the albite and K-feldspar may be dissolving essentially congruently (that is any leached layer would be very shallow). These dislocation dissolution sites are so reactive that by the time these grains get into overlying soil profiles, most of this material has already been removed within several microns of the mineral surface. As these micro- to macro- pores and pits are filled with secondary minerals (mostly clays), and as diffusion rather than fluid flow becomes the dominant transport mechanism, reaction rates will slow down even more. When dissolution rates are measured in the field, this may be precisely the kind of mineral that is being measured, a dramatic contrast to the starting material in many laboratory experiments. [Note that this does not take into account experiments that show that grains collected from soils and dissolved in the lab still show faster dissolution rates then when left in the field and measured in situ. This will be discussed below.]

Plagioclase feldspars. There have been a large number of laboratory-based dissolution studies of plagioclase feldspars all the way along the albite to anorthite join (see Blum, 1994; Blum and Stillings, this volume, and references therein). Most of these studies have not taken into account the microtexture of plagioclases and the affect this may have on dissolution mechanism and kinetics. Labradorite plagioclases are particularly interesting in that they, like a few other compositional ranges within the plagioclase series, are subject to the formation of exsolution lamellae (e.g. Smith, 1983; Ribbe, 1983). These are referred to as Bøggild intergrowths which show exsolved lamellae with compositions ranging from An_{40} to An_{68}. Like so many complex feldspar microtextures, they are best studied with HRTEM. Some of the first hints that plagioclase microtexture was coming into play in laboratory-based dissolution experiments came from Hochella et al. (1988a) and Hochella (1990) who used high resolution scanning Auger spectroscopy and microscopy to show that dissolving labradorite surfaces had highly irregular surface compositions on a micron to submicron scale. However, the specific microtexture of the labradorites used in these studies was not determined. Subsequent to this, Inskeep et al. (1991), using TEM and XPS, clearly showed the microtextural controls of plagioclase (labradorite) dissolution. In this study, mechanically

Figure 9 (opposite page). SEM images of resin-casts derived from weathered alkali feldspar (001) cleavage surfaces equivalent to those in Figure 8. These samples were prepared by impregnating them with epoxy resin, and then etching the feldspar away with HF. The casts thus reveal the internal etch pit form. (a) These etch pit casts show typical complexity of pit structure within the grain. (b) Casts that extended up to the original surface have fallen over, but this image shows the connection between pits in adjacent rows within the grain. (c) Casts of horizontal channels connecting pits intersecting the original surface. Note that these horizontal pits all originate (or terminate) along a single albite lamellae running along the right of the image. (d) Complex sub-surface network of etch pits which penetrate as deep as 15μm. [Used by permission of the editor of *Geochimica et Cosmochimica Acta*, from Lee and Parons (1995), Fig. 9.]

ground An$_{54}$ was dissolved in batch experiments at pH ~ 4 for up to several hundred days. TEM photomicrographs of fine grain edges are striking (Fig. 10). They show exsolution lamellae ranging from 50 to 90 nm in thickness which have been differentially etched. EDX analysis confirmed that the more calcic phases was preferentially etched relative to the sodic phase. This results in a distinctly corrugated surface, with the calcic phase recessed between 50 and 200 nm.

The messages from the Inskeep et al. (1991) work on labradorite are similar to those listed above in connection with the Lee and Parsons (1995) study. Let it just be emphasized here that optical and SEM examination of minerals will frequently not reveal the pertinent microtexture which clearly must affect the results of dissolution experiments based on solution analysis. This is certainly the case in the Inskeep et al. (1991) labradorite study. When the dissolution of this mineral is being tracked by solution analysis only, and a homogeneous sample is assumed, misinterpretation can and have been easily made. For example, in this case, preferential dissolution of the calcic phases would be misinterpreted as incongruent dissolution with a deep Ca-leached layer assuming that the mineral was homogeneous. Further, results from surface sensitive spectroscopies such as XPS will be misleading when used to analyze these surfaces. The results, on average, will be biased towards the more sodic (higher relief) phase. Surface

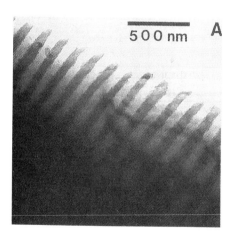 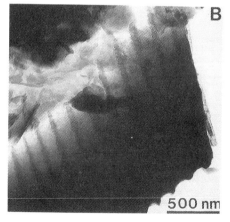

Figure 10. Bright-field TEM micrographs of labradorite (An$_{54}$) grain edges after batch dissolution experiments at a pH of approximately 4 for several hundred days. This plagioclase is exsolved (Bøggild intergrowths), and the more calcic lamellae (as determined by EDX analysis) are clearly dissolving faster than the more sodic lamellae. (A) Calcic lamellae are recessed approximately 140 nm relative to the sodic lamellae. (B) Calcic lamellae are recessed approximately 45 nm relative to the sodic lamellae. [Used by permission of the editor of *Geochimica et Cosmochimica Acta*, from Inskeep et al. (1991), Fig. 3, p. 793.]

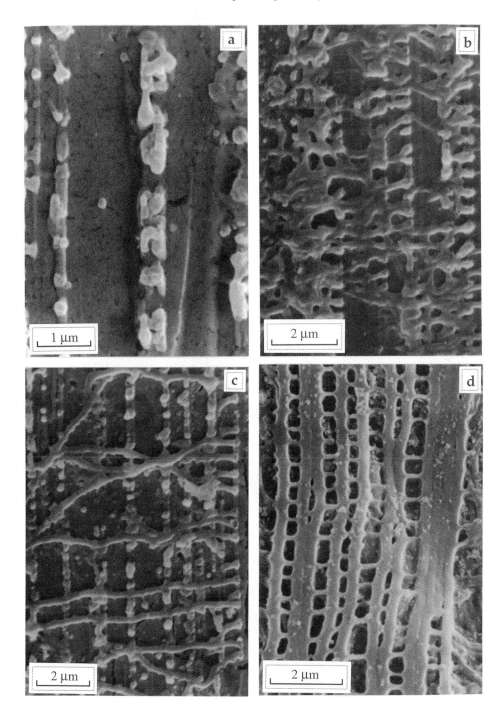

topography must be taken into account to correctly interpret the data. Similar problems will be operative in sputtering depth profiling experiments (see also Hochella et al., 1988b, regarding potential problems with sputtering experiments).

Figure 11. HRTEM image of naturally weathered olivine. Note the pervasive development of channels as narrow as 2 nm. These channels, which parallel (001) olivine, are frequently filled by clay (see Fig. 12).

The incipient stages of silicate mineral alteration

Characterization of silicate minerals showing incipient alteration reveals that dissolution and crystallization of secondary minerals occurs throughout the primary mineral, not just at grain boundaries with other phases. This is especially obvious in olivine, where several studies (Eggleton, 1984; Smith et al., 1987; Banfield et al., 1990) report pervasive development of channels a few to a few tens of nanometer wide (Fig. 11, above). These approximately parallel-sided channels have the potential of increasing the reactive surface area by orders of magnitude. Interfaces between smectite that fills the channels and olivine are semi-coherent, with space between the phases only in the range of a few Angstroms. Consequently, reactive surface areas may greatly exceed areas that would be estimated by conventional measurement procedures (e.g. BET; see section above entitled "Measuring surface areas").

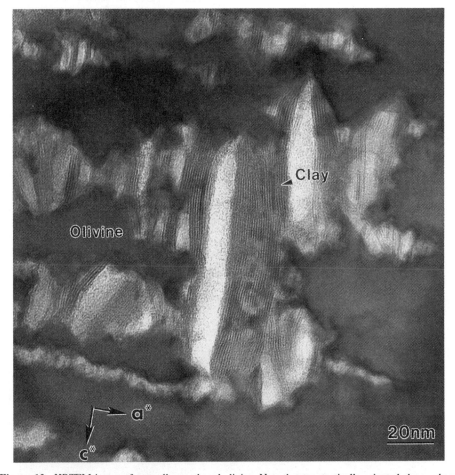

Figure 12. HRTEM image of naturally weathered olivine. Note that topotactically oriented clays, when fully hydrated, essentially fill the volume previously occupied by olivine. [Used by permission of the editor of *Contributions to Mineralogy and Petrology*, from Banfield et al., 1990, Fig. 11, p. 116.]

Smectite formation at reacting surfaces inside silicate minerals is common (but not ubiquitous). Olivine (Fig. 12), pyroxenes (Figs. 13-14), pyroxenoids (Banfield et al.,

1995a), amphiboles (Fig. 15), layer silicates (Fig. 16), and feldspars (Fig. 17) are all extensively replaced by 2:1 layer silicates under weathering conditions. These reactions, where secondary products fill space created by dissolution (called *isovolumetric reactions*), involve fluids with very different chemistries compared to free fluid that would be sampled in a field study. In these confined spaces illustrated in Figures 11-17, the water volume is limited and the ratio of water molecules to surface area is extremely low. Most importantly, the physical properties of water near surfaces differ fundamental-

Figure 13. HRTEM image of the interface between pyroxene (upper right one-third of image) and interstratified topotactically oriented clays (lower right two-thirds) produced by natural weathering. Note the one- to three- unit cell-high steps on the pyroxene surface. From Banfield et al. (in prep.).

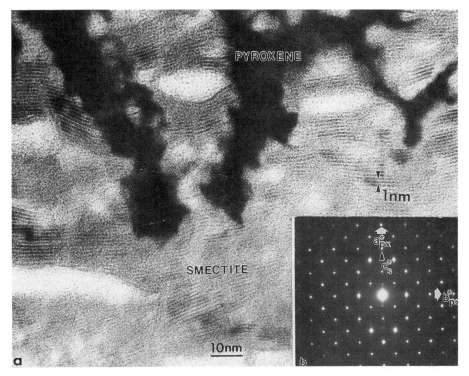

Figure 14. (a) HRTEM image of smectite developed at internal surfaces within naturally weathered pyroxene. The topotactic relationship between pyroxene and clay is apparent from both the image and (b), the inset electron diffraction pattern. [Used by permission of the editor of *Geochimica et Cosmochimica Acta*, from Banfield et al. (1991), Fig. 8, p. 2787.]

ly from those of bulk water as we will describe in detail in the next section. Consequences of these effects will be especially obvious where clay-filled channels insure long diffusion path lengths connecting reaction sites with bulk weathering fluids.

Frequently, 2:1 layer silicates developed in close proximity to dissolving surfaces are smectites with compositions that closely resemble those of the primary minerals (e.g. pyroxene: Banfield et al., 1990; amphibole: Banfield and Barker, 1994). In the case of exsolved amphibole, tens of nanometer-scale variations in primary mineral chemistry are reflected by modulations in the chemistry of the product (Banfield and Barker, 1994). Thus, secondary minerals forming under these conditions differ in composition and structure from those that would crystallize in equilibrium with bulk weathering solutions (e.g. groundwater). Such observations demonstrate the scale of local micro-environmental control of the weathering system.

Differences between bulk fluids and those involved in weathering reactions at internal surfaces can also be inferred based on the extensive mobility of elements normally considered immobile (e.g. Fe, Ti, Zr, and Al). For example, Al is a minor but important constituent of smectites formed by weathering of Al-free olivine (Banfield et al., 1990) and has been added in great abundance to layer silicates adjacent to altered metal sulfides (Al is apparently complexed by sulfate produced by alteration of pyrite; Banfield et al., 1994). Another example involves Fe and Ti which are redistributed during weathering of feldspars, pyroxenes, micas, etc. (e.g. Banfield et al., 1991). From these and many similar observations, it can be inferred that processes such as adsorption of protons onto the mineral surfaces and their incorporation into reaction products (raising pH), redox reactions (locally modifying electrochemical potentials in solutions), reactions that release complexing anions (e.g. SO_4^{2-}), etc., greatly change the relevant aqueous geochemistry.

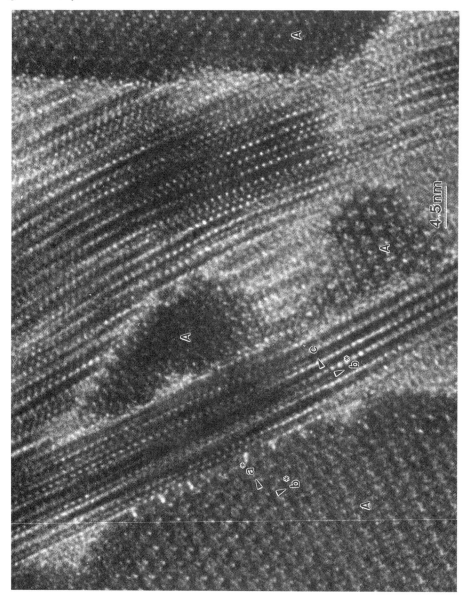

Figure 15. HRTEM image of naturally weathered amphibole (A). Note the topotactic orientation of secondary products and absence of open volume at semi-coherent interfaces between amphibole and smectite (see Banfield and Barker, 1994).

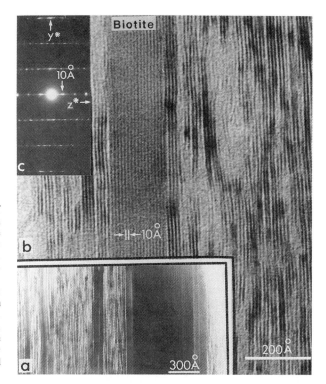

Figure 16. HRTEM image of naturally weathered biotite. In some areas, biotite layers have been converted to vermiculite by a process involving oxidation of Fe, loss of K, and introduction of water to the interlayer. (a) low magnification image; (b) higher magnification image of central area in (a); (c) electron diffraction pattern. [Used by permission of the editor of *Clays and Clay Minerals*, from Banfield and Eggleton (1988), Fig. 5, p. 52.]

In cases where abundant smectite crystallizes at a reaction front, it is often *topotactically* oriented (the crystallographic axes of clay parallel those of the primary silicate or bear a special orientation to them). Clay orientation is determined by similarity in orientation and distribution of structural components of primary and secondary minerals. The grain boundary adopts a structure and orientation that minimizes misfit between structural units, minimizing surface energy. Thus, the consequence of clay formation at mineral surfaces is the creation of interfaces with the *minimum* amount of open space, resulting in porosities just large enough to accommodate water molecules and aqueous metals (Figs. 12, 14, 15). Surface properties (surface free energy) impact the nature of interactions between primary minerals and fluids by reducing atomic-scale porosity at reaction fronts. In effect, the result is to restrict access of bulk water to mineral surfaces, so that secondary mineral growth occurs via a surface- rather than a solution-mediated process.

Topotactic relationships between primary and secondary minerals characterize many reacting systems during retrograde metamorphism and weathering. This results in a semi-coherent solid-solid interface, implying chemical bonding between atoms in the product and reactant. A possible corollary of this (following from the often high degree of continuity between slabs of structure in both minerals) is that these slabs could be inherited, without disaggregation and recrystallization, by the weathering product. Such a hypothesis is difficult to prove and the possibility that the product grew epitactically on the surface (i.e. using the surface as a growth template) cannot be discounted. Certainly, in the case of olivine, extensive structural units shared by smectite do not exist, so that epitactic smectite growth by controlled transfer of atoms across an interface (probably

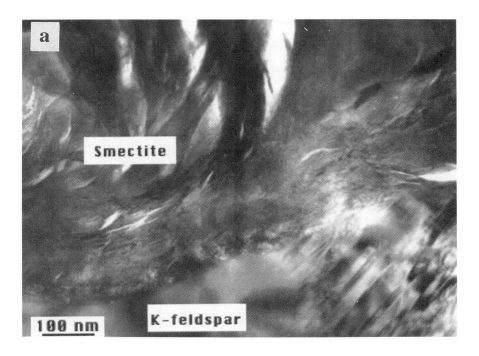

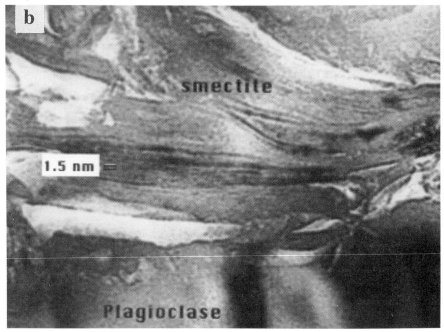

Figure 17. HRTEM images of weathering products developed at internal surfaces within feldspar crystals: cross sections across (a) the interface between K-feldspar and reaction products, and (b) plagioclase feldspar and reaction products (see Banfield and Eggleton, 1990).

with no more than a single hydration sphere) is most likely. However, in layer silicate weathering, evidence for a layer by layer mechanism and extensive similarity between reactant and product structures make it difficult to reject some structural inheritance. A strong case for such a reaction mechanism can be made for higher temperature layer silicate transformation mechanisms (Banfield and Bailey, in press). Because of their intermediate degree of silicate polymerization, amphiboles and pyroxenes probably represent intermediate cases.

Not all silicate weathering reactions result in topotactically oriented secondary minerals that fill the space previously occupied by the primary mineral. In some cases, even in incipient stages, secondary minerals are randomly oriented and occupy only a fraction of the volume created by dissolution. This is the case in plagioclase and K-feldspar weathering (e.g. Eggleton and Buseck, 1980; Banfield and Eggleton, 1990). The absence of topotactic orientation between feldspar and clays has been attributed to lack of structural similarity between these phases.

Minerals that weather to topotactically oriented clays under some conditions alter to porous assemblages of randomly oriented clays and/or oxyhydroxides under others. The degree to which reactions are isovolumetric probably depends upon many factors. Velbel (1993a) suggests that the fundamental determinants are the stoichiometries and molar volumes of reactant minerals and their products. In detail, the nature of the product assemblage must depend upon the rates of a number of processes. The solubility and rate of release of species from mineral surfaces, solubility and rates of secondary mineral crystallization, and rates of transport of constituents to and from reaction sites are probably very important. In the incipient stages of weathering at internal surfaces, rapidly reacting minerals (e.g. olivine) should be isovolumetrically replaced by oriented secondary minerals. Conversely, if minerals dissolve slowly and solutions remain relatively dilute, reaction product growth at mineral surfaces may be suppressed. Any factors (temperature, pH, fluid chemistry, etc.) that affect the relative rates of these processes or products formed could alter the nature of the mineral product interface and thus, the extent to which micro-environmental controls dominate weathering reactions (Banfield et al., in prep.).

CHARACTERISTICS OF
INTERNAL WEATHERING FRONTS IN MINERALS

A common feature to all of the intergranular weathering reactions described in the previous section is the confined space that connects the primary/secondary mineral interface with fluids external to the grain. Chemical components from the primary mineral not needed in the growth of the secondary mineral must be expelled. Likewise, secondary mineral nutrients not in the primary mineral must be supplied. HRTEM images show that these passages likely range down to the nanometer to sub-nanometer size range (e.g. Figs. 11-17). These conduits must be at least 3 Å across to pass a water molecule or partially coordinated aqueous metal, and approximately twice that to pass a fully coordinated aqueous metal. These reaction fronts fall within the definition of internal mineral surfaces as defined above. What is the nature of these interfaces? What are the differences between these interfaces and the water-mineral interface associated with an external surface? What are the properties of the aqueous solutions that exist in nanometer spaces? We will attempt to answer these questions in this section, and we will use the answers in the next section where we explore the consequences of these properties and phenomena to the whole of the weathering process.

Some physical properties of water in confined spaces

Long before many of the present day spectroscopic techniques were known or perfected, some of the physical properties of water in confined spaces could be measured (see, e.g. Derjaguin and Churaev, 1986, and references therein). This was our first clue that water within close proximity of a surface was different than bulk water.

Perhaps the simplest physical property of water to measure in a confined space is viscosity. This can be done by forcing water through very narrow capillaries (down to a few tens of nanometers in diameter) and using Poiseuille's equation. The mean viscosity of a fluid, η_m under such conditions is:

$$\eta_m = \frac{r^2}{8\upsilon} \frac{\Delta P}{L} \tag{4}$$

where r is the capillary radius, ΔP is the pressure difference between both ends of the capillary of length L, and υ is the mean flow rate. Figure 18 shows the viscosity of water (in terms of η_m/η_0, where η_0 is the viscosity of bulk water at the same temperature and pressure as the capillary measurement) versus r in microns for fine quartz capillaries. All data were collected at room temperature and pressure. In this plot, one sees that the viscosity of bulk water is the same as that in a 1 μm diameter quartz capillary. At 0.5 μm, the viscosity is still within one standard deviation of the 1 μm measurement, but the measurements are trending to higher values. As the capillary diameter continues down to the minimum value for these experiments (0.05 μm, or 50 nm), the viscosity increases exponentially. Figure 18 also includes viscosity data for the non-polar liquids benzene and carbon tetrachloride. These liquids clearly behave differently relative to water. Their standard viscosities are maintained through smaller and smaller capillaries, although the viscosity of benzene through the smallest capillary used (0.02 μm) is on the water curve. Although beyond the scope of this chapter, contrast seen here between polar and non-polar liquids may have interesting implications for hydrodynamic studies of organic contaminants within the micropermeable portions of aquifers.

Clay-water systems are also ideal for measuring the physical properties of water near surfaces and in confined spaces. Swelling clays (e.g. montmorillonites) are often used for this type of study, and the properties of the water are measured as the water con-

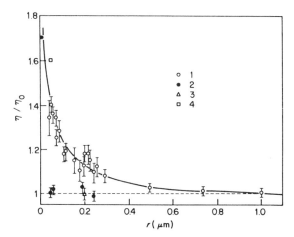

Figure 18. Viscosity of water, in terms of η_m/η_0 (the mean viscosity of water in a capillary relative to bulk water) as a function of capillary radius r. Open circles are for water, filled circles are for benzene, triangles are for CCl$_4$, and the square is for water in porous silica. The non-polar liquids (benzene and CCl$_4$) clearly behave differently from water. However, note the data point for benzene using the smallest capillary, 0.02 μm (20 nm). Modified from Derjaguin and Churaev (1986). First published by B.V. Derjaguin et al. (1974) in *The Surface Forces in Thin Films and the Stability of Colloids.* Nauka, Moscow, p. 90 (in Russian).

tent is varied. In this way, water properties can be measured as a function of interlayer spacing from multiple layers of water in the interlayer all the way down to a monolayer. Philip Low and colleagues performed the pioneering work in this field. They measured specific volume (Anderson and Low, 1958), heat capacity (Oster and Low, 1964), thermal expansion (Clementz and Low, 1976; Ruiz and Low, 1976), heat of compression (Kay and Low, 1975), and viscosity (Low, 1976) of water in the montmorillonite-water system (see also Low, 1979). For comparison with the Derjaguin and Churaev (1986) results from fine capillaries presented above, we will briefly discuss the viscosity results of Low (1976) next, and relate this to the diffusion of species in the near-surface region below.

Low (1976) used several methods to determine the viscosity of water in the Na-montmorillonite system in the presence of various amounts of water. The results are shown in Figure 19, plotted as a function of both d (the distance between montmorillonite layers in Å as determined by X-ray diffraction) and m_w/m_c (the ratio of the mass of water to the mass of clay). Like the data of Derjaguin and Churaev (1986), the viscosity of water increases exponentially as the space in which it must flow becomes confined. The data for the montmorillonite-water system is expressed in the equation:

$$\eta_m/\eta_0 \ = \ \exp\left[\frac{122.8}{(m_w/m_c)T}\right] \tag{5}$$

where T is absolute temperature. According to the plot, as the distance between montmorillonite layers approaches 3 Å (the separation needed for a monolayer of water in the interlayer), the apparent viscosity of the water is approximately one order of magnitude above what it is in the bulk. The reason for this viscosity rise probably results from at least two factors. First, the self-diffusion coefficient of water is reduced in the presence of electrolytes (Wang, 1954; Devell, 1962; Jones et al., 1965), in this case exchangeable Na^+ in the interlayers of the montmorillonites used in these studies. This is because water in the coordination sphere of the electrolyte moves as a unit, albeit a short lived unit due to relatively rapid ligand exchange rates (see, e.g. Sposito, 1984; Casey and Ludwig, this volume). Nevertheless, the overall effect is to slow the general self-diffusion rate of water. This effect, however, is not enough to increase the apparent viscosity of water by as much as that seen in Figure 19. Therefore, the second reason may be more

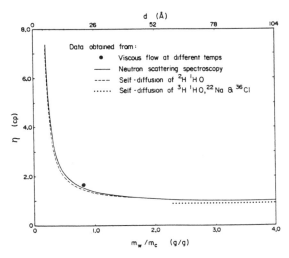

Figure 19. Absolute viscosity of water (in centipoise) plotted as functions of both d (the distance between montmorillonite layers in Å as determined by X-ray diffraction) and m_w/m_c (the ratio of the mass of water to the mass of clay). Data was obtained using four different methods as indicated. [Used by permission of the editor of *Journal of the Soil Science Society of America*, from Low (1976), Fig. 1, p. 502.]

important, and this has to do with the hydrogen bonding interaction between water and the surface of the octahedral layers of sheet silicates. Ravina and Low (1972) have shown a correlation between the *b*-cell dimension of montmorillonite and the self-diffusion of water near its surface, suggesting that the structure of water near the clay surface is responding to the precise crystallography of the octahedral layer through which the viscosity is modified. The structure of water near silicate surfaces will be much more closely scrutinized in the next section, and the reason for a viscosity increase near a surface will be explained further.

Another physical property of water that can be measured in a thin film is the dielectric constant (e.g. Sposito and Prost, 1982; Sposito, 1984), although it is difficult to separate out ionic surface conductance contributions. The dielectric constant is the ratio of the static permittivity to the permittivity of a vacuum. For bulk water, this constant is approximately 80. For a few monolayers of adsorbed water on phyllosilicates, dielectric constant measurements have been quite variable, but the measured constant is always considerably less than 80, and the true value is probably in the neighborhood of 20. This is a clear indication that water molecules are less able to align along an applied electric field when a surface is nearby. As such, ion association with water molecules is favored (e.g. Davies, 1962, and Bolt, 1979) and, at the same time, the development of a diffuse electric double layer near the surface is inhibited.

The findings presented above hint that the molecular mobility of water in confined spaces, or near a surface, is reduced. This is probably the most important underlying property change of water in this state. In the next several subsections, we will continue to revisit this theme while discussing other important details along the way.

The structure of water near silicate surfaces

The atomic structure and dynamics of pure water within a few monolayers of a surface are modified relative to bulk water. The exact form of this modification is dependent on the chemistry and structure of the surface with which it is in contact. In terms of mineral substrates, quartz and kaolinite have been the most often investigated in these studies (kaolinite is the representative phyllosilicate mineral commonly used in these studies because it has no exchangeable cations, and therefore 'pure' water can be studied at its surface). We will use these two mineral surfaces to get an idea of what pure water looks like within a few monolayers of a mineral surface.

Water at the quartz/water interface. When a virgin quartz surface is first exposed, for example immediately after fracture, the first water molecules that arrive at the surface are dissociated according to the general reaction:

$$Si\text{-}O\text{-}Si + H_2O = 2\ SiOH \tag{6}$$

Therefore, the surface is covered with SiOH (silanol) groups. The fact that silicate surface silanol groups form was first proposed long ago (Gaudin and Rizo-Patron, 1942, and references therein), and since has been extensively investigated (e.g. see review by Parks, 1990). Surface silanols are one of the most important surface chemical configuration on silicate surfaces, and can be dominant in establishing surface charge on minerals. These groups are also one of the most important building blocks in surface complexation theory (e.g. see Davis and Kent, 1990, for a recent review of this broad subject).

Surface silanols are ionizable, and they can gain or lose a proton depending on the pH of the overlying aqueous solution. In the presence of excess H^+ (low pH), $Si-OH_2^+$ tend to form, imparting an overall positive electrostatic charge to the surface. In the relative absence of H^+ (high pH), $Si-O^-$ tend to form, giving the surface an overall negative charge. The equilibrium concentration of these groups on a quartz surface is such that the positive and negative species will be in equal abundance at a pH of ~2 (some estimates are as high as pH 2.9). At this pH, the surface will be neutral in charge assuming that there are no other surface-charge determining species in solution. This is often called the point of zero charge or PZC, but it is more appropriately called the pristine point of zero charge or PPZC (e.g. Davis and Kent, 1990).

Recent spectroscopic studies of neighboring water molecules as they interact with the ionizable silanol surface of quartz just described have been enlightening. This is the result of the discovery that optical second harmonic generation (SHG) and sum-frequency generation (SFG) are ideal tools for studying the liquid side of interfaces (e.g. Guyot-Sionnest et al., 1988; 1991). Recently, Du et al. (1994) have used IR-visible SFG to study OH stretching vibrations of water at the fused quartz/water interface. They have determined that at high pH (~12), the water structure becomes highly ordered into an ice-like structure. This is consistent with the fact that the strong negative electrostatic field of the surface at these pHs orients several (at least 3, and perhaps as many as 5 or 6) monolayers of water with their hydrogen dipoles oriented towards the surface. At pH around 2 (the PZNPC of quartz), SFG spectra also indicate that the interfacial water is also ice-like, but the signals from this configuration are much weaker. This is consistent with the likelihood that, in this case, water molecules are oriented with their oxygens towards the surface, and as such hydrogen bonded to surface silanol groups (Klier and Zettlemoyer, 1977). The SFG signal is weaker because only one or two monolayers are oriented in this way due to the short range nature of hydrogen bonding. Further, the Du et al. (1994) study also confirmed using SFG that water molecules have opposite orientations at the two pH extremes used in their study.

At intermediate pH, Du et al. (1994) found that the interfacial water becomes disordered, with the degree of disordering strongly dependent on the surface ionization as outlined above. This observation is apparently due to the competing effects of the surface electrostatic field (a relatively long range influence) and the hydrogen bonding (a relatively short range influence). The final picture here is a fascinating one. From low to high pH, interfacial water goes through an order-disorder-order sequence with a 180° orientation change between the ordered regions, and with the perturbed regions increasing in thickness from a monolayer or two at low pH to as much as 5 or 6 monolayers at high pH.

Water at the kaolinite/water interface. The structure of water at the kaolinite surface is usually studied using halloysite. This mineral, as a disordered form of kaolinite, is found in both dehydrated and hydrated forms with basal spacings of 7 Å and 10 Å, respectively. The 10-Å halloysite has a monolayer of water in the interlayer space between a mostly silica tetrahedral layer on one side and a mostly alumina octahedral layer on the other (e.g. Brindley and Brown, 1980). Given the stereochemical constraints of water in this interlayer, including exactly 4 water molecules per unit cell, hydrogen bonds between the molecules must be about 3 Å long, and IR spectroscopy has confirmed that the hydrogen-bonded network in this interlayer is more water-like than ice-like (Yariv and Shoval, 1975). When water molecules are probed by proton nuclear magnetic resonance (NMR) techniques in this interlayer (Cruz et al., 1978), spectra similar to bulk water are obtained. This indicates that the water molecules are tumbling quickly enough to show no preferred orientation to either the tetrahedral or octahedral walls of the interlayer within the time frame of the NMR measurement. Nevertheless, the NMR results do indicate that molecular (translational) motions are considerably slower than in bulk water, and are in the range of ice. Therefore, in this monolayer of water, we have a compromise between the disorder of liquid water and a rigid configuration as part of a crystal structure.

Little is known about the structure of water on kaolinite surfaces beyond the first monolayer, at least from a spectroscopic point of view. However, various thermodynamic measurements, such as those performed by Jurinak and Volman (1961) and Kohl et al. (1964), indicate that the influence of kaolinite surfaces on surrounding water may extend through about 3 monolayers, or about 10 Å from the surface (see also Fripiat et al., 1982).

The structure of aqueous solutions near silicate surfaces

This section will be broken down into two sections, for quartz and phyllosilicates, as in the previous section. For the phyllosilicates, expandable clays have become the standard by which work on the structure of aqueous solutions near planar silicate/water interfaces is based. For these materials, variable amounts of interlayer water, combined with the presence of exchangeable cations, provides a perfect system for these studies. The majority of this work has been done on smectites, starting with the pioneering research some time ago of Mooney et al. (1952a,b). Therefore, in the phyllosilicate subsection below, we will confine most of our comments to such studies on smectites.

Aqueous solutions at the quartz/water interface. As discussed above, Du et al. (1994) collected SFG spectra from water in contact with quartz while varying pH. They also performed these highly useful measurements with fused quartz in the presence of aqueous solutions (0.1-0.5M NaCl). At a pH of 1.5, the SFG spectrum was no different than the one taken with pure water, even at a concentration of 0.5M. It is not surprising that the background electrolyte does not affect the structure of water near the quartz surface at this pH which is near the PZNPC. With a neutral fused quartz surface, there is no pile-up of counter ions, and therefore the salt does not affect the interfacial system appreciably. At high pH (approximately 12), where the surface is charged strongly negative, Na^+ ions form a diffuse double-layer which helps screen the near surface water molecules from the surface charge. As expected, this results in a lowering of the SFG signal intensity. However, the spectra also indicate an increase in molecular ordering compared to the pure water spectra. This may be due to the fact that, at higher surface charge and a concomitant decrease in the double-layer thickness, water near the interface is now more contained in the coordination spheres of aqueous sodium ions. At intermediate pH (e.g. 5.6), characterized by spectra indicating maximum disorder in the pure water case, the addition of NaCl results in spectra that resemble those collected at low pH around the PZNPC. Water molecules are relatively ordered, apparently because the presence of Na near the negatively charged surface at these pHs has completely upset the delicate balance between the surface charge and hydrogen bonding forces which effectively compete in the pure water case within this pH range.

Aqueous solutions at the smectite/water interface. This interfacial system has been extensively studied (see, e.g. Johnston et al., 1992, and references therein) for the reasons given in the introduction to this section. One of the underlying themes of these studies is that the structure of the water near a smectite surface (usually observed in an expanded interlayer) is influenced most by the presence of the exchangeable cation. A number of studies using various techniques have shown that the water molecules coordinated to these exchangeable metals on smectite surfaces are in less mobile environments relative to those in bulk water (e.g. NMR and ESR used by McBride et al., 1975, McBride, 1982, Brown and Kevan, 1988, Kogelbauer et al., 1989, and Delville et al., 1991; IR used by Russell and Farmer, 1964, and Farmer, 1978, Johnston et al., 1992; and neutron scattering and dielectric permittivity measurements used by Sposito and Prost, 1982). Atomic models for one and two layer hydrate smectites and vermiculites, with monovalent interlayer cations, have come out of these and other studies (Fig. 20). Certainly, at extended swelling, when there are 15 to 20 water molecules or more per exchangeable cation, the spectroscopic signature of this water is similar to that of bulk water. However, with 1 to 3 layers of hydration in interlayer residence (interlayer spacing between approximately 3 and 10 Å), the above mentioned features become apparent. The self-diffusion coefficient of the water in the first coordination shell of monovalent interlayer cations in the hydrated interlayer (encompassing from 3 to 6 water molecules) and divalent interlayer cations (encompassing 6 to 8 water molecules) is considerably smaller (approximately one order of magnitude) than the self-diffusion coefficient for pure water (Sposito and Prost, 1982; Sposito, 1984).

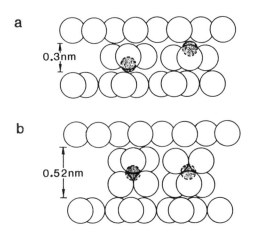

Figure 20. Interlayer atomic structures of (a) one-layer hydrate of Li^+-hectorite (a tri-octahedral smectite) and (b) two-layer hydrate of Na^+-vermiculite. Li^+ and Na^+ are shaded in their respective drawings. The circles within the interlayer each represent a water molecule. The cirlces bordering the top and bottom of the interlayers represent oxygens. [Used by permission of the Soil Science Society of America, from McBride (1989), Fig. 2-4, p. 42.]

In addition to the effects of the exchangeable cations on interlayer water, there is also the influence of the tetrahedral sheets on both sides of the smectite interlayer. As expected, this influence increases as the amount of interlayer water decreases and the interlayer collapses. In the case of the one-layer hydrate and monovalent interlayer cations (Fig. 20a), the water molecules are in an ice-like configuration with hydrogen bonding between interlayer cation complexation shells and the ditrigonal cavities of the bordering tetrahedral layers. In such a configuration, the distance between neighboring water molecules is a typical distance of ~3 Å.

The solvent properties of water in confined spaces

As discussed above, there is a great deal of evidence showing that the structural properties of water are significantly modified within at least 2 to 3 monolayers (about 7 to 10 Å) of a silicate surface. This is in the neighborhood of the semi-coherent spacing between primary and secondary minerals as shown in the TEM images above. Further, atomic structure and physical properties are always interrelated. Therefore, it will be very useful to look at the property modifications of aqueous fluids in these confined spaces. Two of them, viscosity and dielectric permittivity, have already been discussed above.

Below, we will discuss the solvent properties (namely Bronsted acidity and saturation state modification) that will play important and specific roles in characterizing this weathering interface.

Bronsted acidity. Protons can become available in the interlayers of smectites, for example, via the dissociation of coordinating water molecules around interlayer cations. The general reaction describing this can be written as follows (Sposito, 1984):

$$M(H_2O)_n^{m^+} = MOH(H_2O)_{n-1}^{(m-1)^+} + H^+ \tag{7}$$

where M represents an aqueous metal. In aqueous solution, the equilibrium constant for this reaction correlates with the ionic potential of the coordinated metal (e.g. Davies, 1962). As the ionic potential of the metal increases, repulsion of hydrating protons increases and the coordinating water dissociates more readily (e.g. the equilibrium constant for Eqn. 7 increases). This process can also be followed by adding a Bronsted base protonation reaction, such as (Mortland and Raman, 1968):

$$M(H_2O)_n^{m^+} + NH_3 = MOH(H_2O)_{n-1}^{(m-1)^+} + NH_4^+ \tag{8}$$

The important point here is that as a smectite slurry is dried, the reaction in Equation (7) goes to the right. Also, as the ionic potential of M increases, the equilibrium constant for Equation (7) goes up faster compared to cations with lower ionic potential. Therefore, smectitic interlayer cations become more acidic as the interlayer water decreases. This acidity effect is maximized when the interlayer hydration is reduced to a single monolayer.

Support for the above has come from IR, NMR, and conductivity measurements (Calvet, 1975; Fripiat et al., 1965; Fripiat, 1976, and references therein; Poinsignon et al., 1978). NMR in particular has shown that the degree of dissociation in a hydrated interlayer of monolayer thickness is as much as two orders of magnitude more compared to bulk water outside of the clay structure.

Aqueous solution saturation states and nucleation/crystallization. As described in the HRTEM section above, internal weathering fronts in minerals are bound by a dissolving primary mineral and a growing secondary mineral. The dissolving and growing faces of these minerals are within several Ångströms of one another. In this regard, among the many factors that are generally not considered in the crystallization (or dissolution) of a mineral is the space available in which the process must occur. One of the most recent papers that addresses these issues is Putnis et al. (1995), who further emphasize that fluid supersaturation and crystallization are much more satisfactorily described and explained via non-equilibrium rather than equilibrium processes. One of the first to emphasize the ramifications of available crystallization space was Bigg (1953). In this study, the freezing point of water droplets was measured as a function of the volume of the droplet, as well as the degree of undercooling and the rate of cooling. At a constant cooling rate, the amount of undercooling necessary for crystallization to begin (ΔT) was found to be related to the volume of the droplet (v) such that:

$$\Delta T = X - Y(lnv) \tag{9}$$

where X and Y are fitted constants. Therefore, as the volume of the droplet decreases, the amount of undercooling necessary for crystallization goes up. This relationship has since been found in other studies (e.g. see Melia and Moffitt, 1964), as well as by Kubota et al. (1988) who studied the crystallization of potassium nitrate from aqueous solution. Further, in systems slightly more analogous to crystallization of minerals in confined spaces within rocks, Halberstadt et al. (1969) and Henisch (1989) (see also Putnis et al., 1995) have conducted crystallization experiments within gels and gelatins. They found that for gels with smaller pores, where the pore size was varied by preparing gels of different densities, the nucleation probability was decreased while the maximum supersaturation increased. Similar results were obtained when going from silica gel to gelatin, with the pores an order of magnitude smaller in the latter.

Manipulation of Equation (9) gives

$$\exp{(X - \Delta T)} / Y = v \qquad (10)$$

where it is easy to see that changes in ΔT require large (exponential) changes in the crystallization volume. For example, Kubota et al. (1988) found that at a constant cooling rate of 1.9×10^{-3} °C/s, a sample volume of 14 cm^3 required an undercooling of approximately 12°C to nucleate and crystallize potassium nitrate from aqueous solution, while a sample volume of 1 μm^3 (a volume reduction of 10^{12}) was required to increase the undercooling by 32° to ~42°C. Such large variations in crystallization volumes are commonly present within the same rock. Therefore, at a constant temperature, the distribution and size of open space within the rock could be a critical factor in the spacial distribution of the nucleation of secondary minerals.

The reason for the observations pointed out above must be the problem of bringing growth components (nutrients) to the surface of a nucleus quickly enough, that is before the crystal nucleus spontaneously disassembles before reaching a critical size. In confined spaces associated with internal surface area as described in this chapter, transport of nutrients must occur via diffusion operating at reduced rates (see Eqn. 12 below). When coupled to the fact that silicates are sparingly soluble materials (very low solubilities), nucleation will be severely depressed within these confined spaces. Equivalently, very high supersaturation values will be obtainable for silicates in confined spaces.

Another little considered factor that may impact crystallization of gels or liquids in confined spaces is the relatively large contribution of surface free energy to the total structural (configurational) energy of the nanocrystalline nucleation product. For metals, it is known that the melting point is lowered by as much as 50% when particles are of nanometer-scale dimensions. Conversely, significantly greater undercooling is required to cause crystallization. Furthermore, in cases where alternative products have different surface energies, phase stability may be reversed (e.g. Langmuir, 1971; Banfield et al., 1993).

All of these factors may have important consequences within the internal weathering front, as we will describe in the next section.

CONSEQUENCES OF INTERNAL WEATHERING FRONT CHARACTERISTICS ON REACTION MECHANISMS AND RATES

Given the nature of weathering fronts in minerals as discussed at length above, we will now explore the ramifications of such knowledge on factors affecting the mechanisms and rates of silicate weathering processes.

A qualitative model for weathering at internal surfaces

External mineral surfaces are defined above as having hydraulically-driven water moving over their surfaces (if part of an aquifer), or having the potential of supporting such flow (if part of the vadose zone). In this section, we will call these areas, collectively, the external zone. In addition, we defined above an internal mineral surface as not being able to support fluid flow because of highly constricted, dead-end or extremely tortuous channels. These channels would probably include most of the nanometer-scale interfaces between primary and secondary phases as seen in our HRTEM examination of weathering rocks shown above. In these highly constricted spaces and channels in which there is no flow, diffusion is the only means of transporting chemical constituents. We will call all of these diffusion limited channels, collectively, the internal zone.

When weathering reactions occur on internal surfaces, i.e. within the internal zone, chemical constituents releasing from the primary mineral dissolution process and not needed for the secondary mineral crystallization process must be transported by diffusion to flowing water in the external zone for removal. On the other hand, the chemical constituents not available from the breakdown of the primary mineral, but needed for the growth of the secondary phase (including water, hydrogen, or hydroxyls if necessary) must follow the reverse path. Constituent movement by fluid flow to and from the mouth of the internal zone (where it interfaces with the external zone) will be considerably faster than the diffusional transport within the internal zone. Therefore, the flux of material to and from the primary/secondary interface via diffusion will be a critical step in the weathering process. It is this exchange of chemical species that is fundamental to weathering reactions in the internal zone. The rate of such diffusion-mediated exchange will help dictate (and in some cases may control) the rate of the weathering reactions in the internal zone.

A quantitative treatment of this critical diffusional exchange mechanism has been developed by carbonate diagenesis geochemists (e.g. Brand and Veizer, 1980; Pingitore, 1982) into a model called bulk solution disequilibrium. We borrow here from this work, as well as from Gordon (1945) and Garrels et al. (1949), the latter who developed a model for ion diffusion in water-saturated rock intergranular spaces.

The principal equation used to described the solute flux via diffusion between external and internal zones is as follows:

$$\frac{dq}{dt} = \frac{D_q \, (C_i - C_e) \, A}{l} \tag{11}$$

where dq/dt is the flux (quantity per time) of species q per time t, D_q is the diffusion coefficient for species q, C_i is the concentration of the species in the portion of the internal zone where the primary mineral is reacting to the secondary mineral, C_e is the concentration in the external zone (assumed to be homogeneously distributed), A is the effective pore area (i.e. the sum of cross-sectional areas of all diffusional paths between the primary/secondary mineral reaction front and the external zone), and l is the average length of the diffusion paths. This equation assumes that a steady state gradient of species q has already been established between the internal and external zones. It is not appropriate for a new reaction until steady state has been reached.

Bulk diffusion coefficients for aqueous species of interest in silicate weathering reactions can be obtained from, for example, Lerman (1980). These coefficients are all in the neighborhood of 0.5 to 1 cm^2/day. The other terms on the right side of Equation (11) are difficult to quantify and obviously depend on the specific system of interest. The diffusion path length may be from microns to centimeters or more, and the effective pore area can also vary by several orders of magnitude. The concentration gradient, $C_i - C_e$, is perhaps the most interesting of all in that this is what controls, for any particular species, the direction of diffusional transport (into or away from the dissolving/crystallizing surfaces) and at what rate this species will move. For example, the internal weathering front reaction of amphibole to smectite (Fig. 15 and associated text) requires that H_2O is transported into the internal zone, and various proportions of Mg, Fe, Na, Al, and Si are all transported out. If aquifer water flowing in the external zone is relatively depleted in these silicate constituents due to wet, high flow conditions, diffusion transport will be relatively rapid, emphasizing the open system aspects of this arrangement. If water in the external zone is relatively enriched in these silicate constituents due to either low flow or intermittent flow (the latter being the case in the vadose zone), the opposite will be true.

Naturally, a relationship like that shown in Equation (11) is further complicated by temporal changes as a function of reaction progress. Certainly, C_i - C_e, A, and l are subject to changes throughout the course of the weathering process, possibly by orders of magnitude.

There are still additional important factors which will affect the exchange process between the internal and external zones. We saw in Figure 19 that the viscosity of water increases by approximately one order of magnitude as it approaches a single monolayer in the presence of smectite. As described above, this is due to two distinct affects, the first being that water is restricted by participating in the coordination shells of interlayer cations (or, in this case, aqueous metals within the internal reaction front), and the second being that water is translationally restricted due to its interaction with the oxygens of the tetrahedral sheets of the smectite (or, in this case, the surface layers of the primary and secondary minerals). This viscosity increase of water near these surfaces should affect the diffusion rate of species throughout highly confined regions of the internal zone. This is probably most easily described with a modified version of the Stokes-Einstein equation (e.g. Low, 1968), which states that the diffusion of a species in a fluid has an inverse relationship with the viscosity of that fluid as follows:

$$D_q = \frac{\beta c_q T}{\eta} \tag{12}$$

where β is a geometric constant that takes into account the porosity or tortuosity, c_q is a constant that scales with the radius of species q, T is the absolute temperature, and η is the viscosity. Therefore, we would expect species to diffuse at a rate about one order of magnitude slower relative to bulk diffusion when moving along internal weathering fronts as seen, for example, in Figure 15. Combining Equations (11) and (12), we have:

$$\frac{dq}{dt} = \frac{\beta c_q T (C_i - C_e) A}{\eta l} \tag{13}$$

This equation establishes all the principal physical components which must be taken into account when considering factors that will control the flux of species to and from reactive sites within internal weathering fronts.

We also discussed above how the Bronsted acidity of smectite interlayer aquo-complexed metals increases as the interlayer water approaches a monolayer in thickness; see Equations (7) and (8) and associated text. At a monolayer, water dissociation around metals can increase by up to two order of magnitude relative to bulk aqueous solutions. Therefore, within the confined spaces of an internal weathering front, pH levels may be considerably lower (by up to two pH units) relative to fluids in the external zones. Within the weathering front itself, H$^+$ should be relatively abundant to act as a catalyst in dissolution reactions, as well as to participate in hydroxyl groups in the growing clay secondary phases. The Bronsted acidity of the metal complexes at the reaction front should tend to expedite the weathering process, while the diffusion of unneeded constituents away from dissolution sites and nutrients towards growing crystals should tend to limit it.

It is also interesting to consider the surface charge of the minerals on either side of the internal weathering front. We have already briefly discussed the development of surface charge in the section on the quartz-water interface above. Note here that we are not referring to permanent structural charge of a mineral as commonly developed, for example, by phyllosilicates, but to coordinative surface charge, that resulting from charged aqueous species interacting with any mineral surface and balanced by a counter-ion charge in solution (this entire package is known as the electrical double layer). The surface charge in this case is a function of pH. In the simplest case, where H$^+$ and OH$^-$ are the only charge determining species interacting with the surface, the pH at which the surface is neutral (i.e. the pristine point of zero charge or PPZC), is quite different depending on the silicate (Table 1, from Sverjensky, 1994). At pH below the PPZC, the surface is charged positively, and it will attract a diffuse layer or swarm of negatively charged ions (counter-ions) to balance the charge. At pH above the PPZC, the reverse is true. Counter-ions closest to the surface may be bound to the surface as an outer-sphere complex (attached weakly to the surface via an intervening water molecule). With enough space, the counter-ion region in solution may be up to a few nanometers wide, with the counter-ion swarm becoming more and more diffuse as the distance from the surface increases. Within a confined internal weathering front, the charging fields of both mineral surfaces will overlap. The two surfaces will compete for counter-ions if they are the same charge; they will try to exclude the other's counter-ions from entering the confined space between them if the surfaces are oppositely charged. In the latter case, if such counter-ions are needed as nutrients at a secondary mineral surface, the growth rate may be restricted or terminated. In future studies of the effects of cooperating or interfering surface charge in internal weathering fronts, a few additional and important points should be considered. First, H$^+$ and OH$^-$ are not necessarily the only surface charge determining species. Other ions can form strong, inner-sphere complexes with the surface (no intervening water molecules) and play a role in determining its surface charge. Further, hydrogens can substitute for metals on the surface which may change the number of ionizable surface proton sites available. In either case, PZCs will shift, and tabulated PPZCs will not be a reliable guide to surface charge as a function of pH. Second, aqueous species may approach and strongly attach to surfaces as inner-sphere complexes even though they have the same charge as the surface. In this case, the chemical affinity for the species to bond to the surface overcomes the electrostatic repulsion. Finally, any form of the PZC is to some extent crystallographically specific. For example, the PZC of a basal surface of a phyllosilicate should not be the same as on its edge sites. Parks (1990), Davis and Kent (1990) and Sverjensky (1994) are good places to start in attempting to evaluate many of the above factors and concepts.

Finally, we have also considered above aqueous solution saturation states and nucleation probabilities within confined spaces. It is interesting to consider these phenomena in terms of the internal weathering front. The nucleation of a secondary phase at the beginning of the development of an internal weathering front has several serious obstacles to overcome. Nucleation is always suppressed when nutrient transport is restricted to diffusion (e.g. Putnis et al., 1995), as might be the case even at the initial development of an internal weathering front. As discussed above, the reason probably has to do with the supply of growth components to the nucleus at a rate that will allow it to exceed the critical size. In this case, retarded diffusion rates encountered in confined incipient weathering fronts may be rate limiting to nucleation, although the reduction in stability of the nucleation product due to surface free energy contributions and possible metastability of the secondary phase may be equally or even more important. The diffusion problem is exacerbated by the fact that silicates are sparingly soluble, meaning of course that the constituents of the nucleus are relatively dispersed in aqueous solution. A considerable amount of supersaturation will be needed to overcome all of these obstacles and nucleate a secondary phase within the internal zone. In such confined spaces, very high supersaturation may not be a problem. However, one further point should be made. Secondary phase nucleation (and later growth) is clearly augmented by the fact that a semi-coherent boundary separates the primary and secondary minerals. Nucleation barriers may be lowered with the epitaxial growth of the secondary mineral on the surface of the primary phase. The promotion of metastable phases in this case is clear. Structural inheritance from the primary to the secondary mineral (generating the so-called topotactic relationships discussed in the section on HRTEM above) is probably also operative. This inheritance might be as obvious as octahedral and tetrahedral sheets inherited from a sheet silicate (primary) to a clay mineral (secondary). On the other hand, the structural inheritance can be much more subtle, as in the case of olivine weathering to a phyllosilicate.

Table 1. Crystal structure properties and zero point of charge values at 25°C and 1 bar.

PHASE	$\varepsilon_k{}^a$	$s/r_{M-OH}{}^b$	Exptl. pH$_{PPZC}{}^c$	Calc. pH$_{PPZC}$
α-SiO$_2$	4.578	0.3818	2.9	2.91
Al$_2$Si$_2$O$_5$(OH)$_4$	11.8	0.2754	4.5	4.66
α-TiO$_2$	120.91	0.2248	5.8	5.21
Fe$_3$O$_4$	20000	0.1768	6.6	7.10
α-Fe$_2$O$_3$	25.0	0.1645	8.5	8.47
α-Al$_2$O$_3$	10.43	0.1711	9.1	9.37
α-Al(OH)$_3$	8.4	0.1716	10.0	9.84
MgO	9.83	0.1070	12.4	12.24

PHASE	$\varepsilon_k{}^a$	$s/r_{M-OH}{}^b$	Exptl. pH$_{ZPC}{}^c$	Calc. pH$_{PPZC}$
BeO	7.16	0.1890	10.2	9.5
CaO	11.95	0.0976		12.3
NiO	11.9	0.1077	9.85-11.3	11.8
CuO	18.1	0.1686	9.5	8.6
ZnO	8.49	0.1676	8.7-9.3	10.0
FeO(OH)	11.7	0.165	9.0-9.7	9.4
amorphous SiO$_2$	3.807	0.3818	3.5	3.9
β-MnO$_2$	10000	0.2301	4.6-7.3	4.8
SnO$_2$	9.0	0.2177	7.3	7.7
ZrO$_2$	22.0	0.1803		7.9
ThO$_2$	18.9	0.1456	9.0-9.3	9.6
UO$_2$	24.0	0.1480		9.2
CaTiO$_3$	165.0	0.1409		8.8
MgAl$_2$O$_4$	8.3	0.1703		9.9
FeAl$_2$O$_4$	40.0	0.1693		7.9
ZrSiO$_4$	11.5	0.1559		9.8
Zn$_2$SiO$_4$	12.99	0.2389	7.4	6.1
ANDALUSITE	6.9	0.2531	5.2-7.8	6.9
KYANITE	7.8	0.2402	5.2-7.9	7.1
SILLIMANITE	11.0	0.2736	5.6-6.8	4.9
FORSTERITE	7.26	0.1972	10.5	9.1
GROSSULAR	8.53	0.2113		8.1
ALMANDINE	4.3	0.2135		10.4
OH-TOPAZ	5.0	0.2413		8.6
JADEITE	10.0	0.1930		8.5
WOLLASTONITE	8.6	0.2370		7.0
DIOPSIDE	8.02	0.2338		7.3
HEDENBERGITE	17.4	0.2333		5.9
ANTHOPHYLLITE	8.0	0.2497		6.6
TREMOLITE	8.0	0.2402		7.0
TALC	5.8	0.2634		7.0
MUSCOVITE	7.6	0.2539		6.6
PHLOGOPITE	7.6	0.2194		8.0
LOW ALBITE	6.95	0.2920	6.8	5.2
H-ALBITE	6.95	0.3533	2.0	2.6
MICROCLINE	5.48	0.2889		6.1
H-MICROCLINE	5.48	0.3549	2.4	3.3
ANORTHITE	7.14	0.2809		5.6

Applications to stagnant water in the vadose zone

In the previous section, we considered a number of factors that may be fundamental to the reaction progress of an internal weathering front. These fronts have been observed to exist within what we defined as internal zones, that is where the only chemical transport mechanism is diffusion. To this point, we have considered the stagnant (that is, non-flowing) aqueous solutions within these zones to be contained in confined, dead-end or highly tortuous branches off flow paths within a rock aggregate or soil. However, there is abundant evidence that intragranular stagnant fluids exist in the vadose zone of soils (see below). Such regions should be subject to many of the same principles that we have discussed in the previous section. In fact, these stagnant intragranular zones are now being used to rationalize the difference between dissolution rates measured in the field and in the lab (e.g. Walther and Wood, 1986; Schnoor, 1990; Velbel, 1993b; Swoboda-Colberg and Drever, 1993).

Aqueous solution movement through the vadose zone of soils is heterogeneous in both space and time. Evidence for this, from both laboratory and field work, as well as from both experimental and theoretical approaches, has been reviewed by Velbel (1993b). Because of its importance in explaining the difference in laboratory and field-based dissolution rate measurements, it is also a subject of Drever and Clow (this volume). The implications of this phenomenon are clear: when one collects effluent from a watershed to use as a measure of mineral dissolution rate within that area, that water has only been in contact with a portion of the available mineral surface area. Estimates of the mineral surfaces in contact with flowing water range from tenths of a percent to 50% or more (e.g. Sollins and Radulovish, 1988; Germann, 1990; Kung, 1990) depending on a number of factors, including the overall make-up and structure of the soil and its saturation state. This heterogeneous flow, often called fingering, bypass, channelized or macropore flow, also contacts a larger percentage of mineral surfaces present shallow in the soil profile. The flow contact percentage drops off rapidly with depth.

The implications of these observations relative to our previous characterization of internal weathering fronts are interesting to consider. In the vadose zone of a soil profile, at any given time, the majority of the intragranular mineral surface areas are probably not supporting flowing fluids. The fluids in contact with these surfaces are stagnant. Although the spaces in which these stagnant fluids exist will not generally be as confining and restricted as in the case of internal weathering fronts, they are still subject to diffusion-limited transport, Equation (11), and all that this implies. Just like internal weathering fronts, these surfaces do participate in the overall weathering of the soil. Equivalently, the water chemistry of the flowing solutions is modified by the stagnant, or bypassed fluids. However, the exchange between these two regimes is subject to diffusion coefficients, concentration gradients, diffusion path lengths, and cross-sectional areas of diffusional transport. When one considers the vast amounts of internal surface area that can be present within grains (much of which may not even be measurable with BET surface area analysis), and adds to this the intragranular surface area that also supports stagnant fluids in the vadose zone of a soil, it is no surprise that dissolution rates measured in the laboratory exceed those measured in the field by as much as several orders of magnitude.

BEYOND INTERNAL WEATHERING FRONTS:
MORE ADVANCED STAGES OF MINERAL ALTERATION

In many cases, initial isovolumetric weathering reactions will give way to non-isovolumetric reactions as the system matures. Therefore, the environments at the weathering reaction sites (i.e. the micro-environments) do not remain constant, but change as porosity and permeability increase. Evolution in micro-environmental conditions at reaction sites is typically indicated by changes in secondary mineral structure and chemistry. For example, biotites, chlorites, and amphiboles are often replaced sequentially by vermiculite/smectite and kaolinite ± oxyhydroxides. Even in cases of plagioclase and K-feldspars, where smaller volumes of products develop in random orientations, evolution in local chemistry can be inferred from the changeover from smectite-dominated to kaolinite/halloysite-dominated assemblages (Banfield and Eggleton, 1990).

When rock permeability increases sufficiently, through physical, biophysical, and chemical disaggregation, mineral weathering reactions may be strongly impacted by organisms. This interaction may involve dissolved organic chemicals (e.g. oxalic acid, Pittman and Lewan, 1994) or the direct physical contact between complex organic polymers (e.g. polysaccharides) and mineral surfaces (Barker and Banfield, 1995a,b; in prep.). Further, organic coatings (patchy and continuous) are common on mineral grains in soils (e.g. Jones and Uehara, 1973). These organic molecules are associated with bacteria, algae, and fungi, as well as higher organisms, that colonize the external surfaces, cracks, and fractures to depths exceeding several millimeters.

Development of chemically and structurally distinct reactant-product assemblages in inorganically- and organically-dominated micro-environments illustrates that microorganisms may strongly mediate weathering reactions near the Earth's surface (e.g. Barker and Banfield, 1995b, and references therein). Organically-mediated reactions most likely involve interactions between surface functional groups on organic membranes and atoms on the mineral surface, probably modifying the mechanism and rate of release of these atoms from their surface sites. Furthermore, polysaccharide membranes concentrate ions from solution and may modulate the abundance and character of water at mineral surfaces (Barker and Banfield, in prep.).

SUMMARY, CONCLUSIONS, AND FINAL THOUGHTS

Microscopic characterization provides direct information about the physical and chemical characteristics of weathering reactions in nature, including chemical exchange accompanying these reactions and the growth of important secondary minerals. These characterizations also provide insight into why long-term silicate mineral weathering rates may not be easily predicted from experimental data. Combined with the physical nature and properties of mineral surfaces, the fluid covering them, and nanometer to subnanometer conduits between primary and secondary phases lining internal weathering fronts, a great deal can be learned about how minerals, and the rocks and soils that they form, weather in nature. The following points summarize and conclude our findings and thoughts.

(1) Experimental reactions may occur by different mechanisms than those most important in nature. Reactions occurring in the laboratory typically converge on congruent dissolution of the mineral after initially incongruent dissolution, and the formation of secondary phases are typically suppressed. This can also happen in nature, and this type of weathering can be considered as one end member in the range of natural chemical weathering reactions. Much of what we have highlighted in this chapter is toward the other end member, that of direct structural transformation of a primary to a secondary mineral. This process assumes a dramatically different scenario. Although the breakdown of the primary silicate may be congruent, the total weathering reaction is often highly incongruent as secondary phases form in various intimate relationships with primary minerals. The only primary mineral constituents leaving the weathering front and eventually moving into surrounding intragranular fluids are those that are not needed in the production of the secondary phase. In the cases of common ferromagnesian silicates, retreat of the primary mineral surface may involve transfer of atoms from the primary to

the secondary mineral across a semi-coherent interface (no residence time is solution). Atoms may be relocated singly or inherited in sections or slabs (e.g. polymerized silica) directly into the secondary phase. Regardless of whether the transformation occurs atom by atom or in larger pieces, it is probable that only those atoms that leave the primary/secondary intermediate sites become completely solvated, and complete solvation may not occur until the atom has diffused some distance from the these sites due to lack of water molecules or space constraints.

Given the range of silicate minerals encountered in nature and the variability in weathering environments, it is likely that a broad continuum exists between congruent dissolution, as might be observed in a typical laboratory experiment, and direct structural transformation. If this is the case, the whole of laboratory studies in this area of research represent only a small fraction of the dissolution/weathering reactions that occur in nature.

(2) It is well known that in most cases dissolution rates strongly depend on solution conditions. Solution conditions present at the majority of mineral surfaces in natural weathering environments may not be well approximated by bulk fluids associated with weathering profiles. This will be especially true during early stages of weathering, when the physical and chemical characteristics of the fluids participating in reactions associated with internal weathering fronts will be greatly impacted. Within the confined boundaries of these fronts (and relative to bulk aqueous solution properties), viscosities will increase in the neighborhood of an order of magnitude accompanied by a diffusion rate decrease of the same order. Metal-coordinating water molecules will dissociate at a significantly higher rate, driving the pH down, and nucleation of new phases will be dramatically suppressed even at very high supersaturation, making epitaxial nucleation and/or structural inheritance critical for the initial development of secondary phases. Thermodynamic equilibrium is neither implied here nor expected.

The restrictions in the vibrational and translational movement of water molecules within the confined spaces of internal weathering fronts (and by extension, certain intragranular space) may also translate into depressed mineral dissolution rates. It has been shown that the rates of metal release from mineral surfaces scale with the rates of solvent exchange around the corresponding metal ion in solution (Casey and Westrich, 1992; Casey et al., 1993). If this observation has mechanistic significance (see Casey and Ludwig, this volume) rather than simply reflecting a bond strength control, the molecular movement of water molecules may be the key rate controlling step in the dissolution of some minerals (e.g. oxides and orthosilicates, not silicates containing extensive polymerized silicate units; Banfield et al., 1995a). If the restrictions on water discussed above translate into reduction of water exchange around aqueous metal ions within internal weathering fronts, this would result in depressed mineral dissolution rates. A guide to the magnitude of this effect can be found in Casey and Westrich (1992); this suggests that an order of magnitude suppression of the first order rate coefficient for water exchange results in lowering dissolution rates by roughly two orders of magnitude.

(3) The basic end member regimes of rate control during weathering have been described as transport control (transport limited reaction, presumably transport through the host solution only) and surface reaction or interface control (e.g. Berner, 1978, 1981; Velbel, 1993a, and references therein). Berner (1981) compared rates for molecular diffusion in aqueous solution with rates for natural weathering of feldspars and concluded that transport rates cannot be rate limiting. First, it is also important to remember that rates for pyroxene and olivine weathering are several orders of magnitude faster than

those for feldspars. Further, in the early stages of rock weathering, diffusion rates in bulk aqueous solution are probably not appropriate values to use in this comparison considering the HRTEM evidence that documents the development of abundant reaction products at internal surfaces, especially in the initial stages of weathering of Fe-free (e.g. diopside, enstatite) and Fe-bearing silicates. In these cases, solution transport may be dramatically slower than expected due to the exceptionally narrow and tortuous passages to reaction sites, and the modified physical properties of fluids in such spaces. In the early stages of weathering at internal surfaces, rates of solution movement may typically be zero. In this case, the rates of destruction of primary minerals may be transport limited. Transport rates to or from a reaction site will depend on concentration gradients and diffusion coefficients; transport direction will depend on the gradient direction. The flux (quantity per time) of a species to or from a site will depend on the gradient and diffusion coefficient as well as corss-sectional area and length of the diffusion paths according to Equations (11)-(13). As reactions proceed and the system matures, access of solutions to reaction sites will increase as passages widen and connect, and fluid flow begins to occur. In this stage, the rates of transport of constituents into and out from reaction sites are rapid. Under these conditions, rates may be surface reaction limited, more closely resembling those occurring in the laboratory. However, in the vadose zone of soils, the majority of the intragranular mineral surfaces will still only have stagnant solutions in contact with them for significant amounts of time and transport rates will again become critical.

The main point here is that both transport and interface rate-controlled regimes may be important during different stages of weathering of a mineral (especially if the mineral is relatively reactive) because the geochemical environment at reaction sites changes with time.

(4) In more advanced stages of weathering, enhanced access of fluids is probably accompanied by increased impact of organically-derived chemicals, especially where sufficient space is available to accommodate microorganisms. Under these conditions, important interfaces may involve complex polymers such as polysaccharides.

(5) As a first step toward scaling laboratory to field data it is desirable to find a field setting that differs from the laboratory setting in as few ways as possible. When this approach is taken, rates are more comparable than previously suggested. One example of this can be found in the study of Banfield et al. (1995b), where it was found that the rates inferred from retreat of mineral surfaces on dated glaciated surfaces (quartz rate assumed to be zero) generally compare to rates measured in the laboratory. The absence of a discrepancy between laboratory and field rates in this case may be attributed to the simple geometry of the field site (where surface areas are known to within an order of magnitude) and the similarity between field and laboratory geometries (external mineral surfaces exposed to a dilute solution). Another example of laboratory and field rates that agree can be found in Rowe and Brantley (1993). In this study, dissolution rates of andesitic glass, plagioclase, and pyroxene from an andesitic lava and volcano-clastic deposit were calculated from field data. The deposit is exposed to a highly acidic (pH ≈ 2) hydrothermal brine. Field-based estimates of dissolution rates were shown to be generally within one order of magnitude of laboratory rates measured under similar pH and temperatures. Rowe and Brantley (1993) attributed the close agreement between rates measured in the laboratory and field primarily to the low pH. This prevents most secondary mineral formation and considerably simplifies the field calculations. In addition, both the water-rock ratio and the groundwater velocity within this aquifer are high. These aquifer characteristics are closer to laboratory-based dissolution experimental set-ups than is typical.

A word about modeling

Rock-water interaction models, although nearly always based on a number of (over)simplifying assumptions, can have complexities of impressive proportions. These models are built on an experimental or theoretical basis, or more typically on a combination of the two. In the case of weathering, vital model inputs include, but are certainly not limited to, partition coefficients, reactive surface area, flow paths, surface attachment and detachment reactions, and molecular diffusion. In addition, kinetic considerations, dimensional scaling, and temporal scaling (transient versus geologic) are often added to, or built into, these models. Just from a mathematical point-of-view, complexities include having to choose an appropriate dimensionality, weighing the advantages and disadvantages of analytical versus numerical solutions, and being cognizant of computational 'expense'.

Unfortunately, numerical models of natural, open systems, including all rock-water interaction models, can neither be verified or validated (see, e.g. the provocative arguments of Oreskes et al., 1994). If this is the case, what good are models of silicate weathering reactions and kinetics? Will realistic modeling of weathering from the individual mineral scale, to the water-shed scale, and finally to the continental and global scales ever be a reality? Should it be expected to be a reality? Or do attempts at models, which always have a shortcoming, serve most importantly as an organizing of ideas, and to tell us most clearly what we still do not understand? The most obvious example here involves the fact that mineral dissolution rates obtained from laboratory and field data generallly do not agree. Many ideas to explain these discrepancies have been put forth, most of which are worthy of extensive consideration, and some of which may in fact contribute significantly to understanding the problem. Perhaps it is this very process that gives us the most important reason to model, in that these ideas which explain the differences would not have been fully developed (or even considered) without the discrepancy that modeling revealed.

Finally, if one expects to derive a unique model that represents a natural macroscopic observable or process, one must start with microscopic observables, that is directly determine the fundamental process from which a unique model can be derived. If one derives a model from a macroscopic observable (e.g. solution analyses from dissolving minerals), the model by definition must be non-unique for describing a fundamental process. In this case, other models can be found that are just as suitable. Extrapolation (or even interpolation) of these models is rather risky, whereas the same for microscopically-derived models is considerably more sound. This is the best argument for the need to continue to develop microscopic and spectroscopic techniques from which one is able to directly interrogate complex natural processes such as weathering reactions.

ACKNOWLEDGMENTS

The authors are indebted to Art White and Sue Brantley for organizing this volume, and for their assistance and advice throughout this project. As usual, Paul Ribbe somehow accommodated our procrastination, not to mention that *Reviews in Mineralogy* would probably not be in existence without him. MFH wishes to thank current Virginia Tech colleagues Udo Becker, Barbara Bekken, Barry Bickmore, Dirk Bosbach, Jodi Junta-Rosso, Don Rimstidt, and Kevin Rosso for continually sharing and discussing their excellent ideas, and Ian Parsons and Martin Lee for sharing important preprints, Lee and Parsons (1995) and Walker et al. (1995). JFB would like to thank Tony Eggleton for introducing her to mineral weathering as a research topic and for his unique insights, as well as many other collaborators and co-authors, especially David Veblen, Blair Jones, Bill Casey, and Bill Barker. Many thanks also go to George Parks and Pat Brady who supplied us with tabulated silicate PZC values as well as valuable advise on this subject, and to Bill Casey for helping us understand the relationship of his ideas to ours. Research support for this work is as follows: for MFH, National Science Foundation (EAR-910500 and EAR-9305031) and Petroleum Research Fund administered by the American Chemical Society (22892-AC5,2 and 28720-AC2); for JFB, National Science Foundation (EAR-9117386 and EAR-9317082) and Department of Energy (DE-FG02-93ER14328). These funding sources are gratefully acknowledged. This work would not have been possible without them.

REFERENCES

Adamson AW (1982) Physical Chemistry of Surfaces (4th Edn). Wiley, New York, 664 p

Amrhein C and Suarez DL (1992) Some factors affecting the dissolution kinetics of anorthite at 25°C. Geochim Cosmochim Acta 56:1815-1826

Anbeek C (1992) The dependence of dissolution rates on grain size for some fresh and weathered feldspars. Geochim Cosmochim Acta 56:3957-3970

Anbeek C (1993) The effect of natural weathering on dissolution rates. Geochim Cosmochim Acta 57:4963-4975

Anderson DM, Low PF (1958) The density of water adsorbed by lithium-, sodium-, and potassium bentonite. Soil Sci Soc Am Proc 22:99-103

Banfield JF, Bailey SW (in press) Evidence for formation of regularly interstratified serpentine-chlorite minerals by tetrahedral inversion in long-period serpentine polytypes. Am Mineral

Banfield JF, Barker WW (1994) Direct observation of reactant-product interfaces formed in natural weathering of exsolved defective amphibole to smectite: Evidence of episodic, isovolumetric reactions involving structural inheritance. Geochim Cosmochim Acta 58:1419-1429

Banfield JF, Eggleton RA (1988) A transmission electron microscope study of biotite weathering. Clays Clay Minerals 36:47-60

Banfield JF, Eggleton RA (1990) Analytical transmission electron microscope studies of plagioclase, muscovite and K-feldspar weathering. Clays Clay Minerals 38:77-89

Banfield JF, Bailey SW, Barker WW (1994) Polysomatism, polytypism, defect microstructures, and reaction mechanisms in regularly and randomly interstratified serpentine and chlorite. Contrib Mineral Petrol 117:137-150

Banfield JF, Bischoff BL, Anderson MA (1993) TiO$_2$ accessory minerals: Coarsening and transformation kinetics in pure and doped synthetic nanocrystalline materials. Chem Geol 110:211-231.

Banfield JF, Jones BJ, Veblen DR (1991) An AEM-TEM study of weathering and diagenesis, Albert Lake, Oregon (I) Weathering reactions in the volcanics. Geochim Cosmochim Acta 55:2781-2793

Banfield JF, Veblen DR, Jones BF (1990) Transmission electron microscopy of subsolidus oxidation and weathering of olivine. Contrib Mineral Petrol 106:110-123

Banfield JF, Ferruzzi GG, Casey WH, Westrich HR (1995a) HRTEM study comparing naturally and experimentally weathered pyroxenoids. Geochim Cosmochim Acta 59:19-31

Banfield JF, Hemb SB, Casey WH (1995b) Weathering of Ca-Mg-Fe pyroxenes in the laboratory and field and the nature of clay-pyroxene interfaces. V.M. Goldschmidt Conference, Pennsylvania State Univeristy, Program and Abstracts, p. 28

Banfield JF et al. (in prep.) Weathering of Ca-Mg-Fe pyroxenes in the laboratory and field and the nature of clay-pyroxene interfaces. Geochim Cosmochim Acta

Barker WW, Banfield JF (1995a) Biochemical impact on rates and mechanisms of silicate mineral weathering. Geol Soc Am Annual Mtg, New Orleans, LA, in press

Barker WW, Banfield JF (1995b) Biologically- versus inorganically-mediated weathering reactions: Relationships between minerals and extracellular microbial polymers in lithobiontic communities. Chem Geol (in press)

Barker WW, Banfield JF (in prep.) Structure of lithobiontic communities in intimate contact with silicate minerals. Microbial Ecology

Berner RA (1978) Rate control of mineral dissolution under earth surface conditions. Am J Sci 278:1235-1252

Berner RA (1981) Kinetics of weathering and diagenesis. In: Lasaga AC, Kirkpatrick RJ (eds) Kinetics of Geochemical Processes. Rev Mineral 8:111-134

Berner RA, Schott J (1982) Mechanism of pyroxene and amphibole weathering II. Observations of soil grains. Am J Sci 282:1214-1231

Bigg EK (1953) The supercooling of water. Proc Phys Soc (London) 66B:688-694

Blum AE (1994) Feldspars in weathering. In: Parsons I (ed) Feldspars and Their Reactions. NATO Advanced Study Inst, Series C. Kluwer, Dordrecht, Netherlands, 421:595-630

Blum AE, Lasaga AC (1988) Role of surface speciation in the low-temperature dissolution of minerals. Nature 4:431-433

Blum AE, Yund RA, Lasaga AC (1990) The effect of dislocation density on the dissolution rate of quartz. Geochim Cosmochim Acta 54:283-297

Bolt GH (1979) Soil Chemistry. B: Physico-Chemical Models. Elsevier, Amsterdam

Brand U, Veizer J (1980) Chemical diagenesis of a multicomponent carbonate system - 1: Trace elements. J Sed Petrol 50:1219-1236

Brantley SA, Crane SR, Crerar DA, Hellmann R, Stallard R (1986) Dissolution of dislocation etch pits in quartz. Geochim Cosmochim Acta 50:2349-2361

Brindley GW, Brown G (1980) Crystal Structures of Clay Minerals and Their X-ray Identification. Mineralogical Society, London

Brown ZZ, Kevan ZZ (1988) Aqueous coordination and location of Cu^{2+} cations in montmorillonite clay studied by electron spin resonance and electron spin-echo modulation. J Am Chem Soc 110:2743-2748

Brunauer S, Emmett PH, Teller E (1938) Adsorption of gases in multimolecular layers. J Am Chem Soc 60:309-319

Burch TE, Nagy KL, Lasaga AC (1993) Free energy dependence of albite dissolution kinetics at 80°C and pH 8.8. Chem Geol 105:137-162

Buseck PR (ed) (1993) Minerals and Reactions at the Atomic Scale: Transmission Electron Microscopy. Rev Mineral 27. Mineral Soc Am, Washington, DC, 508 p

Calvet R (1975) Dielectric properties of montmorillonites saturated by bivalent cations. Clays Clay Minerals 23:257-265

Casey WH, Westrich HR (1992) Control of dissolution rates of orthosilicate minerals by divalent metal-oxygen bonds. Nature 355:157-159

Casey WH, Carr MJ, Graham RA (1988a) Crystal defects and the dissolution kinetics of rutile. Geochim Cosmochim Acta 52:1545-1556

Casey WH, Westrich HR, Arnold GW (1988b) Surface chemistry of labradorite feldspar reacted with aqueous solutions at pH = 2, 3, and 12. Geochim Cosmochim Acta 53, 821-832

Casey WH, Banfield JF, Westrich HR, McLaughlin L. (1993) What do dissolution experiments tell us about natural weathering? Chem Geol 105:1-15

Clementz DM, Low PF (1976) Thermal expansion of interlayer water in clay systems. I. Effect of water content. Colloid Interface Sci 3:485-502

Cruz MI, Letellier M, Fripiat JJ (1978) NMR study of adsorbed water. II. Molecular motions in the monolayer hydrate of halloysite. J Chem Phys 69:2018-2027

Daubrée M (1867) Expériences sur les déscompositions chimiques provoquées par les actions mécaniques dans divers minéraux tels que le feldspath. Comptes Rendus 55:339-345

Davies CW (1962) Ion Association. Butterworths, London

Davis JA, Kent DB (1990) Surface complexation modeling in aqueous geochemistry. In: Hochella MF Jr, White AF (eds) Mineral-Water Interface Geochemistry. Rev Mineral 23:177-260

Delville A, Grandjean J, Laszlo P (1991) Order acquisition by clay platelets in a magnetic field. NMR study of the structure and microdynamics of the adsorbed water layer. J Phys Chem 95:1383-1392

Derjaguin BV, Churaev NV (1986) Properties of water layers adjacent to interfaces. In: Croxton CA (ed) Fluid Interfacial Phenomena. Wiley, New York, p 663-738

DeSmedt F, Wierenga PJ (1979a) Mass transfer in porous media with immobile water. J Hydrol 41:59-67

DeSmedt F, Wierenga PJ (1979b) A generalized solution for solute flow in soil with mobile and immobile water. Water Resources Res 15:1137-1141

Devell L (1962) Measurements of the self-diffusion of water in pure water, H_2O-D_2O mixtures and solutions of electrolytes. Acta Chem Scand 16:2177-2188

Dibble WE Jr, Tiller WA (1981) Non-equilibrium water/rock interactions - I. Model for interface-controlled reactions. Geochim Cosmochim Acta 45, 79-92

Du Q, Freysz E, Shen YR (1994) Vibrational spectra of water molecules at quartz/water interfaces. Phys Rev Lett 72:238-241

Ebelmen JJ (1847) Recherches sur la décomposition des roches. Annal des Mines 12:627-654

Eggleston CM, Hochella MF Jr, Parks GA (1989) Sample preparation and aging effects on the dissolution rate and surface composition of diopside. Geochim Cosmochim Acta 53, 979-804

Eggleton RA (1984) Formation of iddingsite rims on olivine: A transmission electron microscope study. Clays Clay Minerals 32:1-11

Eggleton RA (1986) The relation between crystal structure and silicate weathering rates. In: Colman SM, Dethier DP (eds): Rates of chemical weathering of rocks and minerals. Academic Press, Orlando, Florida, p 21-40

Eggleton RA, Buseck PR (1980) High resolution electron microscopy of feldspar weathering. Clays Clay Minerals 28:173-178

Enustun BV, Turkevich J (1960) Solubility of fine particles of strontium sulfate. J Am Chem Soc 82:4502-4509

Farmer VC (1978) Water on particle surfaces. In: Greenland DJ, Hayes MHB (eds) Chemistry of Soil Constituents. Wiley, New York

Ferry, JM (1985) Hydrothermal alteration of Tertiary igneous rocks from the Isle of Skye, northwest Scotland. II. Granites. Contrib Mineral Petrol 91:283-304

Fitz Gerald JD, Harrison TM (1993) Argon diffusion domains in K-feldspar I: Microstructures in MH-10. Contrib Mineral Petrol 113:367-380

Folk RL (1955) Note on the significance of "turbid" feldspars. Am Mineral 40:356-357

Fripiat JJ (1976) The NMR study of proton exchange between adsorbed species and oxides and silicate surface. In: Resing HA, Wade CG (eds) Magnetic Resonance in Colloid and Interface Science. Am Chem Soc, Washington, DC

Fripiat JJ, Cases J, Francois M, Letellier M (1982) Thermodynamic and microdynamic behavior of water in clay suspensions and gels. J Colloid Interface Sci 89:378-400

Fripiat JJ, Jelli A, Poncelet G, Andre J (1965) Thermodynamic properties of adsorbed water molecules and electrical conduction in montmorillonites and silicas. J Phys Chem 69:2185-2197

Garrels RM, Dreyer RM, Hewland AL (1949) Diffusion of ions through intergranular spaces in water-saturated rocks. Geol Soc Am Bull 60:1809-1828

Gaudet JP, Jegat H, Vachaud C, Wierenga PJ (1977) Solute transfer, with exchange between mobil and stagnant water, through unsaturated sand. Soil Sci Soc Am J 41:665-671

Gaudin AM, Rizo-Patron A (1942) The mechanism of activation in flotation. Am Inst Mining Metal Engineers Tech Pub 1453:1-9

Germann PF (1990) Preferential flow and the generation of runoff, 1. Boundary layer flow theory. Water Resources Res 26:3055-3063

Goldich SS (1938) A study of rock weathering. J Geol 46:17-58

Gordon AR (1945) The diaphragm cellmethod of studying diffusion. NY Acad Sci Ann 46:285-308

Guthrie GD, Veblen DR (1991) Turbid alkali fledspars from the Isle of Skye, northwest Scotland. Contrib Mineral Petrol 108:298-304

Guyot-Sionnest P, Hunt JH, Shen YR (1988) Sum-frequency vibrational spectroscopy of a Langmuir film: Study of molecular orientation of a two-dimensional system. Phys Rev Lett 59:1597-1600

Guyot-Sionnest P, Superfine R, Hunt JH, Shen YR (1991) Virbrational spectroscopy of a silane monolayer at air/solid and liquid/solid interfaces using sum-frequency generation. Chem Phys Lett 144:1-5

Halberstadt ES, Henisch HK, Nickl J, White EW (1969) Gel structure and crystal nucleation. J Colloid Interface Sci 29:469-471

Helgeson HC, Murphy WM, Aagard P (1984) Thermodynamic and kinetic constraints on reaction rates among minerals and aqueous solutions, II. Rate constants, effective surface area, and the hydrolysis of feldspar. Geochim Cosmochim Acta 48:2405-2432

Henisch HK (1989) Crystals in Gels and Liesegang Rings. Cambridge Univ Press, Cambridge, UK

Hochella MF Jr (1990) Atomic structure, microtopography, composition, and reactivity of mineral surfaces. In: Hochella MF Jr, White AF (eds) Mineral-Water Interface Geochemistry. Rev Mineral 23:87-132

Hochella MF Jr (1994) Mineral surfaces: Their characterization and their chemical, physical, and reactive nature. In: Vaughan DJ, Pattrick RAD (eds) Mineral Surfaces. Mineral Soc Series 5:17-60. Chapman and Hall, London

Hochella MF Jr, White AF (1990) Mineral-water interface geochemistry: An overview. In: Hochella MF Jr, White AF (eds) Mineral-Water Interface Geochemistry. Rev Mineral 23:87-132

Hochella MF Jr, Lindsey JR, Mossotti VG, Eggleston CM (1988b) Sputter depth profiling in mineral-surface analysis. Am Mineral 73:1449-1456

Hochella MF Jr, Ponader HB, Turner AM, Harris DW (1988a) The complexity of mineral dissolution as viewed by high resolution scanning Auger microscopy: Labradorite under hydrothermal conditions. Geochim Cosmochim Acta 52:385-394

Holdren GR Jr, Speyer PM (1985) Reaction rate-surface area relationships during the early stages of weathering: I. Initial observations. Geochim Cosmochim Acta 49:675-681

Holdren GR Jr, Speyer PM (1987) Reaction rate-surface area relationships during the early stages of weathering: II. Data on eight additional feldspars. Geochim Cosmochim Acta 51:2311-2318

Inskeep WP, Nater EA, Bloom PR, Vandervoort DS, Erich MS (1991) Characterization of laboratory weathered labradorite surfaces using X-ray photoelectron spectroscopy and transmission electron microscopy. Geochim Cosmochim Acta 55:787-800

Iwasawa Y, Mason R, Textor M, Somorjai GA (1976) The reactions of carbon monoxide at coordinatively unsaturated sites on a platinum surface. Chem Phys Lett 44:468-470

Johnston CT, Sposito G, Erickson C (1992) Vibrational probe studies of water interactions with montmorillonite. Clays Clay Minerals 40, 722-730

Jones JR, Rowlands DLG, Monk CB (1965) Diffusion coefficient of water in water and in some alkaline earth chloride solutions at 25°C. Trans Faraday Soc:61:1384-1388

Jones RC, Uehara G (1973) Amorphous coatings on mineral surfaces. Soil Sci Soc Am Proc 37:792-798

Junta JL, Hochella MF Jr (1994) Manganese (II) oxidation at mineral surfaces: A microscopic and spectroscopic study. Geochim Cosmochim Acta 58:4985-4999

Junta-Rosso JL (1995) Precipitation at the Mineral-Water Interface: Linking Microscopic to Macroscopic Processes. PhD dissertation, Stanford Univ, Stanford, CA, 164 p

Jurinak JJ, Volman DH (1961) Cation hydration effects on the thermodynamics of water adsorption by kaolinite. J Phys Chem 65:1853-1856

Kay BD, Low PF (1975) Heats of compression of clay-water mixtures. Clays Clay Minerals 23:266-271

Klier K, Zettlemoyer AC (1977) Water at interfaces: Molecular structure and dynamics. J Colloid Interface Sci 58:216-229

Kogelbauer A, Lercher JA, Steinberg KH, Roessner F, Soellner A, Dmitriev RV (1989) Type, stability, and acidity of hydroxyl groups of HNaK-erionites. Zeolites 9:224-230

Kohl RA, Cary JW, Taylor SA (1964) On the interaction of water with a Li-kaolinite surface. J Colloid Sci 19, 699

Kubota N, Fujisawa Y, Tadaki T (1988) Effect of volume on the supercooling temperature for primary nucleation of potassium nitrate from aqueous solution. J Crystal Growth, 89:545-552

Kung, KJS (1990) Preferential flow in a sandy vadose zone. 1. Field observations. Geoderma 46:51-58

Langmuir D (1971) Particle size effect on the reaction goethite = hematite + water. Am J Sci 271:147-156

Lapidus L, Amundson NR (1952) The rate-determining steps in radial adsorption analysis. J Phys Chem 56:373-383

Lasaga AC, Blum AE (1986) Surface chemistry, etch pits, and mineral-water reactions. Geochim Cosmochim Acta 50:2363-2379

Lee MR, Parsons I (1995) Microtextural controls of weathering of perthitic alkali feldspars. Geochim Cosmochim Acta (in press)

Lee MR, Waldron K, Parsons I (1995) Exsolution and alteration microtextures in alkali feldspar phenocrysts from the Shap granite. Mineral Mag 59:63-78

Lerman A (1980) Geochemical Processes. John Wiley & Sons, New York, 481 p

Low PF (1968) Observations on activity and diffusion coefficients in Na-montmorillonite. Isr J Chem 6:325-336

Low PF (1976) Viscosity of Interlayer Water in Montmorillonite. Soil Sci Soc Am J 40:500-504

Low PF (1979) Nature and properties of water in montmorillonite-water systems. Soil Sci Soc Am J 43:651-658

Lowell S, Shields JE (1991) Powder Surface Area and Porosity (3rd Edn). Chapman and Hall, London, 250 p

Matsui I (1973) Catalysis and kinetics of manganous ion oxidation in aqueous solution and adsorbed on the surfaces of solid oxides. PhD disseration, Lehigh Univ, Bethlehem, PA

Melia TP, Moffitt WP (1964) Crystallization from aqueous solution. J Colloid Sci 19:433-447

McBride MB (1982) Hydrolysis and dehydration reactions of exchangeable Cu^{2+} on hectorite. Clays Clay Minerals 30:200-206

McBride MB (1989) Surface chemistry of soil minerals. In: Dixon JB, Weed SB (eds) Minerals in Soil Environments (2nd Edn). Soil Sci Soc Am, Madison, Wisconsin, p 35-88

McBride MB, Pinnavaia TJ, Mortland MM (1975) Electron spin resonance studies of cation orientation in restricted water layers on phyllosilicate (smectite) surfaces. J Phys Chem 79:2430-2435

Michalske TA, Fuller ER (1985) Closure and repropagation of healed cracks in silicate glass. J Am Ceram Soc 68:586-590

Montgomery CW, Brace WF (1975) Micropores in plagioclase. Contrib Mineral Petrol 52:17-28

Mooney RW, Keenan AG, Wood LA (1952a) Adsorption of water vapor by montmorillonite. I. Heat of desorption and application of BET theory. J Am Chem Soc 74:1367-1374

Mooney RW, Keenan AG, Wood LA (1952b) Adsorption of water vapor by montmorillonite. II. Effect of exchangeable ions and lattice swelling as measured by X-ray diffraction. J Am Chem Soc 74:1371-1374

Mortland MM, Raman KV (1968) Surface acidities of smectites in relation to hydration, exchangeable-cation and structure. Clays Clay Minerals 16:393-398

Murphy WM (1989) Dislocations and feldspar dissolution. European J Mineral 1:315-326

Nagy KL, Lasaga AC (1992) Dissolution and precipitation kinetics of gibbsite at 80°C and pH 3: The dependence on solution saturation state. Geochim Cosmochim Acta 56:3093-3111

Nkedi-Kizza P, Biggar JW, van Genuchten MTh, Wierenga PJ, Selim HM, Davidson JM, Nielsen DR (1983) Modeling tritium and chloride 36 transport through an aggregated oxisol. Water Resources Res 19, 691-700

Oelkers EH, Schott J, Devidal JL (1994) The effect of aluminum, pH, and chemical affinity on the rates of aluminosilicate dissolution reactions. Geochim Cosmochim Acta 58:2011-2024

Oreskes N, Shrader-Frechette K, Belitz K (1994) Verification, validation, and confirmation of numerical models in the earth sciences. Science 263:641-646

Oster JD, Low PF (1964) Heat capacities of clay and clay-water mixtures. Soil Sci Soc Am Proc 28:605-609

Page R, Wenk HR (1979) Phyllosilicate alteration of plagioclase studied by transmission electron microscopy. Geology 7:393-397

Parks GA (1984) Surface and interfacial free energies of quartz. J Geophys Res 89:3997-4008

Parks GA (1990) Surface energy and adsorption at mineral/water interfaces: An introduction. In: Hochella MF Jr, White AF (eds) Mineral-Water Interface Geochemistry. Rev Mineral 23:133-175

Parsons I (1978) Feldspars and fluids in cooling plutons. Mineral Mag 42:1-17

Parsons I (1980) Alkali-feldspar and Fe-Ti-oxide exsolution textures as indicators of the distribution of subsolidus effects of magmatic 'water' in the Klokken layered syenite intrusion, South Greenland. Trans Royal Soc Edinburgh Earth Sci 71:1-12

Petrovich R (1981) Kinetics of dissolution of mechanically comminuted rock-forming oxides and silicates - II. Deformation and dissolution of oxides and silicates in the laboratory and at the Earth's surface. Geochim Cosmochim Acta 45:1675-1686

Pingitore NE Jr (1982) The role of diffusion during carbonate diagenesis. J Sed Petrol 52:27-39

Pittman ED, Lewan MD (1994) Organic Acids in Geological Processes. Springer-Verlag, Berlin, 482 p

Poinsignon C, Cases JM, Fripiat JJ (1978) Electrical polarization of water molecules adsorbed by smectites: An infrared study. J Phys Chem 82:1855-1860

Putnis A, Prieto M, Fernandez-Diaz L (1995) Fluid supersaturation and crystallization in porous media. Geol Mag 132:1-13

Ravina I, Low PF (1972) Relation between swelling, water properties and b-dimension in montmorillonite-water systems. Clays Clay Minerals 20:109-123

Ribbe PH (1983) Exsolution textures in ternary and plagioclase feldspars: Interference colors. In: Ribbe PH (ed) Feldspar Mineralogy. Rev Mineral 2:241-270

Rowe GL Jr, Brantley SL (1993) Estimation of the dissolution rates of andesitic glass, plagioclase and pyroxene in a flank aquifer of Poás Volcano, Costa Rica. Chem Geol 105, 71-87

Ruiz HA, Low PF (1976) Thermal expansion of interlayer water in clay systems. II. Effect of clay composition. Colloid Interface Sci 3:503-515

Russell JD, Farmer VC (1964) Infra-red spectroscopic study of the dehydration of montmorillonite and saponite. Clay Mineral Bull 5:443-464

Schnoor JL (1990) Kinetics of chemical weathering: A comparison of laboratory and field weathering rates. In: Stumm W (ed) Aquatic Chemical Kinetics. Wiley, New York, p 475-504

Schott J, Brantley S, Crerar D, Guy C, Borcsik M, Willaime C (1989) Dissolution kinetics of strained calcite. Geochim Cosmochim Acta 53:373-382

Shields JE, Lowell S (1983) A method for the estimation of micropore volume and micropore surface area. Powder Technol 36:1-4

Smith JV (1983) Phase equilibria of plagioclase. In: Ribbe PH (ed) Feldspar Mineralogy (2nd Edn) Rev Mineral 2:223-239

Smith JV (1994) Surface chemistry of feldspars. In: Parsons I (ed) Feldspars and Their Reactions. NATO Advanced Study Inst, Series C. Kluwer, Dordrecht, Netherlands, 421:541-593

Smith KL, Milnes AR, Eggleton RA (1987) Weathering of basalt: Formation of iddingsite. Clays and Clay Minerals 35:418-428

Sollins P, Radulovich R (1988) Effects of soil physical structure on solute transport in a weathered tropical soil. Soil Sci Soc Am J 52:1168-1173

Sposito G (1984) The Surface Chemistry of Soils. Oxford University Press, Oxford, UK, 234 p

Sposito G (1986) Distinguishing adsorption from surface precipitation. In: Davis JA, Hayes KF (eds) Geochemical Processes at Mineral Surfaces. ACS Symp Series 323:217-228. Am Chem Soc, Washington, DC

Sposito G, Prost R (1982) Structure of water adsorbed on smectite. Chem Rev 82:553-573

Stavrinidis B, Holloway DG (1983) Crack healing in glass. Phys Chem Glasses 24:19-25

Sverjensky DA (1994) Zero-point-of-charge prediction from crystal chemistry and solvation theory. Geochim Cosmochim Acta 58:3123-3129

Swoboda-Colberg NG, Drever JI (1993) Mineral dissolution rates in plot-scale field and laboratory experiments. Chem Geol 105:51-69

van Genuchten MTh, Wierenga PJ (1976) Mass transfer studies in sorbing porous media, 1, Analytical solutions. Soil Sci Soc Am J 40:473-480

Vaughan DJ, Pattrick RAD (eds) (1995) Mineral Surfaces. Chapman and Hall, London, 370 p

Velbel MA (1985) Geochemical mass balances and weathering rates in forested watersheds of the southern Blue Ridge. Am J Sci 285, 904-930

Velbel MA (1992) Geochemical mass balances and weathering rates in forested watersheds of the southern Blue Ridge. III. Cation budgets and the weathering rate of amphibole. Am J Sci 292:58-78

Velbel MA (1993a) Formation of protective surface layers during silicate-mineral weathering under well-leached, oxidizing conditions. Am Mineral 78:405-414

Velbel MA (1993b) Constancy of silicate-mineral weathering-rate ratios between natural and experimental weathering: Implications for hydrologic control of differences in absolute rates. Chem Geol 105:89-99

Waldron KA, Parsons I (1992) Feldspar microtextures and multistage thermal history of syenites from the Coldwell Complex, Ontario. Contrib Mineral Petrol 111:222-234

Walker FDL (1990) Ion microprobe study of intragrain micropermeability in alkali feldspars. Contrib Mineral Petrol 106:124-128

Walker FDL, Parsons I, Lee MR (1995) Micropores and micropermeable texture in alkali feldspars: Geochemical and geophysical implications. Mineral Mag 59:507-536

Walther JV, Wood BJ (1986) Mineral-fluid reaction rates. In: Walther JV, Wood BJ (eds) Fluid-Rock Interactions during Metamorphism. Adv Phys Geochem 5:194-211

Wang JH (1954) Effect of ions on the self-diffusion and structure of water in aqueous electrolyte solutions. J Phys Chem 58:686-692

Westall J, Hohl H (1980) A comparison of electrostatic models for the oxide/solution interface. Adv Colloid Interface Sci 12:265-294

White AF, Peterson ML (1990) Role of reactive-surface-area characterization in geochemical kinetic models. In: Melchior DC, Bassett RL (eds) Chemical Modeling of Aqueous Systems II. Am Chem Soc Symp Ser No. 416, 35:461-475

White AF, Blum AE, Schulz MS, Bullen TD, Harden JW, Peterson, M (1995) Chemical weathering of a soil chronosequence on granite alluvium I. Reaction rates based on changes in soil mineralogy. Geochim Cosmochim Acta (in press)

Wilson MJ, McHardy WJ (1980) Experimental etching of a microcline perthite and implications regarding natural weathering. J Microscopy 120:291-302

Wilson MJ, Bain DC, McHardy WJ (1971) Clay mineral formation in deeply weathered boudler conglomerate in north-east Scotland. Clays Clay Minerals 8:435-444

Wollast R, Chou L (1985) Kinetic study of the dissolution of albite with a continuous flow-through fluidized bed reactor. In: Drever JI (ed) The Chemistry of Weathering. D. Reidel, Dordrecht, Netherlands, p 75-96

Worden RH, Walker FDL, Parsons I, Brown WL (1990) Development of microporosity, diffusion channels and deuteric coarsening in perthitic alkali feldspars. Contrib Mineral Petrol 104:507-515

Yariv S, Shoval S (1975) The nature of the interaction between water molecules and kaolin-like layers in hydrated halloysite. Clays Clay Minerals 23:473-474

Chapter 9

CHEMICAL WEATHERING RATES OF SILICATE MINERALS IN SOILS

Art F. White

U. S. Geological Survey
Menlo Park, CA 94025 U.S.A.

INTRODUCTION

Soil can be defined as "a natural body consisting of generally unconsolidated layers or horizons of mineral and/or organic constituents of variable thickness which differ from parent rock in morphological, physical, chemical and mineralogical properties" (Joffe, 1949). As such, the study of soils have long been tied to the nature and extent of chemical weathering. The literature on soil chemical weathering related to secondary mineralogy, chemical equilibrium and ion exchange, in addition to the effects on geomorphology and other physical processes, has been reviewed in a number of books (Lindsay, 1979; Nahon, 1991; Sparks and Suarez, 1991; and Sposito, 1994). The present review discusses the rates of chemical weathering of primary silicate minerals in soil environments. Although soils represent one of the most accessible natural environments, relatively few studies have quantitatively addressed weathering rates of feldspars, amphiboles, pyroxenes, micas and other silicate minerals under such conditions. Estimating weathering rates of silicate minerals in soils is important because such reactions ultimately control rates of soil development and secondary mineral formation. Weathering rates also influence soil buffering capacities related to acidic deposition in watersheds and control the cycling of many inorganic nutrients which are important in soil fertility and carbon cycling.

The chemical weathering of a primary mineral j in a soil can be defined in simplest terms as

$$M_j = k_j \ S \ \Delta t \tag{1}$$

were M_j is the mass loss (mol), k_j is the rate constant (mol cm^{-2} s^{-1}), S is the surface area (cm^2) and Δt the time period (s) during which weathering occurred. These terms, as well others used throughout the paper, are tabulated in Table 1. In weathering studies of natural systems, Equation (1) can be employed as a predictive tool to determine any one of the above parameters provided that the other terms are known. The present paper will discuss methods used to determine mass losses, surface areas and duration of weathering, which in turn, will permit calculation of weathering rate constants for soil environments.

Chemical weathering of primary silicates in soils will be addressed in the present paper from two perspectives. These will be (a) rates determined from solid state element and mineral losses relative to initial or parent material and (b) rates determined from solute fluxes through the soil profile. These parallel approaches require the determination of mass loss M_j based on either mineral or solute compositions. Soil mineralogy represents the residual product of chemical reactions which integrate the weathering rate over the entire period of soil development. In contrast, solute chemistry and fluxes reflect present day weathering under current chemical and hydrologic conditions. Both approaches require the determination of the duration of weathering. In the case of mineral or elemental loses, duration of

Table 1. Summary of parameters referred to in text[1].

Symbol	Definition	Units
a	Rate of etch pit annihilation	s^{-1}
$C_{j,w}$	Concentration of weatherable mineral or element in soil profile	$g\ g^{-1}$
$C_{j,p}$	Concentration of weatherable mineral or element in protolith	$g\ g^{-1}$
$C_{i,w}$	Concentration of inert mineral or element in soil profile	$g\ g^{-1}$
$C_{i,p}$	Concentration of inert mineral or element in protolith	$g\ g^{-1}$
C_{qtz}	Concentration of quartz in soil	$g\ g^{-1}$
C^*	Concentration of aluminosilicates in soil	$g\ g^{-1}$
c_k^t	Concentration of solute species k contributed by dissolution	$mol\ cm^{-3}$
c_k	Solute concentration measured in soil solution	$mol\ cm^{-3}$
c_k^s	Solute concentration incorporated into secondary phases	$mol\ cm^{-3}$
c_k^a	Solute concentration contributed from non-weathering sources	$mol\ cm^{-3}$
D	Mineral grain diameter	cm
E_a	Activation energy	$kJ\ mol^{-1}$
ET	Evapotranspiration	cm
f_m	Mass fraction of mineral in soil	$g\ g^{-1}$
f^ϕ	Mass fraction of mineral in size fraction ϕ	$g\ g^{-1}$
G	Etch pit growth rate	$cm\ s^{-1}$
h_g	Hydraulic head due to gravity flow	$cm\ H_2O$
h_p	Hydraulic head related to matrix potential	$cm\ H_2O$
IAP	Aqueous activity product for soil solution	
K_m	Unsaturated zone hydraulic conductivity	$cm\ s^{-1}$
K	Solubility constant of mineral phase	
k_j	Weathering rate constant for mineral phase j	$mol\ cm^{-2}.s^{-1}$
M	Mass of soil	g
M_j	Mass of mineral phase reacted	mol
$M^\phi_{j,w}$	Mass of residual mineral in soil	g
$M^\phi_{j,p}$	Mass of mineral in protolith	g
$m_{j,flux}$	Mass flux of mineral weathered	$mol\ cm^{-2}$
m	Moisture content of soil	$cm^3 cm^{-3}$
n	Number of etch pits	$cm\ cm^{-2}$
$N_{j,k}$	Stoichiometric ratio of species k in mineral phase j	$mol\ mol^{-1}$
P	Precipitation	cm
Q_w	Watershed mass flux	$g\ m^{-2}s^{-1}$
Q_j	Flux of mineral j	$mol\ cm^2.s$
Q_k	Flux of solute k	$mol\ cm^2.s$
q	Soil water flux	$cm\ s^{-1}$
R	Gas constant	
R_o	Ratio of conservative and weatherable component in soil and parent material	$g\ g^{-1}$
R_s	Ratio of unit soil surface area to total surface area of mineral in soil	$cm^2\ cm^{-2}$
r^o	Initial particle radius	cm
r	Final particle radius resulting from weathering	cm
S	Specific mineral surface area	$cm^2 g^{-1}$
S_i	Internal surface area due to porosity	$cm^2 g^{-1}$
S_s	Unit surface area of soil perpendicular to flow	cm^2
S_t	Total surface area of mineral contained in volume of soil	cm^2
s	Geometric mineral surface area	$cm^2 g^{-1}$

T	Temperature	°K
V_p	Volume of protolith	cm^3
V_w	Volume of soil	cm^3
V_j	Molar volume of mineral phase j	cm^3 mol^{-1}
v_{sp}	Volume fraction of saprolite	cm^3 cm^{-3}
v_i	Volume fraction of inert minerals in soil	cm^3 cm^{-3}
v_{sm}	Volume fraction of secondary minerals produced in soil	cm^3 cm^{-3}
v_p	Volume fraction of soil attributed to porosity	cm^3 cm^{-3}
W	Etch pit diameter	cm
$x_{qtz,w}$	Mass fraction of quartz in soil	g g^{-1}
$x_{j,w}$	Mass fraction of weatherable mineral in soil	g g^{-1}
$x_{qtz,p}$	Mass fraction of quartz in protolith	g g^{-1}
$x_{j,p}$	Mass fraction of weatherable mineral in protolith	g g^{-1}
z	Depth of weathering	cm
α	Mineral sphericity	cm^3 cm^{-3}
ΔG_r	Net free energy of reaction	kJ mol^{-1}
Δt	Time span over which weathering occurs	s
$\epsilon_{j,w}$	Volumetric strain	cm^{3}cm^{-3}
∇H	Hydraulic gradient	cm H$_2$O cm^{-1}
λ	Mineral surface roughness	cm^2 cm^{-2}
λ^*	Aluminosilicate surface roughness	cm^2 cm^{-2}
λ_{qtz}	Quartz surface roughness	cm^2 cm^{-2}
ρ_p	Density of protolith	g cm^{-3}
ρ_w	Density of weathered soil	g cm^{-3}
ρ_j	Density of mineral	g cm^{-3}
τ	Shear modulus	
$\tau_{j,w}$	Transport function describing mass loss from weathering	g g^{-1}
ω_w	Weathering velocity	cm s^{-1}
ϕ	Particle size fraction	
θ	Etch pit slope	degree
φ_m	Mineral molecular weight	mol g^{-1}

[1]Note: To prevent duplication of terms, some original symbols used in cited literature have been changed.

weathering equates to the age of the soil profile. For solutes, the resident time is equivalent to fluid transit time through the soil. Both approaches normalize chemical fluxes to the unit surface area of the mineral phase. A detailed discussion will follow on the definition and role of surface area in weathering rate calculations. Finally, the derived rate constants k_j, based on Equation (1), will be compared for various natural soils and experimental studies. Reasons for observed variations will then be addressed.

WEATHERING ENVIRONMENTS BASED ON SOIL CLASSIFICATION

Weathering rates of primary silicates are strongly influenced by specific soil environments. Due to the complexity of soil weathering and the need to qualitatively describe and compare soil distributions, much effort in soil science has been devoted to methods of soil classification. From the standpoint of weathering, three aspects of soil classifications are relevant.

Soil Profiles: Physical descriptions of soil weathering are dependent on the spatial position or depth in the soil column. Studies often distinguish either explicitly or more often implicitly between a soil profile and a weathering profile (Birkeland, 1984). Where this is done, the soil profile is generally considered to make up the upper part of a much thicker weathering profile. A more inclusive definition of a soil profile is the vertical section that includes all layers that have been pedogenically altered during chemical weathering. Soil horizons defined within such profiles clearly recognize the vertical transition in the degree of weathering from the intensively weathered A and E horizons through the moderately weathered B, to the relatively unaltered C horizons to bedrock (Buol et al., 1989). Such an approach encompasses the entire span of chemical weathering occurring at the earth's surface including both the distribution of primary silicates as well as secondary clay and oxyhydroxide minerals.

Soil Classes: Most of soil science literature is concerned with the upper regions of the soil profile often termed the soil solum. The exact lower boundary of the soil solum is not always defined but is generally shallow (25 to 100 cm) and is the soil profile most influenced by plant roots (Buol and Weed, 1991). Such an emphasis on shallow soil horizons clearly reflects the overriding importance of biological activity in most soil studies as well as the amenability of shallow soils to soil surveys. Modern soil mineral classifications for the major soil orders are based on the description of this shallow zone (Buol et al., 1989). Soil family mineralogy classes are principally oriented toward the distribution of secondary oxyhydroxides and clays in the heavily weathered shallow solum and do not generally include details of the primary silicate distributions more prevalent in deeper soils nor the composition of the underlying bedrock.

Soil Sequences: Soil science attempts to define changes in soil characteristics in terms of specific processes or influences relevant to weathering. Jenny (1980) defined any dependent soil property as a function of a number of independent soil properties or state factors such that

$$Soil\ Property\ =\ f\ (pt,\ cl,\ tp,\ t,\ org) \qquad\qquad (2)$$

where *pt* is the parent material, *cl* is climate, *tp* is the topography, *t* is the age of the soil and *org* is the role of organisms. Soil sequences can be classified based on a single variable if other conditions effecting soil development are subordinate or relatively constant. Such sequences can be defined as follows:

Lithosequences are series of soils developed on different parent materials or rock types but under similar conditions for other state factors. Differences in parent mineralogy may be the dominant factor determining weathering rates in these soil (Dixon and Weed, 1989). As originally proposed by Goldich (1938), the relative reactivity of minerals decreases in the order: carbonates > mafic silicates > feldspars > quartz. Therefore, the proportions of these primary minerals in a soil profile will strongly influence absolute weathering rates.

Climosequences are soils with similar properties that are influenced principally by differences in precipitation and temperature. An example would include different weathering rates of continuous volcanic ash layers in soils deposited under extreme differences in precipitation conditions on Hawaii (Johnsson et al., 1993).

Toposequences refer to soils formed over lateral variations in slope and topography. Soils at bases of hillslopes or on river flood plains accumulate parent material and have higher moisture contents than upland soils which are non-cumulative and have lower moisture conditions. Such sequences may be important in assessing effects of hydrology and physical versus chemical weathering rates on soil development.

Chronosequences are groups of soils for which all soil forming factors except elapsed time of formation are equivalent. The effect of time on chemical weathering rates is one parameter which can not be addressed in laboratory studies and may be significant in explaining many of the observed differences between experimental and natural weathering rates.

Biosequences are soil sequences containing variable biotia such as differences in the extent and type of plant cover. The role of plants in weathering rates is an important issue in soil weathering studies and in deciphering the weathering history of the earth.

Clearly the above soil classifications provide important opportunities to isolate and assess important parameters that influence soil weathering rates.

MINERALOGICAL MASS BALANCES IN SOILS

In the introduction, a soil was defined as being composed of mineral or chemical constituents that differ from the parent material or protolith. As indicated by Equation (1), quantification of the mass difference M_j is required in calculating weathering rates of primary silicates in soils. The following sections discuss the approaches which can been used and the assumptions required to calculate mass losses due to chemical weathering.

Mass Balance Approaches

The simplest approach in determining mineral or chemical losses associated with primary silicates in soils is to determine the ratio of the chemical or mineral concentration j in the weathered soil $C_{j,w}$ to the corresponding concentration in the parent material or protolith $C_{j,p}$ (i.e. $C_{j,w}/C_{j,p}$). However, the mass ratio of mineral or chemical components in soil and parent material is also dependent on concentration and dilution effects caused by the losses and gains of all the other components. The most common method for overcoming this problem is to define the mass ratio in terms of a conservative component i whose absolute mass does not change during weathering of the protolith to form a soil

$$\frac{C_{j,w}}{C_{j,p}} = R_o \left(\frac{C_{i,w}}{C_{i,p}} \right) \tag{3}$$

When $R_o = 1$ the ratio of the weatherable j component equals the ratio of the conservative component i. This signifies that no loss or gain of the j component occurs in the soil and that j also behaves in a conservative fashion during weathering. In contrast, if $R_o = 0$, complete weathering loss of component j occurs. Likewise, the change in volume of a soil relative to the volume of the original parent material can be defined as

$$\frac{V_w}{V_p} = \frac{C_{i,w}}{C_{i,p}} \tag{4}$$

Various forms of Equations (3) and (4) have long been used to estimate the extent of chemical changes in soils (Merrill, 1906; Barth, 1961; Harden, 1987).

A more general approach to soil weathering using the same mass ratios has been developed by Brimhall and Dietrick (1987), Chadwick et al. (1990) and Merritts et al. (1992) where the ratio of the weathered to parent composition is determined by three distinct processes defined on the right hand side of Equation (5).

$$\frac{C_{j,w}}{C_{j,p}} = \frac{\rho_p}{\rho_w} \frac{1}{(\epsilon_{j,w} + 1)} (1 + \tau_{j,w}) \tag{5}$$

Residual enrichment of an element results from density changes ρ_p/ρ_w due to dissolution and removal of mobile elements with a corresponding increase in porosity. Volume changes that may be associated with the density changes are described by the strain factor $1/(\epsilon_{j,w}+1)$ where $\epsilon_{j,w} = V_w/V_p - 1$ (Eqn. 4). Together residual enrichment and strain are described as "closed system" contributions which occur without movement of the component under consideration (i.e. $C_{j,w}$). The mass transport component "tau" $\tau_{j,w}$ is an "open system" contribution that describes mass movement across the sample volume boundaries (Brimhall and Dietrick, 1987).

$\tau_{j,w}$ is computed from density and chemical composition data in combination with volume change derived from the strain calculations

$$\tau_{j,w} = \frac{\rho_w C_{j,w}}{\rho_p C_{j,p}} (\epsilon_{j,w} + 1) - 1 \tag{6}$$

When $\tau_{j,w} = -1$, mass of element j was been completely lost during weathering. If $\tau_{j,w} = 0$, the element is immobile and is only affected by internal closed chemical system processes, specially residual and strain effects caused by changes in bulk density and volume. For a truly immobile element i (i.e. $\tau_{j,w} = 0$), Equation (6) can be rearranged to give

$$\epsilon_{j,w} = \frac{\rho_p C_{i,p}}{\rho_w C_{i,w}} - 1 \tag{7}$$

such that the volumetric strain can be calculated. Substitution of Equations (6) and (7) into Equation (5) produces the expected mass transfer relation $\tau_{j,w} = R_o - 1$ (Eqn. 3).

The mass flux resulting from weathering of a volume of soil V_w can be calculated using the open chemical system transport function $\tau_{j,w}$ such that

$$M_j = \left(\rho_p V_w \frac{C_{j,p}}{100} \right) \tau_{j,w} \tag{8}$$

In turn, the overall flux from a soil profile which has undergone variable amounts of mass transfer as a function of depth z can be calculated as

$$M_j = \left(\rho_p \frac{C_{j,p}}{100} \right) \int_{z=0}^{z=d} \tau_{j,w} dz \tag{9}$$

Mass balances are defined in terms of ratios (Eqns. 3 and 5) and are therefore independent of the units employed. Such mass balance calculations are equally applicable to mineral and elemental soil components (White et al., 1995). Equations (8) and (9) are written in terms of mineral concentrations which leads to a direct correlation with the loss of mineral mass M_j in Equation (1). If the above equations are written in terms of elemental concentrations, then M_j must be calculated from mineral stoichiometries.

Required assumptions for mass balance calculations

Two important assumptions concerning mineral and elemental components are required in the mass balance calculations. The first involves determination of the composition of the parent material $C_{j,p}$ prior to the onset of chemical weathering. For soils developed in situ on parent bedrock, such as saprolites, the potential errors in this determination are confined to local heterogeneities in bedrock compositions.

The estimate of the initial composition becomes more difficult when soils are developed on sedimentary parent materials such as alluvial terraces or loess deposits. One approach would be to use the bedrock composition from the assumed source area as the parent material. However for deposits produced by rivers, significant compositional differences can exist between the bedrock in the source area and the downstream sedimentary deposit. This is caused by weathering in upland watersheds coupled with sorting and winnowing during transport and deposition in the rivers. For such deposits, the parent material is commonly assumed to be the least weathered horizon in the soil profile. In practice, this corresponds to the deepest C horizon soils and in the case of soil chronosequences, the youngest soil profiles (Merritts et al., 1992; White et al., 1995). The assumptions inherent in this approach is that such materials are truly unweathered and that they accurately reflect the composition of initial sediments from which the older and more weathered soils were derived.

The second requirement in the mass balance approach (Eqns. 3 and 5) is that weatherable elements or minerals must be ratioed against an inert component i present in both the parent material and soil. Conservative elements most often employed in mass balance calculations include Zr (Harden, 1987; Chadwick et al., 1990), Ti (Johnsson et al., 1993), and rare earth elements such as Nb (Brimhall and Dietrick, 1987). Considerable disagreement occurs in the literature as to the relative mobility of specific elements under differing weathering regimes. Also as discussed by White et al. (1995), minor elements such as Zr and Ti are often concentrated in the small size-heavy mineral fraction which may be subjected to significant fractionation during sediment transport and deposition.

An alterative approach is to consider relatively inert minerals such as ilmenite (April et al., 1986) and quartz (Sverdrup, 1990; White et al., 1995) as conservative phases. Quartz is often a significant fraction of the parent material and is of comparable particle size and specific gravity to the weatherable aluminosilicate fraction. Thus, quartz is less likely to fractionate in a depositional environment relative to heavy minerals containing elements such as Zr and Ti. However, the rate of quartz weathering in soil environments is open to debate and therefore the extent to which it is conservative relative to other weatherable minerals or elements is open to question. Based on experimental rate data, Lasaga et al. (1994) estimated that the residence time of 1 mm quartz grains at neutral pH to be 34 m yr, a time span 1 to 2 orders of magnitude longer than for feldspar grains of comparable size. White et al. (1995), demonstrated minimal etch pitting and relative constant BET surface areas for quartz in a 3 mA soil chronosequence. However other studies have suggested that significant amounts of quartz can dissolve in some weathering environments (Brantley et al., 1986; Brimhall and Dietrick, 1987) (see Dove, this volume; for a detailed discussion of quartz dissolution).

Use of mass balance equations

The preceding mass balance approaches are now illustrated for deep saprolitic weathering in the Rio Icacos watershed in Luquillo Mountains of Puerto Rico (White et al., 1995) This watershed is situated in a humid tropical rain forest. Bulk soil and bedrock samples were analyzed for major and minor elements. Soil density measurements were obtained using a piston coring device. The density of the granitic rocks was assumed to be $\rho_p = 2.60$. Resulting data, as functions of residual enrichment, ρ_p/ρ_w, strain $\epsilon_{j,w}$, and mass transport $\tau_{j,w}$, as defined in Equation (5), are plotted in Figure 1. In the Rio Icacos soil profile, the density ratio of parent rock to weathered saprolite ρ_p/ρ_w is larger in the shallow A soil horizon (<1 m) due to organic input and bioturbation (Fig. 1A). Deeper in the undisturbed sapolite B horizon, the profile reaches a relative constant density ratio of 2.0 (50% porosity) to a depth of approximately 7 m, which corresponds to the bedrock interface.

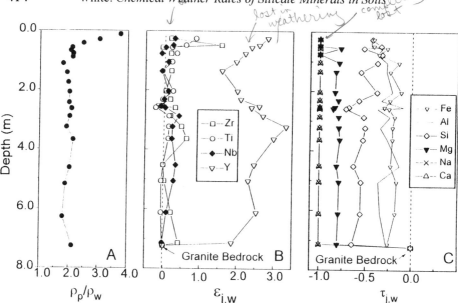

Figure 1. Weathering characteristics as a function of depth in saprolite profile in the Rio Icacos Watershed in Luquillo Mountains of Puerto Rico: (A) bulk density ratios ρ_p/ρ_w, (B) volumetric strain $\epsilon_{j,w}$, and (C) mass transfer $\tau_{j,w}$ calculated from Equations (5-7).

The calculation of the strain factor $\epsilon_{j,w}$,which describes the volume change in the soil ($\epsilon_{j,w} = V_w/V_p - 1$; Eqn. 7), was made using trace elements including Zr, Ti, and the rare earth elements Nb and Y (Fig. 1B). Although such elements are generally considered immobile with respect to weathering, their conservation under any specific geochemical condition is not guaranteed. Therefore, the best strategy is to consider a suite of such elements to establish the extent of volumetric changes undergone by a soil during weathering. As indicated for the Rio Icacos soil, Zr, Ti, and Nb produce consistent estimates for strain $\epsilon_{j,w}$ which center close to zero (Fig. 1B, vertical dashed line). This lack of a significant volume change indicates that weathering is essentially iso-volumetric, which is consistent with a soil porosity of nearly 50%, and with the preservation of primary igneous textures in the saprolite. Iso-volumetric weathering has been reported for other sapprolites based on mass balance calculations (Cleaves, 1993). Use of Y in Equation (7) produces calculated volume increases of up to a factor of three in the saprolite. Such unrealistic increases in volume reflect the loss of Y from the soil during weathering.

The mass transport component $\tau_{j,w}$ (Eqn. 6) is calculated for major cations and SiO_2 in the Luiquillo soil in Figure 1C. Values of -1 for Na and Ca throughout the profile indicate that these elements are completely loss during weathering of the primary silicates, principally from plagioclase and hornblende, and are not retained by secondary clay minerals, principally kaolinite. Approximately 80% of Mg, and 50% of SiO_2 are also lost. The only remaining primary minerals are hydrobiotite, which accounts for the residual Mg, and quartz, which along with secondary kaolinite, which accounts for the remaining SiO_2. Residual Al and Fe are retained in the kaolinite and Fe oxyhyroxide phases. The extent of elemental and mineral losses from weathering in the upper 7 m of the soil profile is constant. Current weathering occurs within a narrow interface (< 10 cm) directly above the underlying quartz diorite and overlying saprolite (Fig. 1).

The element loss from the Rio Icacos soil profile can be calculated by integrating $\tau_{j,w}$ over the depth of the entire profile z using Equation (9) and the composition and density of the bed rock (ρ_p = 2.60).The calculation indicates that for 1 m^2 of soil surface, the molar losses from weathering to a depth of 7 m are Si = $1.1 \cdot 10^5$, Ca = $1.7 \cdot 10^4$ Al = $1.6 \cdot 10^4$, Na = $1.1 \cdot 10^4$, Mg = 6.4×10^3, Fe = $2.3 \cdot 10^3$ and K = $2.0 \cdot 10^1$. Elemental fluxes, when combined with known stoichiometries, can be used to calculated the total mineral mass flux M_j (Eqn. 1).

Elemental losses as functions of time

If the age of a soil horizon is known, mass loss as function of time can be determined from Equations (3) or (5). The most direct application of this approach is to soil chronosequences. The following example demonstrates the usefulness of this approach in interpreting effects of climate on chemical weathering rates for 3 soil chronosequences studied in California. These soils are described briefly as follows:

The *Merced Chronosequence* is comprised of soils developed on alluvial terraces formed from glacially-derived granitic outwash from the Sierra Nevada and deposited along the Merced River in central California (Harden, 1987). The main soil units and estimated ages of these deposits (0.2 to 3000 kA) obtained by ^{14}C, uranium trend analysis, and K-Ar dating methods are listed in Table 2. The present climate is a Mediterranean type with average annual precipitation of 300 mm and an average temperature of 16 °C.

The *Honcut Creek Chronosequence* is a series of soils developed on alluvial outwash terraces from the Sierra Nevada and consisting of meta-volcanics and granodiorite were developed along to the north near the Feather River Canyon. These soils, described in detail by Busacca and Singer (1989), range in age from 0.6 to 1600 kA. The present climate is Mediterranean with an average annual rain fall of 600 mm and an average annual air temperature of 16 °C.

The *Strawberry Rock Chronosequence* is situated near the mouth of the Mattole River in Northern California and is composed of marine terrace deposits composed of arkosic sandstone with some siltstone and shale from the Jurassic Franciscan formation. Soil ages are based on radiocarbon and sea-level stands, and range from 3.6 to 240 kA (Chadwick et al., 1990; Merritts et al., 1992). This site has a seasonal Mediterranean type climate characterized by prolonged cool but dry summers and mild and wet winters with an annual precipitation of 1100 mm and an average annual temperature of 13°C.

The ratios of SiO_2 concentrations in the soil and parent sediments (R_o, Eqn. 3) were chosen as indicators of the extent of weathering within the soil chronosequences. In both the Strawberry Rock and Honcut Creek studies, Zr was assumed to be the conservative component (Busacca and Singer, 1989; Merritts et al., 1992) while quartz was assumed to be chemically conservative in the Merced soils (White et al., 1995). All three studies assumed that the deepest C horizons of the youngest soils were representative of parent chemical compositions. The SiO_2 ratios for the Merced, Honcut Creek and Strawberry Rock soils are approximated by exponential decay functions when plotted against time in Figure 2A (solid lines). Such exponential decay functions have been observed for other chemical properties of chronosequences such as the decrease in base cation saturation with time (Bain et al., 1993). These exponential decreases in weathering rates are attributable to selective removal of the more reactive minerals during the initial stages of weathering and the persistence of less reactive phases during later stages of weathering.

The relative rates of SiO_2 decrease in the three soil chronosequences are significantly different. The Strawberry Rock soils weather faster than the Honcut Creek soils which in turn weather faster than the Merced soils (Fig. 2A). These relative rates are apparently related

Table 2. Soil profiles and ages of the Merced Chronosequence, California.

Soil Unit	Sampled Depth (cm)	Age of surface (kA)
Post-Modesto Deposits	51	0.2 - 3
Modesto Fm., Upper member	254	10
Modesto Fm., Lower member	413	40
Riverbank Fm., Upper unit	400	130
Riverbank Fm., Lower unit	500	250
Turlock Lake Formation	375	600
Laguna Fm., China Hat Unit	350	3000

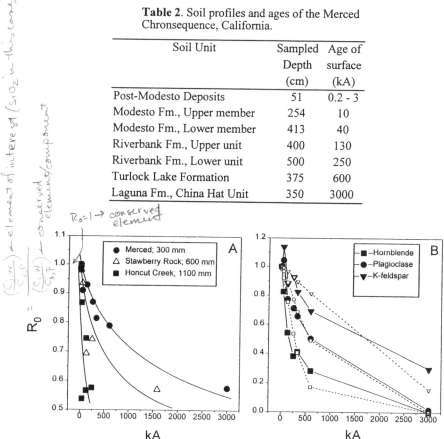

Figure 2. Comparison of mineral weathering rates based on mass ratios (R_o) for: (A) decrease in SiO_2 as a function of time for three chronosequences in California legend (annual precipitation in mm), and (B) decrease in mineral abundances with time for the A horizons in the Merced chronosequence. Dashed lines connect fitted rate constants k_j (Eqns. 21 and 22).

to climatic differences. Although the current mean annual air temperatures for the 3 chronosequences are similar, 14-16 °C, the average annual precipitation varies by almost a factor of 4 from 300 mm for the Merced soils to 1100 mm for the Strawberry Rock soils. A comparable correlation between increasing weathering and precipitation has been recently documented for SiO_2 and Na in watershed studies (White and Blum, 1995b). Comparison of weathering rates of soil profiles based on present climate assumes that similar conditions have persisted in the past. Precipitation in central California reached a long term maximum in the late Pliocene at approximately 1.5 times that of the present day average (Busacca et al., 1989). Past paleoclimatic differences at an individual site therefore appear to be less than present day differences between the 3 sites. Other mass balance studies have investigated differences in weathering rates over time. Cleaves (1993) estimated that the rates of saprolite formation in the Piedmont Province in Maryland may have decreased by 50 to 90% as a result of periglacial conditions during the Pleistocene.

Mineral losses as functions of time

The mass balance calculations use ratios for both weatherable and inert components

Table 3. Comparsion of compositions of parent Tuolume intrusive series and basal C horizons of the 10 kA Merced soils. Units are in wt % unless otherwise indicated.

Unit	Profile[2]	Depth	Qtz.	Plag.	K-feld	Hnbl.	Biot.	Clay	Zr	Ti
Granite[1]			25.4	43.9	22.2	7.0	4.1	0.0	0.016	0.37
Basal Modesto	M12	231 cm	34.6	38.7	15.7	7.0	3.2	0.8	0.017	0.36
Basal Modesto	M31	254 cm	37.2	33.0	12.9	10.7	2.6	3.6	0.033	0.57
Basal Modesto	M46	250 cm	33.6	39.4	10.8	5.7	1.8	8.7	0.021	0.49

[1]Data from Bateman and Chappell, 1989; [2]Units based on classification of Harden, 1987

(Eqns. 3 and 5) and are therefore equally applicable for calculating mineral as well as elemental losses or gains (White et al., 1995). Mineralogical compositions in soils and parent material can be determined using a number of analytical approaches including point counting, chemical modal analysis and quantitative X-ray diffraction techniques. The last two techniques were used to evaluate mineralogical changes in the Merced chronosequence soils (White et al., 1995). As indicated in Table 3, the composition of the assumed parent materials, as represented by the basal C horizons of the youngest 10 kA soils, are significantly depleted in feldspars and enriched in quartz relative to the source granite contained in the Tuolumne intrusive of the Sierra Nevada. Note also that quartz abundances are much more consistent within the deposits than are Zr and Ti contents which are commonly assumed to be conservative elements in mass balance calculations (Eqns. 3 and 5) These differences are attributed to weathering and preferential sorting of heavy minerals prior to deposition in river terrace deposits.

The proportions of K-feldspar, plagioclase and hornblende remaining in the A soil horizons relative to the parent composition were calculated and plotted as a function of soil age in Figure 2B. R_o values, based on quartz as the conservative component, approach unity in the younger soils (± 0.20). During the first 600 kyr, primary silicates decrease exponentially in order hornblende > plagioclase > K-feldspar, which is consistent with commonly observed weathering rates in soils (Dixon and Weed, 1989; Nahon, 1991) and in experimental dissolution studies (Lasaga, 1984). The calculations predict, that after 3000 kyr, all of the primary aluminosilicate minerals are lost except for a residual K-feldspar fraction ($R_o = 0.30$). Optical inspection of residual mineral separates in the China Hat showed no detectable plagioclase or hornblende and only minor amounts of K-feldspar. Decreases in reactive minerals dominate the initial weathering processes and result in rapid SiO_2 loss in younger soils. In the older soils, chemically resistant minerals, principally quartz, are present. The selective decrease in reactive minerals explains the exponential decrease in chemical weathering rates with time.

Other studies have also used changes in mineral abundances to investigate chemical weathering. Mahaney and Halvorson (1986) documented progressive increases in the quartz/feldspar ratios in soils from a 500 kA chronosequence in the Wind River Mountains of Wyoming. April et al. (1986) calculated long term chemical weathering rates based on hornblende depletion in soils in two Adirondack watersheds. In comparing these rates with weathering based on watershed solute discharge, these workers concluded that current rates may be a factor of 3 times greater than long term rates due to the influence of acid deposition. As in the case of elemental loses (Fig. 2A) mineral losses can provide important information on changes in the weathering environment over time (Fig. 2B).

Volumetric analysis and rates of soil formation

Chemical weathering rates in soils can, in some instances, be formulated as the rate of downward movement (velocity) of the chemical weathering front into the regolith. The simplest case is the use mass balances to investigate the iso-volumetric weathering of crystalline bedrock to saprolite (Cleaves, 1993). As indicated for the Rio Icacos soils, weathering of primary minerals occurs predominantly at the base of the saprolite in direct contact with the bedrock (Fig. 1). Mass balance under such conditions requires that the volume of saprolite equals the volume of initial rock ($V_w = V_p$). A unit volume of saprolite v_{sp} therefore can be considered to consist of three components: the volume fraction of secondary minerals produced from weathering v_{sm} (e.g. clay and Fe oxyhydroxides), the volume fraction of stable unweatherable minerals v_i (e.g. quartz) and the volume fraction due to porosity of the saprolite v_p such that

$$v_{sp} = v_{sm} + v_i + v_p \tag{10}$$

The velocity of the weathering front at the saprolite-bedrock interface ω_{sp} (cm s^{-1}) can be related to the mass flux of solute transported from the weathering profile Q_w (g cm^2 s^{-1}) where ρ_p is the density of parent rock (g cm^3)

$$\omega_{sp} = \frac{Q_w}{\rho_p} \cdot \frac{1}{v_{sm}} \tag{11}$$

Q_w is most often determined from net watershed mass fluxes attributed to chemical weathering in the soil profile. Methods for calculating weathering rates in watersheds based on mass balances are discussed in detail by Drever (this volume). Weathering fluxes are determined as the difference between total measured solute discharge in the stream and input and output sources such as wet and dry fall, biological fluxes and ion exchange. Solute fluxes attributed to weathering can be normalized to the geographical surface area of the watershed which in turn can be converted to unit area of soil (g. cm^{-2}).

Cleaves (1993) calculated a range of weathering velocities of ω_{sp} = 3.7 to 29.7 m 10^{-6} yrs for the vertical advancement of the saprolite weathering front into schists in the Pond Branch watershed in the northern Piedmont Province of the USA. Other saprolite weathering rates, including that for granite at the Rio Icacos site, are found to range between 3 and 37 m 10^{-6} yrs (Table 4). Such rates are also comparable to long term denudation rates reported for comparable landscapes assuming long term dynamic equilibrium between rates of saprolitization and rates of physical and chemical erosion. Based on total river discharge, sediment rates and solute fluxes (Wakatsuki and Rasyidin, 1992) estimated a global average soil formation rate of 5.6 m 10^{-6} yrs.

In contrast to iso-volumetric weathering of bedrock to saprolite, soil formation on sedimentary deposits commonly results in significant increases in soil density and decreases in soil volume with time. For example, measured bulk densities ρ_w of the Merced chronosequence (Fig. 3A) increase with age from the Post Modesto (3 kA) through the Turlock Lake (600 kA) soils and then decrease in the China Hat (3000 kA) soils. Density increases in the older soils are related to soil compaction associated with weathering of primary minerals and clay formation. Less compaction in the oldest China Hat soil is probably due to the presence of cobbles which support a less dense soil structure. Higher organic contents and bioturbation create lower bulk densities in the A horizons than the B+C horizons.

Corresponding volume changes for the B+C horizons of Merced soils based on

Table 4. Propagation rates of weathering fronts at the sapolite/bedrock interface (Eqn. 11). Weathering rates are in meters/10^6 years.

Rate	Rock		Reference
3.3-29.7	Schist	Pond Branch Maryland	Cleaves, 1993
4.0	Granite	Occoquan basin, Virgina	Pavich et al., 1986
35.8	Granite	Rio Icacos, Puerto Rico	This paper
37.0	Schist	Coweeta Watershed, North Carolina	Velbel, 1985

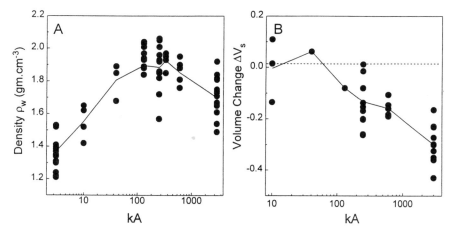

Figure 3. Density ρ_w and volume change ΔV in the B+C horizons of soils in the Merced chronosequence. Lines connect mean average values of soil samples of the same age.

Equation (4) are plotted as functions of time in Figure 3B. The progressive decrease in the volume ratio to 0.7 in the oldest China Hat soils suggest that significant compaction has occurred as a result of weathering even though the bulk densities in the younger soils actually increase (Fig. 3A). Assuming that (1) quartz is conservative, (2) parent mineralogy is equivalent to the average composition listed in Table 3 and (3) ρ is constant for all minerals, the complete weathering of aluminosilicates to kaolinite would produce a volume loss of 30%, which is comparable to the calculated China Hat volume loss (Fig. 3B). The extent of these volumes losses are expected to significantly affect porosity and hydrochemical processes which control weathering in the soils.

The approaches presented in the preceding discussions are important techniques in quantifying mass and volume losses associated with the weathering of primary minerals in soils. When normalized relative to spatial distributions, such mass balances provide information on total mass losses from soil horizons (Eqns. 8 and 9). When coupled with soil ages, mass balances permit comparison of relative weathering rates in different soil profiles and to assess parameters such as climate differences.

SOIL SURFACE AREAS

The calculation of quantitative dissolution rates k_j (Eqn. 1) is dependent on normalizing the reacted mineral mass M_j to the mineral surface area S. As will become apparent in the following discussion, the estimation of surface areas is one of the most difficult problems in quantifying weathering rates in soils and represents one of the sources of greatest discrepancy. The surface area in Equation (1) is related to the density of sites on the solid surface which is exposed to aqueous solution and at which silicate hydrolysis reactions can occur (Helgeson et al., 1984). However, from an operational standpoint, such reactive surface areas in natural weathering studies are almost always assumed to scale directly with measurable physical surface areas ($cm^2 g^{-1}$). The following discussion will therefore focus principally on available data and methods for determining physical surface areas of primary minerals in soils. Other factors, including selective reaction at defect and dislocation sites and physical and hydrologic isolation of mineral surfaces, may decouple physical and reactive surface areas. Additional discussions of the nature of reactive surfaces are presented in a following section of this present paper and elsewhere in this volume (see Lasaga; Blum and Stillings).

Assuming spherical grain geometry for a non-porous mineral, the specific surface area S ($cm^2 g^{-1}$) of mineral grains of constant diameter D (cm) can be described as (Jaycock and Parfitt, 1981; Anbeek, 1992)

$$S = \frac{6}{\rho_j D} \lambda \tag{12}$$

where ρ_j ($g\ cm^{-3}$) is the mineral density and λ ($cm^3.cm^{-3}$) is the roughness factor. λ is the ratio of the measured surface area S to the equivalent geometric surface area s (Helgeson et al., 1984; White and Peterson, 1990; Anbeek, 1992) such that

$$\lambda = S/s \tag{13}$$

Operationally surface roughness is determined from a macroscopic measuring technique used to estimate s (e.g., sieving or light scattering) and a microscopic technique used to measure S (e.g. BET gas sorption). S can be defined by Equation (13) as the product of the spherical geometric surface area s ($cm^2 g^{-1}$), defined by D and ρ_j, and the roughness factor λ (cm ³/cm ³).

Surface areas of silicate minerals in soils

Surface areas of soil components are commonly normalized to soil mass ($m^2 gm^{-1}$ of soil) or as a specific surface area normalized to the unit mass of mineral component ($m^2 g^{-1}$ of mineral). BET surface areas of bulk soils are generally reported to range between 5-70 m^2 per gram of soil (Gallez et al., 1976; Feller et al., 1992). Limited data on the surface area of the primary silicate fraction suggests that this component makes up a minor portion of the bulk surface area. White and Peterson (1990) investigated the relationship between grain size distributions and surface areas of the silt-sand size fractions of several soils. Examples included the relative contributions of Fe-oxyhydroxides, Fe-silicate phases ($\approx$ biotite + hornblende) and non-Fe containing silicates (feldspars + qtz) to total surface areas of a poorly developed soil on Sierra Nevada granite near Lake Tahoe (Fig. 4A) and a deeply weathered granitic saprolite from Montara Mountain near San Francisco (Fig. 4B). Data are plotted as cumulative contributions to the bulk surface area (m^2 per gram of soil) In the younger soil, the surface area of all size fractions are dominated by Fe-oxyhydroxide phases which coat the silicate grains. This contribution is less in the older Montara soil with surface area being dominated by the residual primary Fe-silicate phases. In both soils, the combined surface areas of quartz and feldspar in all size fractions were only a minor component of the total soil surface area.

Soil surface areas generally increase with soil age (Table 5). The BET surface areas of Merced soils increase with age from 4 to 18 m^2/g. This increase correlates closely with the corresponding clay content whose surface areas comprise about 20% of the total surface area in the youngest 3 kA soils, compared to 72% in the oldest 3000 kA soils. Cumulative surface

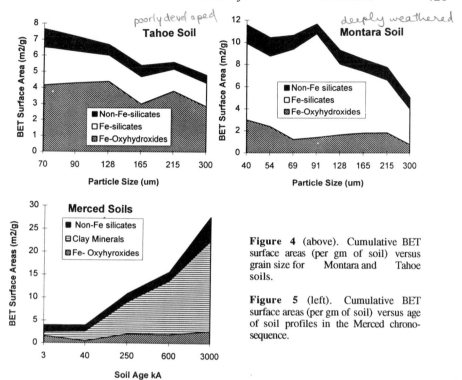

Figure 4 (above). Cumulative BET surface areas (per gm of soil) versus grain size for Montara and Tahoe soils.

Figure 5 (left). Cumulative BET surface areas (per gm of soil) versus age of soil profiles in the Merced chronosequence.

Table 5. BET surface areas of soil and mineral components of the Merced chronosequence ($m^2 g^{-1}$)

Unit	Post Modesto	Modesto	Riverbank	Turlock Lake	China Hat
Age (kA)	3	40	250	600	3000
Profile	PM14	M12	R32	T6	CH2
Depth (cm)	33-99	32-66	152	100-155	90-130
Soil Horizon	C	A	B	B	B
Surface Areas of Soil Components					
Untreated Soil	4.47	4.11	7.51	10.25	17.68
Extractable Fe	1.58	0.49	2.00	1.91	2.67
Clay (<4 um)	0.87	2.10	7.00	11.74	19.80
Silicates (>4 um)	1.52	1.32	1.81	1.92	5.18
Surface Areas of Mineral Separates[1,2]					
Quartz	0.11	0.17	0.23	0.21	0.10
Plagioclase	0.39	0.26	0.46	1.48	na
K-feldspar	0.12	0.26	0.94	0.81	na
Hornblende	na	0.34	0.72	0.67	na

[1] Fe removed by dithionite

[2] Analysis for 500 to 1000 um fraction except hornblende (500-250 um)

areas contributed by primary silicates, clay, and Fe-oxyhydroxides, per gram of soil, are plotted as functions of age in Figure 5 (White et al., 1995). Due to increasing clay content and decreasing abundance, the contribution of primary silicate fraction to the bulk soil surface area decreases from 37% in the youngest soils to 18% in the oldest soils. The preceding data indicate that although the surface areas of the primary silicate fractions are critical in controlling weathering rates, their relative contribution to bulk soil surface area is generally minor.

Very limited information is available on the specific surface area of primary mineral phases in soils. White et al. (1995) documented that the specific surface area of plagioclase grains increases with age from 0.4 $m^2\ g^{-1}$ in the 3 kA Post-Modesto to 1.5 $m^2\ g^{-1}$ in the 600 kA Turlock Lake soils (Table 5). K-feldspar and hornblende exhibit more erratic but significant increases in surface area over the same period (Fig. 6A). In the oldest China Hat soil only residual quartz was abundant enough for BET measurements. This quartz surface area was comparable to quartz in the younger soils (0.1-0.2 $m^2\ g^{-1}$) suggesting significant resistance to chemical weathering in this environment.

Surface area and particle size

Equation (12) predicts that the external surface area is proportional to the geometric surface area. Therefore a plot of *log S* versus *log D* should produce an inverse relationship with a slope of -1 as documented (White and Peterson, 1990) for clays, oxides and freshly crushed silicates (Fig. 7). The parallel offset in the data relative to the geometric surface area *s* is equal to the average surface roughness ($\lambda = 7$) which was independent of particle size. In reviewing surface area for crushed quartz and natural well-rounded quartz grains, Parks (1990) also found surface roughnesses ($\lambda = 2.2$ and 6.2, respectively) that were independent of grain size. In reviewing BET surface area data for crushed feldspar samples employed in dissolution studies Blum (1994) determined a corresponding roughness factors of 9±6. BET surface areas of relatively unweathered minerals are therefore generally proportional to their particle diameter based on an assumed spherical geometry. Likewise the deviation in the magnitude of measured and geometric surface areas over large ranges in particle diameter are relatively constant ($\lambda = 2$ to10).

Surface Areas of Weathered Silicates

BET surface areas of the primary silicate fraction weathered in soils is significantly larger than for the corresponding freshly crushed and unweathered minerals. White and Peterson (1990) reported surface roughnesses of $\lambda = 50$ to 200 for the sand size primary silicate fractions of several soils development on granitic bedrock. Anbeek (1992) reported roughness factors of $\lambda = 130$ to 2600 for silicates from glacial deposits in Switzerland. BET surface area data versus silicate particle size (feldspar + quartz) from 5 soils from the Merced chronosequence are plotted in Figure 8. Aggregated clay particles and Fe-oxyhyroxide coating were removed by pre-treatment. For any size fraction, the BET surface areas of the silicate minerals are much larger ($\lambda = 100$ to 1000) than the corresponding spherical geometric surface area (dashed line, Fig. 8). The weathered surface areas are also significantly larger ($\lambda = 10$ to 100) than BET surface areas of similar size fractions of freshly crushed silicates commonly used in experimental dissolution studies. The surface areas of particles of the same diameter in the Merced soils also generally increase with soil age.

The surface areas for weathered grains have a log slope less than -1 and are clearly non-linear relative to particle size (Fig. 8). The difference between the measured BET surface area and the equivalent geometric surface area decreases with decreasing grain size indicating that the apparent surface roughness also decreases with size. Anbeek et al. (1994)

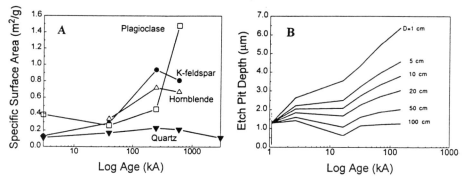

Figure 6. Specific BET surface areas of minerals in Merced chronosequence (A) and etch pit depth of hornblende in the Tobacco Range chronosequences (B) (Hall and Horn, 1995) as functions of soil age. D corresponds to soil depth.

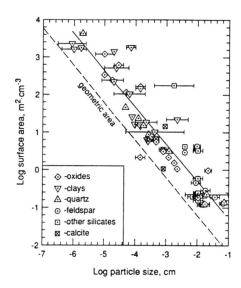

Figure 7. Relationship between reported BET surface areas and grain size for non-weathered minerals (White and Peterson, 1990). Dashed line corresponds to the geometric surface area and solid line to the linear regression fit through the BET data (slope = -0.99, r^2 = 0.86).

Figure 8. Surface area as a function of particle diameter (from White et al., 1995). Geometric areas are calculated for smooth spheres of equivalent diameter (dashed line). Quartz, K-feldspar and albite data are BET measurements of freshly crushed material (Leamnson et al., 1969; Holdren and Speyer, 1987). Curved lines are fits of Equation (16) to the BET measurements of silicate fractions (less clay and Fe oxyhydroxides) of Merced soils of different ages (open symbols).

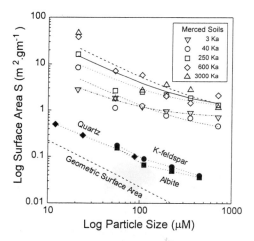

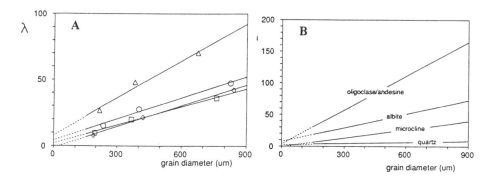

Figure 9. Relationship between surface roughness λ and grain diameter for the bulk silicate fraction (A) and individual minerals (B) from glacial deposits from Switzerland (from Anbeek et al., 1994).

also investigated the relationship between surface roughness and particle sizes for weathered silicates from glacial deposits. As shown in Figure 9 surface roughness for both the bulk samples and individual mineral fractions decrease linearly with particle diameter. Decreases in surface roughness for naturally weathered silicates contrasts with the apparent constant roughness factors for unweathered silicates and oxides as summarized by Parks (1990) and White and Peterson (1990) (Fig. 7).

Surface roughness of weathered silicate grains in soils is related to microscopic morphologic features on grain surfaces not considered in geometric estimates. Pervasive etching and pitting during weathering of primary minerals in soils have been previously observed for pyroxenes and amphiboles (Berner et al., 1980; Velbel, 1989; Cremeens, 1992) feldspars (Berner and Holdren, 1979; Cremeens, 1992; Brantley et al., 1993) and quartz (Brantley et al., 1993). These studies conclude that roughening is due principally to etch pit formation resulting from preferential dissolution from reactive sites with high surface energies at dislocation, defect sites and exsolution lamellae. Several models have been proposed to quantitatively relate surface energies of such sites to the density and morphology of etch pits and dissolution rates (Lasaga; Blum and Stillings; Brantley; this volume).

Grain roughness and pitting have generally been found to increase with soil age. For example, SEM photographs of plagioclase surfaces from the Merced soils (Fig. 10) clearly show more extensive pitting in the 250 kA Riverbank soil relative to the 10 kA Modesto soil. However some grains of the same age and mineralogy were found have nearly pristine surfaces while others were highly weathered. Anisotropic dissolution of different crystallographic faces is commonly observed, and may explain a portion of this variation. This is seen in the 10 k A plagioclase from the Modesto Formation (Fig. 10A) in which the left facing surfaces are much more deeply pitted, and the 250 kA K-feldspar grain (Fig. 10C) in which the (001) cleavage faces (horizontal in the micrograph) are less pitted than surfaces with other orientations. However, heterogeneity is also observed on faces with a nearly constant orientation. This may reflect variations in the composition and/or defect structures of the mineral grains, or large variations in the chemical conditions of weathering micro-environments within the soil.

Efforts have been made to develop quantitative relationships between the extent and morphology of pitting and soil weathering environments using scanning electron and petrographic microscope techniques (Hall and Martin, 1986; Locke, 1986; Cremeens et al., 1992; Hall and Horn, 1993). An example of the relationship between etch pit depth and age of hornblende grains contained in a soil chronosequence developed on glacial moraines in

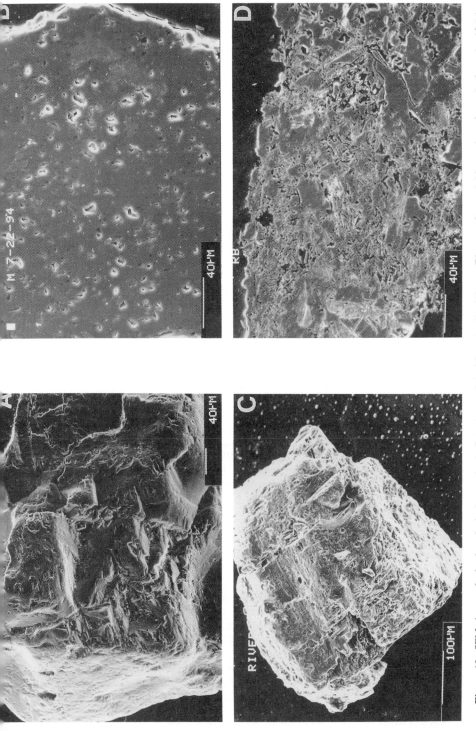

Figure 10. SEM photos of selective plagioclase grains in B horizons of the Merced soils. (A) and (B) are surface and thin section of a grain from the 10 kA Modesto soil. (C) and (D) are surface and thin section of a grain from a 250 kA Riverbank soil.

the Tobacco Root Mountain Range of Montana is shown in Figure 6B (Hall and Martin, 1986). The parent rock types are Precambrian metamorphic rocks and Cretaceous granites. As indicated, the depth of surface pitting of hornblende both increases with soil depth and age. This increase in etch pitting for minerals in the Tobacco Root chronosequence can be qualitatively compared to the increase in surface areas of the primary silicate fraction in the Merced chronosequence for comparable soil ages. This comparison suggests the importance of surface morphology in controlling BET surface areas. However a quantitative relationship between the degree of surface pitting and the surface roughness is not necessarily clear. For example, Anbeek et al. (1994) concluded that surface roughness attributed to etch pits on naturally weathered feldspar surfaces contributed only a fraction of the total measured surface areas.

Internal porosity

Several studies (Wood et al., 1990; Anbeek et al., 1994; and White et al., 1995) have concluded that weathered silicate minerals exhibit significant internal porosity which may be a major contributor to total mineral surface areas. SEM photomicrographs of cross-sectioned mineral grains show internal porosity, even in the youngest soils. However, as shown for examples of plagioclase from the Merced soils, the internal porosity is greater in the 250 kA Riverbank soil relative to the 10 kA Modesto soil. Morphologies range from small isolated equal-dimensional pits (Fig. 10B) to elongated and interconnected pores and cracks (Fig. 10 D). In general, pore diameters increase with soil age even though the absolute number of pores does not change appreciably. In the more highly-weathered grains, the pores enlarge and coalesce, finally forming very open internal structures (Fig. 10D). Pores are generally equally distributed from the rim to the core of grains, and do not form any weathering rim. In general the density of macropores increased in the order hornblende > plagioclase > K-feldspar > quartz which is in general agreement with the weatherability of silicate minerals (Goldich, 1938) and roughness factors calculated by Anbeek et al. (1994).

The distribution of macropores appears to be related to the internal distribution of defects. The density of internal pores counted in a typical cross-section photo of a weathered plagioclase from the Riverbank soil (250 kA) was found to be 500 pores per 5000 μm^2, which is a density of 10^7 cm^2 (White et al., 1995). This pore density is in the range of dislocation densities observed in feldspars from undeformed igneous rocks (Willaime et al., 1979). A surprising amount of porosity generated by these defects may be present within mineral grains before the onset of any soil weathering process. Porosities of up to 4% observed in unweathered plagioclase and K-feldspar have been attributed to exsolution of magmatic water and hydrothermal alternation within plutons (Montgomery and Brace, 1975; Worden et al., 1990). SEM photos produced by these studies show both tubular and equivalent micro-cavities that closely resemble pores permeating the weathered Merced feldspars (Fig. 10).

Calculation of combined surface roughness and internal porosity

For porous media with internal porosity accessible to the mineral surface, the total measured surface area S can be refined from Equation (12) (White et al., 1995),

$$S = \frac{6}{\rho D} \lambda + S_i \tag{14}$$

where S_i (m^2/g) is the portion of the surface area associated internal porosity that is physically connected to the surface.

The total surface area S of very small particles will be dominated by the external surface area which is high relative to the total particle mass. Equation (14) then becomes equivalent to Equation (12) ($S = 6\lambda/\rho D$) and the slope approaches log S/log $D = -1$ for constant λ and ρ (Fig. 7). The total surface areas of very large particles will depend principally on the internal surface area ($S \approx S_i$) which is attributed to the porosity evident in SEM micrographs (Fig. 10). If porosity is evenly distributed throughout the grains, S_i will be independent of grain size and the slope approaches log S/log $D = 0$. For intermediate size particles, a dependence on both external and internal surface areas will predict a non-linear decrease in S with increasing D. Such a condition is evident for the relationship between primary mineral surface areas and particle diameters for the Merced soils which produces non-linear slopes between -1 and 0 (Fig. 8).

Table 6. External roughness λ and internal surface area S_i (m^2/g) for the primary silicate fraction of the Merced soils[1]

Soil Age	λ	λ^*	S_i	S_i^*
3	21	11	0.69	1.32
40	42	36	0.32	0.51
250	83	90	1.05	1.78
600	200	310	0.74	1.32
3000	130	620	0.99	5.81

[1]λ^* and S_i^* are calculated parameters excluding the quartz fraction (Eqn.15).

External roughness λ and internal surface area S_i can be determined by empirically fitting Equation (14) to observed surface area -grain size distributions such as plotted in Figure 8 (White et al., 1995). As shown in Table 6, roughness factors λ for the Merced soils were found to increase consistently with age from $\lambda = 21$ in the 3 kA Post Modesto soil to $\lambda = 200$ in the 600 kA Turlock Lake soil. The higher proportion of quartz in the China Hat soil explains the corresponding decrease in apparent surface roughness ($\lambda = 130$). Quartz BET surface areas are significantly lower than for other minerals and exhibit essentially no increase with soil age (Table 5).

Changes in surface roughnesses of the aluminosilicate fraction λ^* (plagioclase + K-feldspar + hornblende + biotite) were estimated by correcting for dilution by quartz λ_{qtz} (White et al., 1995)

$$\lambda^* = \frac{\lambda - \lambda_{qtz} x_{qtz,w}}{x_{j,w}} \tag{15}$$

$x_{qtz,w}$ and $x_{j,w}$ are the mass fractions of quartz and aluminosilicates in the soil. Resulting values for λ^* show a consistent increase in surface roughness with soil age from $\lambda = 11$ for the 3 kA Post Modesto soil to $\lambda = 620$ for the China Hat soil (Table 6) Surface roughnesses λ^* of the youngest Merced soils approach values of $\lambda = 5$ to 10 reported for freshly crushed silicates (White and Peterson, 1990; Blum, 1994). As shown in Figure (11), the divergence between λ^* (open circles) and λ (closed circles) becomes more pronounced for the older soils as x_{qtz} increases. Surface roughness λ^* of the residual aluminosilicate fraction in the oldest China Hat soils exceeds the range reported previously for weathered feldspar and granitic surfaces (White and Peterson, 1990; Anbeek, 1992) ($\lambda = 50$ to 500). The fitted roughness factors are assumed to be independent of grain size and can therefore be compared to roughness factors determined independently for individual minerals (Fig. 6A). As predicted, both λ and λ^* are bracketed by λ values for plagioclase and quartz which exhibit respectively the largest and smallest increases in surface roughness with time.

Unlike external surface areas, internal surface areas reflecting grain porosity exhibit

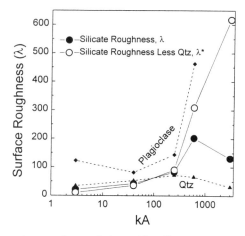

Figure 11. Calculated surface roughness as functions of soil age based on measured BET surface areas and calculations using Equations (14) and (15) (from White et al., 1995). Dashed lines show corresponding surface roughness for the 500 to 1000 μm fractions of quartz and plagioclase.

only a weak correlation with soil age (S_i = 0.69 to 0.99 m^2 /g, Table 6). Corrected S $_i^*$ values for the primary aluminosilicate fraction were calculated by an equation analogous to Equation (15) assuming that the quartz component has negligible internal porosity. Even with this correction, variability in S_i^* is much less than in λ^* (Table 6) suggesting smaller changes in internal relative to external surface area during weathering. SEM photos of sectioned soil grains (Fig. 10) document that porosity increases with soil age and extent of weathering. However a correlation between porosity increases and greater internal surface area is not straight forward. An increase in internal surface area results if pore diameters increase or if the walls of existing pores roughen. However, if weathering causes a coalescence of many small pores into fewer large pores, the resulting internal surface area could decrease. By definition, internal surface area is that component which is independent of particle volume and includes only that internal surface area that is homogeneously distributed throughout a mineral grain. Preferential pore development, enlargement or roughing near grain surfaces would therefore be assigned to surface-dependent roughness.

In summary, weathering rates are directly correlated with the density of reactive sites on the mineral surface which are exposed to aqueous solution. This reactive surface area is assumed to scale with physical surface area which is measurable by either macroscopic or microscopic means. For naturally weathered soil silicates, the surface roughness factor λ, as defined by the ratio of the BET to geometric surface area, varies between $\lambda = 10^2$ to 10^3. This roughness is attributed both to external surface morphology, which dominates in small grain sizes, and to internal porosity, which is dominant in larger grain sizes.

MINERAL DISSOLUTION RATES BASED ON PRIMARY SILICATE LOSSES

Up to this point in the discussion, the rates of chemical weathering have been defined in terms of changes in elemental or mineral ratios (Eqns. 3 and 5) or in terms of mass normalized with respect to a unit surface area of soil (Eqns. 8 and 9). A more rigorous approach is to quantify weathering with respect to rate constants for individual mineral phases normalized to their specific surface areas. This approach has much more general applicability because soil weathering can then be described by the sum of the rates of the individual minerals present. Equation (1) can be rearranged to define the dissolution rate of a primary mineral phase j in a soil in terms of the reaction rate constant k_j (mol cm^2 s^{-1}). The

$$k_j = (\frac{M_j}{\Delta t}) \frac{1}{S} \qquad (16)$$

following sections discuss several approaches which can be used to calculated weathering rate constants for minerals in soil environments based on observed losses of primary silicates during soil formation.

Coupling mineral surface area and dissolution rates

As demonstrated by Lasaga (1984), the simplest approach to relating dissolution rates to changes in surface area is through a simple geometric model. For a spherical mineral grain with no surface roughness, the dissolution rate becomes

$$k_j = \frac{r^o - r}{V_j \, \Delta t} \tag{17}$$

where r^o is the initial particle radius and r is the decreased radius caused by dissolution over time Δt. V_j is the molar volume of mineral phase j. The mean lifetime of a weathering particle can be defined as the time Δt required to reduce r to zero. Based on experimental dissolution rate data (Lasaga, 1984; Lasaga et al., 1994) estimated mean lifetimes of 1 mm particles to range from $7.9 \cdot 10^1$ years for wollastonite to $3.4 \cdot 10^7$ years for quartz.

For natural silicates contained in soils, a range of particle sizes and shapes of a given mineral are actually involved in the weathering process. Sverdrup and Warfvinge (1988), in developing a weathering model for soils in the Gårdsjön watershed (Sverdrup, this volume), estimated the total geometric surface area s as the sum of the surface areas of grains in individual size fractions ϕ defined by particle diameter D'

$$s = \frac{6M}{\rho \alpha} \sum_{\phi=1}^{n} \frac{f^\phi}{D^\phi} \tag{18}$$

where α is the average particle sphericity ($\alpha=1$ for perfect sphere; $\alpha = 0.85$ to -0.90 for particles with cubic shapes) and f is the fraction of total mass M of particle size ϕ.

Sverdup and Warfvinge(1988) assumed the exposed physical surface area S_t of a mineral phase in a unit volume of soil to be,

$$S = s \, \lambda \left(f_m \, \frac{\rho_j}{\rho_w} \, z \right) \tag{19}$$

where f_m is the weight fraction of the mineral relative to the total soil and ρ_j/ρ_w is the density ratio of the mineral to the soil. The surface roughness λ was estimated from the ratio of the BET and geometric surface areas of freshly crushed particles. The weathering rate constant k_j was then be calculated based on mass differences between bedrock and soil using quartz as a conservative component and a mineral surface area calculated by Equation (19)

$$k_j = \rho_p \left(\frac{x_{qtz,w}}{x_{qtz,p}} \, x_{j,p} - x_{j,w} \right) \left(\frac{1}{S \, \Delta t} \right) \tag{20}$$

Note the terms in brackets are equivalent to the mass balance ratio in Equation (3). The overall equation is comparable to Equation (16) which originally defined the weathering rate constant in terms of relative concentrations of weatherable and non-weatherable soil components. Sverdup and Warfvinge (1988) used Equations (18-20) to estimate soil surface areas and weathering rate constants for microcline, plagioclase and hornblende in soils in the Gårdsjön watershed (Table 7).

Table 7. Comparsion of natural and experimental dissolution rates of selected silicate minerals

	Log Rate k_j (mol/cm^2/s)	pH	Environment	Location	Method of Calculation	Surface Area	Age (kA)	Reference
Plagioclase								
Oligoclase	-19.9	4.5-7.0	Soil	Merced, CA	Mineralogy	BET	10-3000	White et al., 1995
Oligoclase	-19.5	4.5-70	Soil	Merced, CA	Solute Flux	BET	10-3000	This paper
Andesine	-18.7	5	Catchment	Filson Ck., MN	Watershed Balance	Geo	10	Siegel & Pfannkuch, 1984
Oligoclase	-18.5	5.8	Catchment	Bear Brook, ME	Watershed Balance	Geo	10	Schnoor, 1990
Labradorite	-17.7	6.0-7.5	Aquifer	Trout Lake, WI	Groundwater	Geo	10	Kenoyer & Bowser, 1992
Oligoclase	-17.4	4.5-7.0	Soil	Merced, CA	Mineralogy	Geo	10-3000	White et al., 1995
Oligoclase	-17.3	2.0-4.5	Soil Solution	Bear Brook, ME	Solute Flux	Geo	10	Swoboda-Colberg & Drever, 1992
Oligoclase	-17.1	5.6-6.1	Soil	Gardsjon, Sweden	PROFILE model	BET	10	Sverdrup, 1990
Oligoclase	-16.5	6.8	Catchment	Coweeta, NC	Watershed Balance	Geo	10	Velbel, 1985
Oligoclase	-16.1	5	Catchment	Hartviko, Czech.	Watershed Balance	Geo	10	Paces, 1983
Albite	-16.1	5	Experiment	NA	Dissolution	BET	Crushed	Knauss & Wolery, 1986
Oligoclase	-16.0	5	Experiment	NA	Dissolution	BET	Crushed	Oxburgh et. al., 1994
Albite	-15.9	5.6	Experiment	NA	Dissolution	BET	Crushed	Chou & Wollast, 1985
K-feldspars								
Orthoclase	-20.5	4.5-7.0	Soil	Merced, CA	Mineralogy	BET	10-3000	White et al., 1995
Orthoclase	-19.7	4.5-7.0	Soil	Merced, CA	Solute Flux	BET	10-3000	This paper
K-feldspar	-18.1		Soil	Loess,Il	Etch Pitting (PSD)	Geo	10-15	Brantley et al. 1993
Orthoclase	-17.8	4.5-7.0	Soil	Merced,CA	Mineralogy	Geo	10-3000	White et al., 1995
K-feldspar	-17.3	5.6-6.1	Soil	Gardsjon, Sweden	PROFILE model	BET	12	Sverdrup, 1990
K-feldspar	-17.3	2.0-4.5	Soil Solution	Bear Brook, ME	Solute Flux	Geo	12	Swoboda-Colberg & Drever, 1992
Microcline	-16.8	5.6	Experiment	NA	Dissolution	BET	Crushed	Schweda, 1989
Hornblende								
Hornblende	-20.1	4.5-7.0	Soil	Merced, CA	Mineralogy	BET	10-3000	White et al., 1995
Hornblende	-18.5	2.0-4.5	Soil Solution	Bear Brook, ME	Solute Flux	Geo	10	Swoboda-Colberg & Drever, 1992
Hornblende	-18.1		Soil	Loess,Il	Etch Pitting (PSD)	Geo	10-15	Brantley et al. 1993
Hornblende	-17.6	5.6-6.1	Soil	Gardsjon, Sweden	PROFILE model	BET	10	Sverdrup, 1990
Hornblende	-17.5	4.5-7.0	Soil	Merced, CA	Mineralogy	Geo	10-3000	White et al., 1995
Hornblende	-15.2	4	Experiment	NA	Dissolution	BET	Crushed	Zhang, et al., 1993
Hornblende	-14.7	5	Experiment	NA	Dissolution	BET	Crushed	Sverdrup, 1990

[handwritten annotation:] vary by ~3 orders of magnitude.

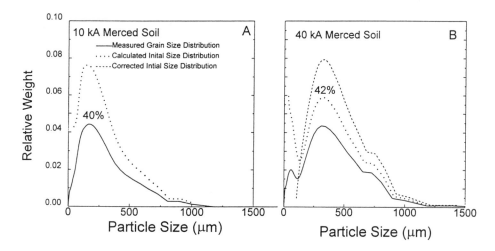

Figure 12. Measured grain size distributions of representative samples of the Merced soils (solid lines) and calculated parent size distributions with (dashed line) and without corrections (dotted lines) for bimodal distribution of fines. Percentages are increases in parent relative to present masses.

Interactive approaches

The above approaches have significant limitations in calculating mineral weathering rates in soils based on mass balances. Although the simple geometric model (Lasaga, 1984) considers the effects of variations in particle mass and size as a function of dissolution, the model did not consider particle size distributions nor the surface roughness of natural mineral grains. The model of Sverdup and Warfvinge (1988) considers grain size distributions but does not consider how particle sizes and surface areas change as a function of mass and volume loss during dissolution. A more inclusive approach to calculating mineral weathering rates in soils is to consider a dynamic model in which a population of grain sizes are considered and for which there is interactive feedback between changes in particle size, surface area, roughness and mass transfer rates (White et al., 1995).

Such a model requires that the initial grain size distribution r^o be known. In soils, such as saprolites, this distribution can be estimated by the mineral size distributions in the parent rock. Under such conditions the surface area and rate constant can be related to changes in grain size distributions in the soil (i e. $r^o - r$, Eqn. 17). For soils developed on sedimentary deposits, the calculation of changing particle diameter with time requires an inverse solution in which the final rather than the initial grain size distribution is known. As discussed by White et al.(1995) this distribution can be reconstructed by incrementally adding mass back to the present grain-size distribution until the original mass is achieved ($M_j = 1$).

Examples of both present day grain sizes (solid lines) and reconstructed parent grain sizes (dotted lines) for 10 kA and 40 kA Merced soils are shown in Figure 12. Initial calculations of grain size changes with time used complete size distributions measured in the present day soil profiles. Results for the youngest soil showed a relatively symmetric increase in the population of grain sizes from the present distribution for r (Fig. 12A, solid line) to that of the initial parent material r^o (Fig. 12A, dotted lines). However the bimodal distribution of measured grain sizes in the 40 kA Merced soil is greatly accentuated in the calculated parent grain sizes (Fig. 12B). Most of the reconstituted mass was preferentially incorporated into very small grain sizes (< 100 μm) due to their large surface area to mass ratios. Such pronounced bimodal distributions in the parent material are almost certainly artifacts of the calculation.

The above results point out the importance of grain size on the relative weathering rates and stability of minerals in soils. Due to their large surface area to mass ratios, small size grains of feldspars and other reactive minerals should be rapidly weathered and will not persist in soil environments. In actuality, the small grains in the bimodal size distributions now present in the 40 kA Modesto soil (Fig. 12B) are probably composed of residual quartz grains and fragments of larger feldspar and hornblende grains decomposed by weathering. In additional calculations performed by White et al. (1995), the small grain size fractions were arbitrarily eliminated by smoothing the measured grain-size distributions. The resulting reconstituted size distribution for the parent material (Fig. 12, dashed line) is no longer bimodal and is centered in the sand size fractions which are more representative of weathering processes in granitic terrains.

Calculation of dissolution rates

An interactive dissolution rate model was developed for the Merced soil data (White et al., 1995) that calculates the reaction rate constant k_j (Eqn. 16) based on changes in specific mineral abundance (Fig. 2B), surface roughness (Fig. 11) and changes in particle sizes with time (Fig. 12). A forward approach was employed in which dissolution of each grain size in a sample was sequentially calculated over a number of time steps n from the age of initial deposition to the present. At each time step, the mineral surface area in each size fraction was calculated from the nominal spherical diameter, the appropriate surface area function ($\lambda = 1$ or λ = variable), and the mass of mineral in the size fraction. The mass of mineral dissolved in each size fraction during the time step is calculated from the surface area of the size fraction and the dissolution rate. The mass dissolved was then subtracted from the mass of the size fraction, and new nominal diameters for the grains within each size fraction were calculated at the end of the time step. The new diameter was used to calculate the surface area for the next time step, and the process is repeated. Thus, the mass remaining in each size fraction $M^{\phi}_{j,w}$ can be calculated as (White et al., 1995)

$$M_{j,w}^{\phi} = M_{j,p}^{\phi} - \sum_{t=1}^{n} M_{t,j,w}^{\phi} \cdot \varphi_m \cdot S_t^{\phi} \cdot k_j \cdot \Delta t \qquad (21)$$

$M^{\phi}_{j,p}$ is the initial mass in the size fraction ϕ, φ_m is the molecular weight of the mineral, and S_t^{ϕ} and $M_{t,j,w}$ are the specific surface area and mass of the mineral remaining for size fraction ϕ at time Δt. The proportion of a mineral remaining in a soil profile today R_o is equal to the sum of the mass remaining in each size fraction ϕ divided by the initial mass of the sample,

$$R_o = \frac{\displaystyle\sum_{\phi=1}^{n} M_{j,w}^{\phi}}{\displaystyle\sum_{\phi=1}^{n} M_{j,p}^{\phi}} \qquad (22)$$

Note that R_o is the same term initially defined in Equation (3) but is recast in term of absolute mineral mass rather than concentrations relative to total soil mass. R_o is also used to define actual variations in mineral masses as a function of time as previously shown for the Merced chronosequence data (Fig. 2B). As written, the forward dissolution model (Eqn. 22) requires a specified dissolution rate k_j to calculate the proportion of mineral masses remaining in the soil as a function of age. In weathering studies, these proportions are known and the dissolution rate constants need to be calculated. This problem was solved by finding the

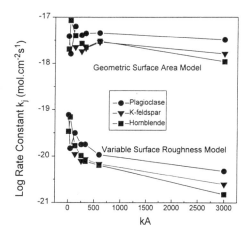

Figure 13. Calculated dissolution rate constants for specific minerals as a function of age in the A horizons of the Merced chronosequence. Rates incorporate geometric surface areas or variable surface roughness (Eqns. 21 and 22).

dissolution rate that yields the closest approximation to the measured mass losses described by R_o (Fig. 2B). Exact solutions of k_j for specific mineral distributions in soils of a given age in the Merced chronosequence were obtained (White et al., 1995). An overall average dissolution rate for each mineral was also determined for the entire chronosequence by minimizing the sum of the squares of the residuals between the predicted and observed mineral abundances. Each soil profile has a different grain-size distribution, which results in a different surface area and a different rate of mass loss. Therefore, this model does not yield a continuous function for mass loss that can be directly interpolated between samples, but rather yields discrete predictions for each individual sample that can be compared with the field observations.

Rates constants calculated from geometric and BET surface area estimates

Geometric and BET surface area measurements are the most commonly used approaches in defining surface areas in weathering rate calculations for natural systems (White and Peterson, 1990). These two scenarios were investigated by White et al. (1995) in relation to surface area effects on weathering rates in the Merced soils. In the first case, geometric surface areas of mineral grains in the model were varied directly with changes in particle diameters and grain-size distributions with time. This approach assumed that the surface roughness was equal to unity ($\lambda=1$) and the grains have no internal porosity ($S_i=0$) (see Eqn. 14). In the second case the BET surface area was interpreted in terms of both an internal surface area, which was assumed constant with time, and an external surface area, which was defined in terms of a roughness factor λ that varied with time (Fig. 11).

Resulting individual dissolution rate constants of plagioclase, K-feldspar and hornblende from the geometric and variable surface roughness models are plotted in Figure 13. The geometric model produces average dissolution rates that are 10^2 to 10^3 orders of magnitude faster than the variable-roughness model. The magnitude of this difference is to a first approximation proportional to the difference between the measured BET and the geometrically-estimated surface areas (Fig. 8). The difference in rates is also dependent on soil age. Based on geometric surface areas, rate constants exhibited no consistent variability with age. In contrast, the variable roughness model predicted that individual dissolution rates decrease by more than an order of magnitude with increasing soil age.

An example of the optimization fit of average dissolution rate constants to mineral distributions in the Merced A horizons is shown in Figure 2. These fits (dashed lines) are based on the variable roughness model for the mineral surfaces. As indicated, the fit does not produce a single smooth function describing the mineral distributions due to differences in particle size distributions in individual soil profiles. Due to the decrease in k_j with time, a single rate constant fitted to the entire chronosequence underestimated weathering rates in the younger soils and overestimate rates in the older soils. This effect is evident in Figure 2 in which predicted mineral proportions (open symbols) are generally greater than observed proportions (closed symbols) in the younger soils and less than the observed mineral proportions in the older soils.

WEATHERING RATES BASED ON ETCH PIT FORMATION

Changes in mineral surface morphology have been employed to directly quantify mineral weathering rates in soils. As previously discussed, systematic increases in the extent and depth of surface pitting have been observed during weathering of silicate minerals in soil environments (Fig. 6). MacInnis and Brantley (1993) and Brantley et al. (1993) (See Brantley, present volume) have recently developed a pit size distribution model (PSD) which considers the population density of etch pits within a given grain size range during dissolution of a mineral phase. The PSD depends upon the number of pits per unit area, n (cm cm^{-2}) and the growth rate, G (cm s^{-1}) of individual pits. The rate of annihilation of pits, a (s^{-1}), is the rate of coalescence of two pits together, or if the pit stops deepening, the rate at which the pit disappears as the flat surface of the crystal recedes.

Assuming the growth rate is time- and size-independent at constant temperature, a population balance on pits of diameter W (cm) can be written in order to derive the steady-state pit size distribution equation. The following PSD equation results when $dn/dt = 0$, assuming that n_o is the density of etch pits at time 0 and τ is the shear modulus

$$n = n_o \exp\left(-\frac{W}{G \cdot \tau}\right) \tag{23}$$

For these assumptions, a plot of $ln\ n$ against W will produce a straight line with slope $-1/G\tau$ and intercept $ln\ n_o$ (Fig. 14). The PSD model defines the characteristic lifetime, τ, as the reciprocal of the annihilation rate.

A linear PSD plot $ln\ n$ vs. W results when there is a balance between nucleation and annihilation of pits, and where there is size-and time-dependent growth rate of pits. At steady state, PSD coefficients do not change with time; however, linear PSDs are also possible for non-steady-state systems. On the other hand, where annihilation or growth is size dependent, the PSD will not be linear: for example, where coalescence preferentially removes the smallest pits, the PSD may become humped (MacInnis and Brantley, 1993).

The contribution of etch pits to the bulk dissolution rate depends upon the number, density and the rate of dissolution of each pit. By integrating over the PSD, and assuming an average geometry for the pits, MacInnis and Brantley (1993) derived a simple expression for the etching rate of a mineral based upon PSD coefficients. For example, for pit geometries of right pyramids, the contribution of etch pits to the bulk dissolution rate equals

$$R_o = n_o \tan\frac{\theta}{Vj} \cdot G(G\tau)^3 \tag{24}$$

where θ is the pit wall slope.

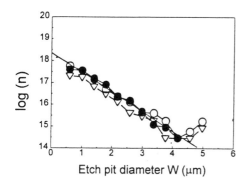

Figure 14. Etch pit size distributions (number of pits n with pit diameter W) for feldspar in Fayatte soil, Illinois. Open circles are upper, solid circles are middle, and triangles are lower soil horizons (from Brantley et al., 1993).

Brantley et al. (1993) analyzed etch pits on loess grains from a 12 kA soil catena from Illinois. For hornblende and potassium feldspar grains, etch pits had not grown to the extent that they had completely coalesced. Where pits have dramatically coalesced, quantification of the pit size distribution is impossible. Figure 14 slows an example of a plot of etch pit width versus number of pits for feldspar from the Fayette soil. Although perfectly linear PSDs were not observed, (typically, distributions exhibited a lack of small pits and an overabundance of large pits), PSD coefficients could be derived for the linear portion of the distribution curves.

By assuming that τ equals the total etching period (12 kA), Brantley et al. (1993) calculated weathering rate constants of log k_j = -18.5 mol cm^{-2} s^{-1} for hornblende grains and k_j = -18.7 2 mol cm^{-2} s^{-1} for K-feldspar. Rates estimated for hornblende were based on consistent crystallographically controlled etch pits, while rates estimated for potassium feldspar were based on irregularly shaped pits. Although little difference in etching rate is observed between soil horizons, the highest etching rates occurred generally in the upper B horizon where the pH was lowest. Etching rates calculated for potassium feldspar were not observed to vary with drainage, while those of hornblende decreased with decreasing drainage. Brantley et al. (1993) argued that decreasing drainage correlates with increasing solute concentration in soil pore waters and increased inhibition of dissolution for the hornblende. Of course, decreased drainage also correlates with lower partial pressures of oxygen in soil waters, which affects hornblende more drastically than K-feldspar.

MINERAL DISSOLUTION RATES BASED ON SOLUTE FLUXES

An alternate approach in determining weathering rates of primary silicates in soils is based on solute fluxes in the soil profile. Conservation requires that the mass of primary mineral phases dissolved in the soil be balanced by the mass increase in secondary minerals minus the loss of mass due to solute fluxes from the profile. The flux Q_k (mol cm^{-2} s^{-1}) for a solute species k through a unit area of a soil profile can be calculated by multiplying the soil water chemical concentration c_k (mol cm^{-3}) by the fluid flux density q (cm.s^{-1}). Note that q is equivalent to the rate of water movement through a unit surface area of the soil profile

$$Q_k = q \cdot c_k \qquad (25)$$

For a number of chemical species n distributed between different mineral phases m, the mass balance relationship in the soil becomes (Velbel, 1986)

$$\sum_{j=1}^{m} Q_j \cdot N_{j,k} = Q_k \qquad k = 1,......n \tag{26}$$

where Q_j (mol cm^{-2} s^{-1}) is the equivalent flux representing dissolution or precipitation of mineral phase j and $N_{j,k}$ is the stoichiometric ratio of species k in mineral phase j.

The solution to the series of linear equations represented by Equation (26) permits the assignment of portions of solute chemistries and fluxes to a specific mineral phase. Solutions based on "balance sheet approaches" have been used by numerous workers studying chemical fluxes in watersheds (Paces, 1983; Siegal and Pfannkuch, 1984; Velbel, 1985). Use of mass balance models which simultaneously solve the matrix of linear equations, however, permit much greater flexibility in the number of aqueous species and primary and secondary minerals considered. BALANCE (Parkhurst et al., 1982) and NETPATH (Plummer et al., 1991) are two widely used codes which can be used to solve Equation (26).

The reaction rate constant described by Equation (16) for a specific mineral phase j can now be rewritten in terms of solute fluxes which are defined by the product of fluid flux and chemical concentrations (Eqn. 25).

$$k_j = (\frac{M_j}{\Delta t})\frac{1}{S_j} = q \ R_s \sum_{k=1}^{n} N_{j,k} \ c_k^t \tag{27}$$

In Equation (27), c_k^t (mol cm$^{-3)}$ is the concentration of species k produced from weathering of mineral phase j. The measured concentration in the soil solution c_k in Equation (25) is related to c_k^t in Equation (27) such that

$$c_k = c_k^t - c_k^s + c_k^a \tag{28}$$

where c_k^s is the concentration of species k incorporated into secondary mineral phases and c_k^a is the concentration contributed from non-weathering processes such as atmospheric deposition. Equation (28) assumes that the fluid mass is constant and not effected by processes such as evapo-transpiration.

The solute flux is defined in right side of Equation (27) in terms of q (cm s^{-1}) which is equal to the volume of water passing through a unit cross-section area of soil perpendicular to the flow direction. In contrast the rate constant k_j and the mass term M_j are normalized relative to a specific mineral surface area. The conversion factor R_s (cm^2 cm^{-2}) is defined as the ratio of unit area of soil surface area S_s (cm^2) divided by the total surface of S_t (cm^2) of mineral phase j contained in the volume of soil. S_t, in turn, is dependent on the specific mineral surface area S (cm^2 gm^{-1}), the depth of the soil zone z the soil density ρ_w, and the mass fraction of mineral j present in the soil f_m. Therefore, for a unit surface area of soil surface (S_s=1 cm^2)

$$R_s = \frac{S_s}{S_t} = \frac{1}{S} \cdot \frac{1}{\left(z \cdot \rho_w \cdot f_m \right)} \tag{29}$$

Based on the preceding analysis (Eqns. 25-29), the mineral weathering rate k_j can be calculated for a mineral phase if the primary and secondary mineral compositions, solute concentrations and soil water fluxes can be determined.

Table 8. Summary of selected soil water chemistries. Numbers are mean average values with numbers in paratheses as stardard deviations. Units except for pH are in umol.l^{-1}.

Reference	Location	Soil type	Method	pH	Ca	Mg	Na	K	Si	Al
Edmeades et al., 1985	New Zealand	Various	2	6.2 (0.3)	794 (293)	237 (74)	786 (392)	641 (294)	346 (89)	37 (28)
Arocena et al., 1994	Alberta	Sandy Luvisols	2	4.9 (0.7)	670 (618)	165 (83)	79 (28)	294 (246)	235 (170)	105 (65)
Manley et al.,1987	Ontario	Podzols	2	5.1 (0.7)	135 (152)	73 (60)	8 (5)	157 (155)	198 (127)	43 (37)
Campbell et al, 1989	England	Sandy loam	2	6.7	669	226	480	452	210	44
Reynolds et al., 1988	Wales	Stagnopodzols	3	4.3 (0.2)	13 (7)	22 (5)	188 (56)	5 (2)	43 (16)	44 (27)
Katz, 1989	Maryland	Loam/saprolite	3	6.9	133	154	96	5	185	nd
Norfleet et al., 1993	South Carolina	Oxidic loams	3	5.3 (0.5)	12 (9)	10 (10)	85 (52)	34 (22)	105 (69)	2.8 (7)
Hughes et al, 1994	Wales	Stagnopodzols	3, 4	4.3 (0.2)	115 (172)	154 (48)	310 (94)	19 (11)	41 (28)	3 (1)
Soulsby and Reynolds, 1992	Wales	Stagnopodzols	3	3.9 (0.2)	27 (9)	75 (26)	350 (76)	20 (21)	69 (31)	185 (125)
Cronan et al., 1990	New York	Spodosols	3	4.5 (0.2)	45 (8)	11 (3)	32 (11)	21 (13)	121 (33)	58 (18)
Cronan et al., 1990	Tennessee	Ultisols	3	4.9 (0.4)	13 (6)	20 (6)	20 (5)	23 (18)	49 (27)	7 (7.8)
Karathanasis, 1991	Kentucky	Undefined	1	6.3 (0.2)	318 (62)	68 (31)	258 (75)	148 (136)	152 (82)	0 (0)
Ugolini et al, 1999	Japan	Spodosols	3	5.7 (0.6)	36 (32)	24 (19)	126 (90)	118 (167)	87 (67)	3 (3)
White et al (unpublished)	Merced, CA	Utisols 10-40 k	3	7.1 (0.7)	364 (166)	166 (87)	194 (71)	146. (105)	720 (204)	2 (5)
White et al (unpublished)	Merced, CA	Utisols 130-330	3	7.0 (0.5)	313 (176)	196 (121)	466 (195)	42 (77)	1150 (570)	4 (8)
White et al (unpublished)	Merced, Ca	Utisols 600 kA	3	7.2 (0.5)	421 (304)	210 (113)	396 (190)	35 (76)	819 (539)	12 (15)
White et al (unpublished)	Merced, CA	Utisols 3000 kA	3	7.5 (0.3)	273 (198)	200 (189)	816 (385)	77 (89)	1116 (570)	6 (80)
White et al (unpublished)	Panola. Georgia	Saprolite	3	5.6 (0.2)	52 (44)	56 (31)	91 (40)	15 (8)	212 (69)	53 (220)
White et al (unpublished)	Puerto Rico	Saprolite	3	5.3 (0.7)	16 (20)	32 (18)	189 (77)	2 (6)	118 (88)	5 (9)

Methods of extraction (1) centifuge, (2) centifuge/displacement, (3) suction cup sampler, (4) zero tension sampler

Soil solution chemistry

As described by Joffe (1933), "soil solutions are the blood circulation of the soil body" and as such are the mobile components controlling, as well as reflecting many soil processes, including weathering, chemical equilibrium, ion exchange and nutrient transport. During periods of active recharge, such as precipitation and snow melt events, soil water can be saturated and sampled using zero tension or free drainage lysimeters. More commonly, soil water is present in unsaturated conditions with capillary tension or matrix potential preventing direct sampling under free flowing conditions. The most commonly used techniques for sampling soil water under unsaturated conditions include laboratory-based extraction using centrifugation and immiscible liquid displacement (Campbell et al., 1989; Norfleet et al, 1993) and in situ field techniques such as suction and porous cup samplers (Huges et al., 1994). Accurate characterization of soil solution chemistry is dependent on extraction procedures which produce the minimal amount of physical disturbance to the soil column and the least impact on complex geochemical conditions in the soil waters.

Chemical composition data of unsaturated soil waters pertinent to silicate weathering studies are relatively limited. Selected mean average soil water chemistries cited in the literature, in addition to unpublished data from the Merced, Rio Icacos, and Panola sites, are presented in Table 8. Also included in Table 8 are locations, soil types and sampling methodologies. Mean average soil pHs generally range from neutral to acidic (7.5 to 3.9). In situ pHs are presumable more acidic due to the loss of CO_2 and possible degradation of organic acids during sampling under atmospheric or sub-atmospheric (suction) conditions. Concentrations of major cations range from several tens to hundreds of µmoles, SiO_2 ranges from hundreds to a thousand µmoles and Al ranges from below detection limits to tens of µmoles.

The standard deviations listed in Table 8 approach the mean average chemical concentrations, indicating significant chemical variability in soil solutions. Such differences are in part attributed to spatial variability which influences the hydrological flow path. Heterogeneity in soils is caused by macropores, duripans and lateral flow associated with hill slope processes. Short term temporal variability stems from periods of active recharge during precipitation and snow melt, which dilutes the soil solution, followed by periods of evapo-transpiration, which tend to concentrate soil solutions.

Longer term chemical variability in soil solutions generally reflects the intensity of soil weathering and the mineral compositions in the soil profile. Solute concentrations often increase with increasing depth in soil profiles, reflecting longer fluid residence times and greater inputs from the weathering of silicate minerals. The effect of soil age on weathering inputs is demonstrated by the bivalent/monovalent chemical ratios in soil solutions of various soil profiles in the Merced chronosequence. As shown in Figure 15, the Ca/Na ratios clearly decrease from approximately 4 in the youngest Modesto soils to less than 0.3 in the oldest China Hat soils. This decrease corresponds to a decrease in the more weatherable plagioclase and hornblende minerals which contain abundant bivalent cations relative to alkali feldspar, which dominants weathering reactions in the older soils. In this case, the evolution of the soil solution chemistry clearly corresponds to the weathering sequence of the minerals in the soils.

Comparison of soil solution and watershed chemistry

Most watershed studies, which have estimated soil weathering rates, assume that surface discharge chemistry reflects the soil water chemistry. However few studies have attempted to make direct chemical comparisons or to address where in the soil profile

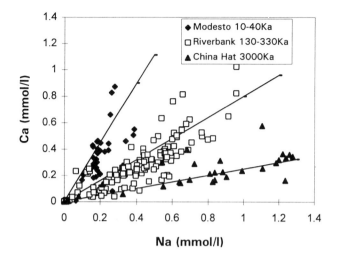

Figure 15. Ca-Na ratios in soil solutions of different ages in the Merced chronosequence. Diagonal lines are regression fits to data.

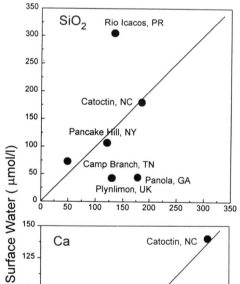

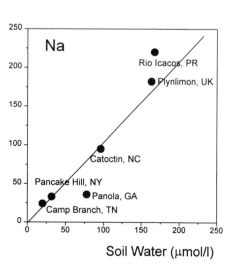

Figure 16. Comparison of average soil solution compositions with corresponding mean annual surface discharge chemistry in selected watersheds.

weathering actually occurs. The average soil solute SiO_2, Ca and Na concentrations reported in Table 8 are compared to mean annual surface water discharge concentrations for corresponding watersheds in Figure 16. The one-to-one solid diagonal lines represent the situation in which the mean soil solution chemistry is directly reflected in the average stream chemistry. As indicated this assumption appears valid for several of the watersheds.

However significant discrepancies between soil and surface chemistries are also apparent. For example, SiO_2 and cation concentrations in Panola watershed are generally lower in the surface water relative to that in the soil water (Fig. 16). This discrepancy is related to significant surface runoff from exposed bedrock in the watershed. Rapid surface runoff, reflecting low dissolved solids, dilutes the surface water discharge relative to the more concentrated weathering input from soil waters. In contrast, average surface discharge water in Rio Icacos watershed is more concentrated than the soil water chemistry. As indicated by the previous discussion of mass balances, weatherable minerals are almost totally depleted in the Rio Icacos saprolite (Fig. 1). Most weathering of granitic mineralogy occurs at a narrow interface between the bedrock and soil zone. Weathering of this interface therefore contributes to higher silica and cation concentrations in the surface water discharge relative to weathering in the overlying sapolite.

Equilibrium controls on soil solution chemistry

The precipitation of secondary minerals, most commonly clays and Al and Fe hydroxides and oxides, is expected to strongly influence the weathering signature contained in soil solutions and to impact mass balance calculations (Eqns 27- 28). For example, SiO_2 is the major chemical species most directly related to weathering reactions. Silica is not contributed from atmospheric precipitation nor is aqueous SiO_2 extensively modified by soil ion exchange processes and biological uptake. However SiO_2 concentrations are strongly affected by precipitation of clay minerals. Kaolinite is the most commonly reported weathering product of silicate rocks in soils. The weathering reaction for plagioclase can be represented as

$$[NaAlSi_3O_8]_{albite} + 1.5\ H_2O + CO_2 \rightarrow 0.5[Al_2SiO_2O_5(OH)_4]_{kaolinite} + Na^+ + HCO_3^- + 2H_4SiO_4$$

The stoichiometry of the above reaction requires the aqueous release of two moles of SiO_2 for every mole incorporated with Al in the solid phases. If gibbsite forms in the soil, 3 moles of SiO_2 will be released and if smectites forms, less than 2 moles of silica will be mobilized. The actual silica concentrations in soil solutions depend strongly on equilibrium with secondary mineral phases.

One common approach to determine mineralogical controls on solution chemistry is the use of activity diagrams (Garrels and Christ, 1965). The above reaction can be characterized by plotting the aqueous activity of H_4SiO_4 in the soil solution versus the ratio of Na^+/H^+. Activity diagrams for the systems $Na_2O-Al_2O_3-SiO_2-H_2O$ in addition to systems $K_2O-Al_2O_3-SiO_2-H_2O$ and $CaO-Al_2O_3-SiO_2-H_2O$ are shown in Figure 17. The soil solution data from the references tabulated in Table 8 are plotted based on the assumption that activities are comparable to chemical concentrations. In terms of SiO_2 phases, soil solutions appear generally unaffected by quartz solubility i.e., chemical concentrations (activities) plot both above and below quartz solubility (Fig. 17, left-hand vertical dashed line). In contrast, maximum aqueous SiO_2 concentrations in soil waters appear to be limited by the precipitation of amorphous silica (Fig. 17, right-hand vertical dashed line) which commonly is associated with the formation of duripans and fragipans in soils (Karathanasis, 1989).

Most of the soil solution data fall within the kaolinite stability field (Fig. 17), a

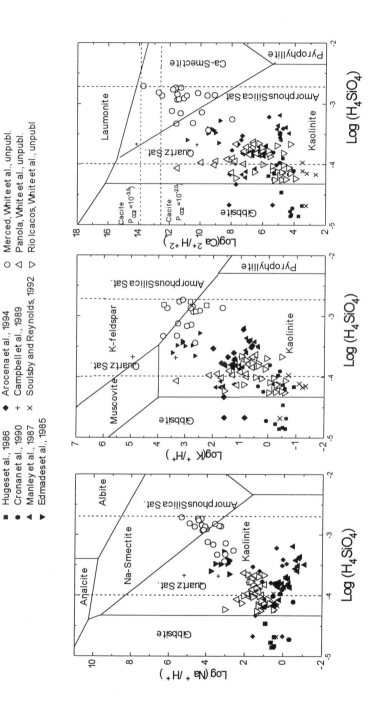

Figure 17. Soil solute concentrations (Table 8) plotted on mineral stability fields for the systems Na2O-Al2O3-SiO2-H2O, K2O-Al2O3-SiO2-H2O and CaO-Al2O3-SiO2-H2O.

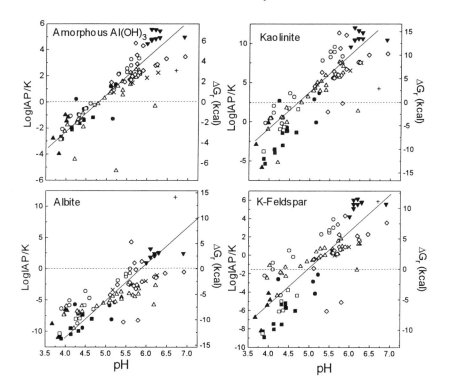

Huges et al., 1986
Cronan et al., 1990
Edmemades et al., 1985
Arocena et al., 1994
Soulsby and Reynolds, 1992
Campbell et al., 1989
Reynolds et al., 1988
Manley et al., 1987
Ugolini et al., 1988
Panola, White, unpublished

Figure 18. Ionic activity products and reaction free energies calculated for individual minerals based on solution soil chemistries cited in Table 8. Calculations assume speciation based on total aqueous Al.

situation previously observed for activity diagrams developed for other soil solution studies (Manley et al., 1987 ; Katz, 1989). Kaolinite is also the most common clay mineral forming in association with common silicate rocks such as granite. The diagrams indicate, however, that some soil solutions are undersaturated with kaolinite. Such a situation is evident in the Rio Icacos soils where kaolinite is actively undergoing dissolution in the upper horizons of the soil profile. Soil waters with high solute concentrations cross over into the Na and Ca smectite stability fields (Fig. 17) suggesting that smectite formation also occurs in some soils. High Ca^{2+}/H^{+2} ratios in some soils also indicate that soil solutions saturate with calcite which is supported by the presence of calciche formation. The activity diagrams also suggest the some soil solutions are approaching saturation with respect to primary silicate minerals such as K-feldspar.

The stability of both secondary and primary mineral phases can also be predicted directly from solubility calculations. The mineral solubility product is defined as K and the corresponding aqueous ionic activity product is defined as IAP. Based on strict thermodynamic interpretation, when $IAP/K = 1$ the free energy of reaction is zero $\Delta G_r = 0$

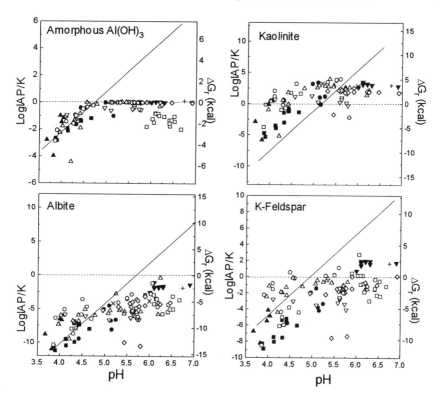

Figure 19. Ionic activity products and reaction free energies calculated for individual minerals based on soil solution chemistries cited in Table 8. Calculations assume that Al concentrations are controlled by gibbsite saturation at measured pH. Symbols as in Figure 18.

and the mineral phase is in equilibrium with the soil solution. When $IAP/K < 1$, then $\Delta G_r < 0$, and the mineral will dissolve. When $IAP/K > 1$, then $\Delta G_r > 0$ and the mineral will precipitate. Under ambient soil solution conditions, the above thermodynamic interpretation is strongly modified by kinetic restraints on the reactions as will discussed in a following section.

Saturation states for soil waters (Table 8) were determined using the SOLMINEQ 88 chemical speciation code (Kharaka et al., 1988). Minerals included in the calculations were gibbsite and kaolinite, secondary phases which commonly precipitate from soil solutions, and albite and K-feldspar, phases that commonly dissolve. IAP/K values based on total Al analysis indicate very wide ranges in solution saturation for the mineral phases (Fig. 18). At low pH, the solutions are generally unsaturated (below the dashed saturation line) and at near neutral pH, the solutions are over saturated (above the dashed saturation line). The maximum calculated IAP values suggest that the soil solutions at near neutral pH exceed gibbsite saturation by 5 orders of magnitude and kaolinite by 15 orders of magnitude.

The plot of pH versus IAP/K for the soil solution data approximate a slope of 1 to 3 (diagonal lines; Fig. 18) even in supersaturated solutions. This slope is proportional to a direct dependence of IAP on H^+ activity if the Al concentration is approximately constant and independent of pH. Precipitation of gibbsite or kaolinite is not reflected in this consistent ratio even though these minerals commonly occur as secondary weathering products in soils.

This discrepancy results from difficulties in the measurement of Al at low concentrations, particularly at near-neutral pH. Dissolved Al in such pore waters are present in a number of forms including inorganic, organic and polymeric species (Huges et al., 1994). Complexation, coupled with common problems associated with sampling, filtration and analysis, often result in significant over-estimation of Al concentrations equilibrated with Al-containing minerals (Hem, 1985).

One approach to correct for analytical measurements and speciation of Al in thermodynamic calculations is to assume that Al is controlled by saturation with a simple $Al(OH)_3$ phase such as gibbsite (Wesolowshi, 1992). This approach is supported by detailed analyses of Al in soil waters which suggest that monomeric inorganic Al concentrations at near-neutral pH is controlled by amorphous or microcrystalline gibbsite saturation (Karathanasis, 1989; Norfleet et al., 1993). Using this approach, the *IAP* values for gibbsite, kaolinite, albite and K-feldspar were recalculated by assuming that the maximum *IAP* for $Al(OH)_3$ is limited by gibbsite saturation at any given pH. This was achieved in the SOLMINEQ.88 program by back-titrating excess Al from the analysis until gibbsite saturation was achieved. These results (Fig. 19) indicate that the increase in *IAP* becomes truncated at gibbsite saturation at pHs above approximately 4.5. These limiting values on Al concentrations produce more realistic results in terms of saturation with respect to the other mineral phases.

Role of solute chemistry and reaction affinity

From elementary thermodynamics, the free energy of the weathering reaction can be defined as

$$\Delta G_r = RT(logIAP - logK) \tag{30}$$

Transition state theory (TST) can be used to relate the kinetic rate k_j to this reaction affinity (Lasaga, 1981) (Also see this volume),

$$k_j = -k_o \left(1 - \exp(\frac{n\Delta G_r}{RT}) \right) \tag{31}$$

where n is the reaction order. In Equation (31), k_j is defined as the net reaction rate between precipitation and dissolution reactions at the mineral surface. For dissolution reactions, values of ΔG_r and k_j are negative and for precipitation reactions ΔG_r and k_j are positive. At $\Delta G = 0$ the rates of dissolution and precipitation are equal. Equation (31) predicts that far from thermodynamic equilibrium, the dissolution or precipitation rate will be independent of ΔG_r and equal to a constant rate $(k_j = k_o)$, whereas at progressively closer to equilibrium, ΔG_r will exert an increasingly stronger effect on the overall reaction rate $(k_j < k_o)$. *IAP/K* values reported in Figure 19 suggest that kaolinite precipitates out of solution at relatively constant supersaturation approximately 100 times above K. This *IAP* corresponds to an average ΔG_r value of approximately 3 kcal. The effect of increasing supersaturation on the increasing rates of kaolinite precipitation was been investigated experimentally at elevated temperatures (80°C) (Nagy et al., 1991; Nagy et al., 1993) (also see Nagy, this volume). Assuming comparable relationships between ΔG_r and rates at lower temperature, these studies suggest that kaolinite may be precipitating out of soil solutions at significantly more rapid rate than if the solutions were less saturated.

The extent of solution supersaturation and therefore the rate of precipitation of kaolinite from soil solutions is controlled in large measure by the rates of Al and SiO_2 release from weathering of primary silicates such as albite and K-feldspar. Based on strictly thermodynamic considerations (Eq. 30), these phases should not dissolve in solutions that exceed their saturation state. This conclusion may be tempered somewhat for soil solutions by kinetic restraints which prevent reprecipitation of these phases from solutions which become supersaturated due to concentration by evapo-transpiration and other processes. As indicated however, soil waters generally remain undersaturated with respect to albite over the entire soil water pH range and saturate with respect to K-feldspar only near neutral pH.

In summary soil solutions reflect both input from weathering processes and may influence weathering rates by controlling both precipitation of secondary minerals and the rates of primary mineral dissolution.

Soil Water Hydrology

The determination of the soil water flux density q is required in order to calculate mineral weathering rates based on soil water chemistry (Eqns. 25 and 26). A general review of unsaturated zone hydrology is beyond the scope of this paper and the reader is referred to texts on the subject such as Hillel (1982). Briefly the flux density per unit area of soil S_s can be defined by the Richards equation (Richards, 1931)

$$q = -K_m \nabla H \tag{32}$$

where K_m is the unsaturated hydraulic conductivity and ∇H is the hydraulic gradient. Equation (32) is equivalent to Darcy's law with the complication that K_m is dependent on degree of saturation or moisture content of the soil (m = volume water/volume soil).

When the soil is saturated, all of the pores are water filled, so that continuity and hence conductivity are maximal. When the soil desaturates, some of the pores become air filled and the conductive portion of the soil cross sectional area decreases correspondingly. The first pores to empty are the largest ones which are the most conductive thus leaving water to flow only in the smaller pores increasing tortuosity. For these reasons, the transition from saturated to unsaturated flow in soils generally entails a steep drop in hydraulic conductivity. An example of the effect of moisture content on the hydraulic conductivity K_m for a silty loam developed on eolian deposits in Central Washington is shown in Figure 20 (Globus and Gee, 1995). As indicated, a decrease in moisture content from 0.27 to 0.05 causes 4 orders of decrease in K_m

In addition to changes in the hydraulic conductivity, the hydraulic gradient ∇H is also dependent on variations in soil saturation. The change in head with soil depth can be described such that

$$\frac{dH}{dz} = \frac{dh_g}{dz} + \frac{dh_p}{dz} \tag{33}$$

h_g is the gravitational head at any point and h_p is the head related to the matrix potential or suction due to the capillary affinity of water of the soil mineral surfaces. Determinations of fluid, and conversely chemical fluxes in soils based on Equations (30) and (31), are difficult for most soil conditions due to the non-linear dependence of conductivity and head potentials on moisture content and matrix potential.

Estimating rates of fluid flow

Several approaches can be used to overcome the complexities of unsaturated flow and

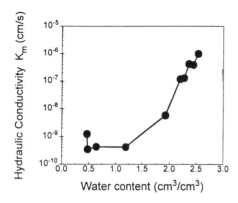

Figure 20. Hydraulic conductivity K_m versus water content for the Palouse silt loam, Central Washington (after Globus and Gee, 1995).

to estimate water fluxes through soils. One approach is to not consider the hydrologic details of soil zone at all and simply calculate soil water fluxes from simple input-output (I/O) balances. As previously mentioned and discussed in greater detail by Drever (this volume) watershed I/O balances is one such approach to calculate both water and solute fluxes through the soil zone. This approach relies on the assumption that discharge from the watershed reflects percolation through the soil zone and that the spatial variability of the watershed soils can be either be characterized or ignored.

The fluid flux can also be calculated from a one dimensional I/O water balance in a soil profile where q is assumed equal to the difference between precipitation P (cm/y) and evapo-transpiration ET (cm/yr) which occur through the upper boundary of the soil profile

$$q = P - ET \tag{34}$$

Because most evapo-transpiration in soils occurs via plant respiration, direct estimates of ET require detailed measurements which are specific to soil and vegetation types. ET can also be indirectly estimated by comparing the concentration c_k^a of a chemical species that is contributed solely from precipitation to its concentration c_k in the soil zone which has undergone concentration by ET. This is equivalent to assuming that c_k^t and c_k^s are zero in Equation (28) and the mass of fluid is no longer conservative. Under steady state conditions this relationship becomes (Erikkson and Kunakasem, 1969)

$$q = P \cdot \frac{c_k^a}{c_k} \tag{35}$$

Finally, estimates of fluid flux can be made based on fluid residence times in the soil profile. Several age dating techniques including ^3H and ^{36}Cl have been successively employed in the age dating relatively young soil waters and in estimating hydraulic conductivity (Allison et al., 1994).

The alternate approach to estimating soil water fluxes is to apply simplifying assumptions to Richards equation (Eqns. 32 and 33) for specific soil environments. Compared to the simple I/O models discussed above (Eqns. 34 and 35), such an approach provides greater information on coupled processes that link soil mineral weathering with soil hydrology. A special case predicted by the Richards equation relates to unsaturated zone flow under conditions of constant matric potential or capillary pressure within a vertical section of a soil profile (i.e $dh_p/dz = 0$) . In such a case $dh_z/dz = -1$ (Eqn. 31) and the flux

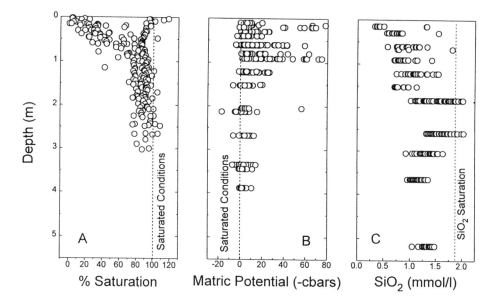

Figure 21. Soil water saturation (A), matrix potential h_p (B), and dissolved SiO_2 (C) in the Riverbank soil profile, Merced, California.

density becomes directly proportional to the hydraulic conductivity

$$q = -K_m \tag{36}$$

One situation in which the flux and hydraulic conductivity may become equal (Eqn. 36) is when an impeding layer is present within a soil column. The effect is to decrease the hydraulic potential in the profile under the impeding layer and to stabilize the development of suction in the subsoil (i.e. $dh_p/dz \rightarrow 0$). This condition, when produced under experimental field conditions, has been proposed as a method for determining unsaturated hydraulic conductivity (Hillel and Gardner, 1970). Such impeding layers are also a common occurrence in many natural soils in the form of caliche layers, duripans and fragipans. Such soil textures are commonly formed by precipitation of silica, iron oxyhydroxides and clay minerals during cyclic wetting and drying of shallow zones of the soil profile (Karathanasis, 1989). A common feature is low hydraulic conductivity within these layers which often results in seasonal saturated or near saturated conditions in overlying portions of the soil profile and relatively steady state unsaturated conditions in the underlying portions.

Example of estimation of soil water flux

In this section, the hydraulic characteristics of the 250 kA Riverbank soil of the Merced chronosequence are used to illustrate methods for estimating hydraulic conductivity K_m and fluid fluxes q. The variations in moisture content (% saturation), matrix potential and SiO_2 concentrations are shown as functions of depth in Figure 21. Soil moisture was calculated from measured gravimetric water content, an average measured soil density of $\rho_w = 1.85$ g cm^{-3} and an assumed mineral density of $\rho_p = 2.65$ g cm^3. In the upper 1 m part of the soil profile soil saturation approaches 100% during wet conditions in the winter and 0% during dry conditions in the summer (Fig. 21A). Below approximately 1 m, the moisture content

Table 9. Solute data for the Riverbank soil profile, Merced, CA.

Element	Precipitation P (mmol l^{-1})	Precipitation Std dev.	Soil Water[1] c_k (mmol l^{-1})	Soil Water Std dev.	Fluid Flux[1] q (cm s^{-1})	Solute Flux[1] Q_k (mol cm^{-2} s^{-1})
Na	0.037	0.002	0.539	0.023	7.6E-08	3.8E-11
K	0.001	0.001	0.032	0.001	7.6E-08	2.4E-12
Mg	0.030	0.012	0.125	0.005	7.6E-08	7.2E-12
Ca	0.018	0.009	0.188	0.005	7.6E-08	1.3E-11
Si	0.006	0.002	1.340	0.024	7.6E-08	1.0E-10

[1] soil water data is at 5.2 m in the soil profile

Table 10. Mineral dissolution and precipitation rates in the upper 5.6 m of the Riverbank soil, CA

Mineral	Formula	Mass (mmol l^{-1})	Fluid Flux q (cm s^{-1})	Mineral flux Q_j (mole cm^{-2} s^{-1})	Mineral rate k_j (mol cm^{-2} s^{-1})
Plagioclase	$Na_{0.684}Ca_{0.316}[Al_{1.316}Si_{2.684}O_8]$	0.566	7.60E-08	4.30E-14	3.47E-20
K-feldspar	$K_{0.0775}Na_{0.225}[AlSi_3O_8]$	0.468	7.60E-08	3.56E-14	2.24E-20
Biotite	$K_{0.958}Na_{0.387}(Mg_{1.686}Fe_{1.1968}(Al_{0.793})[Al_2Si_{6.22}](OH)$	0.063	7.60E-08	4.79E-15	nd[1]
Kaolinite	$[Al_2Si_{6.22}0](OH)_4$	-0.656	7.60E-08	-4.99E-14	nd
Silica	SiO_2	-0.432	7.60E-08	-3.28E-14	nd
Goethite	FeOOH	-0.095	7.60E-08	-7.22E-15	nd

[1] not determined

becomes relatively constant with an average saturation of 85%. The corresponding matrix potentials were measured with nested tensiometers. A matrix potential of zero corresponds to saturated conditions whereas high matrix potentials correspond to low moisture contents. As in the case for moisture content, the upper soil zone exhibits significant variations in matrix potentials between wet and dry seasons, while below 1 m, the matrix potential becomes nearly constant throughout the year with an average potential of approximately 10 cbar (Fig. 21B).

These relatively steady state hydrologic conditions occur below a silica duripan which is pervasive between 1 to 2 m depth in the older soils of the Merced chronosequence. The formation of this layer corresponds to high SiO_2 solute concentrations which at these depths approach silica gel saturation (Fig. 21C, dashed vertical line). Resulting SiO_2 gel precipitation is achieved by concentration through evapo-transpiration of soluble SiO_2 originally derived from silicate weathering. With time, the formation of such duripans at shallow depths become self perpetuating, whereby evapo-transpiration produces additional silica precipitation which in turn leads to lower permeability which furthers the effects of evapo-transpiration.

The above observations (Fig. 21) suggest the rate of soil water movement is controlled principally by rates of slow infiltration through the duripan. Above this zone, the hydraulic head is strongly influenced by cyclic seasonal rainfall and ET, while below this interface, the head is controlled principally by gravimetric flow (dh_t/dz =0, Eqn. 33). Under such conditions, the soil water flux density can be approximated by a constant conductivity ($q = K_m$, Eqn. 36). A single saturated conductivity measurement made on in situ core from the Riverbank duripan, using a biased falling head setup, produced a value of $1 \cdot 10^{-7}$ cm s^{-1}. This is comparable to the saturated conductivity reported by Globus and Gee (1995) for the Palouse soil (Fig. 20) and is higher than expected for less saturated soil conditions.

An alternative approach is to calculate the fluid flux q through the profile based on Cl concentrations. Under such steady state conditions, q and therefore k_s can be calculated from Equation (35). Precipitation chemistry was analyzed for samples periodically taken from open precipitation collectors which represented both wet and dry fall inputs. The average atmospheric Cl concentration over a 4 year time period was $c_{Cl}{}^a = 25 \pm 18$ µmol l^1 . Cl distributions in the soil pore waters exhibit a similar trend as SiO_2 (Fig. 21C). In the deeper unsaturated zone below the duripan, seasonal variations damped out. At a maximum measured depth of 5.6 m, the average soil water Cl concentration is $c_{Cl} = 308 \pm 109$ µmol l^{-1}. Based on these estimates, approximately 8% of the precipitation was recharged through the 5.6 m of soil while 92% was lost to *ET*. As previously indicated, annual mean precipitation at the Merced site is P = 30 cm yr^{-1}. Substituting these values into Equation (35) results in a fluid flux through the system of $q = 2.4$ cm. yr^{-1} . This corresponds to a unsaturated hydraulic conductivity $K_m = 7 \cdot 10^{-8}$ cm s^{-1} (Eqn. 36). As expected, this value is somewhat lower than the saturated hydraulic conductivity measured experimentally.

Soil mineral weathering rates based on solute fluxes

The determinations of the chemical concentrations and fluid fluxes allows us to now calculate solute fluxes through the Riverbank soil profile (Eqns. 27 and 29). Table 9 lists average chemical concentrations for major cations and silica measured over a 4 year period in precipitation and at 5.6 m depth in the Riverbank soil. The difference in these concentrations is assumed to correspond to the chemical concentration contributed from chemical weathering (Eqn. 28; $c_k - c_k{}^s$). The standard deviations of the soil water solutes are small (0.01 to 0.024 mol l^{-1}). This suggests, as shown in the examples for SiO_2 (Fig. 21C), that chemical fluxes in the deeper unsaturated zone are relatively constant over time. The solute fluxes calculated by Equation (27) were found to range between $Q_k = 2 \cdot 10^{-12}$ mol cm^{-2} s^{-1} for K to $Q_k = 1.0 \cdot 10^{-10}$ mol cm^{-2} s^{-1} for SiO$_2$.

The solute concentrations for precipitation and soil water (Table 9) were used as input to the NETPATH program to calculate corresponding masses of mineral dissolved in the first 5.6 m of the soil zone. Mineral stoichiometries used in the model were previously determined for the Merced soils (White et al., 1995). Attempts to constrain the model using K were unsuccessful apparently due to the fact that dissolved K concentrations do no realistically reflect weathering reactions. K in the Merced soils is probably dominated by ion exchange processes and influenced by nutrient cycling in the vegetative cover.

NETPATH output, in terms of reacted mineral masses and weathering fluxes are tabulated in Table 10. The mass flux of each mineral phase per unit of surface area of soil can be calculated in a manner analogous to that for solute fluxes (Eqn. 25) by multiplying mineral concentrations times the fluid flux density. Positive values correspond to mineral dissolution and negative values to mineral precipitation. Approximately 60% of the original mass of plagioclase and 18% of the K-feldspar have dissolved between Riverbank time and the present (250 kyr) These results are in agreement with the loss of these phases previously calculated from solid state mass balances (Fig. 2B). The NETPATH output predicts only minor dissolution of biotite which is in agreement with its observed resistance to weathering in the soil profiles. The model was found not to converge with calculations involving hornblende apparently due to errors in estimating mineral stoichiometry and/or problems with K balances. The precipitated secondary mineral phases (Table 10, negative values) are kaolinite, amorphous silica, and goethite which is consistent with the observed secondary mineralogy in the soil sequence.

The final step in calculating the mineral dissolution rate k_j is to normalize the mineral flux Q_k to the mineral surface S rather than the soil surface area S_s. From Equations (26-29)

$$k_j = R_o \cdot Q_j \qquad (37)$$

R_o was determined by Equation (29), where the specific surface areas of plagioclase and K-feldspar were 0.46 and 0.94 m^2 g^{-1} (Table 5), and the respective mineral wt fractions in the Riverbank soil are $f_m = 0.26$ and 0.16 (Fig. 2B). The total soil depth is $z = 5.6$ m and the soil density is $\rho_w = 1.85$ g cm^{-3}. Combining these parameters in R_o with the mineral flux values in Table 10 resulted in mineral dissolution rate constants log $k_j = 19.5$ mol cm^{-2} s^{-1} for plagioclase and log $k_j = -20.5$ mol cm^{-2}.s^{-1} for K-feldspar (Eqn. 37). The corresponding rates previously derived from the mineral mass balance model (Eqns. 21 and 22) were log $k_j = -19.9$ mol cm^{-2} s^{-1} for plagioclase and -19.7 mol.cm^{-2}s^{-1} for K-feldspar (Table 7). The degree of similarity in the rates is remarkable considering that the number of assumptions used in both models. This similarity also suggests that current weathering rates, characterized by solute fluxes, must be comparable to average rates impacting weathering over the 250 kyr of soil development.

INTERPRETATION OF WEATHERING RATE CONSTANTS IN SOILS

The following discussion will compare the weathering rate constants derived from examples described in the previous sections with k_j values reported in the literature for soils and for selected experimental studies for minerals common to soils. Table 7 tabulates weathering rate constants k_j in increasing order of reaction for plagioclase, K-feldspar and hornblende. These minerals are common in soils developed on crystalline bedrock such as granite. Rates were determined by various methods including mineralogical mass balances (White et al., 1995), etch pit distributions (PSD model, Brantley et al., 1993), soil solute fluxes (Swoboda-Colberg and Drever, 1992; present paper), watershed solute balances (Paces, 1983; Velbel, 1985; Siegal and Pfannkuch, 1984; and Schnoor, 1990) and the integrated PROFILE model (Sverdrup, 1990) (see Sverdrup, this volume). Also included are selected rates based on experimental mineral dissolution. The rate constants range from log k_j = -19.9 to -15.9 mol cm^{-2} s^{-1} for plagioclase, from log k_j =-20.5 to -16.8 mol cm^{-2} s^{-1} for K-feldspar and from log kj =-20.1 and -15.2 mol cm^{-2} s^{-1} for hornblende.

An important observation apparent from the tabulation is that the differences in rate constants for individual minerals greatly exceeds any apparent difference in weathering rates for different minerals. Blum (1994) previously noted that the experimental dissolution rates of albite and K-feldspar in the acid and neutral pH range are indistinguishable. Comparable differences between mineral phases in natural systems are also generally small (Velbel, 1993). In the Merced soils, plagioclase weathers only 2.5 times faster and hornblende only 5.5 times faster than K-feldspar. The time-dependent distribution of residual minerals in the Merced soils, however, puts very tight controls on the magnitude of these rate constants. A 50% increase or decrease in observed residual mineral abundances, such as reported in Figure 2B, changes the calculated reaction rate by only a factor of ±2. Such small differences in rates clearly fall below the resolution of the range in rate constants reported in Table 7.

In general, calculated rates for natural systems are significantly slower than for the experimental studies. Several researchers have previously documented that natural weathering rates appear one to three orders of magnitude lower than experimentally predicted rates (Claassen and White, 1979; Paces, 1983; Swoboda-Colberg and Drever, 1992; Velbel, 1993). Differences in rates can be contrasted by comparing residual mineral abundances in the Merced soils (Fig. 2B) with abundances predicted by the simple geometric model expressed by Equation (17) (Lasaga, 1984). An experimental k_j=-16.0 mol cm^{-2} s^{-1} for albite (Table 7) predicts a residence time of 575k yr for a 1 mm diameter grain. Yet >50% of a comparable size fraction of plagioclase persists in the Turlock Lake soil after 600 kA (Fig. 2B). An experimental k_j = -16.5 mol cm^{-2}.s^{-1} predicts that K-feldspar would persist for 920 kyr and yet >20% of the K-feldspar remains in the China Hat soil after 3000 kA (Fig. 2B). These discrepancies between experimental and natural rates are based on extremely conservative estimates. If the average roughness for crushed feldspar used in the dissolution experiments were introduced into the geometric model ($\lambda \approx 8$; Blum, 1994), the residence times for the feldspars would be reduced to 35 to 70 kA. If the surface areas comparable to actual weathered silicates were used (λ >100), the model would produce extremely unrealistic residence times of <10 kA.

These apparent discrepancies in magnitude of the rate constants signify either that significant variability must exist in the approaches and methodologies used to calculate the rate constants and/or that significant variability must exist in fundamental processes which control dissolution kinetics in natural and experimental systems. The following sections present a brief review of probable causes of such variability. Additional discussions on the relative rates of experimental and natural systems are given by Sverdrup (this volume).

Surface area

Variation in the rate constants tabulated in Table 7 can be attributed, in part, to differences in normalizing the solid or solute flux relative to unit surface area of mineral surface (Eqn. 1). As indicated in Table 7, rate constants for weathering in natural systems have employed both geometric or BET estimates of surface areas. Indeed there has been a tendency in the literature to use these two approaches interchangeably in comparing weathering rates. Due to natural surface roughness, BET surface areas are generally 2 to 3 orders of magnitude greater than geometric estimates of surface area. As indicated for the weathering study on the Merced chronosequence, the geometric surface area model produces dissolution rates that are 10^2 to 10^3 orders of magnitude faster than the variable-roughness model (Fig. 13).

In contrast to natural systems, experimental studies have almost exclusively utilized BET surface area measurements. Therefore, differences in surface area estimates do not explain why the experimental rates are significantly faster than natural rates. In fact if many of the reported natural rates, based on geometric surface areas estimates, were renormalized using realistic values of surface roughness, the apparent discrepancies with experimental rates would even be greater. Therefore, other processes, representing fundamental controls on chemical weathering rates, must be also responsible for the discrepancies apparent in Table 7.

Soil age

Several lines of evidence suggest that mineral weathering rates decrease with time. In terms of a total soil, exponential decay in rates (Fig. 2) can be attributed to decreases in weatherable aluminosilicates relative to residual phases such as quartz. Slower weathering rates in the Merced chronosequence can be attributed to the greater age of most of these soil profiles relative to most other studies which have considered weathering in areas glaciated during the Pleistocene (~10 kA). However, other evidence indicates that the dissolution rate constants for individual aluminosilicate minerals may also decrease with time The Merced study is the first time that weathering rates for individual minerals have been documented to decrease with time (White et al., 1995). As shown in Figure 13, rate constants for plagioclase, K-feldspar and hornblende calculated from BET surface areas decrease by more than an order of magnitude over 3 million years of weathering.

Decreases in weathering rate constants with time can be explained, in part, by the decoupling of the direct proportionality between physical surface areas and the density of reaction sites with increasing time. On an atomistic level, heterogeneity in surface energies is expected to decrease as reaction sites at compositional impurities, dislocations and other crystallographic defects are selectively reacted (Brantley et al., 1986; Blum and Lasaga, 1987). Therefore, the reactivity per unit of physical surface area may decrease. Rates for minerals with high-energy heterogeneities are expected to be initially very rapid as reactive material is removed from etch pits. Rates will then decrease significantly after these sites are depleted and less reactive sites dominate. In contrast, minerals with initially less surface energy heterogeneity would be expected to exhibit a less pronounced decrease in dissolution rate with time. Larger decreases in reaction rates with time for minerals with higher potential surface heterogeneity are suggested for the Merced soils (White et al., 1995). In the A horizon, for example, hornblende weathering rates appears to decrease by a factor of about 45 between 10 kA and 3000 kA, plagioclase by a factor of ~16 and K-feldspar by ~4.5, the same order as decreasing etch pits abundance observed by SEM (Fig. 10).

BET measurements on primary silicates from soils have generally been performed on clean surfaces from which adhering clay and Fe oxyhydroxide layers were removed (White and Peterson, 1990; Anbeek, 1992; White et al., 1995). Increasing concentrations of these secondary phases in soils of increasing age may also exert a negative impact on dissolution rates by effectively shielding chemical interaction between the physical mineral surface and soil waters. This scenario is supported by the observation that with increasing age, the measured BET surface areas of the untreated bulk soils become progressively less than the sum of the separated Fe hydroxides, clays, and primary silicate components (White et al., 1995). This implies that in untreated soils, components such as Fe oxyhydroxides, are occluding the gas absorption and by inference, aqueous interaction in the soil environment.

Solute chemistry

Soil solute chemistry is both produced by and may influence weathering reactions in soils. Solution pH has been shown to be a significant factor in controlling experimental silicate dissolution rates (see Blum and Stillings and Brantley, this volume). This is because both H^+ and OH^- are important participants in dissolution mechanisms, and fluctuate more widely in concentration than any other natural solute. Except in the case of strong acidification [Swoboda-Colberg and Drever, 1992], the reported pH of soil and catchment waters generally range between 4 to 8. In reviewing dissolution of plagioclase and K-feldspar, Blum (1994) (also see Blum and Stillings, this volume) showed dissolution rates are at minimum constant values between pH 5 and 8. Below approximately pH 5, the rate constant for feldspars becomes proportional to $[H^+]^{0.5}$ This data suggest that weathering rates change by a factor of 5 between pH 4 and 5, a pH range typical of acid soils. A variation in solution input pH of 4.5 to 4.0 during artificial acidification of soils at Bear Brook, Maine (Swoboda-Colberg and Drever, 1992). produced a factor of 3 increase in rate constant ($k_j =$ 7.8×10^{-18} to 2.0×10^{-17} mol.cm^{-2}.s^{-1}) for reactive minerals dominated by plagioclase and mica. However for soils in the more neutral pH range, weathering rates are not particularly sensitive to pH changes.

As previously discussed (Eqn. 31), soil solute compositions impact soil weathering rates via reaction affinity. Burch et al. (1993) investigated the effects of approach to equilibrium from undersaturation on the rates of albite dissolution at pH 8.8 and 80 °C. At $\Delta G_r \geq -10$ kcal, dissolution rates were independent of saturation state. The rates of dissolution decreased rapidly in the range of -10 to -7 kcal and reached a plateau from ΔG_r of -7 to close to equilibrium. Based on Figure 19, ΔGr for most soil solutions at pH ≥ 5 are in excess of -10 kcal with respect to albite (Fig. 19). Assuming that the experimental ranges (Burch et al., 1993) can be extrapolated to ambient soil temperatures, the saturation state of most soils waters should strongly affect albite dissolution rates. Although no experimental data is available on the effects of ΔG_r on K-feldspar dissolution, soil solutions are generally closer to saturation than for albite (Fig. 19) and therefore the K-feldspar rates may be even more retarded in soil solutions. This effect has been suggested by Brantley et al. (1995) as the reason for why K-feldspar apparently weathers at a slower rate in field conditions relative to albite.

In addition, thermodynamic saturation, the concentration of specific aqueous species in soil solutions may also impact mineral dissolution rates. Oelkers et al. (1994) showed at elevated temperatures (150°C) increasing Al concentrations decreased reaction rates far from equilibrium. This effect was attributed to the presence of a reactive surface precursor that does not have the same Al:Si stoichiometry as the original silicate mineral. Studies have also documented that the addition of aqueous SiO_2 retards the dissolution of kaolinite in alkaline solutions far from equilibrium (Devidal et al., 1992). This is likely the result of adsorption of aqueous silica which adds cross links to the silica network forming silanol groups. As suggested by the above discussion, specific soil water chemistry can influence weathering rates of minerals in complex ways. For example, although decreasing pH will accelerate weathering rates, concurrent increases in Al release will tend to concurrently retard dissolution. As pointed by Sverdrup (1990) (also see Sverdrup, this volume), an rigorous approach to quantitatively modeling weathering reactions requires the ability to simultaneous consider this solution effects.

Hydrologic heterogeneity

Heterogeneity in soil zone hydrology impacts soil solution chemistry, fluid residence

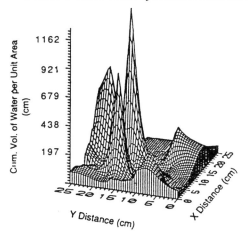

Figure 22. Soil hydrologic heterogeneity as demonstrated by variations in cumulative volume of water that has flowed through sections of an *in situ* soil column (after Andreini and Steenhuis, 1993).

times, and effective mineral surface areas exposed to weathering reactions. Formation of duripans or fragipans, which significantly retard the movement of water and solutes through the soil zone, is reviewed by Smeck and Ciolkosz (1989) and Nettleton (1991). The effect of SiO_2 fragipan formation on soil moisture content, matrix potential and solute chemistry is demonstrated for Riverbank soil profile (Fig. 21).

In contrast, preferential flow can be generated by soil macropores generated from plant roots, burrowing animals, soil cracks and sedimentological heterogeneity (Gish and Shirmohammadi; 1991). An example of preferential flow due to macropores is shown in Figure 22 for the rates of fluid flux through the cross sectional area of a large scale field lysimeter (Andreni and Steenus, 1993). Such preferential flow, which can increase fluid fluxes by an order of magnitude, will also decrease the fluid residence time and thus decrease soil solution concentrations and reaction affinities.

If a large fraction of the mobile fluid interacts with a small proportion of the potentially available mineral surface area , rates calculated from total soil surface area will significantly underestimate the natural weathering rate (Velbel, 1993). In addition, if soil lyimeters predominantly sample matrix pore water, the corresponding high solute concentrations will overestimate solute fluxes which in reality are dominated by lower solute concentrations associated with macropore flow. Several researchers suggested soil heterogeneity is responsible for the apparent lower weathering rates in natural systems relative to experimental studies (Swoboda-Colberg and Drever, 1992; Velbel, 1993).

Role of biological activity

The role of plants and associated mirco-biota in chemical weathering has received intense recent interest in relation weathering and the earths' climate (Berner, this volume). Most data collected to date on the effects of plants on silicate weathering have been descriptive and generally inconclusive (Huang and Schnitzer, 1986). Potential influences of plants on chemical weathering in soils include effects of pH, organic ligands and issues related to physical and hydrologic processes, such as erosion, moisture retention and transpiration (Drever,1994).

Plant root respiration impacts soil water pH through the release of CO_2. P_{CO_2} is commonly 1 to 2 orders of magnitude higher than atmospheric CO_2 (P_{CO2} =$10^{-3.5}$ atm). Plant CO_2 respiration is the main source of H^+ which drives the silicate hydrolysis reaction responsible for soil weathering. However due to associated buffering reactions, even very high P_{CO2} values will not generate soil pHs in excess of 4.5. As previously indicated, reaction rates at or above these pHs probably do not vary by more than a factor of 2 or 3. Organic acids produced by biological activity can also depress soil solution pH. However, these species are not present in most soil solutions at concentrations sufficient to significantly lower the pH to a degree which can accelerate weathering reactions (Drever, 1994).

Organic ligands produced by biological activity influence soil weathering rates by complexing with the species such as Al on silicate surfaces and thus facilitate breaking of the structural bonds and accelerating weathering. In contrast, the sorption of non-complexing organics on to mineral surfaces may decrease dissolution rates by effectively shielding the reactive surface from proton-induced reactions. Most experimental studies to date have focused on dissolution involving strong complexing ligands such as oxalate at concentrations significantly higher than that measured in soil waters. Even these results suggest that complexation does not increase silicate dissolution rates by more than a factor of 2 to 3 (Welch and Ullman, 1993).

Field data, in contrast, suggests that vegetation can significantly affect chemical weathering. Conifer forest plantations impact soil acidity and the release of Al and other species from soil exchange and weathering of soils (Huges et al., 1994). In a recent paper, Cochran and Berner (1995) (see Berner, this volume) showed that vascular plants augment chemical weathering by at least an order of magnitude relative to abiotic conditions. These and other results suggest that such accelerated weathering occurs in micro-environments associated with roots and fungal hyphae where pH and organic ligands may be significantly more concentrated than measured in soil water.

Climate

The origin of the term "weathering" implies that chemical weathering in soils is strongly affected by climate, principally by moisture and temperature. Moisture is influenced by the total amount, intensity, and seasonality of precipitation, humidity, evapo-transpiration, runoff and infiltration. Thermal effects include average air temperatures, seasonal temperature variations, and thermal gradients in soils. Changes in these climatic parameters are expected to impact chemical weathering.

The most simplistic description of the effect of temperature on chemical reaction rates is the Arrhenius relationship. The ratio of rate constants k_j and k_j^o at temperatures T and T° (°K), respectively, can be predicted by the expression

$$\frac{k_j}{k_j^o} = \exp\left[\frac{E_a}{R}\left(\frac{1}{T^o} - \frac{1}{T}\right)\right] \tag{38}$$

where R is the gas constant factor and E_a (kJ mol^{-1}) is the reaction activation energy.

Figure 23 indicates that increases in activation energy increases the temperature effect on weathering rates. Experimental studies generally report activation energies for silicate minerals ranging between 30 and 90 kJ mol^{-1} (Knauss and Wolery, 1986; Sverdrup, 1990; Brady and Carroll, 1994) (see present volume). As pointed out by Velbel (1990), the exponential character of the Arrhenius equation can produce strong non-linear effects on

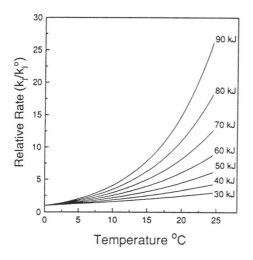

Figure 23. Relationship between ratios of reaction rate constants k_j and k_j^o and temperatures $T^o = 0°C$ and T for activation energies ΔG_r between 30 and 90 kJ.

weathering due to temperature variations in shallow soils. For example, short periods of relatively warm temperatures will have a disproportionate effect on weathering rates relative to long periods of cool temperatures.

Water content also has long been considered an important parameter in soil development but has only recently been considered in terms of weathering mechanisms (Swoboda-Colberg and Drever, 1992; Velbel, 1993). Increasing precipitation, moisture, and solvent throughput can affect the unsaturated hydrology of the soil zone, and accelerate weathering rates by increasing the wetted surface areas of minerals. In addition, as soils become wetter, stagnant pore waters, that are immobile under drier conditions, become hydrologically connected and can potentially activate weatherable mineral surfaces. Areas of high precipitation also generally correlate with areas of greater vegetative cover which can influenced weathering processes as previous discussed. In contrast, low rates of precipitation and/or high rates of *ET* will produce more concentrated soil solutions that will be thermodynamically closer to mineral equilibrium than in wetter soils. High solute concentrations may also promote the effects of Al and other species that have been experimentally shown to inhibit feldspar dissolution

Based on watershed fluxes, linear relationships between weathering and annual average precipitation and runoff have been developed (Dethier, 1986). Recently White and Blum (1995b) proposed a coupled relationship

$$Q_w = (a_1 * P) \exp\left[-\frac{E_a}{R}\left(\frac{1}{T} - \frac{1}{T_o}\right)\right] \tag{39}$$

in which the chemical flux Q_w varied as the product of a linear precipitation term P and an exponential temperature term (Eqn. 38). Equation (39) predicts that the effect of precipitation on weathering is much greater in regions of higher temperatures than in regions of lower temperatures. Conversely, the temperature function is dependent on precipitation. For weathering exposed to low precipitation, the apparent exponential effect of temperature on weathering will be quite low. For wet regions, the exponential influence of temperature on weathering becomes very pronounced. The net effect of the above relationship is to reinforce weathering in areas of both high temperature and precipitation as in the tropics and to decrease weathering fluxes in watersheds with low temperature and precipitation such as in the arctic. At present, this model has been calibrated on a watershed scale and not yet been validated in terms of actual weathering rates in soil environments.

CONCLUSION

Although the subject of chemical weathering in soils has been the focus of extensive research over the last century, surprisingly meager information is available on quantitative weathering rates of primary silicates in soil environments. This paper documents approaches employed to obtain such rates, summarizes available data, and identifies chemical, physical, and hydrologic processes responsible for the observed variability.

Silicate weathering rates are addressed from several perspectives. Changes in mineral masses during soil formation require determination of initial and final compositions of both weatherable silicates and inert elements or minerals in the soil. Approaches are demonstrated for calculating weathering velocities in saprolite profiles and in documenting the effects of climate on rates of soil development. Use of soil solute fluxes in calculating mineral weathering rates involves solving mass balances which relate the array of solute species to dissolution and precipitation of specific mineral phases. Solute mass fluxes require the determination of the fluid flux through the soil, based on either estimates of hydraulic conductivity or on input/output balances for conservative tracers such as chloride.

Calculation of weathering rates requires normalization of mass balances to specific mineral surface areas. Approaches to surface area characterization include macroscopic geometric estimates and microscopic BET measurement. Surface areas determined from these approaches progressively diverge as mineral surfaces roughen and porosity increases with weathering intensity.

Weathering rates, derived from mineral mass balances and solute fluxes, are compared for a Merced soil profile, and found to be similar. This agreement is remarkable considering the number of assumptions used and the fact that the mineral balance approach integrates weathering over the entire time of soil development while the solute approach is based on fluid residence times of several years. However reported rates of specific minerals weathering under both natural and experimental conditions where found to vary between 3 and 4 orders of magnitude. This range is greater than difference in rates between minerals. Such variations are caused, in part, by differing approaches used to calculate rate constants, particularly in regard to surface areas.

Other reasons for variations in rate constants are attributable to fundamental processes which impact weathering reactions in the natural environment. These include the effects of soil age on mineral surface reactivity, the role of hydrologic heterogeneity on fluid residence time and reactive mineral surfaces, the effects of soil solution pH, speciation and reaction affinities, and finally the impact of vegetation and climate. In summary this paper outlines the quantitative framework by which soil weathering rates can be calculated and compared. Clearly much additional work is required to understand the geochemical processes that impact the results generated by these approaches.

ACKNOWLEDGMENTS

Recognition is given to numerous colleagues who provided important insights into the role of chemical weathering in soil environments. These include Alex Blum, Tom Bullen, Marjorie Schulz, Jennifer Hardin and David Stonestrom at the U.S. Geological Survey and

Susan Brantley at Pennsylvania State University. Additional field data and support was provided by Matt Larsen in San Juan, Puerto Rico and Tom Huntington in Atlanta, Georgia. This effort was supported by the Global Change Program at the U.S. Geological Survey.

REFERENCES

Allison GB, Gee GW, Tyler SW (1994) Vadose zone techniques for estimating groundwater recharge in arid and semiarid regions. Soil Science Am J 58:6-14

Anbeek C (1992) Surface roughness of minerals and implications for dissolution studies. Geochim. Cosmochim. Acta 56:1461-1469

Anbeek C Van Breemen N, Meijer EL, Van Der Plas L (1994) The dissolution of naturally weathered feldspar and quartz. Geochim Cosmochim Acta 58:4601-461

Andreneni MS and Steenhuis TS (1993) Role of soil heterogeneity in macropore flow. J Hydrol 143:79

April R, Newton R, Coles LT (1986) Chemical weathering in two Adirondack watersheds: Past and present-day rates. Geol Soc Am Bull 97:1232-1238

Arocena JM, Pawluk S, Dudas MJ (1994) Mineral transformations in some sandy soils from Alberta, Canada. Geoderma 61:17-38

Bain DC, Mellor A, Roberston-Rintoul MS, Buck, ET (1993) Variations in weathering processes and rates with time in a chronosequence of soils from Glen Feshie, Scotland. Geoderma 57:275-293

Bateman PC, and Chappell BR. (1979) Crystallization, fractionation, and solidification of the Tuolumne Intrusive Series, Yosemite National Park, California Geol Soc Am Bull 90: 465-482

Barth TF (1961) Abundance of the elements, aerial averages and geochemical cycles. Geochim Cosmochim Acta 23:1-8

Berner RA, Holdren GR (1979) Mechanism of feldspar weathering- II. Observations of feldspar from soils. Geochim Cosmochim Acta 43:1173-1186

Berner RA, Sjoberg EL, Vebel MA, Kromand DM (1980) Dissolution of pyroxenes and amphiboles during weathering. Science 207:1205-1206

Birkeland PW (1984) Soils and Geomorphology, Oxford University Press, 372 p

Blum AE (1994) Feldspars in weathering. In: Parsons, I (ed) Feldspars and Their Reactions. Kluwer-Academic Press, Netherlands p 595-629

Blum AE, Lasaga AC (1987) Monte Carlo simulations of surface reaction rate laws. In: Stumm W (ed) Chemical Processes at the Particle-Surface Interface. Wiley and Sons, New York, p 255-291

Brady PV, Carroll SA (1994) Direct effects of CO_2 and temperature on silicate weathering: possible implications for climate control. Geochim Cosmochim Acta 58:1853-1863

Brantley SL, Crane SR, Hellimann R, Stallard R (1986) Dissolution at dislocation etch pits in quartz. Geochim Cosmochim Acta 50:2349-2361

Brantley SL, Blai AC, Cremeens DL, MacInnis I, Darmody RG (1993) Natural etching rates of feldspar and hornblennde Aquatic Sci 55:262-272

Brantley SL, Richards P, Murphy S (1995) Soil porewater chemistry and the relative rate of weathering of feldspar and quartz. Geochim. Cosmochim. Acta (in review)

Brimhall G, Dietrich WE (1987) Constitutive mass balance relations between chemical composition, volume, density, porosity, and strain in metasomatic hydrochemical systems: Results on weathering and pedogenesis. Geochim Cosmochim Acta 51: 567-587

Buol W. Hole FD, McCracken RJ (1989) Soil Genesis and Classification. Iowa State University Press, Ames Iowa, 453 p

Buol S, Weed SB (1991) Saprolite-soil transformations in the Piedmont of North Carolina. Geoderma 51:15-28

Burch TE, Nagy KL, Lasaga AC (1993) Free energy dependence of albite dissolution kinetics at 80°C, and pH 8.8. Chem Geol 105:137-162

Busacca AJ, Singer MJ (1989) Pedogensis of a chronoseqence in the Sacramento Valley, California, U. S. A., II. Elemental chemistry of silt fractions. Geoderma 44: 43-75

Busacca AJ, Singer MJ, Verosub KL (1989) Late Cenozoic stratigraphy of the Feather River, Yuba Rivers area, California with a section on soil development in mixed alluvium at Honocut Creek. U. S. Geol Surv Professional Paper 1590:131 p

Campbell DJ, Kinniburgh, DG, Beckett PHT (1989) The soil solution chemistry of some Oxfordshire soils: temporal and spatial variability. J Soil Science 40:321-339

Chadwick OA, Brimhall GH, Hendricks DM (1990) From black box to a grey box: a mass balance interpretation of pedogensis. Geomorphology 3:369-390

Claassen HC, White AF (1979) Application of geochemical kinetic data to groundwater systems, Part I. A tuffaceous-rock system in southern Nevada. In: Jenne E (ed)   Am Chem Soc Symp Ser 93:447-473

Cleaves ET (1993) Climatic impact on isometric weathering of a coarse-grained schist in the northern Piedmont Province of central Atlantic states. Geomorphology 8:191-198

Chou L, Wollast, R (1984) Study of the weathering of albite at room temperature and pressure with a fluidized bed reactor. Geochim Cosmochim Acta 48:2205-2217

Cochran MF, Berner RA (1995) Promotion of chemical weathering by higher plants: field observations on Hawaiian basalts. Chem Geol (in press)

Cremeens DL, Darmody RG, Norton LD (1992) Etch-pit size and shape distribution on orthoclase and pyriboles in a loess catena. Geochim Cosmochim Acta 56: 3423-3434

Cronan C, Driscoll C, Newton RM, Kelly M, Schofield CL, Bartlett RJ, April R (1990) A comparative analysis of aluminum biogeochemistry in a Northeastern and a Southeastern forested watershed. Water Resources Res 26:1413-1430

Dethier DP (1986) Weathering rates and the chemical flux from catchments in the Pacific Northwest, U.S.A. In: Coleman S, Dethier DP (eds) Rates of Chemical Weathering of Rocks and Minerals. Academic Press, Orlando, p 503-530

Devidal JL, Dandur JL, Schott J (1992) Dissolution and precipitation kinetics of kaolinite as a function of chemical affinity (T = 150°C, pH 2 and 7.8) In: Kharaka Y, Maste A (eds) Water Rock Interaction 7, Balkema. p 93-96

Dixon JB, Weed JB (1989) Minerals in Soil Environments. Soil Sci Soc Am Ser 1 893 p

Drever JI (1994) The effect of land plants on weathering rates of silicate minerals. Geochim Cosmochim Acta 58:2325-2332

Edmeades DC, Wheeler DM, Clinton OE (1985) The chemical composition and ionic strength of soil solutions from New Zealand top soils. Aust J Soil Res 23:151-165

Erikkson E, Kunakasem V (1969) Chloride concentrations in groundwater, recharge rate and rate of deposition of chloride in the Israel coastal plain. J Hydrology 7:178-197

Feller E, Schoulle E, Thomas F, Roiller J, Herbillon A (1992) N_2 BET surface areas of some low activity soils and their relationships with secondary constituents and organic matter. Soil Sci 153:279-299

Flury M, Fluhler H, Jury WA, Leuberger J (1994) Susceptibility of soils to preferential flow of water: A field study. Water Resources Res 30:1945-1954

Gallez A, Juo ASR, Herbillion AJ (1976) Surface charge characteristics of selected soils in the tropics. Soil Sci Soc Am J. 40:601-608

Garrels RM, Christ CL (1965) Solutions, Minerals, and Equilibrium. Freeman, Cooper, San Francisco, 395 p

Gish TJ, Shirmohammadi A (1991) Preferential Flow. Proc. Nat. Symp, Chicago, IL, Am Soc of Agricultural Engineers, 525 p

Globus AM, Gee, GW (1995) Method to estimate water diffusivity and hydraulic conductivity of moderately dry soil. Soil Sci Soc Am J 59:684-689

Goldich SS (1938) A study in rock-weathering. J Geol 46:17-53

Hall RD, Horn LL (1993) Rates of hornblende etching in soils in glacial deposits of the northern Rocky Mountains (Wyoming-Montana, USA): Influence of climate and characteristics of the parent material. Chem Geol 105:17-19

Hall RD, Martin RE (1986).The etching of hornblende grains in the matrix of alpine tills and periglacial deposits. In: Colman S, Dethier D (eds.) Rates of Chemical Weathering of Rocks and Minerals. Academic Press, Orlando, p 101-128

Harden JW (1987) Soils developed in granitic alluvium near Merced, California. U.S. Geo Surv Bull 1590-A, 129 p

Helgeson HC, Murphy WM, Aagard P (1984) Thermodynamic and kinetic constraints on reaction rates among minerals and aqueous solutions, II. Rate constants and effective surface area, and the hydrolysis of feldspar. Geochim Cosmochim Acta 48:2405-2432

Hem JD (1985) Study and interpretation of the chemical characteristics of natural water. U.S. Geol Surv Water Supply Paper 2254:264 p

Hillel D (1982) Introduction to Soil Physics., Academic Press San Diego, 329 p

Hillel D, Gardner WR (1970) Measurement of unsaturated conductivity diffusivity by infiltration through an impeding layer. Soil Sci 109:149-154

Holdren Jr GR, Speyer P (1987). Reaction rate-surface area relationships during the early stages of weathering. II. Data on eight additional feldspars. Geochim Cosmochim Acta 51:2311-2318

Huang PM, Schnitzer M. (1986) Interactions of Soil Minerals with Natural Organics and Microbes. Soil Sci Am Spec Publ.17, 529 p

Huges S, Norris DA, Reynolds B, Williams TG (1994) Effects of forest age on surface drainage and soil solution aluminum chemistry in stagnopodzols in Wales. Water, Air, Soil Pollut 77:115-139

Jaycock MJ Parfitt GD (1981) Chemistry of Interfaces, Ellis Horwood, New York, 329 p

Jenny H (1980) Factors of Soil Formation. McGraw-Hill, New York, 239 p

Joffe JS (1933) Lysimeter studies. 2. The movement and translocation of soil constituents in the soil profile. Soil Sci 35:239-257

Joffe JS (1949) Pedology. Pedological Publications, New Brunswick, N J 2nd Ed, 239 p

Johnsson MJ, Ellen SD, McKittrick MA (1993) Intensity and duration of chemical weathering: An example from clays of southeastern Koolau Mountains, Oahu, Hawaii. Geol Soc Am Spec Paper 284:147-169

Karathanasis AD (1989) Solution chemistry of fragipan formation, thermodynamic approach to understanding fragipan formation. In: Smeck NE, Ciolkosz EJ (eds) Fragipans: Their Occurrence, Classification and Genesis. Soil Sci Am. Spec. Publ 24:113-141

Katz BG.(1989) Influence of mineral weathering reactions on the chemical composition of soil water, springs, and groundwater, Catoctin Mountains, Maryland. Hydrol Processes 3:185-202

Kenoyer GJ, Bowser C (1992) Groundwater evolution in a sandy silicate aquifer in Northern Wisconsin 1. Patterns and rates of change. Water Resources Res 28:579-589

Kharaka YK, Gunter WD, Aggarwal PK, Perkins E, Debraal JD (1988) SOLMINEQ 88: A computer program for geochemical modeling of water-rock interactions. U. S. Geol Surv Water Resources Invest Rept 88-4227, 419 p

Knauss, KG, Wolery TJ (1986).Dependence of albite dissolution kinetics on pH and time at 25°C and 70°C. Geochim Cosmochim Acta 50:2481-2497

Lasaga AC (1981) Transition state theory. In: Lasaga, AC, Kirkpatrick, RJ (eds.) Kinetics of Geochemical Processes. Reviews Mineral 8:135-169 p

Lasaga AC (1984) Chemical kinetics of water-rock interaction. Geophys Res 89:4009-4025

Lasaga AC, Soler JM, Ganor J, Burch TE, Nagy KL (1994) Chemical weathering and global chemical cycles. Geochim Cosmochim Acta 58:2361-2386

Leamnson RN, Thomas Jr. J, Ehrlinger HP (1969) A study of the surface areas of particulate microcrystalline silica and silica sand. Illinois State Geol Surv Cir 444:12 p

Lindsay W (1979). Chemical Equilibria in Soils. John Wiley and Sons, New York, 395 p

Locke WW (1986) Rates of hornblende etching in soils on glacial deposits, Baffin Island, Canada. In: Coleman S, Dethier, D (eds.) Rates of Chemical Weathering of Rates and Minerals. Academic Press, Orlando, p 129-146

MacInnis IN, Brantley SL (1993) Development of etch pit size distributions on dissolving minerals. Chem Geol 105:31-49

Mahaney WC, Halvorson DL (1986) Rates of mineral weathering in the Wind River Mountains, western Wyoming. In: Coleman, S, Dethier, D.(eds.) Rates of Chemical Weathering of Rates and Minerals. Academic Press, Orlando, p 147-168

Manley EP, Chesworth W, Evans LJ (1987) The solution chemistry of podzolic soils from the eastern Canadian shield: A thermodynamic interpretation of the mineral phases controlling soluble Al^{3+} and H_4SiO_4. J Soil Sci 38:39-51

Merrill G.P (1906). A Treatise on Rocks, Rock Weathering and Soils. MacMillian, New York, 389 p

Merritts DJ, Chadwick OA, Hendricks DM, Brimhall GH, Lewis CJ (1992) The mass balance of soil evolution on late Quaternary marine terraces, Northern California. Geol Soc Am Bull 104:1456-147

Montgomery CW, Brace WF (1975) Micropores in plagioclase. Contrib Mineral Petrol.52:17-28.

Nagy KL, Lasaga AC (1993) Simultaneous precipitation kinetics of kaolinite and gibbsite at 80°C and pH 3. Geochim Cosmochim Acta 57:4329-4335

Nagy KL, Blum AE, Lasaga AC (1991) Dissolution and precipitation kinetics of kaolinite at 80°C and pH 3: The dependence on the saturation state. Am J Sci 291:649-686

Nahon DB (1991) Introduction to the Petrology of Soils and Chemical Weathering. John Wiley and Sons, New York, 429 p

Nettleton WD (1991) Occurrence, Characteristics and Genesis of Carbonate, Gypsum, and Silica Accumulations in Soils. Soil Sci Soc Am. Spec Publ 26, 149 p.

Norfleet ML, Karathanasis AD, Smith BR (1993) Soil solution composition relative to mineral distribution in the Blue Ridge Mountain soils. Soil Sci Amer J 57:1375-1380

Oelkers EH, Schott J, Devidal JL (1994) The effect of aluminum, pH and chemical affinity on the rates of alumino-silicate dissolution reactions. Geochim Cosmochim Acta 58:2011-2024

Oxburgh R, Drever, JI, Sun YT (1994) Mechanism of plagioclase dissolution in acid solutions at 25°C. Geochim Cosmochim Acta 58:661-669

Paces T (1983) Rate constants of dissolution derived from the measurement of mass balance in hydrologic catchments. Geochim Cosmochim Acta 47: 1855-1863

Parkhurst DL, Plummer LN, Thorsten DC (1982) BALANCE-A computer program for calculating mass transfer for geochemical reactions in groundwater. U.S. Geol Surv Water Resources Invest 82-14, 29 p

Parks GA (1990) Surface energy and adsorption at mineral-water interfaces: An introduction. In: Hochella Jr MF, White AF (eds) Mineral-Water Interface Geochemistry, Reviews   Mineral 23:133-169

Pavich MJ, Brown L, Harden J, Klein J, Middleton R (1986) [10]Be distribution in soils from Merced River terraces, California. Geochim Cosmochim Acta 50:1727-1735

Plummer LN, Presterson EC, Parkhurst DL (1991) An interactive code (NETPATH) for modeling   NET geochemical reactions along a flow path. U.S. Geol Surv Water Resources Invest Rept 91-4078, 277 p

Reynolds B, Neal C, Hornung M, Hughs S, Stevens PA (1988) Impact of afforestation on the soil solution chemistry of stagnopodzols in Mid-Wales. Water, Air, Soil Pollut 38:55-70

Richards LA (1931) Capillary conduction of liquids in porous media. Physics 1: 318-333

Schnoor JL (1990) Kinetics of chemical weathering: A comparison of laboratory and field rates. In: Stumm W (ed) Aquatic Chemical Kinetics, John Wiley and Sons, New York,   p 475-50

Schweda P (1989) Kinetics of alkali feldspar dissolution at low temperature Proc. 6th Internat. Symp. Water/Rock Interaction, A.A. Balkema, Rotterdam, p 609-612

Siegal DI, Pfannkuch HO (1984) Silicate dissolution influence on Filson Creek chemistry, northeastern Minnesota. Geol Soc Amer Bult 95:144-1453

Smeck NE, Ciolkosz EJ (1989) Fragipans: Their Occurrence, Classification and Genesis. Soil Sci Amer Spec Publ 24, 153 p

Soulsby C, Reynolds B (1992) Modeling hydrological processes and aluminum leaching in an acid soil at Lyn Brianne, Mid-Wales. J Hydrology 138:409-429

Sparks DL, Suarez DL (1991) Rates of Soil Chemical Processes. Soil Sci Am Spec Publ. 27, 329 p

Sposito G (1994) Chemical Equilibria and Kinetics in Soils. Oxford Press, New York, 423p

Sverdup H (1990) The Kinetics of Base Cation Release Due to Chemical Weathering. Lund University Press, Lund, Sweden 346 p

Sverdup H, Warfvinge F (1988) Weathering of primary silicate minerals in the natural soil environment in relation to a chemical weathering model. Water, Air, Soil Pollut.38:387-408

Swoboda-Colberg NG, Drever JI (1992) Mineral dissolution rates: A comparison of laboratory and field studies Proc. 7th Int. Symp Water Rock Interaction. Park City, Utah, Balkema, Rotterdam, p 115-117

Ugolini FC, Dahlgren R, Shoji S, Ito T (1988) An example of podzolization as revealed by soil solution studies, southern Hakkoda, Northeastern Japan. Soil Sci 145:111-125

Velbel MA (1985) Geochemical mass balances and weathering rates in forested watersheds of the southern Blue Ridge. Am J Sci 285:904-930

Velbel MA (1986) The mathematical basis for·determining rates of geochemical and geomorphic processes in small forested watersheds by mass balance: Examples and implications. In: Colman S, Dethier D (eds) Rates of Chemical Weathering of Rocks and Minerals. Academic Press, Orlando, p 439-448

Velbel MA (1989) Weathering of hornblende to ferruginous products by a dissolution-reprecipitation mechanism: petrography and stoichiometry. Clay Clay Minerals 37:515- 524

Velbel MA (1990) Influence of temperature and mineral surface characteristics on feldspar weathering rates in natural and artificial systems: a first approximation. Water Resources Res 25:3049-3053

Velbel MA (1993) Constancy of silicate-mineral weathering-ratios between natural and e x p e r i m e n t a l weathering: Implications for hydrologic control of differences in absolute rate. Chem Geol 105:89-99

Wakatsuki T, Rasyidin A (1992) Rates of weathering and soil formation. Geoderma 52:251-263

Welch SA, Ullman WJ (1993) The effect of organic acids on plagioclase dissolution rates and stoichiometry. Geochim Cosmochim Acta 57:225-2736

Wesolowski D J (1992) Aluminum speciation and equilibrium in aqueous solution: I. the solubility of gibbsite in the system NA-K-Cl-OH-Al(OH)$_4$ from 0 to 100°C. Geochim Cosmochim Acta 56:1065-1091

White AF, Peterson ML (1990) Role of reactive surface area characterization in geochemical models. Chemical modeling of aqueous systems II, In: Melchior DC, Bassett RL (eds) Am Chem Soc Symp Ser 416, p 461-475

White AF, Blum AE (1995a) Climatological influences on chemical weathering in watersheds; application of mass balance approaches. In: Trudgill S (ed) Solute Modeling in Catchment Systems. John Wiley, New York, (in press)

White AF, Blum AE (1995b) Effects of climate on chemical weathering rates in watersheds. Geochim Cosmochim Acta 59:1729-1747

White AF, Blum AE, Schulz MS, Bullen TD, Harden JW, Peterson ML (1995) Chemical weathering of a soil chronosequence on granitic alluvium 1: Reaction rates based on changes in soil mineralogy. Geochim Cosmochim Acta (in press)

Willaime C, Christie JM, Kovacs MP (1979) Electron microscope study of plastic defects in experimentally deformed alkali feldspars. Bull. Soc fr Mineral Cristallogr 100:263-271

Wood WW, Kraemer TF, Hearn PP (1990) Intergranular diffusion: An important mechanism influencing solute transport in clastic aquifers. Science 247:1596-1571

Worden RH, Walker FD, Parsons I, Brown WL (1990) Development of microporosity, diffusion channels and deuteric coarsening in perthitic alkali feldspars. Contrib Mineral Petrol 104:507-515

Zhang H, Bloom PR, Nater EA (1993) Changes in surface area and dissolution rates during hornblende dissolution at pH 4.0. Geochim Cosmochim Acta 57:1681-1689

Chapter 10

WEATHERING RATES IN CATCHMENTS

J. I. Drever

Department of Geology & Geophysics
University of Wyoming
Laramie, WY 82071 U.S.A.

D. W. Clow

U.S. Geological Survey
MS 415, Denver Federal Center
Lakewood, CO 80225 U.S.A.

INTRODUCTION

Weathering is a very general term for the processes by which rocks undergo physical and chemical alteration at the earth's surface. In this paper we shall be concerned only with chemical weathering and, more specifically, with the production of solutes. For the purpose of this paper we shall define weathering rate as the rate of production of solutes by alteration of minerals in a catchment. The solutes may be transported away in solution, taken up by the biomass, or adsorbed by organic or inorganic substrates in the catchment. We shall exclude cation exchange from our definition of weathering, although the boundary between ion exchange and alteration is somewhat arbitrary. We are proposing this definition strictly for the present discussion. Other definitions of weathering may be appropriate in other contexts.

The discussion in this chapter will focus on weathering of silicate minerals. The mechanisms controlling rates of silicate weathering on the catchment scale are much less well understood than those controlling carbonate dissolution. Much of the work on silicate weathering rates in the field has been motivated by the importance of weathering rates for predicting the long-term effects of acid deposition on surface waters. Weathering is also important in models of long-term controls on the carbon dioxide content of the atmosphere and hence global temperature (e.g. Berner, 1992, 1994; Brady, 1991). Weathering removes CO_2 from the atmosphere, so there is considerable interest in understanding how the rate of weathering responds to a change in environmental conditions.

Weathering rate is generally normalized to unit area of catchment and is expressed in units of (k)moles ha^{-1} y^{-1} or (k)equivalents ha^{-1} y^{-1}. Mineral dissolution rates in laboratory experiments are generally normalized to unit area of mineral surface area and expressed in units of (p)moles m^{-2} s^{-1}. In this paper we shall always use moles ha^{-1} y^{-1} for rates normalized to land surface area and moles m^{-2} s^{-1} for areas normalized to mineral surface area. Estimating mineral surface area on a catchment scale is difficult; the topic will be discussed below.

MEASUREMENT OF WEATHERING RATES

By solute budgets

Solute budgets are established by measuring the inputs to a catchment from the atmosphere and the output from the catchment, usually in the form of surface runoff. The

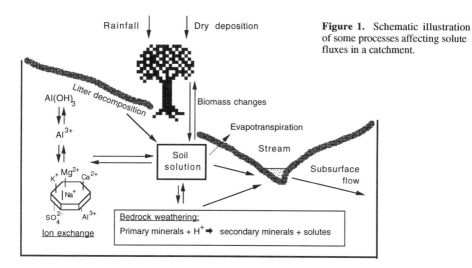

Figure 1. Schematic illustration of some processes affecting solute fluxes in a catchment.

difference between the flux in and the flux out for each element represents the generation (or removal) of the element by processes within the catchment. Chemical weathering, however, is not the only process affecting solute budgets (Fig. 1). Even if we restrict our discussion to the major inorganic solutes, uptake by plants, and cation exchange may significantly influence element fluxes. These processes will be discussed in more detail below. We can interpret the fluxes for a pristine catchment in terms of a mass balance equation:

solutes in outflow = solutes from atmosphere + solutes from weathering

$$\pm \text{ solutes from change in biomass} \pm \text{change in exchange pool.} \quad (1)$$

In writing this equation we have already made the assumption that our time-scale is long enough that changes in water storage can be ignored. The question of time-scale is very important in catchment-scale studies: it is rarely possible to construct a meaningful catchment budget for a time-scale of less than a year. On the shorter time-scale, biomass effects tend to be large, whereas they tend to average out over an annual cycle. We shall discuss the individual terms in the above equation in some detail because they are often ignored. Ignoring them can introduce significant errors.

Solutes in outflow. The information needed, in principle, is the volume and chemical composition of water leaving the catchment. It is usually assumed, with varying degrees of justification, that discharge via groundwater is negligible compared to surface runoff. The volume of surface runoff is commonly monitored continuously, ideally at a weir constructed for the purpose. Water chemistry is commonly measured at some fixed time interval, typically weekly. Interpolated values from these weekly samples are coupled with the continuous discharge record to calculate a continuous record of the flux of solutes leaving the catchment. One problem with this approach is that short-term events such as floods (when high discharge occurs) tend to be missed in the chemical sampling, whereas base flow periods (associated with low discharge) tend to be over-sampled. If the chemistry of the stream varies with discharge, as is normally the case (e.g. Miller and Drever, 1977; Hooper and Aulenbach, 1993; Huntington et al., 1994), there will be a systematic error in estimating the annual solute flux. This error can be corrected to some extent (e.g. White and Blum, 1995), but it remains a source of uncertainty.

Solutes from the atmosphere. Atmospheric deposition occurs in the form of rain and snowfall, condensation from mist or fog, and dry deposition. Dry deposition in turn consists of solid particles and gases such as SO_2 and NO_x that are taken up directly by surfaces, particularly vegetation or moist foliage. The term *occult deposition* is sometimes used to cover dry deposition and deposition from mist or fog. Occult deposition is difficult to quantify. One approach, which is useful in catchments with a reasonably uniform forest cover, is to measure *throughfall*, which is rain collected after it has passed through the forest canopy, and *stemflow*, which is water running down the outside of tree trunks. The solutes in throughfall and stemflow can be attributed to:

1. Solutes in the incoming precipitation.
2. Solutes from dry deposition and deposition from fog on the foliage of the trees.
3. Solutes translocated from the soil by the trees and then leached from the foliage.

Table 1. Mean concentrations of rainfall and throughfall, and fluxes of elements from the atmosphere into an old spruce stand, Vosges Mountains, France (Probst et al., 1990).

	NH_4	Na	K	Mg	Ca	H^+	Cl	NO_3	SO_4
Concentrations (µeq/L)									
Bulk precipitation	19.1	10.0	2.8	4.5	11.9	33.9	12.5	24.1	41.5
Throughfall	36.9	46.4	52.7	17.8	65.5	114.8	63.4	78.3	185.0
Fluxes (mole/ha/y)									
Bulk precipitation	270	142	39	32	84	480	177	340	290
Throughfall	385	484	550	93	342	1197	661	817	966
Difference	115	342	511	61	258	717	484	477	676
Occult deposition[*]	115	342	102	31	206	1282	484	477	676

[*]Difference corrected for elements translocated through the trees.

It is possible to distinguish *approximately* between occult deposition and translocation on the basis of tree physiology (Matzner, 1986), but ambiguities remain, particularly for trace elements. As an example, Table 1 shows the composition of open-field rainfall and throughfall in an old spruce stand in the Vosges Mountains of eastern France. The corresponding fluxes in units of moles ha[-1] y[-1] are also shown. The occult deposition flux represents the difference between rainfall and throughfall, corrected for elements translocated through the trees. It is clear from Table 1 that occult deposition may be a major input into a catchment. It is quite sensitive to the amount and type of vegetation present. Conifers are much more effective in trapping atmospheric solutes that deciduous trees, and trees are more effective than grassland. Significant dry deposition occurs even on rock outcrops (Peters, 1989; Clow and Mast, 1995). It is very difficult to measure dry deposition for a catchment containing different vegetation types.

Dry deposition can also be estimated by chloride balance (White and Blum, 1995). If it is assumed that chloride is a conservative tracer and has no bedrock source, the flux of chloride leaving a catchment will be equal (over sufficient time) to the flux entering. If the flux leaving is greater than the measured input from precipitation, the difference is a measure of chloride entering the catchment as dry deposition. If the ratios of other ions to chloride in dry deposition are known or can be estimated, then the inputs of these other ions can be calculated. These ratios cannot be measured directly; they are typically assumed to be equal to the ratios in wet deposition or, in coastal areas, in sea water. A problem with this method is that chloride concentrations tend to be quite variable, and they are very susceptible to contamination. Also, the ratios of other species, particularly nitrogen and sulfur compounds, to chloride in dry deposition are not well constrained.

Changes in the exchange pool. Soils contain exchangeable cations and anions that are in equilibrium with the soil solution. As the composition of the soil solution changes, ions will be exchanged between the solid phase and solution (see Sverdrup, this volume). If the soil solution composition does not change with time, adsorbed ions will not change either and ion exchange will make no net contribution to the solute budget. In the short term, changes in solution composition occur as a result of precipitation events, evaporation, and growth cycles of plants. It is often assumed that these changes average out over an annual cycle so that, on a time scale longer than a year they cause no net change to the exchange pool and hence no net contribution to solute flux. This assumes that year-to year variations in solution chemistry (caused, for example by alternating wet and drought years) are insignificant.

If there is a permanent change in the chemistry of atmospheric input, this will cause a unidirectional change in the exchange pool. If, for example, the concentration of H^+ in precipitation increased, the H^+ (or Al^{3+} generated from dissolution of $Al(OH)_3$ in response to the input of H^+) would displace other cations (commonly mostly Ca^{2+}) from exchange sites, and these cations would appear in the outflow from the catchment (e.g. Reuss and Johnson, 1986). If the composition of precipitation remained constant, the composition of the exchange pool would gradually adjust to the new input and a new steady state, in which there was no net change in the exchange pool over time, would be established. The time taken by the exchange pool to adjust to a change in atmospheric input is generally quite long—on a time-scale of decades—because the reservoir of exchangeable ions in the soil is generally large compared to the annual input from the atmosphere.

If a catchment does not receive a large input of anthropogenic acidity from the atmosphere, it is generally assumed that the exchange pool is in steady state, and is not a net contributor of solutes.

Changes in the biomass. As plants grow, they extract inorganic nutrients from the soil solution and incorporate them into plant tissue. The stoichiometry is approximately (Schnoor and Stumm, 1985):

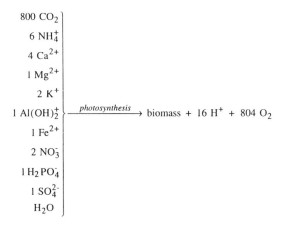

$$\left.\begin{array}{r} 800 \text{ CO}_2 \\ 6 \text{ NH}_4^+ \\ 4 \text{ Ca}^{2+} \\ 1 \text{ Mg}^{2+} \\ 2 \text{ K}^+ \\ 1 \text{ Al(OH)}_2^+ \\ 1 \text{ Fe}^{2+} \\ 2 \text{ NO}_3^- \\ 1 \text{ H}_2\text{PO}_4^- \\ 1 \text{ SO}_4^{2-} \\ \text{H}_2\text{O} \end{array}\right\} \xrightarrow{\text{photosynthesis}} \text{biomass} + 16 \text{ H}^+ + 804 \text{ O}_2$$

Plant growth affects the budgets of the major cations and protons and will affect both the exchange pool and the net output from a catchment. When plants die and decompose, the process is reversed and the elements are returned to the soil.

If a forest (or grassland) is in steady state, that is to say the growth of new vegetation is exactly balanced by the death and decay of old vegetation, the biomass will be neither a net source nor a sink in the mass balance equation. However, in forested catchments, the biomass is rarely in a steady state. Even without human intervention such as tree cutting or planting, forests typically go through cycles of gradual biomass increase interrupted by catastrophic events, such as fire or disease, that result in rapid biomass loss (Vitousek and Reiners, 1975). It is only on a very long time scale, a scale of centuries, that the biomass of a forest can be considered in steady state. Hubbard Brook, New Hampshire, in the northeastern United States, is an example of a catchment where the biomass term is large compared to the weathering term (Fig. 2). According to the data of Likens et al. (1977), the net uptake of Ca in the form of biomass increment and forest floor (organic) increment was 45% of the amount released by weathering. For potassium, the biomass increment was 86% of the amount released by weathering: the net outflow of K from the catchment (runoff – precipitation) was only 14% of the K calculated to be released by chemical weathering. More recent measurements have decreased the biomass uptake term relative to the weathering term, but the conclusion remains: in forested catchments with silicate bedrocks, the biomass uptake term is likely to be important. On the other hand, in high elevation catchments where the vegetation is sparse, such as Loch Vale, Colorado, the biomass term is small compared to the weathering term (Baron, 1992).

The biomass term in Equation (1) can be measured directly by measuring the rate of growth of trees and the chemical composition of different tissues. This requires long-term measurements over several years, implying a significant investment of time and resources (e.g. Likens et al., 1977). Even if the long-term average biomass uptake rate is known, the rate may vary from year to year depending on weather conditions, particularly the availability of moisture.

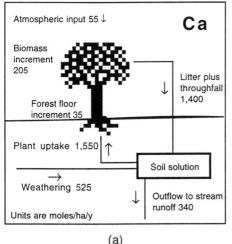
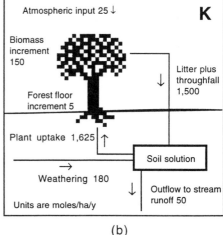

(a)   (b)

Figure 2. Calcium (a) and potassium (b) fluxes at Hubbard Brook Experimental Forest, New Hampshire, USA. Data from Likens et al. (1977). (After Drever, 1988).

The term "biomass" as used here includes non-living organic matter on the ground surface and in the soil. This organic matter contains cations such as calcium (and, particularly, nitrogen species), so changes in its decomposition rate may also cause a net flux of inorganic solutes.

Chemical weathering. There is no simple direct way of measuring the contribution of chemical weathering to the solutes measured at the output of a catchment. The most common approach is to assign values to (or simply ignore) the biomass change and ion exchange terms in Equation (1) and determine the weathering term by difference. The utility of this approach depends very much on the catchment. In catchments where the bedrock weathers rapidly, the weathering term is much larger than the other two, so there is not a problem. Where atmospheric deposition has not been significantly affected by pollution, neglecting the ion exchange term should not introduce much error, and in catchments that are largely unvegetated, the biomass term should be negligible. Unfortunately, the catchments that have been studied in the greatest detail (motivated by research on acid deposition) are those that are characterized by low weathering rates, the presence of forests (often complicated by changes in land use in historic times), and significant anthropogenic inputs of acid (Wright, 1988). In these situations, the weathering term may be small compared to the other terms in Equation (1) and it cannot be determined with much confidence.

An alternative approach is look at the budget of an element that is not significantly affected by ion exchange or biomass uptake, such as sodium or silicon (Stauffer and Wittchen, 1991; White and Blum, 1995). Although sodium is involved in ion exchange reactions, in the forest soils of the northeastern United States it forms a very small fraction of the exchange pool. If the flux of sodium from weathering is known (output – atmospheric input), and the stoichiometry of weathering is known, then the fluxes of the other cations, notably Ca^{2+} and Mg^{2+}, can be calculated by multiplying the sodium flux by the appropriate stoichiometric ratio. There are two problems with this approach: in coastal areas with low weathering rates, the flux of Na from weathering may be small compared to the flux from the atmosphere, making it difficult to measure accurately. The other problem is that there is no easy way of measuring the stoichiometry of weathering. Human inputs such as road salt or fertilizer would, of course, make the method unworkable. Silica can be used instead of sodium. The advantage is that the input-output budget of silica can be established with more certainty than that of sodium (usually the atmospheric input is negligible and the flux from weathering relatively large). The disadvantage is that silica is partially (and variably) retained in secondary minerals such as kaolinite and smectite and may be affected by adsorption-desorption reactions and plant uptake. The problem of variable retention of silica can be illustrated by considering the weathering of albite to different secondary phases. If albite weathers to gibbsite

$$NaAlSi_3O_8 + H^+ + 7\,H_2O = Al(OH)_3 + Na^+ + 3\,H_4SiO_4$$

The ratio of moles of albite weathered to moles of Si released will be 1:3. If albite weathers to kaolinite:

$$NaAlSi_3O_8 + H^+ + 4.5\,H_2O = 0.5\,Al_2Si_2O_5(OH)_4 + Na^+ + 2\,H_4SiO_4$$

the ratio will be 1:2, and if it weathers to a smectite (for example, beidellite)

$$2.3\,NaAlSi_3O_8 + 2\,H^+ + 6.4\,H_2O =$$
$$Na_{0.3}Al_2(Si_{3.7}Al_{0.3})O_{10}(OH)_2 + 2\,Na^+ + 3.2\,H_4SiO_4$$

the ratio will be approximately 1:1.4. Thus the silica flux is not a simple direct indicator of the rate of weathering of primary minerals. Information about the types of secondary minerals that are forming must be coupled with data concerning which primary minerals are weathering.

Several researchers have used Sr-isotopes to characterize the relative contributions of cation-exchange and weathering in the output budgets of catchments (Aberg et al., 1989; Miller et al., 1993). If it is assumed that Ca^{2+} behaves exactly as Sr^{2+}, then the Ca^{2+} in the outflow can be assigned to the two sources in the same ratio as for Sr. Since Ca^{2+} is the most important cation released by weathering, establishing a budget for Ca^{2+} is almost equivalent to establishing a budget for weathering as a whole. Sr-isotopes have also been used to identify the relative weathering rates of minerals in catchments (Blum et al., 1994). Thus, they can provide an independent check on results from mass-balance calculations, or can be used where requisite data for mass-balance calculations is lacking. This is often the case because detailed sampling and discharge measurements for at least one and preferably more annual hydrologic cycles are required to construct a solute budget with confidence.

Another approach to establishing a long term weathering budget for a catchment is to measure the accumulation of soil and the change in composition that occurs as bedrock is converted to soil. This approach is discussed by White (this volume).

RATES NORMALIZED TO MINERAL SURFACE AREA

In order to compare weathering rates in the field with mineral dissolution rates in laboratory experiments, we need to know the surface area of minerals exposed to weathering in the field. This raises two issues: the issue of surface area itself, and the issue of contact between mineral surfaces and infiltrating meteoric water. Several different approaches have been used (discussed below). Most comparisons are based on *geometric surface area*. Geometric surface area is calculated by assuming (or measuring) a grain-size distribution and assuming a simple shape (sphere, cube) for minerals in the field. Surface measurements in laboratory experiments (Lasaga, this volume) are based on BET measurements (Brunauer et al., 1938). In the BET method, the surface area is measured by measuring the amount of a gas (commonly nitrogen or krypton) required to form a monolayer on the mineral surface. BET surface areas are always larger than geometric surface areas: the ratio of BET area to geometric area is referred to as the *surface roughness*. Surface roughness in fresh minerals generally increases with increasing grain-size (Anbeek, 1992a). It is higher in weathered minerals than in fresh minerals (Anbeek, 1992b) and it changes with time during a dissolution experiment (Anbeek, 1993; Anbeek et al., 1994; Murphy, 1993). The conversion between BET area and geometric surface area is thus not simple. Although the total dissolution rate of fresh minerals is proportional to BET area, it is not strictly proportional to geometric surface area because surface roughness is itself a function of grain size.

It is generally not possible to measure meaningful BET areas on field samples. Soils consist of a mixture of minerals, primary and secondary, and it is difficult to apportion the total area among the different phases. Any clay minerals or oxides present will distort the results by contributing a large amount of surface area. This topic is discussed in more detail by White (this volume).

Catchments in the Czech Republic

Paces (1983) compared the rate of weathering of oligoclase in two catchments in (then) Czechoslovakia underlain by granitic gneiss to the laboratory dissolution rate of oligoclase measured by Busenberg and Clemency (1976). The rocks affected by weathering consisted of about 1 m of regolith/alluvium and 7 m of fractured bedrock. The field rate was based on the flux of Na over a five-year period, and surface area in the field was calculated by assuming a joint pattern in the bedrock calculated from its hydrologic properties. With these assumptions, Paces calculated an albite dissolution rate in the field of 5.2×10^{-15} to 6.8×10^{-13} moles m^{-2} s^{-1}. He compared this number to a literature value (Busenberg and Clemency, 1976) of 1.7×10^{-12} moles m^{-2} s^{-1}. The laboratory value corrected to the lower temperature of the field conditions was 7.9×10^{-13} moles m^{-2} s^{-1}. Thus the corrected laboratory rate corresponds to the upper limit of the estimated field rate. The "best estimate" of the field rate was about an order of magnitude slower. On the other hand, the laboratory rate was measured under a CO_2 pressure of 1 atm., which probably represented a lower pH than the field. Paces attributed the slower rate in the field to either the presence of protective secondary phases covering the feldspar surface or "aging" of the surface. Aging would represent the progressive removal over time of high-energy sites in the crystal surface such as dislocations.

It could certainly be argued that, given the uncertainties, the field and laboratory rates are indistinguishable. We believe, however, that Paces' estimate of surface area based on assumed planar fractures was extremely conservative, and so the difference is probably real.

Coweeta Basin, North Carolina

The Coweeta Hydrologic Laboratory consists of a series of forested catchments underlain by metamorphic silicate rocks. The area had been the subject of ecological research for many years, so well-constrained input-output budgets were available. Rainfall is high (around 200 cm y^{-1}) and the mean annual temperature about 13°C. The major chemical weathering process is the conversion of fresh rock to saprolite along a well-defined weathering front (Velbel, 1985). The principal reactions, as determined by optical microscopy, are alteration of biotite to vermiculite, dissolution of almandine garnet with precipitation of goethite and gibbsite, and alteration of plagioclase feldspar to gibbsite and kaolinite. Solutes in the runoff could be assigned to these specific reactions (and a biomass uptake term). Velbel calculated weathering rates for the individual minerals of approximately 200 to 500 moles ha^{-1} y^{-1}. Rates for individual minerals were independent of discharge (flushing rate), suggesting that chemical reaction and not transport in solution was the rate-limiting process.

In order to compare the field rates to laboratory rates, Velbel assumed a grain-size of 1 mm--garnet as spheres, plagioclase as cubes, and biotite as cylinders. One mm was chosen on the basis of petrographic observation. The total mineral area accessible to solution was then calculated from the modal abundance of each mineral and the thickness (average 6.1 m) of the saprolite in the catchments. Dissolution rates calculated on these assumptions are shown in Table 2. On the basis of the literature values for mineral dissolution rates available to him at the time, Velbel concluded that the rate for oligoclase was perhaps a factor of 10 faster in the lab, the rate for garnet a factor of 3, and the rate for biotite a factor of about 2000. However, when more recent values for the dissolution rates of oligoclase and biotite are used in the calculation (Table 2, final column), the discrepancy for oligoclase becomes a factor of 2, and that for biotite a factor of about 8. These discrepancies would become even smaller if the laboratory rates were corrected to the field temperature of 13°C. Within the uncertainties of the data, the field rates and laboratory rates are indistinguishable. On the other hand, the comparison of a geometric surface area (the field) with a BET area (laboratory experiments) probably introduces a bias of about a factor of 10. If the field and laboratory rates had been normalized in the same way, the discrepancies

Table 2. Weathering rates (moles m^{-2} s^{-1}) of minerals in the regolith of the Coweeta catchments compared to laboratory dissolution rates from the literature (after Velbel, 1985).

Mineral	Field rate	Laboratory rate[1]	Laboratory rate[2]
Oligoclase	8.9×10^{-13}	3×10^{-12} to 2×10^{-11}	1.8×10^{-12}
Garnet	3.8×10^{-12}	1×10^{-11}	
Biotite	1.2×10^{-13}	2×10^{-10}	1×10^{-12}

[1] Literature rates reported by Velbel (1985).

[2] More recent rates: oligoclase from Mast and Drever, 1987; biotite from Acker and Bricker (1992).

would have been about a factor of 10 greater. Velbel's (1985) results suggest that the field rate was about an order of magnitude slower than the laboratory rate, but that number is not well constrained.

Bear Brooks, Maine

The Bear Brooks catchments in Maine is the site of a series of experiments designed to investigate the effect of acidic deposition from the atmosphere on surface water chemistry (Norton et al., 1995). As part of the project, a series of small (2 m^2) plots were irrigated with dilute HCl (pH 2, 2.5, 3) over a period of three years (Swoboda-Colberg and Drever, 1993). The percolating water was collected at 25 and 50 cm depths and its composition was used to calculate the amount of weathering occurring in the top 25 and 50 cm of the soil. The mineralogy and grain-size distribution of the soil was measured in great detail to provide a measurement of the geometric surface areas of primary minerals in the soil. The dissolution rate of the 75 to 150 μm fraction of the soil was measured in flow-through (fluidized bed) reactors for comparison. The pH values in the reactor experiments corresponded to those of the solutions collected from the soils at 25 and 50 cm (pH 4-4.5).

From these experiments, Swoboda-Colberg and Drever (1993) concluded that dissolution rates in the field plots were about a factor of 200 slower than rates in the fluidized-bed reactors. The comparison was based on the assumption that dissolution rate scales with geometric surface area, but this should not introduce a large error because the grain-size was similar in field and laboratory experiments. Also, the dissolution rate of plagioclase feldspar in the reactor experiments with minerals from the soil was similar to that of crushed, fresh oligoclase under the same conditions (Swoboda-Colberg and Drever, 1989). The difference between the field rate and the laboratory rate was thus not a consequence of "aging" of mineral surfaces. Swoboda-Colberg and Drever preferred a hydrologic explanation--solutions did not percolate uniformly through the soil and so only a small fraction of the minerals in the soil were in effective contact with the percolating solution.

The natural mineral weathering rate in the whole Bear Brooks catchment (Norton et al., 1992; Drever et al., 1995) is about a factor of three slower than in the small plot experiments. The difference reflects the higher water flux and lower pH in the small plot experiments. Thus the weathering rate for the catchment as a whole was at least two orders of magnitude slower than would be predicted from laboratory experiments using minerals from the site in fluidized-bed flow-through reactors.

Rocky Mountain National Park, Colorado

Clow (1992) measured weathering rates of soil minerals in a very small alpine catchment on granitic bedrock in Loch Vale, Rocky Mountain National Park. The catchment consisted of a 15.5 m² pocket of soil on a 40 m² outcrop. Mean soil depth was 0.3 m, and vegetation consisted of sparse short grasses and herbs. Weathering rates were measured using input-output budgets, with natural rainfall' and irrigation with distilled water as inputs. The irrigation water served to maintain the soil in a continuously wet condition, and kept the volume of wetted soil constant. Outflow was confined to a narrow slot in the bedrock. The sparseness of the vegetation indicated that the biomass term could be ignored in the mass-balance. Cation exchange was assumed negligible for the main solutes used to calculate weathering rates, which were Na, Si, and K.

The small size of the catchment allowed a very well-constrained estimate to be made of mineral surface area in the zone of weathering. This was done by measuring the grain-size distributions for the three soil horizons present, determining the percentages of the main primary minerals in each of four size-fractions of each soil horizon, and measuring the specific surface area of the primary minerals by BET analysis. This provided an estimate of the surface area of the primary minerals in a unit volume of soil, which was multiplied by the volume of wetted soil in the catchment to obtain the total surface area of each of the primary minerals.

Weathering rates of soil minerals collected at the field site were also measured in the laboratory using saturated columns and fluidized-bed reactors. By using similar materials in the field and laboratory experiments it was possible to compare directly weathering rates and try to bridge the gap between lab and field measurements. Fluidized-bed reactors were employed in some of the experiments to allow comparisons with previously published work, and because it is easy to keep solutions undersaturated with respect to secondary minerals. Saturated columns were used in other experiments because they provide flow rates more similar to those encountered in nature. The main disadvantage of the columns is that it is slightly more difficult to limit solute concentrations below saturation with respect to secondary minerals.

Weathering rates in the field were dependent on flow conditions through the soil (Table 3). Average daily outflow from the catchment was 34 l/d in 1990 compared to 154 l/d in 1991 due to greater irrigation inputs in 1991. Volume-weighted mean concentrations declined only slightly in 1991, and weathering rates increased almost in direct proportion to the increased water flux. These results suggest that solute release was stimulated in some manner by the higher flow rates in 1991. The soil was continuously wet during both summers of the study, but was saturated only during the second summer. Clow (1992) hypothesized that the higher flow in the second year may have stimulated flow in micropores and cracks in mineral grains where water normally moved very slowly. This in turn could have increased the amount of mineral surface area contributing solutes to soil solutions that made their way to the outflow of the catchment.

Weathering rates measured in the laboratory in reactor experiments were much higher than rates measured in the field (Table 3). However, weathering rates measured in column experiments were somewhat more comparable to field rates. The oligoclase dissolution rate in the field under high-flow conditions (1991) was roughly two-thirds of the rate measured

Table 3. Mineral weathering rates in laboratory and field experiments in Rocky Mountain National Park. Units are picomoles m^2 s. The 106 to 208 μm size-fraction was used in all lab experiments. Results are expressed as mean ±95% confidence interval (after Clow, 1992).

	Oligoclase	Biotite	Microcline
Field			
low-flow (34 L/D, 1990)	0.09	0.002	0.006
high-flow (154 L/D, 1991)	0.32	0.012	0.020
Column			
unfractionated, treated soil	0.22±.04	0.06±.01	
crushed granite	0.22±.03	0.03±.01	
Reactor			
feldspar concentrate	1.70±.20		0.58±.14
biotite concentrate		0.12±.02	
unfractionated, treated soil	1.84±.48	0.09±.02	
unfractionated, untreated soil	1.93±.31	0.09±.04	

in column experiments. The biotite dissolution rate in the field in 1991 was one-fifth of the rate measured in the column experiments. Much of the difference in rates obtained in column experiments compared to the field could be due to the difference in temperatures in the lab and field. Mean daily soil temperatures in the field ranged from 7 to 11°C during the study compared to 19±1.5°C in the laboratory. Assuming an activation energy of 7.1 kcal mol^{-1} (Paces, 1983), the difference in rates attributable to lower temperatures in the field compared to the lab is about 50%.

There was a large difference in weathering rates measured in column and reactor experiments (Table 3). Oligoclase dissolution rates in column experiments were one-tenth of the rates measured in reactors, and biotite rates in columns were one-half of the rates in reactors. Flow rates were similar in both types of apparatus. None of the solutions in the laboratory experiments were supersaturated with respect to pertinent secondary minerals during the period of steady-state solute release, so precipitation of secondary phases in the columns is unlikely to have caused the difference in rates. Fluidized-bed reactors provide a fundamentally different flow regime than saturated columns. In the reactors, excellent contact is maintained between mineral surfaces and solution because the mineral grains are kept in suspension by a jet of the reacting fluid. In columns, where mineral grains are at rest, it is conceivable that fluids may move relatively slowly along some paths, in a situation which is analogous to nature. The exact mechanism by which weathering rates are reduced under these flow conditions is open to debate, and will be discussed in more detail below.

The minerals used in these experiments differed from those used in most previous studies in that the grains in this study were extracted from natural soils, and thus had oxide coatings and organic material attached to them. One other important difference was that in most previous laboratory studies, fresh mineral grains were crushed, ground, and sieved prior to use. Even after rigorous cleaning procedures, ultrafine particles and high-energy sites might remain which would yield artificially high dissolution rates (Schnoor, 1990). Clow (1992) sought to test the importance of these differences in mineral surface characteristics by running a series of dissolution experiments with a variety of pretreatments. Weathering rates of soil minerals and crushed, ground granite from the field site were measured in columns to test the effect of grinding on mineral dissolution. To test the effect of coatings, the material used in the reactor experiments discussed previously was split into a subsample and treated to remove organic coatings (30% hydrogen peroxide, Kunze and Rich, 1959) and iron oxide coatings (sodium citrate/sodium dithionite, Mehra and Jackson, 1960). The dissolution rate of minerals from the soil in column experiments were indistinguishable from that of the same size-fraction of freshly-crushed granite bedrock, indicating that grinding had little effect on dissolution rates (Table 3). Pretreatment of soil minerals to remove organic matter and iron oxides also had no appreciable effect on mineral dissolution rates (Table 3). These results indicated that differences in mineral surface characteristics were not important controls on dissolution rates in the Rocky Mountain National Park study.

Causes of discrepancies between field and laboratory rates

Dissolution rates of minerals in the field in the Bear Brooks and Rocky Mountain National Park sites are clearly slower than dissolution rates of the same minerals in fluidized-bed flow-through reactors conducted at the same pH. The studies from Coweeta and the Czech Republic suggest similar discrepancies, but there are large uncertainties in the calculation, and the discrepancies may not be real. From this rather meager data base it would be rash to say that the discrepancy is universal, and even if it is universal, to ascribe it to a single universal cause. It is useful, however, to consider possible causes, provided the limited nature of the data base is kept in mind. The primary reasons that have been presented are:

1. *"Aging" of mineral surfaces and formation of protective secondary layers* (e.g. Paces, 1983; Velbel, 1985). The idea is that dissolution takes place preferentially at high-energy lattice defects (Lasaga, this volume). Over time, these high-energy sites would disappear, and dissolution rate would decrease. Alternatively, as dissolution of the primary minerals proceeded, secondary material (amorphous Al or Fe oxides?) would be deposited on the mineral surface and decrease contact between the mineral and solution.

    This mechanism clearly does not explain the results from Bear Brooks or Rocky Mountain National Park. In both studies, the minerals used in the laboratory experiments were obtained from the soils in the field and should thus already be "weathered". Furthermore, these "weathered" minerals dissolved at the same rate as "fresh" minerals or minerals in freshly-crushed bedrock. It is possible, however, that this mechanism is significant in more mature soils; it is also possible that formation of secondary minerals has a significant effect on soil hydrology (see below).

2. *The presence of some solute, particularly Al, which inhibits the dissolution of silicate minerals* (Chou and Wollast, 1985; Amrhein and Suarez, 1992; Oelkers et al., 1994; Gautier et al., 1994) . The role of Al in silicate dissolution is complex, and is complicated in nature by the common correlation between Al concentration and pH. However, it is clearly not a factor in the Rocky Mountain National Park study. There, Al concentrations (and Al^{3+} activities) are generally lower in the field than in the laboratory experiments (Fig. 3).

3. *Solutions in the field approach saturation with respect to primary minerals* (e.g. Burch et al., 1993; Lasaga et al., 1994). As chemical reactions approach equilibrium the rate of the back reaction, here reprecipitation of the original solid phase, become finite. Thus the net rate of reaction decreases as equilibrium is approached and ultimately becomes zero at equilibrium, where, by definition, the rate of the back

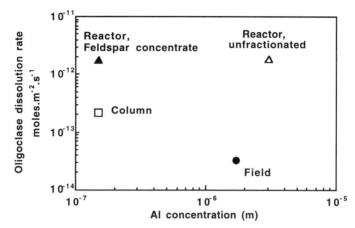

Figure 3. Rates of oligoclase dissolution in field, column, and fluidized-bed reactor experiments compared to total dissolved Al concentrations, Rocky Mountain National Park site. Data from Clow (1992).

reaction exactly equals the rate of the forward reaction. The question is how close to equilibrium the reaction has to be before this effect becomes significant. This is a controversial question which has not been fully resolved in the literature. Transition state theory gives, *for a simple, elementary reaction*, (Aagaard and Helgeson, 1982; Velbel, 1989):

$$\text{Rate} = k_a[1 - \exp(\Delta G_r/\sigma RT)]$$

where k_a is the apparent forward rate constant, σ is a "stoichiometric number" (the ratio of the rate of destruction of the activated complex to the rate of dissolution of the mineral; if the activated complex has the same formula as the mineral, σ is equal to 1), R is the gas constant and T temperature (K). ΔG_r is the departure from equilibrium, or *chemical affinity*, in units of kJ or kcal per mole. This relationship is shown in Figure 4. Chemical affinity is related to the activity product for the reaction by

$$\Delta G_r = RT \, ln \, (Q/K_{eq})$$

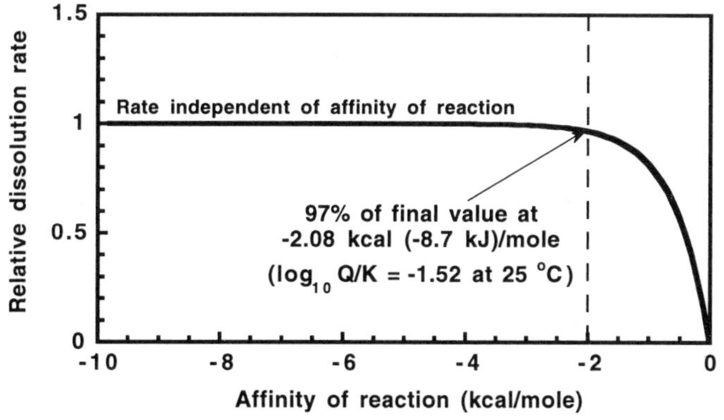

Figure 4. Theoretical relative dissolution rate of a mineral as a function of chemical affinity (degree of undersaturation) at 25°C. The rate is essentially independent of chemical affinity when the departure from equilibrium is greater than about 2 kcal/mole. σ is assumed to be 1

where Q is the activity product and K_{eq} the equilibrium constant. Thus for congruent dissolution of albite in acid solution:

$$NaAlSi_3O_8 + 4H^+ + 4H_2O = Na^+ + Al^{3+} + 3H_4SiO_4$$

the activity product, Q, is given by:

$$Q = \frac{[Na^+][Al^{3+}][H_4SiO_4]^3}{[H^+]^4}$$

where [ ] represent activities in solution. For undersaturated solutions, the value of Q is less than the equilibrium constant, K_{eq} and the affinity of reaction is negative. Degree of undersaturation can also be expressed as the saturation index, SI, where:

$$SI = \log_{10}(Q/K_{eq})$$

The term $\exp(\Delta G_r/\sigma RT)$ represents the rate of the "back reaction"—in this case reprecipitation of the primary mineral. It will be effectively negligible compared to the rate of the forward reaction if the rate of the backward reaction is less than 3% (the number is an arbitrary choice) of the forward rate. It will be 3% of the forward reaction if

$\exp(\Delta G_r/\sigma RT) = 0.03$, yielding $\Delta G_r/RT = -3.51$ (assuming $\sigma = 1$),

that is, if $\Delta G_r = -2.08$ kcal (-8.7 kJ) at 25 °C, or SI = -1.52.

If the departure from equilibrium is greater than this, the back reaction should be negligible and rate should be independent of affinity of reaction. Most solutions in the weathering environment should have saturation indices much more negative than -1.52 with respect to most important primary minerals other than potassium feldspars (Velbel, 1989) so, according to this model, chemical affinity should have no effect on the dissolution rates of most primary minerals in the weathering environment.

The problem with this model is that it assumes that mineral dissolution behaves as a simple elementary reaction. In experiments conducted by Burch et al. (1993) at 80°C and pH 8.8, the dissolution rate of albite seemed to be reduced by chemical affinity effects at a saturation index of about -5, much higher than the theoretical value of -1.52. On the other hand, in experiments by Oelkers et al. (1994) and Gautier et al. (1994) at 150°C and pH 9, the dissolution rate of albite was a strong function of aluminum concentration but there was no apparent "anomalous" dependence on chemical affinity. There are really almost no data on the effect of chemical affinity on dissolution rates of feldspars at 25°C and neutral to mildly acid pH. It is quite probable that chemical affinity effects are significant at greater degrees of undersaturation than indicated by Figure 4.

Solutions from the field at the Rocky Mountain National Park site had saturation indices of approximately –4.3 for albite (a surrogate for oligoclase in the rock) and –43 for phlogopite (a surrogate for biotite in the rock). It is conceivable, on the basis of results of Burch et al. (1983), that chemical affinity effects are decreasing the rate of oligoclase dissolution. On the other hand, the dissolution rate of oligoclase in the column experiments was a factor of 8 lower than in the fluidized bed reactor (Fig. 3) and that difference cannot be explained by chemical affinity effects. It is most unlikely that they are affecting the dissolution rate of biotite.

The weathering rate of potassium feldspars in the field, on the other hand, are probably limited by chemical affinity effects (or even supersaturation). The dissolution rates of K-feldspar and albite are similar in laboratory experiments at mildly acid pH (Blum, 1994), but K-feldspar generally weathers much more slowly than albite in the field. K-feldspar is less soluble than albite, so solutions approach saturation with respect to it before they approach saturation with respect to albite. The weathering rate of limestone in the field is clearly limited by approach to saturation and not by the dissolution rate of calcite far from equilibrium.

4. *Apparent rates in the field are slow because the estimate of mineral surface area in contact with percolating water is unrealistically high* (e.g. Schnoor, 1990; Swoboda-Colberg and Drever, 1993). In calculating the weathering rates at Coweeta, Bear Brooks, and Rocky Mountain National Park, it was assumed that all minerals in the soil or regolith were in contact with the percolating water and, conversely, that the water in contact with all mineral surfaces had the composition of either runoff from the site or solutions collected by tension lysimeters. This is obviously unrealistic (e.g. Hornberger et al., 1990). When a soil is saturated or nearly saturated with water, water movement will take place predominantly through a few relatively large channels ("macropores"), whereas water present in smaller channels ("micropores") will be relatively immobile. The concept of micropores can be extended to include water in fractures or submicroscopic channels within mineral grains (Banfield and Barker, 1994). Water in micropores will be chemically different from water in macropores. Its longer residence time will lead to higher pH values and a closer approach to equilibrium, both of which could cause decreased weathering rate. The "weathering rate" calculated from the flux out of the system would thus be determined largely by transport of solutes between micropores and macropores rather than directly by mineral dissolution rates. The difference in rates between column experiments and fluidized-bed reactor experiments for the minerals from the Rocky Mountain National Park site discussed above probably reflects greater solution-water contact in the fluidized-bed reactors. The situation is more complex in the vadose zone because the population of pores transmitting water varies with the degree of water saturation, but the same general principle should apply.

Drever et al. (1994) visualized the control in terms of two end-members. Under rapid flow conditions, solutions would move largely through high-permeability channels and weathering rate would be determined by reactions in the large channels (representing a small fraction of the total surface area) and by diffusional transport between the large channels and the small pores of the soil matrix. The other end-member would be represented by slow flow conditions. Concentrations would be more spatially uniform but, because of the implied long contact time, solutions would be more concentrated and closer to saturation with respect to primary phases. Weathering rate would be limited by the affinity of reaction or a related chemical effect.

RATES NORMALIZED TO LAND SURFACE AREA

In this section we shall discuss the effect of climate, lithology, and topography on weathering rates in catchments. We are concerned here only with solute production per unit area of land surface, and not with mineral surface area. The ideal data set for such a discussion would be input-output budgets for a large population of catchments encompassing all permutations of rock type, climate, topography, soil development, and vegetation. The data base that exists is very limited, so any generalizations must be regarded as tentative.

Effect of climate

White and Blum (1995) compiled input-output budgets from every gauged catchment on granitoid rocks in the world for which data were available. They restricted their study to granitoid rocks in order to minimize variability due to differences in lithology. They ignored biomass uptake and cation exchange, but focused their discussion on fluxes of sodium and silica, which are relatively unaffected by these processes. Their conclusions were:

1. Net sodium flux (corrected for atmospheric input) and silica flux could be described by an equation:

$$Q_{i,w} = (a_i * P)\exp\left[-\frac{E_a}{R}\left(\frac{1}{T} - \frac{1}{T_o}\right)\right]$$

$Q_{i,w}$ is the flux of element i from weathering (moles ha^{-1} y^{-1}), a_i and E_a are constants (fitting parameters), P is annual precipitation (mm), R is the gas constant, T the mean annual temperature of the catchment (K), and T_o a "reference temperature", taken to be 278 K or 5°C. Thus at a given temperature, the flux from weathering was proportional to precipitation. At a given precipitation, the flux increased exponentially with increasing temperature. E_a is analogous to an activation energy for a single chemical reaction. For SiO_2, a had a value of 0.046 and E_a 59.4 kJ mol^{-1}; for Na the values were 0.097 and 62.5 kJ mol^{-1} respectively.

The fact that weathering flux is linearly proportional to precipitation (i.e. water flux) suggests that weathering rate is not controlled directly by mineral dissolution rates far from equilibrium, but by either a chemical affinity effect or a transport process (cf. Schnoor, 1990). It is interesting to note that this relationship is the opposite to that observed by Velbel (1985) at Coweeta. In that catchment, the solute flux was independent of discharge, which is what would be expected from surface reaction control.

2. Although silica and sodium fluxes could be related to precipitation and temperature, no such relationship could be established for Ca, Mg, and K. White and Blum implied that "noise" from variations in mineralogy, biomass uptake, and/or cation exchange was swamping the "signal" from precipitation and temperature. This lack of correlation is somewhat disturbing, because Ca is generally the most important cation in the total weathering flux.

Effect of lithology

Meybeck (1986, 1987) measured solute fluxes and calculated chemical denudation rates in more than 200 small, nonpolluted, monolithologic catchments in France. Amiotte Suchet and Probst (1993) retabulated his results and calculated the weathering rate (determined as flux of CO_2 from the atmosphere) for each rock type. The data could be described well by the relationship:

$$F_{CO2} = a.Q_{water}$$

where F_{CO2} represents the flux of CO_2 consumed by weathering mmole km^{-2} s^{-1}, Q_{water} is the runoff (1 km^{-2} s^{-1}), and a is a fitting parameter, which reflects the "weatherability" of the particular rock type. The constant a can be normalized to a value of 1 for plutonic and metamorphic rocks (Table 4). It then provides an index for the relative weathering rate of

Table 4. Correlation between CO_2 consumption by weathering and runoff for monolithologic catchments in France (after Amiotte-Suchet and Probst, 1993).

Rock Type	N[1]	a (see text)	Correl. coeff.	Relative Rate CO_2 cons.	Solute flux[2]
Plutonic and metamorphic (granite, gneiss, schist)	41	0.095	0.92	1.0	1.0
Felsic volcanic rocks (rhyolite, andesite, trachite etc)	22	0.222	0.98	2.3	
Basalt	18	0.479	0.98	5.0	
Sandstone, arkose, graywacke	47	0.152	0.71	1.5	1.3
Argillaceous rocks (clays, shales, slate)	34	0.627	0.95	6.6	2.5
Carbonate rocks (limestone, dolomite, chalk, marl)	19	1.586	0.98	16.7	12.0
Evaporites	9	0.293	0.99	3.1	40-80

[1] Number of catchments.   [2] Solute flux (from Meybeck, 1986, 1987) includes solutes (e.g. Cl^-, SO_4^{2-}) not directly related to CO_2 consumption.

the minerals in different lithologies (in the climatic environment(s) of France). It is interesting to compare these numbers to the relative rates of mineral dissolution far from equilibrium (Table 5). The differences in rates for the catchments (less than a factor of 20) are much less than the differences in mineral dissolution rate (a factor of around 10^6 for calcite vs. sanidine; a factor of perhaps 50 to 100 for the minerals of a basalt vs. those of a granite). This suggests that differences in mineral dissolution rates far from equilibrium are not the only reason for the different rates observed in catchments.

Like White and Blum (1995), Amiotte Suchet and Probst found a linear dependence of weathering rate on discharge. They obtained good correlations without including a temperature term, perhaps because the temperature range of the catchments in France was relatively small.

Stallard and Edmond (1983, 1987) and Stallard (1985) showed a similar overall dependence of chemical denudation rate (assumed to be proportional to solute concentration in runoff) and lithology in the Amazon basin. Highest concentrations were associated with evaporites, intermediate concentrations with limestones, and lowest concentrations with silicate rocks.

Table 5. Relative rates of dissolution of different minerals in laboratory experiments at pH 5 far from equilibrium.

Mineral	Rate rate for albite
Quartz[a]	0.02
Muscovite[a]	0.22
Biotite[d]	0.6
Microcline[a]	0.6
Sanidine[a]	2
Albite	1
Oligoclase[b]	1
Andesine[b]	7
Bytownite[b]	15
Enstatite[a]	57
Diopside[a]	85
Forsterite[a]	250
Dolomite[c]	360000
Calcite[c]	6000000

[a] from Lasaga et al. (1994);
[b] from Oxburgh et al. (1994)
[c] from Wollast (1990)
[d] from Acker and Bricker (1992)

Relief

The relationship between chemical weathering rate and relief is controversial. In the Amazon basin, there is a clear relationship: chemical weathering increases with increasing relief, such that most (about 86%) of the solutes delivered to the ocean by the Amazon come from the Andes mountains, which make up about 12% of the total basin area (Gibbs, 1967; Stallard and Edmond, 1983). The problem is that in the Amazon basin, lithology and relief are highly correlated. Outcrops of limestone and evaporites are present in the Andes, whereas the rest of the Amazon basin is underlain by silicate rocks (including alluvium). The question is thus whether the weathering rates of silicate rocks are a function of relief, or whether the high rate simply reflects the presence of easily weatherable rocks. Stallard (1985) presented a model in which weathering rate was limited by chemical processes where erosion was active (a weathering-limited system) and by the accumulation of secondary products where erosion is minimal (a transport-limited system). Weathering rate should thus reflect erosion rate, which should be a function of relief. Stallard (1985) and Edmond et al. (1995) imply that silicate weathering rate in the Andes does in fact increase with increasing relief. Probst et al. (1994) disagree, and contend that the apparent effect of elevation is simply an effect of lithology.

White and Blum (1995) saw no relationship between weathering rate and relief in the catchments they studied. This may be because their population of catchments did not include any areas of really low relief such as the Amazon basin, or it may be that any effect was lost in the "noise" of their data.

In any event, we would not expect weathering rate to increase as a simple monotonic function of relief. According to Stallard's (1985) model, weathering initially increases as soil thickness increases, only decreasing when a very thick layer of weathered material has accumulated. Thus at very high rates of erosion, weathering should be slow because of the absence of soil. It should go through a maximum at some intermediate erosion rate/soil thickness, and then decrease again at very low erosion rates. Even this discussion assumes that soil thickness represents some sort of steady-state, which may be unrealistic. The data of Drever and Zobrist (1992) show a clear increase in weathering rate associated with increasing soil thickness in a region of active physical erosion.

SUMMARY

In order to compare weathering rates on the catchment scale with mineral dissolution rates measured in laboratory experiments, we need an estimate of the mineral surface area exposed to weathering in the catchment. Two studies of silicate weathering rate on the scale of a whole catchment (Paces, 1983; Velbel, 1985) suggest that rates in nature are slower than would be predicted from laboratory experiments, but their estimates of surface area are not well constrained. The difference was documented unambiguously in two much smaller-scale studies (Swoboda-Colberg and Drever, 1993; Clow, 1992). The reason for the discrepancy is probably both preferential flow through high-permeability channels and a decrease in rate as saturation is approached. The relationship between dissolution rate and saturation state represents a big gap in our knowledge.

On a catchment scale, the weathering rates of different lithologies correspond qualitatively to differences in dissolution rates measured in the laboratory. The differences are, however, less dramatic than would be predicted from the laboratory dissolution rates. On the basis of laboratory dissolution rates far from equilibrium alone, limestone should weather approximately 10^6 times as fast as granite, and basalt perhaps 50 times faster. The observed differences are not nearly as great as this, indicating that processes other than dissolution rate far from equilibrium have a major influence on weathering rates in the field.

REFERENCES

Aagaard P, Helgeson HC (1982) Thermodynamic and kinetic constraints on reaction rates among minerals and aqueous solutions, I. Theoretical considerations. Am J Sci 282:237-285

Aberg G, Jacks G, Hamilton PJ (1989) Weathering rates and $^{87}Sr/^{86}Sr$ ratios: an isotopic approach. J Hydrol 109:65-78

Acker JG, Bricker OP (1992) The influence of pH on biotite dissolution and alteration kinetics at low temperature. Geochim Cosmochim Acta 56:3073-3092

Amiotte Suchet P, Probst JL (1993) Flux de CO_2 consommé par altération chimique continentale: influences du drainage et de la lithologie. CR Acad Sci Paris 317 (II):615-622

Amrhein C, Suarez DL (1992) Some factors affecting the dissolution kinetics of anorthite at 25°C. Geochim Cosmochim Acta 56:1815-1826

Anbeek C (1992a) Surface roughness of minerals and implications for dissolution studies. Geochim Cosmochim Acta 56:1461-1469

Anbeek C (1992b) The dependence of dissolution rates on grain size for some fresh and weathered feldspars. Geochim Cosmochim Acta 56:3957-3970

Anbeek C (1993) The effect of natural weathering on dissolution rates. Geochim Cosmochim Acta 57:4963-4975

Anbeek C, Van Breemen N, Meijer EJ, Van Der Plas L (1994) The dissolution of naturally weathered feldspar and quartz. Geochim Cosmochim Acta 58:4601-4613

Banfield JF, Barker WW (1994) Direct observation of reactant-product interfaces formed in natural weathering of exsolved, defective amphibole to smectite: Evidence for episodic, isovolumetric reactions involving structural inheritance. Geochim Cosmochim Acta 58:4601-4613

Baron J (1992) The Biogeochemistry of a Subalpine Ecosystem: Loch Vale Watershed. Ecological Series 90, Springer-Verlag, New York

Berner RA (1992) Weathering, plants, and the long term carbon cycle. Geochim Cosmochim Acta 56:3225-3231

Berner RA (1994) Geocarb II. A revised model of atmospheric CO_2 over Phanerozoic time. Am J Sci 294:56-91

Blum AE (1994) Feldspars in weathering. In: Parsons I (ed) Feldspars and Their Reactions. NATO Advanced Study Inst, Series C 421:595-629, Kluwer: Dordrecht, Netherlands

Blum JD, Erel Y, Brown K (1994) $^{87}Sr/^{86}Sr$ ratios of Sierra Nevada stream waters: implications for relative mineral weathering rates. Geochim Cosmochim Acta 58:5019-5025

Brady PV (1991) The effect of silicate weathering on global temperature and atmospheric CO_2. J Geophys Res 96:18101-18106

Brunauer S, Emmett PH, Teller E (1938) Adsorption of gases in multimolecular layers. J Am Chem Soc 60:309-319

Burch TE, Nagy KL, Lasaga AC (1993) Free energy dependence of albite dissolution kinetics at 80°C and pH 8.8. Chemical Geology 105:137-162

Busenberg E, Clemency CV (1976) The dissolution kinetics of feldspars at 25°C and 1 atm CO_2 partial pressure. Geochim Cosmochim Acta 40:41-49

Chou L, Wollast R (1985) Steady-state kinetics and dissolution mechanisms of albite. Am J Sci 285:963-993

Clow DW (1992) Weathering rates from field and laboratory experiments on naturally weathered soils. Unpublished PhD dissertation, Univ Wyoming, Laramie, WY, USA

Clow DW, Drever JI (1991) A study of the weathering rate of of soil in an alpine catchment using field and laboratory experiments. Geol Soc Am Abstr Progr, San Diego, p A318.

Clow DW, Mast MA (1995) Composition of precipitation, bulk deposition, and runoff at a granitic bedrock catchment in the Loch Vale Watershed, Colorado. In: Biogeochemistry of Seasonally Snow-Covered Catchments. Tonnessan KA, Williams MW, Tranter M (eds), Int'l Assoc of Hydrological Sciences Publ 228:235-242

Drever JI (1988) The Geochemistry of Natural Waters. 2nd Ed. Prentice Hall: Englewood Cliffs, NJ

Drever JI, Zobrist J (1992) Chemical weathering of silicate rocks as a function of elevation in the southern Swiss Alps. Geochim Cosmochim Acta 56:3209-3216

Drever JI, Murphy KM, Clow DW (1994) Field weathering rates versus laboratory dissolution rates: an update. Min Mag 58A:239-240

Drever JI, Swoboda-Colberg, N, Kahl JS, Schnoor JL (1995) Mineral weathering rates at the Bear Brooks catchment, Maine, USA, from laboratory experiments, plot-scale acidification experiments, and catchment fluxes. Acid Reign '95? 5th Int'l Conf on Acidic Deposition, Gothenburg, Sweden, Abstract

Edmond JM, Palmer MR, Measures CI, Grant B (1995) The fluvial geochemistry and denudation rate of the Guayana Shield in Venezuela, Colombia and Brazil. Geochim Cosmochim Acta (in press)

Gautier J-M, Oelkers EH, Schott J (1994) Experimental study of K-feldspar dissolution rates as a function of chemical affinity at 150°C and pH 9. Geochim Cosmochim Acta 58:4549-4560

Gibbs RJ (1967) The geochemistry of the Amazon River system: I. The factors that control the salinity and the composition and concentration of the suspended solids. Geol Soc Am Bull 78:1203-1232

Hooper RH, Aulenbach BT (1993) The role of sampling frequency in determining water-quality trends. EOS Trans Am Geophys Union 74:279

Hornberger GM, Beven KJ, Germann PF (1990) Inferences about solute transport in macroporous forest soils from time series models. Geoderma 46:249-262

Huntington, RG, Hooper RP, Aulenbach BT (1994) Hydrologic processes controlling sulfate mobility in a small forested watershed. Water Resources Res 30:283-295

Kunze GW, Rich CI (1959) Mineralogical methods. In: Certain Properties of Selected Southeastern United States Soils and Mineralogical Procedures for their Study. Rich CI, Seatz LF, Kunze GW (eds) Southern Coop. Series Bull 61:135-146

Lasaga AC, Soler JM, Ganor J, Burch TE, Nagy KL (1994) Chemical weathering rate laws and global geochemical cycles. Geochim Cosmochim Acta 58:2361-2386

Likens GE, Bormann FH, Pierce RS, Eaton JS, Johnson NM (1977) Biogeochemistry of a Forested Ecosystem. New York: Springer-Verlag

Mast MA, Drever JI (1987) The effect of oxalate on the disolution rates of oligoclase and tremolite. Geochim Cosmochim Acta 51:2559-2568

Matzner E (1986) Deposition/canopy interactions in two forest ecosystems of Northwest Germany. In: HW Georgii (ed), Atmospheric Pollutants in Forested Area. Dordrecht, Netherlands: Reidel, p 247-462

Mehra OP, Jackson, ML (1960) Iron oxide removal from soils and clays by a dithionite-citrate system buffered with sodium bicarbonate. Clays and Clay Minerals, Proc. Nat'l Conf 7:317-327

Meybeck M (1986) Composition chimique des ruisseaux non pollués de France. Sci Géol. Bull 39:3-77

Meybeck M (1987) Global chemical weathering of surficial rocks estimated from river dissolved load. Am J Sci 287:401-428

Miller EK, Blum JD, Friedland AJ (1993) Determination of soil exchangeable-cation loss and weathering rates using Sr isotopes. Nature 362:438-441

Miller WR, Drever JI (1977) Water chemistry of a stream following a storm, Absaroka Mountains, Wyoming. Geol Soc Am Bull 88:286-290

Murphy KM (1993) Kinetics of albite dissolution:the effect of grain size. Unpublished MS thesis, Univ Wyoming, Laramie, WY

Norton SA, Kahl JS, Fernandez IJ, Rustad LE, Scofield P, Haines TA (1995) Response of the West Bear Brook Watershed, Maine to the addition of $(NH_4)_2SO_4$—Three-year results. Forest Ecology and Management (in press)

Norton SA, Wright RF, Kahl JS, Scofield JP (1992) The MAGIC simulation of surface water acidification at, and first year results from, the Bear Brook Watershed Manipulation, Maine, USA. Environ Pollution 77:279-286

Oelkers EH, Schott J, Devidal J-L (1994) The effect of aluminum, pH, and chemical affinity on the rates of aluminosilicate dissolution reactions. Geochim Cosmochim Acta 58:2011-2024

Oxburgh R, Drever JI, Sun Y-T (1994) Mechanism of plagioclase dissolution in acid solution at 25°C. Geochim Cosmochim Acta 58:661-669

Paces T (1983) Rate constants of dissolution derived from the measurements of mass balance in hydrological catchments. Geochim Cosmochim Acta 47:1855-1863

Peters NE (1989) Atmospheric deposition of sulfur to a granite outcrop in the piedmont of Georgia, USA. Int'l Assoc of Hydrological Sciences Publ 179:173-180

Probst A, Dambrine E, Viville D, Fritz B (1990) Influence of acid atmospheric inputs on surface water chemistry and mineral fluxes in a declining spruce stand within a small granitic catchment (Vosges Massif, France). J Hydrol 116:101-124

Probst JL, Mortatti J, Tardy Y (1994) Carbon river fluxes and weathering CO_2 consumption in the Congo and Amazon river basins. Applied Geochem 9:1-13

Reuss JO, Johnson DW (1986) Acid Deposition and the Acidification of Soils and Waters. Ecological Studies 59. New York: Springer-Verlag

Schnoor JL (1990) Kinetics of chemical weathering: A comparison of laboratory and field weathering rates. In: Stumm W (ed) Aquatic Chemical Kinetics. New York: Wiley, p 475-504

Schnoor JL, Stumm W (1985) Acidification of aquatic and terrestrial systems. In: Stumm W (ed) Chemical Processes in Lakes. New York: Wiley, p 311-338

Stallard RF (1985) River chemistry, geology, geomorphology, and soils in the Amazon and Orinoco basins. In: Drever JI (ed) The Chemistry of Weathering. Dordrecht, Netherlands: Reidel, p 239-316

Stallard RF, Edmond JM (1983) Geochemistry of the Amazon: 2. The influence of the geology and weathering environment on the dissolved load. J Geophys Res 88:9671-9688

Stallard RF, Edmond JM (1987) Geochemistry of the Amazon: 3. Weathering chemistry and limits to dissolved inputs. J Geophys Res 92:8293-8302

Stauffer RE, Wittchen BD (1991) Effect of silicate weathering on water chemistry in forested, upland, felsic terrane of the USA. Geochim Cosmochim Acta 55:3253-3271

Swoboda-Colberg NG, Drever JI (1989) Mineral weathering rates in acid-sensitive catchments: Extrapolation of laboratory experiments to the field. In: Miles DL (ed) Water-Rock Interaction. Rotterdam: Balkema, WRI-6:211-214

Swoboda-Colberg NG, Drever JI (1993) Mineral dissolution rates in plot-scale field and laboratory experiments. Chemical Geol 105:51-69

Velbel MA (1985) Geochemical mass balances and weathering rates in forested watersheds of the southern Blue Ridge. Am J Sci 285:904-930

Velbel MA (1989) Effect of chemical affinity on feldspar hydrolysis rates in two natural weathering systems. Chemical Geol 78:245-253

Vitousek PM, Reiners WA (1975) Ecosystem succession and nutrient retention: A hypothesis. Bioscience 25:376-381

White AF, Blum A (1995) Effects of climate on chemical weathering in watersheds. Geochim Cosmochim Acta (in press)

Wollast R (1990) Rate and mechanism of dissolution of carbonates in the system $CaCO_3$–$MgCO_3$. In: Stumm W (ed) Aquatic Chemical Kinetics. New York: Wiley-Interscience, p 431-445

Wright RF (1988) Influence of acid rain on weathering rates. In: Lerman A, Meybeck M (eds) Physical and Chemical Weathering in Geochemical Cycles. NATO ASI Series 251:181-196 Kluwer: Dordrecht, Netherlands

Chapter 11

ESTIMATING FIELD WEATHERING RATES USING LABORATORY KINETICS

Harald Sverdrup and Per Warfvinge

Department of Chemical Engineering II
University of Lund, Box 124
S-221 00, Lund, Sweden

INTRODUCTION

The understanding of chemical weathering is of importance for analyzing and understanding a large number of important environmental issues. Acidification of soils and waters are closely connected to the geochemistry of weathering. Weathering of soils and rocks represents the only self-repairing mechanism for acidified ecosystems. Weathering of silicate minerals seem to play an important role for the understanding of what causes global climatic changes. Ecological change is closely connected to changes in local chemical and climatic conditions, both in terms of major and minor elements. The integrated result of local effects over large regions, regulate these large cycles. With tools available based on new geochemical understanding, such ecological changes can now begin to be quantitatively addressed, and their causes and effects properly evaluated. Assessments of long term sustainability of biomass production in forestry and agriculture, water quality assessment or atomic waste storage safety require the weathering rate to be derived from geological and mineralogical properties of the system.

The importance of models in geochemistry

Models are especially important in research, not because they produce results of their own right, but because they allow complex and non-linear systems to be investigated and data from such systems to be interpreted. With models, the interaction of several simultaneous processes in a single experiment can be studied. Basically all models serve one or both of two purposes:

* Test the synthesized understanding of a system, based on mathematical representaion of its subsystems and the proposed coupling of subsystems.
* Predict what will happen in the future, based on the capability to explain how and why things have worked in the past.

When the researcher is forced to form equations and parameterize the coefficents, then his formal understanding of the ecosystem is put to test. Thus the model can be seen as the integrated bearer of the modeler's knowledge and understanding of the system. There are no "maybes" in modelling, as all parameters are assigned quantitative values according to unique and precise rules.

Good and bad models. A model is any consequence or interpretation taken from a set of observations or experience. A good model is one that adheres to these rules:

* The model must be transparent. It must be possible to inspect and understand the rules and principles the model is using.
* It must be possible to test the model. It must work on inputs that can be defined and determined, and it must yield outputs that can be observed.

The model can be a mental understanding of a mechanism, system, pattern or principle, and it can be substantiated as an equation or a set of equations or rules. If the principles and rules are many, then it is practical to let a computer program keep track of all connections and accounting. Goodness or badness of a model does not have anything to do with the adequacy of the principles inside the model. If the model is good, then we can verify or falsify the performance of the model with a specific principle incorporated.

Soil and catchment models

The wide range of catchment solute models and the different ways of modeling different processes, shows that there is a significant element of empiricism even in models that claim to be based on fundamental physical and chemical principles. This is true with respect to the mathematical formulation of individual processes, but especially the selection of parameter values that tie the magnitude of one chemical component or chemical process to another. The wide range of catchment solute models have emerged for several reasons. Important factors that have led to this variety are:

1. Management objective
2. Dependent and independent variables chosen
3. Availability of data allowing a model to be validated
4. Differences in catchment characteristics
5. Scientific background of the modeler

Virtually all models described here were developed with financial support from organizations with clearly defined management responsibilities. For example, the models dealing with soil and water acidification were created with support from environmental ministries to provide a scientific platform to assess the effects of long range air-transported pollutants on terrestrial and aquatic ecosystem. With this objective, the independent variables in such models always become the rate of deposition of acidifying substances, and the dependent variables would be the biologically most relevant quantity, such as stream pH and/or Al-concentration. Examples of such models are the American MAGIC model (Modelling Acid Groundwater In Catchments, Cosby et al., 1985), the Swedish SAFE model (Simulation Acidification in Forested Ecosystems, Warfinge et al., 1992) and the Dutch RESAM model (Research Acidification Model, de Vries and Kros, 1989).

To guarantee usefulness, catchment solute models have to be validated. This calls for a good data set with which the predictive capacity can be assessed. Indeed, many models were developed in close connection with a certain set of data. For example, the development of the ILWAS (Integrated Lake Water Acidification Study) model (Chen et al., 1983) relied heavily on the data that was generated within the catchment studies at Woods Lake and Panther Lake in the Adirondack Mountains. With the high level of ambition within the field project, it was feasible to design a model containing a large number of processes (and parameters), and still maintain the possibility to parameterize and calibrate the model.

The MAGIC model was developed with access to the 1000 lakes survey in Norway and a survey of 4000 streams, and the whole structure of the model, use and calibration, was designed to make maximum use of available data. The simple way MAGIC was built allows both a modeling on a catchment scale, and on a regional scale with available data.

ILWAS was also successfully calibrated to its research catchment, but the regional application was fraught with difficulty, since the amount of input data to such a complicated model was much larger (Chen et al., 1983). Many assumptions had to be made, several of which cannot really be checked, and thus, even if the model may by some be perceived as being more "right", it may still produce results with more uncertainty. Both the ILWAS and MAGIC modeling exercises share a common objective, to assess the impact of acid deposition on runoff chemistry, but the resulting models and the outputs are different.

But doesn't the diversity of models reflect a waste of effort ? Do we need so many models addressing the same issues? It is important to recognize that in all multi-disciplinary fields of work, such as ecosystem research, different views will always be represented. Different individuals will, literally, direct their consciousness to different parts of the ecosystem and search for explanations to various phenomena within their own area of specialization. As long as individuals are different, so their models will be.

Process-oriented models. The dominant class of catchment solute model is the process-oriented model. This class of model ties an input to the catchment, i.e. atmospheric input, fertilization, etc, to a set of output responses, i.e fluxes of components in a stream leaving a catchment through chemical and physical processes. The process-oriented models thus take their starting point in the basic properties of the catchment and the laws of the sub-systems.

The basic assumption of all process-oriented catchment models is that the law of continuity is applicable on a catchment scale. In words, the law of continuity states that what is produced of element i within a control volume must either leave the volume or be retained within it, i.e. either diverge or accumulate. In mathematical terms, this is written as

$$\frac{\partial C_i}{\partial t} = r_i - \nabla \cdot N_i \tag{1}$$

where C_i denotes the total concentration of component i, r_i the rate of production of i per unit volume and time, while N_i is a molar flux per unit area of the control volume. Most often this basic law is called mass balance, and models based on this law of mass conservation are called biogeochemical models.

In principle, we can calculate the concentration and fluxes of any solute with this equation. In practice, however, it is impossible to model every chemical aspect of every possible solute component. It is therefore necessary to restrict the level of ambition to what is absolutely necessary to meet the overall objectives of the modeling work.

We can also see that the law of continuity is quite clear about which possibilities that the modeler has access to for simplifications. Without simplifications, modelling becomes difficult and the models very difficult to apply to experimental conditions. The possibilities for variation are:

1. Spatial resolution and temporal resolution
2. Components included
3. Chemical reactions included

Although the equation of continuity can be solved analytically for a number of special cases, even inclusion of very simple chemistry leads to considerable mathematical difficulties. Therefore, no 'real' biogeochemical systems can be modelled at a catchment scale based on analytical solutions of the law of continuity. Inevitably, the law of continuity must be discretized, i.e. the catchment must be divided into one or more discrete compartments, and the time span of the simulation must be divided into an appropriate number of time-steps. It is also necessary to make decisions on which components are necessary, considering interactions between components and the rates of different reactions.

Dynamic versus static models. One important divide in model structure is that between dynamic models and static models. While the dynamic model is designed to predict the change in state variable as a function of time, the static model calculates the steady-state conditions for a set of boundary conditions. The dynamic models thus include both sides of Equation (1), while static models set the left hand side of Equation (1) $\equiv 0$. One example of a process-oriented static catchment model is PROFILE (Warfvinge and Sverdrup, 1992; Sverdrup and Warfvinge, 1993).

Several models neglect temporal variations on a time scale of an order of magnitude less than the other important forcing functions in the catchment. For example, the seasonal variations in nutrient uptake and biomass degradation is neglected in most catchment models that use annual average values for atmospheric inputs of nutrient and eutrofying/acidifying substances. This is also common for hydrological variations in catchments. Presently most attention is given to dynamic model approaches, but the catchment models developed by Kirchner (1990) and Hooper et al. (1990) represent an interesting line of development in static modelling. Also, it is noteworthy that the static model PROFILE has been more widely used for its final objective, environmental management within the context of critical loads of acid deposition than any dynamic model so far. The Henriksen steady-state water chemistry model (Sverdrup et al., 1990) has also been very successful in providing a tool for environmental policy decisions, by giving input to the new sulphur emissions reduction protocol signed in Oslo in June 1994 under the UN/ECE convention on long range transport of airborne pollutants (LRTAP).

Spatial and temporal resolution of dynamic models. One of the most striking differences between dynamic catchment solute models is the differences in temporal and spatial resolution (see Table 1). The models calculate solute chemistry with a temporal resolution ranging from hours to years, while the spatial resolution varies between 150 km by 150 km to a few m^2.

Table 1. Classification of temporal and spatial resolution of some models.

[More information on the models may be found in Chen et al., 1983; Gehrini et al., 1985; Christophersen et al., 1982; Cosby et al., 1985; Henriksen et al., 1988; Holmberg et al., 1989; Hooper et al., 1990; Kirchner, 1990; Sverdrup et al., 1995; Warfvinge and Sverdrup, 1992; Warfvinge et al., 1992.]

Spatial resolution	Temporal resolution		
	Year	Intermediate	Day/Hour
1-D, Catchment	MAGIC, SMART	BIRKENES EMMA	
2-D	NUCSAM, SAFE		MACRO, SOIL PULSE
3-D			TOPMODEL

The key questions that determine the resolution of process-oriented catchment solute models is how the catchment integrates the chemistry and physics of its subsystems, and further, how the biological component affected by the discharge quality (such as fish in a stream) is integrated by solute chemistry. Obviously, nature does integrate properties on different scales. One example is the formation of distinct soil horizons during podzol-isation. The horizons are chemically (macroscopically) homogeneous and often form a very well defined boundary to the next layer. With respect to vertical discretisation, the soil horizon appears to be a good starting point. For simulation of short term variations in solute chemistry, applications of end-member mixing analysis (Hooper et al., 1990) supports this view, both with respect to chemistry and hydrology. For long term simulations of acidification of catchments with a lateral flow components, distributed (soil layers or catchment subareas) models indeed yield different results than spatially lumped models (Wright et al., 1991).

In a comparative study, some catchment scale models such as MAGIC, SAFE and SMART were used to calculate the stream water chemistry at three Scandinavian sites for different loadings of acidifying atmospheric input (Warfvinge et al., 1992). The overall objective was to determine the acid load that would not cause damage to fish within a certain time frame. All models worked with an annual time-resolution in input (deposition, hydrology), so the models could only predict annual mean values of stream-chemistry. Yet, it is known that one important cause of damage to fish populations, is the occurrence of short-term acid episodes during the hatching period. The data available to relate the stream-chemistry to damage on fish populations also showed the risk for damage and extinction of fish populations as a function of annual mean water chemistry (Henriksen et al., 1990), expressed as Acid Neutralizing Capacity (ANC). This biological data did therefore integrate the temporal variability effects of annual mean level of acidity on the biota. Therefore the catchment solute models matched the data available and were appropriate for the predictive purpose.

In this context, it is appropriate to mention the distinction *between temporal resolution* and *numerical time-step*. While the former refers to the time-interval between input/output data, the latter refers to the number of subdivisions in this time-interval that is necessary to solve the underlying equations without errors. The time-step is therefore always shorter than the time-resolution.

The law of continuity (Eqn. 1) allows the modeler to involve any number of components (solutes) in a model, but the challenge is to find the minimum amount of components while still obtaining the link between system input and desired output. As the number of components (state variables) increase, so does the number of mass balance equations, the number of parameters and the structural complexity of the model.

Chemical weathering.   One key process in catchments is chemical weathering. While cation exchange is a process with a limited pool available, chemical weathering of primary minerals is an almost inexhaustible source of cations, such as Na, K, Mg and Ca, and a sink for acidity. In the long-term perspective, weathering is a key process in ecosystems because it provides the nutrient and acts a buffering mechanism. All prediction regarding effects of acidification on solute chemistry with process-oriented models do therefore include weathering as an important process. To illustrate the importance of reasonable estimates of weathering rates for over-all model performance, let us consider a catchment where the weathering rate is 100 m^2 yr^{-1} and the runoff is 0.4 m yr^{-1}: differences in model predictions of 50% would result in a difference of 125 meq m^{-3} in runoff alkalinity, a significant difference indeed.

The mechanism behind chemical weathering is that a dissolution reaction takes place at the mineral-solute interface, and that the chemical composition at the interface determines the weathering rate. The historic problem has been that there hasn't been rate laws available that have made it possible to link solute concentration to mineral dissolution, and no scheme for how the mineral phase should be characterized.

Despite the historic lack of scientific support, ecosystem and catchment modellers have been forced to assess weathering rates. Three main approaches have been taken:

1. Assignment of a constant weathering rate, determined either by calibration or by using an estimate from the actual catchment or some neighboring site.
2. Assigning of a standard rate, either arbitrary or determined as a function of soil properties, modified by the the solution pH
3. Development and application of a geochemical weathering model.

The first two approaches can be summarized in the following simple rate equation:

$$R_W = kA \cdot [\text{H}^+]^n \tag{2}$$

where W is the weathering rate, kA is a calibrated rate constant and n is the reaction order with respect to H^+. Using a constant weathering rate corresponds to $n = 0$ and $kA = R_w$ and this is the submodel used in MAGIC (Cosby et al., 1985), BIRKENES (Christophersen et al., 1982), NAP (Oene, 1992) and MIDAS (Holmberg et al., 1989) and many others. In ILWAS (Gherini et al., 1985) and RESAM (de Vries and Kros, 1989), $n = 0.5$ is an option, in ETD (Schoor et al., 1982) and a few other models, $n = 1.0$. The Dutch SMART model, which normally use a basic weathering rate rate inferred from soil type, total analysis or bedrock geology (de Vries et al., 1993), is intermediate, because it can be set to scale up the rate proportionally using $n = 0.5$. but also $n = 0$ is an option. Geochemical properties are partly considered, but the process is not really modelled.

The MAGIC model has found wide application. In this model the weathering rate for each individual base cation is adjusted, using runoff chemistry as objective. The obtained rate is interpreted as a weathering rate but is really a residual term for all sinks or sources not included in the model. As a tool to determine the weathering rate, the MAGIC model can be seen as an enhanced budget-study type of tool. The model estimates the ion exchange component, and an approximate weathering rate is obtained in the process. The model makes no connection to any geochemical property of the soil. The one-layer approach also limits the possibility to take any detailed geochemical properties or ion exchange properties and their variation in the soil profile into account.

A more detailed geochemical approach, as implemented in SAFE and PROFILE, treats weathering is a function of solute composition (Al, base cations, DOC, DIC), and geochemical properties such as mineralogy and soil texture. With the consideration of geochemical properties comes the need for a layered approach. Geochemical properties such as texture and mineralogy may vary significantly within the soil profile. With $n = 0.5$, the weathering rate increases 3-fold for a change in 1 unit of pH, a change that is in the range of actual decline in pH due to acidification over a period of 50 years (Falkengren-Grerup, 1987). The change in weathering rate is 2-fold per pH-unit if $n = 0.3$. The geochemical model SAFE includes rate equations for the actual dissolution reaction with $n = 0.5$, but due to the chemical interactions in the model, the geochemical model suggests even less change between pH 4.5 and 5.5 (40%) under field conditions than an $n = 0.3$ model.

We can see that in a vertically lumped model, an "average" H^+-concentration will be used to drive chemical weathering. Therefore, the higher the reaction order n is, and the greater the pH gradient through the soil is, the more difficult is it to conceptualize the effect of spatial lumping on the calculated weathering rate. We can thus conclude that models with a high value of n will have larger behavioral complexity than models with lower reaction order for chemical weathering.

Other model type. Besides the process-oriented models, there is another important class of models. Many of these models are holistic, which means that ecosystem units with fixed properties are defined. An example of such modeling is the holistic model introduced by Kirchner (1990). In its simplest form, the proposed model predicts variations in the concentration of one solute, as a function of the concentration of another. The only assumption is that heterogeneous equilibrium rules in the catchment, and that the parameters governing these equilibria are constant. One application has been to predict the concentrations of base cations in drainage water as a function of the acid anion concentrations, which, in a sense, reflect the conditions at the upstream boundary of the catchment.

Another holistic model is the end-member approach (EMMA) to predict stream-water variations in chemistry on a short-term (weekly) basis. EMMA is based on the hypothesis that stream-water is a mixture of water from different distinct source with different, time-invariant composition. An underlying idea is that variations in hydrological flow paths are responsible for these chemical variations. Indeed, examples show that the back-calculated hydrograph produced by EMMA may agree with the actual hydrograph (Hooper et al., 1990). Thus, EMMA is dynamic with respect to hydrology but is static with respect to the internal catchment chemistry. The dynamics in stream water chemistry is entirely due to the dynamics in hydrological pathway.

The same basic idea was implemented by Bergström et al. (1990) in PULSE. There, the hydrochemical properties of runoff from different soil compartments was determined by a fitted, time-variant function, rather than experimentally confined end-members. Since this category of model is based on time-invariant end-member composition, they can only be applied to short term variations, where the hydrological variations are greater than the changes in chemical properties of the catchment. They can thus not be applied to make prognosis regarding long-term trends, only short term-variations in stream water acidity.

THEORETICAL AND EXPERIMENTAL KINETICS

Introduction

In the following presentation we define chemical weathering the release of base cations due to chemical dissolution from specific minerals in the soil matrix, and the neutralization or production of alkalinity connected to this process. This definition is separate from the chemical dissolution of bulk rock or mineral mass, which is also called chemical weathering. We also strictly separate this from the denudation of base cations from ion exchange complexes in the soil which is termed leaching or cation exchange, as well as the degradation of different types of organic matter. We also disregard any type of mechanical degradation or physical weathering of soil minerals.

The capability to predict field weathering rates arose from a systematic and specific way of interpreting the laboratory kinetics. This may differ somewhat from some of the approaches presented in other chapters of this book. This introduction will explain how we determined the kinetic coefficients from laboratory experiments, and explain why some of our coefficients have slightly different values than those usually found in the literature. The basis for making use of laboratory experiments is to evaluate them in terms of a specific and consistent theory. Applying a model permits data from experiments of very different design to be used together in such a way that the effects of differences in conditions and properties can be accounted for. Thus very different experiments can be normalized to a common platform for comparison. This favors experimental studies where one factor at a time is studied under great care and control. Evaluation of laboratory experiments without models is difficult, and lead to oversimplification and confusion. In order to extract the kinetic coefficients from laboratory data, a clear understanding of important mechanisms affecting mineral dissolution and the quantitative kinetic expressions connected with each mechanism is needed.

A brief history of weathering kinetics

The long years of confusion. Research on the kinetics of weathering has gone on since the beginning of this century, and for decades most results were confusing, laboratory results were not always consistent, and the rates could not be translated to field observations (Paces, 1983; Velbel, 1986; Sverdrup, 1990). Experimental weathering of minerals was considered to be a three-stage process. First, cations would be leached from the fresh mineral, and secondly, the non-steady-state dissolution through a growing layer of secondary product mineral would occur; it was called parabolic kinetics. Finally, the mineral would reach a pseudo-steady-state rate called long term linear kinetics. It has been shown that parabolic kinetics were caused by experimental artifacts (Berner, 1978; Holdren and Berner, 1979; Dibble and Tiller, 1980). Deficiencies in sample preparation and improper characterization of the exposed surface area of the mineral powders used in experiments muddled earlier studies significantly (Sverdrup, 1990). This caused confusion when comparisons with field rates were made, and the impression of a large discrepancy between laboratory and field rates (Paces, 1983; Velbel, 1986).

Many early studies were hampered by the lack of a multiple-reaction perspective, rendering many experimental designs inadequate. Ignoring the effect of CO_2 or an organic acid may under certain circumstances make a difference on the rate of more than one order of magnitude. Organic ligands in buffers used to keep the pH value constant during experiments, react with minerals. If different buffers are used in a series of dissolution experiments, then each organic ligand will react differently with the mineral, and no clear interpretation can be made of the results.

Early rate equations. Aagaard and Helgeson (1982), Helgeson et al. (1984), Lasaga (1981, 1984) and Murphy and Helgeson (1987) formalized the transition theory for dissolution of one mineral occurring through one elementary reaction:

$$R_W = k \cdot A_W \cdot \prod_{i=1}^{n} a_i^{-n_{i,j}} \cdot \left(1 - e^{\left(-\frac{AA}{\sigma \cdot RT}\right)}\right) \tag{3}$$

R_W is the amount of ion released due to weathering of the mineral, k the specific rate coefficient, a the activity of the i-th species, n_{ij}, the reaction coefficient of the i-th reactant species in the j-th reversible reaction corresponding to the formation of one kmole of activated complex on the surface of the mineral. AA is the chemical affinity, σ a stoichiometric number, R the universal gas constant and T the temperature. A_W is the surface area of the mineral exposed to the solution.

Helgeson et al. (1984) assumed the formation of only one type of complex at a time at low pH-values below pH 2.9, between pH 2.9 and pH 8 and above pH 8, resulting in a simple expression for albite dissolution:

$$r = k_H \cdot a_{H^+}^{1.0} + k_{H_2O} + k_{OH} \cdot a_{OH^-}^{0.4} \tag{4}$$

r_W is the weathering rate per unit surface area of the mineral. Using the theory developed by Aagaard and Helgeson (1982) and Helgeson et al. (1984), Murphy (1985) went on to determine rate coefficients for albite, olivine, wollastonite, diopside and a few other minerals. Due to the limited amount of data used and differences in experimental conditions, the coefficients are not applicable to field conditions.

Results by Chou and Wollast (1985) have shown that the expression derived by Helgeson et al. (1984) for albite, assumed a too high reaction order with respect to the hydrogen ion. The transition state theory combined with surface coordination chemistry was applied in the work of Chou and Wollast (1985) based on a larger basis of consistent experimental data:

$$r_W = k_H \cdot \left(\frac{a_{H^+}^n}{a_{Al^{3+}}^y} \right) + k_{H_2O} + k_{OH} \cdot \left(\frac{a_{OH^-}^w}{a_{Al(OH)_4^-}^y} \right) \tag{5}$$

where r_W is the weathering rate expressed as the ion release rate. In this text, this generally is expressed as release of Ca, Mg, K and Na. Chou and Wollast (1985) determined the reaction orders experimentally; $n = 0.5$, $w = 0.25$ and $y = 0.35$. They also deduced that the last term accounts for the alkaline range is an aspect of the hydrogen ion reaction.

We consider the studies of Chou and Wollast (1984, 1985) as the most important steps towards deriving a theory that would work under field conditions. They were the first to fully realize the importance of distinguishing several simultaneous weathering rate reactions for silicate minerals, perhaps inspired by earlier work of Plummer et al. (1973) on calcite. With them came for the first time experimental laboratory data that had been produced with a multiple reaction view.

Chemical dissolution reactions

A number of weathering reactions have been experimentally identified to take place in parallel at the mineral surface (Sverdrup, 1990). Five simultaneous reaction systems are considered:

* The reaction with the hydrogen ion
* The reaction with water
* The reaction with the hydroxyl ion
* The reaction with carbon dioxide
* The reaction with strongly complexing organic acids

The total base cation release rate by chemical weathering will be the sum of the rate of all parallel simultaneous processes regardless of the molecular mechanism, minus the rate of precipitation with secondary solid phases. (Plummer et al., 1976; Lasaga, 1981; Aagaard and Helgeson, 1982; Helgeson et al., 1984). The forward reaction rate can under certain circumstances be affected by dissolved species which interfere with the formation mechanism of the activated surface complexes. The initial stage of mineral dissolution is also characterized by non-stoichiometric dissolution where the molar composition of the reaction products is different from the molar composition of the parent mineral. This applies to almost all minerals. During long term steady state dissolution, most mineral dissolution is stoichiometric, and any deviation from this, usually indicates that some secondary solid phase is simultaneously precipitated from the solution. Under certain circumstances, sheet silicates may dissolve partially or non-stoichiometrically and form secondary mineral residues.

The Transition State Theory applied to weathering

The transition state theory was defined by Eyring (1935) and Wynne-Jones and Eyring (1935), stating that the rate of a chemical reaction is controlled by the decomposition of an activated complex. The theory of Eyring applies to chemical reactions in general, and the principle may be applied to solid-liquid reactions such as chemical reactions with minerals.

Consider a reaction between the solid mineral A, and the dissolved species B, which will react to form a surface complex $\Psi*$ at the mineral surface which is a part of the solid solution at the solid-liquid interface, and two dissolved reaction products D and E, according to the stoichiometry of the elementary reaction controlling the concentration of the activated complex

$$f \cdot A(solid) + p \cdot B \rightleftharpoons s \cdot \psi^*(solid) + q \cdot D + v \cdot E \tag{I}$$

f, p, s, q, v are stoichiometric coefficients of the reaction. The formation of the activated complex is assumed to take place from a surface complex, and the activated complex is assumed to be in equilibrium with and bear similar configuration to the surface complex. The surface complex is assumed to form a regular solid solution with the reacting solid. The activated complex is considered to be in or near equilibrium with the reactants, and the decay of the complex is considered formally to be reversible. The theory assumes the decay to be reversible, but for most primary minerals, elevated temperature and pressure required for their formation, and the rate formation of the activated complex from the reaction products will be small enough to justify the assumption that the decay is virtually irreversible.

$$\psi^* \rightarrow \text{reaction products} \tag{II}$$

The net weathering rate of mineral per unit surface area and time is proportional to the concentrations of the complex [$\Psi*$]

$$r = k' \cdot [\psi^*] \tag{6}$$

The reaction products from creating the activated surface complex must be desorbed from the mineral, and may be in equilibrium with the absorption site and the activated surface complex. The presence of the reaction products at the surface can interfere with the adsorption of reactants and the decay mechanism of the activated surface complex. The dissolution process has been illustrated in Figure 1.

It is probable that the reactive sites have a range of energy levels, and that the energetically most favorable sites get occupied by sorbed species first. This is equivalent to a logarithmic change in the adsorption enthalpy as adsorption progresses, which leads to a Freundlich adsorption isotherm for the surface concentration of reactant (Swalin, 1972). The isotherm exponent has the same value for both the adsorbing and the desorbing species, being a function of the ion exchange matrix only (Swalin, 1972; Sverdrup, 1990). When the Freundlich adsorption isotherm is applied to ion exchange between dissolved species in aqueous solution and a surface, then the Rothmund-Kornfeld type of equilibrium equation can be derived for Reaction (I) (Bolt, 1982, Bruggenwert and Kamphorst, 1982:

$$K_{Eq} = \left(\frac{a^s_{\psi^*}(s)}{a^f_A(s)} \right) \cdot \left(\frac{a^q_D \cdot a^v_E}{a^p_B} \right)^{1/N} \tag{7}$$

where $a_i(s)$ is the activity of species i in the regular solid solution at the mineral surface, $\Psi*$ represents the activated surface complex, A the solid mineral. The concentration of activated compölex can be expressed as:

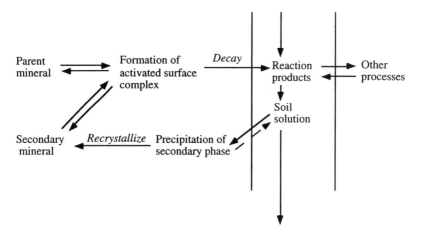

Figure 1. Schematic view of the weathering process. Reactants form an activated unstable surface complex at the mineral surface, reaction products in the soil solution affect the formation process. The formation process is reversible. The complex decays irreversibly, proportionally to the number of complexes. Reaction products may interfere with the formation process for the complex. Secondary phases may be formed from the soil solution. Once the precipitates have formed irreversibly, they can only dissolve through the kinetic process involving an activated surface complex.

$$[\psi^*] = \left(\frac{a_A^f(s)}{\gamma_{\psi^*}}\right) \cdot K_{\gamma_{\psi^*}} \left(\frac{a_B^p}{a_D^q \cdot a_E^v}\right)^{1/N} \tag{8}$$

The activity of a species i in a solution is defined as

$$a_i = \gamma_i \cdot [C]_i \tag{9}$$

where a is the activity of species i, $[C]$ the molar concentration and γ the activity coefficient. (s) indicates the concentration at the surface in the phase where the reactions take place. For any component i of the solid solution, the activity coefficient γ_i is defined as (Swalin, 1972):

$$\gamma_i(s) = e^{\frac{\Delta H_E}{RT} \cdot (1 - X_i)^2} \tag{10}$$

and the partition coefficient N for the solid solution (Swalin, 1972) can be approximated with a series truncated to:

$$N = 1 - \frac{\Delta H_E}{RT} \tag{11}$$

ΔH_E is the excess enthalpy of mixing of the solid solution including all species involved at the surface. For an ideal solution the excess energy is zero, and N has the value one. γ_{ψ^*} is the activity coefficient for the activated surface complex in the solid solution. X_i is the mole fraction of species i in the solid solution at the preferred site on the mineral surface. N is related to the exponent of the Freundlich adsorption isotherm and a measure of the non-ideality of the solid solution at the surface between the parent surface material, the activated surface complex and its eventual precursors, and the surface adsorbed reactant. The observed rates of weathering may be interpreted to imply that the concentration of activated complex at the mineral surface is very small, but this does not as such imply ideal solid solution behavior, even if it is usually assumed that dilute species show ideal behavior. The activated complex may also be in near equilibrium with a surface complex of higher concentration, acting as precursor for the activated complex, and the more abundant surface complex may be responsible for the observed non-ideal behavior.

Several reactions occur at the mineral surface simultaneously, each individual elementary reaction depending on individual activated surface complexes formed at the mineral surface. The total rate of base cation release to the liquid solution from the mineral, is taken to be the sum of all forward rates minus the rate of all backward reactions, including precipitation of secondary solid phases (Chou and Wollast, 1985; Grandstaff, 1977, 1980, 1986; Lagache, 1965, 1970; Busenberg and Clemency, 1976). This is based on the assumption that each individual activated surface complex occupy only a small fraction of the possible active sites for complex formation on the mineral surface. Backward reactions like precipitation of the parent primary mineral are not likely to occur at ambient temperatures found in the natural soil, because most of the minerals considered here require significantly elevated temperatures and pressure for their formation from solution or mineral formation through rock metamorphosis. The backward dissolution reaction must be virtually zero for thermodynamic reasons. In relation to base cation release, one or more secondary minerals can be precipitated from the solution. At the surface, we have a surface saturation index Ω for the parent mineral a and the activated surface complex:

$$X_a + X_{\psi^*} = \Omega \tag{12}$$

and for all components at the surface:

$$X_a + \overbrace{\sum_i X_{\psi_i^*}}^{activated} + \overbrace{\sum_i X_i}^{inactivated} = 1 \tag{13}$$

where the first sum (*activated*) is the sum of the fraction of surface sites occupied by activated surface complexes formed by reacting component, and the second sum (*inactivated*) is the adsorbed fraction of other ligands adsorbed to the surface which do not react at any significant rate. For some minerals, some types of organic ligands can create inactive complexes at the surface. Such inactive complexes do not dissolve the mineral, but they occupy reactive sites. Such ligands will slow down dissolution.

If the complex and the parent mineral surface are the only species present in a significant proportion at the surface, then $\Omega = 1$. The concentration of activated surface complex is related to the surficial mole fraction by the relation:

$$[\psi^*] = X_{\psi^*} \cdot \left(\frac{\rho \cdot l \cdot \sigma}{MW} \right) \tag{14}$$

where ρ is the density of the mineral, l is the thickness of the reacting layer at the surface, and MW the moleweight of the mineral. σ is the relative fraction of the mineral surface being covered with preferred sites for weathering reactions and formation of activated surface complexes.

A general expression for any stoichiometry can be derived if Equation (8) is combined with Equation (14), this is then inserted in Equation (12), and the whole expression rearranged to be solved for the fraction activated surface occupied by activated surface complex X_{ψ^*}

$$(\Omega - X_{\psi^*})^f \cdot K_{\gamma_{\psi^*}} - (X_{\psi^*})^s \cdot \left(\frac{a_D^q \cdot a_E^v}{a_B^p}\right)^{1/N} = 0 \tag{15}$$

For any values for f and s and values for the fraction of activated surface complex at the reaction site, X_{ψ^*}, that satisfy the equation must be found in order to determine the kinetic expression.

It can be shown that the general expression may under circumstances where the reactions take place in dilute solutions, be approximated with a simple expression where the parameter X_{ψ^*} will disappear as compared to the total mineral compostion fraction at the surface Ω, and which is valid for all values of f and s, as long as the fraction of activated surface complex at the preferred surface sites is small as compared to the reacting mineral component in the solid solution:

$$(\Omega - X_{\psi^*})^f \to 1 \tag{16}$$

It is evident from Equation \ref{omega}, that the surface mole fraction of activated surface complex is (Sverdrup, 1990):

$$X_{\psi^*} = (K_{\psi^*} \cdot \frac{a_B^p}{a_D^q \cdot a_E^v})^{1/(N \cdot s)} \tag{17}$$

The forward rate expression can be obtained by inserting the expression derived for the complex concentration (Eqn. 15) in the Eyring equations (Eqns. 6 and 7):

$$r_w = \sum_{i=1}^{reactions} k_i' \cdot (K_{\psi_i^*} \cdot \frac{a_B^{p_i}}{a_D^{q_i} \cdot a_E^{v_i}})^{1/(N \cdot s)} \cdot \left(\frac{\rho \cdot \sigma \cdot l}{MW \cdot \gamma_{\psi^*}}\right) \tag{18}$$

where k_i is the rate coefficient of the i-th forward reaction. K_{ψ^*} is the equilibrium constant for the equilibrium between the reactants and the activated surface complex in the i-th reaction. This corresponds to the actual conditions in most cases of weathering in the natural soil environment where reactant concentrations are generally relatively low. A general observation from experimental studies is that the condition above is valid for solution reactant concentrations less than 3×10^{-3} kmol/m³ (Sverdrup, 1990). At very high reactant concentration all the preferred sites on the mineral surface may become occupied by activated surface complexes, X_{ψ^*} will approach unity, and the rate will approach a constant value.

The reaction between H^+***-ions and a mineral.*** For a large number of minerals experiments show that the reaction with the hydrogen ion dominates at low pH values, we will use this reaction as an example to derive the kinetic expression. When the reaction between a mineral and the hydrogen ion is considered we will have a stoichiometry such that:

$$f \cdot mineral + p \cdot H^+ \rightleftharpoons q \cdot Al^{3+} + v \cdot BC^{2+} + s \cdot activated\ complex \tag{III}$$

For K-feldspar, one proposal for the formation of the activated complex is as given below, somewhat modified based on Chou and Wollast (1985) and Sverdrup (1990). First most, but not all alkali is leached out from the lattice:

$$KAlSi_3O_8 + (1 - \frac{v}{2}) \cdot H^+ \rightleftharpoons (1 - \frac{v}{2}) \cdot K^+ + H_{(1-\frac{v}{2})}K_{\frac{v}{2}}AlSi_3O_8 \tag{IV}$$

Some K is required to remain in the lattice to explain the dependence of the retardation on K or Na. In the next step the activated and negatively charged surface complex is formed:

$$f \cdot H_{(1-\frac{x}{2})}K_{\frac{x}{2}}AlSi_3O_8 + p \cdot H^+ \rightleftharpoons q \cdot Al^{3+} + v \cdot K^{2+} + s \cdot H_{f+p}Si_{f\cdot3}O_{f\cdot8}^{(3q-p)-} \qquad (V)$$

For K-feldspar, Chou and Wollast (1985) and Sverdrup and Warfvinge (1993) propose the reaction:

$$KAlSi_3O_8 + 0.75 \cdot H^+ \rightleftharpoons 0.75 \cdot K^+ + H_{0.75}K_{0.25}AlSi_3O_8 \qquad (VI)$$

First K is leached from the lattice, next, Al and K are lost in the step that sets up the complex in the activated state:

$$2 \cdot H_{0.75}K_{0.25}AlSi_3O_8 + 5.5 \cdot H^+ \rightleftharpoons 2 \cdot Al^{3+} + 0.5 \cdot K^{2+} + 1 \cdot H_7Si_6O_{16}^- \qquad (VII)$$

Thus we propose $f = 2$ and $s = 1$. According to this particular stoichiometry, the rate dependence on K is 1/4 of that on Al. Since $n = p/N$, and N is counted per unit mineral, then it becomes evident that $N = 4.5$ to 5.5. Thus at molecular level, the reaction order n is 5.5. Due to non-ideality of the surface this become a macroscopic reaction order of 0.5 to 0.6. This consept can be compared to the interpretation by Stumm et al. (1987) where the number of H^+-ions reacting in the rate limiting reaction step, $p,$ would be equal to the oxidation state of the center of complexation at the surface. For corundum Stumm et al. (1987) find a reaction order of 6.

Many authors have tried to work out the exact composition of the activated complex, its composition is important for the exact interpretation of the molecular mechanism, but this is very difficult. For example for K-feldspar several different complexes have been proposed, which all lead to similar rate expressions. Most of the proposed complexes imply that the base cation first detaches from the surface, then Al is stretched out of the lattice in an activated complex with the reactant. This is a very unstable crystalline formation that will break to pieces with time and the complex decays. It is by no means certain that the complex really looks exactly as we have illustrated. However, these constructs can be seen as operational models, simpler descriptions that still retain some of the important properties of the real activated complexes. These model constructs are used to predict the consequence of various reactants on mineral dissolution rates.

The expression for the concentration of activated surface complex becomes:

$$X_{\psi^*} = k'_{H+}\left(\frac{K_{\psi^*,H+} \cdot a_{H+}^p}{a_{BC2+}^v \cdot a_{Al3+}^q}\right)^{1/N} \qquad (19)$$

It can then be shown (Eqn. 15) that the reaction rate with the hydrogen ion in dilute solutions for any value of f and $s = 1$ becomes (Sverdrup, 1990):

$$r_{H+} = k'_{H+}\left(\frac{K_{\psi^*,H+} \cdot a_{H+}^p}{a_{BC2+}^v \cdot a_{Al3+}^q}\right)^{1/N} \cdot \frac{\rho \cdot l \cdot \sigma}{MW \cdot \gamma_{\psi^*}} \qquad (20)$$

p, q and x are the stoichiometric numbers of the reactants involved in the formation of the activated surface complex. Protonation of the surface may cause v to be less than the indicated value, at full protonation, $v = 0$. Under the conditions normally prevailing in soils the concentration of reactant species will be low, $\Omega = 1$, and the rate expression can be simplified to:

$$r_{H+} = k_{H+} \frac{a_{H+}^n}{a_{BC^{2+}}^x \cdot a_{Al^{3+}}^y} \tag{21}$$

where the reaction orders are given by the ratio between the stoichiometric number and the partition coefficient yield per unit of a parent mineral reacting. In assessing experimental rates the partition coefficient must first be estimated from the experiments, assuming that the proposed stoichiometry is right; $N = p/n \cdot 1/f = 5.5$. The partition coefficent N is then used to estimate what the other reaction orders for Al and BC should be; for base cations we have $x = v/N = 0.25/2.5 = 0.1$, for Al we have $y = q/N = 1/5.5 = 0.18$. Experimental data indicate that $n = 0.5$ for K-feldspars, $v = 0.1$ to 0.2, and $y = 0.2$ to 0.4. For other minerals where only the reaction order for the H^+-ion is available, this may be used to estimate the approximate reaction order with respect to BC and Al.

The molecular mechanism involved when Al and BC affect the rate is seen from Reaction (V) and Equations (7), (17) and (18). The presence of Al or BC change the conditions for forming the activated surface complex. It is evident from Equation (17) that higher concentrations of either Al or BC will cause a lower equilibrium concentration of activated surface complex at the surface, and the rate is proportional to the surface concentration of complex. Under conditions where the reactant concentration is very high, such as at pH below 1 or 2, the analytical solution derived from Equation (15) implies a weak or no dependence of the rate on pH. We then can have:

$$r_{H^+} = k'_{H+} \frac{a_{H+}^n}{a_{H+}^n + a_{BC^{2+}}^x \cdot a_{Al^{3+}}^y} \tag{22}$$

This is confirmed by results for several minerals including albite, garnet, staurolite, disthene, pyroxenes, hornblende, and chlorite (see Chou and Wollast, 1985; Sverdrup, 1990). Thus the derived expressions describe both high and low concentration effects on dissolution kinetics.

Chou and Wollast (1985) hypothesized that the rate and reaction order of the OH^- reaction for albite feldspar can be deduced from the kinetics of the reaction with H^+, as a result of pH dependent speciation of aluminium. Their conclusion is that the OH^- reaction is an aspect of the H^+ reaction. This appears to be the case for many minerals containing Al. For minerals containing no or little Al such as certain nesosilicates and mafic minerals, independent activated complexes involving the OH^- ion may be formed (Sverdrup, 1990).

Organic acids.   Grandstaff (1977, 1980, 1986) studied the dissolution of a for-steritic olivine in different organic acids. For olivine, organic acids reacted with the mineral and increased the rate. There is a separate reaction between the organic ligand and the mineral, which must be separated from the reaction between H^+ from the acid which is another type of mechanism. Further studies have shown that organic acids have a larger effect on some minerals than others. Some minerals, like feldspars, are relatively slowly dissolving in organic acids.

Grandstaff (1980, 1986) described rate of reaction as being dependent on the adsorption of a bidentate ligand to two adjacent adsorption sites on the mineral with a Langmuir isotherm and the rate depend on the surface saturation with reactant:

$$r_{org} = k'_{org} \cdot \left(\frac{a_{org}}{1 + K_{org} \cdot a_{org}} \right)^{0.5} \tag{23}$$

assuming the activated surface complex to be proportional to the concentration of surface adsorbed polydentate ligand. Similar expressions have been found for feldspars, basaltic glass, quartz, etc. The mechanism implies that when enough organic acid is present to saturate the surface, the rate will not increase further. This saturation of the surface may block the reactive sites, and therefore the rate of the other reactions may decrease. The overall result may be that the total rate does not increase very much, maybe even not at all. It can therefore not always be assumed that the presence of organic acids will automatically increase the weathering rate. There are examples available where the presence of organic acids will lower the rate (Sverdrup, 1990).

The transition state theory rate expression. It follows from what we wrote earlier that the total base cation release rate caused by silicate dissolution consists of the rate contributions from each chemical reaction:

$$r_W = r_{H+} + r_{H_2O} + r_{OH-} + r_{CO_2} + r_{Org} \tag{24}$$

For each of the reactions included in Equation (24), an expression based on the transition state theory may be worked out. It can be shown by applying the transition state theory to each of the reactions discussed above that the rate expression for dissolution of silicate minerals in dilute solutions has the following form (Sverdrup, 1990; Sverdrup and Warfvinge, 1993):

$$r = k'_{H+} \left(\frac{a^n_{H+}}{a^x_{BC^{2+}} \cdot a^y_{Al^{3+}}} \right) + k^*_{H_2O} \left(\frac{a^{nn}_{H_2O}}{a^{vv}_{BC^{2+}} \cdot a^{yy}_{Al^{3+}}} \right) + k'_{OH-} \left(\frac{a^{n-y}_{OH-}}{a^w_{BC^{2+}} \cdot a^y_{Al^{3+}}} \right)$$
$$+ k_{CO_2} \cdot P^{nCO_2}_{CO_2} + \overset{Org.acids}{\underset{i}{\sum}} k_{Org_i} \cdot \left(\frac{K_{org_i} \cdot a^{nOrg_i}_{org_i}}{1 + K_{org_i} \cdot a^{nOrg_i}_{org_i}} \right) \tag{25}$$

The observed reaction order with respect to a reactant species, is related to the ratio between the stoichiometric number ($p,q,...$) of that species in the reaction, and the partition coefficient N. The partition coefficient describes the non-ideality of the surface adsorption of reactants. Thus if the stoichiometry of the reaction is known and the reaction order n observed from dissolution experiments, then the partition coefficient N and the degree of non-ideality of the dissolution process can be determined.

In the soil model described later the kinetic expression for the reactions were modified somewhat for the model to give realistic predictions. In Equation (19), an extremely low Al concentration could potentially increase the rate to infinity. This is obviously not the case in natural systems. If the soil solution concentration of Al become very low, then the Al concentration at the surface will be controlled by the surface release rate of Al through dissolution, and there will be a threshhold below which Al in the bulk of the solution has no longer any effect on the rate. The rate equation was rewritten as:

$$k'_{H+} \cdot \left(\frac{a^n_{H+}}{a^y_{Al^{3+}} \cdot a^x_{BC^{2+}}} \right) \rightarrow k'_{H+} \cdot \frac{a^n_{H+}}{(C_{Al} + a_{Al^{3+}})^y \cdot (C_{BC} + a_{BC^{2+}})^x} \tag{26}$$

C_{Al} is the Al concentration that can be experimentally determined (Sverdrup, 1990), below which the solution Al concentration do not affect the dissolution rate. The value for it has been taken from experimental studies (mostly Lennart Sjöberg's private letters, but see also Sverdrup, 1990 for other data). The value of C_{Al} may be different under natural field conditions, when diffusion in heterogeneous media is included. The same mechanism of a lowest concentration of influence on the rate would apply to other cations (Ca, Mg, K, Na, Fe, Si,..) released from the mineral during dissolution. The surface Al concentration should not be confused with that on the ion exchange matrix. In most soils, even those with only a few percent organic matter, most of the ion exchange capacity exist on the organic matter. Per weight organic matter has 5 times the capacity of clay minerals. For ion exchange, silicate minerals play a minor role as exchange matrices.

Combination of Equations (25) and (26) modifies the rate equation to the one used for evaluation available laboratory experiments:

$$
\begin{aligned}
r = & \; k'_{H^+} \cdot a^n_{H^+} \cdot \frac{1}{(a_{BC^{2+}} + C_{BC,H})^{x_{BC}}} \cdot \frac{1}{(a_{Al^{3+}} + C_{Al,H})^{y_{Al}}} \\
& + k'_{H_2O} \cdot \frac{1}{(a_{BC^{2+}} + C_{BC,H_2O})^{z_{BC}}} \cdot \frac{1}{(a_{Al^{3+}} + C_{Al,H_2O})^{z_{Al}}} \\
& + \sum_{i=1}^{org.acids} k_{org_i} \cdot a^{n_{CO_2}}_{R_i^-} \cdot \frac{1}{(1 + K_{org_i} \cdot a_{R_i^-})^{n_{org}}} + k_{CO_2} \cdot P^{n_{CO_2}}_{CO_2} \\
& + k'_{OH^-} \cdot a^h_{OH^-} \cdot \frac{1}{(a_{BC^{2+}} + C_{BC,OH})^w} \cdot \frac{1}{(a_{Al^{3+}} + C_{Al,OH})^q}
\end{aligned}
\tag{27}
$$

The organic acid reactions are operationally often lumped into one rate expression.

Extension of the theoretical expression to field conditions. The synthesis of data and theory lead to a kinetic rate law for release of Ca, Mg, Na and K from weathering of silicate minerals (Sverdrup, 1990). This expression based on concentrations instead of activities was incorporated in SAFE and PROFILE is the same as Equation (26), but somewhat rearranged:

$$
r_j = k_{H^+} \cdot [H^+]^n \cdot \frac{1}{f_H} + k_{H_2O} \cdot \frac{1}{f_{H_2O}} + k_{CO_2} \cdot P^m_{CO_2} \cdot \frac{1}{f_{CO_2}} + k_{org} \cdot [R^-]^{0.5} \cdot \frac{1}{f_{org}}
\tag{28}
$$

where r_j is the dissolution rate of mineral j, and the retardation factors are given by:

$$
f_H = (1 + \frac{[BC]}{C_{BC,H}})^{x_{BC}} \cdot (1 + \frac{[Al^{3+}]}{C_{Al,H}})^{y_{Al}}
\tag{29}
$$

$$
f_{H_2O} = (1 + \frac{[BC]}{C_{BC,H_2O}})^{z_{BC}} \cdot (1 + \frac{[Al^{3+}]}{C_{Al,H_2O}})^{z_{Al}}
\tag{30}
$$

$$
f_{org} = (1 + \frac{[R^-]}{C_R})^{0.5}
\tag{31}
$$

$$
f_{CO_2} = 1
\tag{32}
$$

The symbols of the equation are:

r	=	Reaction rate	kmol m^{-2} s^{-1}
k_{H^+}	=	Rate coefficient in the H$^+$ reaction	
k_{H_2O}	=	Rate coefficient in the H$_2$O reaction	
k_{CO_2}	=	Rate coefficient in the CO$_2$ reaction	
k_{org}	=	Rate coefficient in organic acid reactions	
C_{Al}	=	Aluminium saturation constant	μm l^{-1}
C_{BC}	=	Base cation saturation constant	μm l^{-1}
C_R	=	Organic reaction saturation constant	μm l^{-1}
x	=	Base cation reaction order in the H$^+$ reaction	
u	=	Al reaction order in the H$_2$O reaction	

m	=	Reaction order in the CO_2 reaction	
$[H^+]$	=	Hydrogen ion concentration	kmol m^{-3}
$[BC]$	=	Base cation concentration	kmol m^{-3}
$[R]$	=	Free dissociated organic ligand concentration	kmol m^{-3}
P_{CO_2}	=	Partial pressure CO_2 in the soil solution	atm

The reaction with OH has been omitted as it is generally unimportant at pH values below 7. The retardation factors have been derived by a rearrangement of the rate equation based on the transition state theory given in Equation (27), and subsequent separation of the retarding part of the expressions.

It is unknown if the reaction with CO_2 or organic acids are influenced by the concentration of Aland BC. We think such effects should be expected and sought for experimentally. The retarding expressions should be expected to have a form similar to what has been found for the H$^+$-reaction. The OH-reaction and a precipitation reaction to form secondary minerals, have only recently been incorporated in specialized versions of SAFE used for calculation of long term soil development (Rietz, 1995) and nuclear waste storage.

In the reaction with organic acid, only dissociated acid ligand is considered to react (Grandstaff, 1980; Sverdrup, 1990). The amount of freely dissociated organic ligands will depend on the degree of protonation and the degree of complexation with Al and base cations from the solution. In the presence of high concentrations of these ions or in acid solutions, the concentration of freely dissociated ligand may be very low despite high total organic acid concentration.

The total rate for the soil horizon is obtained by repeating the calculation additively for all minerals present:

$$R_W = \sum_{j=1}^{minerals} r_j \cdot A_W \cdot x_j \cdot \theta \cdot Z \qquad (33)$$

where:

A_W	=	Exposed surface area of soil minerals	m^2 ha^{-1}
x_j	=	The surface area fraction of soil mineral j	
r_j	=	The reaction rate of mineral j according to Eq. 28	keq ha^{-1}yr^{-1}
θ	=	The soil moisture saturation	
Z	=	Soil layer thickness	m

The total exposed area can be estimated using the formula:

$$A_W = (8.0 \cdot x_{clay} + 2.2 \cdot x_{silt} + 0.3 \cdot x_{sand} + 0 \cdot x_{coarse}) \cdot \frac{\rho}{\rho_0} \qquad (34)$$

where:

x_{clay}	=	Less than 2 micron size fraction
x_{silt}	=	2-60 micron size fraction
x_{sand}	=	60-250 micron size fraction
x_{coarse}	=	250- micron size fraction
ρ	=	Soil density, kg m^{-3}
ρ_0	=	Reference density, kg m^{-3}
A_W	=	Surface area, m^2 m^{-3}soil

The rate is proportional to the exposed surface area of the mineral, A_W. Only a small part of this surface is actually involved as active reaction sites, but since there is no simple method for determining the surface area of the active sites only, using the total mineral surface seem to be the best alternative. The relative ratio between the BET surface and the effective surface of the mineral being active sites then become a property of the mineral, and is included in the experimentally laboratory determined rate coefficients.

Determining the surface area on soil samples. The formula has been derived empirically from Swedish soil samples (Melkerud, 1990, Sverdrup, 1990). A series of forest soil samples were analyzed for particle size and BET surface. Use of the formula in other countries have shown that it is valid elsewhere in Europe. In a first series of experiments the forest soil samples were treated in several steps to take away precipitates and secondary minerals, in order to estimate the surface area of the minerals that really participate in weathering. In a second series, all organic matter was removed with concentrated hydrogen-peroxide at 8°C, and partitioned into different size fractions. On each particle size fraction with known granulometry, the BET surface was determined in five point determinations.

The results of our study showed that BET method will overestimate the surface area of natural bulk forest soil samples, and that the given formula give a better estimate of the available surface relevant to weathering. It also showed that for the soil samples used, a simple formula assigning a fixed area to the clay, silt and sand fractions would give an acceptable estimate of the mineral surface. An important result is that it is possible to correlate this to the standard forestry field classification of texture.

Soil wetness and surface area. The reactions only will take place on wetted surfaces, and the degree of surface wetting is taken to be proportional to the soil moisture saturation. All surfaces to participate in the reactions must be wetted, but there must also be sufficient soil solution present for the weathering process to respond to and exchange mass with other soil processes. At lower degrees of soil moisture saturation, it would seem reasonable to expect the reactivity of the exposed mineral surface to decrease. This has been verified in both laboratory experiments and under field conditions for calcite (Sverdrup and Warfvinge, 1985; Warfvinge, 1988).

The soil moisture saturation is calculated by combining the densities of the solid (2700 kg/m^3), wet (1000 kg/m^3), gaseous phase (0 kg/m^3), and the bulk soil density ρ_{soil} with the volumetric water content Θ, in an expression for the ratio between void space and water content (Sverdrup, 1990; Warfvinge and Sverdrup, 1991; Sverdrup and Warfvinge, 1993):

$$\theta = \frac{\rho_{mineral} \cdot \Theta}{\rho_{mineral} + \rho_{water} \cdot \Theta - \rho_{soil}} \tag{35}$$

$\rho_{mineral}$	=	Mineral density	kg m^{-3}
ρ_{water}	=	Water density	kg m^{-3}
ρ_{soil}	=	Soil bulk density	kg m^{-3}
Θ	=	The soil moisture content	ton H$_2$O m^{3}soil
θ	=	The soil moisture saturation	

Considering the temperature. The weathering rate in the natural soil is calculated with rate coefficients taken from laboratory studies carried out at 25°C, and adjusted to soil temperatures using an Arrhenius relationship:

$$\ln\left(\frac{r_1}{r_2}\right) = EA \cdot \left(\frac{1}{T_2} - \frac{1}{T_1}\right) \tag{36}$$

where:

EA	=	Arrhenius activation energy	°Kelvin
T	=	The absolute temperature	°Kelvin
r	=	Chemical weathering rate in a specific reaction	kmol $m^{-2}s^{-1}$

The weathering rate in the natural soil is calculated with rate coefficients taken from laboratory studies carried out at 25°C, and adjusted to soil temperatures using the Arrhenius relationship (Sverdrup, 1990).

How rate coefficients were determined from data

The rate coefficients used in the PROFILE model and its dynamic counterpart SAFE, were determined from the experimental data, either reported in the literature, or in case of missing information, from our own experimental data. The basic principle in our reevaluation of the available data was very strict data quality control, screening of data and use of the same theory across all experiments.

Screening of kinetic data was our first step. Since most researchers had ignored the possibility for several simultaneous reactions, all experiments had to be resorted and re-evaluated with respect to this. Undefined experimental chemical conditions, use of mineral samples contaminated with ultra-fines, or failure to achieve steady state dissolution, were reasons for elimination of an experiment. Of ~200 reported experiments reviewed, we could learn something from almost all of them, but only ~50 studies provided useful kinetic data of high quality.

The assessment would not have been possible without the complementary information taken from unpublished material made available to us by colleagues (Sjöberg, Schweda at Stockholm University) and our own experiments. Our own experiments included experiments on olivine, forsterite, grossular garnet, almandine garnet, andradite garnet, anortite, epidote, zoisite, staurolite, cordierite, augite, wollastonite, spodumene, jadeite, glaucophane, hornblende, calcite, dolomite, magnesite and biotite, in the pH range from 1.5 to 7, usually at 20°C. For feldspars, special attention was paid to the experiments of Chou and Wollast (1985), Sjöberg (1989; unpublished), and Schweda (1989).

The reaction with H^+*-ion.* For most of the minerals studied here, we get a range where the rate is clearly pH-dependent, a section of the curve where there is little or no dependence of the rate on pH and at high pH, a pH dependent section. An example is shown in Figure 2 for K-feldspar, albite, labradorite and olivine. This is used to determine the rate order of the major reactant according to

$$r = k_{H+}^* \cdot [H^+]^n \tag{37}$$

In a first step, the traditionally coefficient $k^*{}_{H+}$ can be determined. The slope of a plot of reactant concentration keeping other reactant concentrations constant versus rate give the different reaction orders. Data from L. Sjöberg at Stockholm University (mostly un-published) but also our own results were used (Sverdrup, 1990). A similar procedure is applied to other reactions between the minerals and different types of solutions. When the reaction orders have been established, then this is used to determine the full rate coefficient used in the full expression:

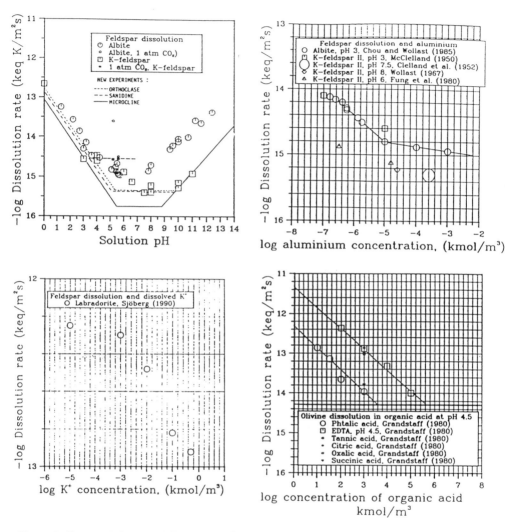

Figure 2. The dissolution rate of K-feldspar at 25°C in aqueous solution. The dotted line represents the dissolution rate of albite in the absence of organic ligands. 2-The dependence of the dissolution rate of feldspar on the concentration of Al in the solution. 3-The effect of different concentrations of potassium in the solution on the dissolution rate of labradorite at 25°C and pH 1. From experiments by Dr. L. Sjöberg at the Institue of Geology, Stockholm University. 4-The dependence of the dissolution rate of olivine at 25°C in different aqueous solutions containing organic acids. For mafic minerals such reactions may be important, but there are large differences between different mineral types.

$$r = k_{H^+} \cdot [H^+]^n \cdot \frac{1}{(1 + \frac{[Al^{3+}]}{C_{Al}})^y} \cdot \frac{1}{(1 + \frac{[BC]}{C_{BC}})^x} \qquad (38)$$

This can be rearranged to an expression for the rate coefficient:

$$k_{H^+} = (1 + \frac{[Al^{3+}]}{C_{Al}})^y \cdot (1 + \frac{[BC]}{C_{BC}})^x \cdot \frac{r}{[H^+]^n} \qquad (39)$$

The relation between the coefficient $k^*{}_{H_+}$ ignoring the retarding factors and the one including them, k_{H^+} is derived by combining Equations (39) and (37). The procedure for the kinetic coefficients for the reaction with mineral and water was similar, except that less data were available. The procedure was tried on data from almost 40 different minerals with varying degree of success (see Sverdrup, 1990). The best data in general, were available for feldspars and olivine. Data for the organic acid reaction were available for olivine and quartz, with more limited experiments for such minerals as feldspars, diopside, enstatite, biotite, muscovite and hornblende (Fig. 2). For the CO_2 reaction, useful data were only available for feldspars; some stray experiments were found for the same minerals as was found for the organic acid reaction

The reaction with organic acid and CO_2. For the reaction rate in relation to organic ligands, a plot of the rate at constant pH versus the -log of the concentration of the organic ligand will yield the rate coefficient as intercept and the reaction order as slope. For the reaction with organic ligands, the available data always gave reaction order 0.5.

For many minerals, very few experiments under CO_2 or in organic acids are available. Often only one single experiment is available. Then a reaction order of 0.5 was assumed and a rate coefficient was determined. The flat pH-independent section of the dissolution curve was used to determine the rate of reaction between the mineral and water, but only in the absence of organic ligands and elevated CO_2.

For the reaction with CO_2, a reaction order of 0.5 has been determined for dolomite, calcite and feldspars (Sverdrup, 1990, using data from Lagache, 1965; Plummer, 1973; Sverdrup et al., 1981; Sverdrup and Warfvinge, 1986). Because no other reaction orders were available, and the values appeared to be consistent, the same reaction order was a priori assumed for the other minerals. If a multiple of reactions were acting under the experimental conditions recorded, then the effects of the different reactions must be properly separated in order to quantify the rate of a particular reaction. As an example, if an experiment occurred in the presence of an organic ligand and CO_2, below pH 7, the rate of the reaction between the organic ligand and the minerals can be approximated from

$$r_{org} = r - r_{H+} - r_{H_2O} - r_{CO_2} \tag{40}$$

according to the principle that the total dissolution rate is the sum of the rates of the participating reactions. If any of the terms on the right side in such an equation cannot be determined, then the net rate cannot be separated from the effect of other reactions. A detailed description of the methodology for the evaluation is given by Lasaga (1981), Sverdrup et al. (1981, 1986) and Sverdrup (1990).

The kinetic parameters. Shown in Table 2 are the rate coefficients, reaction orders, and product inhibition limit concentrations for the H^+-, water-, CO_2- and organic acid-reactions for the minerals included in PROFILE; the coefficients relate to the production of base cations expressed as keq m^{-2} s^{-1} at 8°C.

In the table, the names of minerals respresent mineral types, grouped by their dissolution kinetic properties. Thus, the name on the list represents a group of minerals with similar stoichiometry and similar dissolution kinetic coefficients. Muscovite includes muscovite, secondary illite, biotite include biotite, glauconite, primary illite and phlogopite, hornblende includes hornblende, tremolite and glaucophane, pyroxene includes minerals of the pyroxene group and talc, it has the dissolution properties of augite. Garnet includes rate coefficients for all nesosilicates. In this group we have also included anorthite of high purity since it dissolves with the same rate and reaction order. Vermiculite includes proper

Table 2. Approximate laboratory rate coefficients for the chemical weathering rate for minerals expressed as the flux of base cations (Ca, Mg, Na, K) related to a total rate expressed as keq m^{-2} s^{-1} at 8°C, applied in the PROFILE model. Values within brackets are interpolations and estimates from Madelung site energies and other crystallographic data. (Modified after Sverdrup, 1990).

Mineral	pk_H	n_H	C_{Al}	y_{Al}	C_{BC}	x_{BC}	pk_{H_2O}	z_{Al}	z_{BC}	pk_{CO_2}	n_{CO_2}	pk_{org}	C_R
K–Feldspar	14.7	0.5	4	0.4	500	0.15	17.5	0.14	0.15	16.8	0.6	15.0	(5)
Plagioclase	14.6	0.5	4	0.4	500	0.2	17.2	0.14	0.15	15.9	0.6	14.7	(5)
Albite	14.5	0.5	4	0.4	500	0.2	16.7	0.14	0.15	15.9	0.6	14.7	5
Hornblende	13.3	0.7	30	0.4	200	0.3	15.9	0.3	0.3	(15.9)	(0.6)	14.4	(5)
Pyroxene	12.3	0.7	500	0.2	200	0.3	17.5	0.1	0.3	15.8	0.6	14.4	(5)
Epidote	14.0	0.5	500	0.3	200	0.2	17.7	0.2	0.2	(16.2)	0.6	(14.4)	(5)
Garnet	12.4	1.0	300	0.4	500	0.2	16.9	0.2	0.2	(15.8)	(0.6)	(14.7)	(50)
Biotite	14.8	0.6	10	0.3	(500)	0.2	16.7	0.2	0.2	15.8	(0.5)	14.8	(50)
Muscovite	15.2	0.5	4	0.4	(500)	0.1	17.5	0.2	0.1	16.5	(0.5)	15.3	(5)
Fe-Chlorite	14.8	0.7	(50)	(0.2)	(200)	(0.2)	(17.0)	(0.1)	(0.1)	16.2	(0.5)	15.0	(5)
Mg-Chlorite	14.3	0.7	(50)	(0.2)	(200)	(0.2)	(16.7)	(0.1)	(0.1)	15.8	(0.5)	14.5	(5)
Fe-Vermiculite	15.2	0.6	4	0.4	500	0.2	17.6	0.1	(0.1)	(16.5)	(0.5)	(15.6)	(5)
Mg-Vermiculite	14.8	0.6	4	0.4	500	0.2	17.2	0.1	(0.1)	(16.2)	(0.5)	(15.2)	(5)
Apatite	12.8	0.7	100	–	300	0.4	15.8	–	0.2	15.8	0.6	(19.5)	(5)
Kaolinite	15.1	0.7	4	0.4	500	0.4	17.6	0.2	0.2	(16.5)	(0.5)	19.5	(5)
Calcite	13.6	1.0	5000	0.4	1000	0.4	15.2	–	0.2	13.2	1.0	13.2	(5)

Table 3. Observed and estimated temperature dependence factors (EA) used to adjust the rate coefficients to any temperature T (in Kelvins), in PROFILE. Data from Sverdrup (1990). Values within brackets are estimates.

Mineral	Arrhenius factor			
	H^+	H_2O	CO_2	Org
K–Feldspar	3500	2000	(1700)	(1200)
Plagioclase	4200	2500	(1700)	(1200)
Albite	3800	2500	1700	1200
Hornblende	(4300)	3800	(1700)	(2000)
Pyroxene	2700	3800	(1700)	(2000)
Epidote	4350	(3800)	(1700)	(2000)
Garnet	2500	3500	(1700)	(1800)
Biotite	4500	3800	(1700)	(2000)
Muscovite	4500	(3800)	(1700)	(2000)
Chlorite	4500	(3500)	(1700)	(1800)
Vermiculite	4300	3800	(1700)	(2000)
Apatite	3500	4000	(1700)	(2200)
Kaolinite	5310	3580	(1700)	(2000)
Calcite	444	4000	2180	2200

vermiculite, secondary illite, smectite, bentonite and montmorillonite. Values in brackets are interpolations and estimates from crystallographic data. Explanations of where the data come from, who did the experiments, and descriptions of our own experiments make a very long story and can be found in Sverdrup (1990).

Table 3 shows observed and estimated temperature dependence factors (EA) used to adjust the rate coefficients to any temperature T in PROFILE. The data were taken from the literature review by Sverdrup (1990).

MODELING KINETICS OF FIELD WEATHERING

Introduction

The laboratory kinetic coefficients have been integrated into a biogeochemical model, PROFILE. PROFILE and it's dynamic version SAFE have been built for calculation of soil and surface water runoff calculation[1].

[1] SAFE = Simulating Acidification in Forested Ecosystems; PROFILE is the steady state version of SAFE. Both models and a users manual is available free of cost from the authors. The PROFILE model exists in a site-by-site version and a regional version that will interface with a database. The SAFE model is available in a site-by-site version. The models are available for MacIntosh and PC-Windows. The SAFE and PROFILE models are also available through the World Wide Web on a FTP server at the Department of Chemical Engineering, Lund University, Sweden.

PROFILE and SAFE model description

SAFE has been developed with the objective of studying the effects of acid deposition on soils and groundwater. It calculates the values of different chemical state variables as a function of time. It can therefore be used to study the process of acidification and recovery, as effected by deposition rates, soil parameters and hydrological variations. PROFILE is a steady state version of SAFE. It bypasses the changes in soil state over time, and calculates the final steady state directly. It is used to calculated the initial state from which SAFE is started.

Processes included. SAFE and PROFILE are based on a conceptual model of a forest soil; the soil may represent a profile or the whole catchment. SAFE includes the following chemical subsystems (See Fig. 1):

Deposition, leaching and accumulation of dissolved chemical components
Chemical weathering reactions of soil minerals with the soil solution
Cation exchange reactions
The net result of reactions of N-compounds; complete nitrification
Internal cycling of N in the canopy, such as canopy exchange, litterfall and net mineralization
Internal cycling of base cations in the canopy, such as canopy exchange, litterfall and net mineralization
Biological net uptake of base cations and nitrogen
Solution equilibrium reactions involving CO_2, Al and organic acids

Naturally, these processes only represent a selection of chemical reactions in the soil. Among processes that have not been been included are sulphate adsorption, a series of reactions that may change the CEC of the soil matrix, store sulphur irreversibly or affect the Acid Neutralizing Capacity (ANC) balance in certain soils. All processes included in the model have been subject to necessary simplification in some respect.

The SAFE model is split into different compartments to represent the natural vertical differences in soils, which result in a marked variation in chemical properties between soil layers. The apparent increase in spatial resolution as compared to one-box models has been carefully evaluated to see if data can be obtained for each horizon, rather than the soil profile as a whole.

Mass balance equations. The change in soil solution chemistry and the subsequent change in the distribution of elements on the cation exchange matrix is calculated by means of conservation equations. The cations Mg, Ca, and K are grouped together on an equivalent basis of as a divalent component, "Base cations" or BC. These are physiologically active with plants, whereas Na is generally not. We have therefore ignored Na in BC:

$$[BC^{2+}] = [Ca^{2+}] + [Mg^{2+}] + [K^+] \tag{41}$$

In SAFE, Na is considered as a tracer with respect to plant uptake and ion exchange. The hydrogen ion is treated as dependent on the variable acid neutralizing capacity; *ANC*. It is defined in terms of molar concentrations as:

$$[ANC] = 2[CO_3^{2-}]+[HCO_3^-]+[OH^-]+[R^-]-[H^+]-3[Al^{3+}]-2[AlOH^{2+}]-[Al(OH)_2^+] \tag{42}$$

Each soil horizon is assumed to be homogeneous, there are no macroscopic concentration gradients in the individual soil layers. The conservation equations, derived as differential mass balances, and constituting the framework for the mathematical model, are:

$$\frac{d[ANC]}{dt} = \frac{1}{z \cdot \Theta}(Q_0[ANC]_0 - (Q + Z\frac{d\Theta}{dt})[ANC]) + R_W + r_{ex} + U_{BC} + r_N \tag{43}$$

$$\frac{d[BC^{2+}]}{dt} = \frac{1}{Z\Theta}(Q_0[BC^{2+}]_0 - (Q + z \cdot \frac{d\Theta}{dt}) \cdot [BC^{2+}]) + R_{W_{BC}} + r_{ex} + U_{BC} \tag{44}$$

$$\frac{d[Na^+]}{dt} = \frac{1}{z \cdot \Theta}(Q_0[Na^+]_0 - (Q + Z\frac{d\Theta}{dt}) \cdot [Na^+]) + R_{W_{Na}} \tag{45}$$

$$\frac{d[NH_4^+]}{dt} = \frac{1}{z \cdot \Theta}(Q_0[NH_4^+]_0 - (Q + Z\frac{d\Theta}{dt}) \cdot [NH_4^+]) - r_{nitr} - U_{NH_4} \tag{46}$$

$$\frac{d[Cl^-]}{dt} = \frac{1}{z \cdot \Theta}(Q_0[Cl^-]_0 - (Q + Z\frac{d\Theta}{dt}) \cdot [Cl^-]) \tag{47}$$

$$\frac{d[SO_4^{2-}]}{dt} = \frac{1}{z \cdot \Theta}(Q_0[SO_4^{2-}]_0 - (Q + Z\frac{d\Theta}{dt}) \cdot [SO_4^{2-}]) + R_{ads} \tag{48}$$

$$\frac{d[NO_3^-]}{dt} = \frac{1}{z \cdot \Theta}(Q_0[NO_3^-]_0 - (Q + Z\frac{d\Theta}{dt}) \cdot [NO_3^-]) + r_{nitr} - r_{denitr} - U_{NO_3} \tag{49}$$

$$\tag{50}$$

where Z is soil layer thickness, Q is water flow rate, Θ is volumetric water content per meter of soil depth. r_{ex} is the ion exchange rate, which is defined later. U_{BC} is the plant base cation uptake rate and r_N is the ANC net production from transformations in the nitrogen cycle. Weathering enters into the equations as the term $R_{W_{ANC}}$. $R_{W_{ANC}}$ is the ANC production rate, which is assumed to be equal to the release rate for base cations R_W.

$$R_{W_{ANC}} = R_W = R_{W_{BC}} + R_{W_{Na}} \tag{51}$$

In PROFILE there is a separate mass balance for each base cation. In SAFE Ca, Mg, K and Na are lumped into BC. The weathering rate also enters into the mass balance for base cations, ANC and through limitations as uptake. Chloride, sulphate and sodium are assumed to flow straight through without taking part in any retention processes. At present sulphate adsorption has been set to zero. There are also mass balances for nitrate and ammonium. The absence of a subscript means that a parameter refers to conditions in that particular soil layer, and consequently to the leachates from that layer, while 0 denotes concentrations in the inflow to the layer. Q is the flow intensity in $m^3 \, m^{-2} \, s^{-1}$, through the layer. If the hydrological data fed to the model suggest variations in soil moisture content, Θ, changes in solution chemistry will occur due to dilution/concentration. This is quantified in the equations by the $Z \cdot d\Theta/dt$ term.

Thus, the essence of the whole modeling process is to find expressions that quantify each of these terms. The chemical components of the soil solution equilibrium are also specified. The quantities $R_{W_{ANC}}, r_{ex}$, U_{BC} and r_N refer to the rate by which ANC is supplied to, or withdrawn from the soil solution by weathering ($R_{W_{ANC}}$), cation-exchange (r_{ex}), base cation uptake (U_{BC}) and nitrogen reactions (r_N). These terms are expressed in keq m^{-3} yr^{-1} rather than a molar basis in order to avoid changing units during calculations. The interactions between the phases and the reactants in the system always occur through the soil solution.

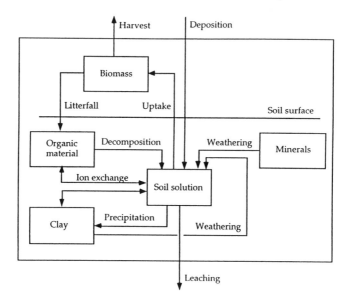

Figure 3. The weathering process must considered in the context of other important soil processes. In addition to weathering, uptake by plants, decomposition of organic matter and ion exchange are particularly important processes for general soil chemistry. This is the only way a weathering model based on laboratory kinetics and laboratory rate coefficients can be applied to field conditions. Coupling of the individual processes in the steady state model PROFILE and the dynamic model SAFE occur via the soil solution. Ion exchange is not a steady state process and is excluded in PROFILE.

The pH–ANC relationship.   In the model, buffering in the liquid phase is controlled by the CO_2-carbonate system, the acid-base reactions of a model organic acid, and an idealized aluminium sub-model based on the gibbsite polynomial expression. The latter is based on the assumption that the concentration of Al-species is governed by the dissolution and precipitation of a solid gibbsite phase, $Al(OH)_3$.

In the organic soil layers, the buffering reaction of organic components may be important. In SAFE, dissolved organic carbon (DOC) is modelled as a monovalent organic acid. The dissociation of the acid functional groups of the DOC is quantified using the Oliver equation:

$$[R^-] = \frac{K_{Oliv} \cdot [DOC] \cdot \alpha - [AlR]}{K_{Oliv} + [H^+]} \tag{52}$$

where DOC is dissolved organic content in mg/l and the coefficient is given by:

$$K_{Oliv} = 10^{-0.96+0.9-0.039 \cdot pH^2} \tag{53}$$

The Al term generally is omitted, but the complexed form can be estimated using a simpler version of Tipping equations:

$$[AlR] = 2.09 \cdot 10^{-7} \cdot [DOC] \cdot \frac{[Al^{3+}]^{0.72}}{[H^+]^{1.054}} \tag{54}$$

In general, the pH corresponding to a certain ANC is governed by the carbonate and the organic matter buffer systems at positive values for ANC, i.e. at pH-levels above 5, while the aluminum buffering system is effective at lower pH values. The major parameter for quantifying the importance of the aluminium buffering is the gibbsite solubility constant, K_G. The reason for selecting the gibbsite dissolution/precipitation reaction as the foundation for modeling aluminum equilibrium is primarily that reasonable titration curves can be derived with this aquatic chemistry model, rather than a conviction that gibbsite is the actual solid phase present in the terrestrial system. In the model, the gibbsite approximation may be substituted with a model based on weathering of minerals, ion exchange and kinetics of precipitation.

Cation exchange. Traditionally, cation-exchange has been modeled as an equilibrium reaction. In soil systems that are subject to episodic variations in flow intensity and soil solution composition, however, there is no a priori justification for assuming equilibrium between the liquid phase and the exchanger matrix. When soil is limed, for example, the exchange matrix may be far from equilibrium with the soil solution. If equilibrium is erroneously assumed in a hydro-chemical model, an excessively strong buffering mechanism is introduced, short term dynamics are lost, and long term trends become incorrect.

As the intrinsic rate of cation exchange is considered to be infinite, the exchange reaction between exchangeable calcium and acidity may be modeled as a perfectly reversible chemical reaction only limited in rate by the transport of base cations between the bulk of the soil solution and active sites on the exchanger surface, i.e. diffusion. The pseudo-steady state rate equation for such a process can be derived from Fick's first law. The change in exchangeable bases, the fraction base cations on the cation exchange matrix, $\bar{X}_{BC}$ is proportional to the flux of base cations to the exchanger, which is driven by the concentration difference between the surface of the exchange complex and the concentration in the bulk of the soil solution. The rate equation is (Warfvinge, 1988):

$$\frac{d\bar{X}_{BC}}{dt} = k_x \cdot ([BC^{2+}]_{sol} - [BC^{2+}]_{ex}) \tag{55}$$

and the ion exchange rate is

$$r_{ex} = CEC \cdot \frac{d\bar{X}_{BC}}{dt} \tag{56}$$

Subscript *ex* indicates that the conditions on the exchanger surface are considered. k_x is a mass transfer coefficient. The above equation thus states that the flux of base cations to the exchange surface is proportional to the concentration gradient between the solution and the surface. The base cation concentration at the liquid-solid interface $[BC]_{ex}$ is given by an equilibrium model, the Gapon exchange equation:

$$[BC]_{ex} = \frac{[H^+]^2_{ex} \cdot \bar{X}^2_{BC}}{K^2_{H/BC} \cdot (1 - \bar{X}_{BC})^2} \tag{57}$$

where $K_{H/BC}$ is the Gapon selectivity coefficient for H/BC exchange. Calculating $[BC^{2+}]_{ex}$ will require computation of pH at the exchanger phase surface, and one equation in addition to an exchange equilibrium equation. The additional relationship required can be derived from the charge balance evaluated at the surface as well as for the soil solution. When this is combined with the Gapon selectivity equation for heterovalent exchange, it becomes possible to solve for $[H^+]_{ex}$. This procedure has been described in detail by Warfvinge and Sverdrup (1992).

If exchangeable aluminum is dominating the exchanger, however, the interface concentrations are calculated from an Al-Ca exchange isotherm. One may combine this expression with the assumption that Al^{3+} is given by a simple gibbsite equilibrium. It then becomes evident that it does not matter whether H-Ca or Al-Ca cation exchange is assumed, provided that the Gapon exchange equation is applied.

An increase in base saturation corresponds to withdrawal of acid neutralizing capacity from the soil solution, the rate of which is calculated as being proportional to the change in base saturation multiplied by the exchange capacity.

Nutrient uptake reactions.   PROFILE relies on input data that specify the magnitude of base cation and N uptake in the different soil layers. Uptake of N and base cations couple to the mass balances (Eqns. 44 and 45) through the terms R_N and R_{BC} Nitrification is modelled by a kinetic rate expression. SAFE includes corresponding differential mass balances for nitrate and ammonium. These are necessary to calculate the nitrification rate in the different soil layers. The complete set of equations used are given in Warfvinge and Sverdrup (1992). The input of ANC from N-reactions is

$$r_N = U_{NO_3^-} - U_{NH_4^+} - 2 \cdot r_{nitr} \tag{58}$$

where U_{NO_3} and $U_{NH_4}{}^+$ are uptake rates read in as input data, and r_{nitr} is the rate of nitrification. The nitrification rate is described with a Michaelis-Menten expression (Sverdrup et al., 1990; Warfvinge and Sverdrup, 1992):

$$r_{nitr} = k_{nitr} \cdot \frac{[NH_4^+]}{K_{nitr} + [NH_4^+]} \tag{59}$$

Uptake is assumed to be preferential for NH_4, and NO_3 uptake occurs only after all NH_4^- has been consumed by uptake or nitrification.

It is assumed that there is no leaching of ammonium from the organic layer, N-deposition is assumed to be nitrate or to be converted to nitrate instantly and all N-uptake generates alkalinity. It should be pointed out that the reaction path that ammonium takes in the organic layer is unimportant; the key assumption is that ammonium leaches from the top layer. The biology of the catchment is in most cases very important for the cycling of nutrients, such as base cations, phosphorus and nitrogen. It is also of great importance for the balance of acidity in a catchment; acidity may be a strong driving force in many soil processes. Major biological processes in this context are the growth of plants, decomposition of organic matter, nitrification and denitrification.

Weathering limits forest growth.   Trees take up base cations from the soil in order to create biomass. Accordingly, the rate of uptake will be proportional to the rate of growth. During calculations, two conditions must be fulfilled. First some base cations will escape uptake due to plant physiological limitations (2 meq/m³; Sverdrup and Warfvinge, 1993). This minimum leaching can however not be larger than what is available from weathering and atmospheric depositionBC_D:

$$BC_{min} = Q \cdot [BC]_{min} \tag{60}$$

Q is percolation, [BC] is the limiting concentration for uptake, provided there is sufficient quantities of base cations:

$$BC_{min} > R_{W_{BC}} + BC_D \rightarrow BC_{min} = R_{W_{BC}} + BC_D \tag{61}$$

The other condition is that uptake cannot be larger than what is available for uptake:

$$U_{BC} > R_{W_{BC}} + BC_D - BC_{min} \rightarrow U_{BC} = R_{W_{BC}} + BC_D - BC_{min} \tag{62}$$

U_{BC} is plant uptake of base cations release from weathering (i.e. Ca, Mg, K) BC_D is the atmospheric deposition of these ions, BC_{min} is the amount that always leaks out of the soil and Q is the runoff.

In the model, growth is subject to nutrient restrictions. Growth is assumed to be proportional to uptake of the limiting nutrient, according to "Liebig's law." Nitrogen and base cation uptake is coupled, and either nitrogen or base cation deficit will reduce total growth. The long term sustainable growth is based on available nutrients.

Limits for uptake are determined by the availability of nutrients either from deposition (N) or deposition and weathering (BC, P). Thus the "Liebig's law function is expressed as the ratio between nutrient demand to sustain growth according to the yield model, and what is available in the soil:

$$f(Nutr) = \min_{i=nutrients} \sum_{j=1}^{4 \, layers} \frac{W_{j,i} + BC_{D_{j,i}}}{G_{max} \cdot x_i} \tag{63}$$

x_i is the content of the nutrient i in the tree. If the amount required by the plant is less than the amount available, then the function has the value 1, and this nutrient will not limit growth. We allow both soil chemistry and supply to restrict growth, but net growth is maximized to present net uptake (Warfvinge et al., 1992) in the steady state PROFILE calculations. In the PROFILE and SAFE models, base cation uptake occurs from each soil layer, but uptake is stopped if the base cation soil solution concentration falls below 7 µeq l^{-1} for Mg and Ca and 1 µeq l^{-1} for K, in PROFILE, in SAFE when the base cation concentration falls below 15 µeq l^{-1} (Arovaara and Ilvesniemi, 1990, Sverdrup and Warfvinge, 1993b). The minimum leaching is introduced because experiments show that there is a lower limit for uptake. It can also be interpreted as taking into account that the roots do not perfectly penetrate the soil, and some nutrients will always escape uptake.

Chemical feedback from soil chemistry. Trees satisfy their need for nutrients from the soil, in proportion to their availability in the soil. The transport process from the bulk of the solution to the root surface is governed by solute transport mechanisms such as solute flow, mass diffusion and ion exchange (Nye and Tinker, 1977, Cronan, 1991). It is known from laboratory studies that Al in the soil solution will be able to disturb the growth of trees (Sverdrup et al., 1992; Sverdrup and Warfvinge, 1993). Trees incorporate base cations in wood and leaves and take them up after they have become available by adsorbtion to the outside of the root. In acid soils large amounts of Al^{3+} and H^+ may absorb to the root, competing with Ca, Mg and K. This may decrease uptake of Ca, Mg and K. This effect can be expressed in terms of the (Ca+Mg+K)/Al ratio (Sverdrup and Warfvinge, 1993). If uptake is less, then the base cations will stay in the soil solution, and the concentration will be changed by the fact that the trees have decreased the growth rate or stopped growing. Decomposition or nitrification are affected by chemistry and affect chemistry in the same way. Uptake reduction due to soil acidity alone in a specific soil layer is calculated as relative growth (G), using the following equation (Sverdrup and Warfvinge, 1993):

$$f(BC/Al) = \frac{[BC]^n}{[BC]^n + KR \cdot [Al^{3+}]^m} \tag{64}$$

[BC] is the concentration, and KR is the response coefficient as defined by Sverdrup and Warfvinge (1993), for spruce n = m = 1 for pine and deciduous trees n = 3, m = 2.

Integrated feedback from soil chemistry and weathering. The effect of soil acidity and availability of mineral nutrients are multiplicative, in PROFILE, we apply:

$$G = G_{max} \cdot \sum_{i=1}^{layers} u_i \cdot f(BC/Al)_i \cdot f(Nutr)_i \qquad (65)$$

where u_i is the fraction of total uptake assigned to layer i, G_{max} is uptake at the site with no acidification or nutrient limit, f(Nutr) expresses the action of Liebig's law and f(BC/Al) the action of soil acidity, G is actual uptake at the site. The two effects may act independently, but both can decrease uptake from a soil layer. G_{max} is growth without limiting effects caused by soil acidification. The reference level for growth in our calculations is the growth calculated using standard yield tables for growth modified for site characteristics.

Spatial structure. The model is structured into a series of layers, each modelled as a stirred tank. The mass balances are solved for each soil layer, solute equilibria, kinetic rates and weathering. Input data are assembled for the model according to layers. This can be done based on available soil chemistry data for different depths in the soil profile. The obvious principle is to divide the soil into compartments that correspond to the natural soil stratification, as the soil horizons are the largest chemically isotropic elements in the system.

Hydrological structure. Water is supplied to the model at the top (Fig. 4). In each layer water is divided among (1) evapotranspiration, (2) horizontal flow and the residual which continues on to the next layer. Data on hydrology is calculated using a separate hydrological model or taken from data (SAFE), or by using long term averages (PROFILE). Each layer starts with a specified soil moisture content. This is usually kept constant, but when available, can be read as input data (SAFE).

Computer appearance. PROFILE (Fig. 5) is controlled from two decks ofcards, in a program called HyperCard. Figure 6 shows the run card for PROFILE, which also carries information concerning climate and major inputs to the soil PROFILE. Figure 7 shows the card for entering the inputs related to the vegetation. Here uptake of nutrients and cycling between soil and canopy is given. The feedback response between growth and soil chemistry is also controlled from this card. Figure 8 show one of the layer cards. Each layer has one such card. For each layer, mineralogy, mineral surface area as reflected by texture, soil moisture, etc. must be entered. Finally, a card (Fig. 9) can be filled out to estimate steady state stream water chemistry for the catchment. This is only reasonable if the total soil depth of all layers correspond to the average soil depth of the catchment.

PROFILE has a separate stack for controlling the kinetics of weathering in PROFILE (Fig. 4 and Figs. 6-9). For each mineral there is one input card. This can be used to assign new kinetic coefficients, reaction orders, mineral stoichiometries and temperature dependencies to the model. This way new kinetic information can be integrated by any user, without recompilation of the whole model code (Fig. 10). The outputs are presented on a number of cards, Figures 11-14. The first card shows the steady state soil solution concentrations. The second card shows the concentration of individual ions. The third card shows the response of the vegetation. The responses are divided between response to soil acidity on uptake and a sustainability balance with respect to available Ca, Mg and K. The fourth card shows the amount of ions produced by chemical weathering, layer by layer, and ion by ion. The ions produced by chemical weathering are Ca, Mg, K,

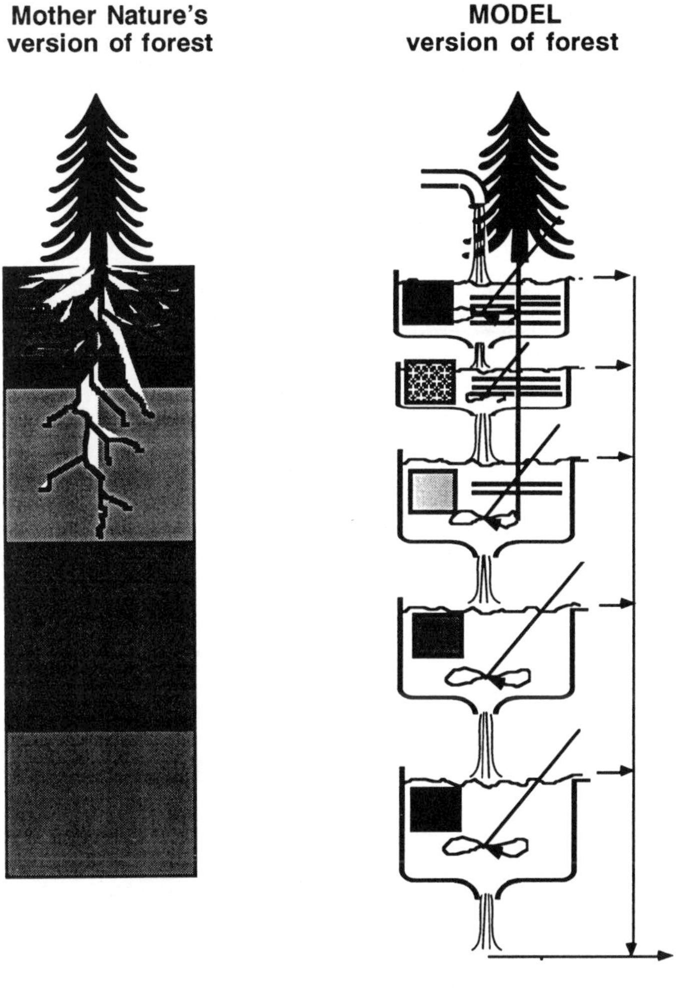

Mother Nature's version of forest

MODEL version of forest

Figure 4. PROFILE and SAFE are multi-layer models of a mostly vertically draining soil profile. Interflow can be specified. The processes listed interact in each layer.

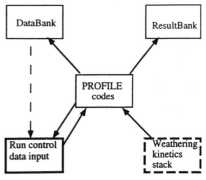

Figure 5. Basic structure of PROFILE on the screen. The model is run from a stack where input of data is made. Optional is to control kinetics from the mineral and kinetic stack.

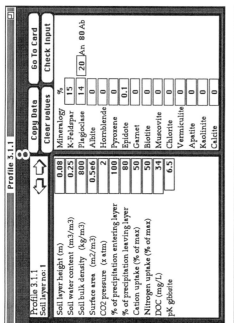
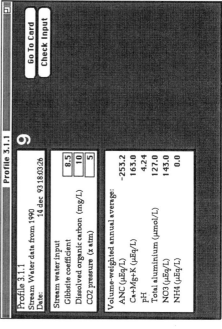
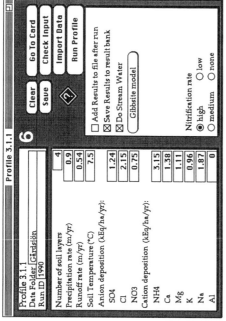
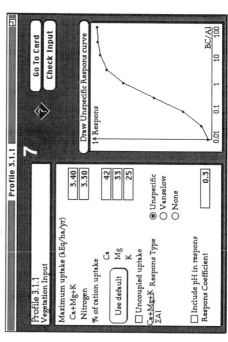

Na, Al, Si and P. The amounts of Al and Si are the total amounts produced by congruent dissolution at the mineral surface and not what appears in the soil solution. For most conditions, a major part of the released Al and Si will precipitate to form secondary phases.

Figures 6-9 are on the previous page.

Figure 6. The run ID input card, the first card in the PROFILE stack. Data from the Swedish research catchment F1 at Lake Gårdsjön have been used

Figure 7. The vegetation input card. Data from the Swedish research catchment F1 at Lake Gårdsjön have been used.

Figure 8. A layer input card. There is one input card for each layer in PROFILE. Data from the Swedish research catchment F1 at Lake Gårdsjön have been used.

Figure 9. The stream water card. This card contain both input and output. Data from the Swedish research catchment F1 at Lake Gårdsjön have been used.

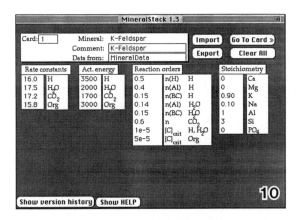

Figure 10. A mineral kinetics input card. There is a separate stack for controlling the kinetics of weathering in PROFILE. For each mineral there is one input card. This can be used to assign new kinetic coefficients, reaction orders, mineral stoichiometries and temperature dependencies to the model. This way new or alternative kinetic information can be integrated by any user, without recompilation of the whole model code. Kinetic coefficients can be found in the stack for K-feldspar, oligoclase, albite, hornblende, pyroxene, epidote, garnet, biotite, muscovite, chlorite, vermiculite, apatite and calcite. Additional kinetic information is found in the book by Sverdrup (1990) for anorthoclase, anortite, labradorite, slawsonite, forsterite, fayalite, disthene, staurolite, nepheline, zoisite, cordierite, tourmaline, bronzite, enstatite, diopside, wollastonite, jadeite, spodumene, leucite, tremolite, riebeckite, glaucophane, phlogopite, serpentine, chrysotile, glauconite, and obsidian in addition to the minerals found in Table 1.

Figures 12-14 are on the following page.

Figure 11. The first output card in the PROFILE model. This show the concentrations in the soil solution at steady state with the deposition in the input.

Figure 12. The second output card in PROFILE. Here base cations are given as individual ions.

Figure 13. On this card, results on tree growth are shown. The uptake response will only be valid for nitrogen if uptake is specified as uncoupled.

Figure 14. The calculated weathering rates is available for individual base ions released in congruent dissolution reaction, before any precipitation.

Profile 3.1.1 — [11]

Profile 3.1.1
Soil Output from: 1990
Date: 14 dec 93 18:03:26

Depth	pH		ANC	BC	Al	Si	NO3	NH4
m	solution	atm-eq	µEq/l		µmol/l		µEq/l	
Dep	7.78	7.78	481.	245.	0.	0.	83.	350.
0.040	4.05	4.05	-89.	160.	14.	3.	241.	71.
0.155	3.94	3.94	-261.	84.	234.	20.	207.	0.
0.345	4.22	4.22	-290.	87.	131.	104.	154.	0.
0.575	4.57	4.58	-253.	104.	167.	184.	143.	0.

Profile 3.1.1 — [12]

Profile 3.1.1
Detailed Soil Output from: 1990
Date: 14 dec 93 18:03:26

Depth	Ca	Mg	K	Na	Cl	SO4	Al-org	Al-inorg
m	µEq/l						µmol/l	
0.040	93.	77.	75.	260.	299.	172.	12.	3.
0.155	39.	34.	48.	350.	398.	230.	180.	54.
0.345	50.	45.	39.	360.	398.	230.	48.	82.
0.575	67.	51.	45.	377.	398.	230.	76.	90.

Profile 3.1.1 — [13]

Profile 3.1.1
Uptake Output from: 1990
Date: 14 dec 93 18:03:26

Depth	Uptake		N Uptake response			BC/Al	Al precip.	Denitrif.
m	N	BC	Al	deficiency	total			kEq/ha.yr
0.040	1.643	1.692	0.995 * 1.000 = 0.995			65.864	–	0.011
0.155	1.114	1.148	0.844 * 1.000 = 0.844			1.620	–	0.014
0.345	0.260	0.268	0.789 * 1.000 = 0.789			1.120	–	0.028
0.575	0.000	0.000	–			1.325	–	0.059
Totals:	3.017	3.108	0.914 * 1.000 = 0.914					0.111

Acidity produced by N-uptake and N-reactions   3.172   kEq/ha.yr
Atmospheric deposition of acidity   -4.330   kEq/ha.yr
Total potential acidity   1.970   kEq/ha.yr
Alkalinity leaching (acidity leaching if negative) -1.367   kEq/ha.yr

Profile 3.1.1 — [14]

Profile 3.1.1
Weathering output from: 1990
Date: 14 dec 93 18:03:26

Depth	Weathering (kEq/ha.yr)							
m	BC+Na	Ca	Mg	K	Na	Al	Si	PO4
0.040	0.009	0.002	0.000	0.002	0.004	0.027	0.089	0.000
0.155	0.054	0.021	0.011	0.006	0.016	0.115	0.350	0.004
0.345	0.394	0.173	0.146	0.022	0.053	0.722	1.816	0.013
0.575	0.245	0.093	0.033	0.029	0.090	0.534	1.715	0.029
Totals:	0.702	0.289	0.190	0.059	0.164	1.398	3.970	0.047

THE GÅRDSJÖN CASE STUDY

Site characteristics

The Gårdsjön area is a small watershed with several lakes, located on the Fenno-Scandian granitic shield on the Swedish West Coast, near the city of Göteborg. The soils in the area are shallow orthic podzols. Lake and soil acidification are intensively studied in the area in a large multidisiplinary project, and it has been determined that the lake acidified in the early 1960s (Renberg et al., 1985). Since 1970 the lake and its watershed have stayed in a stable acid condition. The acidity of the lake is but a mirror of the chemical conditions in the soil of the watershed, as the runoff water composition is largely formed by chemical reactions occurring when the water passes through the soil matrix. As a part of the lake and soil acidification study, the mineralogy of the soil as well as the bedrock were determined along with params concerning the soil texture and chemical status at a number of sites in the watershed (Melkerud, 1983; Olsson et al., 1985). Soil thickness has been estimated in F1 to 0.55 to 0.65 m, maybe even less.

The input data requirements for a model calculation of the weathering rate in the three watersheds at Gårdsjön are different from the requirements of a budget calculation, but the required information for the watersheds are available in the literature (Olsson et al., 1985, Melkerud, 1983, Hultberg, 1985). The soil matrix particle size distribution was determined at different levels in the soil profile at three locations, in all 26 samples (Melkerud, 1983), and from this material the surface area was calculated as indicated earlier. See Tables 4-6 for data.

Table 4. Approximate soil and bedrock composition in three catchments on the north slope of Gårdsjön (Adapted from Melkerud, 1983) estimated using a combination of total analysis and XRD analysis.

Mineral	Bedrock composition %	Depth 2 meter %	Depth 0.7 meter %	Till composition F1 %	F2 %	F3 %
K-feldspar	20.0	14.0	19.0	15.0	15.0	12.0
Plagioclase	32.3	27.0	16.0	14.0	12.0	12.0
Hornblende	2.6	2.0	1.0	0.5	0.5	
Epidote	4.5	2.0	2.0	0.1	0.1	1.1
Biotite	13.0	1.0	1.0	1.2	1.0	-
Garnet	0.1	0.1	-	0.1	0.1	-
Chlorite	0.4	0.4	-	0.4	0.4	-
Vermiculite	-	1.1	5.0	5.0	5.0	8.0
Apatite	0.1	0.1	0.1	0.1	0.1	0.1
Undefined	-	12.7	5.0	5.0	5.0	5.0
Quartz	27.0	40.0	28.0	28.0		

The soil solution have pH 3.9 to 4.3 in the upper 0.30 m of the soil profile and pH 4.3 to 4.5 in the 0.3 to 0.6 m layer and pH 4.5 to 5.0 in the lowest layers based on observations made in 1991 reported by Hultberg (1995). In 1981 the soil pH was slightly higher Melkerud (1983). The till mineral composition has been estimated in the subcatchments F1, F2 and F3, located on the south shore of the lake, using data from total analysis and XRD analysis (Melkerud, 1983). The soil moisture was set at 0.15 m^3 m^{-3} in

Table 5. Minerals present in small amounts in the bulk forest soil.*

	Light fraction >2700 kg/m³	Heavy fraction <2700 kg/m³
Sand-silt	I	II
<2μm	Feldspars Quartz Muscovite Biotite	Pyroxene Amphiboles Epidotes Biotite Nesosilicates
Clay	III	IV
>2μm	Vermiculite Montmorillonite Quartz	Chlorite

* Minerals such as hornblende, pyroxene or apatite, may be difficult to determine in a bulk sample. A special procedure for soil mineralogy analysis has been developed together with the Czech and Swedish Geological Surveys. The minerals may be divided into four classes, depending on relative abundance in different texture classes and their specific gravities.

the top of the soil profile increasing to 0.35 at 0.6 m depth in G1. The average depth of the soil layer is 0.43 m in G1, 0.63 m in F1, it has been guessed that it is 0.6-0.8 m in F2 and F3. No soil mineralogy is available for the roofed catchment, G1, but it was assumed that the soil mineralogy was the same as in catchment F1.

The net amount of biomass cation accumulation was estimated as the difference between the measured storage in tree trunks and substantial roots (Hultberg, 1985; Sverdrup et al., 1990). The weathering of sodium was calculated after the subtraction of marine salt from seaspray.

Method

In order to be able to validate the PROFILE approach against field data, other independent estimates of the weathering rate was made using the best data available. Such estimates are difficult to make, and uncertainties may be large. Therefore a number of methods were used.

Results

Modelling the field weathering rate with PROFILE. The present base cation release rate was calculated for the Gårdsjön watersheds using the the model described, based on observed particle size distributions, soil pH values, soil texture, water content and weathering rate coefficients determined in laboratory studies. PROFILE was used with data for the soils in catchment G1, F1, F2 and F3 in the Gårdsjön catchment, obtained from Olsson et al. (1985), shown in Tables 7 and 8. The weathering rate per mineral is shown in Table 9. It can be seen that K-feldspar, plagioclase, hornblende, epidote, vermiculite and apatite account for all the weathering in the profile. Of this the feldspars account for 60% and hornblende and epidote account for 20%.

The calculated rate values correlate well with the observed values, and the observed difference is within the uncertainty of the values. The model has been established as a tool for calculating critical loads for acid deposition. This has lead to its testing against data in several European Countries.

Table 6. SAFE and PROFILE input data for F1 subcatchment at Lake Gårdsjön.

Parameter	Unit	1	2	3	4
Morphological characterization		O	A/E	B	B/C
Soil layer thickness	m	0.09	0.05	0.3	0.23
Moisture content	$m^3\,m^{-3}$	0.15	0.25	0.3	0.35
Soil bulk density	$kg\,m^{-3}$	300	800	700	1250
Specific surface area	$m^2\,m^{-3} * 10^{-6}$	0.48	1.2	1.11	1.98
Cation exchange capacity (CEC)	$keq\,kg^{-1} \cdot 10^{-6}$	352	22	61	13
CO_2 pressure	times ambient	2	5	20	30
Dissolved organic carbon	$mg\,l^{-1}$	70	70	10	5
log Gibbsite eq. constant	$kmol^2\,m^{-3}$	6.5	7.5	8.5	9.5
Inflow	% of precipitation	100	64	54	54
Percolation	% of precipitation	64	54	54	54
Mg+Ca+K uptake	% of total max	50	40	10	0
N uptake	% of total max	50	40	10	0
Initial pH		4.3	4.4	5.1	5.3
Initial BC concentration	$\mu mol(+)\,m^{-3}$	200	80	80	100
Mineral	% of total				
K-feldspar		15	15	18	19
Oligoclase		14	14	15	16
Albite		0	0	0	0
Hornblende		0.1	0.5	1.5	1.5
Pyroxene		0	0	0	0
Epidote		0.1	0.5	0.75	1.0
Garnet		0	0.1	0.1	0.1
Biotite		0	0.5	0.5	0.5
Muscovite		0	0	0	0
Chlorite		0	0.4	0.4	0.4
Vermiculite		0	3	15	5
Apatite		0	0.1	0.2	0.3
Kaolinite		0	0	5	0
Calcite		0	0	0	0
BC uptake	$kmol\,ha^{-1}yr^{-1}$	0.63			
N uptake	$kmol\,ha^{-1}yr^{-1}$	0.83			

Table 7. The weathering rate in keq ha^{-1} yr^{-1} for catchment F1 calculated with PROFILE.

Depth meter	Ca	Mg	K	Na	BC	P
			$keq\,ha^{-1}yr^{-1}$			
0-0.05	0.002	0.000	0.002	0.003	0.007	0.000
0.05-0.11	0.009	0.003	0.006	0.009	0.027	0.001
0.11-0.37	0.049	0.028	0.034	0.039	0.151	0.009
0.37-0.59	0.107	0.043	0.052	0.078	0.280	0.021
0.59-0.64	0.041	0.014	0.017	0.030	0.101	0.009
Sum	0.209	0.088	0.111	0.158	0.566	0.040

Table 8. The weathering rate in keq ha^{-1} yr^{-1} for catchment G1 calculated with PROFILE.

Depth meter	Ca	Mg	K	Na	Layer sum	P
			keq ha^{-1}yr^{-1}			
0-0.05	0.003	0.000	0.003	0.004	0.009	0.000
0.05-0.11	0.008	0.002	0.006	0.008	0.024	0.001
0.11-0.33	0.046	0.027	0.033	0.038	0.144	0.008
0.33-0.46	0.072	0.030	0.036	0.054	0.191	0.013
Sum	0.129	0.059	0.077	0.103	0.369	0.022

Table 9. The weathering rate for each mineral in the soil in keq ha^{-1} yr^{-1} for catchment G1, as calculated with PROFILE.

Mineral	Weathering in layer, keq ha^{-1}yr^{-1}					%
	1	2	3	4	Sum	of total
K-feldspar	0.003	0.006	0.025	0.029	0.062	16.8 %
Plagioclase	0.006	0.014	0.060	0.082	0.161	43.6 %
Hornblende	-	0.001	0.015	0.018	0.034	9.2 %
Epidote	-	0.001	0.009	0.028	0.038	10.3 %
Garnet	-	-	0.001	-	0.001	0.3 %
Biotite	-	-	0.002	0.002	0.004	1.1 %
Chlorite	-	0.001	0.004	0.004	0.009	2.4 %
Vermiculite	-	0.001	0.017	0.007	0.024	6.5 %
Apatite	-	-	0.013	0.023	0.035	9.5 %
Sum	0.009	0.024	0.144	0.191	0.369	100 %

Budget study method. There are a number of studies where the base cation release rates due to chemical weathering have been estimated from catchment studies. Until recently, only a few studies have had access to datasets that permit the base cation transport rates to be separated into rates of chemical weathering and depletion of the exchange pool in the soil, often laboratory rates were also not adequate (Arrhenius, 1954; Paces, 1983a,b; 1986; Cronan, 1984; Velbel, 1985; Olsson et al., 1985; Fölster, 1985; Matzner, 1987; Jacks and Åberg, 1987). A mass balance for a base cation keeps account of all sources and sinks considered in the system. The mass balance equation can be rearranged to an expression for the weathering rate for each base cation:

$$weathering = leaching + uptake - deposition - base\ sat.\ decrease \qquad (66)$$

In most budget calculations, the base cation flux from the change in base saturation cannot be separated from the flux of base cations from weathering. If it is assumed that the catchment is already at steady state, then the ion exchange term is set to zero. The basic assumptions of steady state must be carefully tested in each case in order to use the methodology for determining the weathering rate.

The base cation release rate due to chemical weathering of silicate minerals was calculated for the Gårdsjön watersheds using data available in the literature (Hultberg, 1985; Melkerud, 1983; Olsson et al., 1985) and new data from the monitoring program. The results are very uncertain, and very different values are obtained, depending on which year the data was taken. Table 10 shows the results for 1981, Table 11 shows the results for 1991-1993. The high rates for Mg in 1981 may be caused by acidity leaching of Mg from the ion exchange complex.

Table 10 (left). Results from a budget study. The weathering plus ion exchange rate for the three subcatchments in the Gårdsjön (Sverdrup and Warfvinge, 1988) using input data from Hultberg (1985) for 1981.

Table 11 (right). Results from a budget study. The weathering plus ion exchange rate for the using input data for 1991-1993 for F1 and G1.

Catchment	Ca	Mg	K	Na	Sum
		keq ha^{-1}yr^{-1}			
F1	0.21	0.42	0.11	0.01	0.75
F2	0.18	0.21	0.07	0.02	0.48
F3	0.19	0.19	0.07	0.02	0.47
Average	0.19	0.27	0.08	0.02	0.56

Catchment	Ca	Mg	K	Na	Sum
		keq ha^{-1}yr^{-1}			
F1	0.25	0.16	0.10	0.13	0.54
G1	0.13	0.10	0.05	0.08	0.36

At Gårdsjön, base cation uptake was estimated to 0.6-0.7 keq ha^{-1} yr^{-1} .Leaching in 1991 is estimated to be in the range 0.65-0.75 keq ha^{-1} yr^{-1} for Ca+Mg+K, for Na it is 2.2-2.4 keq ha^{-1} yr^{-1}. The amounts of cations which may potentially be removed from the soil by biological processes will be dependent on whether there is a net accumulation of biomass in the system or a net breakdown of biomass. If the canopy is harvested, the removal becomes irreversible.

Nilsson (1985) determined the weathering rate from an aluminium budget in 1985 to be approximately 0.6 keq ha^{-1} yr^{-1} .Aluminium is involved in many reactions in the soil which are not well understood, and the number is very approximate. Hultberg (1985) determined weathering plus ion exchange plus net mineralization to 1.20 keq ha^{-1} yr^{-1} for 1981. Lundström (1991) estimated the same sum to be 0.85 keq ha^{-1} yr^{-1}.

Using input values for 1991-1993 yield similar weathering rates for Ca, lower for Mg, higher for Na and a very uncertain value for K. The throughfall value for K is higher than the deposition value due to canopy leaching. Deposition of K was set at 0.2 keq ha^{-1} yr^{-1}. In F1, atmospheric inputs are a little lower than at G1. This could imply that the reservoir of K is being depleted, since the roof was installed, the number is very uncertain. In 1991-1993 the acid deposition load was approximately 40% lower than in 1981. This could explain the lower Mg values, the acid rain is no longer depleting the ion exchange reservoir to the same extent.

Strontium isotope method. Åberg and Jacks (1985), Jacks and Åberg (1987) and Wickman and Jacks (1991) used a modified budget approach where the difference in the ratio between the two naturally occurring strontium isotopes (Sr^{87}/Sr^{86}) in the deposition and the bedrock, were used to determine the fraction of the cation export coming from the weathering of soil minerals. The approach assumes the isotopic ratio in the part originating from chemical weathering to be equal to the ratio measured in a citrate or salisylic acid leachate. They calculated the weathering rate from the relation:

$$R_W = x_{Ca} \cdot Ca_D \cdot \left(\frac{Sr_L - Sr_D}{Sr_W - Sr_L} \right) \tag{67}$$

where:

R_W =   Base cation release by weathering   keq ha^{-1}yr^{-1}
Ca_D =   Calcium cation deposition rate   keq ha^{-1}yr^{-1}
x_{Ca} =   Ca fraction of weathering release
Sr_D =   The strontium isotope ratio in the deposition
Sr_L =   The strontium isotope ratio in the runoff
Sr_W =   The strontium isotope ratio in the mineral matrix

The method is based on the assumptions that the strontium isotope ratio in the exchange pool is the same as in the runoff, and that the weathering of strontium from the mineral matrix is linearly correlated with the weathering rate of calcium. The method is unique in the respect that when combined with a budget study it allow the weathering rate to be separated from the leaching from the ion exchange complex.

The water samples used for the method were collected in April 1992. This was just after sprinkling had commenced in under the roof. Calcium deposition was 0.13 keq Ca ha^{-1} yr^{-1} April 1991 to March 1992, the next hydrological year April 1992 to March 1993, 0.09 keq Ca ha^{-1} yr^{-1} was distributed under the roof. Salisylic acid was used for soil extraction, a check indicated that this did not extract all exchangeable Sr. Thus the ratio estimated will be an underestimate. The ratio measured for the extract is assumed to be approximately the same as in the weatherable minerals releasing Sr and Ca. The results obtained for G1 were:

Deposition	Sr^{87}/Sr^{86}	=	0.7097
Exchangeable	Sr^{87}/Sr^{86}	=	0.7270
Runoff	Sr^{87}/Sr^{86}	=	0.7097

Using a deposition of 0.09 keq Ca ha^{-1} yr^{-1} and a mineral Sr ratio of 0.7270, give a weathering rate of 0.12 keq Ca ha^{-1} yr^{-1}. If a deposition of 0.13 keq Ca ha^{-1} yr^{-1} and a ratio of 0.7255 is used, the calculations give a weathering rate of 0.22 keq Ca ha^{-1} yr^{-1} .Using a ratio of 0.7270 and the average deposition, we get 0.147 keq Ca ha^{-1} yr^{-1}. In the PROFILE calculations, the ratio of Ca weathering to total base cation weathering is 2.7. In the budget calculations the ratio of Ca weathering to total base cation weathering is 3.07 during 1981 and 2.46 for 1991, the average is 2.76. The value 2.7 was used to scale up from Ca weathering to total weathering. The obtained value is 0.39 keq ha^{-1} yr^{-1} for G1, with an uncertainty range from 0.32 keq ha^{-1} yr^{-1} to 0.57 keq ha^{-1} yr^{-1}. Using a C-layer sample of soil water at approximately 0.6 m depth yield a weathering rate of 0.26 keq ha^{-1} yr^{-1}, corresponding to 0.59 keq ha^{-1} yr^{-1}. This could serve as an approximate value for F1 since that catchment has an average soil depth of 0.63 m.

Table 12. Total analysis for subcatchments F1 and G1 at Lake Gårdsjön.

Site	CaO	MgO	Na$_2$O	K$_2$O	P$_2$O$_5$	SiO$_2$	Al$_2$O$_3$
F1; 21:1	2.11	0.88	2.8	2.76	0.06	73.3	13.9
G1; 23:1	2.46	1.57	2.9	2.61	0.10	69.4	15.6

Total analysis correlation. The correlation is based on the fact that the base cations magnesium and calcium are mainly associated to easily weatherable minerals (hornblende and epidote are rich in Mg and Ca), whereas sodium and potassium are associated with very slowly dissolving minerals such as feldspars. [Feldspars are rich in Na and K, but poor in Mg and Ca.] It also uses the fact that the weathering rate is dependent on the temperature. The empirical correlation (derived using data from Olsson and Melkerud, 1990), applicable to soils of granitic origin, is expressed keq ha^{-1} yr^{-1} (Sverdrup, 1990, Sverdrup et al., 1990):

$$R_{W_{Ca}} = Z_{soil} \cdot (1.0 \cdot 10^{-4} \cdot \%CaO \cdot (768 + 104 \cdot T) - 0.08) \qquad (68)$$

$$R_{W_{Mg}} = Z_{soil} \cdot (2.12 \cdot 10^{-4} \cdot \%MgO \cdot (768 + 104 \cdot T) - 0.09) \qquad (69)$$

$$R_{W_{Na}} = Z_{soil} \cdot 0.37 \cdot 10^{-4} \cdot \%Na_2O \cdot (768 + 104 \cdot T) \qquad (70)$$

$$R_{W_K} = Z_{soil} \cdot 0.21 \cdot 10^{-4} \cdot \%K_2O \cdot (768 + 104 \cdot T) \qquad (71)$$

where the symbols are:

T	=	Average annual temperature	^oC
%CaO,...	=	Cation content expressed as oxide in the soil	%
Z_{soil}	=	Soil depth	meter

Another empirical expression available for the total weathering rate is (Sverdrup et al., 1990; Olsson and Melkerud, 1990):

$$R_W = Z_{soil} \cdot (10^{-4} \cdot (768 + 104 \cdot T) \cdot (\%CaO + \%MgO + \%Na_2O + \%K_2O) - 0.37) \quad (72)$$

A similar expression is obtained by adding the Equations (68-71) for individual base cations. The expressions are calibrated for Swedish podzolic soils. The base cation content is based on an analysis of a totally dissolve soil sample. The correlation above were based on the historical weathering rate in Sweden, partly as recorded by Olsson and Melkerud (1990), and derived from depletion studies.

The total analysis correlation was applied to catchments F1, F2, F3 and G1. For F1, 0.53 keq ha^{-1} yr^{-1} was obtained. Data corresponding to Melkerud's sample 21:1 was used for F1, and 23:1 was used for G1. In G1, soil density may be somewhat less than in F1, since in a thinner soil the organic horizons make up a larger part of the whole. We are assuming average catchment soil density for G1 to be 870 kg m^{-3} and F1 1000 kg m^{-3}. The results have been compared to other results in in Table 16. For F2 and F3, 0. keq ha^{-1} yr^{-1} and 0.61 keq ha^{-1} yr^{-1} was obtained. If soil depth is less than estimated in 1983, then these rates must be changed correspondingly.

Table 13 (left). Results from total analysis correlation. The weathering for F1 and G1.

Table 14 (right). Results from the simple total analysis correlation. The total base cation weathering for F1 and G1 using the lumped type expressions.

Catchment	Ca	Mg	K	Na	Sum
		keq ha^{-1}yr^{-1}			
F1	0.15	0.14	0.05	0.10	0.44
G1	0.12	0.17	0.03	0.06	0.38

Catchment	Weathering keq ha^{-1}yr^{-1}
F1	0.55-0.52
G1	0.43-0.41

Table 15. The dissolved amounts of mineral in the upper m of soil matrix, since the glaciation in the soil from catchment F3, called Folkes Vik, in the Gårdsjön area. For catchment F1 it would be approximately the same. ME is the equivalents base cations per kg mineral.

| Mineral | ME | Mineralogy | | | | Weathering rate | | |
| | | Gneiss bedrock IDQ 211 % | Granite bedrock IDQ 43 % | Profile bottom F1 % | Average soil mineral % | Gneiss ref. | Granite ref. | Bottom ref. |
						keq ha^{-1}yr^{-1}		
Microcline	279	19.5	35.4	14	15	0.12	0.58	-
Plagioclase	260	32.8	29.8	27	14	0.52	0.44	0.35
Hornblende	70	2.6	-	2	0.5	0.22	-	0.22
Epidote	145	4.2	-	2	0.1	0.20	-	0.09
Biotite	142	12.9	1.1	1	1.2	0.59	-	-
Weathering						1.64	1.02	0.66

The historic weathering rate. The investigations by Olsson et al. (1985) in the Gårdsjön acidification research area may be used to estimate the average weathering rate of the past. The till in Scandinavia is generally of a geologically recent origin, most of the mineral matrix was formed during the last period of glaciation. One approach may be to assume that at the end of the glaciation, the mineral matrix of the soil consisted of freshly ground material, which only has been exposed to weathering since then. If the composition of the parent minerals is known, this may be used together with the present composition of the soil mineral matrix to estimate the approximate weathering rate during the last 12,500 years. The mineral composition of the soil and the different parent bedrock materials is shown in Table 15. The dissolved mass of each mineral is calculated from the difference between the parent material composition, either at the bottom of the profile or bedrock and the present day till mineral composition. The formula used was

$$R_W = \sum_j^{minerals} \frac{(C_{REF,j} - C_{soil,j}) \cdot \rho_{SOIL} \cdot Z}{100 \cdot ME} \tag{73}$$

$C_{REF,j}$	= Fraction of mineral in reference	
$C_{soil,j}$	= Fraction of mineral in reference mater	
ρ_{SOIL}	= Soil density	kg m^{-3}
ME	= Equivalent amount of base cation per mineral mass	keq kg^{-1}
t	= time since the material was formed	
Z	= Soil depth	m

The rates obtained are shown in Table 16.

The present matrix may be a mixture of minerals weathered before the glaciation and after the glaciation. For the first case where we use the bedrock values we get a weathering rate of 1.31-0.81 keq ha^{-1} yr^{-1} for subcatchment F3, assuming average soil depth to be 0.6 m and t to 12,500. Using the values from the bottom of the profile we get a past weathering rate in the range 0.53 keq ha^{-1} yr^{-1}. The range in rates arise from uncertainties in the estimates of mineralogy and soil density. The average historic rate was estimated using SAFE to be 0.74 keq keq ha^{-1} yr^{-1}, but that may be a low estimate.

Table 16. Summary of base cation release rates due to chemical weathering at different catchments at Lake Gårdsjön, Sweden, using different methods. It can be seen that the different methods give approximately the same results.

F1	F2	F3	G1	Method
	keq ha^{-1}yr^{-1}			
0.54	0.61	0.59	0.36	Budget study
0.57	0.60	0.67	0.37	Laboratory kinetics, PROFILE
0.53	0.60	0.61	0.38	Total analysis correlation
0.59	-	-	0.39	Sr isotope ratio method
0.56	0.60	0.64	0.38	Average present value
1.02-1.64	-	0.48-1.2	0.67-1.08	Historic rate, bedrock mineralogy reference
0.44-0.66	-	-	0.30-0.43	Historic rate, C-layer mineralogy reference
0.74	-	- 1.08	0.30	Historic rate, laboratory kinetics, SAFE
0.47	0.47	- 0.48	0.37	Historic conditions, present soil, PROFILE

Instead of using individual minerals as the conservative substance, quartz, zircon or rutile may be used as the conservative tracer. They are relatively resistant to chemical dissolution (Nickel, 1973; Olsson et al., 1985). However, Ti may be incorporated in biotite, pyroxenes and amphiboles in a less resistant form, which may offset the methodology. Another confounding factor occur when the soil is layered with different geological origins of the layers. This approach has several inherent weaknesses. The degree of mixing between old soils and newly formed soils is unknown. The soil may have been formed in a stepwise manner, creating layers of different origin.

Historic weathering as reconstructed with the SAFE model. The SAFE model was applied to the Gårdsjön catchment and run for 12,500 years that has elapsed since the last glaciation. The results have been shown in Figure 15. The model was set up to subtract each year's weathering of each mineral from the amount calculated for the previous year's amount of each mineral. At the same time the particle size was recalculated, as to a shrinking sphere. This means that the particle size distrubution change over time, depending on how the different minerals were initially distributed over the different particle size fractions. The model calculates the development of soil chemistry, soil mineralogy and soil texture over time, and partly reconstructs soil development. Both the consumption of primary minerals and the production of weathering residues and secondary minerals are considered.

The reconstruction of the weathering rate at Lake Gårdsjön starts with a mineral matrix composed of crushed bedrock material, of the same texture as the soil at 2 m depth. The model calculates present weathering, 12,500 years after ice-out, to 0.45 keq ha^{-1} yr^{-1}. The average for the whole time-period is 0.74 keq ha^{-1} yr^{-1}. At present, the model underpredicts present soil texture, yielding a slightly too low present weathering rate. This is still very close to the observed present value 0.58 keq ha^{-1} yr^{-1} as estimated by other methods. At the same time, the calculated present soil content of K-feldspar, plagioclase, hornblende and vermiculite are closely reconstructed by the model (Rietz, 1995). This is an indication that laboratory kinetics does indeed apply to field conditions. The calculated pH curves could be verified against reconstructions of lake pH from paleolimnological

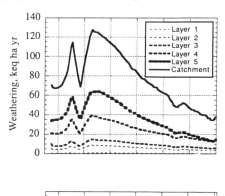

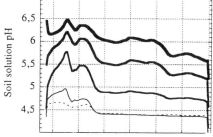

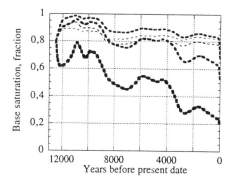

Figue 15. Reconstructed historical weathering using SAFE at the Lake Gårdsjön catchment. Soil chemistry like soil solution pH and base saturation also vary with time at Lake Gårdsjön. The change over time is most pronounced in the B- and C-layers of the soil. The lines starting with the thinnest and increasing in thickness with layer depth represents the the O-, A, B-, B/C- and C-layers.

sediment investigations (See Rietz, 1995). The diagrams also show how the soil is depleted of base cations mainly in the B- and C-layers, mostly due to the action of the vegetation. Root transport base cations up from the deeper layers to be deposited on the upper layers by litterfall. Only in the deeper layer are they not replenished. It becomes evident that the development of vegetation is very important for soil chemistry development at the site, but temperature and moisture is more important for variations in the weathering rate. Very recently the soil chemistry change very rapidly, due to acid deposition effects.

Comparing methods for estimating the weathering rate. As discussed above, there are several different weathering rate estimates available for the catchments in the Gårdsjön area. A summary of the weathering rates obtained are shown in Table 16, above. The rates appear to be consistent between measurements even when different methods have been used. The budget method seem to give less accuracy, since ion exchange may still be playing a role in the catchment.

MODEL VALIDATION

The performance of the PROFILE model in weathering rate calculations was tested at a number of sites in Europe and North America. Some of the sites were used for blind tests; Aubure, Alt'a Mharcaidh, X-14, Fårahall, Rock Carving (first run is used, no input adjustments, no weathering estimates were known until afterwards).

PROFILE was been tested in 15 different catchments in Europe and North America, and has been able to reproduce the observed weathering rates under very different conditions better than approximately ±10% in the range from 0.007 to 18. keq ha^{-1} yr^{-1}, as partly evident from Figures 16 and 17. The model, the tests of it, and its use have been described in detail in Warfvinge and Sverdrup (1993) and Sverdrup and Warfvinge (1993). Input data forms and the PROFILE model were sent out to colleagues throughout Europe. They tested the model on their data, and returned to us their independent estimate of the weathering rate together with the rate calculated with PROFILE for the same site. The data received were used to draw the diagram. The mineralogy at the sites is listed in Table 1.

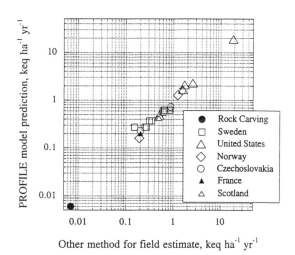

Figure 16. The weathering rates calculated with PROFILE as compared to other field estimates. In total the accuracies of the weathering rate calculations are of the order of ±10% on the rate in the range from 0.007 to 20 keq ha^{-1} yr^{-1}.

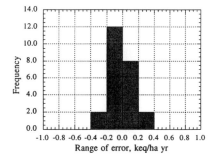

Figure 17. The distribution of errors in the weathering rates estimated with PROFILE.

The sites exhibited a large variability, from weathering in a pre-historic rock-carving, to epidote-rich soils in northern Maryland, the temperatures varied from +1.8°C in Swedish Lappland to +13.3°C in Maryland.

The rock-carving is located on the Swedish west coast, it dates from the Middle Bronze Age (1200 BC). The carving studied was investigated for increased weathering due to exposure to rain of increasing acidity. The carving is located on a flat, polished, southwest sloping side of exposed bedrock, and it depicts ships, people, cupmarks and different objects and symbols of the Bronze Age cult. The rock is a gneiss, composed of feldspar, plagioclase and biotite, the mineralogy is found under "Swedish bedrock" in Table 1.

The weathering rates of the the Maryland sites were estimated from mass balances and the assumption that the sites were at steady state with the present deposition input of acidity and base cations. To evaluate the mass balances, the effects of agricultural liming in the catchments were estimated; this effect adds to the uncertainty in the determinations. The weathering rates are in large excess of the present acid deposition rate. That the model also predicts the weathering rate of these soils with good success, show that the model is valid for both young glaciated soils as well as old unglaciated soils.

Table 17. Texture of soils in sites used in the test of the model.
S = Sweden; N= Norway; F= France; UK= United Kingdom;
CS = Czech Republic; NY = New York, USA; MD = Maryland, USA.

Location	O-layer	E-layer	B-layer	C-layer
	10^6 m^2/m^3			
Svartberget (S)	0.03	0.8	0.9	1.2
Rock carving (S)	0.01	-	-	-
Risfallet (S)	0.04	0.2	0.66	0.66
Fårahall (S)	0.47	0.71	0.88	0.9
Höylandet (N)	1.7	1.7	2.1	2.2
Nordmoen (N)	0.13	1.3	1.0	1.0
Aubure (F)	0.01	1.9	2.6	3.1
Mharcaidh (UK)	0.5	0.7	0.64	0.56
X-14, Most (CZ)	0.5	2.0	2.0	1.5
Woods Lake (NY)	0.5	1.5	2.0	1.5
Queen Anne (MD)	0.5	3.7	5.2	3.0
Anne Arundel (MD)	1.0	4.0	5.8	5.0
Washington (MD)	0.5	3.9	4.3	2.7
Wicomico (MD)	0.3	3.0	4.4	2.8

The soils in Maryland are old soils, normally not subjected to glaciation for the last 800,000 years. The soils there are generally low in all types of primary minerals, typically feldspars make up 1 to 5% of the weight, mafic minerals from 0.05 to 1% and the rest is quartz and insoluble residues, something which could perhaps be expected such for old soils. The weathering rates are still comparable to those observed for Sweden where the feldspars make up 20 to 50% of the weight, heavy minerals 2 to 6% and the rest quartz. This is due to the basic difference in soil texture. For Maryland the surface area range from 1.5 to 7×10^6 m^2 m^{-3}, the annual average temperature range from 7° to 13°C, in Sweden, soil texture vary from 0.2 to 2.5×10^6 m^2 m^{-3} and annual average temperature 1° to 7°C, which tends to compensate for the effect on the weathering rate caused by the difference in mineralogy.

Table 18. Climatic and vegetation input data for sites used in the test of the model. They span a large variety of climates, textures, pollution climate, soil chemistry and vegetation. S = Sweden; N = Norway; F = France; UK = United Kingdom; CZ = Czech Republic; NY = New York, USA; MD = Maryland, USA.

Location Temp.	Precip.	Runoff	S	NO$_3$	NH$_4$	BC	BC	N	Temp
	m yr^{-1}	keq ha^{-1}yr^{-1}	Deposition keq ha^{-1}yr^{-1} °C				Uptake		°C
Svartberget (S)	0.72	0.37	0.37	0.13	0.13	0.13	0.08	0.15	1.80
Rock carving (S)	0.90	0.55	1.50	0.60	0.60	0.75	0.50	0.70	6.00
Risfallet (S)	0.55	0.22	0.58	0.25	0.22	0.14	0.20	0.50	4.80
Fårahall (S)	1.05	0.45	1.82	0.71	0.56	0.67	0.40	0.5	6.50
Höylandet (N)	1.60	1.30	0.17	0.12	0.14	0.83	0.15	0.25	1.50
Nordmoen (N)	0.78	0.41	0.80	0.31	0.31	0.19	0.40	0.61	4.30
Aubure (F)	0.86	0.66	1.44	1.98	1.60	2.10	1.58	2.56	6.00
Mharcaidh (UK)	0.92	0.86	0.70	0.16	0.10	0.55	0.00	0.00	5.50
X-14, Most (CZ)	1.00	0.43	7.50	1.00	1.00	0.70	0.30	0.20	10.00
Woods lake (NY)	1.20	0.75	1.40	0.28	0.21	0.31	0.5	0.55	6.50
Queen Anne (MD)	0.97	0.35	0.89	0.31	0.18	0.24	0.10	0.20	13.3
Anne Arundel (MD)	0.97	0.35	0.85	0.32	0.19	0.22	0.61	1.1	11.0
Washington (MD)	0.98	0.33	1.01	0.35	0.17	0.14	0.88	0.65	11.4
Wicomico (MD)	0.90	0.35	0.69	0.26	0.21	0.27	0.78	0.35	12.4

The conclusion drawn from this test is that laboratory kinetics are applicable to field conditions, and that the field weathering rate can be calculated with very good accuracy. There is apparently full consistency between laboratory kinetics and observed field rates. The mineralogy of the soil can vary quite significantly in the landscape. The mineralogy of the soil is a function of the soil forming process. In Scandinavia this was driven by the last glaciation. These soils contain ground bedrock, but not necessarily from the bedrock under that site, and not of the same weight composition as the bedrock of origin. The mineralogy may be a mixture of ground material from bedrock in a wide circumference, selectively weathered to change the mineralogical composition.

Large mineralogical surveys have shown that the mineralogical composition of soils is quite different from the underlying bedrock. Many minerals are present in podzolic soil profiles, however the majority of the podzolic soils on the Fenno-Scandian Shield contain mainly minerals of granitic origin where the dominant minerals are quartz, mica and feldspars. Several secondary minerals occur, mainly clays such as illite, smectite and kaolinite. Amphiboles and pyroxenes are not present as major constituents, except in small confined areas only, and generally only as accessory minerals. Small amounts of hydroxides, oxides and sulfides of iron and manganese are generally present in the soil profile. Typical soil mineralogies have been listed in Table 1.

Regional applications. In several countries (Sweden, Norway, Switzerland, Russia, Spain, Slovakia) the PROFILE model has been used to create regional weathering rate maps, based on regional surveys of the necessary mineralogical information. Such maps are valuable for assessing soil resistance to acid deposition For critical loads see Sverdrup et al. (1990; Hettelingh et al. (1991); Sverdrup and de Vries (1994); and Sverdrup and Warfvinge (1995), or for estimation of the nutrient supply capacity of the soil. Figure 18 shows the weathering rate map for the upper 0 to 0.5 m of Swedish forest soils, based on calculations with PROFILE in 1804 sites where mineralogy was available.

Table 19. Mineralogy of the soils at different locations in wt % of the bulk soil. All soils have significantly different mineralogy from the parent bedrock material. All the sites have bedrock below which contain approximately 40 wt % plagioclase, 35 wt % K-feldspar, 5 to10 wt % dark minerals, and the remainder quartz. The bedrock in Maryland (MD) is made up of schist with 10 to 20 wt % feldspar, 1 to 2 wt % dark minerals and quartz. The first group of soils listed are only 6000 to 12000 years old, and are the result of the last glaciation. The middle group are soils that were only partly glaciated, and have been weathered for a longer period of time, maybe 15,000 to 25,000 years. The Maryland soils are up to 800,000 years old, two of them are located on old volcanic bedrock.

Legend:

Kf = K-feldspar; Pl = Plagioclase; Al = Albite; Ho = Hornblende; Py = Pyroxene; Ep = Epidote;
Ga = Garnet; Bi = Biotite; Mu = Muscovite; Ch = Chlorite; Ve = Vermiculite; Ap = Apatite;
S = Sweden; N = Norway; F = France; UK = United Kingdom;
CZ = Czech Republic; NY = New York, USA; MD = Maryland, USA.

Location	Kf	Pl	Al	Ho	Py	Ep	Ga	Bi	Mu	Ch	Ve	Ap
Gårdsjön F1 (S)	18	12	-	0.5	-	0.5	0.1	0.5	-	0.4	5	0.35
Svartberget (S)	7.6	16	-	7.7	-	2	-	-	-	2	4	0.4
Risfallet (S)	25	26	-	3.5	-	-	-	-	-	0.5	-	0.1
Fårahall (S)	29	28	-	4	-	-	-	3	-	-	5	0.3
Nordmoen (N)	7.7	12.8	-	4.0	-	-	-	-	21.3	15.4	10	0.1
Mharcaidh (UK)	25	40	-	-	-	-	-	-	-	-	5	-
Woods Lake (NY)	31	10	-	2	0.4	0.5	0.6	3.6	-	-	-	-
Höylandet, (N)	3.5	20	-	0.6	8.6	2.9	-	-	-	3.5	-	-
Aubure (F)	8.0	-	4.0	-	-	-	-	-	4.0	-	-	0.1
X-14, Most (CZ)	9	20	-	-	-	-	-	2.5	5	2.5	7	-
Anne Arundel (MD)	2.1	2.9	-	0.01	-	0.02	-	-	12.4	-	9.5	-
Queen Anne (MD)	6.0	7.0	-	0.12	0.03	0.01	-	-	7.4	-	7.6	-
Washington (MD)	2.2	14	-	0.25	-	10.2	-	-	4.7	-	16.7	-
Wicomico (MD)	4.1	3.6	-	0.06	0.02	0.01	-	-	7.2	-	10.0	-
Tanum bedrock (S)	30	59	-	-	-	-	-	-	1	-	-	-
Gårdsjön bedrock (S)	33	20	-	3.5	-	4.2	-	4.9	-	0.4	-	0.4

DISCUSSION

The important minerals

The weathering rate in nature is determined by the abundance of weatherable minerals and their texture in the soil. The PROFILE model represents the first tool available for estimation of the weathering rate from independent data on geochemistry and soil conditions. Accordingly, an understanding of the soil mineralogy and texture distribution on a regional scale is essential for understanding the conditions for weathering available in the region. A few minerals in the soil are of major controlling importance for the weathering rate in the field:

Plagioclase   K-feldspar   Hornblende   Epidote   Apatite

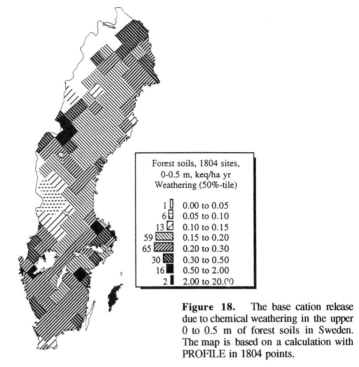

Forest soils, 1804 sites,
0-0.5 m, keq/ha yr
Weathering (50%-tile)

1	0.00 to 0.05
6	0.05 to 0.10
13	0.10 to 0.15
59	0.15 to 0.20
65	0.20 to 0.30
30	0.30 to 0.50
16	0.50 to 2.00
2	2.00 to 20.00

Figure 18. The base cation release due to chemical weathering in the upper 0 to 0.5 m of forest soils in Sweden. The map is based on a calculation with PROFILE in 1804 points.

In non-glaciated, old soils primary minerals such as pyroxene and biotite may sometimes play a minor role, secondary minerals of some importance are vermiculite, chlorite and illite. It is our experience from surveying the mineralogy in both glaciated regions (Denmark, Sweden, Norway, Switzerland) and unglaciated regions (Maryland, Spain) that epidote is present in very many locations, but almost always in small amounts.

A simple model based on mineralogy

We may use the model to calculate the weathering rate per weight unit mineral in a typical soil. Table 20 shows the approximate weathering rate from a specific mineral assuming the same soil properties as in our G1 example, soil moisture of 0.25 $m^3\,m^{-3}$ of soil, T = 6.5°C and the specific total mineral surface is $1.3 \times 10^6\ m^2\,m^{-3}$ of soil. The list corresponds to a Goldich series of weatherability. The values in the table can be used by adjusting for differences in temperature, soil moisture and soil texture to roughly estimate the weathering rate from mineralogy.

$$R_W = \sum_i^{minerals} r_i \cdot \Theta \cdot Z \cdot e^{-0.13 \cdot TX} \cdot 8 \cdot 10^{(3500/279.5 - 3500/(273+T))} \qquad (74)$$

where

z is actual soil depth in m

Θ is soil water content in tonnes $H_2O\ m^{-3}$ soil

TX is Swedish Forest Inventory field texture class on a scale from 1 (coarse, $0.3 \times 10^6\ m^2\,m^{-3}$) to 9 (very fine clay, $15 \times 10^6\ m^2\,m^{-3}$), Gårdsjön is TX = 4 ($1.5 \times 10^6\ m^2\,m^{-3}$), r_i is the weathering rate of each individual mineral as listed in Table 20, and T is temperature in °C.

Table 20. The weathering rate per mineral in keq ha⁻¹ yr⁻¹ for 0.5 m of soil found in catchment G1, assuming the soil to constitute 100% mineral, calculated with PROFILE. (Soil texture is 1.3 mill. m² m⁻³, corresponding to the Swedish Forestry field texture class 4, and soil moisture is 0.25 m³ m⁻³). The minerals have been organized according to weatherability as predicted by PROFILE.

Mineral	Ca+Mg+Na+K release in keq ha^{-1}yr^{-1} per 100% and 0.5 m soil depth
Calcite	82.00
Apatite	17.00
Epidote	8.80
Hornblende	2.63
Pyroxene	2.06
Albite	1.68
Plagioclase	1.34
Garnet	1.16
Chlorite	1.05
Biotite	0.93
Vermiculite	0.56
K-feldspar	0.48
Muscovite	0.30

The difference disappeared!

In summary, that after going through all the steps described in this study, the often stated discrepancy between field and laboratory weathering rates disappeared. The Laboratory rate coefficients (Table 2) go into the calculations, and what comes out are values very close to observed field rates.

The earlier discrepancy observed between the face value of the laboratory rate and the field rate may add up as follows in Table 21. The most important factor for explaining the discrepancy is the occurrence of product inhibition of Al and base cations under field conditions, the partial wetting of mineral surfaces in the soil and the generally lower temperature as compared to laboratory conditions. Many earlier attempts have also stranded because of faulty laboratory rate coefficients from the very start.

Table 21. The effect of different conditions on the weathering rate in the laboratory as compared to apparent field rates. Most often factors overlooked are the the partial wetting, temperature difference and the effect of product inhibition.

Factor	Effect on weathering rate	Cumulative effect on weathering rate
Effect of product inhibition in field	2-5 times increase	2-5
Temperature difference field/lab	2-10 times increase	4-50
Partial wetting of field minerals	4-10 times increase	16-500

It is very important to realize that unless a multi-reaction approach is used when evaluation laboratory studies is done and unless an integrated model is used for transferring laboratory *kinetics* to field conditions, it is not even worth the attempt to compare. Ignoring other soil processes like ion exchange, soil solution composition, uptake, etc., will not do. Comparing a laboratory rate to a field rate without further thought *must* yield a large difference, 2 to 4 orders of magnitude!

Table 22. The effect of different laboratory experimental conditions on the weathering rate in the laboratory as compared to apparent field rates. Most often factors overlooked are the the partial wetting, temperature difference, the effect of product inhibition and the effect of buffers. Further confusion was added by using incorrect laboratory coefficients.

Factor	Effect on weathering rate	Cumulative effect on weathering rate
Effect of buffers in lab	2-5 times increase	2-5
Freshly ground minerals used in lab	2-10 times increase	4-50
CO_2 overpressure in experiment	2-15 times increase	8-750

We have been able to show that the different weathering reactions depend on state variables such as soil solution concentrations of Al, Ca or pH earlier in this study. These solution concentration effects can be summarized as in Table 23.

Table 23. Summary of the effects of different soil solution constituents on weathering rates.

Substance	Effect
H^+	Increase
Al^{3+}	Decrease
Ca^{2+}, Mg^{2+}, Na^+, K^+	Decrease
CO_2	Increase
Organic acids	Increase

Quantitatively the effects are different for different types of minerals (Sverdrup, 1990), minerals rich in aluminium react stronger to dissolved Al than those poor in Al, and the same applies to base cations. However these factors will all interact in the natural environment, and the net result of for instance the pH dependence may not be the same as in the laboratory. In the field the effect of pH may be pronounced, but this may also be overwhelmed by the effects of other factors such as Al or organic acids. Under certain circumstances where much aluminium or base cations are produced in the soil, this may overrule the effect of pH and the weathering rate may decrease.

The importance of product inhibition under field conditions. The calculated weathering rate in the 0.25 to 0.75 soil layer at Fårahall during 1800 to 2100 A.D. is shown in Figure 19. During this period, soil pH drops from 6.0 to 4.0, Al increases from 0 to 300 μeq l^{-1}, and the weathering rates increase from 0.66 keq ha^{-1} yr^{-1} in 1800 to 0.90

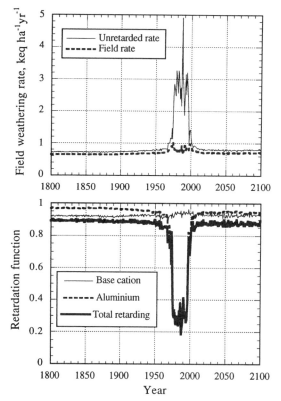

Figure 19. The calculated weathering rate in the 0.25 to 0.75 soil layer at Fårahall during the years 1800 to 2100. During this period soil pH drops from 6.0 to 4.0, Al increase from 0 to 300 μeq l^{-1}, and weathering rate increase from 0.63 keq ha^{-1} yr^{-1} in 1800 to an average of 0.90 keq ha^{-1} yr^{-1} in 1995. The weathering rate is shown as calculated for field conditions with SAFE, and as without the retarding functions for Al and BC applied. The lower diagram show the actual contribution of Al and BC to the total retarding of the rate. At most the rate is retarded by 70 % by the presence of Al and 10% by BC.

keq ha^{-1} yr^{-1} in 1995. The weathering rate is shown as calculated for field conditions, and without the retarding functions applied. The lower diagram show the value of the retarding functions in the kinetic equation. The retarded rate may appear almost independent on soil pH despite soil acidification. The increase due to soil acidification predicted by the model is modest, an increase by 30 to 40% increase in the rate from 0.66 keq ha^{-1} yr^{-1} to 0.90 at maximum soil acidity, something that appear reasonable with respect to field observations. Paces (1983) observed a 20 to 40% increase which he ascribed to the effect of soil acidification. Figure 20 shows a comparison of the calculated field rate with the rate without Al retardation, (2) without Al retardation and with full wetting and (3) with no Al retardation, full wetting and at laboratory temperature (25°C). The observed dependence of the weathering rate unretarded by Al and BC concentrations is similar to laboratory experiments, whereas the rate calculated for field conditions show a different and smaller dependency on pH.

Figure 21 shows the calculated field weathering rate plotted versus the calculated laboratory weathering rate, corresponding to the rate without application of the BC and Al retardation, with full wetting and at laboratory temperature. The diagram suggests that the laboratory rate should in this case be 100 times larger than the field rate.

The whole question of the dependence of the weathering rate on soil pH must be carefully handled when such a relationship is sought for in field data. The soil pH is a result of the balance between sources of acidity and sources of alkalinity in the soil, where weathering is a major part of the alkalinity source. Accordingly the pH of the soil may be

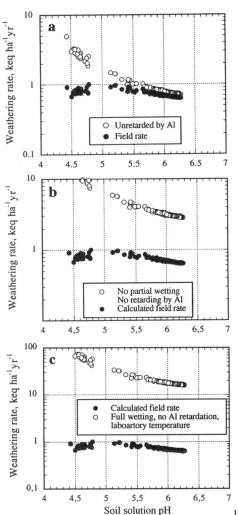

Figure 20. (a) Comparison of the calculated field rate with the rate without Al-retardation, (b) without Al-retardation and with full wetting, and (c) with no Al-retardation, full wetting and 25°C. The observed dependence of the weathering rate unretarded by Al and BC concentrations is similar to lab experiments, whereas the rate calculated for field conditions shows a different and smaller dependency on pH. Further compensating for experimental bias from CO_2 or buffers would increase the difference by another order of magnitude.

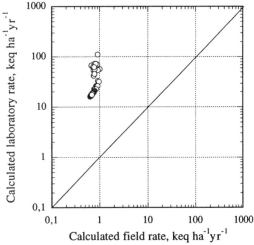

Figure 21. Calculated field vs. calculated laboratory weathering rates, corresponding to the rate without application of BC- and Al-retardation, with full wetting and at laboratory temperature.

dependent on the weathering rate as well as the weathering rate depending on the soil pH. This implies that if the observed field rate, incorporating different textures and mineralogies are plotted versus soil pH, we would find an increase in weathering with an increase in pH. This is a function of how weathering drives ANC and thereby pH. From this we can conclude that it is not possible to unscramble the information contained in field weathering rates into conclusions over the dependency of mineral weathering kinetics on soil solution composition such as H^+ or organic acid concentrations. Temperature variations within each year, and between years, was ignored in order to have a simpler case (SAFE can operate with variable temperature within and between years). In the same diagram the weathering rate with no effect of retarding functions has been drawn. It can be seen that the retarding factors are a very important part of determining the field rate, and that the base cation retarding function has little effects at Fårahall, where the base cation concentration is in the interval 150 to 250 μeq l^{-1}. The limiting concentration for BC-retarding effect is ~200 μeq l^{-1} for feldspars. The Al-retarding function appears to be important in this case.

Taking away the difference step-by-step. Figure 20 is a plot of the calculated weathering rates with and without retarding factors plotted against soil solution pH. It can be sen that the rate without the effects of retarding functions shows the same dependence on solution pH as observed in laboratory experiments, but that this is not evident after including retardation functions in the calculated field weathering rate.

Multiplying the difference factors for the typical case would, when compared at face value, give a ratio between experimental laboratory rate and field rate of 128 to 375,000, depending on the conditions (2 to 5 orders of magnitude). Tables 21 and 22 (above) show the additive effect of not considering the difference between field and laboratory conditions and the effect of inadequate experimental conditions in the laboratory. It can be seen that in most cases, a large difference build up. This is also what is observed when laboratory rates and field rates are compared at face value. Often the rates are 1 to 2 orders of magnitude off, which easily can be explained by overlooking one or two of the factors listed above. These differences have been adequately accounted for in the model and the interpretation of the laboratory data (Sverdrup, 1990). Having done that, there is no longer any significant difference between the laboratory rate and the field rate

Conclusions

It may be concluded from the analysis of laboratory kinetics, field weathering data and PROFILE model applications to European and American catchments:

* There is full consistency between field rate and laboratory weathering rates, when the differences in chemical and physical conditions are accounted for.

* Under field conditions, product inhibition by Al and base cations, is a key factor for modifying the weathering rate.

* Plagioclase, K-feldspar, hornblende and epidote account for 85% of the weathering rate observed in most soils of granitic origin.

* The PROFILE model can be used to calculate the field weathering rate within ±10% from independent geophysical and geochemical input data.

* Process-oriented models are necessary for interpreting the significance of observed field weathering rates in relation to different hypothesizes concerning single weathering processes and mechanisms.

REFERENCES

Aagaard P, Helgesson H (1982) Thermodynamic and kinetic constraints on reaction rates among minerals and aqueous solutions: I. Theoretical considerations. Am J Sci 282:237-285

Åberg G, Jacks G (1985) Estimation of weathering rate by Sr(87)/Sr(86) ratios. Geol Föreningens Stockholms Förhandlinar 107:289-290

Alcamo J, Amann M, Hettelingh J, Holmberg M, Hordijk L, Kämäri J, Jauppi P, Mäkälä A (1987) Acidification in Europe: A simulation model for evaluating control strategies. Ambio 16:232-244

April R, Newton R, Coles L (1986) Chemical weathering in two Adirondack watersheds: Past and present-day rates. Bull Geol Soc Am 97:1232-1238

Arovaara H, Ilvesniemi H (1990) The effect of soluble inorganic aluminum and nutrient imbalances on *Pinus sylvestris* and *Picea abies* seedlings. In: Kauppi P (ed) Acidification in Finland. Springer-Verlag, Berlin, p 715-733

Bain DC, Mellor A, Wilsson M (1990) Nature and origin of an aluminous vermiculite weathering product in acid soils from upland catchments in Scotland. Clay Minerals 25:467-475

Bergström S, Lindström G (1990) A review of models for analysis of ground-water and surface water acidification. Swedish Meteorological and Hydrological Inst, Norrkoping

Berner RA (1978) Rate control of mineral dissolution under earth surface conditions. Am J Sci 278:1235-1252.

Bolt GH (1982) Thermodynamics of ion exchange. In: Bolt GH (ed) Soil Chemistry. B Physio-chemical Models, p 27-46

Bonneau M (1991) Effects of air pollution via the soil. In: Landemann G (ed) French Research into Forest Decline—DEFORPA programme. ENGREF, Nancy, France, p 87-100

Bruggenwert MGM, Kamphorts A (1982) Survey of experimental information on cation exchange in soil systems. In Bolt GH (ed) Soil Chemistry. B Physio-chemical Models, p 141-175

Busenberg E, Clemency C (1975) The dissolution of feldspars at 25°C and at 1 atm CO_2 partial pressure. Geochim Cosmochim Acta 40:41-46

Chen C, Gherini S, Hudson R, Dean S (1983) The integrated lake-watershed acidification study. Final Report EPRI EA-3221, Electrical Power Research Inst, Palo Alto, California

Chou L, Wollast R (1985) Steady state kinetics and dissolution mechanisms of albite. Am J Sci 285:963-993

Christoffersen N, Seip H, Wright R (1982) A model for stream water chemistry at Birkenes, Norway. Water Resource Res 18:977-996

Cosby B, Wright RF, Hornberger G, Galloway J (1985) Modeling the effects of acid deposition: Assessment of a lumped parameter model for soil water and stream water chemistry. Water Resource Res 21:51-63

Cosby J, Hornberger, G, Wright, R (1985) Estimating time delays and extent of regional de-acidification in southern Norway in response to several deposition scenarios. In: Kämäri J, Jenkins A, Brakke, DF, Wright, RF, Norton SA (eds) Regional Acidification Models. Springer-Verlag, Berlin, p 151-166

Cronan CS (1985) Chemical weathering and solution chemistry in acid forest soils: Differential influence of soil type, biotic processes and H^+ deposition. In: Drever J (ed) The Chemistry of Weathering. D. Reidel, Dordrecht, Netherlands, p 175-195

Eggleston CM, Hochella M Jr, Parks GA (1989) Sample preparation and aging effects on the dissolution rate and surface composition of deposed. Geochim Cosmochim Acta 53:797-804

Eyring H (1935) The activated complex in chemical reactions. J Chem Phys 3:107-115

Falkengren-Grerup U (1987) Long-term changes in pH of forest soils in southern Sweden. Environmental Pollution 43:79-90

Falkengren-Grerup U, Eriksson H (1990) Changes in soil, vegetation and forest yield between 1947 and 1988 in beech and oak sites of southern Sweden. Forest Ecology and Management 38:37-53.

Falkengren-Grerup U, Linnermark N, Tyler G (1987) Changes in soil acidity and cation pools of south Sweden soils between 1949 and 1985. Chemosphere 16:2239-2248.

Furrer G, Stumm W (1984) The coordination chemistry of weathering: I. Dissolution kinetics of δ-Al_2O_3 and BeO. Geochim Cosmochim Acta 50:1847-1860

Gherini S, Mok J, Hudson RJM, Davis GF, Chen CW, Goldstein RA (1985) The ILWAS model, formulation and application. Water, Air, Soil Pollution 26:425-459

Goldich SS (1938) A study in rock Weathering. J Geol 46:17-58

Grandstaff D (1986) The dissolution rate of forsteritic olivine from Hawaiian beach sand. In: Coleman S, Dethier D (eds) Rates of Chemical Weathering of Rocks and Minerals, p 41-59. Academic Press, New York

Grandstaff DE (1977) Some kinetics of bronzite orthopyroxene dissolution. Geochim Cosmochim Acta 41:1097-1103

Grandstaff DE (1980) The dissolution rate of forsteritic olivine from Hawaiian beach sand. In: 3rd Int'l

540   Sverdrup & Warfvinge: *Estimating Weathering Rates*

Water-Rock Interaction Symp, p 72-74. Geochemical Cosmochemical Soc, Alberta Research Council

Helgeson H, Murphy W, Aagaard P (1984) Thermodynamics and kinetic constraints on reaction rates among minerals and aqueous solutions: II. Rate constants, effective surface area and the hydrolysis of feldspar. Geochim Cosmochim Acta 48:2405-2432

Henriksen A, Lien L, Traaen TS, Sevaldrud IS, Brakke DF (1988) Ambio 17:259-266

Hettelingh J, Downing R, de Smet P (1991) Mapping Critical Loads for Europe. Coordination Center for Effects, RIVM. CCE Tech Report No. 1, RIVM Report No. 259101001

Holmberg M, Hari P, Nissinen A (1989) Model of ion dynamics and acidification of soil: Application to historical soil chemistry data of Sweden. In: Kämäri J, Brakke, DF, Jenkins A, Norton SA, Wright RF (eds) Regional Acidification Models: Geographical Extent and Time Development, p 229-240. Kluwer Academic Publishers, Dordrecht, Netherlands

Hooper R, Christophersen N, Peters N (1990) Modeling streamwater as a mixture of soil water end-members—An application to Panola Mountain catchment. J Hydrol 116:321-343

Hultberg H (1985) Budgets of base cations, chloride, nitrogen and sulphur in the acid Lake Gårdsjön. Ecological Bull 37:133-157

Jacks G, Åberg G, Hamilton PJ (1989) Calcium budgets for catchments as interpreted by strontium isotopes. Nordic Hydrology 20:85-96

Jönsson C, Warfvinge P, Sverdrup H (1994) Uncertainty in prediction of weathering rate and environmental stress factors with the PROFILE model. Water, Air, Soil Pollution 78:37-55

Kirchner JW (1990) Heterogeneous geochemistry of catchment acidification. Geochim Cosmochim Acta 56:2311-2327

Knauss K, Wolery TJ (1989) Muscovite dissolution kinetics as a function of pH and time at 70°. Geochim Cosmochim Acta 53:1493-1502

Lagache M (1965) Contribution a l'etude de la dissolution des feldspates, dans l'eau, entre 100 et 200°C, sous diverse pressions de CO_2, et application a la synthese des mineraux argileux. Bull Soc fr Minéral Crist 88:233-253

Lagache M (1976) New data on the kinetics of the dissolution of the alkali feldspar at 200°C in CO_2 charged water. Geochim Cosmochim Acta 40:157-161

Lasaga A (1981) Transition state theory. In: Lasaga A, Kirkpatrick R (eds) Kinetics of Geochemical Processes, 8:135-170

Lasaga A (1984) Chemical kinetics of water-rock interactions. J Phys Res 89:4009-4025

Lasaga A, Blum AE (1986) Surface chemistry, etch pits and mineral-water reactions. Geochim Cosmochim Acta 50:2363-2379

Levenspiel O (1980) The coming of age of chemical reaction engineering. Chem Eng Sci 35:1821-1839

Melkerud P (1983) Quaternary deposits and bedrock outcrops in an area around Gårdsjön, southwest Sweden, with physical, mineralogical geochemical investigation. Reports on Forest Ecology and Forest Soils No. 40. Swedish Univ Agricultural Sciences, Uppsala, Sweden

Nickel E (1973) Experimental dissolution of light and heavy minerals in comparison with weathering and intrastitial solution. Contrib Sedimentology 1:1-68

Oene H (1992) Acid deposition and forest imbalances: A modeling approach. Water, Air, Soil Pollution 63:33-50

Olsson B, Hallbäcken L, Johansson S, Melkerud PA, Nilsson I, Nilsson T (1985) The Lake Gårdsjön area—Physiogeographical and biological features. Ecological Bull 37:10-28

Olsson M, Melkerud PA (1990) Determination of weathering rates based on geochemical properties of the soil. In: Pulkinene E (ed) Proc Conference on Environmental Geochemistry in Northern Europe. Symp Geol Surv Finland, 34:45-61

Paçes T (1973) Steady-state kinetics and equilibrium between ground water and granitic rock. Geochim Cosmochim Acta 37:2641-2663

Paçes T (1978) Reversible control of aqueous aluminum and silica during the irreversible evolution of natural waters. Geochim Cosmochim Acta 42:1487-1493

Paçes T (1983) Rate constants of dissolution derived from the measurements of mass balances in catchments. Geochim Cosmochim Acta 47:1855-1863

Paçes T (1985) Sources of acidification in central Europe estimated from elemental budgets in small basins. Nature 315:31-36

Plummer NL, Wigley TML, Parkhurst DL (1978) The kinetics of calcite dissolution in CO_2-water systems at 5°C to 60°C and 0.0 to 1.0 atm CO_2. Am J Sci 278:179-216

Renberg I, Hellbert T, Nilsson M (1985) Effects of acidification on diatom communities as revealed by analyses of lake sediments. In: Andersson F, Olsson B (eds) Lake Gårdsjön—An acid lake and its catchment. Ecological Bull 37:219-223

Rietz F (1995) Modeling mineral weathering and soil chemistry during post-glacial period. Reports in Ecology and Environmental Engineering, 1, Dept Chemical Engineering, Lund Univ, ISRN LUTKDH/TKKT-3004-SE

Sverdrup H (1990) The Kinetics of Base Cation Release Due to Chemical Weathering. Lund Univ Press

Sverdrup H, Bjerle I (1986) Dissolution of calcite and other related minerals in acidic aqueous solution in a pH-stat. Vatten 38:59-73

Sverdrup H, Warfvinge P (1988a) Assessment of critical loads of acid deposition on forest soils. In: Nilsson J (ed) Critical Loads for Sulphur and Nitrogen. Nordic Council of Ministers and United Nations Economic Commission for Europe (ECE), p 81-130

Sverdrup H, Warfvinge P (1988b) Chemical weathering of minerals in the Gårdsjön catchment in related to a model based on laboratory rate coefficients. In: Nilsson J (ed) Critical Loads for Sulphur and Nitrogen. Nordic Council of Ministers and United Nations Economic Commission for Europe (ECE), p 131-150

Sverdrup H, Warfvinge P (1985) Upplösning av kalksten och andra neutralisasjonsmedel i mark. Naturvårdsverket, Rapport 3311

Sverdrup H, Warfvinge P (1988c) Weathering of primary silicate minerals in the natural soil environment in relation to a chemical weathering model. Water, Air, Soil Pollution 38:387-408

Sverdrup H, Warfvinge P (1991) On the geochemistry of chemical weathering. In: Rosén K (ed) Chemical Weathering under Field Conditions. Report on Forest Ecology and Forest Soils 63:79-118. Swedish Univ Agricultural Sciences, Uppsala, Sweden

Sverdrup H, Warfvinge P (1993) Calculating field weathering rates using a mechanistic geochemical model—PROFILE. J Appl Geochem 8:273-283

Sverdrup H, Warfvinge P, Bjerle I (1986) Experimental determination of the kinetic expression for the dissolution of K-slag, a tricalcium silicate, in acidic aqueous solution at 25°C and in equilibrium with air, using the results of pH-stat experiments. Vatten 42:210-217

Sverdrup H, Warfvinge P, Blake L, Goulding K (1995b) Modeling recent and historic soil data from the Rothamsted Experimental Station, England, using SAFE, Agriculture, Ecosystems and Pollution 52:223-259

Sverdrup H, Warfvinge P, Janicki A (1992) Mapping critical loads and steady state stream chemistry in the state of Maryland. Environmental Pollution 77:195-203

Sverdrup H, Warfvinge P, Johansson M (1992) Critical loads to forest soils in the Nordic countries. Ambio 5:348-355

Sverdrup HU, de Vries W, Henriksen A (1990) Mapping Critical Loads. Nordic Council of Ministers, Geneva. Miljörapport 1990:15, Nord 1990:98

Swalin R (1972) Thermodynamics of Solids. John Wiley & Sons, New York

Tamm C, Hallbäcken L (1985) Changes in soil pH over a 50-year period under different canopies in SW Sweden. In: Martin HC (ed) Int'l Conference on Acid Precipitation, Muskoka, Canada, Part 2. Riedel, Dordrecht, Netherlands, p 337-341

Velbel M (1985) Geochemical mass balances and weathering rates in forested watersheds in Southern Blue Ridge. Am J Sci 285:904-930

Velbel M (1986) The mathematical basis for determining rates of geochemical and geomorphic processes in small forested watersheds by mass balance. Examples and implications. In: Coleman S, Dethier D (eds) Rates of Chemical Weathering of Rocks and Minerals, p 439-451. Academic Press, New York

Vries W, Kros J (1989) The long term impact of acid deposition on the aluminum chemistry of an acid forest soil. In: Kämäri J, Jenkins A, Brakke DF, Wright RF, Norton SA (eds) Regional Acidification Models, p 231-255

Warfvinge P (1988) Modeling acidification mitigation in watersheds. PhD dissertation, Inst Technology, Lund Univ, Lund, Sweden

Warfvinge P, Falkengren-Grerup U, Sverdrup H (1993) Modeling long-term base cation supply to acidified forest stands. Environ Pollution 8, in press

Warfvinge P, Holmberg M, Posch M, Wright R (1992d) The use of dynamic models to set target loads. Ambio 5:369-376

Warfvinge P, Sverdrup (1992a) Calculating critical loads of acid deposition with PROFILE—A steady-state soil chemistry model. Water, Air, Soil Pollution 63:119-143

Warfvinge P, Sverdrup H (1992b) Hydrochemical modeling. In: Warfvinge P, Sandén P (eds) Modeling Acidification of Groundwater. SMHI, Norrköping, Sweden

Warfvinge P, Sverdrup H (1992c) Scenarios for acidification of groundwater. In: Warfvinge P, Sandén P (eds) Modeling Acidification of Groundwater. SMHI, Norrköping, Sweden

Wickman T, Jacks G (1992) Strontium isotopes in weathering budgeting. In: Kharaka Y, Mast A (eds) Water-Rock Interaction, p 611-614. AA Balkema, Rotterdam

Wynne-Jones WFK, Eyring H (1935) The absolute rate of reaction in condensed phases. J Chem Phys 3:492-502

Chapter 12

RELATING CHEMICAL AND PHYSICAL EROSION

R. F. Stallard

U.S. Geological Survey
3215 Marine Street
Boulder, CO 80303 U.S.A.

INTRODUCTION AND BACKGROUND

Two great classes of material move through river systems—solutes and solids. This discussion is dedicated to analyzing the processes that relate solute transport to solid transport under natural conditions and under the influence of human activities. This topic has received surprisingly little recent attention from geochemists, despite its importance in discussions a quarter of a century ago. The concept was embodied in the title of one of the great geochemical texts, Garrels and Mackenzie's (1971a) "Evolution of Sedimentary Rocks," in which the chemical evolution of the major elements of the Earth's crust through geologic time was examined. Many geomorphologists working at that time were concerned with the degree to which tectonic processes and human activities had affected the discharge of sediments from rivers into the oceans (Schumm, 1963; Judson, 1968; Holeman, 1968; Meade, 1969; Ahnert, 1970). Geochemists took an active part in the interpretation of these observations (Garrels and Mackenzie, 1971b). Among geologists and geomorphologists the interest has continued (Stallard, 1995).

Subsequent time saw the enthusiastic embrace of small-watershed studies by ecologists and geochemists. Such studies permitted the examination of the complex array of processes that interact to control the composition and discharge of substances exported from small watersheds. Sediment discharge was never a major focus of these studies, because of two factors. First, unlike solutes, for which the bulk of transport is during low to normal flow, most solid transport is at high to extremely high flow. High flows arise during "events"—typically, major storms and snow melt. Event sampling is often difficult, if not dangerous, and automatic samplers are best used. Second, sediments are difficult to sample representatively.

Funding for sediment research is largely derived from the needs to mitigate sedimentation or erosion problems such as the reduction of reservoir capacity to unacceptable agricultural-soil loss. The watersheds examined tend to be larger than those studied by landscape ecologists and geochemists. As such, they differ from the small watersheds in one fundamental aspect—considerable quantities of sediment are stored in larger watersheds. Sites of storage include the base of hillslopes (colluvium), within channels and floodplains (alluvium), and in lakes and reservoirs (lacustrine sediments). For solutes, storage is typically not so important.

Sediments and the geologic record

Geologic history is recorded in sedimentary rocks. The record takes three principal forms—fossils, chemical sediments, and clastic sediments. Fossils embody considerable information about past environments, especially if one can develop hypotheses based on modern analogues and population models. Fossils work best for the Phanerozoic, or roughly the last 550 million years. Chemical sediments record information about the composition of the water in which they were deposited. The information is in the form of the mineralogy of the sediment and of the isotopes and trace elements incorporated into the sediment. This record gets obscured by chemical changes after deposition—diagenesis. With increasing geologic age, the diagenetic overprint becomes a more serious impediment to interpreting the

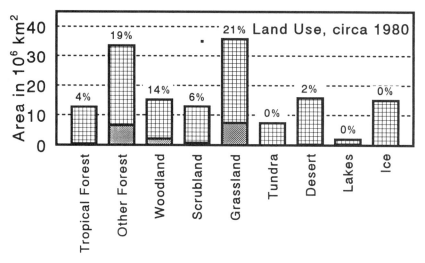

Figure 1. Histograms of different land/vegetation associations on the Earth's surface, circa 1980. The lower, shaded, part of each bar indicates the area of each land/vegetation type that was under cultivation or urbanization, but not grazing. The upper, gridded, part is either natural or grazed. The percentage of modified land is indicated above each bar. Data are from Matthews (1983).

record. Chemical sediments are typically deposited from more concentrated solutions such as seawater and saline lakes. Carbonate, sulfate, and chloride minerals form the bulk of these sediments. The most important chemical sediments of freshwater environments are organic materials—peats and coal. Clastic sediments are formed from deposits of solid particles commonly transported by water, wind, and glaciers. Most sedimentary rocks are clastic. These record information about the erosional environment in their composition and their depositional environment in their texture and style of deposition. The interpretation of the clastic record is complicated by hydrologic sorting, intermediate storage during transport, and diagenesis.

The present is unusual

The present is unusual. If we are to use the present as a key to the past, we must disentangle features particular to the present from those applicable to the past—even immediately prior to technological time starting 7,000 years BP. Although early civilizations affected erosion in the Mediterranean region and southern Asia for several millennia (Judson, 1986; Lowdermilk, 1975; Hillel, 1991), the dramatic growth of the human population in the past two centuries has resulted in a marked alteration of the appearance of the Earth's surface (Fig. 1, Hooke, 1994). This was preceded by two to three million years of major glaciations. Such episodes of glaciation appear to have affected the Earth only three or four times during the entire Phanerozoic. Finally, we have been affected by two particularly major orogenies, the Himalayan and Andean uplifts, and the closing of the Isthmus of Panama within the past few tens of millions of years.

The erosion of solids is seldom just a physical process. Most rock is a hard, cohesive material in which mineral grains are bound together either through intimate chemical contact or by means of a mineral cement between grains. Only a few geologic materials are so weakly bound that they crumble when exposed to the actions of rain and wind. These include some unconsolidated mudrocks, loess, volcanic ash, colluvium, alluvium, lacustrine sediments, beach sands, and soil. Unconsolidated coarse material may not erode, because wind and flowing water may not be able to lift individual particles. Some erosional processes act largely through physical phenomena. Glaciation, wind erosion, channel erosion, frost weathering, rock slides, and debris flows are examples of physical phenomena that disaggregate or move large quantities of material.

In many environments, once vegetation becomes established, even on the most unconsolidated geologic material, rates of physical erosion diminish dramatically. Paucity of vegetation appears to be required for high erosion rates. For example, the Yellow River of China transports $1.1 \cdot 10^9$ tonnes yr^{-1} of sediment to the ocean (Milliman et al., 1987). The average concentration of suspended sediment is 18 kg m^{-3}. Rapid erosion of the loess plateaus of interior China generates the sediment. This erosion is not a natural state of affairs. Milliman et al. (1987) reconstruct sediment discharge for the Yellow River using 5,000 years of written records, stratigraphic work on the river floodplain, and cores from the South China Sea. From 10,000 years ago to 7,000 years ago discharges were 20 percent modern. From 7,000 years ago to 3,000 years ago rates dropped to 10 percent modern as the loess plateau became heavily vegetated. Early human occupation and perhaps climate change increased discharge to 35 percent modern by 200 BC. Discharge then rose rapidly to 80 percent modern, and remained high until the Mongol invasion. A drop in sediment discharge to 50 percent modern followed. At 600 AD, intensive agriculture was reestablished and sediment discharge rose to the modern rate.

PROVIDING A FRAMEWORK: STEADY-STATE EROSION

To erode, most geologic materials must disaggregate into particles fine enough to be moved by water or wind. Wetting alone may weaken poorly consolidated sediments, because it causes clay minerals to swell. Many rocks will not erode, however, unless the rock is first chemically degraded. During chemical weathering, minerals within the rock react with water and water-borne ions to produce dissolved and solid weathering products. Dissolved products are usually removed as river-borne solutes. Solid products are occasionally removed rapidly, but more often they remain in a developing soil. Eventually, a combination of physical and biological processes moves the solids downslope and out of the local weathering environment. Chemical processes determine the rate at which weathered solids are produced. Physical and biological processes, by removing partially weathered material from the local environment, determine how long and to what extent the weathering proceeds. Moreover, the development of soil controls the access of water and reactive ions to reactive minerals within the soil and bedrock. We see a complex interplay of chemical, physical, and biological processes as a prerequisite to generating weathering products and controlling the partitioning of elements among the various phases produced during weathering.

Working hypotheses

Steady-state erosion is closely linked to the concept of an equilibrium landscape (Stallard, 1995). The core idea is that the geomorphic properties of the landscape are not changing substantially through time. The appearance may change—rivers may shift; there may be debris flows; forest fires may burn—but in a statistical sense, average measures of landscape form hold constant. Of course, erosion is an ongoing irreversible process. Unless uplift, driven by tectonic processes, balances erosion, we know that the statistical properties of the landscape must change. Moreover, erosion is a slow process compared with rates of climate change, and we know that many features of contemporary landscapes are inherited from previous climate regimes. Thus, a steady-state landscape is an idealized condition. Nevertheless, erosion under steady-state conditions has characterizable properties that provide a useful frame of reference for studying the earth today.

In this discussion, I examine the implications of a simple working hypothesis—that the production of fine, loose solid materials for physical erosion requires certain chemical reactions. Steady-state erosion implies a chemical mass balance. The sum of the fluxes of all solid and dissolved weathering products leaving a landscape should equal the mass of bedrock being degraded during weathering. In other words, given rates of chemical weathering and the composition of weathering products, one can predict steady-state rates of physical weathering.

Collection of representative samples

The measurement of material fluxes from a landscape is a complex endeavor. Typically, material discharges from river catchments are measured. These discharges are divided by the area of the catchment to calculate yields that have units of mass area^{-1} time^{-1}. The term "denudation rate" is used to describe these yields, rather than "erosion rate." Denudation rates embody the idea that eroded material may be stored and released from storage as it moves down hillslopes and through river systems. Ideally, three components are measured—the dissolved load, the suspended load, and bedload. The best measurement site is a cross section in which the channel is well defined, has a stable bed, and is laterally well mixed. The last condition demands that sampling not be immediately downstream of tributaries.

For the bulk of material moving through rivers, the distinction between solutes and solids is clear, and an element is either an ion in solution or part of a crystal lattice. The elements that form the major solute ions (Na^+, K^+, Mg^{2+}, Ca^{2+}, HCO_3^-, CO_3^{2-}, Cl^-, SO_4^{2-}) are in this category. For some elements, no clear distinctions exist between solute and solid. Instead, there is a gradual transition from truly dissolved forms that can be characterized through physical chemistry, to large floppy molecules or especially fine colloids, on to unambiguous particles. Major elements in this category include silicon (at high concentrations), carbon (in organic molecules), iron, and aluminum.

Sediment sampling has become quite sophisticated (Nordin, 1985). The near-surface "grab" samples typically used for solutes undersample coarser sediment. Today, perhaps the best sampling procedure, the "equal-transit-rate, depth-integration method," uses bottles or bags having "isokinetic" nozzles (Meade and Stevens, 1990; Moody and Troutman, 1992). The nozzle has an opening with a slightly expanded taper designed to admit water at a velocity that is the same as the ambient flow velocity. This minimizes the shear that promotes size segregation of particles entering the sampler. The sample enters either a vented bottle or a loose bag designed to minimize the static pressure gradient through the nozzle. The sampler fills at a rate that is proportional to the flow velocity. Samples are collected across the channel at equal-width increments so that the effects of lateral inhomogeneities are reduced. Each width increment is called a "vertical." The sampler is lowered and raised at a constant rate, the same at each vertical. This serves to reduce the effects of vertical inhomogeneities, especially those caused by the settling of coarser particles. In the end, the sample is a velocity-weighted average of the water moving through the cross section—a so called "representative sample."

The dissolved load is the simplest discharge to measure. Settling is not a problem, and well-mixed cross sections can be sampled at a single point. Concentrations of solutes decrease with increasing discharge. Consequently, the bulk of solute transport typically is during times of low to intermediate water discharge. Rivers in aridlands (little low to intermediate flow) and carbonate terrains (concentrations do not drop strongly with increasing discharge) are important exceptions.

Suspended sediment from an isokinetic sampler includes solids of all sizes—colloids (50± nm to 0.2± μm), clay (0.2± to 2 μm), silt (2 to 63 μm), and sand (63 to 2000 μm). In practice, a sieve is used to separate sand from the sample before filtration. The separation between truly dissolved material from particles is defined operationally. Ideally, molecular filtration would be used (Koehnken and Stallard, 1988), but vacuum and pressure filtration are often more practical. I use 0.2 μm or 0.45 μm filters that have been "pre-clogged" by filtering and discarding a volume of the sample. The process is slow, but the composition of the filtrate is similar to samples filtered through a molecular system. Concentrations of suspended solids increase with discharge, and typically the bulk of solid transport is during episodes of high water discharge—"events." Determination of solid concentration is operational as well. For example, the concentration of bedrock-derived material is best determined after heating the sample to 900°C to convert the sample back to elemental oxides.

The difficulties in accurately determining bed-material transport are a significant source of error in studies of mass-discharge in rivers. Coarse material moves by rolling and bouncing along of the river bed. Often movement is confined to narrow zones, and it is especially episodic. Bed-material transport is measured using a separate class of samplers that collect material moving on the bed. Very complicated schemes, such as cross-channel conveyor belts, have been used to sample the entire bed-material transport of a river (Nordin, 1985). These procedures are utterly impractical because of their cost. Instead, formulas developed on the few rivers that have been studied in detail are applied to rivers undergoing study. These formulas (Stevens and Yang, 1989) show that periods of high discharge are especially important to bed material transport. Bed-material transport varies from null to more than half the total sediment transport, depending on the river. To estimate bed-material transport, a crude convention of using 10 percent of the suspended load transport has evolved.

Caveats and corrections

The determination of solute discharge attributable to chemical weathering requires corrections. Solutes transported through the atmosphere or derived from human activities may be important. The correction for atmospheric solutes is commonly called a "cyclic-salt correction." The name derives from the assumption that on the scale of the continents, one must correct for salts that cycle from the ocean onto land only to return to the ocean in river water. For smaller individual catchments, solutes in rain are not necessarily from the ocean. Instead, adjacent lands may be a source of solutes as may be the watershed itself. Determination of the atmospheric flux of solutes moving into and out of a watershed is problematic. An estimate based on use of solute ratios in rain relative to a reasonably conservative constituent such as chloride is an approach that is often successful. In continental interiors, calcium, sulfate, nitrate, and ammonia dominate the atmospheric inputs, whereas near the coast, atmospheric solutes look like seasalt with added sulfate, nitrate, ammonia, and sometimes calcium.

Humans transport large quantities of chemicals into populated watersheds. Domestic activities, fertilizer use, road salting, and industrial wastes all contribute. Most of the inorganic solutes contributed by human activities are simple salts that can be easily subtracted from chemical budgets. In some regions, good relationships exist between watershed population and solute discharge (Ceasar et al., 1976). Thus, human loading can be calculated from a per person contribution for each solute and census information (Meybeck, 1979). Little evidence exists that human conversion of land from a natural to a developed state strongly affects rates of chemical weathering of silicate rocks (Stallard, 1995). Much of chemical weathering is deeper in the soil profile than are the direct effects of clearing. Acid rain and the effects of heavy irrigation and water diversion are exceptions to this condition. Sodium and silica, which are not strongly accumulated in biological or ion-exchange reservoirs, seem to be especially insensitive to the effects of the destruction of biomass and loss of soil.

Steady-state denudation definitions

The instantaneous mass discharge of any substance, D, is the product of water discharge, Q (units = volume time^{-1}), and the average cross-section concentration, [D] (units = mass volume^{-1}). The average discharge is best estimated by integrating several years of instantaneous discharge. Average total mass discharge of river-borne material, q_t (units = mass time^{-1}), is the sum of solute (dissolved), q_d, and solid, q_s, discharges. At a steady state the total of mass material transported out of the catchment should be the mass of material derived from bedrock, q_b.

$$q_t = q_d + q_s \underset{\text{state}}{\overset{\text{steady}}{=}} q_b \tag{1}$$

Similarly, the discharge of any element, D, can be partitioned between solute and solid transport. D is a concentration (units = mass fraction) relative to the total discharge of all similar (solute, solid, total) bedrock-derived material being transported. Thus D_d = [D]÷[total dissolved bedrock]. In this discussion, concentrations are percent oxides. Thus, D_d is a percent oxide concentration, and the sum of all D_d is 100 percent. An implicit assumption is that carbonates, sulfates, and reduced iron and sulfur compounds are not complicating factors in the watersheds being modeled:

$$D_t q_t = D_d q_d + D_s q_s \underset{\text{state}}{\overset{\text{steady}}{=}} D_b q_b \tag{2}$$

The partitioning of elements among dissolved and solid phases during weathering controls the composition of the solutes, river-borne solids, and soils. The ratio of bedrock that dissolves to total bedrock that is weathered, W, is a measure of the "intensity" of chemical weathering. Similarly, for a given element, we can define a W_D, which varies between zero (for an element that does not dissolve) and one (for an element that dissolves completely). I was in part responsible for applying the name "intensity" to W_D (Murnane and Stallard, 1990) but in retrospect "efficiency" may have been better:

$$W = \frac{q_d}{q_t} \; \text{(a)}, \quad \text{and} \quad W_D = \frac{D_d q_d}{D_t q_t} = \frac{D_d}{D_t} W \; \text{(b)}. \tag{3}$$

A major problem in the application of the above equation is that measurement errors are large. The measurement error of water discharge going into the estimate is typically about 10 percent. The error from interannual variability of discharge may be even larger. In addition, the difficulty of measuring solid transport, especially bed-material transport, contributes additional error. Because particles undergo sorting based on their size, density and propensity for sticking to other particles and surfaces, it is quite difficult even to get a representative sample of solid material.

Elemental ratios in a given sample are easy to measure. For any given sample, the ratio of discharge of two elements in a sample is based solely on analytical errors. Discharge errors do not contribute. The solid and solution ratios of two elements, C and D, are related through the degree of leaching, W_C, of one of the elements:

$$\frac{D_t}{C_t} = W_C \frac{D_d}{C_d} + (1 - W_C) \frac{D_s}{C_s} \quad \overset{\text{steady}}{\underset{\text{state}}{=}} \quad \frac{D_b}{C_b}. \tag{4}$$

The above equation can be rearranged to solve for W and W_C. Use of subscript b, for bedrock assumes the existence of steady state.

$$W_C(C,D) = \left(\frac{D_b}{C_b} - \frac{D_s}{C_s}\right) \div \left(\frac{D_d}{C_d} - \frac{D_s}{C_s}\right) \quad \text{and} \quad W(C,D) = \frac{C_b}{C_d} W_C(C,D) \tag{5}$$

In this form, $W_C(C,D)$ is expressed as ratios that one can hope to measure. Two special "end-member" cases are worth examining. In one, D is insoluble, in the other, D is totally soluble.

Zirconium is a good example of an insoluble element. The solubility of crystalline zircon (as opposed to metamict zircon) is virtually immeasurable at room temperature. If we suppose that D is insoluble, then

$$D_t q_t = D_s q_s. \tag{6}$$

W_C can be expressed in a form involving only solid phases:

$$W_C = 1 - \frac{D_b}{C_b} \cdot \frac{C_s}{D_s} \tag{7}$$

The fundamental problem in applying this equation is that hydrodynamic sorting makes it very difficult to estimate elemental ratios in the solid load. In particular, zircons are hard and dense and sort into a "heavy mineral" fraction that does not move with the bulk of the solid load.

Philosophically, Equation (7) is interesting. A valid argument is that at steady state, processes that remove solids in turn control the composition of solutes that are eroded. It is physical transport that removes reactive solids from watersheds, and physical processes enhance or impede the access of water to reactive solids. To understand solute transport, one must disentangle the relative role of chemical, biological, and physical processes in exposing solids to water, isolating solids from water, and removing solids from a watershed before they completely weather.

In the second end-member case, D is totally soluble, and the above equation simplifies into a form involving only bedrock and solute ratios.

$$W_C = \frac{C_d}{D_d} \cdot \frac{D_b}{C_b} \tag{8}$$

In geologically simple catchments, one can reasonably determine solute and bedrock ratios, and Equation (8) can be used for soluble components to estimate bedrock weathering rates.

Hillslope processes

To identify phenomena that control partitioning of elements among solid and solute phases, emphasis must be given to those processes that mobilize solids and solutes such that they may be moved out of the catchment. Unlike solids, solutes can be, and are typically, transported from deep in the soil profile by moving water. Solutes may be stored for short periods in biomass and in soils as exchangeable ions and organic complexes. Biomass storage can vary considerably, depending on such factors as seasonal growth, burning and regrowth, blight, and harvesting. Ion exchange in soils is affected by vegetation, wetness, and human factors such as acid precipitation. Under steady state only a small part of these reservoirs is exported from a catchment as solids. Often, a tight cycle is established in which nutrients are absorbed by vegetation to be released by decay only to be reabsorbed, once again. In dry environments, solutes can be reprecipitated as salts in soil Stallard(1992, 1995).

Erosion regimes

The concept of erosion regimes classifies the geomorphic processes that control the transport of weathered material down hillslopes (Stallard, 1985). It can be used to characterize soil genesis on hillslopes and to understand the composition and yield of dissolved and solid erosion products transported by rivers.

Erosion on hillslopes is characterized as a continuum between weathering limited and transport limited. In weathering-limited erosion, the capacity of transport processes to move material down hillslopes exceeds the generation rate of loose material by chemical and physical weathering. Under weathering-limited conditions, the capacity of transport processes exceeds the rate of generation of loose material. Under weathering-limited conditions, without vegetation, little fine loose material would remain on a slope. Vegetation anchors soil and allows it to develop. The development continues until mass-wasting processes strip off non-cohesive soil. These processes include landslides, erosion following tree throws, and deforestation (including natural deforestation by fires and blight). When vegetation is effective in anchoring loose material, soils would be expected to be thicker.

Given enough time with a stable climate, a dynamic equilibrium may be reached on the hillslope. Mass is lost from the ground surface by physical erosion and from throughout the soil profile by chemical erosion. This lost mass is balanced by material newly derived from the weathering of bedrock. If mass wasting is relatively steady, the appearance of a soil profile would not change significantly through time. By analogy, the soil would appear much like an assembly line running in reverse. Steady-state soils would have no clearly defined age because there is no obvious time when unweathered bedrock was presented at the ground surface. Similarly, the soil surface has no special meaning. Material at the surface has been exposed by transport processes and will be removed by the same processes. This is a very Huttonian view of soil: "no evidence of a beginning no prospect of an end."

Under the contrasting regime, transport-limited erosion, the rate of generation of weathered material from bedrock exceeds the capacity of physical transport processes to remove that material. Soils continue to develop until homeostatic mechanisms (development of considerable thickness or formation of impermeable horizons) reduce the rate of weathering by restricting access of water to fresh bedrock. Transport-limited erosion typically occurs in nearly flat landscapes—alluvial terraces and erosion surfaces. Erosion surfaces can be very ancient (Stallard, 1988). For example, in South America and Africa the oldest surfaces may be 150 million years old. Exploratory studies of bauxite on these surfaces show that the grade of bauxite increases with increasing age of the soil, indicating continued development of the soil. Clearly, this is not a steady state.

In transport-limited regimes the concept of soil age has significance. For alluvial soils, the clock begins with the deposition of the alluvium. The starting date is more ambiguous for erosion surfaces—it is presumably the time when erosion went from weathering limited to transport limited (with landform evolution as formulated by King, not Davis—see Stallard, 1988). With a sufficiently old erosion surface, this date becomes reasonably well defined.

For solids, we have to concern ourselves with the details of the processes that mobilize particles such that they can be transported down hillslopes into rivers. The range of processes that anchor or mobilize particles is truly vast, and the role of organisms is often critical.

Roles of organisms in physical erosion

Many observers visualize physical erosion, especially in vegetated regions, as a surface process. As bedrock weathers, some primary minerals dissolve completely; others break down into secondary solid phases and solutes, while others fail to react. The solids remain in place, and solutes are transported away from the site of reaction by flowing soil waters. With time, unreacted minerals start to break down, and secondary phases reacted further. Meanwhile, physical erosion removes solids from the top of the profile. As physical erosion lowers the surface, chemical weathering lowers the base of the profile. With a steady state, the distance between the surface and the base would remain fixed. This "conveyer-belt" model is unrealistic. Bioturbation frequently completely reworks the top meter of the soil profile, and energetic processes, such as debris flows and gully formation, excavate deeply into soils.

In most environments, the top meter of soil is very dynamic, with vigorous bioturbation by roots and fungi are important. These penetrate into, grow in, and chemically interact with the soil. Once roots decay, voids are formed that often fill with soils. Soils are overturned when trees topple (tree throws). Earthworms eat soil to digest organic debris. Many animals burrow into soils for shelter and to establish colonies. Burrows form voids that eventually refill. Typically, organisms have an upper size limit for material that they can excavate. Coarse material that is excluded settles to the bottom of the bioturbated layer forming stone lines. On slopes, tree throws and animal excavations leave loose material on the surface. Based on my observations, this material easily washes down the slope.

Soil bioturbation is rapid. For example, Piperno (1990) observes that throughout a summit plateau in Barro Colorado Island in Panama artifacts from a human occupation 400 to 600 years ago are buried under 10 to 20 cm of soil. The soil is a normal tropical soil with a 7- to 10-cm thick A horizon. Because this plateau is the height of land, with no upslope sources of sediment, biological overturn has probably brought deep soil to the surface at a rate of about 1 cm in 20 to 40 years. The most likely processes are tree throws (Putz and Milton, 1982) and excavations by leaf-cutter ants. Similarly, in tropical soils of the Luquillo mountains of Puerto Rico, bioturbation homogenizes in situ-produced ^{10}Be in the top half meter of the soil profile (Brown et al., 1995).

Deep erosion

Several erosional phenomena excavate deeply, stripping through the soft, less cohesive parts of the soil profile. Vegetation appears to protect the soil surface from physical erosion. In forests and grasslands, ground cover and root systems anchor the loose material. Rapid physical erosion caused by sheet wash and gully formation appears to follow the removal of this cover by natural or by human activities. In watersheds where human activities have removed the cover (see discussion in Meade et al., 1990), rates of physical denudation typically increase 10- to 100-fold. Increased denudation does not imply an increase in chemical weathering or supply of material. Instead, erosion is excavating soils that have developed for thousands to millions of years. Deep physical erosion can also operate in naturally vegetated settings. It is often an episodic process that requires the removal of

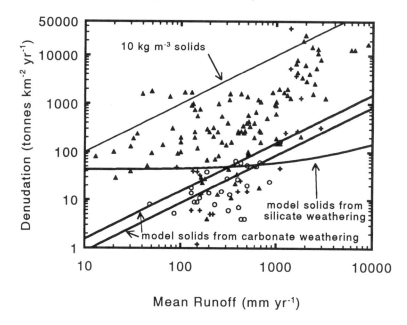

Figure 2. Sediment yields compared with runoff for rivers that drain into the ocean throughout the world. Data are from Milliman and Syvitski (1992). Symbols combine categories from Milliman and Syvitski (1992); open circles—coastal plain (0-100 m headwaters) and lowland (100-500 m); pluses—upland (500-1,000 m); solid triangles—mountains (1,000-3,000 m) and high mountains (> 3,000 m). The model lines for model steady-state solid yields for carbonate and silicate weathering are derived in the text. The silicate curve uses the data set presented in Figure 3.

vegetation by phenomena such as fire and blight, or it may require the attainment of a threshold of slope instability to initiate erosion. Debris flows are in the latter category.

Debris flows that involve surface horizons are quite common in regions that experience intense rainfall and earthquakes. Larsen and Simon (1993) summarize landslides throughout the humid tropics. Landslides are triggered when rainfall exceeds about 200 mm. Larsen and Torres (1995) have studied landslides in eastern Puerto Rico using several decades of aerial photography. Under natural conditions on slopes that are greater than 12 degrees, landslides strip about 1 percent of the land surface per century. The average slide is about 200 m². The authors note that rates of sliding dramatically increase when humans clear land for agriculture, road building, and construction projects. These observations were supported by a study of in situ-produced [10]Be in quartz transported by a river in the study region (Brown et al., 1995). Coarse quartz in the river load had to be derived from deep in the soil profile, whereas the fine quartz had to be largely from the bioturbated zone. Strong earthquakes also generate landslides that strip much of the soils off of mountainous landscapes (Keefer, 1984).

Stallard (1985) examined the role of deep erosion in controlling the chemistry of rivers. Sodium is enriched relative to potassium and calcium is enriched relative to magnesium in rivers that drain mountainous regions. The explanation is that the bulk of sodium and calcium is removed as solutes, whereas potassium and magnesium are removed as solids. Potassium minerals (feldspars and micas) and magnesium-bearing phyllosilicates persist until much shallower parts of soil profiles. Surface soils, however, typically do not have an abundance of potassium and magnesium-bearing minerals. Erosional processes must excavate more

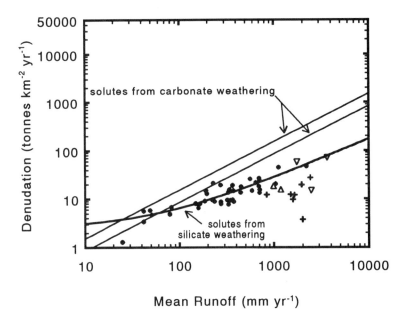

Figure 3. Solute yields for rivers that drain crystalline rocks. Solid circles—data from Kenyan rivers that drain crystalline basement from Dunne (1978); pluses—rivers that drain South American shields (Lewis et al., 1987; Lewis and Saunders, 1989a, 1990; Weibezahn, 1990); empty up triangles—large, geologically complex, mostly siliceous neotropical rivers (Stallard, 1980; Lewis and Saunders, 1989b); empty down triangles—Luquillo Mountains of Puerto Rico (McDowell and Asbury, 1994). The river with especially low yields in this figure and Figure 7 is from the Atabapo River in Venezuela, a river that drains a shield peneplain.

deeply to remove these phases. The composition of solid phases transported within the Orinoco River system is consistent with the greater stability of potassium-bearing and magnesium-bearing solids (Stallard et al., 1990; Johnsson et al., 1991).

Exceptional erosion

Geologic settings exist for which these weathering-regime models should not apply. These are environments in which chemical weathering is not required to generate solid phases. Examples include unconsolidated deposits of all types—alluvial sediments, sand dunes, loess, volcanic ash, and young marine sediments exposed by rapid uplift. In these environments, the development of a vegetation cover or a cohesive clay soil appears to limit erosion.

Milliman and Syvitski (1992) presented a global summary of the solid yields of rivers that enter the ocean. They observe good correlations between basin elevation and solid yield and between basin size and solid yield. The correlation between solid yield and relief is particularly strong if one distinguishes among rivers that drain old and new orogenic belts (Pinet and Souriau, 1988). The relation between runoff and solid yield is poor (Fig. 2). This contrasts strongly with solutes, which show good correlations between runoff and yield (Fig. 3; Stallard, 1995; Bluth and Kump, 1994).

The points that represent the greatest yields in Figure 2 presumably represent erosion of poorly consolidated materials minimally protected by vegetation. A line corresponding to a suspended solid concentration of 10 kg m^{-3} is shown. Lack of vegetation protection may be

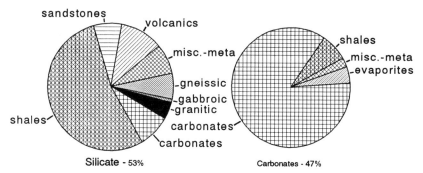

Figure 4. Relative removal of carbon dioxide from the atmosphere by the chemical weathering of various rock types. Estimates are from Meybeck (1987, Table 6, "atm. CO_2"). The left pie, representing silicate weathering, is 53 percent of the total. The right pie, carbonate weathering, is 47 percent of the total.

caused by human activities, harshness of climate, or inability for an adequate cover to be established. If we assume a bedrock density of 2.6 tonnes m^{-3}, the greatest solid yields in Figure 2 correspond to denudation rates of about 0.1 m yr^{-1}. Sustaining such rates for long periods would be difficult.

Glaciers are powerful producers of sediment; however, chemical weathering appears to have little significance in the production of this sediment. Glacial erosion is most effective when the base of the glacier is at the melting point of water—"warm-based" glaciers (Sugden, 1978). The style and efficacy of the erosion process depend on whether the glacier has incorporated debris, on whether the base is warming and melting or cooling and freezing, and on the speed of the glacial flow. In contrast, cold-based glaciers are weak erosion agents. In some areas of the Laurentide Ice Sheet, pre-glacial soils and tree trunks were preserved under cold-based ice. Bell and Laine (1985) estimate that the Canadian Shield was lowered about 120 m during the Pleistocene. During the same period, 3 My, the present-day erosion on the Shield could lower the shield only about 40 m. The latter rate may be high because of the abundance of fresh glacial debris on the shield. Available data indicate that before glaciations the shield was deeply weathered (Stallard, 1992).

Clastics and carbonates

As stated above, the steady-state assumption implies that the combined discharge of bedrock-derived solutes and solids should equal the mass of bedrock weathered in a catchment. This should apply to each element. This proposition becomes interesting when some chemical reaction or suite of chemical reactions is necessary to generate loose material for physical erosion. We will first illustrate this with carbonate-cemented clastic rocks.

Carbonate weathering removes roughly half the carbon dioxide from the atmosphere by chemical weathering (Fig. 4). This carbon dioxide is removed as a fast cycle, because it is restored to the atmosphere when carbonates reprecipitate in the ocean. Most carbonates have little associated clastic material or biogenic opal (Blatt et al., 1980), because the turbidity associated with clastic sedimentation suppresses the biogenic activity required for carbonate deposition. A upper limit for clastics in typical carbonates is about 5 percent. Occasionally, carbonate cemented silicious sediments are deposited. Sedimentation in upwelling zones next to active orogenic belts is an example of such a setting. The siliceous components of the sediments are released as the carbonates dissolve. If the terrain is flat, siliceous soils develop over the carbonates, while on slopes, physical erosion limits soils to a thin veneer.

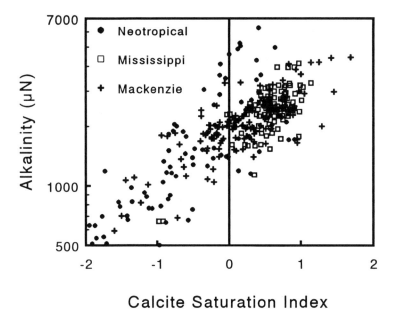

Calcite Saturation Index

Figure 5. Alkalinity compared with the saturation index for calcite. Symbols: solid circles— neotropical rivers (mostly from Amazon and Orinoco basins, rivers in Puerto Rico and Panama—Stallard and Edmond, 1983, and Stallard, unpublished); open squares— Mississippi basin (Garbarino and Taylor, unpublished); pluses—Mackenzie system (Reeder et al., 1972).

Only simple assumptions are required to calculate physical denudation rates for an environment in which carbonate dissolution controls the release of clastic material. The dissolution of calcium carbonates is quite rapid, and saturation with respect to local carbon-dioxide vapor pressure and temperature is typically established (Stallard and Edmond, 1983). In this discussion, it is assumed that the alkalinity in rivers near carbonate saturation represents a good general estimate of dissolved carbonate minerals. Solid yields can then be calculated from the mass concentration of siliceous material in the bedrock. When expressed in units of mass per volume, the solution concentration of dissolved $CaCO_3$, $[CaCO_3]$, times the runoff, R, gives the solute yield, Y_s:

$$Y_s(\text{g m}^{-2}\text{ yr}^{-1} \text{ or tonnes km}^{-2}\text{ yr}^{-1}) = [CaCO_3](\text{g m}^{-3}){\cdot}R(\text{m yr}^{-1}) \qquad (9)$$

When only calcite dissolution is contributing to solutes, the concentration of dissolved calcite is the alkalinity expressed as g m^{-3} $CaCO_3$. Usually, such units are ugly, because the weathering of silicates and dolomite may also contribute to alkalinity, but for this exercise, it is a practical approach.

The solid yield, Y_d, can then be calculated from the weight percent silicate concentration, C_{sil}. For convenience, it is assumed that the silicates do not undergo significant chemical weathering:

$$Y_d(\text{tonnes km}^{-2}\text{ yr-1}) = Y_s(\text{tonnes km}^{-2}\text{ yr-1}){\cdot}\frac{C_{sil}}{100 - C_{sil}} \qquad (10)$$

When carbonates are abundant in a watershed, their chemical weathering products dominate solute chemistry (Stallard, 1995). The saturation index, $SI_{calcite}$, is calculated from the solution activities of calcium ($a_{Ca^{2+}}$), carbonate ions ($a_{CO_3^{2-}}$), and the solubility constant for calcite (K_{sp}):

$$SI_{calcite} = \log\left(\frac{a_{Ca^{2+}} \cdot a_{CO_3^{2-}}}{K_{sp(calcite)}} \right) \tag{11}$$

The saturation index is greater than one when water is supersaturated with respect to calcite. When alkalinity is compared with the saturation index for calcite for rivers ranging from the tropics to the Arctic (Fig. 5), several features are noted. There is a tight cluster of slightly supersaturated samples, with only a few samples being greatly supersaturated. The supersaturation is thought to be caused by slight degassing of carbon dioxide from waters as they exit the weathering environment and enter surface waters. The cluster starts at a saturation index of -0.5. The geometric mean and standard deviation for alkalinity in this cluster is 2300±700 μN, or 115±35 g m^{-3} CaCO$_3$. The ranges of P_{CO_2} and temperature that correspond to this range of alkalinity is indicated as bold entries in Table 1. This provides sufficient information to apply Equations (9) and (10) to determine solid yields for rivers.

Table 1. Alkalinity[1] in μN (equals Ca^{2+} in μN) for samples equilibrated with calcite at the indicated temperatures and CO_2 vapor pressure ratios, sample / atmosphere.

Temperature, Celsius	Ratio of sample CO_2 vapor pressure to that of the atmosphere (350 ppmv)						
	10	21.5	46.4	100	215	464	1000
0	1487	**1920**	**2479**	3201	4134	5338	6894
5	1381	**1783**	**2302**	**2972**	3838	4957	6402
10	1280	1652	**2133**	**2754**	3556	4593	5931
15	1184	1527	**1972**	**2546**	3288	4246	5483
20	1092	1409	**1819**	**2349**	**3033**	3917	5058
25	1006	1298	1675	**2163**	**2792**	3606	4657
30	924.5	1193	1539	**1987**	**2566**	3314	4280
35	848.4	1094	1412	**1823**	**2354**	**3040**	3926

[1]Calculations were done using thermodynamic constants from Tables 3.1, 4.7, 4.8, 4.9, 5.1.III of Stumm and Morgan (1981), without an ionic strength correction.

[2]Numbers in bold are in the range of samples from neotropical rivers (mostly the Amazon and Orinoco systems), the Mississippi basin, and the Mackenzie basin.

The curves that predict solid yields for the erosion of carbonate rocks having 50 percent siliceous clastics (Fig. 2, labeled "model solids from carbonate weathering") separate fields defining mountainous (Solid up triangles) and flatland (open circles) river basins, as defined by Milliman and Syvitski (1992). Such a siliceous carbonate is unusual, but the separation of mountainous and flatland regions by a curve describing a steady-state erosion process has significance.

Weathering-limited erosion operates in mountainous regions. Under a steady-state, solid erosion rates would be controlled by the rate at which the carbonate matrix dissolves, and points corresponding to the weathering of clastic-rich carbonates should plot along the curve. Human activities might accelerate physical erosion and increase physical denudation rates, causing points to scatter above the steady-state curve. Rates of physical erosion of unconsolidated material are not limited by chemical weathering processes. Solids yields for such rivers should also plot above curves defining steady-state erosion of consolidated bedrock.

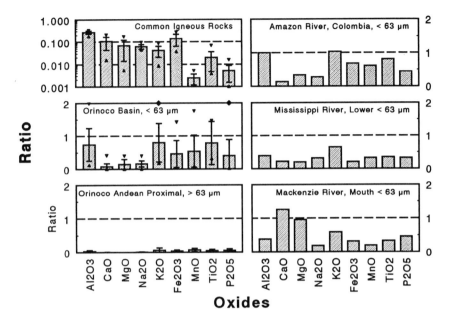

Figure 6. Ratios used in deriving equations described in the text. (Upper left panel) Ratios of elemental oxide concentrations for common igneous rocks to SiO_2. Analyses are from Le Maitre (1976). (Remaining panels) Ratio $(D_s/C_s)/(D_b/C_b)$, where C is SiO_2 and D represents other oxides. For < 63 μm sediments from throughout the Orinoco basin, and the Amazon, Mississippi, and Mackenzie Rivers, analyses are from Koehnken (1990). Analyses for sands near the Andes Mountains are from Johnsson et al. (1992).

In flatlands, transport-limited erosion should predominate. Physical erosion would be limited by two factors. Instead of being transported through rivers, solid weathering products would remain in developing soils. Developing soils would in turn limit the access of water to the carbonate minerals and the rate of chemical weathering would decrease. With less carbonate dissolution solid generation would further decrease. Eventually a steady state may be reached, but rates would be less than the steady-state curve for transport-limited conditions. Thus, the spread of data below the theoretical line is expected. The effect of accelerated erosion caused by human activities would be limited by storage of sediment in low-order rivers near the site of erosion.

Solid yields and silicate rocks

The bulk of clastic material transported by rivers is derived from the weathering of silicate rocks. Minerals in silicate rocks weather to solutes and secondary phases. At some point the original rock is sufficiently degraded that it can be mobilized by physical processes, assuming these are active. In the following discussion we examine steady-state physical erosion of igneous and metamorphic rocks assuming that erosion is weathering limited and that some unspecified silicate weathering reaction is required for the rocks to be softened for physical erosion.

Usually, leaching of soils and river-borne solids increases with increasing runoff. Two classic studies of soils in Hawaii (Sherman, 1952) and California (Barshad, 1966) describe general relations between clay minerals in soils and precipitation. Smectites predominate for precipitation less than 250 to 1,000 mm yr^{-1}, usually associated with carbonate accumulation—"caliche." For precipitation between 250 and 2,000 mm yr[-1] in California, kaolinite combines with illite (sialic rocks, precipitation < 1,000 mm yr[-1]) and vermiculite. In Hawaii, where bedrock is mostly basaltic lava, kaolinite and sesquioxides are important in the same range. Sesquioxides become important in all settings when precipitation exceeds 2,000 mm yr[-1]. At high precipitation, there are alternative scenarios depending on its seasonality. Hawaiian soils are ferruginous under seasonal conditions and aluminous in more uniformly wet environments. For siliceous bedrock in South America (Stallard, 1988), kaolinite is replaced by quartz when conditions are uniformly wet. Under more seasonal wetness, sesquioxides predominate. Differentiation, based on seasonality of wetness, may be related to the role of organic matter in mobilizing aluminum and iron in soil profiles (Stallard, 1988). With wet conditions, iron and aluminum are maintained as solutes by organic complexation or colloid stabilization. In lowland areas, waters are frequently undersaturated with respect to quartz, whereas in uplands, supersaturation is typical. For the major solute cations, these mineralogical trends suggest retention of Ca^{2+} in caliche under dry conditions and as an exchangeable cation where 2:1 clays are present. Magnesium would be retained as a lattice atom and an exchangeable cation in 2:1 clays and to a lesser extent in caliche. Potassium would be retained on interlayers in illites and vermiculites. Sodium would be weakly retained as an exchangeable cation on 2:1 clays, but should be considered mobile under conditions typical of most soils. The composition of clays in river-borne sediments is similar to those in the soils that the rivers drain.

The composition of river-borne sediments confirms the general mobility of sodium. The ratios of elemental oxides to silica in river-borne sediments (Fig. 6) show that the ratios of all oxides, except sodium are high in sediments in at least one large river. For example, the ratio of CaO/SiO_2 is high in sediments from the Mackenzie River because of abundant calcite. If we return to Equations (5) and (7), we note that sodium and silicon would be an ideal pair of elements for calculating $W(C,D)$ and $W_C(C,D)$.

$$W(Si, Na) = \frac{Si_b}{Si_d} \cdot W_{Si}(Si, Na) \approx \frac{59 \pm 16}{Si_d} \cdot \frac{0.052 \pm 0.015}{(Na_d \div Si_d) - 0.010 \pm 0.005} \qquad (12)$$

The constants in the above formulation are derived from the average of Na_2O/SiO_2 ratios for all suspended sediments from the Orinoco River system (Fig. 6). These samples represent rivers that drain a very broad range of bedrock geology. In addition, river-bed sands from the same locations (Fig. 6) show a still lower Na_2O/SiO_2 ratio for sand-size sediment. The average bedrock SiO_2 concentration of bedrock is based on the average and standard deviation for common igneous rocks (rock data are from categories 6-20 and 22-28 of Le Maitre, 1976). Note that Si_d and Na_d are corrected for atmospheric inputs and are expressed as oxide weight percent of total solute oxides. The above equation should have broad application to many geologic settings. It can be applied rigorously to rivers that drain common igneous and metamorphic rocks. In clastic sedimentary rocks, except rock-salt, Si_b tends to be greater than for igneous rock, whereas Na_b tends to be lower.

Information on the solute concentration of all solutes as a function of runoff is needed to calculate $Si_d(R)$ and $Na_d(R)$. Dunne (1978) provides a set of equations (Table 2) that accompany the data in Figure 3. These relate bedrock-derived solute concentrations to runoff. A fundamental problem with these equations is that the trend at high runoff violates features seen in high-runoff rivers draining igneous rocks in upland areas. The ratio of each major cation to silica is close to that estimated assuming the weathering of bedrock to kaolinite and iron sesquioxides, with quartz remaining stable (Stallard, 1995). Sodium and calcium are most thoroughly leached. The equations of Dunne (1978) depart significantly from this scenario, particularly for sodium.

Table 2. Solute concentration in tropical rivers on siliceous rock: solute concentration compared with runoff.

Element	Dunne (1978) Equations[1]: Coefficient units μM	Exponent	Equations[2] as modified in this paper Coefficient units μM	Exponent	Ratio[3]	Conversion from μM to g m^{-3} oxide Oxide	Conversion factor
Na	16529	-0.74	47.482	-0.813	0.236	Na$_2$O	0.03099
K	1990	-0.54	6.372	-0.672	0.105	K$_2$O	0.0471
Mg	954	-0.43	3.053	-0.601	0.2	MgO	0.04031
Ca	1427	-0.41	4.516	-0.589	0.224	CaO	0.05608
Si	864	-0.16	864	-0.16	0	SiO$_2$	0.06009

[1]Element=Coefficient·Runoff$^{(Exponent)}$

[2]Element=(Coefficient·Runoff$^{(Exponent)}$+Ratio)·Si$_{Dunne}$ (Runoff)

[3]Ratio is (Element÷Si for weathering average common igneous rock to kaolinite and iron sesquioxides, with quartz stable). Model conditions are assumed to apply to a runoff of about 4,000 mm, as is discussed in the text.

Dunne's (1978) equations can be easily modified to be consistent with the assumption that at high runoff, bedrock weathers to kaolinite and iron sesquioxides, with quartz being stable. For this example, I averaged common igneous rocks from Le Maitre (1976). The average closely resembles tonalite. The quartz concentration were averaged from the CIPW norms calculated by Le Maitre. The cation-to-Si(OH)$_4$ ratios are calculated for the model weathering scenario, assuming model conditions approximately met at 4,000 mm runoff. This runoff is the maximum for data in Figure 3, and based on my experience, model conditions are fulfilled.. In the following equation, coefficients A and B were determined by fitting model equations to solutions of Dunne's (1978) equations using least squares:

$$\log\left(\frac{Cation_{Dunne}(R=25..2500)}{Silica_{Dunne}(R=25..2500)} - \frac{Cation_{Dunne}(4000)}{Silica_{Dunne}(4000)} \right) = A + B \cdot \log(R) \qquad (13)$$

Because this is an exponential fit, the model ratio approached asymptotically.

Predicted solid yields

Equation (13) and Table 2 describing solute concentration as a function of runoff provide all the information needed to calculate solid yields from solute yields. Equation (12) gives $W_{Si}(Si_d(R),Na_d(R))$ and $W(Si_d(R),Na_d(R))$. Solute yield can be determined by summing all the equations for solute concentration as a function of runoff, and then multiplying by runoff. In doing this calculation, expression of solute concentrations as oxides in solution is essential, because the cations and silicon exist as oxides in the bedrock that is weathering. The steady-state solute yield, $Y_d(R)$, can be used to calculate a steady-state solid yield, $Y_s(R)$:

$$W(R) = \frac{Si_b}{Si_d(R)} \cdot W_{Si}(Si_d(R),Na_d(R)) \quad \rightarrow \quad Y_s(R) = Y_d(R) \frac{1-W(R)}{W(R)} \qquad (14)$$

The model curve for steady-state silicate weathering is shown in Figure 2 (labeled "model solids from silicate weathering"). Note that it appears to separate the flatland and mountain rivers as well as does the carbonate curve (labeled "model solids from carbonate weathering"). Dunne (1979) also published solid yields for the watersheds from which this model was derived (Fig. 7, "model solids from silicate weathering, 4000 mm"). Dunne (1979) distinguished among rivers where land was forested, where forests exceeded cultivation,

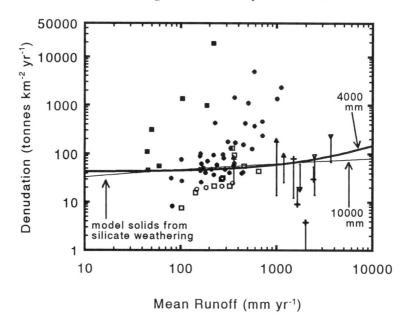

Figure 7. Solid yields for rivers in Figure 3. Data for Kenyan rivers (Dunne, 1979): open circles—forested catchments; open squares—forest > agriculture; solid circles—agriculture > forest; solid squares—grazing. Remaining symbols are the same as in Figure 3. The small dots connected to symbols by lines represent model predictions for individual rivers. The 4,000 mm and 10,000 mm versions of the model are discussed in the text.

where cultivation exceeded forests, and where land was grazed. Note that the steady-state line separates the first two types of land use (minimum human impact) from the other types of land use (maximum human impact). Data for other tropical watersheds are included as well. These watersheds have been minimally affected by human activities. Model solid yields calculated from observed solute yields for each of these rivers are also shown. These also plot along or below the model line. Those rivers that plot well below the model line drain extensive flat areas. As argued before, the inability to erode solid weathering products leads to increased soil thickness and subsequently to diminished chemical weathering rates and concomitant production of solid weathering products. Even where humans enhance physical erosion in relatively flat watersheds, the solids remain trapped as colluvium at the base of hillslopes or as alluvium in river channels.

This model curve should not be applied loosely. The model is sensitive to the choice 4,000 mm as the runoff for meeting model conditions (Fig. 7, "model solids from silicate weathering, 4000 mm"). Y_s (10000) is about one half, if 10,000 mm is chosen, instead (labeled "model solids from silicate weathering, 10000 mm"). In part, this instability reflects contradicting assumptions in Equation (12) and the fits to Dunne's (1978) model. In Equation (12), sodium is assumed to be in the solid load, whereas the fit assumes completely leaching of sodium. The sediment Na_2O/SiO_2 ratio used in Equation (12) is from rivers having about 1,000 mm runoff. When Equation (2) is applied to each element, solid load concentrations are slightly negative for Ca and K at high runoff. This can be controlled through use of a more complex regression procedure than the one used here. This is planned.

Process-related assumptions must also be considered. Stallard (1995) notes that silicate weathering in island-arc terrains may be acting several-fold faster than on cratonic terrains or

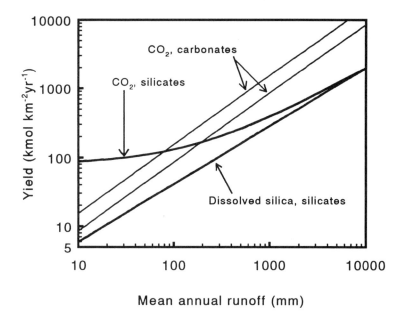

Figure 8. Yields of atmospheric carbon dioxide fixed by chemical weathering as a function of runoff calculated using the carbonate and silicate weathering models presented in this paper. Carbon dioxide fixation was set to half of the alkalinity for carbonate weathering and to the alkalinity for silicate weathering. When carbonates reprecipitate in the ocean, the carbon dioxide fixed by carbonate weathering returns to the atmosphere, whereas roughly half the carbon dioxide fixed by silicate weathering returns.

continental mountain belts. Several factors may be involved including finer grain size and the presence of glasses in the island-arc rocks. An additional factor may be volatiles, perhaps derived from seawater-magma interactions. The Kenyan setting of Dunne (1978, 1979) has been affected by rift-related volcanism. The model curve may overpredict solid yields for cratonic terrains. Temperature effects may also be important, with lower silicate-weathering rates and higher carbonate-weathering rates for a given runoff at lower temperatures. Demonstrations of temperature effects are equivocal. Where freezing and frozen water are factors, chemical weathering may not be the primary process that breaks apart rocks. This is clearly the case for glacial erosion, where sediment yields are very high. The exposure of so much fresh rock by physical processes appears to drive chemical weathering at rates that are as high as in the humid tropics, but the reactions are completely different (Axtmann and Stallard, 1995).

SUMMARY CONCLUSIONS

The models used above not only allow prediction of solid yields as a function of runoff, but they also allow comparative assessment of carbonate and silicate weathering and implications for landscape development. The comparison here is through the fixation of carbon dioxide as a function of runoff (Fig. 8). This builds on Figure 3 that compares the importance of various rock types globally in carbon dioxide fixation. We see that at low runoff, silicate weathering is more effective. The importance of these two classes of rocks reverses, so that at high runoff carbonates weather much more rapidly. This is consistent with the apparent importance of carbonate rocks as major landform elements in drier regions and their virtual absence in old tropical landscapes.

The release of solute and solid products for carbonate weathering is far more sensitive to runoff than is the case for silicate weathering (Figs. 2, 3, and 8). In particular, the generation of solids by silicate weathering is quite insensitive to runoff. This is consistent with a long history of observation (see summaries in Stallard, 1995).

The concept of steady-state erosion is a convenient device to compare measured rates of physical denudation with a model. Models can be derived from solute chemistry for the river system and a knowledge of bedrock chemistry. As developed here, the models are both robust and general. This is because use of siliceous-clastic concentrations in carbonates and sodium and silicon in silicate rocks appears to introduce manageable errors (see Eqn. 12 and model-measurement tie lines in Fig. 7).

We can only speculate what the Earth was like in the geologic past. Perhaps data gathered from rivers on a quiescent Earth, lacking recent super orogenies, continental glaciations, and human activities, would plot near or below a steady-state curve. The insensitivity to runoff variations of the solid yield for weathering-limited erosion of siliceous terrains variation is significant. This model implies that sedimentation rates would not vary strongly in response to climate change; instead, climatic effects would be reflected in sediment composition. The opposite is implied for erosion of carbonate-cemented clastic rocks.

The use of steady-state physical erosion models has direct application to studies of contemporary land-use change. Measurements of physical erosion seldom predate the changes in land use. Because land-use change does not appear to strongly affect the yield of dissolved sodium and silica derived from bedrock, their respective yields can be used to model the rate at which bedrock is degrading. This gives a probable predevelopment erosion rate. In watersheds from flat regions, solid yields calculated from steady-state erosion models allow a determination of whether soils are thickening or whether sediments are being stored in river systems. Under such circumstances, the model should significantly over-predict solid yields. If model and measure solid yields agree, and denudation rates are especially low, then the landscape may have reached steady-state erosion under transport-limited conditions.

REFERENCES

Ahnert F (1970) Functional relationships between denudation, relief and uplift in large mid-latitude drainage basins. Am J Sci 268:243-263

Axtmann EV, Stallard RF (1995) Chemical weathering in the South Cascade Glacier basin, comparison of subglacial and extra-glacial weathering. in Tonnessen KA, Williams MW, Tranter M (eds) Biogeochemistry of Seasonally Snow-Covered Catchments, Int'l Assoc Hydrological Sciences Publ 228:431-439

Barshad I (1966) The effect of a variation in precipitation on the nature of clay mineral formation in soils from acid and basic igneous rocks. in: Proc Int'l Clay Conf., Jerusalem, Israel. 1:167-173

Bell M, Laine EP (1985) Erosion of the Laurentide region of North America by glacial and glaciofluvial processes. Quarternary Res 23:154-174

Blatt H, Middleton GV, Murray RC (1980) Origin of Sedimentary Rocks. Prentice-Hall, Englewood Cliffs, NJ, 782 p

Bluth GJS, Kump LR (1994) Lithologic and climatologic controls of river chemistry. Geochim Cosmochim Acta 58:2341-2359

Brown ET, Stallard RF, Larsen MC, Raisbeck GM, Yiou F (1995) Denudation rates determined from the accumulation of in situ-produced ^{10}Be in the Luquillo Experimental Forest, Puerto Rico. Earth Planet Sci Lett 129:193-202

Ceasar J, Collier R, Edmond J, Frey F, Matisoff G, Ng A, Stallard RF (1976) Chemical dynamics of a polluted watershed, the Merrimack River in northern New England. Env Sci Tech 10:697-704

Dunne T (1978) Rates of chemical denudation of silicate rocks in tropical catchments. Nature 274:244-246

Dunne T (1979) Sediment yields and land use in tropical catchments. J Hydrology 42:281-300

Garrels RM, Mackenzie FT (1971a) Evolution of Sedimentary Rocks. New York, New York: W. W. Norton, 397 p

Garrels RM, Mackenzie FT (1971b) Gregor's denudation of the continents. Nature 231:382-383

Garwood NC, Janos DP, Brokaw N (1979) Earthquake-caused landslides: A major disturbance to tropical forests. Science 205:997-999

Hillel DJ (1992) Out of the Earth, Civilization and the Life of the Soil. New York, N.Y.: The Free Press, 321p

Holeman JN (1968) The sediment yield of major rivers of the world. Water Resources Res 4:737-747

Hooke RL (1994) On the efficacy of humans as geomorphic agents. GSA Today 4:218,224-225

Johnsson MJ, Stallard RF, Lundberg N (1991) Controls on the composition of fluvial sands from a tropical weathering environment: Sands of the Orinoco River drainage basin, Venezuela and Colombia. Geol Soc Am Bull 103:1622-1647

Judson S (1968) Erosion of the land, or what's happening to our continents? Am Sci 56:356-374

Keefer DK (1984) Landslides caused by earthquakes. Geol Soc Am Bull 95:406-421

Koehnken Hernández L, Stallard RF (1988) Sediment sampling through ultrafiltration. J Sed Petrol 58:758- 759

Koehnken L (1990) The composition of fine-grained weathering products in a large tropical river system, and the transport of metals in fine-grained sediments in a temperate estuary. PhD Dissertation, Princeton University, Princeton, NJ, 246 p

Larsen MC, Simon A (1993) A rainfall intensity-duration threshold for landslides in a humid tropical environment, Puerto Rico. Geografiska Annaler 75A:13-23

Larsen MC, Torres Sánchez AJ (1995) Geologic relations of landslide distribution and assessment of landslide hazards in the Blanco, Cibuco, and Coamo basins, Puerto Rico. U S Geol Surv Water Resources Investigations Rep 95-4029, 123 p

Le Maitre RW (1976) The chemical variability of some common igneous rocks. J Petrology 17:589-637

Lewis WM Jr, Hamilton SK, Jones SL, Runnels DD (1987) Major element chemistry, weathering, and element yields for the Caura River drainage, Venezuela. Biogeochemistry 4:159-181

Lewis WM Jr, Saunders JF III (1989a) Concentration and transport of dissolved and suspended substances in the Orinoco River. Biogeochemistry 7:203-240

Lewis WM Jr, Saunders JF III (1989b) Transport of major solutes and the relationship between solute concentrations and discharge in the Apure River, Venezuela. Biogeochemistry 8:101-113

Lewis WM Jr, Saunders JF III (1990) Chemistry and element transport by the Orinoco main stem and lower tributaries. In: Weibezahn FH, Alvarez H, Lewis WM Jr (eds) El Río Orinoco como Ecosistema. Caracas, Venezuela: Impresos Rubel, 211-239

Lowdermilk WC (1953) Conquest of the land through 7,000 years. US Dept Agri Soil Conserv Serv, Agri Information Bull 99, 30 p

Matthews E (1983) Global vegetation and land use: New high-resolution data bases for climate studies. J Climate Appl Met 22:474-487

McDowell WH, Asbury CE (1994) Export of carbon, nitrogen, and major ions from three tropical montane watersheds. Limnol Ocean 39:111-125

Meade RH (1969) Errors in using modern stream-load data to estimate natural rates of denudation. Geol Soc Am Bull 80:1265-1274

Meade RH, Stevens HH (1990) Strategies and equipment for sampling suspended sediment and associated toxic chemicals in large rivers -- with emphasis on the Mississippi River. Sci Total Environ 97/98:125-135

Meade RH, Yuzyk TR, Day TJ (1990) Movement and storage of sediment in rivers of the United States and Canada. In: Wolman MG, Riggs HC (eds) Surface Water Hydrology, The Geology of North Amica O-1. Boulder, Colorado: Geol Soc Am. 255-280

Meybeck M (1979) Concentrations des eaux fluviales en éléments majeurs et apports en solution aux océans. Revue de Géologie Dynamique et de Géographie Physique 21:215-246

Meybeck M (1987) Global chemical weathering of superficial rocks estimated from river dissolved loads. Am J Sci 287:401-428

Milliman JD, Qin YS, Ren ME, Saito Y (1987) Man's influence on the erosion and transport of sediment by Asian rivers: the Yellow River (Huanghe) example. J Geol 95:751-762

Milliman JD, Syvitski JPM (1992) Geomorphic / tectonic control of sediment discharge to the ocean: The importance of small mountainous rivers. J Geol 100:525-544

Moody JA, Troutman BM (1992) Evaluation of the depth-integrated method of measuring water discharge in large rivers. J Hydrol (Amsterdam) 135:201-236

Murnane RJ, Stallard RF (1990) Germanium and silicon in rivers of the Orinoco drainage basin, Venezuela and Colombia. Nature 344:749-752

Nordin CF Jr (1985) The sediment load of rivers. In: Rodda JC (ed) Facets of Hydrology. New York, New York: John Wiley 183-204

Piperno D. (1990) Fitolitos, Arqueología y Cambios Prehistóricos de la Vegetación en un Lote de Cincuenta Hectáreas de la Isla de Barro Colorado. In: Leigh EG Jr, Rand AS, Windsor DM (eds) Ecología de un Bosque Tropical: Ciclos Estacionales y Cambios a Largo Plazo. Balboa, República de Panamá: Smithsonian Tropical Research Institute 153-156

Putz FE, Katharine M (1982) Tree mortality rates on Barro Colorado Island. In: Leigh EG Jr, Rand AS, Windsor DM (eds) The Ecology of a Tropical Forest Seasonal Rhythms and Long-term Changes. Washington, DC: Smithsonian Institution Press, 95-100

Reeder SW, Hitchon B, Levinson AA (1972) Hydrochemistry of the surface waters of the Mackenzie River drainage basin, Canada - I. Factors controlling the inorganic composition. Geochim Cosmochim Acta 36:825-865

Schumm SA (1962) The disparity between present rates of denudation and orogeny. U S Geol Surv Prof Paper 454-H 13 p

Sherman GD (1952) The genesis and morphology of the aluminum-rich laterite clays. In: Frederickson AF (ed) Problems of Clay and Laterite Genesis. Am Inst Mining Metal Engineers: New York, 154-161

Stallard RF (1980) Major element geochemistry of the Amazon River system. PhD dissertation, Mass Inst Tech - Woods Hole Ocean. Inst Joint Prog in Oceanography. WHOI-80-29. Woods Hole, MA: Woods Hole Oceanogr Inst 366 p

Stallard RF (1985) River chemistry, geology, geomorphology, and soils in the Amazon and Orinoco basins. In: Drever JI (ed) The Chemistry of Weathering. Dordrecht, Holland: D. Reidel, 293-316

Stallard RF (1988) Weathering and erosion in the humid tropics. In: Lerman A, Meybeck M (eds) Physical and Chemical Weathering in Geochemical Cycles. Dordrecht, Holland: Kluwer Academic Publishers, 225-246

Stallard RF (1992) Tectonic Processes, continental freeboard, and the rate-controlling step for continental denudation. In: Butcher SS, Charleson RJ, Orions GH, Wolfe GV (eds) Global Biogeochemical Cycles. San Diego, California: Academic Press, 93-121

Stallard RF (1995) Tectonic, environmental, and human aspects of weathering and erosion: A global review using a steady-state perspective. Ann Rev Earth Planet Sci 12:11-39

Stallard RF, Edmond JM (1983) Geochemistry of the Amazon 2: The influence of the geology and weathering environment on the dissolved load. J Geophys Res 88:9671- 9688

Stallard RF, Koehnken L, Johnsson MJ (1991) Weathering processes and the composition of inorganic material transported through the Orinoco River system, Venezuela and Colombia. Geoderma 51:133-165

Stevens HH, Yang CT (1989) Summary and use of selected fluvial sediment-discharge formulas. U S Geol Surv Water Resources Investigations Rep 89-4026, 62 p

Stumm W, Morgan JJ (1981) Aquatic Chemistry. New York, NY, John Wiley, 780 p

Sugden DE (1978) Glacial erosion by the Laurentide Ice Sheet. J Glaciology 20:367-391

Weibezahn FH (1990) Hidroquimica y solidos suspendidos en el alto y medio Orinoco. In: Weibezahn FH, Alvarez H, Lewis WM Jr (eds) El Río Orinoco como Ecosistema. Caracas, Venezuela: Impresos Rubel, 151-210

Chapter 13

CHEMICAL WEATHERING AND ITS EFFECT ON ATMOSPHERIC CO_2 AND CLIMATE

Robert A. Berner

Department of Geology and Geophysics
Yale University
New Haven, CT 06520 U.S.A.

INTRODUCTION

It is well established that climate exerts an important influence on chemical weathering, but this statement can be inverted; it is also true that chemical weathering exerts an important influence on climate. Weathering affects climate by way of the atmospheric greenhouse effect. During the weathering of silicate and carbonate minerals atmospheric CO_2 is taken up and converted to dissolved HCO_3^- in natural waters. The HCO_3^-, after delivery to the oceans by rivers, can be stored there or removed from the oceans in the form of carbonate minerals or organic matter in sediments. Either way there is a net loss of atmospheric CO_2. Since CO_2 is a greenhouse gas, any change in its concentration, due to changes in its rate of uptake by weathering, affects its ability to absorb infrared radiation and warm the earth's surface. [For a summary discussion of CO_2 and the greenhouse effect consult IPCC, 1990.] It is the purpose of the present paper to document how CO_2 is converted to HCO_3^- by weathering, to show how various factors affect the rate of this process, and to show how atmospheric CO_2 levels could have changed during the geologic past as a result of changes in the rate of weathering.

Let us trace a carbon atom as it makes its way from the atmosphere to the oceans via the land (see Fig. 1). First of all the carbon is taken up via photosynthesis by terrestrial vegetation. The fixed carbon is then transferred to the soil as organic acids and CO_2 by roots and associated microflora, and by the decay of dead roots and litter (leaves, branches, dead trees, etc.). The carbon-containing acids attack silicate and carbonate minerals resulting in the formation of bicarbonate and organic anions in soil solution, with the organic anions soon thereafter oxidized to bicarbonate. These steps can be combined in terms of the overall weathering reactions:

For carbonates:

$$CO_2 + H_2O + CaCO_3 \rightarrow Ca^{++} + 2\,HCO_3^- \tag{1}$$

$$2\,CO_2 + 2\,H_2O + CaMg(CO_3)_2 \rightarrow Ca^{++} + Mg^{++} + 4\,HCO_3^- \tag{2}$$

For silicates (generalized and simplified as wollastonite and enstatite):

$$2\,CO_2 + 3\,H_2O + CaSiO_3 \rightarrow Ca^{++} + 2\,HCO_3^- + H_4SiO_4 \tag{3}$$

$$2\,CO_2 + 3\,H_2O + MgSiO_3 \rightarrow Mg^{++} + 2\,HCO_3^- + H_4SiO_4 \tag{4}$$

These reactions are accompanied by a separate weathering reaction involving the release of CO_2 via the oxidation of organic matter (kerogen, coal, oil) contained in old sedimentary rocks. The overall reaction, with organic matter simplified as CH_2O, can be represented as:

$$CH_2O + O_2 \rightarrow CO_2 + H_2O \tag{5}$$

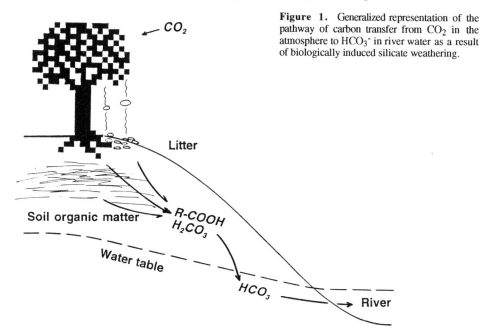

Figure 1. Generalized representation of the pathway of carbon transfer from CO_2 in the atmosphere to HCO_3^- in river water as a result of biologically induced silicate weathering.

This reaction shows that weathering can be a source, as well as a sink, for atmospheric CO_2. [An extreme example of Reaction (5) is the burning of fossil fuels which represents greatly accelerated weathering by humans.] Unfortunately the natural weathering of ancient organic matter is poorly understood (e.g. Hedges, 1993) and will not be discussed here. This is not too problematic because mass balance calculations (Berner, 1991) indicate that the release of CO_2 via kerogen oxidation is quantitatively less important than its uptake via silicate and carbonate weathering.

The HCO_3^- formed by silicate and carbonate weathering is transported from soil and ground waters to rivers and by rivers to the sea. Over time scales of 10^2 to 10^5 years this dissolved carbon can be stored in seawater, but over longer, multimillion year periods carbon storage cannot continue because seawater can hold just so much dissolved HCO_3^- (and accompanying CO_3^{-2}) before inorganic $CaCO_3$ precipitation occurs. [The same can be said about carbon storage in biota over millions of years since the earth can sustain just so much living matter.] Thus, oceanic storage of CO_2 in, and release from, the oceans is an important aspect of the carbon cycle on the hundred-to-thousand year time scale (e.g. Opdyke and Walker, 1992; Sarmiento, 1993) but not over many millions of years when oceanic inputs must be balanced by outputs (Berner, 1991). Here we distinguish the short term (10^3 to 10^5 year) carbon cycle where carbon storage and release involves transfers between the atmosphere, oceans, and the biosphere, from the much longer multimillion year *geochemical* carbon cycle where storage and release is only to and from rocks.

Since on a multimillion year time scale carbon cannot be stored in the oceans, we need to consider the means by which it is removed from seawater. Some carbon may be removed by CO_2 release during the formation of new silicates (reverse weathering—see Mackenzie and Garrels, 1966), but by far most is removed by the burial of $CaCO_3$ as skeletal remains and CH_2O as organic matter in bottom sediments (Berner and Berner, 1987). Extra Ca^{++}, beyond that contributed by rivers, is added to the oceans by exchange for Mg^{++} during basalt-seawater reaction (e.g. Holland, 1978) which enables the removal, as CaCO$_3$, of the additional HCO$_3^-$ derived from the weathering of Mg silicates (Reaction 4). Representative reactions are:

For carbonate burial :

$$Ca^{++} + 2\,HCO_3^- \rightarrow CaCO_3 + CO_2 + H_2O \tag{6}$$

For organic matter burial:

$$CO_2 + H_2O \rightarrow CH_2O + O_2 \tag{7}$$

Note that from Reaction (6) one mole of CO$_2$ is liberated for each mole of CaCO$_3$ buried and that Reaction (6) is the reverse of Reaction (1), which represents CaCO$_3$ weathering. As a result, if bicarbonate is not stored in the ocean, there is no net effect on atmospheric CO$_2$ of carbonate weathering combined with carbonate burial. This means that carbonate weathering is not an important factor in the control of atmospheric CO$_2$ on the multimillion year time scale. It is only on the thousand year time scale that HCO$_3^-$ storage in the oceans can result from an excess of carbonate weathering over burial. Since we will only be discussing the long term geochemical carbon cycle here, we will ignore carbonate weathering as a factor affecting atmospheric CO$_2$ and climate.

The above reactions can be summarized in terms of succinct overall reactions, first elucidated by Högbom (1894) (see Berner, 1995) and later expressed explicitly by Urey (1952):

$$CO_2 + CaSiO_3 \rightarrow CaCO_3 + SiO_2 \tag{8}$$

$$CO_2 + MgSiO_3 \rightarrow MgCO_3 + SiO_2 \tag{9}$$

Note that here, in contrast to the weathering of Ca- and Mg-carbonates, the weathering of Ca- and Mg-silicates results in the net removal of CO$_2$ from the atmosphere. It is the weathering of silicates, and not carbonates, that exerts a major long term control on atmospheric CO$_2$. Also, only the weathering of Ca- and Mg-silicates is important. This is because Na and K added to the oceans by silicate weathering on the continents are not removed as carbonates, by reactions analogous to (8) and (9) above, because of the great solubility of Na- and K-carbonate minerals. Most likely Na and K added by silicate weathering are removed by silicate formation (reverse weathering or basalt-seawater reaction) in the oceans with the return of CO$_2$, originally consumed by continental weathering, to the atmosphere.

WEATHERING AS A FEEDBACK CONTROL ON CO$_2$

The carbon dioxide level of the atmosphere on the million year time scale is affected by processes other than weathering. Both carbonates and organic matter are deeply buried and thermally decomposed at depth by diagenesis, metamorphism, and magmatism giving rise to degassing of CO$_2$ to the atmosphere. Additional CO$_2$ is supplied by the volcanic release of carbon stored for long periods in the mantle. Thus, atmospheric CO$_2$ is affected by changes in the rate of global degassing as well as by changes in the rate of weathering of Ca and Mg silicates. It is the balance between degassing and weathering that controls the level of CO$_2$ over geologic time (e.g. Walker et al., 1981; Berner et al., 1983). Note that the Högbom-Urey Reactions (8) and (9) above, can also be read from right-to-left, in which case they reflect carbonate decomposition and CO$_2$ degassing.

There is a need for a negative feedback control on atmospheric CO$_2$ level. If degassing became excessive it is possible that in the past there could have been a runaway greenhouse, resulting in very high temperatures as exist today on the planet Venus. However, such a situation (or its opposite a runaway icehouse due to a severe drop in degassing) has not existed on Earth for at least the past 3 billion years as attested to by a continuous record over this period of both life and an ocean consisting of liquid water. To prevent the development of a runaway greenhouse or icehouse a negative feedback mechanism must have existed. Furthermore, it is believed that the level of solar radiation 3 billion years ago was 20 to 30% less than that today and that it has been increasing linearly with time since then (Caldeira and Kasting, 1992). Calculations indicate that with such a reduced solar input, at the present level of atmospheric CO_2, the oceans would freeze over and there would be a *global* icehouse condition for much of geologic time (Kasting and Ackerman, 1986). Again this has not happened and there must have been a control mechanism that allowed for an enhanced greenhouse effect due (presumably) to higher CO_2 levels to compensate for lower solar input.

Chemical weathering is an ideal candidate for the needed negative feedback control mechanism. As CO_2 increases there is global warming due to the atmospheric greenhouse effect, but the rate of CO_2 uptake by silicate weathering also increases with increasing temperature, and this increased weathering rate provides negative feedback. Conversely, as CO_2 decreases, there is global cooling and the rate of CO_2 uptake by weathering decreases. Similar reasoning can be applied to solar radiation. As radiation increases with time, temperature and weathering rate increase so that CO_2 drops bringing about a counter-balancing effect. This temperature/weathering control mechanism first elucidated by Walker et al. (1981) is now accepted by most workers as the principal feedback control on global climate on the million year time scale.

Incorporation of the temperature feedback mechanism into global carbon cycle modeling (e.g. Berner, 1994) is presently done based on an empirical expression fitted to the results of general circulation models (GCMs) of the relation between global mean temperature and both CO_2 and solar radiation (e.g. Marshall et al., 1994; Manabe and Bryan, 1985).The expression employed (Berner, 1994) is:

$$T(t) - T(o) = \Gamma \, ln \, RCO_2 - W_s \, (t/570) \tag{10}$$

where: T = global mean surface temperature
RCO_2 = ratio of atmospheric CO_2 mass at time t to that at present
t = time before present in millions of years
o = present
Γ = empirical CO_2 greenhouse parameter
W_s = empirical solar radiation parameter

Use of this expression to calculate the feedback effect of temperature on weathering is discussed below under the subject of climate and weathering.

There are other possible feedback controls. Experiments indicate that plant growth, under conditions of abundant nutrients, water and light, accelerates with higher levels of CO_2 (for a summary consult Bazzaz, 1990). If enhanced growth involves enhanced silicate weathering in order to gain plant nutrients, then higher CO_2 could bring about accelerated weathering via a purely biological mechanism in addition to the temperature mechansim discussed above. This biological feedback mechanism was introduced by Volk (1989) and has been incorporated into the GEOCARB models (Berner, 1991; 1994). The GEOCARB formulation is:

$$f_B(CO_2) \, _{(plants)} = [\, (\Pi CO_2)_t / (\Pi CO_2)_o \,]^{0.4} \tag{11}$$

$$\Pi(CO_2)_t / \Pi(CO_2)_o = 2 \, RCO_2 / (1 + RCO_2) \tag{12}$$

where: $f_B(CO_2)$ = dimensionless feedback factor which is multiplied times the present weathering flux to obtain weathering fluxes for times in the past when RCO_2 was different from one. The subscript (plants) refers to the effect of plants alone which is multiplied by the temperature effect on weathering $f_B(CO_2)_{(temp)}$ to give the complete expression for $f_B(CO_2)$ (see below). $\Pi(CO_2)$ = land plant productivity

The expression for productivity (12) uses a standard Monod formulation and the exponent 0.4 is used to express the fact that not all plants respond to changes in atmospheric CO_2 due to limitation of growth by nutrients, water, or light.

Other feedbacks have been suggested, but they are less tenable. One is decreased CO_2 removal via less marine organic matter burial to accompany increased CO_2 removal via accelerated weathering due to mountain uplift (Raymo and Ruddiman, 1992), but there is no known reason why decreased organic burial should result from increased weathering. If anything, increased weathering should deliver more nutrients, such as phosphorus, to the oceans which should bring about increased biological production and, therefore, *increased* organic matter burial. Both Raymo and Ruddiman (1992) and Edmond (1993) have suggested that increased physical weathering accompanying colder climates should lead to increased CO_2 uptake via chemical weathering. Although this may help to trigger glaciations, this cannot persist for long because it constitutes positive feedback which would ultimately lead to a runaway icehouse. Others (Staudigel et al., 1989; Francois and Walker, 1992) have suggested submarine basalt weathering as a major feedback control on CO_2. However, there are both theoretical considerations (Caldeira, 1995) and experimental evidence (P. Brady, pers. comm.) for little or no CO_2 dependence of the low temperature reaction of basalt with CO_2 in seawater. Further, the field evidence for global CO_2 uptake via basalt weathering itself is limited to just two drill cores in buried Cretaceous seafloor rocks (Staudigel et al., 1989; Spivack and Staudigel, 1994).

FACTORS AFFECTING WEATHERING OVER GEOLOGIC TIME

In this section brief discussions of the major factors that affect weathering are presented, with emphasis on factors that are global in scope. They are: continental land area and lithology of exposed rocks, mountain uplift, global climate, and land vegetation. Each subject is discussed here in terms of how these factors may have varied over geologic time as incorporated into the Phanerozoic carbon cycle model (GEOCARB II) of the author (Berner, 1994). Because it is a subject of current active research by the author, special emphasis also will be on the role of land vegetation.

Continental land area and lithology

It has been amply demonstrated that over Phanerozoic time (past 600 million years) global sea level has risen and fallen on the scale of a few hundred meters giving rise to considerable variation in the land area of the continents. At first sight it would seem that the rate of uptake of CO_2 via silicate weathering might have varied over the same period in direct proportion to the area of land exposed to the atmosphere. However, those areas affected by hundred meter changes in sea level, i.e. coastal plains bordering the sea, often undergo little weathering because of the development there of thick covers of highly weathered, and therefore less reactive, material such as clay minerals (Stallard, 1992). By this reasoning covering of low-lying coastal lands by the sea, or their exposure by a drop in sea level, should not make much difference to the rate of global silicate weathering (Berner, 1994). This of course is true only where the lowlands are underlain by heavily weathered material. Coastal lowlands at the foot of high mountains undergoing glaciation and intense physical erosion may be underlain by a high proportion of unweathered primary silicates, such as feldspars, delivered rapidly to the lowlands. In this case exposure of the lowlands due to a drop in sea level could contribute appreciably to global weathering (Lasaga et al., 1994). Until lowlands of this type can be quantified for the geologic past, it will be assumed in the model of the present paper that silicate weathering is not appreciably affected by changes in sea level or land area.

Although carbonate weathering does not directly influence atmospheric CO$_2$, as discussed earlier, it is still must be included in mass balance expressions for carbon input to and output from the atmosphere/ocean system as part of overall carbon cycle modeling. The treatment of carbonate weathering in terms of land area is different from that for silicate weathering. The dissolution of carbonates occurs under both high relief and low relief conditions. For example, intense carbonate dissolution, with the formation of karst features, occurs beneath the state of Florida which is exceedingly flat and near sea level. Because of the non-dependence of carbonate weathering on relief or elevation, it is assumed to be directly proportional to the area of carbonates exposed on land. The area of land underlain by carbonate over Phanerozoic time has been calculated from paleolithologic maps by Bluth and Kump (1991) and their results are incorporated in the GEOCARB modeling via the dimensionless expression f_{LA} (t) (Berner, 1994):

$$f_{LA}(t) = \text{area of carbonates on land at time t / area of carbonates at present} \qquad (13)$$

Mountain uplift

The importance of mountain uplift to the weathering of silicates has been emphasized recently (Raymo, 1991; Francois and Walker, 1992; Raymo and Ruddiman, 1992). The idea is that uplift results in rugged relief, and cold temperatures at high elevations. The rugged relief enhances physical erosion and the removal of protective covers of highly weathered clay residues allowing greater exposure of primary silicates to chemical weathering. Cold temperatures result in greater physical weathering due both to freeze-thaw and to grinding where glaciers are present. The enhanced physical weathering, by granulation, exposes more surface area of the primary minerals to weathering solutions. In addition, mountains can bring about enhanced rainfall due to orographic effects resulting in greater flushing of rocks by water. All these factors should have brought about greater weathering of silicate minerals during geologic periods when the extent of high mountains was globally more important. An example is the late Cenozoic when the uplift of the Himalayan/Tibetan system occurred (Raymo and Ruddiman, 1992).

The author has no quarrel with the importance of mountain formation to silicate weathering and has in fact incorporated the effect into carbon cycle modeling. However, I do differ with the workers cited above in that I believe that other factors are also important in weathering and that mountain uplift is not always the primary cause of changes with time in Phanerozoic atmospheric CO$_2$. In the latest model, GEOCARB II (Berner, 1994), mountain uplift is parameterized in terms of the ^{87}Sr/^{86}Sr of seawater. It is assumed that more radiogenic ^{87}Sr is delivered to the oceans when there is greater exposure of deeply buried old highly radiogenic rocks, or rocks containing remobilized radiogenic Sr (Edmond, 1992), by mountain uplift into the zone of weathering. In the GEOCARB II model correction for changes in ^{87}Sr/^{86}Sr of the oceans with time due to basalt-seawater reaction is made and the expression used to represent the mountain uplift effect on weathering is:

$$f_R(t) = 1 - L\ [(R_{ocb}(t) - R_{ocm}(t))\ /\ (R_{ocb}(t) - 0.7000)] \tag{14}$$

where: $f_R(t)$ = dimensionless weathering factor which is multiplied times the present weathering flux to obtain weathering fluxes for times in the past when the $^{87}Sr/^{86}Sr$ of the oceans was different from that at present.

$R_{ocm}(t)$ = $^{87}Sr/^{86}Sr$ of the oceans at a time t as recorded in the geologic record

$R_{ocb}(t)$ = $^{87}Sr/^{86}Sr$ calculated for the oceans if the only control on the isotope ratio was basalt-seawater reaction

L = multiplying factor for sensitivity analysis.

Unfortunately the interpretation of Sr isotopic data in terms of mountain uplift and weathering rate is equivocal (Richter et al., 1992; Edmond, 1992). An increase of the $^{87}Sr/^{86}Sr$ of the oceans can be due to either increased weathering of average continental rocks, to a constant rate of global weathering but of unusually radiogenic rocks, or to changes in the rate of exchange of Sr isotopes during basalt-seawater reaction not accounted for in Equation (14). In the future it is hoped that a new independent method of deducing paleotopography and its effect on silicate weathering rate will be developed. One approach might be the construction of paleogeographic maps where mountain chains and their relative height can be deduced from the collisional history of the continents.

Climate

Most workers agree that climate exerts an important influence on chemical weathering. That weathering is a function of climate is revealed by the incorporation of the word "weather" within it. Climate exerts its influence directly in terms of temperature and precipitation, and indirectly in terms of vegetation, runoff, and glacial/periglacial phenomena. Because the development of terrestrial vegetation is a product of biological evolution, in addition to climate, this topic is discussed separately in the present paper.

Temperature is important in weathering, first of all because minerals dissolve faster as temperature increases (cf. Lasaga et al., 1994). An average temperature coefficient (equiv-alent to an Arrhenius activation energy of 15 kcal/mol) for the dissolution of common silicate minerals (e.g. Brady, 1991; Lasaga et al., 1994) has been used in carbon cycle modeling (Berner, 1994) and is in agreement with results from natural watersheds (Velbel, 1994; White and Blum, 1995). White and Blum (1995) determined, for a large number of purely granitic watersheds, the rate of silica and sodium release via chemical weathering as a function of precipitation (rainfall) and temperature. A summary of their results is shown in Figure 2. General agreement with the formulation of the GEOCARB II model indicates that the use of laboratory derived temperature coefficients by this model is justified. This suggests that surface reaction controlled dissolution of primary minerals may be the rate limiting step in silicate weathering.

Temperature has another, indirect effect, because globally it correlates with precipitation. A warmer global climate in the past should have resulted in greater evaporation from the oceans and consequently greater precipitation on the continents. If greater precipitation leads to greater runoff, then there should have been enhanced weathering. The response of silicate weathering to increased runoff, however, is not linear because of dilution effects (Holland, 1978). Silicate dissolution does not keep pace with increased flushing of rocks so that the riverine concentration of dissolved ions liberated by weathering decreases as river runoff increases. A survey of recent field studies suggests that the silicate dissolution flux, because of dilution, should follow only the 0.65 power of runoff (Berner, 1994). The carbonate dissolution flux, by comparison, should undergo

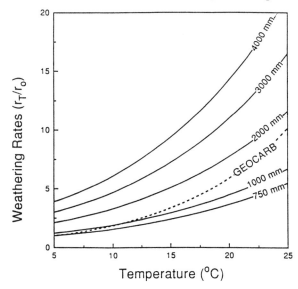

Figure 2. Observed dependence of silicate weathering rate as a function of mean annual temperature and precipitation. Curves summarize hydrochemical data on a large number of small watersheds underlain only by granite. The climate dependence used in GEOCARB modeling is shown for comparison (adapted from White and Blum, 1995).

less dilution with runoff because ground waters, due to rapid rates of dissolution, are generally fairly close to saturation with CaCO$_3$ (Langmuir, 1971; Harmon et al., 1975).

Based on the temperature effects on both the rate of mineral dissolution and runoff the temperature feedback expression $f_B(T)$ for silicate weathering is (Berner, 1994):

$$f_B(T) = [J(T) / J(T_0)] \times [R(T) / R(T_0)]^{0.65} \qquad (15)$$

where: $J(T)$ = the global weathering dissolution flux at some past time with global mean surface temperature T

$R(T)$ = global runoff at some past time with global mean surface temperature T

T_0 = present global mean surface temperature and the subscript (o) refers to the present.

Explicit expressions for $J(T)$ and $R(T)$ as a function of T are obtained by use of the GEOCARB II model. From the standard Arrhenius formulation of dissolution rate we have the relation:

$$J(T) / J(T_0) = \exp \{[\Delta E/R][(T - T_0) / T T_0]\} \qquad (16)$$

where: ΔE = activation energy; R = the gas constant (1.99 cal deg^{-1} mol^{-1}); and T is in Kelvins. Substituting E = 15 kcal/mol and simplifying, we obtain approximately:

$$J(T) / J(T_0) = \exp [0.090 (T - T_0)] \qquad (17)$$

The relation for runoff as a function of temperature (Berner, Lasaga, and Garrels, 1983) is derived from GCM modeling (Manabe and Stouffer, 1980; Manabe and Bryan,

1985). (Changes in runoff due to factors other than temperature are included in a separate dimensionless parameter $f_D(t)$—see below). The derived relation is:

$$R(T_0) / R(T_0) = 1 + 0.038 (T - T_0) \qquad (18)$$

Combining Equations (10), (15), (17) and (18), one obtains the CO_2-temperature feedback factor $f_B(CO_2)_{(temp)}$ for silicate weathering:

$$f_B(CO_2)_{(temp)} = \exp [- 0.09 \, W_s(t / 570)] \times (RCO_2)^{0.09\Gamma}$$

$$\times [1 + 0.038 \, \Gamma \, ln \, (RCO_2) - 0.038 \, W_s \, (t / 570)]^{0.65} \qquad (19)$$

This is the dimensionless factor which expresses the effect of CO_2, by way of the greenhouse effect, on weathering. When multiplied by Equation (11) for $f_B(CO_2)_{(plants)}$ one obtains the total feedback effect of CO_2 on weathering for periods of time since the advent of vascular land plants.

Global rainfall is a function of many factors besides global mean temperature. Both rainfall and runoff depend on the size, location, and topography of a given land mass (Crowley and North, 1991). Large continents located along subtropical latitudes can be very dry in their interiors and undergo intense monsoonal activity as was the case for the supercontinent of Pangea during the Permian and Triassic Periods (Kutzbach and Gallimore, 1989). Coastal mountain belts, if located facing into prevailing onshore winds, can experience extensive precipitation on their windward sides due to orography. Small equatorial land masses can be very wet. At any rate, it is obvious that greater rainfall should in general bring about greater weathering and for periods of time when continents were smaller and located at wetter latitudes weathering should have been enhanced (providing all other factors remained the same). The GEOCARB modeling currently employs the global runoff calculations of Otto-Bliesner (1995) for principal periods of the Phanerozoic in terms of the runoff factor $f_D(t)$:

$$f_D(t) = \text{global runoff at time } t \, / \text{ global runoff at present (for constant } CO_2) \qquad (20)$$

For silicate weathering $f_D(t)$ is raised to the 0.65 power to express the effects of dilution as discussed above. Note that $f_D(t)$ is calculated only for the effects on runoff of changing geography, and not for changes in atmospheric CO_2 or global temperature which is covered separately by Equation (18).

Large scale glacial and peri-glacial activity associated with times of global cooling (e.g. the Pleistocene Epoch) should bring about enhanced grinding of rocks to produce more surface area for contact with weathering solutions. This has led some workers (e.g. Raymo and Ruddiman, 1992; Edmond, 1993) to suggest that global weathering is faster when the earth is colder. However, as stated above this situation cannot persist globally for a long time or it would lead eventually to a runaway ice house.

Edmond (1993) has cited the chemical composition of the Aldan River in Siberia as evidence that cold climates at high latitudes favor rapid silicate weathering due to freeze/thaw action. However, he did not find that high dissolved Ca^{++} and HCO_3^- concentrations were accompanied by appreciable dissolved silica or clay mineral weathering products as would be expected for silicate weathering. These observations suggest carbonate weathering rather than silicate weathering, which is in agreement with the presence of abundant limestones in the Aldan drainage basin (Gordeev and Siderov, 1993). Furthermore the idea that global cooling should lead to increased weathering at high latitudes is not in accord with climatological considerations. Model calculations indicate that greenhouse induced global warming should result in even greater warming at high latitudes with a concomitant increase in rainfall (Manabe and Stauffer, 1993). During the Cretaceous period when high latitudes were unusually warm, heavily vegetated, wet and frost-free in many areas, conditions must have favored greater weathering than at present where there is greater aridity and the ground is covered with snow most of the year and frozen at depth as permafrost. Finally, the observation that present day clay mineral formation by weathering is greater at lower latitudes (Biscaye, 1965) is a simple example that warmer, and not colder, climates lead to greater silicate weathering.

Vegetation

The role of vegetation in weathering is usually included under a discussion of climate and weathering. However, vascular land plants did not exist before about 400 million years ago and they have undergone evolution and many changes of habitat since that time. Prior to 400 million years the land surface may have been covered with primitive algae or lichens (Wright, 1985; Schwartzman and Volk, 1989), but the effectiveness of these organisms compared to higher plants in affecting the rate of weathering is debatable, as will be pointed out below. Because vascular plants exert a major influence on silicate weathering, their evolution must have had an effect on the level of atmospheric CO_2 and it is the purpose of this section to demonstrate this. [Parameterization of the effects of plants in the GEOCARB model is discussed in the following section.]

Vascular plants affect the rate of weathering in numerous ways. Some of the more important ones are:

1. Rootlets (+ symbiotic microflora) with high surface area secrete organic acids/ chelates which attack minerals in order to gain nutrients.

2. Organic litter decomposes to H_2CO_3 and organic acids providing additional acid for weathering.

3. On a regional scale plants recirculate water via transpiration followed by rainfall and thereby increase water/mineral contact time. There is greater rainfall in forested regions than there would be in the absence of the trees.

4. Plants anchor clay-rich soil against erosion allowing retention of water and continued weathering of primary minerals between rainfall events.

Plants accelerate weathering in order to obtain nutrient elements from rocks (Berner, 1992b). As part of this process the secretion of organic acids by plant roots and their associated mycorrhizae (micro-organisms attached to rootlets) is of special interest. A number of studies (e.g. Cromack et al., 1979; Koch and Matzner, 1993; Griffiths et al., 1994) have demonstrated enhanced concentrations of organic acids within short distances of mycorrhizae and plant roots (the rhizosphere) which can lead to localized weathering (April and Keller, 1990). This is due largely to the secretion by the mycorrhizae of various low molecular weight acids such as oxalic acid or citric acid. These acids, besides providing extra H^+ for the attack on minerals, also are able to complex and solubilize (chelate) otherwise insoluble elements such as Fe and Al. This helps to further break down those minerals containing Al and Fe. Because of an enhanced acid concentration within the microenvironment of the rhizosphere, one can be misled as to the chemical composition of the soil solutions that actually attack minerals. Laboratory experiments based on the use of organic chelators at concentrations found in average soil waters, as sampled by lysimeters, ignore the intense microenvironment of the rhizosphere, and as a result can come to erroneous conclusions regarding the importance of organic acids in weathering. This has been emphasized recently by Drever (1994).

An example of the effectiveness of rhizosphere solutions in attacking silicate minerals, based on the research of M.F. Cochran, is shown in Figure 3. Here plagioclase phenocrysts from a young Hawaiian basalt have been completely dissolved away by biological exudates, leaving molds of the phenocrysts in the immediate vicinity of plant

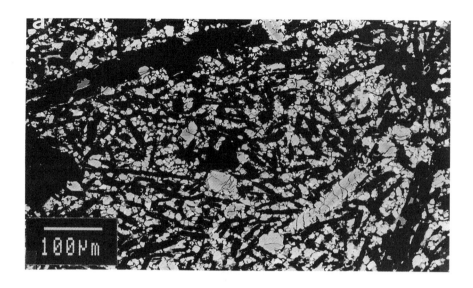

Figure 3. Evidence of plant induced weathering. (a) Molds of pre-existing plagioclase phenocrysts from a young Hawaiian basalt that have been dissolved away by the action of adjacent plant rootlets of the Ohia tree *Metrosideros*. (b) Plagioclase phenocryst from a young Hawaiian basalt showing etching from the activity of adjacent plant rootlets—note the access "tunnel" for aggressive solutions leading to the phenocryst (after Cochran, 1995).

rootlets (Fig. 3A). By contrast, away from rootlets there is no evidence for such intense weathering. The plant in this case (the Ohia tree *Metrosideros*) has dissolved away anorthitic plagioclase in order to extract the important nutrient calcium. One can actually observe widening of minute cracks by biological fluids in order to gain access to the Ca-rich minerals (Fig. 3B). By measuring the degree of porosity, due to the removal of minerals and volcanic glass, Cochran has been able to quantify the rate of initial biochemically induced weathering of Hawaiian basalts ranging in age from 100 to 5000 years (Cochran, 1995).

In contrast to the dissolution found around the roots of higher plants, we have found little or no evidence of dissolution under the lichen *Stereocaulon vulcani* developed on the same young basalts (Cochran, 1995; Cochran and Berner, 1993), This is in conflict with the earlier work of Jackson and Keller (1970) who have maintained that this lichen weathers Hawaiian basalts rapidly, even over time spans as short as 50 years. The material described by Jackson as the alteration residue of the underlying basalt, we have shown is, instead, largely dust trapped by the sticky, lipid-rich lichen thalli and altered in situ. This finding of much less weathering by *S. vulcani* is important in that Schwartzman and Volk (1989) have stated that, based on Jackson's findings, the role of lichens in weathering is immense. They used Jackson and Keller's results to calculate that silicate weathering during the Precambrian was greatly accelerated by lichens and that this helped to bring atmospheric CO₂ levels down to the point where earth surface temperatures became low enough to support higher forms of life. We believe this conclusion is based on insufficient evidence.

It is still possible that other lichens or primitive micro-organism could have been effective weathering agents prior to the rise of vascular land plants as has been maintained by others (e.g. Retallack, 1990; Schwartzman and Volk, 1989; Jackson, 1993; Horodyski and Knauth, 1994). However, we suggest that such weathering could not have been as important as that which accompanied the rise of vascular plants. There is definite evidence of limited leaching immediately underneath lichens (e.g. Wilson and Jones, 1983; Thorseth et al., 1992) but because of the very slow growth of these organisms (Ahmadjian, 1993) their effect over time becomes negligible compared to that of higher plants. The reason that higher plants are more effective is simple. Underneath even small crop plants the interfacial area between minerals and roots, rootlets, and root hairs is immense with values of 1000 m² per square meter of land surface (Wild, 1993). For lichens, on the other hand, the area of rock/biota interface is only about 1 m² per square meter of rock (land). Combined with the much slower growth rate (and, therefore, slower nutrient storage) by lichens compared to higher plants, this difference in interfacial area must mean that rooted higher plants are much more effective in attacking rocks and weathering them than are lichens.

A knowledge of the quantitative effect of plants on the rate of silicate weathering is necessary if one is to obtain any estimate of how the rise and spread of vascular plants on land could have affected levels of atmospheric CO₂ by altering rates of global weathering. Before the Devonian Period there were some vascular plants but they were small and confined to the vicinity of watercourses (Stewart, 1983). It was only during the Devonian that plants developed deep roots and spread to upland areas where they could be effective in altering rates of weathering (Algeo et al., 1995). If major vascular plants (e.g. trees) are much more important in affecting weathering than what may have existed earlier (e.g. algae and lichens), then their population of the continents during the Devonian could had a major impact, not only on atmospheric CO₂, but also on geochemical cycles in general and even on marine ecology (Algeo et al., 1995).

It is difficult to estimate the quantitative effect of plants on the rate of weathering. However, some initial estimates have been made. Drever and Zobrist (1992) have shown that the stream flux of dissolved HCO_3^-, resulting from CO_2 uptake during silicate weathering in the southern Swiss Alps, is a strong function of elevation. The weathering flux at 300 m elevation is about 25 times higher than it is at 2400 meters. The two elevations differ in both mean annual temperature and in vegetation, but not in bedrock lithology. The low area is heavily forested by deciduous trees whereas the high area is above tree line and is essentially barren with lichen-encrusted rocks. If correction is made for the effect of temperature difference on the rate of mineral dissolution, which is a factor of about three based on the adiabatic lapse rate plus an activation energy of 15 kcal/mole, then the residual effect is a factor of about eight. If this difference is due mainly to the presence of the trees at the low elevation, then this data suggests a large effect (8×) of vascular plants on the rate of silicate weathering. This conclusion, however, is tentative because of the possibility that differences in hydrology and/or microclimatology between the two elevations could explain the results.

Another study whose results can be applied to the problem of the effect of plants on weathering was conducted by Arthur and Fahey (1993) at the Loch Vale Watershed of the Rocky Mountain National Park located at an elevation of between 3100 and 4000 meters. Here a small forest of Engelmann spruce is located in the lower elevations of the watershed which otherwise consists of bare bedrock, talus, and permanent snowfields. They found that the cationic denudation rate per unit area of the forested area was 3.5 times higher than that for the surrounding barren area.

A better estimate of the quantitative role of plants in weathering could be gained by performing controlled experiments where, climate, hydrology etc. were held constant. Fortunately such an experiment exists and is currently underway at the Hubbard Brook Experimental Forest Station in New Hampshire (Bormann et al., 1987). In 1983 three small equisized plots were constructed by filling 1.5-meter depressions with the same feldspathic glacial sand and lining the bottom of each with impermeable plastic sheeting to allow flow of all water passing through each plot to an exit pipe located below ground where the water could be sampled. One plot was planted with red pine trees, the second with two species of grass, and the third left fallow. [The fallow field has unavoidably developed a cover of lichens and other primitive plants which is an advantage in that this makes it a reasonable stand-in for the ancient land surface.] For the past 12 years the exit waters have been sampled, and chemical analyses for all major elements have been performed recently by our laboratory. Other workers in the meantime have been measuring the uptake of major elements by exchangeable phases and by the growing biota.

Results for dissolved HCO_3^- as a function of time for the Hubbard Brook experiment, taken from the work of Cochran (1995), are shown in Figure 4. Here measured HCO_3^- concentrations are normalized to measured Cl^- concentrations in order to correct for changes in the concentration of all ions and in outflow rate due to evapotranspiration. As can be seen there are large seasonal variations. The release of bicarbonate from the tree-lined plot, as compared to the barren plot, is very large during periods of active growth and intense snowmelt, as occurs in the spring. Also, the acceleration of weathering by the pine trees is much greater than that by the grasses (not shown). A mean value for the total period of the relative bicarbonate flux for the trees relative to the barren plot is about a factor of four. This must be a minimum value because of the storage of nutrients in the actively growing trees. In a natural forest, trees eventually die, due to drought, fires, windstorms, landslides, etc., and release nutrient cations (Bormann and Likens, 1981), such as K^+ or Ca^{++}, which can be accompanied by additional HCO_3^-. It is the sum of HCO_3^- loss by

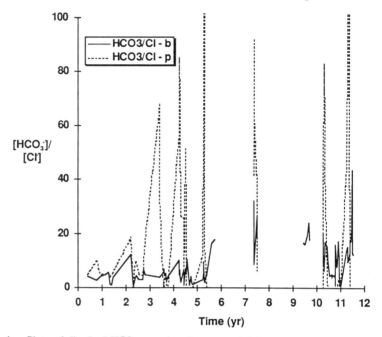

Figure 4. Plots of dissolved HCO₃⁻ vs. time for waters draining each of two small experimental weathering plots ("sandboxes') at Hubbard Brook, New Hampshire. All HCO₃⁻ concentrations have been normalized to dissolved Cl⁻ to correct for differences in evapotranspiration. p = plot with pine trees; b = "bare" plot populated by algae and lichens (after Cochran, 1995).

streams, plus that released with cation nutrients that were previously stored in growing plants, that constitutes the total time-integrated effect of plants on weathering rate (e.g. Taylor and Velbel, 1991). Once data on nutrient storage in the pine trees at Hubbard Brook is obtained, we will be able to derive a more accurate idea of the total effect of plants on the rate of uptake of CO_2 and its conversion to HCO_3^- during the weathering of silicate minerals.

SENSITIVITY OF THE GEOCARB CARBON CYCLE MODEL TO SOME WEATHERING PARAMETERS

In this section results, in terms of plots of carbon dioxide concentration vs. time, for different values for critical weathering parameters are presented based on the latest version of the GEOCARB geochemical carbon cycle model for Phanerozoic time (Berner, 1994 modified to introduce the rise of vascular land plants between 380 and 350 my BP—see Algeo et al., 1995). This model is explained in detail in previous publications (Berner, 1991; 1994) and the parameterization of weathering in the model is discussed above, so that only its essence will be summarized here.

The GEOCARB modeling assumes that, at each million year time step, the rate of input of carbon dioxide to the earth surface (the atmosphere/ocean/biosphere reservoir), by thermal degassing and the weathering of old carbonates and organic matter, is equal to its removal by weathering and the burial of new carbonates and organic matter in sediments. In other words, as input and output rates change over time it is assumed that a new steady state is re-established during each million year period. From the dynamics of the long term carbon cycle this is justified (Kump, 1989; Lasaga, 1989). A diagrammatic representation

Figure 5. Diagrammatic representation of the geochemical carbon cycle as treated by the GEOCARB modeling (Berner, 1991; 1994). F_{wc} = weathering of carbonates; F_{mc} = degassing from thermal carbonate decomposition; F_{wg} = weathering of organic matter; F_{mg} = degassing from thermal organic matter decomposition; F_{bc} = burial of carbonates in sediments; F_{bg} = burial of organic matter in sediments. On a multimillion year time scale carbon cannot be stored in the ocean/atmosphere reservoir, so that at each time in the past the total inputs to the ocean + atmosphere (+ biosphere) must be very close to that for total outputs. [In other words, $F_{wc} + F_{mc} + F_{wg} + F_{mg} = F_{bc} + F_{bg}$.]

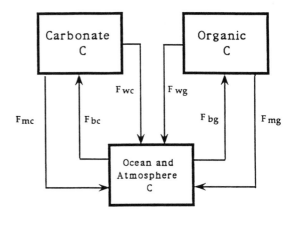

of the model is shown in Figure 5. Mass balance expressions are constructed for both total carbon and ^{13}C using the ^{13}C values for limestones over the Phanerozoic. The weathering flux terms for carbonates and silicates take into consideration changes with time by multiplying present weathering rates by time-dependent dimensionless weathering parameters such as those shown in Equations (11) through (20). The factors consider continental land area, surface lithology of rocks undergoing weathering, continental relief due to mountain uplift, global river runoff, the rise and evolution of land plants, slow constant increase in solar radiation, and changes in atmospheric CO$_2$ as they act as a negative feedback and affect global mean surface temperature and the rate of growth (weathering) by land plants. Degassing changes with time are parameterized by variations in the rate of seafloor spreading via plate tectonics and by fluctuations in the amounts of carbonates (deep sea vs. shallow water) available for thermal decomposition upon burial. Paleolevels of CO$_2$ are computed, using an inversion procedure, from values obtained for the feedback expression $f_B(CO_2)$ for silicate weathering at each million year time step.

To illustrate the effects of changing the values of input parameters for weathering, two sensitivity runs are presented here. [Many additional sensitivity runs can be found in the original paper—Berner, 1994.] The first, Figure 6, shows the effects on the rate of weathering of varying the quantitative importance of the rise of vascular land plants during the Devonian. The dimensionless parameter RCO$_2$ represents the ratio of the mass of CO$_2$ in the atmosphere at some time t to that at present. The dimensionless parameter $f_E(t)$ represents the ratio of the rate of weathering under existing biota at a time t to that under present day plants (with all other factors held constatnt). Values of $f_E(t)$ attached to the curves represent those assumed for the pre-vascular plant situation, and the range chosen is based on our Hubbard Brook experiments and the study of Drever and Zobrist (1992). Note that a knowledge of the quantitative effect of plants on weathering is critical in calculating values of CO$_2$. This is a principal reason why we are presently concentrating our research on this subject.

In Figure 7 is shown the sensitivity of CO$_2$ to the GCM climate model used for converting atmospheric CO$_2$ concentration and solar radiation level to global mean temperature. The empirical expression fitted to the results of general circulation modeling used for this purpose (Berner, 1994) is stated above as Equation (10). Three formulations are used to obtain values of Γ and W_S in Equation (10). One is based on the results of Marshall et al. (1994), another on those of Manabe and Bryan (1985) and S. Manabe (pers.

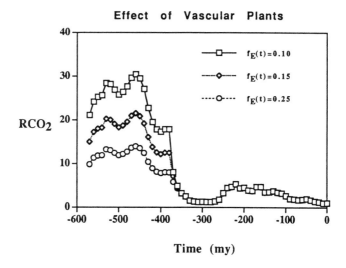

Figure 6. Plot of CO_2 vs. time, calculated according the GEOCARB II model, showing sensitivity to the parameter $f_E(t)$ for the period prior to the rise of deeply rooted vascular plants in upland areas. RCO_2 represents the mass of CO_2 in the atmosphere at some time t divided by that today. The parameter $f_E(t)$ is set equal to one at present; thus, a value of $f_E(t) = 0.10$ means that present day plants accelerate weathering by a factor of ten over that for the primitive vegetation of the early Paleozoic. Plot is for values of $\Gamma = 6$, $W_S = 12.9$ (see caption to Fig. 7).

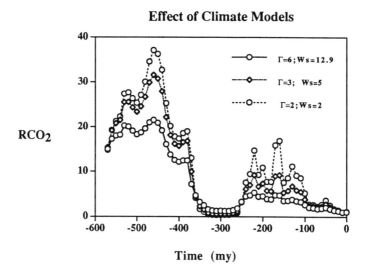

Figure 7. Plot of CO_2 vs. time, calculated according to the GEOCARB II model, showing sensitivity to the climate feedback parameters Γ and W_S. RCO_2 represents the mass of CO_2 in the atmosphere at some time t divided by that today. The parameters Γ and W_S represent the sensitivity of global mean temeperature to atmospheric CO_2 level and solar radiation respectively based on an empirical equation fitted to the results of general circulation modeling (see text). The highest values are based on the work of Marshall et al. (1994), intermediate values on that of Manabe and Bryan (1985) and Manabe (pers. comm.), and the lowest values are minima allowable for physically possible values of RCO_2.

comm.), and the third using the lowest values of Γ and W_S permissible without producing negative RCO$_2$ values. The amplitude of variations in CO$_2$ with time depends on which GCM formulation is used, but the general shape of the curve is not altered. Also, similar results are obtained if different formulation are used for the response of plant-induced weathering to CO$_2$ level. Thus, the *shape* of the CO$_2$ curve is robust relative to whichever feedback is employed; this is an important result. It is in agreement with earlier work (Berner, 1991; 1994) that has shown that variation of *all* modeling parameters within geologically reasonable limits, either those associated weathering or those reflecting degassing, results in the same general curve shape.

SUMMARY: WEATHERING, CLIMATE, AND CO$_2$

There is no doubt that atmospheric CO$_2$ has varied appreciably over Phanerozoic time and a large part of the variation has been due to changes in the rate of weathering. Variations with time in continental land area, elevation, and surface lithology and in global climate, vegetation, and the amount of solar radiation all have influenced weathering and consequently the level of atmospheric CO$_2$. The values of CO$_2$ vs. time calculated by the GEOCARB model and shown in Figures 6 and 7 are in rough agreement with independent estimates (see Berner, 1992a) in that all methods show at each time about the same value of RCO$_2$ (within a factor of two). Also, the modeling agrees with observed climates if the atmospheric greenhouse effect is a major control of climate (Crowley and Baum, 1992). High CO$_2$ levels occur during periods of global warming and low levels during glacial times. [The Ordovician glaciation is a special case—see Crowley and Baum, 1995.]

Modeling results show that the dramatic drop in CO$_2$ during the Devonian-Carboniferous was mainly the result of an increase in weathering intensity accompanying the rise and evolution of vascular land plants. The newly risen plants accelerated weathering, but to keep the CO$_2$ removal flux equal to the CO$_2$ degassing input flux, the level of CO$_2$ dropped, thus counterbalancing the plant accelerating effect and providing the necessary feedback to stabilize the system. [Additional CO$_2$ drop was provided by the enhanced burial of vascular-plant derived organic matter—see Berner, 1991.] The CO$_2$ drop led to global cooling and the onset of the Permo-Carboniferous glaciation, the longest and most extensive glacial period of the entire Phanerozoic. Since these results hinge on assumed values for the quantitative effects of plants on weathering, they show that it is imperative that a better knowledge of the role of plants in weathering be obtained. This is an outstanding example of how weathering may be important to global climate and shows why a better understanding of weathering, as discussed by the many papers in the present volume, has major application to other aspects of earth science.

ACKNOWLEDGMENTS

This work has been supported by grants from the National Science Foundation (EAR-9117099 and EAR-9417325). The writer thanks M. Ford Cochran for many helpful discussions.

REFERENCES

Ahmadjian V (1993) The lichen symbiosis. Wiley, New York, 250 p
Algeo TJ, Berner RA, Maynard JB, Scheckler SE (1995) Late Devonian oceanic anoxic events and biotic crises: "rooted" in the evolution of vascular land plants? GSA Today 5:45, 5:64-66
April R, Keller D (1990) Mineralogy of the rhizosphere in forest soils of the eastern United States. Biogeochemistry 9:1-18
Arthur MA, Fahey TJ (1993) Controls on soil solution chemistry in a subalpine forest in north-central Colorado. Soil Science Soc Am J 57:1122-1130

Bazzaz FA (1990) The response of natural ecosystems to the rising global CO_2 levels. Ann Rev Evolution and Systematics 21:167-196

Berner EK, Berner RA (1987) The Global Water Cycle:Geochemistry and Environment. Prentice-Hall, Engleood Cliffs, NJ, 397 p

Berner RA (1991) A model for atmospheric CO_2 over Phanerozoic time Am J Sci 291:339-376

Berner RA (1992a) Paleo-CO_2 and climate. Nature 358:114

Berner RA (1992b) Weathering, plants, and the long-term carbon cycle. Geochim Cosmochim Acta 56:3225-3231

Berner RA (1994) GEOCARB II: A revised model for atmospheric CO_2 over Phanerozoic time. Am J Sci 294: 56-91

Berner RA (1995) A.G. Hogbom and the development of the geochemical carbon cycle. Am J Sci 295:491-495

Berner RA, Lasaga AC, Garrels RM (1983) The carbonate-silicate geochemical cycle and its effect on atmospheric carbon dioxide over the past 100 million years. Am J Sci 283:641-683

Biscaye PE (1963) Mineralogy and sedimentation of recent deep-sea clay in the Atlantic Ocean. Geol Soc Am Bull 76:803-832

Bluth GJS, Kump, L R (1991) Phanerozoic paleogeology. Am J Sci 291:284-308

Bormann FH, Bowden WB, Pierce RS, Hamburg SP, Voigt GK, Ingersoll RC, Likens, GE (1987) The Hubbard Brook sandbox experiment. In: Restoration Ecology. Jordan R, Gilpin ME, Aber JD (eds) Cambridge Univ Press, p 251-256

Bormann FH, Likens GE (1981) Pattern and process in a forested ecosystem. Springer-Verlag, New York, 253 p

Brady PV (1991) The effect of silicate weathering on global temperature and atmospheric CO_2. J Geophys Res 96:18, 101-18, 106

Caldeira K (1995) Long-term contol of atmospheric carbon dioxide: low-temoerature seafloor alteration or terrestrial silicate-rock weathering. Am J Sci 295 (in press)

Caldeira K, Kasting JF (1992) The life span of the biosphere revisited. Nature 360:721-723

Cochran MF (1995) Weathering, plants and the geochemical carbon cycle. PhD Dissertation, Yale Univ, New Haven, CT

Cochran MF, Berner RA (1993) Reply to the comments on "Weathering, plants and the long-term carbon cycle. Geochim Cosmochim Acta 57:2147-2148

Cromack K, Sollins P, Graustein WC, Speidel K, Todd AW, Spycher G, Li CY, Todd RL (1979) Calcium oxalate accumulation and soil weathering in mats of the hypogeous fungus *Hysterangium Crassum* . Soil Biol and Biochem 11:463-468

Crowley TJ, Baum SK (1992) Modeling late Paleozoic glaciation. Geology 20:507-510

Crowley TJ, Baum SK(1995) Reconciling late Ordovician (440 Ma) glaciation with very high (14×) CO_2 levels. J Geophys Res 100:1093-1101

Crowley TJ, North GR (1991) Paleoclimatology. Oxford University Press, New York, 339 p

Drever JI (1994) The effect of land plants on weathering rates of silicate minerals. Geochim Cosmochim Acta 58:2325-2332

Drever JI, Zobrist J (1992) Chemical weathering of silicate rocks as a function of elevation in the southern Swiss Alps. Geochim Cosmochim Acta 56:3209-3216

Edmond JM (1992) Himalayan tectonics, weathering processes and the strontium isotope record in marine limestones. Science 258:1594-1597

Edmond JM (1993) Weathering processes on hot and cold cratons: the Guyana and Aldan Shields. Geol Soc Am Ann Meeting Abstracts, 414-415

Francois LM, Walker JCG (1992) Modeling the Phanerozoic carbon cycle and climate: constraints from the $^{87}Sr/^{86}Sr$ isotopic ratio of seawater. Am J Sci 292:81-135

Gordeev VV, Siderov IS (1993). Concentrations of major elements and their outflow into the Laptev Sea by the Leana River. Marine Chem 43:33-45

Griffiths RP, Baham JE, Caldwell BA (1994) Soil solution chemistry of ectomycorrhizal mats in forest soil. Soil Biol and Biochem 26:331-337

Harmon RS, White WB, Drake JJ, Hess JW (1975) Regional hydrochemistry of North American carbonate terrains. Water Resources Res 11:963-967

Hedges J (1993) Global biogeochemical cycles: progess and problems. Marine Chem 39: 67-93

Högbom AG (1894) On the probability of secular variations of atmospheric carbon dioxide (in Swedish) Svensk Chemisk Tidsskrift 6:169-176

Holland HD (1978) The Chemistry of the Atmosphere and Oceans. Wiley, New York, 351 p

Horodyski RJ, Knauth LP (1994) Life on land in the Precambrian. Science 263, 494-498

IPCC (Intergovernmental Panel on Climate Change) (1990) The IPCC assessment. Cambridge Univ Press, 200 p

Jackson TA (1993) Comment on "Weathering, plants and the long-term carbon cycle". Geochim Cosmochim Acta 57,:2141-2144

Jackson TA, Keller WD (1970) A comparative study of the role of lichens and "inorganic" processes in the chemical weathering of recent Hawaiian lava flows. Am J Sci 269:446-466

Kasting JF, Ackerman TP (1986) Climatic consequences of very high carbon dioxide levels in the earth's early atmosphere. Science 234:1383-1385

Koch AS, Matzner E (1993) Heterogeneity of soil and soil solution chemistry under Norway spruce) and European beech as influenced by distance from the stem base. Plant and Soil 151:227-237

Kump LR (1989) Alternative modeling approaches to the geochemical cycles of carbon, sulfur and strontium isotopes. Am J Sci 289:390-410

Kutzbach JE, Gallimore RG (1989) Pangean climates: megamonsoons and megacontinents. J Geophys Res 94:3341-3357

Langmuir D (1971) The geochemiistry of some carbonate ground waters in central Pennsylvania. Geochim Cosmochim Acta 35:1023-1046

Lasaga AC (1989) A new approach to isotopic modelling of the variation of atmospheric oxygen theough the Phanerozoic. Am J Sci 289:411-435

Lasaga AC, Soler JM, Ganor J, Burch TE, Nagy K (1994) Chemical weathering rate laws and global geochemical cycles. Geochim Cosmochim Acta 58:2361-2386

Mackenzie, FT and Garrels, RM (1966) Chemical mass balance between rivers and oceans. Am J Sci 264:507-525

Manabe S, Bryan K (1985) CO$_2$-induced change in a coupled ocean-atmosphere model and its paleoclimatic implications. J Geophys Res 90:11,689-11,707

Manabe S, Stauffer RJ (1993) Century-scale effects of increased atmospheric CO$_2$ on the ocean-atmosphere system. Nature 364:215-218

Marshall S, Oglesby RJ, Larson, JW, Saltzman B (1994) A comparison of GCM sensitivity to changes in CO$_2$ and solar luminosity. Geophys Res Lett 21:2487-2490

Opdyke BN, Walker JCG (1992) Return of the coral reef hypothesis: basin to shelf partitioning of CaCO$_3$ and its effect on atmospheric CO$_2$. Geology 20:733-736

Otto-Bliesner B (1995) Continental drift, runoff and weathering feedbacks: implications from climate model experiments. J Geophys Res (in press)

Raymo ME (1991) Geochemical evidence supporting T.C. Chamberlin's theory of glaciation. Geology 19:344-347

Raymo ME, Ruddiman WF (1992) Tectonic forcing of late Cenozoic climate. Nature 359:117-122

Retallack GJ (1990) Soils of the Past. Unwin-Hyman, Boston, 520 p

Richter FM, Rowley DB, DePaolo DJ (1992) Sr isotope evolution of seawater: the role of tectonics. Earth Planet Sci Lett 109:11-23

Sarmiento JL (1993) Oceanic carbon cycle Chemical and Engineering News 71:30-43

Spivack AJ, Staudigel, H (1994) Low-temperature alteration of the upper oceanic crust and the alkalinity budget of seawater. Chem Geol 115:239-247

Stallard RF (1992) Tectonic processes, continental freeboard, and the rate-controlling step for continental denudation. In SS Butcher, RJ Charlson, GH Orians, GV Wolfe (eds) Global biogeochemical cycles. Academic Press, New York, p 93-121

Staudigel H, Hart SR, Schminke HU, Smith BM (1989) Cretaceous ocean crust at DSDP Sites 417 and 418: Carbon uptake from weathering vs. loss by outgasing. Geochim Cosmochim Acta 53:3091-3094

Stewart WN (1983) Paleobotany and the evolution of plants. Cambridge Univ Press, Cambridge, 405 p

Taylor AB, Velbel MA (1991) Geochemical mass balances and weathering rates in forested water-sheds of the southern Blue Ridge II Effects of botanical uptake terms. Geoderma 51:29-50

Thorseth H, Furnes H, Heldal M (1992) The importance of microbiological activity in the alteration of natural basaltic glass. Geochim Cosmochim Acta 56:845-850

Urey HC (1952) The Planets: Their Origin and Development. Yale Univ Press, New Haven, 245 p

Volk T (1987) Feedbacks between weathering and atmospheric CO$_2$ over the last 100 million years. Am J Sci 287:763-779

Volk T (1989) Rise of angiosperms as a factor in long-term climatic cooling. Geology 17:107-110

Walker JCG, Hays PB, Kasting JF (1981) A negative feedback mechanism for the long-term stabilization of Earth's surface temperature. J Geophys Res 86:9776-9782

White AF, Blum AE (1995) Effects of climate on chemical weathering in watersheds. Geochim Cosmochim Acta 59:1729-1748

Wild A (1993) Soils and the Environment: An Introduction. Cambridge Univ Press, Cambridge, 287 p

Wilson MJ, Jones D (1983) Lichen weathering of minerals: implication for pedogenesis. In: RCL Wilson (ed) Residual deposits surface related weathering processes and materials. Blackwells, Oxford, p 5-12

Wright VP (1985) The precursor environment for vasacular plant colonization. Phil Trans Royal Soc London Ser B 309:143-145